BIOFLUID DYNAMICS OF HUMAN BODY SYSTEMS

Expanded and Revised Edition

BIOFLUID DYNAMICS OF HUMAN BODY SYSTEMS

Expanded and Revised Edition

Edited by

Megh R. Goyal, PhD, PE

Arka Bhowmik, PhD

Anamika Chauhan, PhD

APPLE ACADEMIC PRESS

First edition published 2025

Apple Academic Press Inc.
1265 Goldenrod Circle, NE,
Palm Bay, FL 32905 USA

760 Laurentian Drive, Unit 19,
Burlington, ON L7N 0A4, CANADA

CRC Press
2385 NW Executive Center Drive,
Suite 320, Boca Raton FL 33431

4 Park Square, Milton Park,
Abingdon, Oxon, OX14 4RN UK

Library and Archives Canada Cataloguing in Publication

CIP data on file with Canada Library and Archives

Library of Congress Cataloging-in-Publication Data

CIP data on file with US Library of Congress

ISBN: 978-1-77491-870-8 (hbk)
ISBN: 978-1-77491-871-5 (pbk)
ISBN: 978-1-00360-431-0 (ebk)

DEDICATION

To our esteemed mentor

Professor Megh R. Goyal

Senior Technical Editor-in-Chief (Agricultural and Biological Engineering)
Apple Academic Press Inc. (Copartner CRC Press/Taylor and Francis Group)

For your leadership, motivation, and guidance have been instrumental in shaping several books on food science and engineering, microirrigation engineering, bioengineering, soil and water engineering, and plant science health.

For your invaluable contributions to the field of science and technology research.

For being a guiding light, providing us with profound impact both personally and academically, throughout our professional careers.

For your mentorship at every stage of our professional and personal development is truly priceless.

For your unwavering support and dedication.

About Senior Technical Editor-in-Chief

Megh R. Goyal, PhD, PE

Retired Professor in Agricultural and Biomedical Engineering, University of Puerto Rico, Mayaguez Campus; Senior Acquisitions Editor, Biomedical Engineering and Agricultural Science, Apple Academic Press, Inc.

Megh R. Goyal, PhD, PE, is currently a Retired Professor of Agricultural and Biomedical Engineering from the General Engineering Department at the College of Engineering at the University of Puerto Rico–Mayaguez Campus (UPRM); and Senior Acquisitions Editor and Senior Technical Editor-in-Chief for Agricultural and Biomedical Engineering for Apple Academic Press Inc. During his professional career, he has worked as a Soil Conservation Inspector; Research Assistant at Haryana Agricultural University and Ohio State University; Research Agricultural Engineer/Professor at the Department of Agricultural Engineering of UPRM; and Professor of Agricultural and Biomedical Engineering in the General Engineering Department of UPRM. He spent a one-year sabbatical leave in 2002–2003 at the Biomedical Engineering Department of Florida International University, Miami, USA.

Dr. Goyal was the first agricultural engineer to receive the professional license in agricultural engineering from the College of Engineers and Surveyors of Puerto Rico. In 2005, he was proclaimed the "Father of Irrigation Engineering in Puerto Rico for the Twentieth Century" by the American Society of Agricultural and Biological Engineers, Puerto Rico Section, for his pioneering work on micro irrigation, evapotranspiration, agroclimatology, and soil and water engineering.

During his professional career of 52 years, he has received many awards, including Scientist of the Year, Membership Grand Prize for the American Society of Agricultural Engineers Campaign, Felix Castro Rodriguez Academic Excellence Award, Man of Drip Irrigation by the Mayor of Municipalities of Mayaguez/Caguas/Ponce and Senate/Secretary of Agriculture of ELA, Puerto Rico, and many others. He has been recognized as one of the experts "who rendered meritorious service for the development of

[the] irrigation sector in India" by the Water Technology Centre of Tamil Nadu Agricultural University in Coimbatore, India, and ASABE (American Society of Agricultural and Biological Engineers), which was bestowed on him the 2018 Netafim Microirrigation Award.

Dr. Goyal has authored more than 200 journal articles and edited more than 120 books. AAP has published many of his books, including *Management of Drip/Trickle or Micro Irrigation; Evapotranspiration: Principles and Applications for Water Management;* ten-volume set on *Research Advances in Sustainable Micro Irrigation.* He has also authored the textbooks *Elements of Agroclimatology* (Spanish) by UNISARC, Colombia, and two *Bibliographies on Drip Irrigation.* Dr. Goyal has also developed several book series with AAP, including Innovations in Agricultural & Biological Engineering (with over 60 titles in the series to date), Innovations and Challenges in Micro Irrigation; and Innovations in Plant Science for Better Health: From Soil to Fork.

Dr. Goyal received his BSc degree in Engineering from Punjab Agricultural University, Ludhiana, India, and his MSc and PhD degrees from the Ohio State University, Columbus, Ohio, USA. He also earned a Master of Divinity degree from the Puerto Rico Evangelical Seminary, Hato Rey, Puerto Rico, USA.

About the Editors

Arka Bhowmik, PhD
Research Associate, Department of Radiology, Memorial Sloan Kettering Cancer Center, New York, USA

Arka Bhomik, PhD, is a mechanical/biomedical engineer and is currently working as a Research Associate in the Department of Radiology at Memorial Sloan Kettering Cancer Center, New York, USA. He formerly worked as a postdoctoral fellow at the School of Medical Science and Technology at the Indian Institute of Technology Kharagpur, India, and at the University of California, Los Angeles (USA). He was also an adhoc faculty in mechanical engineering at the National Institute of Technology Calicut, Kerala, India.

Since 2010, he has carried out scientific research in applications of mechanical engineering in biological systems, such as radiative bioheat transfer, mathematical modeling of heat and mass transfer, design and optimization of photobioreactor for microalgae biomass production, development of medical devices for diagnosis and therapy of diseases, planning models for treatment of superficial and deep microwave hyperthermia, and microfluidics. His primary interest is in developing a fundamental understanding of the light and heat transport processes in healthy and diseased biological cells and tissues and to articulate the acquired knowledge for development of medical tools and performance of such tools for diseases diagnosis and therapy.

He has been awarded as an outstanding reviewer by three flagship journals, and one of his research projects has earned him a best paper award at an international conference. He has to his credit 11 research articles in international journals, an invited editorial, two Indian patents, and a book chapter.

Dr. Bhowmik received his BE. degree from Amravati University, Maharashtra, India; his MSc. (Research) degree from University of Sheffield, Sheffield, UK; and his PhD degree from the Indian Institute of Technology Ropar, Punjab, India.

Anamika Chauhan, PhD
Assistant Professor, Department of Home Science
Chanam Lal Mahavidhyalya Haridwar
(Sri Dev Suman University, Tehri, Uttarakhand),
Uttarakhand, India

Anamika Chauhan, PhD, is an Assistant Professor in the Department of Home Science at Chanam Lal Madhavidhyalya Haridwar, Sri Dev Suman University, Uttarakhand, India, where she is also a member of the Board of Studies. She is UGC-NET, USET, MPhil (Home Science) qualified. She was awarded a Young Scientist Award by the Uttarakhand State Council for Science and Technology and a Teacher's Excellence Award by Motherhood University, Uttarakhand, as well as a Dr. Sarvepalli Radhakrishnan Global Educator Award (2022) by the International Council for Education, Research and Training. She holds one patent ("Gold Nanoparticles with Albizia L Ebbeckl. Benth Leaves Extract for Dermal Cancer") to her credit. She is a lifetime member of the Indian Science Congress Association Kolkata, Scientific and Technical Research Association (STRA), and Home Science Association of India. She has published several research papers in peer-reviewed national and international journals, eight edited books, and a textbook. She has also been associated with workshops and conferences. Her research interest and areas range across the right to education, nutrition, and human health. Dr. Chauhan obtained her master's degree and PhD from HNB Garhwal Srinagar, Central University, Uttarakhand, India. She also has a master's degree in English, Education, and Political Science.

Contents

Contributors

Arka Bhowmik
Research Associate, Department of Radiology, Memorial Sloan Kettering Cancer Center, New York, NY, USA

Anamika Chauhan
Assistant Professor, Department of Home Science, Chaman Lal Mahavidhyalya, Landhora, Haridwar, Uttarakhand, India

Megh R. Goyal
Senior Technical Editor-in-Chief, Apple Academic Press Inc., Rincon, Puerto Rico, USA

Ramjee Repaka
Associate Professor, Department of Mechanical Engineering & Department of Biomedical Engineering, Indian Institute of Technology – Ropar, Rupnagar, Punjab, India

Manpreet Singh
Research Assistant, Department of Mechanical Engineering, College of Engineering and Information Technology, University of Maryland, Baltimore County, Baltimore, Maryland, MD, USA

Rupesh Singh
Director of Animal Imaging Core and Research Associate · Cleveland Clinic, Cole Eye Institute, Cleveland, Ohio, USA

LIST OF FORMER CONTRIBUTING STUDENTS OF PROF. MEGH R GOYAL

(Compiled from each chapter in the First Edition)

[University of Puerto Rico–Mayagüez Campus (UPRM)[*], & Florida International University (FIU)–Miami–USA[**]]

Amaury Pita[**]
Angel E. Rodríguez Burgos[*]
Annette Romero[*]
Bayron Ortiz[*]
Brijesh Kadam[**], MD
Carlos A. Cruz[*]
Carlos Ortiz[*]
Carlos Vazquez[*]
Danny Gonzalez[**]
Delhy Arias[**]
Giovanny Soto[*]
Guillermo Colón[*]
Hector Peña Santiago[*]
Hong Guo[**]
Human Arjomandi[**]
Jackeline Martínez[*]
Joel Blanco[*]
Jorge Alvarado Colon[*]
Jorge Rodríguez[*]
Jose Cesar[*]
Juan Flores Lozada[*]
Julian Bedoya[**]
Luis de Jesus Diaz[*]
Mahwish Ahmed[**]
Martha Vallejo[**]
Melissa Pérez González[**]
Nelly Arboleda[**]
Noel Faría González[**]
Punkaj Gupta[**], MD
Rafael Oliver–Vidaud[**]
Rajendra Boldhane[**], MD
Reyinald Garcia Montañez[*]
Ricardo Baez[*]
Roberto La Torre[*]
Rosaura De Zayas[**]
Siobhain L. Gallocher[**]
Stephanie S. Barcelona[**]
Tejas Choksi[**]
Víctor Guerra[*]
Xavier Piñero[*]
Yarimar Padua Rosado[*]

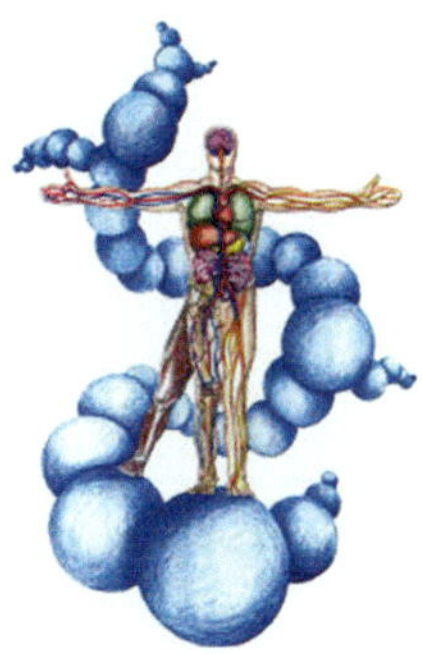

Courtsey: Students of Prof. Megh R. Goyal

Abbreviations

AC	alternating current
AFI	amniotic fluid index
AHL	ascending Henle's loop
AME	absorptive-mediated transport
ATP	adenosine triphosphate
AVAs	arterio-venous anastomoses
BBB	blood–brain barrier
BL	basal lamina
BOLD	blood oxygenation level dependent
BPG	bisphosphoglycerate
BPH	benign prostatic hyperplasia
BPT	Buckingham π theorem
BSA	body surface area
BUN	blood urea nitrogen
CB	central blood
CBF	cerebral blood flow
CBV	cerebral blood volume
CCK	cholecystokinin
CEM	cumulative equivalent minutes
CFD	computational fluid dynamics
CFTR	cystic fibrosis transmembrane conductance regulator
CHO	Chinese hamster ovary cells
CL	clearance
CM	cervical mucus
$CMRO_2$	cerebral metabolic rate of oxygen
CNS	central nervous system
CO	cardiac output
CPP	cerebral perfusion pressure
CSF	cerebrospinal fluid
CT	collecting duct
CT	computerized tomography
CTA	CT angiography
CTS	carpal tunnel syndrome
CV	control volume

CVP	central venous pressure
CW	continuous wave
DC	direct current
DHL	descending Henle's loop
DIVA	discrete vascular
DJD	degenerative joint disease
DMSA	dimercaptosuccinic acid
DPL	dual-phase-lag
DPP	dynamic pressure profile
DT	distal tubule
DV	ductus venosus
DWI	diffusion-weighted image
EC	endothelial cell
ECA	extra cranial connections
ECF	extracellular fluid
ECG	electrocardiogram
EEG	electroencephalogram
ECL	enterochromaffin
ELISA	enzyme-linked immunosorbent assay
EMG	electromyography
EMS	electrical muscle stimulator
ERCP	endoscopic retrograde cholangio pancreatography
eRPF	effective renal plasma flow
ESR	erythrocyte sedimentation rate
FAIR	fluid attenuated inversion recovery
FBG	fiber Bragg grating
FBS	fasting blood sugar
Fd	frequency shift
FDA	Food and Drug Administration
FDG	fluorodeoxyglucose
FDM	finite difference method
FEG	frequency encoded gradient
FID	free induction decay
FIU	Florida International University
FLM	force, length, and mass
fMRI	functional MRI
FRC	functional residual capacity
FUO	fevers of undetermined origin
GABA	gamma-aminobutyric acid

GERD	gastroesophageal reflux diseases
GI	gastrointestinal
GTP	glutamyl transpeptidase
HDL	high-density lipoprotein
HIFU	high-frequency focused ultrasound
HPV	human papillomavirus
HVPG	high-voltage pulsed galvanic
HSP	heat shock proteins
HU	Hounsfield unit
HVAC	heating, ventilation, and air conditioning
HVT	half value thickness
IB	indirect bilirubin
IC	interstitial cystitis
ICC	intracranial connections
ICP	intracranial pressure
IF	input function
IgSF	immunoglobulin superfamily
IIF	intracellular ice formation
IM	inner medulla
IMO	implantable membrane oxygenator
ISD	intrinsic sphincter defect
IUI	intrauterine insemination
IV	intravenous
IVF	in vitro fertilization
IVT	illini variable temperature
JAM	junctional adhesion molecules
KDMC	King-Drew Medical Center
LA	latex agglutination
LDL	low-density lipoprotein
LES	lower esophageal sphincter
MAP	mean arterial pressure
MCV	mean corpuscular volume
MENS	microcurrent electrical nerve stimulation
MFTT	magnetic fluid thermal therapy
MH	malignant hyperthermia
MI	medulla interstitial
MMC	migrating motility complex
MR	magnetic resonance
MRI	magnetic resonance imaging

MRP	multidrug resistance associated protein
MRS	MR spectroscopy
MSE	mean squared error
MUCP	maximum urethral closure pressure
MUP	maximum urethral pressure
MW	microwave
NAA	N-acetylaspartic acid
NAF	nipple aspirate fluid
NASA	National Aeronautics and Space Administration
NCS	nerve conduction studies
NDF	net driving force
NMI	nuclear medicine imaging
NMR	nuclear magnetic resonance
NS	Navier–Stokes
NTC	negative temperature coefficient
NU	nucleus
OCT	optical coherence tomography
OCTA	OCT angiography
ODE	ordinary differential equations
OEF	oxygen extraction fraction
OM	outer medulla
P	pericyte
PAP	prostatic acid phosphatase
PCR	polymerase chain reaction
PD	proton density
PDE	partial differential equations
PEG	phase–encoded gradient
PET	position emission tomography
PI	propidium iodide
PMC	pleural mesothelial cells
PMN	polymorphonucleocyte
POIS	post-orgasmic illness syndrome
PRF	proton resonance frequency
PRUV	post-void residual urine volume
PSA	prostate-specific antigen
PTC	positive temperature coefficient
PTS	peptide transport system
PVR	post-void residual
RBC	red blood cells

Re	Reynolds number
RECG	reo-encephalography
RF	radio frequency
RME	receptor-mediated endocytosis
ROI	region-of-interest
Rp	pulmonary resistance
RPF	renal plasma flow
RTDs	resistive temperature detectors
RTE	radiative transfer equation
SAP	systemic arterial pressure
SA	sinoatrial
SAR	specific absorption rate
SBS	Shaken baby syndrome
SE	spin echo
SG	specific gravity
SIDS	sudden infant death syndrome
SLD	super luminescent diode
SP	short pulse
SPECT	single-photon emission computed tomography
SSG	slice selection gradient
SVR	systemic vascular resistance
TBI	traumatic brain injury
TBW	total body water
TE	time of echo
TENS	transcutaneous electrical nerve stimulation
TJ	tight junction
TSH	thyroid-stimulating hormone
UPP	urethral pressure profile
US	ultrasound
UTI	urinary tract infection
UV	umbilical vein
VLDL	very low-density lipoprotein
VTF	Vogel–Tammann–Fulcher
WBCs	white blood cells
WHO	World Health Organization

Foreword

The human body is a complex functional system that thrives due to the functional nature of the circulatory system, their chemical compositions, and the transportation process at the physiological levels (such as: molecular, cellular, tissue, and organs). Understanding the physical behavior of human body systems in the area of biofluid mechanics, dynamics, and bioheat transport has led to remarkable research and development in biomedical engineering. I am delighted to know that Dr. Bhowmik and Dr. Chauhan have reedited all the chapters in the first edition of *Biofluid Dynamics of Human Body Systems,* which was originally authored by Prof. Megh R. Goyal.

I believe that the contents of this revised book will be helpful as a reference manual for students and researchers in bioengineering. This book combines fundamental and applied research topics of fluid dynamics and heat transfer in biological systems. The content provides an understanding of transport processes and biofluid mechanics strategies for disease diagnosis and therapy. This book also includes a chapter on working principles of commonly used medical devices, which makes it a complete guide for engineering students.

In reality, everybody is curious about the functionality of the human body circulatory system as it relates to all systems, be it the coronary system, respiratory system, cerebral hemodynamics, digestive and urinary systems, and drug transport in our body and also since the majority of present diseases arise due to dysfunction in these systems. I believe this book will enhance a proper understanding of biofluid and bioheat transport in various organ systems in our body, and provide a fundamental base for students and researchers to develop new methods and technologies to tackle diseases. I wish success for the authors, editors, and publisher of this textbook.

—Ramjee Repaka, PhD
Associate Professor, Department of Mechanical Engineering
Associate faculty, Department of Biomedical Engineering,
Indian Institute of Technology–Ropar,
Rupnagar, Punjab, India

Preface to Revised Edition

I have been teaching fluid mechanics to undergraduate and graduate students since 1971. The subject of fluid mechanics is my hobby and passion. I decided to apply principles of fluid mechanics to human body systems when my respiratory system collapsed in 1989 and I had three strokes in 2002 and my vagus nerve failed in 1999. Currently, my pulmonary system works only 27% with severe neuropathy. I thought why this would happen to a human body and I found out that a river of life flows inside my body and the fluids give life to my human body systems. This "River of Life" carries different kind of fluids to give life to all cellular organisms of our body. Also, our professional life is like a river. Let us all keep our spirits high so that the River of Life will always be dynamic. **My hypothesis is that my body system with adequate fluid quantity and properties will contribute to low incidence of disease conditions.** This is how idea for this book was born.

In July of 2001, students of my course on–INGE4015 Fluid Mechanics–conducted the first congress on "Biofluid Dynamics of Human Body Systems." The purpose of this congress was to show and learn how the biofluids in our body systems function and help to maintain a great physical state. In this congress, student groups discussed biofluid dynamics of: brain system; ear/throat/nose system; circulatory system; reproductive system; digestive system; respiratory system; urinary system; arthritis; instrumentation and measurements for the human body; challenges in biofluid dynamics of human body systems. In July of 2002, students conducted the second congress on "Biofluid Dynamics and Engineering of Artificial Organs." The purpose of this congress was to learn biofluid dynamics of bioheat transfer in a human body, biomass transfer in a human body, biofluid mechanics of artificial heart, biofluid mechanics of artificial lung, and biofluid mechanics of artificial kidneys.

In April of 2003, students of my course on–BME4999 Human Body Systems at FIU–conducted the third congress on "Biofluid Dynamics of Human Body Systems" to discuss: circulatory system; artificial heart; cardiovascular bypass surgery; respiratory system; kinetics of drug transport; digestive system; maternal fetal system; stenting of intracranial arteries as endovascular bypass of berry aneurysms (Baruch Barry Lieber, PhD, PE, University of Miami).

In July of 2004, we conducted the fourth congress on "Biofluid Dynamics of Human Body Systems" to discuss: our body fluids; properties of body fluids; ear/nose and throat system; ballooning and stenting; urinary system; artificial kidney; advances in human body fluids; dimensional analysis; biomass transfer; body pain; instrumentation and measurements. It has been an excellent learning experience for me. I thank all of the students at the University of Puerto Rico-Mayagüez and at Florida International University who have enriched my knowledge in biomedical engineering.

Fluid mechanics is a branch of engineering mechanics that deals with the response of liquids and gases at rest or in motion. Biofluid mechanics deals with living organisms: agricultural plants, animals, and humans. In this book, we consider only human body systems. One may be interested in variation of velocity, pressure, density, volume, momentum, heat, mass, nutrient, or other parameters of the biofluids. In this book, we will consider only continuous fluids (i.e., there is no discontinuity at any point). Biofluid mechanics and computational fluid dynamics (CFD) are one of the specialty areas of biomedical engineering. Biofluid mechanics applies classical mechanics (fluids at rest or in motion, thermodynamics, and continuum mechanics) to biological or medical problems. It includes the study of motion flow within our body and in medical devices, and transport of chemical constituents across biological and synthetic media and membranes. Progress in biofluid mechanics has led to the development of the artificial heart and heart valves, stents, as well as a better understanding of the function of the heart and lungs, blood vessels, and capillaries. Medical imaging combines knowledge of a unique physical phenomenon (sound, radiation, magnetism, etc.) with high-speed electronic data processing, analysis, and display to generate an image and study motion of the fluids in our body.

The mission of this compendium is to serve as a textbook or a reference manual for graduate and undergraduate students of biomedical engineering, biotechnology, nanotechnology, nursing, and medicine and health sciences. We hope that it will be a valuable reference for professionals that work with medicine and health sciences, for nursing institutes, and other agencies that work with human health.

There are two books on biofluid dynamics published by CRC Press: (1) *Biofluid Dynamics: Principles and Selected Applications* (April 26 of 2006). The book has two main parts: theory, comprising the first two chapters; and applications, constituting the remainder of the book. Specifically, the author reviews the fundamentals of physical and related biological transport phenomena, such as mass, momentum, and heat transfer in biomedical systems, and highlights complementary topics such as two-phase flow, biomechanics, and fluid-structure interaction; and (2) *Biofluid Mechanics: The Human Circulation* (2nd Edition, 2012, p. 451) edited by Krishnan B. Chandran, Stanley E. Rittgers, Ajit P. Yoganathan, published by CRC Press. The contents are PART I: Fluid and Solid Mechanics and Cardiovascular Physiology: Fundamentals of Fluid Mechanics; Introduction to Solid Mechanics; Cardiovascular Physiology. PART II: Biomechanics of the Human Circulation: Rheology of Blood and Vascular Mechanics; Static and Steady Flow Models; Unsteady Flow and Non-uniform Geometric Models; Native Heart Valves. PART III: Cardiovascular Implants, Biomechanical Measurements, and Computational Simulations: Prosthetic Heart Valve Dynamics; Vascular Therapeutic Techniques; Fluid Dynamic Measurement Techniques; Computational Fluid Dynamic Analysis of the Human Circulation; Index. I have thoroughly read both the books.

This expanded and revised edition complements the above two books and is unique because it is complete, simple, one-stop manual worldwide on Fluid Mechanics of Human Body Systems. It includes basic principles and applications, types and mechanics, flow dynamics through human body systems. It offers basic principles/ knowledge/ techniques of fluid mechanics to human body that are necessary to understand before designing/developing and evaluating a medical device. This book is a must for physicians, scientists, educators, and students.

This book, *Biofluid Dynamics of Human Body Systems-Expanded and Revised Edition,* consists of 15 chapters that include information on physical and chemical properties of fluids in the human body; dimensional analysis; Navier–Stokes equation for modeling biofluid dynamics; bioheat transfer in the human body; bioheat transfer models: theory and applications in bioengineering; biomass transfer in the human body; biofluid dynamics of human body systems: circulatory, respiratory, brain, urinary, digestive, maternal fetal systems; kinetics of drug delivery in the human body; instrumentation and measurements of human body fluids; medical imaging of the human body fluids: principle and diagnostics. Each chapter includes an introduction, body of the chapter, conclusions, summary, keywords, and a bibliography.

Twelve out of fifteen chapters of the book are from the first edition and are based on research and teaching materials and special projects by students in the courses on fluid mechanics at University of Puerto Rico-Mayaguez Campus and Florida International University. The contribution of the students at the University of Puerto Rico-Mayaguez (UPRM) and Florida

International University (FIU) has been most valuable in the compilation of these twelve chapters. Their names are mentioned in each chapter.

The three additional chapters have been diligently edited by Dr. Arka Bhowmik (Mechanical & Biomedical Engineer), and are: (1) **Chapter 3** on "*Navier-Stokes equation for modeling biofluid dynamics*" is an important aspect, and the simplified examples of these equations are a unique feature of this revised version; (2) **Chapter 5** on "*Bioheat Transfer Models: Theory and Applications in Bioengineering*" includes new transport models to regulate and monitor heat by various modes of transfer and includes analysis in tissues and whole-body. The bioheat transport process is related to the biofluid dynamics for biological systems and can regulate and/or change the temperature by collective efforts of heat, biofluid flow and mass transfer processes. Further, the heat and mass transfer processes are complex in a biological system and are yet to be understood properly. This is mainly due to various physiological processes, viz., variation in tissue thermophysical and optical properties, blood circulation, oxygen distribution, metabolism, neuro-regulation, vasomotion, etc. This new edition illustrates the role of such biophysical processes on heat transfer and has expanded the understanding of mathematical models, which are of particular interest in various medical applications, viz., diagnosis, therapy, and organ preservation sciences. This chapter also introduces some of the applications and instruments used for research purposes and provides a database for various thermophysical properties. Additionally, the chapter includes numerical problems and provides detailed analytical and numerical solutions of transient bioheat equation; and (3) **Chapter 15** on "*Medical Imaging of Human Body Fluids: Principles and Diagnostics*" focuses on the operating principle of commonly used image acquisition systems such as computed tomography, nuclear medicine imaging, magnetic resonance imaging, ultrasound imaging and optical coherence tomography. This chapter presents a simplified illustration of basic image reconstruction methods or algorithms used by medical imaging devices which are often difficult to understand for readers. Also, it provides a concentration to the physics of imaging methodology, followed by its application in structural and functional imaging of human diseases. This chapter also focuses on the applications of imaging the dynamic nature of blood, which is essential for the human system to live and function and provides us with an essential clue for underlying pathology or treatment efficacy. Additionally, this new chapter also includes numerical problems and practice MATLAB script snippets along with a special section on compartment analysis for pharmacokinetics modeling in nuclear medicine.

The intention of my team is to improve this book in the next edition, and we encourage book readers to offer useful suggestions.

As educators, we would pass a message of wisdom to the world: "*Permit that our Almighty God, our Creator, provider of all an excellent Teacher, keep us healthy with His Grace—; and Get married to our profession—*".

This book would not have been written without the valuable cooperation of a group of engineers and physicians from the University of Puerto Rico-Mayagüez Campus (UPRM) and Florida International University (FIU),

Dr. Arka Bhowmik and Dr. Anamika Chauhan have joined my team for the organization, editing of galleys, and final format of this expanded edition. My team owe special gratitude to Professors Anthony McGoron at FIU and Ramjee Repaka at the Indian Institute of Technology (IIT), Ropar for their contribution and suggestions; as well as to our friends and families, who encouraged my team for the importance of clear goals, hard work, and standing for the right belief. This lesson has guided us in everyday activities. Lastly, my team wish to thank them all for their understanding and patience throughout this book project.

Finally, a river of thanks flows from my heart and soul to my wife, Subhadra D. Goyal, for the understanding and collaboration of sharing the responsibility, time, and devotion necessary to prepare the first and this edition. She has always motivated me to live longer and to live happier to serve the world community. I dedicate this book to "those who want to live happily." Good health brings happiness to one and all.

—Megh R. Goyal, PhD, PE
Senior Technical Editor-in-Chief

Acknowledgments

We would like to extend our appreciation to:

Ramjee Repaka, PhD, for playing an advisory and a helpful role toward the development of this revised version and providing valuable suggestions to enrich the contents of this book. We also acknowledge him for providing necessary plagiarism checks for each chapter. We also thank him for editing and providing foreword for this book.

All cooperating authors and former students of Professor Goyal (University of Puerto Rico–Mayaguez Campus and Florida International University–Miami) for their contribution to the first edition (2013) and this revised edition. Their input has been most valuable.

Anamika Chauhan, PhD, for editing and addition/or replacement of images and updating all references in each chapter from the first edition; and for her contribution to the chapter on *Human Maternal Fetal System*.

C. David Kreger, PhD (http://www.modernhumanorigins.com/anth501.html) for his permission to use his copyrighted text in Chapter 4 on Bioheat of this book (such reprint permission has been filed with Edition 2013 of this book by the Apple Academic Press).

We thank Saadatmand, M. J. Abdekhodaie, M. R. Pishvaie, and F. Farhadi for their contribution to Chapter 10.

We appreciate William E. Prentice for "Therapeutic Modalities in Rehabilitation" in chapter 14.

Miguel A. Muñoz Muñoz, PhD, (Ex-President of University of Puerto Rico–USA); Saravana Kumar Jaganathan, PhD, (Former Dean of Research and Professor at PSNA College of Engineering and Technology); Anthony McGoron, PhD, Richard Schoephoerster, PhD, and James E. Moore Jr., PhD (Florida International University, Miami–USA); Ruben E. Diaz, Ricky Valentin, Ismael Pagan Trinidad, Carlos Rinaldi Ramos, Madeline Torres Lugo, Eduardo Juan, Ramón Vásquez Espinosa, and Ivette Ríos Lamberty (University of Puerto Rico–Mayaguez Campus) for their guidance and review of chapters in the first edition.

Physicians Jesús Manuel Román Velez, MD (Internal Medicine and Pulmonary Mechanics), Norberto Baez, MD (Orthopedist); Carlos Reinaldo Garcia, MD (Heart Surgeon) for their availability to edit, review, and contribute to chapters of their expertise.

Our colleagues throughout the world in the educational community for their suggestions, comments, and professional inputs. We extend our appreciation to these educators for trusting our editorial skills. Their cooperation is valuable for the completion of a large portion of this book.

Our friends and families for encouraging us on the importance of clear goals, hard work, and standing for the right belief. This lesson has guided us in everyday activities. Lastly, we wish to thank them all for their understanding and patience throughout this project.

—*Editors*

Reviews of the First Edition (2013)

"I am enthusiastic to know that Apple Academic Press Inc. and Dr. Goyal (whom I know since 1979) have put together this textbook on *Biofluid Dynamics of Human Body Systems*. The reading of this book has enriched my knowledge and has convinced me that liquids play tremendous role in our health. The book is not only simple but also thorough. The book is the most fascinating and complete manual on how human body fluids flow to carry necessary nutrients and chemical agents to each cell. Our body is a marvelous living robot that will not survive more than three days without water. Dr. Goyal, master on fluid dynamics, has applied principles of engineering mechanics to human body fluids in a simple understandable language and has been able to communicate to the reader. I hope that this manual will improve the quality of life."

—**Miguel A. Muñoz Muñoz, PhD**
President of University of Puerto Rico–USA
Professor-cum-Researcher and Ex-Chairman of Department,
University of Puerto Rico–Mayaguez Campus.

"I believe this innovative book will not only aid student community but also researchers coming from the Asia-Pacific region, who are eagerly waiting for such a manual. Dr. Goyal's vast experience in *Biomechanics* is greatly reflected in the book chapters. Since I belong to the same fraternity, I wish success for both author and publisher to bring this exciting textbook."

—**Saravana Kumar Jaganathan, PhD**
Dean of Research and Professor
Former Head at Department of Biomedical Engineering
PSNA College of Engineering and Technology
Dindugal, Palani Highway, Tamil Nadu, India

"All chapters in this book are from a practical but less complicated and technical perspective. The book should serve as a textbook or as a valuable reference with practical applications and values of human physiological processes that will be useful to students, instructors, researchers, clinicians, scientists, bioengineers, and device designers alike. Dr. Goyal is uniquely qualified to present these important topics in this book. I learned a great deal during his yearlong visit in 2003 to Biomedical Engineering Department at Florida International University."

—Anthony McGoron, PhD
Associate Dean and Professor
Biomedical Engineering Department
Florida International University, Miami, USA

"I hope that this concise textbook on human body fluids will be appealing and valuable to the health science communities. It is unique, user-friendly, and introductory focusing on essential aspects of "*Biofluid Dynamics of Human Body Systems.*" I would like to see more books on such topics.

—Jesús Manuel Román Vélez, MD
Practicing Physician, Specialist in Internal Medicine, and Pulmonary Mechanics Director of Intensive Care Unit at Hospital Perea–Mayaguez
Fellow of American Society of Pulmonary Medicine
Mayaguez–PR–USA

CHAPTER 1

Introduction and Properties of Biofluids in the Human Body Systems*

MEGH R. GOYAL, RICARDO BAEZ, ANNETTE ROMERO, and GIOVANNY SOTO

1.1 INTRODUCTION

Fluid mechanics is a branch of engineering mechanics that deals with the response to motion and resting fluids, such as gas and liquid. Biofluid mechanics deals with living organisms: Agricultural plants, animals, and humans. In this book, we will consider only the human body systems. One is interested in the variation of velocity, pressure, density, volume, momentum, heat, mass, nutrient, or other parameters throughout the biofluid. In the human body, we will consider only continuous fluids (i.e., there is no discontinuity at any point, and a continuity equation is applicable). Biofluid mechanics and computational fluid dynamics (CFD) are special fields of biomedical studies, such as biomechanics, biomass transport, bioheat transfer, drug transport phenomena, clinical engineering, cancer engineering, physiologic modeling, rehabilitation engineering, medical imaging, bioelectricity, biomaterials, bioinstrumentation, prosthetic devices and artificial organs, biotechnology/bionanotechnology, biosensors, biomodeling, bioinformatics, etc.

Biofluid dynamics ranges from a theoretical to nonexperimental understanding of state-of-the-art applications. It encompasses research, development, implementation, and operation. Experts in biofluid dynamics can provide the tools and techniques to make our "human body systems" more

*This chapter has been modified and edited from the review article prepared by the students [Ricardo Baez, Annette Romero, and Giovanny Soto] of Professor Megh R. Goyal for the course on Fluid Mechanics, INGE 4015 (Course Instructor: Megh R. Goyal, PhD, P.E.), Retired Professor in Biomedical Engineering, General Engineering Department, University of Puerto Rico – Mayaguez Campus.

Biofluid Dynamics of Human Body Systems: Expanded and Revised Edition.
Megh R. Goyal, Arka Bhowmik, & Anamika Chauhan (Eds.)

effective, efficient, and healthy. Our hypothesis in this book is based on the fact that a specific human body system can stay healthy if the supply of biofluids in this system is within normal limits.

According to the Biomedical Engineering Society (<bmes.org>), medical and biological scientists can view biofluid mechanics as classical mechanics (fluids at rest or in motion, thermodynamics, continuum mechanics, and Navier–Stokes' Equation with specific boundary conditions applicable for the flow of biofluid. This field deals with biofluid transport in synthetic media and biological systems such as flow in our body and in medical devices, cancer tissues, and membranes. Progress in biofluid mechanics helped in the creation of artificial heart, heart valves, and stents. It helped to develop an understanding of functions of transport phenomenon in capillaries and vessels, lung as well as heart. This is possible due to fast image reconstruction, and medical development in high-quality image acquisition systems that use acoustic, X-ray radiation, and magnetic field to study the motion of the fluids in our body.

Any living human system contains different body fluids which performs an essential role for sound body function. The amounts of these fluids in a person's body depend on age, health condition, daily work activities, etc. Blood plasma is one of the liquids in the blood that carries various blood cells, proteins, hormones, and other essential substances across the body parts. Fluids such as lymphatic fluid and mucus play an important role in defending the body from diseases. Similarly, saliva and gastric juices aid in the proper process of digestion. Other essential body fluids include cerebrospinal fluid (CSF), interstitial fluid, tears, sweat, semen and vaginal secretions, etc.

Body fluids differ from one another because of the composition of their solutes. Regardless these all have a common characteristic: These are all formed mainly by water. Despite their solid resemblance, animals are 70–90% water. In fact, the body is composed of 70% water. Of this 70%, approximately 50% is made up of cellular water, 15% is intestinal fluids, and 5% is plasma found in the blood. Body fluids contain various diluted organic and inorganic substances. These fluids move freely in our body systems: Circulatory; respiratory; digestive; urinary; brain; reproductive; ear/throat/nose; arthritis and pain; etc.

Our heart is a pumping station for 5200 mL of blood daily, which is about 8–10% of our total body weight. Our heart has four valves that control the direction of blood flow through the heart, permitting forward and preventing backflow. The artificial heart and circulatory assist devices are utilized in humans as a bridge to transplantation.

Pulmonary ventilation results from tidal expansion and relaxation of the lung. Respiratory muscles serve as mechanical pumps. Our right lung accounts for 55% and the left lung for 45% of total gas volume. Gas-filled cavities (250–350 million) have a total alveolar surface area of 50–100 m^2 and are responsible for gas exchange. Up to the 15th generation, the bulk flow of gases through the airway tract takes place between mouth and nose. Beyond the 15th generation, gas diffusion is more important. Artificial ventilators or respirators help to artificially ventilate the lungs of diseased patients. These ventilators are classified as negative-pressure and positive-pressure types. The therapist selects the oxygen content and the pressure at which the breath is to be delivered. The attending nurse can control the volume and the rate of respiration.

Our kidney is a waste factory for eliminating the excess electrolytes, the water-soluble nitrogenous end-products of protein metabolism; the main balance of electrolytes; contributing to obligatory water loss and discharge of excess water in the urine and maintaining acid–base balance in body fluids and tissues. To fulfill these functions, our kidneys process plasma water. Renal failure requires artificial machines to fulfill these functions. These machines are used as a bridge to kidney transplantation, thus saving the life of a patient.

Biomass transfer in our human body provides significant constraints on the metabolic rates and anatomy of living organisms, from the organization of organ networks to internal cellular structures. Convective transport dominates the major blood vessels and airways but becomes comparable to diffusion and reaction in the functional units surrounding capillaries and sinusoids.

At the cellular and subcellular levels, electrical effects and a wide variety of carrier transport processes complicate diffusion. This interacts in complex ways with enzymatic and genetic reactions. The human body is composed of 70% of fluids, which help in the thermoregulatory processes for maintaining a constant temperature of 37.4°C. In our body, net bioheat conducted out must be equal to the heat generated within our body minus the change in the energy storage in our body. Measuring the body temperature is a simple, objective, and accurate indicator of the physiologic state of a human body. We are warm-blooded living beings because we can maintain constant body temperature regardless of the environment. Heat is produced through the metabolism of food and body secretions and during physical work. Body heat is lost through radiation, convection, conduction, evaporation, and urine/feces losses. Sweat glands help in perspiration. A person at rest may lose about 900 mL of water daily.

Therefore, to have in-depth knowledge of conventional fluid mechanics, it is more appropriate to study our own body. We have an ocean within our body. Fluids include liquids and gases. Our 12 body systems need or produce a diversity of fluids to carry out different functions. Also, artificial organs (such as kidneys, lungs, heart, etc.) use fluids that can be biological or nonbiological. For example, some artificial heart uses hydraulic fluid to work.

This review chapter includes a brief description of the properties of fluids in the human body systems.

1.2 PROPERTIES OF FLUIDS[1-5]

The fluid property that provides resistance to the motion of the fluid is referred to as Viscosity. As viscosity increases the fluid motion hinders, whereas less viscous fluid can move under a small force. The energy absorbed by fluid in motion depends on the strength of viscosity. The viscosity (symbol μ) is a ratio of stress that arises due to shearing two fluid layers (F/A) to the differential change in velocity also referred to as gradient of velocity ($\Delta v_x/\Delta z$ or dv_x/dz) in a fluid and is defined by Newton's law of viscosity.

$$\mu = \left(\frac{F}{A}\right) \Big/ \left(\frac{\Delta v_x}{\Delta z}\right) \quad \text{or} \quad \mu = \left(\frac{F}{A}\right) \Big/ \left(\frac{dv_x}{dz}\right) \tag{1.1}$$

Newton's law of viscosity states that the shear stress developed by the moving fluid is proportional to the gradient of velocity, which reflects a similar law of motion by Newton ($F = ma$).

$$\frac{F}{A} = \mu \frac{\Delta v_x}{\Delta z} \quad \text{or} \quad \frac{F}{A} = \mu \frac{dv_x}{dz}$$

$$F = m\frac{\Delta v}{\Delta t} \quad \text{or} \quad F = m\frac{dv}{dt} \quad \text{or} \quad F = ma \tag{1.2}$$

Viscosity in eq 1.1 is also called absolute viscosity or dynamic viscosity. The common viscosity measurement instruments are falling ball viscometer, calibrated glass capillary viscometer, capillary tube viscometer, rotating drum viscometer, and Saybolt viscometer. The SI units of viscosity are Pascal-second (Pa·s). Another common unit is the dynes second per square centimeter (dynes-sec/cm^2), which is called poise (P) after Jean Louis Poiseuille (1799–1869). Conversion factors are 1 pascal × second = 10 poise = 1000 millipascal × second and 1 centipoise = 1 millipascal × second.

"Kinematic viscosity" is given by the division of fluid viscosity by fluid density.

$$\upsilon = \frac{\mu}{\rho} \tag{1.3}$$

"Fluid volume" refers to the total space contained within a three-dimensional system. Volume is expressed in cubic feet, cubic inches, or cubic meters. Some special geometric figures have special volume formulas. For a parallelepiped or parallelopiped:

l = length, w = width, h = height

A = area = $l \times w$

$$V = \text{volume} = h \times \text{A} \tag{1.4}$$

"Density" is a measure of how much mass is contained in each unit volume (density = mass/volume) expressed by kg/m^3. In eq 1.5: ρ = density, m = mass and V = volume.

$$\rho = m/V \tag{1.5}$$

Mass is a measure of how much fluid is there in an object, and density is a measure of how tightly that fluid is packed together. Specific volume is a reciprocal of the density. A floating object in water is less dense than water, whereas a dense object fails to float in water. A metallic anchor being a dense and heavy object sinks and the inner tube floats in water.

For example, a coin floats on mercury (Figure 1.1). A coin is less dense than mercury which makes it float.

"Specific heat capacity" refers to the amount of heat energy required to result in an increase in temperature. The specific heat capacity of water is quite large which makes it useful as a coolant in many applications. It plays a significant role in maintaining homeostasis by keeping the body temperature at an optimal level so that enzymes and proteins work effectively to catalyze vital reactions in our bodies. Since our body is made mainly of water, it takes significant energy to make small changes to our body's core temperature and it is a key reason for water being so critical to living beings.

"Specific weight" (eq. 1.6) is defined as the weight of a body per unit volume. The SI units of specific weight are kN/m^3. The imperial English units are lb/ft^3.

$$\gamma = \rho g \tag{1.6}$$

Where: ρ = density, γ = specific weight, and g = acceleration of gravity.

FIGURE 1.1 A floating coin on mercury.

Source: Reprinted from Ref. [6]. https://creativecommons.org/licenses/by-sa/4.0/

"Specific gravity" (eq 1.7) is dimensionless and is a ratio of the density of the fluid to the density of water at a specified temperature. Specific gravity can be expressed as

$$SG = \rho / \rho_{H_2O} \tag{1.7}$$

Where: SG = specific gravity, ρ = density of the fluid, and ρ_{H_2O} = density of water.

"Surface tension" (σ) originates on the surface of the fluid due to the cohesive force between liquid molecules and can be defined as the force of attraction along a line of unit length perpendicular to the surface. Away from the surface inside bulk fluid, molecules are subjected to equal cohesive force in every direction leading to net zero force or no surface tension. In the case of the molecules at the surface, they experience a cohesive force acting inward due to other liquid molecules having no cohesive force by the surface to have a balance of forces, therefore the surface molecules are subjected to a net inward force. Surface tension is measured in Newton per meter (Nm^{-1}). After dimensional analysis, one can find the units of surface tension (Nm^{-1}) are equivalent to joules per square meter (Jm^{-2}). This indicates that surface tension can also be viewed as energy at the surface, popularly known as surface energy.

The "bulk modulus of elasticity" (eq 1.8) characterizes fluids' ability to compress, that is, the ease with which one can compress a unit of the fluid volume by pressure applied. Note that volume decreases with pressure increase and vice versa. This also leads to density increase or decrease. The SI unit of the bulk modulus elasticity is N/m^2 (Pa). The imperial (BG) unit is lb/in^2 (psi).

$$E_v = -\frac{dp}{dV/V} \quad or \quad E_v = \frac{dp}{d\rho/\rho} \tag{1.8}$$

where E_v = bulk modulus for elasticity, dp = differential change in pressure, dV = differential change in volume, V = volume, $d\rho$ = differential change in density, and ρ = density of a fluid.

1.3 PRESSURE

The normal force experienced by a fluid elements surface is defined as pressure. Pressure is often measured as the excess of that pressure above atmospheric pressure, sometimes called "gauge pressure instead of absolute pressure." Atmospheric pressure is measured using a mercury barometer and the gauge pressure is measured by Bordon gauge or by vacuum gauge. The complete vacuum is a reference pressure line. The standard atmosphere (atm) is a unit of pressure and 1 atm = 101,325 Pa = 101.325 kPa = 1013.25 hPa = 760 mmHg.

The gauge pressure is defined as follows:

$$P_a = P_g + P_{atm} \tag{1.9}$$

where P_a = Absolute pressure, P_g = gauge pressure and P_{atm} = atmospheric pressure. If the pressure of a system is below atmospheric, it is called a vacuum pressure or suction. The SI unit of pressure is pascal and 1 Pa = 1 N/m^2. The standard atmospheric pressure is 1.01325 bar or 101325 Pa or 101.3 kPa. In hydraulics, "m of water" is common, and 1 m of water = 9806.65 Pa. Many gauges use mercury (Hg) as the measurement medium. Table 1.1 shows pressure units with their conversion factors, which are summarized as follows:

1 mmHg = 133.4 Pa	1 inch of H_2O = 249.08 Pa	1 ksi = 1000
1 dynes/cm^2 = 0.1 Pa	1 inch of Hg = 3386.4 Pa	psi = 68.947 MPa
kg$_f$/cm^2 = kp/cm^2 = 1 Pa	1 psi = lb$_f$/in^2 = 6894.7 Pa	
	1 atm. = 760 mmHg	

TABLE 1.1 Pressure Units and Conversion Factors.[1–4]

	Pascal, Pa	Bar	N/mm^2	kp/m^2	kp/cm^2 (atm. tech.)	atm
1 Pa (N/m^2) =	1	10^{-5}	10^{-6}	0.102	0.102×10^{-4}	0.987×10^{-5}
1 Bar (daN/cm^2) =	10^5	1	0.1	10200	1.02	0.987
1 N/mm^2 =	10^6	10	1	1.02×10^5	10.2	9.87
1 kp/m^2 =	9.81	9.81×10^{-5}	9.81×10^{-6}	1	10^{-4}	0.968×10^{-4}
1 kp/cm^2 (atm. tech.) =	98100	0.981	0.0981	10000	1	0.968
1 atm =	101325	1.013	0.1013	10330	1.033	1

TABLE 1.2 Typical Pressures in Our Body.[1]

	kPa	mmHg
Arterial blood pressure		
Maximum (systole)	13–18	100–140
Minimum (diastole)	8–12	60–90
Venous blood pressure	0.4–0.9	3–7
Great veins	<0.1	<1
Capillary blood pressure		
Arterial end	4	30
Venous end	1.3	10

The manometer measures the pressure difference between two vertical points. A pitot tube or Cobra or Kiel probes can be connected to a manometer to measure the pressure of a moving fluid. The measurement can be stagnation or static pressure that depends on the location of inlet holes on the probe. Pressure due to the density and depth of a fluid ($p = \rho gh$) is static pressure. Stagnation pressure is the pressure due to the velocity of a fluid and is defined to include static pressure. The hydrostatic law of variation of pressure is defined as the difference in fluid height in the manometer as below:

$$dp = -\rho g dz \tag{1.10}$$

Pressure is an important physiological parameter for the human body—blood pressure defines how well the human cardiovascular system is working. Table 1.2 shows typical values of pressures in our body. Terms related to our body pressure are defined in the following subsections.

1.3.1 BLOOD PRESSURE

Blood pressure is the pressure experienced by artery walls due to blood flow. As the pressure exerted by blood on the walls elevates, the delicate inner tissue layer constituted by endothelial cell lines in the artery walls wears thin and may become leaky. Any form of obstruction such as fat and cholesterol deposits reduces the blood flow, narrowing the arteries, increases the wall shear stress, increases the blood pressure which is damaging to vessels. Arteriosclerosis speeds up coronary artery degradation due to increased blood pressure than without arteriosclerosis. Elevated pressure can lead to kidney failure and heart failure also establishes stroke.

Blood flow in the arterial system depends on contraction and relaxation of the heart which gives rise to systolic and diastolic pressure. Systolic pressure is the pressure experienced by the walls of an artery due to contraction and diastolic is the pressure experienced by a wall when the heart relaxes. For adults, 19 kPa systolic and 12 kPa diastolic are considered normal, and values offset from these are unhealthy conditions. It is indicated as a ratio of systolic pressure to diastolic pressure. For a healthy person, this ratio is 120/80.[1,7–9]

1.3.2 VENOUS PRESSURE

Venous pressure[10] is the average pressure of blood in the venous compartment. Sometimes it is called central venous pressure (CVP) which is the pressure of blood in the thoracic vena cava close to the right atrium. The respiratory activity, cardiac output (CO), contraction of muscles, sympathetic vasoconstrictor tone, and hydrostatic forces due to gravity are dependent on CVP. These physiological responses change either venous blood volume or venous compliance (C_v) as shown in eq 1.11:

$$\Delta P_v \propto \frac{\Delta V}{C_v} \tag{1.11}$$

Here, P_v depends on the value of venous blood volume and venous compliance. Moreover, P_v can increase due to decrease in venous compliance, due to sympathetic activation of veins. Other factors that can bring changes to CVP are as follows:

- Reduction in the output of the heart is related to either on loss of inotropy (e.g., in ventricular failure) which causes a halt of blood inside venous circulation causing venous volume to increase or due to decreased heart rate. This means a small fraction is pumped into the arterial circulation causing an increase in CVP.

- Renal failure or activation of the renin–angiotensin–aldosterone system and will increase the venous pressure.
- Venous constriction elicited by a decrease in venous compliance due to vasoconstrictor substances circulation (e.g., catecholamines, angiotensin II) or sympathetic activation of veins, thereby increasing venous pressure.
- CVP may also increase by change in the volume of blood in the thoracic venous compartment arising due to standing posture to supine posture in a person.
- Venous volume and CVP may increase due to an arterial dilation caused by sympathetic tone withdrawal; or induced by vasodilator drugs resulting in increased flow of blood to venous compartments during normal function of the heart. Whereas, failure to arterial dilation in ventricular causes CVP to decrease. This leads to an increase in ventricular stroke volume. The afterload of failure of arterial dilation in ventricular influences the ventricular stroke volume.
- An increase in CVP can occur due to force expiration, in particular against a high resistance (as occurs with a Valsalva maneuver) caused by thoracic vena cava external compression due to a rise in intrapleural pressure.
- Muscle contraction causes compression in veins of the abdomen and limbs that decreases compliance causing an increase in thoracic compartment blood volume.

1.3.3 CAPILLARY HYDROSTATIC PRESSURE (PC)

"Capillary hydrostatic pressure" (P_C) causes the capillary fluid to exit and maintains high pressure at the arteriolar end as compared to the venular end. This pressure could drop along the capillary length (axial pressure gradient) by 15–30 mmHg in different organs. The axial pressure gradient causes filtration and reabsorption between the arteriolar and venular end of the capillary, respectively. The average value of pressure depends on pressures (P_A and P_V) at both ends of the capillary; and post-to-precapillary resistance ratio (R_V/R_A). The low resistance of the venous part ensures P_v changes to transmit back to the capillary, whereas high resistance of the arterial part causes poor transmission in downstream for change in P_A. Hence, P_C is readily dependent on changes in P_V compared to P_A.

1.3.4 TISSUE (INTERSTITIAL) HYDROSTATIC PRESSURE (P_T)

"Tissue (interstitial) hydrostatic pressure" (P_T) is determined based on the value of compliance of the tissue and the interstitial fluid volume. Normally, P_T can be either close to zero or in some tissues slightly subatmospheric while in other tissues, it could be slightly positive. Tissue compliance generally appears to be low because of small increases in tissue volume during enhanced filtration or lymphatic blockage that led to dramatic increases in P_T. The increased interstitial fluid volume and P_T reduces the gradient in capillary directing a less filtration.[1,7]

1.3.5 CAPILLARY PLASMA ONCOTIC PRESSURE (PC)

"Capillary Plasma Oncotic Pressure" (P_C): Often the capillary barrier is easily permeated by ions, however, the osmotic pressure acting on the capillary walls is exerted by plasma proteins, which are relatively impermeable. Therefore, often "osmotic" pressure is viewed as colloid osmotic pressure or oncotic pressure, since exerted by colloids. The blood plasma protein albumin contributes about 70% of the oncotic pressure. This pressure is typically 25–30 mmHg.

The increase in concentration of proteins in filtered fluids particularly high net filtration capillaries results in increasing the oncotic pressure along the length of the capillary. The large protein molecules are not permeable which gives rise to oncotic pressure measured across a semipermeable membrane. The permeability of proteins across endothelium is finite for capillaries. Also, permeability relies on capillary type and protein characteristics (size, shape, and charge). As a result, true oncotic pressure experienced at capillary walls is lower than the total concentration of proteins. The permeability of proteins on oncotic pressure can be computed by knowledge of capillary wall reflection coefficients. If $s = 1$, that means impermeable capillary for proteins and $s = 0$ refers to proteins that are free to cross the capillary wall. Continuous capillaries exhibit high values (i.e., $s > 0.9$), whereas discontinuous capillaries (very "leaky" to proteins) exhibit low s value. In the latter case, the net driving force (NDF) is less governed by tissue and plasma oncotic pressures.[1,3,7,9]

1.3.6 TISSUE (INTERSTITIAL) ONCOTIC PRESSURE (P_T)

"Tissue (interstitial) oncotic pressure" (P_T) of the interstitial fluid will depend upon the reflection coefficient of the capillary wall and concentration of interstitial protein. The value of interstitial oncotic pressure is higher for the barrier to protein transport. It is also dependent on the filtered fluid into the interstitium. For instance, the decrease in interstitial protein concentration and reduction of the oncotic pressure is governed by higher affinity to capillary filtration. This effect of filtration on protein concentration also helps to restrict capillary filtration. In general, tissue oncotic pressure is about 5 mmHg, often lower than plasma oncotic pressure.[1,7]

1.3.7 MEAN ARTERIAL PRESSURE (MAP)

Figure 1.2 shows a variation in mean arterial pressure in relation to systolic and diastolic pressure for different size blood vessels of a typical systemic circulation. The pressure is generated due to the pumping of blood by the heart when blood is subjected to resistance to systemic circulation. This pressure is pulsatile waves due to the cardiac pumping effect. In a simplified framework of nonpulsating flow, the MAP depends on systemic vascular resistance (SVR), CO, and CVP as shown in eqs 1.12 and 1.13:

$$\text{MAP} = (\text{CO} \times \text{SVR}) + \text{CVP} \tag{1.12}$$

Here, CVP is close to 0 mmHg, which leads to

$$\text{MAP} = \text{CO} \times \text{SVR} \tag{1.13}$$

In other words, MAP is affected by CO or SVR changes. If the CO and SVR change in a reciprocal manner, then proportionately MAP remains unchanged. These variables in eqs 1.11 to 1.13 are interdependent to each other. The function curves of cardiac and systemic can reveal interdependency. The function of cardiac is pulsatile hence cannot be computed from CO and SVR instead of direct measurement of arterial pressure. The rate of change of aortic pressure indicates geometric mean pressure is lower than the arithmetic average of systolic and diastolic pressures. Therefore, MAP could be evaluated under normal conditions as follows:

$$\text{MAP} \cong P_{dias} + \frac{1}{3}\left(P_{sys} - P_{dias}\right) \tag{1.14}$$

Whereas MAP is considered the arithmetic average of systolic and diastolic pressure at elevated heart rates due to modification of arterial pressure pulse, changes to narrow. In other words, digital methods are used to compute MAP for better accuracy.

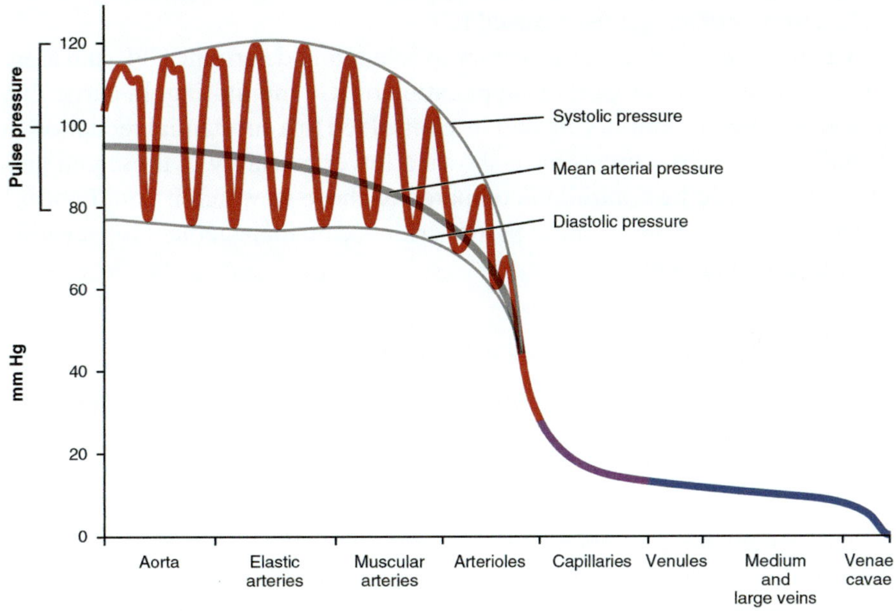

FIGURE 1.2 Diagram of mean arterial pressure in relation to systolic and diastolic pressure in blood vessels.

Source: Reprinted from Ref. [11]. https://creativecommons.org/licenses/by-sa/4.0/

1.3.8 INTRACRANIAL PRESSURE (ICP)

The volume of the cranial vault is constant since it is a closed system with a fixed volume. This volume is described by the Monro–Kellie doctrine as follows:

$$V_{\text{intracranial}}\left(\text{constant}\right) = V_{\text{brain}} + V_{\text{CSF}} + V_{\text{blood}} + V_{\text{mass lesion}} \tag{1.15}$$

The pressure elevates to a high value after the cranial vault is filled. The rise in ICP could lead to an interruption of cerebral of blood flow by reducing the cerebral perfusion pressure (CPP). As an intracranial mass lesion or edematous brain expands, some compensation is possible because CSF and blood move into the spinal canal and extracranial vasculature, respectively. Beyond this point, further compensation is impossible and ICP rises dramatically, and the normal ICP is 0–10 mmHg. There is no defined set point, at which treatment for intracranial hypertension should be initiated, but levels above 20 mmHg are usually treated. However, it is probably more important to maintain adequate CPP. In a hypotensive patient, even a small increase in ICP may be harmful. Alternatively, an elevated mean arterial pressure (MAP) may protect against a raised ICP.

The ICP measurement is necessary to accurately determine CPP, and there is much evidence to support the application of ICP monitoring. In particular, the introduction of monitoring and its control led to a substantial reduction in mortality. Similarly, studies have shown a lower mortality in those patients, whose ICP could be controlled compared to those in which it could not be. ICP monitoring is now a central part of critical care management for severely brain-injured patient.[1,7]

1.3.9 CEREBRAL PERFUSION PRESSURE (CPP)

Cerebral perfusion pressure (CPP) is the difference between the MAP and the ICP.

$$CPP = MAP - ICP \tag{1.16}$$

This parameter corresponds to the pressure gradient driving the cerebral blood flow (CBF), which carries oxygen and metabolites to each neuronal cell. To ascertain a constant flow of blood to different brain cells, the normal brain autoregulates the resistance of the blood vessels by varying its diameter even in case of an unfavorable pressure gradient or CPP. However, the mechanism behind the tight maintenance of blood and oxygen supply to neuronal cells is yet to be understood well, but recent studies have indicated neuronal nitric oxide influences cerebral blood vessel constriction and relaxation. In fact, in case of cerebral blood supply dysfunction including head trauma, the observed homeostatic mechanism in the normal brain alters (with an increase in cerebral vascular resistance) and the brain cells become vulnerable to ischemia or plausible cell necrosis due to limited oxygen and nutrients exchange. This is a serious medical condition since the brain cell operates in a complex manner and demands significantly larger blood flow compared to other cells for its proper function.[1,7]

1.4 FLUIDS IN THE CIRCULATORY SYSTEM

The human body cell's function is carried out by indirect contact with the external environment by transport of fluids and signals. In fact, this transport service for the cells is achieved by the circulatory system. Blood and lymph are two different fluid exchanges take place in the circulatory system. A lymphatic system is composed of lymph, lymph nodes, and lymph vessels, whereas the cardiovascular system is composed of blood, heart, and blood vessels. The definition of circulatory system is the combination of cardiovascular and lymphatic systems (Figure 1.3).

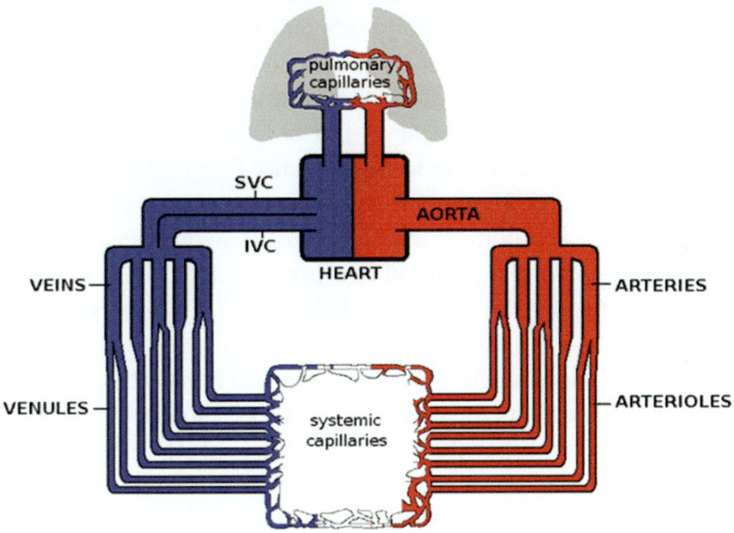

FIGURE 1.3 Simplified models of human circulatory system.

Source: Reprinted from Ref. [12]. https://creativecommons.org/licenses/by-sa/4.0/

TABLE 1.3 Properties of Adult Human Blood.[1,7–9]

Property	Units	Value	
		Range	Normal
Density	g/cm^3	1.050–1.064	1.057
Viscosity	cP	2.18–3.59	3.0
pH	–	7.35–7.45	7.4
Specific gravity	–	1.0501–1.0619	1.057
Refractive index	–	16.2–18.5	17.4
Specific heat	g/cal	–	0.92
Surface tension	dynes/cm	55.5–61.2	–
Colloid osmotic pressure: plasma	mmHg	280–480	330
Specific weight	N/m^3	–	10,369

The blood forms 8% of our body weight and in an adult human it is about 5 liters, its normal pH range is 7.35–7.45 (Table 1.3). A given volume of blood is composed of 45% blood cells and of 55% plasma, and the liquid portion of blood is 92% water (Table 1.4). Blood has a viscosity that is 3–5 times thicker than water. Blood is a nonhomogeneous, ionic, anisotropic, composite fluid, which is formed of a mixture of blood cells and plasma. Due to the high molecular weight of long-chain polymers and the presence of rouleaux and aggregates, blood exhibits a non-Newtonian and viscoelastic flow.

In contrast, for Newtonian fluid stress strain rate curve is linear with viscosity being the slope of the curve. For non-Newtonian fluids, the stress strain rate curve is nonlinear with viscosity is not constant, that is, an apparent viscosity being a slope of a tangent drawn on the curve. Blood can behave both as Newtonian and non-Newtonian fluid depending on its shear rate. Apparent viscosity is high for blood at a low value of shear rate. However, only individual cells subsist and blood behaves as if it is a Newtonian fluid at shear rates greater than 100 s^{-1}.

The velocity of blood at the center of the vessel is greater than the velocity near the walls. This affects the distribution of red blood cells (RBCs) in the circulatory system. It implies that the distribution of RBCs is not uniform so there are more in the center than at the edges. The effect of this is of 2-folds: (1) First, as blood enters a small vessel branching from the side of a main vessel, the hematocrit (percentage of RBCs in the blood) will be less than that of the main vessel; (2) Second, extremities have a greater percentage of RBC's than blood from the heart. The plasma protein and hematocrit can bring significant variation the in rheology of blood. Increasing the hematocrit increases the viscosity of blood. Appendix I in this chapter gives allometric relations of some hemodynamic parameters. Appendix II shows the effects of elevation on blood properties.[1–4,7]

A similar observation arises due to an increase in the concentration of fibrinogen. Fibrinogen is primarily responsible for the non-Newtonian fluid behavior of blood. Other plasma proteins, such as albumin, do not contribute to this non-Newtonian behavior.[1–4,7] It is observed that an increase in shear rate lowers viscosity, whereas, elevation in hematocrit level leads to an increase in viscosity. Viscosity decreases with increasing shear rate but also elevates due to a fall in temperature.

1.5 BLOOD AS A NON-NEWTONIAN FLUID[1–5,7–9]

Blood is a nonhomogeneous, anisotropic, ionized fluid containing a suspension of viscoelastic particles, and it displays non-Newtonian,

TABLE 1.4 Constituents and Functions of Blood Cells.[1,13]

Constituents		Main functions
		Formed elements (37–54%)
Erythrocytes (RBC) (99.9%)		CO_2 and O_2 transport
Leukocytes (WBC) (<0.1%)	Neutrophil (50–70%)	Phagocytosis of bacteria
	Eosinophil (2–4%)	Defense against parasitic helminths; modulation of inflammatory processes
	Basophil (<1%)	Release of histamine and other inflammation mediators
	Monocyte (2–8%)	Generation of mononuclear–phagocyte system cells in tissues.
	Lymphocytes (20–30%)	T cells kill virus-infected cells; B cells generate antibody-producing terminal cells; Natural killer cells kill tumor and virus-infected cells
Thrombocytes (Platelets) (<0.1%)		Clotting of blood
		Plasma (46–63%)
Plasma proteins (7%)	Albumins (60%)	Contributes to the osmotic pressure of plasma; transports fatty acids, thyroid hormones, and some steroid hormones.
	Globulins (35%)	Immunoglobulins attack foreign proteins and pathogens; transport globulins bind ions, hormones, and other compounds that might be lost at kidneys or have very low solubility in water.
	Fibrinogen (4%)	Converted to fibrin, which provide the basic framework for a blood clot.
	Regulatory proteins (<1%)	Catalyze specific biochemical reactions and regulate activities of other cells.
Other solutes (1%)	Electrolytes	Maintain normal extracellular fluid ion composition essential for cellular activities; contribute to the osmotic pressure of body fluids.
	Organic nutrients	Needed for ATP production, growth, and cell maintenance.
	Organic wastes	Carried to sites of breakdown or excretion.
Water (92%)	Transports organic and inorganic molecules, formed elements, and heat.	

viscoelastic deformation characteristics that can be modeled by a Power law (eq 1.17).

$$\tau = k\left(\frac{du}{dy}\right)^n + \tau_y \tag{1.17}$$

In eq 1.17, τ is the time dependent stress, *du*/*dy* is the time-dependent strain, K is a material-dependent constitutive parameter which reduces to the dynamic coefficient of viscosity for a Newtonian fluid, *n* is the material-dependent nonlinearity index (i.e., >0), and τ_y is the yield stress. For blood, τ_y in dynes per square centimeter can be calculated using eq 1.17.

$$\sqrt{\tau_y} = (H - 0.1)(C_F + 0.5) \tag{1.18}$$

In eq 1.18, H is the hematocrit expressed as a fraction, and i is the plasma concentration of fibrinogen expressed in g per 100 mL. The typical value for *k* and *n* are 0.01352 Pa·s and 0.7844, respectively. The most contributing factor in making blood non-Newtonian is the aggregation of RBCs. Human RBCs have the propensity of aggregating also referred to as rouleaux, which largely depends on fibrinogen and globulin concentration in plasma. Casson's equation (eq 1.19) is another method to describe the shear stress of whole blood.

$$\sqrt{\tau} = \sqrt{\tau_y} + \eta\sqrt{\frac{du}{dy}} \tag{1.19}$$

As RBCs form rouleaux, they will tumble while flowing in large vessels. The tumbling disturbs the flow and requires the consumption of energy, thus increasing blood viscosity at low shear. As the shear rate increases, blood aggregates tend to be broken up resulting in a drop in blood viscosity. In short, rouleaux formation increases blood viscosity, whereas breaking up rouleaux decreases blood viscosity.

The blood behaves like a non-Newtonian at low shear rates (called Casson fluid), however, as the shear rate increases, the blood begins to act more and more like a Newtonian fluid. The Newtonian fluid displays a straight line in stress v/s strain curve, whereas the non-Newtonian fluid follows a nonlinear curve in stress v/s strain curve.[1,3,7]

1.6 FLUIDS IN THE RESPIRATORY SYSTEM

The process by which air is brought into the lungs during inhalation and into close contact with the blood is respiration and it carries oxygenated blood to different body parts. At the same time, the blood gives up carbon dioxide, which is carried out of the lungs during exhalation.[1,7] The respiratory system includes nose, mouth, trachea, lungs, and diaphragm.

1.6.1 OXYGEN

Oxygen is a colorless, odorless, and tasteless gas.[2–4] It is slightly soluble in water only and is denser than the air. Normal atmospheric oxygen (O_2) has a molecular weight of 32. When cooled below its boiling point, oxygen becomes a pale blue liquid; and the liquid solidifies on cooling further. Oxidation is a chemical process during which oxygen binds directly with another element to form oxides, such as many acids and hydroxides, carbohydrates, proteins, fats and oils, alcohols, cellulose, and numerous other compounds, such as carbonates; chlorates; nitrates and nitrites; phosphates and phosphites; and sulfates and sulfites, etc. The respiration by living systems is an oxidation process.[1,7–9] The physical properties of oxygen and carbon dioxide are indicated in Table 1.5.

1.6.2 CARBON DIOXIDE

The physical properties of carbon dioxide can be found in Table 1.5. The carbon dioxide (CO_2) is a colorless, odorless, and tasteless gas that is about 1.5 times denser than the air.[1–4,7] Carbon dioxide is stable, inert, and nontoxic in normal conditions. It is a fairly stable compound but decomposes at very high temperatures into carbon and oxygen and is slightly soluble in water. Although it is not a poison, it can cause death by suffocation if inhaled in large amounts. One volume of CO_2 dissolving in an equal volume of water at room temperature and pressure results in carbonic acid. The gas is easily liquefied by compression and cooling. If liquid carbon dioxide is quickly decompressed, it rapidly expands removing enough heat so that it cools into solid carbon dioxide.

TABLE 1.5 Physical Properties of Oxygen and Carbon Dioxide at 70°F and 1 atm.[2–4]

Parameter	Oxygen, O_2	Carbon dioxide, CO_2
Boiling point temperature	–297.37°F (–183.0°C)	109.2°F (–79°C)
Critical point pressure	731.4 psia	1070.6 psia
Critical point temperature	–181.4°F	87.9°F
Gas density at 70°F and 1 atm	0.08281 lb/ft^3	0.1144 lb/ft^3
Latent heat	91.4 BTU/lb.	247 BTU/lb.
Liquid density at 70°F and 1 atm	71.23 lb./ft^3	47.64 lb/ft^3
Melting point temperature	–361.8°F	–69.9°F
Molecular weight	32.00 lb/mol	44.01 lb/mol
Specific gravity at 70°F and 1 atm	1.1	1.555
Specific heat at 70°F and 1 atm	7.03 BTU/lbmol-deg F	8.92 BTU/lbmol-deg F
Specific volume at 70°F and 1 atm	12.1 ft^3/lb (0.755m^3/kg)	8.74 ft^3/lb (0.42 m^3/kg)
Viscosity at 70°F and 1 atm	0.0192×10^{-3} Pa-s	0.0138×10^{-3} Pa-s

Carbon dioxide constitutes 1% of the volume of dry air in the atmosphere.[2–4] The natural balance of carbon dioxide in the atmosphere is increasing from its stable level, and this extra carbon dioxide will fuel the greenhouse effect, warm the atmosphere, and further disrupt the natural carbon dioxide cycle.[2–4] For commercial use, it is available as a liquid under high pressure in steel cylinders, as a low-temperature liquid at lower pressures, and as a solid dry ice.

1.6.3 WATER VAPOR

"Water vapor" is water in gaseous form, which forms either through evaporation of liquid water or sublimation of solid ice. Due to condensation and deposition, vapor form converts to water or ice. Above 100°C (212°F) and below sea-level pressure, most water molecules in a closed system of pure water assume the form of vapor.

Humidity is the amount of water vapor in the air and is expressed as absolute humidity or relative humidity.[2–4] Absolute humidity is the mass of water vapor divided by a unit volume of air (grams of water/cm^3 of air). Relative humidity *is* the amount of water vapor present in the air divided by the maximum amount that the air could be held at that temperature. The relative humidity is expressed as a percentage.[7] The relative humidity is 100% if the air is saturated with water vapor and zero percent if no water vapor is present in the air at all. Table 1.6 shows the specific weight of water vapor at different temperatures. The viscosity of water at 100°C is 0.125 cP.

TABLE 1.6 Absolute Pressure and Specific Weight of Water at Temperatures from 212 to 440°F.[2–4]

Temperature (°F)	Specific weight (lbm/ft^3)	Absolute pressure (psia)	Temperature (°F)	Specific weight (lbf/ft^3)	Absolute pressure (psia)
212	59.81	14.70	320	56.66	57.56
220	59.63	17.19	340	55.96	67.01
230	59.38	20.78	360	55.22	89.66
250	58.82	24.97	380	54.47	118.01
270	58.24	35.43	400	53.65	153.03
280	57.94	41.86	420	52.80	195.31
290	67.64	49.20	440	51.92	308.59

1.7 BIOFLUIDS IN THE URINARY SYSTEM

The volume and composition of body fluids within the normal limits are maintained by the urinary system.[1,2,7,9] This is attained by expelling the waste liquid in the form of urine accumulated in the body because of cellular metabolism. Urine is a principal liquid in the urinary system that includes kidneys, ureters, bladder, urethra, etc.

1.7.1 URINE

Urine is the body's liquid waste that originated as a byproduct from the kidney and is expelled from our body by a process of urination. Urinary excretion maintains homeostasis of the body's liquids; however, many species also use it for olfactory communication. The fluid collects generally from the blood or interstitial fluid. The composition of urine is adjusted by returning the necessary substances back to the body fluid and the byproduct with excess toxic substances is released from the body.

Urine is produced by a process of filtration, reabsorption, and tubular secretion. Urine contains a significant amount of urea. These chemicals can be detected and analyzed by urine analysis (Table 1.7). In the case of kidney or urinary tract infection (UTI), the urine will contain bacteria, but otherwise, urine is virtually sterile and is nearly odorless when it leaves the body. However, the bacteria that contaminate the urine are responsible for converting chemicals in the urine and give rise to the distinctive smell of urine.

1.8 THE HUMAN REPRODUCTIVE SYSTEM[1,5,7–9,16,17]

1.8.1 SEMINAL FLUID

The seminal fluid (or "semen" from the male person) is a mixture of glandular secretions. It is composed of 60% from the seminal vesicles, 30% from the prostate gland, 5% from the epididymis, and 5% from the bulbourethral glands. The average ejaculate produces from 2 to 6 mL of seminal fluid. The concentration of spermatozoa per milliliter of semen reaches over 50 million which helps to neutralize the acidic environment inside the vagina by giving semen a slightly alkaline pH of >7. When semen enters the vagina, special enzymes initially clot the fluid. After 15–30 min, the semen re-liquefies. The semen also includes water and mucus, which help to lubricate and to provide a liquid medium for the spermatozoa.

The minerals and nutrients in the semen nourish the spermatozoa as they travel in the vaginal canal, and these include ascorbic acid, citric acid, and

TABLE 1.7 Chemical Properties of Urine.[8,9,13]

Test description	Units	Range	Test description	Units	Range
Albumin serum	gr/dL	2.50–5.00	Mean corpuscular volume (MCV)	fl	84.00–96.00
Albumin/Globulin ratio	N/A	0.80–2.20	Mean corpuscular hemoglobin (MCH)	Pg	28.00–34.00
Alkaline phosphatase	U/L	30.00–115.00	MCH concentration (MCHC)	Percent	32.00–36.00
Anion gap	N/A	8.00–22.00	Monocyte	–	4.00–8.00
Bili total	mg/dL	0.20–1.20	Neutrophil	–	36.00–66.00
Blood urea nitrogen (BUN)	mg/dL	8.00–30.00	Platelet count	10^3 mm^3	140.00–440.00
BUN/Creatinine ratio	N/A	5.00–35.00	Potassium serum	mmol/L	3.50–5.00
Calcium	mg/dL	8.50–10.50	RBC	10^6 mm cu	4.00–5.00
Carbon dioxide	mmol/L	20.00–35.00	Sgot–Ast	U/L	0.00–41.00
Chloride	mmol/L	95.00–110.00	Sgpt–Alt	U/L	0.00–32.00
Cholesterol	mg/dL	120.00–200	Sodium serum	mmol/L	135.00–150.00
Cholesterol/HDL ratio	N/A	–	T4-Total	ug/dl	4.70–13.00
Creatinine	mg/dL	0.10–2.00	Total protein	gr/dl	6.00–8.00
Fasting blood sugar (FBS)	mg/dL	70.00–120.00	Triglycerides	mg/dl	44.00–200.00
Globulin	mg/dL	2.00–3.80	Thyroid-stimulating hormone (TSH)	uUl/mL	0.35–5.50
High-density lipoprotein (HDL)	mg/dL	35.00–85.00	Very-low-density lipoprotein (VLDL)	mg/dL	5.00–40.00
Hematocrit	percent	36.00–48.00	White blood cell (WBC)	10^3 mm^3	4.80–10.80
Hemoglobin	g/dL	12.00–16.00	Specific gravity	–	1.002–1.028
Low-density lipoprotein (LDL)	mg/dL	0.00–129.00	pH	–	6
LDL/HDL ratio	N/A	–	Lymphocyte	–	22.00–40.00

fructose. Hormones in the semen serve other functions. For instance, the hormone prostaglandins aid in sperm transport and in contracting the smooth muscle of the vagina. Table 1.8 lists several properties of semen. The normal values of semen are shown in Table 1.9.

1.8.2 AMNIOTIC FLUID

The female vagina secret fluids for lubrication and protection.[1,7–9,16–18] The metabolic rate of resident bacteria in the vagina is an acidic environment with a pH of 5.0. This acidic environment restricts the growth of many pathogens and decreases the motility of sperm in the vagina. Some of the fluids present in the vagina originate from the mucus formed in the cervix. When a woman is pregnant, the cervix forms a mucus plug serving as a physical barrier between the developing fetus and the vagina. These cervical plugs contain antimicrobial properties to kill bacteria, and contain immune system proteins to fight infection. Upon labor, the mucus plug breaks down to lubricate the vagina when the baby is ready to be delivered.

TABLE 1.8 Physical and Chemical Properties of Semen.[1,7–9,18]

Property	Units	Value
	Physical properties	
Conductivity	Mho	$(88–107) \times 10^{-4}$
Freezing point depression	°C	0.55–0.58
pH	–	7.1–7.5
Specific gravity	–	1.031–1.039
Spermatozoa	10^6/mL	50–100
Volume of ejaculate	ML	2–6
	Chemical properties	
Water	%	91.8
Calcium	mEq/L	43.0
Chloride	mEq/L	43.0
Sodium	mEq/L	117.0
Potassium	mEq/L	23.0
Zinc	mg/100 mL	14.0
	Acids	
Ascorbic acid Citric acid	mg/100 mL mg/100 mL	12.8 480.0
	Carbohydrates	
Fructose	mg/100 mL	290.0
Creatine	mg/100 mL	20.0

The amniotic fluid absorbs any shocks from any light stimulus. The physical properties of amniotic fluid (Table 1.10) are like water. The fluid volume can range from 400–1200 mL depending on the stage of pregnancy. The pH of the fluid is slightly alkaline ranging from 7.12 to 7.38, and it is slightly acidic ranging from 6.49 to 7.12 in the latter term of pregnancy. This occurs as a result of a buildup of uric acid and other wastes in the fluid.

TABLE 1.9 Normal Values of Semen Components.[1,7–9,18]

Property	Normal values
Morphology	30% or more with normal morphology.
Motility	50% or more with forward progression; or 25% or more with rapid progression within 60 min after collection.
pH	7.2–7.8
Sperm Concentration	$\geq 20 \times 10^6$ spermatozoa per mL
Total sperm count	$\geq 40 \times 10^6$ spermatozoa per mL
Vitality	75% or more live
Volume	2.0 mL
WBCs	Fewer than 1×10^6 per mL
	Immunological tests
Immunobead test	Fewer than 20% spermatozoa with adherent particles.
MAR test	Fewer than 10% spermatozoa with adherent particles
	Seminal plasma biochemical analysis
	Epididymal markers
α-glucosidase (neutral)	≥20 mU per ejaculate
Carnitine	0.8–2.9 mmole per ejaculate
	Prostate markers
Acid phosphatase (total)	≥200 U per ejaculate
Citric acid (total)	≥52 mmole per ejaculate
Zinc (total)	≥2.4 mmole per ejaculate
	Seminal vesicle marker
Fructose (total)	≥13 mmole per ejaculate

TABLE 1.10 Physical Properties and Chemical Composition of Amniotic Fluid.[1,7,16–18]

Properties	Units	Value	Properties	Units	Value
Physical properties			**Chemical composition**		
Volume	mL	400–1200	Water	%	98.5
Appearance	–	Clear	Calcium	mg/100 mL	7.2
Osmotic pressure	mOsm	265.5	Chloride	mg/100 mL	365

TABLE 1.10 *(Continued)*

Properties	Units	Value	Properties	Units	Value
P_{CO2}	mmHg	33-55	Fructose	mg/100 mL	1–2
Specific gravity	–	1.006–1.008	Glucose	mg/100 mL	4
			Iron	mg/100 mL	0.07
			Potassium	mg/100 mL	19
			Sodium	mg/100 mL	290
			Urea	mg/100 mL	20
			Uric acid	mg/100 mL	4.4

1.8.3 HUMAN MILK

The mammary glands have 15–20 lobes of glandular tissue.[1,7–9,16–18] The labels contain lactiferous ducts that converge toward the nipple. These ducts dilate just before they reach the lactiferous sinus, and then constrict again before passing out of the nipple through 15–20 openings. These glands are modified sweat glands that produce and secrete milk during the lactation process to feed the newborn.

During pregnancy, high blood estrogen and progesterone levels stimulate lactation. The corpus luteum produces these hormones during early pregnancy and the placenta takes over later. The hormones stimulate the ducts and glands in the breasts, enlarging the breasts. Milk is the natural secretion of the breast glands (Figures 1.4 and 1.5). There are different kinds of milk, but maternal milk is the best (Table 1.11). The density of the milk ranges from 1.027 to 1.033 kg/m^3 at 20°C. The milk plays a fundamental part in the nutrition, development, growth, and defense of the baby. The composition of this milk is as follows:

0.65% Minerals (Ca, P, citrate, Mg, K, Na, Zn, Cl, Fe, Cu, sulfate, bicarbonate).

0.18% Acids (citrate, formate, acetate, lactate, and oxalate); enzymes, gases, and vitamins.

87.3% Water.

3.9% Milk fat.

8.8% Solids-not fat.

3.25% Protein (¾ casein).

4.6% Lactose.

TABLE 1.11 Chemical Composition of Human Milk.[8,9,18]

Nutrients	Units	Human		Nutrients	Units	Human	
		Urban	Rural			Urban	Rural
Calories	–	586	554	Iron	Mg	3100	570
Calcium	mg	1130	250	Vitamin C	mcg	10	22
Carbohydrates	G	35	60	Thiamin	mcg	500	160
Fat	G	34	30	Niacin	mcg	1200	1550
Proteins	G	35	11	Riboflavin	mcg	1000	270

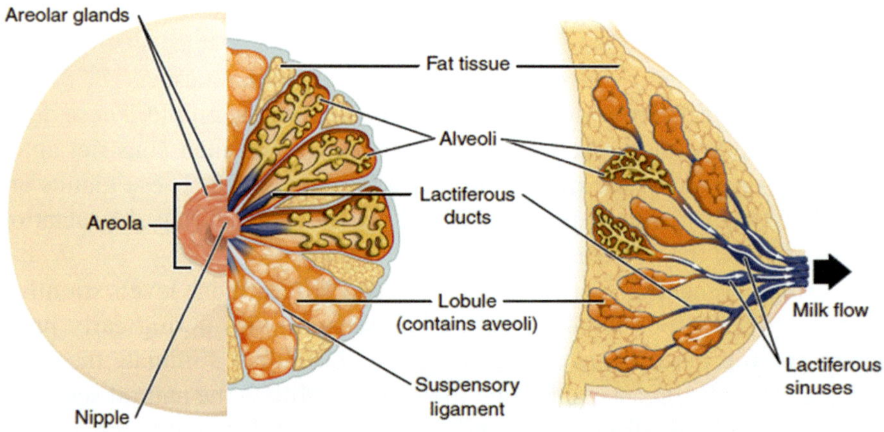

FIGURE 1.4 The pictorial representation of female mammary glands.

Source: Reprinted from Ref. [19]. https://creativecommons.org/licenses/by-sa/4.0/

1.9 FLUIDS IN THE HUMAN BRAIN[1,3,5,7–9,16–18]

1.9.1 CEREBROSPINAL FLUID

Cerebrospinal fluid (CSF) is a clear, colorless liquid that surrounds the brain and spinal cord and fills the spaces in them.[16,17] It helps to support the brain, acts as a lubricant, maintains pressure in the skull, and cushions shocks. Common diagnostic tests on CSF include protein and glucose levels, cell counts and differential, microscopic examination, and culture. Additional tests, such as lumber puncture, opening pressure, supernatant color, latex agglutination, and polymerase chain reaction (PCR) also may be performed. These tests allow the physician to diagnose a variety of diseases.

1.9.1.1 OPENING PRESSURE

To measure CSF opening pressure, the patient must be in the lateral decubitus position with the legs and neck in a neutral position. The meniscus will fluctuate between 2 and 5 mm with the patient's pulse and between 4 and 10 mm with respirations. The patient should be advised not to strain, because straining can increase the opening pressure, and is cautioned not to hyperventilate that will lower the opening pressure. Normal opening pressure ranges from 10 to 100 mm H_2O in young children, from 60 to 200 mm H_2O after 8years of age, and up to 250 mm H_2O in obese patients. Intracranial hypotension is defined as an opening pressure of less than 60 mm H_2O. In an elevated opening pressure, CSF should be removed slowly, and the pressure is monitored during the procedure. No additional CSF should be removed once the pressure reaches 50% of the opening pressure.

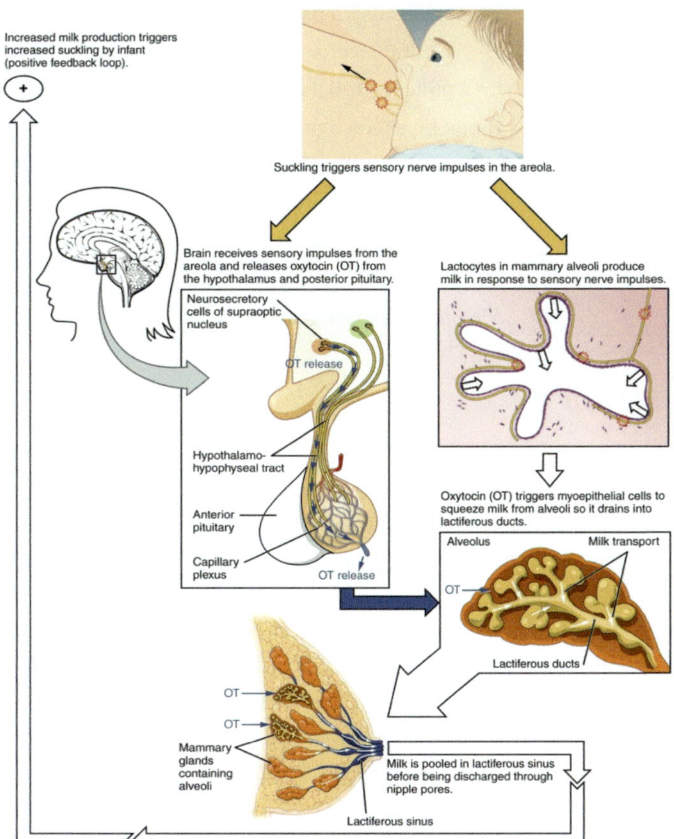

FIGURE 1.5 Positive feedback loop for human breast milk.

Source: Reprinted from Ref. [20]. https://creativecommons.org/licenses/by/4.0/

1.9.1.2 SUPERNATANT COLOR

Normally, CSF is a clear fluid. However, the turbid appearance of CSF is primarily due to its composition, that is, 200 WBCs per mm^3 or 400 RBCs per mm^3. Xanthochromia gives a yellow, orange, or pink shade to the CSF; and it is found in greater than 90% of patients within 12 h of the beginning of subarachnoid hemorrhage (a serious medical condition of bleeding in the brain) and in patients with serum bilirubin levels between 171 and 256.5 μmol/ L. Discoloration begins after RBCs have been in spinal fluid for about 2 h, and such condition lasts for 2–4 weeks (Table 1.12).

TABLE 1.12 Cerebral Spinal Fluid (CSF).[1,16–18]

Color of CFS supernatant	Conditions or causes
Yellow	Blood breakdown products
	Hyperbilirubinemia
	CSF protein ≥150 mg/dL (1.5 g/ L) > 100,000 rRBCs per mm^3
Orange	Blood breakdown products
	High carotenoid ingestion
Pink	Blood breakdown products
Green	Hyperbilirubinemia
	Purulent CSF
Brown	Meningeal melanomatosis

1.9.1.3 CELL COUNT

Normally, CSF may contain up to 5 WBCs per mm^3 in adults and 20 WBCs per mm^3 in newborns. Eighty-seven percent of patients with bacterial meningitis have a WBC count higher than 1000/mm^3, while 99% shows in more than 100/mm^3. Having < 100 WBCs per mm^3 is more common in patients with viral meningitis. WBC counts may also elevate after a seizure, in intracerebral hemorrhage, in malignancy cases, and in a variety of inflammatory conditions (Table 1.13). Peripheral blood in the CSF after a "traumatic tap" will result in an artificial increase in WBCs by one WBC for every 500–1000 RBCs in the CSF. A traumatic tap occurs in approximately 20% of lumbar punctures. The common practice is to measure cell counts in three consecutive tubes of CSF. If the number of RBCs is relatively constant, then it is assumed that the blood is caused by an intracranial hemorrhage. A falling count is attributed to a traumatic tap. The three-tube method, however, is not always reliable. If a traumatic tap occurs within 12 h of a suspected subarachnoid hemorrhage, it is reasonable to repeat the lumbar puncture one interspace to try and obtain clear CSF.

TABLE 1.13 Typical CSF Findings in Various Types of Meningitis.[16–18]

Test	Bacterial	Viral	Fungal	Tubercular
Opening pressure	Elevated	Usually Normal	Variable	Variable
WBC count	≥1000 per mm^3	<100/mm^3	Variable	Variable
Cell differential	Predominance of PMNs*	Predominance of lymphocytes	Predominance of lymphocytes	Predominance of lymphocytes
Protein	Mild to marked elevation	Normal to elevated	Elevated	Elevated
CSF-to-serum glucose ratio	Normal to marked decrease	Usually Normal	Low	Low

*Polymorphonucleocyte.

1.9.1.4 *CELL DIFFERENTIAL*

The WBC count seen in normal adult CSF is comprised of approximately 70% lymphocytes and 30% monocytes. Occasionally, a solitary eosinophil or polymorphonucleocyte (PMN) will be seen in normal CSF. Several PMNs in a neonatal patient's CSF are not unusual. Most patients with Guillain–Barré syndrome will have 10 or fewer monocytes per mm^3 and a minority of patients will have 11–50 monocytes per mm^3. In clinical diagnosis, the differential count of cells alone cannot differentiate between bacterial and nonbacterial meningitis. Lymphocytosis is seen in viral, fungal, and tuberculosis infections of the central nervous system (CNS), although a predominance of PMNs may be present in the early stages of these infections. CSF in bacterial meningitis is typically dominated by the presence of PMNs. However, more than 10% of bacterial meningitis cases will show a lymphocytic predominance, when there are fewer than 1000 WBCs per mm^3 (Table 1.13).

1.9.1.5 *MICROSCOPIC EXAMINATION*

Gram stain is positive in 60–80% of untreated cases of bacterial meningitis and in 40–60% of partially treated cases. Acid-fast staining should be done if tuberculosis is clinically suspected. Only 37% of initial smears will be positive for acid-fast bacilli. This result can be increased to 87% if four smears are done. Sensitivity also can be increased by examining the CSF sediment.

1.9.1.6 PROTEIN LEVEL

Newborn patients have up to 150 mg/ dL (1.5 g/L) of protein. The adult range of 18–58 mg/dL (0.18–0.58 g/L) is reached between 6 and 12 months of age. In the case of infections, intracranial hemorrhages, and multiple sclerosis, the protein found in CSF can be elevated, Guillain Barré syndrome, malignancies, some endocrine abnormalities, certain medication use, and a variety of inflammatory conditions. False alarm of protein concentration increase can also take place due to the presence of RBCs in a traumatic tap situation. This can be corrected by subtracting 1 mg/dL (0.01 g/ L) of protein for every 1000 RBCs per mm^3. Low CSF protein levels can occur in conditions, such as repeated lumbar puncture or a chronic leak, in which CSF is lost at a higher-than-normal rate. Low CSF protein levels also are seen in some children between the ages of 6 months and 2 years, under acute water intoxication, and in a minority of patients with idiopathic intracranial hypertension.

1.9.1.7 GLUCOSE LEVEL

As a general rule, CSF glucose is about two-thirds of the serum glucose measured during the preceding 2 to 4 h in a normal adult. This ratio decreases with increasing serum glucose levels. CSF glucose levels generally do not go above 300 mg/dL (16.7 mmol/L) regardless of serum levels. Glucose in the CSF of neonates varies much more than in adults, and the CSF-to-serum ratio is generally higher than in adults. CNS infections can cause low CSF glucose levels, although glucose levels are usually normal in viral infections.

1.9.1.8 CULTURE

Antibiotic treatment prior to lumbar puncture can decrease the sensitivity of culture, especially when given intravenously or intramuscularly. Enterovirus, the leading cause of viral meningitis, can be recovered in 40–80% of cases. Culture for herpes simplex virus is 80–90% sensitive but can take 5–7 days to become positive. Mycobacterium tuberculosis culture is positive 56% of the time on the first sample of CSF and improved to 83% of the time if four separate samples are cultured.

1.9.1.9 LATEX AGGLUTINATION (LA)

Latex Agglutination (LA) allows rapid detection of bacterial antigens in CSF. LA for *Haemophilus influenzae* has a sensitivity of 60–100%, but is much lower for other bacteria. The specificity for LA is very low. Because false positives lead to unnecessary treatment, LA is not routinely used today.

1.9.1.10 POLYMERASE CHAIN REACTION (PCR)

Polymerase Chain Reaction (PCR) of the CSF has a sensitivity of 95–100%, and a sensitivity of 100% for herpes simplex virus type 1, Epstein-Barr virus, and enterovirus. PCR is faster and more sensitive than culture for enterovirus meningitis. When PCR is positive for enterovirus, it allows earlier hospital discharge and less intervention. PCR is the most sensitive means of diagnosing cytomegalovirus (CMV) infections of the CNS, and it has been suggested that PCR should replace brain biopsy as the gold standard for herpes encephalitis. PCR has a sensitivity of 54–100% and a specificity of 94–100% for tuberculosis meningitis. PCR is sensitive for acute neurosyphilis but not for more chronic forms. PCR also is being studied as a diagnostic tool for bacterial meningitis and other infections of the CNS.

1.10 THE HUMAN DIGESTIVE SYSTEM[1,3,5,7–9,16–18]

The human digestive tract [or gastrointestinal (GI) tract] consists of a long continuous tube that extends from the mouth to the anus (Figure 1.6). It includes the mouth, pharynx, esophagus, stomach, small intestine, and large intestine. The tongue and teeth are accessory structures located in the mouth. The salivary glands, liver, gallbladder, and pancreas are major accessory organs that secrete fluids (Table 1.14) into the digestive tract. Food undergoes digestion, absorption, and elimination processes in our body. Digestion and absorption occur in the digestive tract. The digestive system prepares nutrients for utilization by body cells through gastric secretions and five activities, such as ingestion; mechanical digestion; chemical digestion; movements; and absorption.

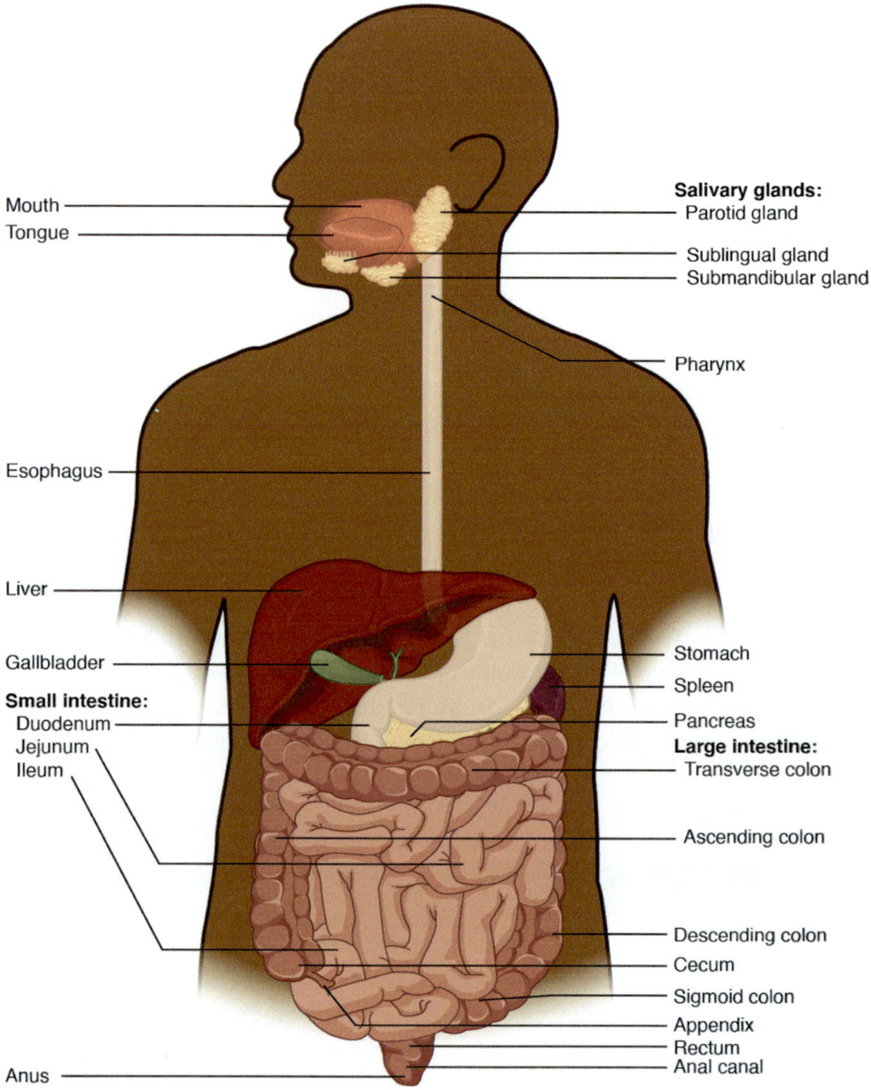

FIGURE 1.6 Human gastrointestinal (GI) tract.

Source: Reprinted from Ref. [21]. https://creativecommons.org/licenses/by/4.0/

1.10.1 GASTRIC SECRETIONS

The secretions of the exocrine gastric glands (mucous, parietal, and chief cells) are found in gastric juice. The products of the endocrine cells are secreted directly into the bloodstream and are not part of the gastric juice.[1,7,16–18] The endocrine cells secrete the hormone gastrin, which is used in the regulation of gastric activity. From the aspect of regulatory factors, the secretion of gastric juice (Table 1.14) and its amount may differ. Regulation of gastric secretions may be divided into three phases: (1) Thoughts and smells of food start the cephalic phase of gastric secretion; (2) the presence of food in the stomach initiates the gastric phase; and (3) the presence of acid chyme in the small intestine begins the intestinal phase.

TABLE 1.14 Secretion Volumes of Fluids in the GI Tract (see Fig. 1.9).[16–18]

GI secretion	Daily volume (mL)	pH	GI secretion	Daily volume (mL)	pH
		–			–
Saliva	1500	6.0–7.0	Succus entericus	2000	7.8–8.0
Gastric secretion	2000	1.0–3.5	Brunner's gland secretion	50	8.0–8.9
Pancreatic secretion	1500	8.0–8.3	Large intestinal secretion	60	7.5–8.0
Bile	500	7.8			
Total secreted volume in the GI tract, mL =				**7610**	–

Note: Secretion values are approximate in Figure 1.8 and this table; and these will vary depending on age, sex, ethnic group, exercise, amount of fluid intake, and health conditions.

1.10.2 SECRETIONS IN THE SMALL INTESTINE

Relaxation of the pyloric sphincter allows chyme to pass from the stomach into the small intestine. Exocrine cells in the mucosa of the small intestine secrete mucus, peptides, sucrose, maltose, lactose, lipase, and enterokinase. Endocrine cells secrete cholecystokinin and secretion. There is a local direct reflex action in response to chemical and mechanical irritation due to the presence of chyme and in response to distention of the intestinal wall.

1.10.3 LARGE INTESTINE

The large intestine consists of the colon, rectum, and anal canal. The mucosa has a large number of goblet cells but does not have any villi. Unlike the small intestine, digestive enzymes are not produced by the large intestine. Chemical digestion is completed in the small intestine before the chyme reaches the large intestine. Functions of the large intestine include the absorption of water and electrolytes and the elimination of feces.

1.10.4 FLUIDS IN THE ACCESSORY ORGANS OF THE DIGESTIVE SYSTEM

The salivary glands, liver, gallbladder, and pancreas (Figure 1.8) are considered accessory organs.

a. **Salivary glands**: Three pairs of major salivary glands and numerous smaller ones secrete saliva into the oral cavity, where it is mixed with food during mastication. Saliva contains water, numerous proteins and digestive enzymes, and inorganic salts. Saliva is a somewhat viscous fluid and is secreted by the surrounding salivary glands. Once in the mouth, it mixes with the food, bacteria, dead cells, and WBCs. Saliva has multiple functions, such as: moistening and lubricating the mouth to help with speech and swallowing; beginning the digestive process with enzymes to break down complex carbohydrates into sugars, removing organic debris; and it also plays an important part in the sequence of dehydration, thirst, and potation.
b. **Liver**: The common hepatic artery delivers freshly oxygenated blood to the liver. Most of the functions of the liver are performed by hepatocytes, whereas the phagocytes Kupffer cells are responsible for cleansing the blood.
c. **The gallbladder** is a storage reservoir for bile which is a yellowish-green fluid produced by liver cells. The main components of bile are water, bile salts, bile pigments, and cholesterol. Salts in the bile act as emulsifying agents in the digestion and absorption of fats. Cholesterol and bile pigments from the breakdown of hemoglobin are excreted from the body in the bile.
d. **Pancreas**: The endocrine portion of the pancreas consists of the scattered islets of Langerhans, which secrete the hormones insulin and glucagon into the blood. The exocrine portion consists of pancreatic acinar cells that secrete digestive enzymes into tiny ducts interwoven between the cells. Pancreatic enzymes include amylase, trypsin, peptides, and lipase. Pancreatic secretions are controlled by the hormones, such as, secretion and cholecystokinin.

1.11 DERMAL SECRETIONS[1,5,7,16–18]

Dermal fluid secretion takes place is in the form sweat by two types of sweat glands: (1) The **eccrine glands** help our bodies to regulate temperature by secreting water (with a mix of inorganic salts) whenever our internal temperature rises. As this water evaporates from the skin, some of the excess heat is dissipated into the atmosphere. During sex, the increased respiration and circulation, along with some perhaps vigorous physical activity, usually causes a rise in internal body temperature and therefore produces sweat from the eccrine glands. (2) The apocrine glands secrete a fatty kind of sweat into tiny tubes in the skin. In response to various strong emotional stimuli (anger, stress, lust, etc.), the walls of the tubes contract, pushing the fatty substance up to the surface of the skin, where bacteria break it down into a variety of aromatic fatty acids. Some of these fatty acids smell good to humans while others are considered unpleasant, even offensive.

Sweat is composed of electrolytes and water, these electrolytes are sodium, chloride, and potassium. The increase in core temperature sensed by the hypothalamus causes an increase in blood flow to the skin, thus stimulating the sweat glands. The result is an increase in the rate of water and electrolytes lost in the form of sweat. During low-to-moderate-intensity exercise of less than 1 h, there are minimal electrolyte losses because the body reabsorbs most of the electrolytes from the sweat. However, during moderate- to high-intensity exercise of greater than 1 h, the electrolyte loss in sweat becomes significant and the sweat rate is too fast for re-absorption of electrolytes.

TABLE 1.15 Water Content of Organs in the Human Body.[1,7,16,17]

Organ	Water content (Wt. %)
Blood	83
Brain	74.84
Heart	79.21
Intestines	74.54
Kidneys	82.68
Liver	68.25
Lungs	78.96
Muscles	75.67
Skeleton	22.04
Skin	72.03
Spleen	75.77
Entire body	63

1.12 BODY WATER

Water and its dissolved constituents make up the bulk of our body and determine the nature of nearly every physiological process.[1,7,16,17] The 70 kg standard male contains 42 liters of water. The 55% of body weight of an adult female is composed of water. Neonates hold more water in their body as compared to adults, about 75–80% water with similar proportions of extracellular fluid (ECF) than adults. In fact, the interstitial fluid is three times higher than in an adult at birth. By the age of 12 months, the proportion of water decreases to 60% similar to the amount found in adults. Total body defined in percentage of weight of body reduces with progress in age. By the age close to 60 years, total body water (TBW) reduces to 50% of total body weight, especially due to adipose tissue in males. The TBW/total body weight is low in obese adults. Majority of tissues comprises 70–80% of water. However, considerate exception is 93% water in plasma (and 7% "plasma solids"), fat 10–15% water and bone 20% water. Table 1.15 shows the water content of organs in our body.

TABLE 1.16 Body Fluid Compartments for a 70 kg Male.[1,7,16,17]

Component	Percent of body weight	Percent of total body water	Volume (Liters)
Bone water	4.5	7.5	3.2
Dense CT water	4.5	7.5	3.2
ECF	27	45	19
ICF	33	55	23
ISF	12.0	20.0	8.4
Plasma	4.5	7.5	3.2
TBW	60	100	42
Transcellular	1.5	2.5	1.0

TABLE 1.17 Average Daily Water Balance.[1,16–18]

Water intake (approximate) (mL)		Water excretion (approximate) (mL)	
Drinking water	1200	Urine	1400
Water content of food	1000	Insensible water loss	700
Water of oxidation	300	Sweat	200
		Stool	200
Total	**2500**	Total	**2500**

The plentiful tissues and organs contain water in the body. It can be divided into intracellular fluid (ICF ~ 23 liters) and extracellular fluid (ECF ~ 19 liters) defined by whether water is inside the cell and outside the cell, respectively. Body water for different fluid compartments is listed in

Table 1.16. The steady-state values (time average values) for water intake, production, and excretion for an adult under average conditions are shown in Table 1.17. Water of oxidation refers to the water produced by metabolism of food (e.g., glucose). The insensible water loss refers to losses by diffusion through the skin (as opposed to water exuded by the sweat glands) and by evaporation from the lungs (expired air is normally saturated with water at body temperature). Table 1.18 indicates the effects of hot weather and heavy exercise on water loss (which must ultimately be matched by water input, mainly by drinking). Note the shift of water from urine to sweat under the hot exercise conditions.

TABLE 1.18 Daily Loss of Water (in mL).[1,7,16,17]

Part	Normal temperature	Hot weather	Prolonged heavy exercise
Feces	200	200	200
Lungs	350	250	650
Skin	350	350	350
Sweat	200	1400	5000
Urine	1400	1200	500
Total	**2500**	**3400**	**6700**

The water loss via the lungs does not change much, dropping some in hot weather (inspired air is usually more humid, and the expired air may be less than saturated), and rising with exercise due to increase in breathing rate. Changes in total body water[18] with increasing weight and age are shown in Figure 1.10. The regression equation for increasing human body weight is as follows:

$$\log(\text{total body water}) = 2.91696 - 0.11128\ \log(\text{body weight}) \quad (1.20)$$

Effects of age on total body water (TBW) are shown in the following regression equation:

$$\log(\text{total body water}) = 2.77903 - 0.00143 \times \text{age in years} \quad (1.21)$$

1.13 PURE WATER

Bulk modulus of elasticity of pure water at standard conditions is 2.15 GPa. Effects of temperature on properties of water are shown in Tables 1.19 and 1.20. For detailed study, the reader should refer to books on Fluid Mechanics and Physics.[1–4,7]

TABLE 1.19 Effect of Temperature on Density, Specific Weight and Specific Volume of Pure Water.

Temperature	Density	Specific weight	Specific volume
	P	γ	v
°C	kg/m^3	kN/m^3	m^3/kg
0	999.9	9806	0.00100
5	1000	9807	0.00100
10	999.7	9804	0.00100
20	998.2	9789	0.00100
30	995.7	9765	0.00100
40	992.2	9731	0.00101
50	988.1	9690	0.00101
60	983.2	9642	0.00102
70	977.8	9589	0.00102
80	971.8	9530	0.00103

Note: $\gamma = \rho\, g$; $v = (1/\rho)$

1.14 ARTHRITIS VERSUS SYNOVIAL FLUID[16,17]

Arthritis is a disease referred to a collection of ailments that distress the muscle and joints. Every joint is surrounded by a lubricating or defensive capsule that act a cushion and aid in motion. Cartilage and a slippery smooth substance cover most joints to assure uniform fluid motion of the joint. The rate or the propensity of RBCs to cling together and sediments in test tube is referred to erythrocyte sedimentation rate (ESR). The infection increases the sedimentation rate. Rheumatologists also aimed to identify antibodies (developed by body to fight against disease) in blood or synovial fluid for infection. These are reflected in the report by existence of antibodies that originates due rheumatoid arthritis, or lupus specific antinuclear antibodies.

TABLE 1.20 Effect of Temperature on Absolute Viscosity of Pure Water.[2–4]

Temperature °C	Viscosity mPa·s
10	1.3059
20	1.0016
30	0.79722
50	0.54652
70	0.40355
90	0.31417

Synovial fluid is a dialysate of plasma. The synovial fluid provides nutritive support, lubrication, and "cushioning" of articular cartilage. Concentrations of synovial fluid, electrolytes and nonelectrolytes, such as glucose and urea, are similar to those of blood. The cells of the synovial membrane add hyaluronate, a glycosaminoglycan that contributes to the viscosity of the fluid. The volume, color, turbidity, viscosity, mucin (hyaluronic acid) concentration, and protein content of synovial fluid can help to categorize an arthropathy as hemorrhagic, degenerative, or inflammatory. In a degenerative joint disease (DJD), joint fluid can help to distinguish degenerative disease from the other categories. Synovial fluid is a transparent, viscid fluid secreted by the synovial membrane that are exists in tendon sheaths, joint cavities, and bursae. Because of its high concentration of hyaluronic acid, the synovial fluid is very viscous.

The viscosity of the fluid can be assessed by allowing a drop of synovial fluid to fall slowly from the tip of a needle, or by placing a drop between the thumb and forefinger (of a gloved hand) and slowly separating them. The quantity and quality of mucin (and therefore hyaluronic acid) can be assessed using a mucin clot test. In case of swollen joint, the viscosity of the fluid is reduced due to depletion of hyaluronic acid. A drop of fluid from an osteoarthritic joint may form a string of 1–2 cm long. The reduction in hyaluronic acid may also not be detectable. Diff-Quik is a rapid test to evaluate synovial fluid samples. Nucleated cell counts are variable from joint to joint, even in the same object.

1.15 FLUIDS FOR THE ARTIFICIAL HEART

The artificial heart AbioCor uses specifically silicone hydraulic fluid to work. This fluid is nontoxic. MASIL® SF polydimethyl siloxane fluids are available in varying viscosities (Table 1.21) to meet the demands of hundreds of applications.[1,7–9]

TABLE 1.21 Physical Properties of MASIL® SF.

Type of MASIL® SF	Viscosity/ temperature coefficient	Specific gravity @ 25°C	Refractive index @ 25°C	Pour point (°C)	Flash point (°C) PMCC	Volatility (% wt loss, 24 h 150°C)
0.65	0.32	0.760	1.3750	–68	–1	—
5	0.53	0.916	1.3970	–84	280	90
100	0.59	0.968	1.4030	–55	461	0.5
350 FG	0.60	0.973	1.4032	–50	500	0.5
500	0.60	0.973	1.4033	–50	500	0.5
1,000	0.60	0.974	1.4035	–50	500	0.5
10,000	0.60	0.975	1.4035	–47	500	2.0
60,000	0.60	0.977	1.4035	–44	500	2.0
10^6	0.60	0.979	1.4035	–25	500	2.0

MASIL® SF silicone fluids exhibit excellent thermal and oxidation stability and are chemically inert/nonreactive with synthetic natural fibers, rubbers, and organic coatings. They have low viscosity/temperature coefficient ratios, excellent stability to mechanical shear stress, and are noncorrosive and nontoxic. They provide high surface activity to formulate products because of their low surface tension. They quickly spread and provide high surface activity to formulate products. The more viscous products have exceptionally high flash points and auto-ignition temperatures. They could display less volatility or pour points, and remain liquid even at very high viscosities (Table 1.21).

a. **Solubilities:** MASIL® SF is completely miscible in aliphatic and aromatic solvents, including halogenated solvents.
b. **Applications:** MASIL® SF silicone fluids can be used as heat transfer and hydraulic fluids, defoamers, lubricants, and release aids. Products with viscosities of 5000 cSt are used as fiber and thread lubricants, mechanical lubricants, mold release agents, and calendar lubricants.

1.16 FLUIDS FOR ARTIFICIAL KIDNEY

The removal of cellular waste and water is accomplished by kidneys. Hormones such as renin, erythropoietin, and Vitamin D are secreted by kidneys. It receives about 20% of the blood from the heart with each beat. The blood flow rate through both kidneys is approximately 1.2 L/min. The normal kidney maintains balance in total body water by removal of excess water via urine. Despite large intakes of salt and water, almost no changes in blood volume or concentration occur.

1.17 OTHER FLUIDS[2,3,5,8,16,17]

1.17.1 NOSE

The nose warms, cleans, and humidifies the air breathe with us. In addition, it assists us to smell. A normal daily produce of about 1 L daily of mucus, which aids in trapping the dust, and keeps the respiratory tract clean and moist. Nearly all of the mucus ends up in our stomach. The mucus is produced by tiny mucus-secreting glands that are scattered throughout the nasal and sinus cavities. The cilia also referred to as tiny hairs that cover the surface allows brushing of particles. The following conditions may cause an increase in thin mucus:

- Allergies.
- Certain foods or spices.
- Cold temperature.
- Drug side-effects (particularly high blood pressure medications).
- Pregnancy or hormonal changes.
- Structural problems (deviated septum, large turbinates).
- Vasomotor rhinitis (an abnormal regulatory problem with the nose).
- Viruses.

Decreasing the fluid content of the mucus usually thickens the secretions leading to the impression of increased mucus. The following may cause thickened secretions:

- Advanced age: The mucus membrane lining the nose can shrink with age leading to a reduced volume of secretions.
- Drug side-effects (antihistamines).
- Environmental irritants (tobacco smoke, smog).
- Foreign bodies (especially if the drainage is from one side).
- Hormonal problems.
- Low humidity.
- Sinus or nasal infections.
- Structural problems (deviated septum, enlarged turbinates, enlarged adenoids).

1.18 THROAT

The throat consists of the pharynx, epiglottis, and larynx. Food travels through the pharynx on the way to the esophagus and stomach. Air passes through the pharynx to the trachea and lungs. When we swallow, food and liquid chewed or manipulated in the mouth are propelled backward by the tongue. It then moves through the pharynx and onto the esophagus and stomach. During this automatic swallowing response, the epiglottis covers the larynx to prevent food from entering the lungs.

1.19 TEARS

The eye is protected by fluids in the eyes or tears. Tears are classified as (1) reflex tears that are generated by stimuli such as irritation, bright light, yawning, and (2) basal tears, which are the nonstimulated secretion of the tear glands. Tears ensure the health of the corneal through actions such as diluting, flushing, and neutralizing foreign bodies and chemicals. Fluid from the conjunctival sac is slightly opalescent as it includes all the products of the conjunctival secretions, including mucus and epithelial debris. The physical and chemical properties of tears are shown in Table 1.22.

1.20 BIOHEAT TRANSFER[1,7]

The human body is composed of 70% of fluids, which is responsible for maintaining a constant body temperature of 37.4°C by various thermoregulatory processes. Our body's heat balance can be defined as net bioheat transferred to the surroundings must be equal to the heat generated within our body minus the change in the heat energy stored in our body. Measuring the body temperature is an accurate indicator of the physiologic state of a human body. We are warm-blooded living beings because we can maintain constant body temperature regardless of the environment. Heat is produced through the metabolism of food and body secretions and during physical work. Body heat is lost through radiation, convection, evaporation, and urine/feces losses. Sweat glands help in perspiration. A person at rest may lose 900 mL of water daily. Readers can refer Chapters 4 and 5 for more details on bioheat transfer and modeling techniques in bioheat transfer.

1.20.1 FEVER[8,9,16]

A 1° rise in body temperature from normal 37°C is identified by fever. Temperature elevation due to infections of minor nature could lead to mild or short-term elevations. However, strong infection can lead to fever temperatures of 103°F and above. To balance fluid loss by perspiration through skin, it is advised to consume 1–2 quarts fluid. This topic was also discussed in Chapter 4.

1.20.2 BODY TEMPERATURE

The core body temperature of a human is 37°C (or 98.6°F), but it may fluctuate in the range of 36.4°C and 37.2°C (or 97.5° to 99°F). If body temperature exceeds the normal range, the cellular function may become impaired or the cells may damage by themselves; if it drops below a certain limit, the cell function and the metabolization rate of foodstuffs reduces. The body temperature is regulated by the rate at which heat is radiated from the skin and by the evaporation of water vapor. Perspiring (evaporation through pores in the skin) and panting (evaporation through pores in the mouth) are common temperature regulators in warm-blooded animals. These processes are controlled involuntarily by the brain.

1.21 NUMERICAL EXERCISES

The difference in pressure, Δp, due to stenosis also viewed as a blockage in the artery is provided by the equation[1–4,7]

$$\Delta p = K_v \frac{\mu V}{D} + K_u \left(\frac{A_0}{A_1} - 1\right)^2 \rho V^2$$

where V, μ, ρ, D, A_0, and A_1 are blood velocity, viscosity, density, artery diameter, area of the unobstructed artery, and area of the stenosis, respectively. Determine the dimensions of the constants K_v and K_u. Specify with reason whether or not the above equation is valid in any system of units.

$$\Delta p = K_v \frac{\mu V}{D} + K_u \left(\frac{A_0}{A_1} - 1\right)^2 \rho V^2$$

$$\left[FL^{-2}\right] = [k_v]\left[\left(\frac{FT}{L^2}\right)\left(\frac{L}{T}\right)\left(\frac{1}{L}\right)\right] + [k_u]\left[\left(\frac{L}{L}\right)^2 - 1\right]\left[\frac{FT^2}{L^4}\right]\left[\frac{L}{T}\right]^2$$

$$\left[FL^{-2}\right] = [k_v]\left[FL^{-2}\right] + [k_u]\left[FL^{-2}\right]$$

Since each term must have the same dimensions, k_v and k_u are dimensionless. Thus, the equation is a general homogenous equation with consistent system of units. Yes, this equation would be valid in any system of units.

Blood pressure is shown by ratio of peak systolic pressure and minimum diastolic pressure, which are measured by help of mercury manometer. A normal ratio for humans is given by 120/70 mmHg. (a) Give the value of the same in Pascals. (b) An inflated car tire is 120 mmHg pressure, is it sufficient in psi for driving?

TABLE 1.22 Physical and Chemical Properties of Tears.[18]

Property	Subjects		No. of tests	Method	Value	Remarks
	Age years	–				
				Physical properties		
Conductivity	–	50	–	–	0.90% NaCl	–
Freezing point depression	Adult	3	3	Bechmann thermometer	0.551°C	Samples were collected 3 times daily in Wright capsules and pooled
pH	Adult	77	77	Beckman pH meter	7.4 (7.3–7.7)	Lacrimation induced by instilling distilled water
	–	–	–	Brown's colorimetric	7.35	Lacrimation artificially induced
	–	50	–	–	7.4	–
Refractive index	22–40	16	40	Refractometer	1.3369 (1.3361–1.3379)	No artificial lacrimator used
Vapor pressure	35	5, 5	10	Krogh's modification of the A.V. Hill method	0.90 (0.87–0.92) percent NaCl	Lachrymation is stimulated by placing a capillary tube along the inner side of the lower lid
	–	50	–	–	0.93	–
Volume	–	5	–	–	(0.50–0.67) g/16 h	During waking hours
	12–29	–	–	–	13 mm/5 min	Schirmer test (measures linear distance tears are absorbed on a strip of filter paper when one end is placed in conjunctival sac)
	15–29	–	–	–	20 mm/5 min	
				General chemical components		
Ash	Adult	2	2	Gravimetric	1.05 g/100 mL	Samples were collected 3 times daily in Wright capsules and pooled
Solids, total	Adult	2	2	Gravimetric	1.8 g percent	
Water	Adult	2	2	Gravimetric	98.2 g percent	

TABLE 1.22 *(Continued)*

Property	Subjects		No. of tests	Method	Value	Remarks
	Age years	–				
Electrolytes						
Bicarbonate					26 mEq/L	
Chloride	Young	2	19	Modification of differential potentiometric titrations of MacInnes and Dole	128.0 (117.6–138.4) mEq/L	Lacrimation induced by sliced onions
	8–12	32	32	Keys titration	135 (101.8–168.2) mEq/L	Lacrimation induced by sliced onions
	–	20	2	Sendroy technique	135.5 (134.5–136.5) mEq/L	No artificial lacrimator used
Potassium	Young	2	36	Perkin–Elmer photometer	14.9 (7.7–22.1) mEq/L	Lacrimation induced by sliced onions
	8–12	32	32	Flame photometry	29.0 (22.8–35.2) mEq/L	Lacrimation induced by sliced onions
Sodium	Young	2	42	Perkin—Elmer photometer	146 (126–166) mEq/L	Lacrimation induced by sliced onions
	8–12	32	32	Flame photometry	142.0 (108.4–175.6) mEq/L	Lacrimation induced by sliced onions
Nitrogenous substances						
Protein, total	5–64 yr	30	30	Nephelometry	0.360 (0.136–0.592) g/100 mL	Varies with the rate of tear flow
	Adult	3	3	Calculated from nitrogen content	0.669 g/100 mL	Varies with the rate of tear flow; samples collected 3 times daily in Wright capsules and pooled
Albumin	Adult	3	3	Hawk and Bergein	0.394 g/100 mL	Samples were collected 3 times daily in Wright capsules and pooled
Globulin	Adult	3	3	Calculated by subtracting albumin from total protein	0.275 g/100 mL	–

TABLE 1.22 *(Continued)*

Property	Subjects		No. of tests	Method	Value	Remarks
	Age years	–				
Ammonia	–	–	–	–	0.005 g/100 mL	Samples were collected 3 times daily in Wright capsules and pooled
Urea	Adult	2	2	Marshall urease	0.03 g/100 mL	Samples were collected 3 times daily in Wright capsules and pooled
	–	2	36	Micromodification of Archibald calorimetric	1.02 g/100 mL plasma concentration	Lacrimation induced by sliced onions
Nitrogen Total N	Adult	6	6	Folin microKjeldahl	0.158 g/100 mL	Samples were collected 3 times daily in Wright capsules and pooled
Nonprotein N	Adult	4	4	Folin microKjeldahl	0.051 g/100 mL	
Carbohydrates						
Reducing substances (as glucose)	–	12		–	6.06 mg/100 mL	Greatly reduced by prolonged tearing
Glucose	–	12	24	Fermentation	2.5 mg/100 mL	Lacrimation induced by tear gas
Miscellaneous organic acids, vitamins, enzymes						
Citric acid	–	1, 1	6	Enzymatic	0.5–0.7 mg/100 mL	No artificial lacrimator used
Vitamin C (ascorbic acid)	–	–	–	–	0.14 mg/100 mL	Precorneal and preconjunctival film
Lysozyme	–	17	35	Viscosimetric	1438 units^2/mL	–

$$p = \delta h$$

$$\text{a) For 120 mm Hg}: p = \left(133 \times 10^3 \frac{N}{m^3}\right)(0.120\ m) = 16.00\,\text{KPa}$$

$$\text{For 120 mm Hg}: p = \left(133 \times 10^3 \frac{N}{m^3}\right)(0.070\ m) = \ 9.31\,\text{KPa}$$

$$\text{b) For 120 mm Hg}: p = \left(16.0 \times 10^3 \frac{N}{m^3}\right)\left(1.450 \times 10^{-4} \frac{lb/in^2}{N/m^2}\right) = \ 2.32\,\text{psi}$$

or 5.35 ft

A cuff is used in the arm to measure blood pressure, the pressure of cuff is the measure of brachial arterial blood pressure measured by mercury manometer. The peak value of systolic pressure is 120 mmHg. What is the reason that water in not used in manometer instead of mercury for pressure measurement? Provide the answer with the necessary calculations and explanation.

$$p = \delta h$$

$$\text{For 120 mm Hg}: \ p = \delta h$$

$$= \left(133 \times 10^3 \frac{N}{m^3}\right)(0.120\ m)$$

$$= 16.00\,\text{KPa}$$

To obtain this pressure with water column :

$$h_{H_2O} = \left(\frac{16.0 \times 10^3 \frac{N}{m^3}}{9.80 \times 10^3 \frac{N}{m^3}}\right) = \ 1.63 \quad m$$

Thus, if water were used in the manometer, the required column heights would be too high and impractical. Water manometer is not therefore recommended.

1.22 SUMMARY

In this chapter, several biofluids are presented with important properties: specific weight, density, specific gravity, pH, and flow rate. In the circulatory system, the principal fluid is the blood in bringing oxygen to the body. In the respiratory system, the oxygen and the carbon dioxide are the most important fluids because these are inhaled and exhaled gases. The urine is present only in the urinary system and its function is to clear the cellular waste accumulated by the body. The brain includes CSF which behaves as a lubricant, ensures desired pressure, and protects from shocks. The sweat is a dermal secretion, and it helps our bodies to regulate its temperature.

KEYWORDS

- **amniotic fluid**
- **bioheat transfer**
- **blood pressure**
- **blood vessel**
- **body temperature**
- **body water**
- **CSF**

REFERENCES

1. Bronzino, J. D.; Peterson, D. R.; Eds. *The Biomedical Engineering Handbook*, Vol. 1–4; CRC Press LLC (Taylor & Francis Group): Boca Raton, FL, 2015; p 5422.
2. Fox, R. W.; McDonald, A. T.; Mitchell, J. W. *Fox and McDonald's Introduction to Fluid Mechanics*; John Wiley & Sons, 2020.
3. Gerhart, P. M.; Gerhart, A. L.; Hochstein, J. I. *Munson, Young and Okiishi's Fundamentals of Fluid Mechanics*; John Wiley & Sons, 2016.
4. Halliday, D.; Resnick, R.; Walker, J. *Fundamentals of Physics Extended*; John Wiley & Sons, 2021.
5. Goyal, M. R. Introduction and Properties of Biofluids in the Human Body Systems. In *Biofluid Mechanics of Human Body Systems*, 2nd ed.; CRC and Apple Academic Press, 2013; pp 1–62.
6. Alby, *A pound coin floating in mercury due to surface tension.* https://en.wikipedia.org/wiki/File:Pound-coin-floating-in-mercury.jpg, (accessed 03/10/2023)
7. Cooney, D. *Biomedical Engineering Principles*; Marcel Dekker, 1976.
8. Goyal, M. R. *Class Notes: Fluid Mechanics*. University of Puerto Rico: Mayaguez; 1997–2012.
9. Goyal, M. R. Proceedings of Congress on: Biofluid Dynamics of the Human Body Systems. *http://faculty.eng.fiu.edu/~mgoyal*; 2003.
10. Wikipedia. *Central Venous Pressure* [Online], 2023. https://en.wikipedia.org/wiki/Central_venous_pressure. (accessed 03/08/2023).

11. OpenStaxCollege. Mean arterial pressure in relation to systolic and diastolic pressure in blood vessels. https://commons.wikimedia.org/wiki/File:2109_Systemic_Blood_Pressure.jpg, (accessed 03/10/2023)
12. Gccwang. Diagram of the circulatory system. https://commons.wikimedia.org/w/index.php?curid=35020064, (accessed 03/10/2023).
13. *Urine and Blood Analysis*; Clinical Laboratory Baco: Mayaguez, Puerto Rico.
14. Pardee, W. D. *Microsoft® Encarta® Encyclopedia*; Microsoft Corporation, 2002. (accessed 03/08/2006).
15. Stuart, I. F.; Rompolski, K. *ISE Human Physiology (ISE HED APPLIED BIOLOGY)*; McGraw Hill, 2021.
16. Altman, P. *Blood and Other Body Fluids*; Federation of American Societies for Experimental Biology, 1961.
17. OpenStaxCollege. *Illustration from Anatomy & Physiology, Connexions*. https://commons.wikimedia.org/wiki/File:Figure_28_02_09.jpg, (accessed 03/10/2023)
18. OpenStax. *Anatomy and Physiology*. https://openstax.org/books/anatomy-and-physiology/pages/28-6-lactation#tbl-ch29_03, (accessed 03/10/2023)
19. OpenStaxCollege. *Illustration from Anatomy & Physiology, Connexions*. https://commons.wikimedia.org/wiki/File:Figure_28_02_09.jpg, (accessed 03/10/2023)
20. OpenStax. *Anatomy and Physiology*. https://openstax.org/books/anatomy-and-physiology/pages/28-6-lactation#tbl-ch29_03, (accessed 03/10/2023).
21. HarvardEdu. *Allometric Relations for Some Hemodynamic Parameters* [Online]. https://bionumbers.hms.harvard.edu/files/Allometric%20relations%20of%20some%20hemodynamic%20parameters.pdf (accessed 03/10/2023).

APPENDIX I

Allometric Relations for Some Hemodynamic Parameters
(*Source:* Adapted from Harvard.edu, 2023).[24]

Parameter	Y	A	b	Y^*	Units	Y =
Arterial pressure	P	$1.17*10^5$	0.033	134	(dynes/cm^2)	aW^b
Arterial pressure	–	–	–	100.8	mmHg	ρgh
Body surface area	–	–	–	1.73	m^2	
Cardiac output	–	–	–	4.62	l/min	VA
Heart rate	f_h	3.60	–0.27	1.17	(s^{-1})	aW^b
Heart weight	W_h	0.0066	0.98	0.40	(kg)	aW^b
Length of aorta	L	17.5	0.31	63.8	(cm)	aW^b
Metabolic rate	MR	$3.41*10^7$	0.734	73.1	(ergs/sec)	aW^b
Pulse velocity	c	446.0	0.0	446	(cm/s)	aW^b
Radius of aorta	r	0.205	0.36	0.92	(cm)	aW^b
Stroke volume	V_s	0.66	1.05	53.30	(mL)	aW^b
Total peripheral resistance	–	–	–	0.13	dynes-seg	

*Values were calculated for an adult with Weight (W) = 65 kg and height = 167 cm.

APPENDIX II

Effects of Elevation on Blood Properties

Altitude m	Country	Place	No. of subjects	Erytrocytes count million/mm^3 of blood	Blood hemoglobin concentration g/100 mL blood	Mean corpuscular hemoglobin
<395	United States	Kansas	350	5.11	15.0	29.4
		Kansas City	100	4.84	15.1	31.2
		New Orleans	115	5.26	15.6	29.7
			100	5.85	15.9	27.2
		Omaha	100	4.69	15.0	32.0
		Portland, Ore	259	5.42	15.8	29.2
	Argentina	Buenos Aires	50	5.50	15.4	28.0
			50	5.30	14.8	27.9
	Denmark	Copenhagen	60	5.07	15.0	29.6
	Germany	Giessen	40	4.96	16.0	32.3
		Jena	52	5.06	16.0	31.6
	Hawaii	Honolulu	137	5.08	15.1	29.7
	India	Bombay	121	5.11	15.4	30.1
		Calcutta	50	5.36	14.8	27.6
	Norway	Oslo	50	5.52	16.2	29.3
	Peru	Lima	100	5.26	15.7	29.8
			175	5.14 (4.46–5.82)	16.0 (14.4–17.6)	31.1
			14	5.00 (4.5–5.6)	15.1 (13.4–16.2)	30.2

APPENDIX II *(Continued)*

Altitude m	Country	Place	No. of subjects	Erytrocytes count million/mm^3 of blood	Blood hemoglobin concentration g/100 mL blood	Mean corpuscular hemoglobin
395	Argentina	Tucumán	153	5.31	16.1	30.3
500	Canada	Saskatchewan	20	5.52	15.6	28.3
	Switzerland	Zurich	139	5.00	15.0	30.0
1524	United States	Denver	40	5.42 (4.83–6.07)	16.5 (15.0–18.3)	30.4
			40	4.63 (4.41–5.00)	14.5 (12.7–15.7)	31.1
1768	South Africa	Johannesburg	60	5.99	14.7	24.5
1830–1890	India	Coonoor and Wellington	80	5.33 (3.95–6.04)	15.9 (13.1–20.3)	29.8
2300	India	Ootacamud	20	5.38 (4.75–6.02)	15.8 (14.8–16.8)	29.4
	Mexico	Mexico City	23, 21	5.26 (4.37–6.14)	14.2 (13.1–15.3)	27.0
			100	5.38 (4.53–6.17)	17.7 (14.4–20.1)	32.9
			100	5.01 (4.27–6.01)	15.2 (12.8–17.7)	30.3
3730	Peru	Oroya	40	5.67 (4.89–6.45)	18.8 (15.9–21.7)	33.2
4540	Argentina	Mina Aguilar	81	6.46 (5.07–9.43)	19.4 (15.7–24.9)	30.0
	Peru	Morococha	32	6.15 (5.01–7.29)	20.8 (17.3–24.2)	33.8
			11	6.70 (5.30–9.30)	19.3 (17.4–24.0)	28.8
5340	Chile	Quilchua	6	7.37	22.6	30.7

CHAPTER 2

Applications of Dimensional Analysis in Bioengineering*

MEGH R. GOYAL, CARLOS ORTIZ, JORGE RODRÍGUEZ, and CARLOS VAZQUEZ

2.1 INTRODUCTION

In practice, many fluid mechanics problems can be tackled by employing numerical solution of mathematical equations that demonstrates the fluid behavior or by approximate analytical solutions for a simplified version of the problem, but there are critical problems in real world that demands experimental observations or justifications for obtaining a reliable solution. In fact, the behavior of fluid (gas or liquid) is a collective response of probabilistic nature of molecules, which is too complex to be solved accurately using any mathematical version of the problem too date, yet at a scale quite larger than molecular scale, the mathematical interpretation often provides an insight on the collective behavior with some degree of accuracy. To overcome the associated error and/or to introduce the empirical error parameters in mathematical relations or predictive models, the input from experimental data is a key approach in many fluid mechanics experiments. Especially, the goal of any scientist/experimentalist is to make the results or empirical relations as simple as possible that could easily apply in many scenarios. To achieve this goal, the concept of dimensional analysis or similitude or similarity conditions is often employed in classical fluid

* This chapter has been modified from the review article prepared by students [Carlos Ortiz, Jorge Rodríguez and Carlos Vazquez] of Professor Goyal for the course on Fluid Mechanics, INGE 4015. Course instructor: Megh R. Goyal, PhD, PE, retired professor in Biomedical Engineering, General Engineering Department, University of Puerto Rico—Mayaguez.

Biofluid Dynamics of Human Body Systems: Expanded and Revised Edition.
Megh R. Goyal, Arka Bhowmik, & Anamika Chauhan (Eds.)

mechanics experiments, wherein measurements made on one system in the laboratory can be used to describe the behavior of other similar systems outside the laboratory.

This chapter focuses on the principles of dimensional analysis in the biological systems.

2.2 DIMENSIONS AND UNITS[1]

2.2.1 SCOPE OF DIMENSIONAL ANALYSIS

Dimensional analysis is a procedure of determining information of physical phenomenon that can be fully described by a dimensionally correct equation among system variables.[2,3] The dimensional analysis is a generalized method. With little effort, a partial solution to nearly any problem can be obtained. On the other hand, a complete solution is not possible under certain situations by dimensional reasoning alone.

The dimensional analysis can reduce the number of variables that are expressed in the form of dimensionless groups in the system. Evidently, reducing variables number while describing the phenomenon improves the information that is obtained from a few experiments and the dimensionless groups defining the system. It is consistent for all units. In practice, often resultant curve between the dependent and independent dimensionless groups can be defined by equations that are obtained by regression analysis.

2.2.2 FORCE AND MASS SYSTEM OF UNITS[1,4–8]

According to Newton's second law of motion, force is given as $F = ma$, which can also be written as $W = mg$ for a freely falling body with weight being W, m being mass and g being gravity. When $m = 1$ is set to 1, we can find that weight of unit mass is equal to g units of force. There is different system of units which are discussed under.

2.2.2.1 CGS SYSTEM

The letters CGS denote "centimeter," "gram," and "second," respectively. In CGS system, unit of mass is gram (= 0.001 kg). Mass is accurately measured by a balance-type scale. For consistency with Newton's Second Law of

Motion, the unit of force in the CGS system is the force (in dyne), which is needed to move one-gram mass at an acceleration of one cm/s^2. Similarly, one-gram of mass exerts nearly 981 dynes force. Therefore, work in CGS unit is defined by unit "dyne centimeter (= one erg)."

2.2.2.2 MKS MASS SYSTEM

The letters MKS denote "meter," "kilogram," and "second," respectively. The mass is defined in the units of kilogram (kg) in MKS system. In fact, we can define the force in unit called Newton (= 1000 dynes = 0.225 lb) that is needed to move one kg of mass at an acceleration of one m/s^2. Similarly, work is defined by unit Newton–meter (= one Joule). Whereas, power can be defined by unit watt (= Joule/second).

2.2.2.3 MKS FORCE SYSTEM

In the MKS force system, the kilogram force is a unit of force, which is 980,665 dynes also designated as weight of 1 kg mass under acceleration due to gravity. Therefore, based on $F = ma$, the unit of mass in the MRS force system is a "kilogram second^2 per meter" (kg s^2/m). The equation $W = mg$ shows that one kg s^2/m is a mass that weighs about 9.81 kg force. Table 2.1 indicates parameters in the System of International (S.I.) units.

TABLE 2.1 List of Parameters that is Used for Defining a Physical Process in International System of Units.

Physical quantity with symbols	S.I. Units with symbols	Alternate units with symbols
Displacement (*d*)	Meter (m)	Centimeter (cm) foot (ft)
Mass (*m*)	Kilogram (kg)	Gram (g)
Time (*t*)	Second (s)	Hour (h) Mean solar day (dy) Solar year (yr)
Current (*I*)	Ampere (A)	Statampere (statA) Abampere (abA)
Temperature (*T*)	Kelvin (K)	Degree celsius (°C) Degree fahrenheit (°F) Degree Rankine (R)
Amount of substance (*N*)	Mole (mol)	–
Luminous intensity (*B* or *L*)	Candela (cd)	–

2.2.2.4 BRITISH MASS SYSTEM

In this system, mass is defined by unit pound (= 0.4536 kg mass). According to $F = ma$, the unit of force is a force that is needed to move one-pound mass at an acceleration of one foot per square second. In fact, the force is also defined by unit pound. Based on $g = 32.174$ ft/s^2 and $W = mg$, the weight 32.2 pounds is equivalent to one-pound mass on the earth.

2.2.2.5 AMERICAN ENGINEERING SYSTEM (US CUSTOMARY UNITS)

Engineers in United States defines force by unit pound (= 0.4536 kg force). Based on $F = ma$, the unit of mass is then a "pound square second per foot" (lb s^2/ft). This unit of mass is called a "slug," which is equivalent to 32.2 pounds in earth. Similarly, spring scale measurement of one pound quantity of matter is referred to as "pound of matter."

For example, in hydraulics, "head" is an energy (= foot pounds of energy per pound of fluid) called as energy head (ft or m). Lord Rayleigh remarked the on the omni-presence of the effect of gravity in measurement of every physical quantity measured in earth. An example to illustrate Rayleigh's point is the expression of Bernoulli's equation in terms of heads. For a horizontal vessel between two points (inlet and exit), the energy equation is expressed as:

$$\frac{P_1}{\rho} + \frac{V_1^2}{2} + g\,z_1 = \frac{P_2}{\rho} + \frac{V_2^2}{2} + g\,z_2 \tag{2.1}$$

where each term is called pressure head, velocity head, and static head, respectively.

For a static fluid, fluid velocity is zero. Therefore, eq 2.1 reduces to: $P_1 = P_2 + \rho gh$. Each term in eq 2.1 can be interpreted as energy per unit volume of fluid. Now, dividing by density (ρ), both sides led to the definition of energy per unit fluid mass.

2.2.3 PHYSICAL DIMENSIONS

When the units of force, length, and time are given, the unit of mass is uniquely determined by Newton's Second Law of Motion, $F = ma$. In this case, the system of measurement is called a "force system." J. Clerk Maxwell employed symbols such as [F], [M], [L], [T], and [θ] to denote force, mass, length, time, and temperature, respectively. He formed products of powers of these symbols, which he called as "dimensions." These are the only characteristic dimensions based on which dimensional analysis is developed.

The dimensions of a physical parameter follow from its definition or from a physical law. For example, a velocity (u) is a change in distance (x) in a unit time (t), i.e., $u = dx/dt$. Therefore, the dimensions of velocity are [L/T] or [LT^{-1}]. Similarly, the dimensions of acceleration are [L/T^2] or [LT^{-2}]. Based on these dimensions, units of velocity are: feet/second, centimeters/second, etc.; and units of acceleration are: feet per second,^{2} centimeters per second,^{2} miles per hour,^{2} etc.

Because the dimensions of force and acceleration are [F] and [LT^{-2}], respectively, hence, $F = ma$ under force system will give us the dimensions of mass as [FT^2L^{-1}]. On the other hand, force has the dimensions of [MLT^{-2}] in the mass-system. To convert dimensions from the force system to the mass system, the dimensional term [MLT^{-2}] is substituted for [F]. The dimensional relationships among [F], [M], [L], and [T] indicate the relationships among the corresponding units, e.g.,

$$1\,lb = 1\,slug\frac{ft}{s^2} \quad 1\,slug = 1\,lb\frac{s^2}{ft} \quad 1\,dyne = 1\,g\frac{cm}{s^2} \quad 1\,g = 1\,dyne\frac{s^2}{cm}$$

Here, the pound is considered to be a unit of force, and unit of mass is gram. The dimensions of density (mass/volume) are [ML^{-3}] or [$FL^{-4}T^2$]. Therefore, units of mass density are also expressed as slug/ft^3 or lbs^2/ft^4. Pressure and stress can be expressed as force/area that have the dimensions [FL^{-2}] or [$ML^{-1}T^{-2}$]. Based on Newton's Law of Viscosity, dynamic viscosity (μ) has dimensions as [$FL^{-2}T$] or [$ML^{-1}T^{-1}$]. Then, the kinematic viscosity ($\nu = \mu/\rho$) will have dimensions of [L^2T^{-1}]. The symbol [θ], denoting temperature, is regarded as one of the fundamental dimensions. The dimension of an arbitrary variable z is denoted by [z]. If z is dimensionless, this fact may be denoted by [z] = [1]. The dimension of an integral, $\iiint (y\,dx)$, is [y dx], [y] [dx], [L] [L], or L^2. Table 2.2 shows dimensions and units of selected physical parameters.

2.3 CONVERSION OF UNITS OF MEASUREMENT

Dimensional expressions enable us to convert units readily from one system of measurement to another. For example, we want to covert an acceleration of 20 cm/s^2 to miles per hour.[2] Set 1 cm = 6.2137 × 10^{-6} mi and 1 sec = 2.77 × 10^{-4} h. Then, the following example can be used in an alternate case:

$$20\left[\frac{cm}{s^2}\right]\left[\frac{6.213\times10^{-6}\,miles}{1\,cm}\right]\left[\frac{1\,s}{2.777\times10^{-4}\,hr}\right]^2 = 1610\frac{miles}{hr^2}$$

2.4 DIMENSIONAL ANALYSIS

In Bioengineering, dimensional analysis[1,3] reduces total number of problem variables to the smallest number of essential dimensionless groups. These dimensional parameters are expressed in fundamental units "length," time, and mass (MLT), or in terms of force, length, and mass (FLM). These parameters are defined by units that are powers of fundamental units, such that, in dimensional analysis:

- Addition and substruction apply to only quantities of same dimension.
- The left-hand and right-hand side of the equation should have same dimensions.
- Two dimensionless groups can be multiplied or divided to give us a new group.
- Any dimensionless group can be inverted to give us a new group.
- The constants of exponential, trigonometric, and logarithmic functions are dimensionless numbers.

2.4.1 BUCKINGHAM Π (PI)-THEOREM

Buckingham π-Theorem (BPT) states that "*If an equation is dimensionally homogeneous, it can be reduced to a relationship among a complete set of dimensionless products*."[3,9] This theorem provides a way to simplify the complex system to number of important dimensionless parameters. Based on this theorem, every physical equation with n-variables can be rewritten in the form generalized predication equation that includes n–m dimensionless parameters, such that quantity of fundamental dimension is denoted by m. Even if the form of the equation is not known, it provides an approach to compute dimensionless parameters.

For example, the variables involved for a moving object in the fluid are: u, ρ, v, A, and F, which are velocity, density, viscosity, frontal area, and force, respectively. The π-theorem helps us to reduce five variables to two dimensionless groups (drag coefficient and Reynolds number). Here, $n = 5$

and $m = 3$ number of dimensions. We can write a general unknown function, "*f*," as follows:

$$f(F, u, A, \rho, \upsilon) = 0 \tag{2.2}$$

TABLE 2.2 Dimensions of Selected Physical Parameters.[1]

Physical parameter	FLT system	MLT system
Acceleration	LT^{-2}	LT^{-2}
Angle	$F^{\circ}L^{\circ}T^{\circ}$	$M^{\circ}L^{\circ}T^{\circ}$
Angular acceleration	T^{-2}	T^{-2}
Angular velocity	T^{-1}	T^{-1}
Area	L^{2}	L^{2}
Density	$FL^{-4}T^{-2}$	ML^{-3}
Energy	FL	$ML^{2}T^{-2}$
Force	F	MLT^{-2}
Frequency	T^{-1}	T^{-1}
Heat	FL	$ML^{2}T^{-2}$
Length	L	L
Mass	$FL^{-1}T^{2}$	M
Modulus of elasticity	FL^{-2}	$ML^{-1}T^{-2}$
Moment of a force	FL	$ML^{2}T^{-2}$
Moment of inertia (area)	L^{4}	L^{4}
Moment of inertia (mass)	FLT^{2}	ML^{2}
Momentum	FT	MLT^{-1}
Power	FLT^{-1}	$ML^{2}T^{-3}$
Pressure	FL^{-2}	$ML^{-1}T^{-2}$
Specific heat	$L^{2}T^{-2}\,\theta^{-1}$	$L^{2}T^{-2}\,\theta^{-1}$
Specific weight	FL^{-3}	$ML^{-2}T^{-2}$
Strain	$F^{\circ}L^{\circ}T^{\circ}$	$M^{\circ}L^{\circ}T^{\circ}$
Stress	FL^{-2}	$ML^{-1}T^{-2}$
Surface tension	FL^{-1}	MT^{-2}
Temperature	Θ	θ
Time	T	T
Torque	FL	$ML^{2}T^{-2}$
Velocity	LT^{-1}	LT^{-1}
Viscosity (dynamic)	$FL^{-2}T$	$ML^{-1}T^{-1}$
Viscosity (kinematic)	$L^{-2}T^{-1}$	$L^{2}T^{-1}$
Volume	L^{3}	L^{3}
Work	FL	$ML^{2}T^{-2}$

*In SI system of units, F = N, M = Kg, L = m, T = s, θ = K.

It can be seen that the equation is set to zero in any system of units. Hence, a relation can be expressed by arranging the "f" in terms of $n - m = 5 - 3 = 2$, dimensionless groups. Buckingham π theorem states that there will be only two such groups in this example. This π – theorem has the following restrictions:

- The list of independent dimensions should not contain redundant dimensions.
- If two dimensions occur only in a specific ratio, then they are not independent and must be treated as one dimension.

Let us consider that certain force depends on velocity, density, viscosity, and two spatial lengths. The question is "How many numbers of non-dimensional groups needed for this problem?". We demonstrate three choices which will give the same results.

1. If we choose our dimensions (FLT) as force, length, and time, then we must express the density (ML^{-3}) in dimensions of force x $(\text{time})^2/(\text{length})^4$ and the viscosity in $(\text{force} \times \text{time})/(\text{length})$.[2] In this case, $n + 1 = 6$ and $k = 3$. Is it possible to identify parameters that cannot be integrated in dimensionless group? The answer is force, one of two lengths, and the density cannot be combined; and these dimensions are independent. We obtain that the number of dimensionless groups are $6 - 3 = 3$.
2. In case of dimensions (MLT) is mass–length–time, then we must express the force in dimensions of mass $\times$ $\text{length}/(\text{time})^2$ and the viscosity as mass divided product of length and time. Therefore, $n + 1 = 6$, and $k = 3$. The three dimensions are independent, and we can observe that the number of dimensionless groups are $6 - 3 = 3$.
3. We may also include our dimensions (FMLT) as force, mass, length, and time. In this case, we must add g_c to our list of parameters. Now we can select a length, force, density, and g_c as four parameters, which cannot be combined into any dimensionless group. Therefore, all four dimensions are independent, and the obtained number of dimensionless groups are $7 - 4 = 3$.

The following procedure provides an algebraic method for selecting groups:

1. Select k variables, which cannot be combined into any dimensionless group. These are called repeating variables.

2. Then the combination of the repeating variables with any of the non-repeating variables can form a dimensionless group and we shall make up n + 1 – k such groups from the variables. This can often be done by inspection or it can be done by obtaining a solution of equations that lead to the required non-dimensional groups.

In the above example, we presented one of the methods. In this choice of dimensions, the independent dimensions (FLT) are length, force, and time. Next, the below table for variable can be constructed with their dimensions.

Step 1: Form a table of physical parameters for the system, e.g., above example:

Variable, *n*	Description	Dimensions [FLT]
A	Force, F_1	F
B_1	Velocity, V	L/T
B_2	Density, ρ	FT^2/L^4
B_3	Viscosity, μ	FT/L^3
B_4	Length, L_1	L
B_5	Length, L_2	L

A = Dependent variable; B_1 *..... B_5 = are independent variables; n = number of variables, m = number of dimensions.*

$n - m = 6 - 3 = 3$, according to π-theorem

Step 2: We will use "Method of repeating variables."

Select any three repeating variables. Let us select force (A), density (B_2), and length (B_4) as repeating variables.

Step 3: Apply π – theorem to decide how many groups are needed; $n - m = 6 - 3 = 3$.

Step 4: Select variables to form three groups by using variables, e.g., A, B_2, B_4, B_1; A, B_2, B_4, B_3; and A, B_3, B_4, B_5.

Step 5: Form the groups using Power Method as follows:

$$\pi_1 = (A)^a (B_2)^b (B_4)^c (B_1)^d \tag{2.3}$$

Step 6: Find values of exponents *a* to *d*. Because the groups are non-dimensional and can be written as:

$$F^0L^0T^0 = (F)^a \left(\frac{FT^2}{L^4}\right)^b (L)^c (L/T)^d \tag{2.4}$$

We have three algebraic equations by equating exponents of F, L, and T on both sides. This can be true only if the following three simultaneous equations are satisfied:

$$\begin{aligned} &\text{Equation for length:} && 0 = -4b + c + d \\ &\text{Equation for time:} && 0 = 2b - d \\ &\text{Equation for force:} && 0 = a + b \end{aligned} \tag{2.5}$$

Solving three equations for *a*, *b*, and *c* (one can solve for any three constants), we get:

$$a = -\,d/2,\ b = d/2 \text{ and } c = d.$$

Now, we may select an arbitrarily value for *d*, e.g., we select $d = 2$ as the lowest value of *d*, which avoids fractional exponents. Then our three dimensionless groups from eq 2.3 by inserting "*a, b, c,*" we get:

$$\pi_1 = (A)^a (B_2)^b (B_4)^c (B_1)^d$$

$$\pi_1 = \frac{\rho L^2 V^2}{F_1} \quad \pi_2 = \frac{\mu^2}{F_1 \rho} \quad \text{and} \quad \pi_3 = \frac{L_1}{L_2} \tag{2.6}$$

where: π_3 is the ratio of the two lengths; and π_2 can be shown to be equal to [2/(Reynolds number × pressure coefficient)].

Examining this set of π-terms, most experienced bioengineers will select dimensionless groups as: $[\pi_3/\pi_2]$ and $[(\pi_1/\pi_2)^{0.5}]$, i.e., the length ratio, the pressure coefficient, and Reynolds number. The justification for this procedure is that a different selection of the repeating variables will give us a different set of π-terms. However, the different sets are all products or ratios of the first set. The dimensions of physical parameters are shown in Table 2.2. Biofluid mechanics are a combination of theoretical and experimental results.

2.5 STANDARD DIMENSIONLESS PARAMETERS[3,10]

Standard π groups can be interpreted as a ratio of typical values of different physical parameters (force or energy). Table 2.3 indicates different π-groups.

1. The ratio of inertial force to viscous force is defined as Reynolds number (Re):

$$\text{Re} = \frac{\rho V D}{\mu}$$

2. Pressure coefficient (C_p) is a ratio of pressure force of inertial force:

$$C_p = \frac{\Delta p}{\frac{1}{2}\rho V^2}$$

3. Froude number (Fr) is a ratio of inertial force to gravitational force:

$$Fr = \frac{V^2}{2gh}$$

4. Strouhal number is a ratio of flow time to oscillation time:

$$St = \frac{\omega L}{V} = \frac{(L/V)}{(I/\omega)}$$

2.6 THERMODYNAMICS

The various laws governing the energy transformation and conservation of energy is referred to as thermodynamics. There are seven forms of key energy in the universe: kinetic, potential, heat, electrical, chemical, electromagnetic (or radiant energy), and nuclear energy. The thermodynamic laws are true for energy transformation and conversion of any form. Except for special cases where matter can be transformed to energy or vice-versa. The electromagnetic energy is a unique form of energy that exists without matter. In fact, all other forms of energy originate from matter or transport through matter only. The conversion process of energy or the concept of thermodynamics for efficient conversion of energy is used in day-to-day applications for instance conversion process of kinetic energy in the reciprocating engine to heat energy of combustion process, or the conversion process of kinetic energy of hydro- or wind- power plant to electrical energy for power generation. Thermodynamics also allow us to identify the propensity of any chemical reactions to work or not and the direction of reaction, but it does not allow us to predict the rate of reaction. The principles of thermodynamics laws is also found in human body system such as efficient energy conversion by cells from consumed food to bioheat and muscular work, or biomass transport to organelles in the form of nutrient and oxygen, via., by blood transport. It is governed by four basic laws:

- **Zeroth Law:** It states that in case of three body system, all are in thermal equilibrium with each other if either of the two system is in thermal equilibrium with third system.

TABLE 2.3 Standard Dimensionless Groups.[1]

Parameter	Definition	Qualitative ratios	Importance
Biot number	$B_i = hL/K$	Resistance offered to thermal conduction in the body to resistance offered to thermal convection at the surface of the body.	Conduction flows involving surface convection effects.
Cavitation number (Euler number)	$E_u = [\Delta P]/[\rho V^2]$	Pressure force on internal force.	Flows in which pressure drop is significant.
Eckert number	$E_c = [U^2]/C_P\, T_0$	Kinetic energy to enthalpy.	Dissipation.
Froude number	$F_r = [V^2]/gL$	Inertial force on gravity force.	Flow with a free surface.
Grashof number	$G_r = \beta\Delta\theta g L^3 \rho^2 c_p/[\mu^2]$	Buoyant force on viscous force.	Free convection flow.
Mach number	$M_a = V/a$	Inertial force of compressible flow.	Compressible flow.
Nusselt number	$Nu_L = hL/K$	Temperature gradient at the surface.	Measure of the convection coefficient.
Prandtl number	$P_r = C_p\mu/\kappa$	Momentum to thermal diffusivities.	Define fluid property.
Rayleigh number	$Ra_L = \beta\Delta\theta g L^3 \rho^2/[\mu\kappa]$	Product of Grashof and Prandtl number.	Free convection flow.
Reynolds number	$R_e = [\rho v D]/\mu$	Inertial force on viscous force.	Always, forced convective flow.
Roughness ratio	ε/L	Wall roughness to body length.	Turbulent, rough walls.
Specific heat ratio	$\gamma = [C_p]/[C_V]$	Ratio of enthalpy to internal energy.	Compressible flow.
Strouhal number	$S_t = [\omega L]/V$	Centrifugal force on internal force.	Flow with a steady component that repeats itself periodically.
Temperature ratio	$T_r = [\theta]/[\theta_a]$	Body temperature on ambient temperature.	N/A
Weber number	$W_e = [V^2 L\rho]/\sigma$	Internal force on surface tension force.	Surface tension influences the flow with an interface.

- **1st Law:** Energy can neither be created nor destroyed, it can only be changed. The sum of heat flowing into a system and work done by the system is zero.
- **2nd Law:** It states that it is impossible to obtain a process so that the unique effect is the subtraction of a positive heat from a reservoir and the production of a positive work. The entropy of a closed system never decreases.
- **3rd Law:** All processes cease as temperature approaches zero. As temperature goes to 0, the entropy of a system approaches a constant value.

Let us assume $U, T, S, P, V, \rho, F, H, G, \mu_1$, and N is internal energy, temperature, entropy, pressure, volume, density, Helmholtz free energy, enthalpy, Gibbs free energy, chemical potential, and particle number, respectively.

Helmholtz energy and the Gibbs energy: $G = U - TS + PV$

Enthalpy: $H = PV + U$

The state for atom/molecule is defined by measurable parameters such as temperature, pressure, and chemical potentials, except in case of phase change and spontaneous symmetry breaking in the ordered phase.

2.6.1 THERMODYNAMIC SYSTEM

It defined region of universe covered by a control volume. In fact, a boundary, either real or imaginary, separates it from the universe. The region outside the system boundaries is identified as surroundings. Hence, a thermodynamic system can be defined as follows:

- **Isolated systems** are systems that neither exchange heat nor matter with its surroundings.
- **Closed systems** are systems which can exchange heat but not matter with its surroundings.
- **Open systems** are system that can exchange both heat and matter with surroundings.

2.6.2 THERMODYNAMIC STATE

Thermodynamic state is defined by a set of measurable quantities. In order to achieve the thermodynamic state, the value of these measurable quantities need to have same or unique value. As the state of the system changes, the value of these quantities (also referred to as properties) varies. It is important to note, that the property's value is independent of path by which the system arrives to the specific state. The Gibbs phase rule provides a relation to determine the number of properties desired to define the state of a specific system.

2.7 NON-DIMENSIONAL AND SIMILITUDE MODELING OF CARDIOVASCULAR SYSTEM

Thomas H. Dawson (Engineering Professor in Biomechanics at Department of Naval Architecture and Ocean Engineering—United States Naval Academy, US Government) published his first book titled *Engineering Design of the Cardiovascular System of Mammals* (1991, Prentice Hall).[11] Since then, he and other investigators[1,2,4–7,10–17] have used non-dimensional and similitude modeling to investigate *Allometric Relations and Scaling Laws for the Cardiovascular System of Mammals.* Investigators concluded that the heart weight in an adult mammal bears a constant proportion to its body weight. Heat flux distributions for a carotid bifurcation with severe stenosis has also been simulated.[11] The study reviewed the dimensional analysis method for the cardiovascular system of mammals; and he has suggested therapeutic drug dosage requirements for children and adults based on these models. Dawson[11] indicated following generalized equation for liquid flow (Q_b) with five dimensionless groups:

$$\left(\frac{Q_b}{w\alpha^2 l}\right) = f\left(\left[\frac{\rho L_a w^2 \alpha^2 L}{r_a^2 E}\right], \left[\frac{\mu L_c w^2 \alpha^2 L}{n_c r_c^4 E}\right], \left[\frac{\mu L_v w^2 \alpha^2 L}{r_v^4 E}\right], \left[\frac{F_v}{hlE}\right]\right) \quad (2.7)$$

Where, number of arterial and venous vessels are constant and each of the same size.

Q_b = cardiac output of blood
w = heart rate
l = ventricle length
a = arterial vessel
ρ = blood density
f = generalized function
E = modulus of elasticity
$w^2\alpha^2 l_o/hl$ = ventricle volume
F_o/hl = Contractile stress
r_a = radius of arterial vessel
r_c = radius of capillary vessel
r_v = radius of venous vessel
L_a = length of arterial vessel
L_c = length of capillary vessel
L_v = length of venous vessel
μ = blood viscosity
n_c = number of capillary vessels
F_o = ventricle wall force with periodic contractions; Note, F_o is same as F_v

2.8 NUMERICAL EXAMPLES[1]

2.8.1 *EXAMPLE 1:*

Investigate the pressure drop across a round pipe (see Figure 2.1).

$$gh_f \sim \frac{p_1 - p_2}{\rho}$$

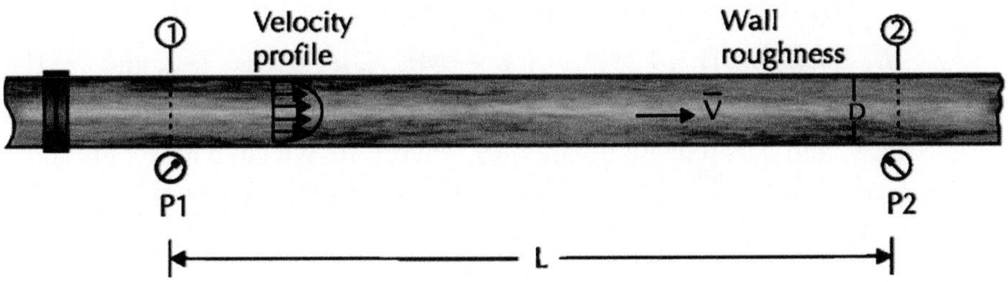

FIGURE 2.1 Pressure drop across a round pipe.

Source: Adapted from Ref. [1].

Physical variables that define this system in Figure 2.1 are: pipe length (L), pipe diameter ($D = 2\ r$), fluid velocity (V), surface wall roughness (ε), fluid density (ρ), fluid viscosity (μ), and the pressure drop ($P_2 - P_1 = gh_f$, where g and h_f are gravity and friction loss). Therefore, we can write the governing equation as:

$$gh_f = f\left(L, D, V, \rho, \mu, \varepsilon\right) \tag{2.8}$$

The objective of this experiment is to find the function f, where f need not be an analytical function. If we vary each parameter for 10 data points, we require 10^6 experiments including the experiments with fluid with different values of ρ and μ. Also, anyone using the data would have to make a six-parameter interpolation. To reduce the number of parameters, we take advantage of Dimensional Analysis. Any dimensionally homogeneous equation or process can be written in the nondimensional form. We follow the following steps:

1. Important parameters of the system are:
   Flow properties: V, gh_f
   Geometry: L, D, ε
   Fluid properties: ρ, μ

2. Determine the number of dimensionless groups as follows:
   - Select as repeating variables "K" of the dimensional parameters with all of the fundamental dimensions.
   - Do not use the dependent parameter (gh_f) as a repeating variable.
   - Do not use a parameter that may be an important only temporarily: μ, ε.
   - Select parameters, whose dimensions are as close to "pure" as possible.

   The independent dimensions of parameters are as follows:

   $$L = [L], D = [L], V = [LT^{-1}], \rho = [ML^{-3}], \mu = [ML^{-1}T^{-1}], \text{ and } \varepsilon = [L]$$

   The Buckingham π theorem (BTP) yields K = 7-3 = 4 and we choose D, V, and ρ as repeating variables. Therefore, we have four π groups.
3. Create a dimensionless form of remaining dimensional parameters using the repeating variables:
   - Take the remaining parameters and combine them with repeating variables (g_f, L, μ, ε) to form the four π groups. Start with $\pi_1 = (gh_f)\rho^a V^b D^c$. Find a, b, and c so that π_1 is dimensionless:

     $$[\pi_1] = [M]^o[L]^o[T]^o = [L^2T^{-2}]\,[ML^{-3}]\,[LT^{-1}]^b[L]^c$$
     $$= [M^a]\,[L]^{2-3a+b+c}\,[T]^{-2-b}$$

     Next, equating M, L, T on both sides leads us to three simultaneous equation having three variables to determine:

     $$a = 0,\ 2 + b = 0, \text{ and } 2 - 3a + b + c = 0$$

     Solving these equations for three unknowns, we get values of exponents a = 0, h = −2, and c = 0. This will lead to $\pi_1 = (gh_f)/V^2$. In a similar approach, we can determine π_2 as

     $$[M]^o[L]^o[T]^o = L\rho^a V^b D^c = [M]^a[L]^{1-3a+b+c}[T]^{-b}$$

     Next, equating the exponents of M, L, T on both sides lead us to a =0, b = 0, and 1–3a +b + c = 0.

     This means a = 0, b = 0, c = –1.

     Substituting these values of exponents "a, b, c" we obtain our next dimensionless group $\pi_2 = L/D$ and $\pi_3 = \mu\rho^a V^b/D^c$ will yield a = −1, b = −1, and c = −1, which yields: $\pi_3 = \mu/\rho VD$ and $\pi_4 = \varepsilon/D$.
   - Now, the generalized equation $\pi_1 = f(\pi_2, \pi_3, \pi_4)$ or

     $$\frac{gh_f}{V^2} = f\left(\frac{L}{D}, \frac{\mu}{\rho VD}, \frac{\varepsilon}{D}\right)$$

4. Next, rearrange the π groups to correspond with customary usage of dimensionless parameters. It is to be noted this method will not provide a unique set of π terms. Therefore, we are allowed to multiply π terms by a constant, raise π terms to any positive or negative power, and multiply any power of a π term with another term. This gives us

$$\frac{gh_f}{\frac{1}{2}V^2} = 2\left(\frac{gh_f}{V^2}\right) \text{ and } \frac{\rho VD}{\mu} = \left(\frac{\mu}{\rho VD}\right)^{-1}$$

$$\text{and } 2\left(\frac{gh_f}{V^2}\right) = \phi\left(\frac{L}{D}\right)^{n2} \cdot \left(\frac{\rho VD}{\mu}\right)^{n3} \cdot \left(\frac{\varepsilon}{D}\right)^{n4} \qquad (2.9)$$

Here, a general form of equation for the system is given by eq 2.9, where the constants (φ, n_2, n_3, and n_4) must be determined.

2.8.2 *EXAMPLE 2*

Blood flow in a human artery[18] can be assumed as flow in a horizontal small diameter tube, which is a function of:

- The pressure loss per unit length (Δp_l)
- Volume rate of flow (Q),
- The diameter (D)
- The blood viscosity (μ).

For a series of test runs for $D = 2$ mm and $\mu = 0.004$ N-s/m^2, following data of Q and Δp were measured (see Table 2.4) along the length, $L = 300$ mm. The test run input physical parameters are defined in Table 2.5. The heat loss is due to friction between the blood and thin vessel. The objective of the problem is to predict the effects of these variables on head loss using dimensional analysis. We are given with following parameters $D = 0.002$ m; $\mu = 0.004$ N-s/m^2; $L = 0.3$ m; $\gamma = \rho g = 10{,}369$ N/m^3; and A = tube area = $(\pi/4)\,D^2 = 3.14 \times 10^{-6}$ m^2. Note, the values of Q and Δp for 5 test runs can be found in Table 2.4, and hydrostatic laws of variation of pressure is given as "$\Delta p = \gamma \Delta h$."

- The head loss depends directly on these parameters (see Table 2.5). In addition, the head loss decreases with an increase in density and viscosity.[6,7]

TABLE 2.4 Test Run.

Test #	Q (m³/s)	ΔP (N/m²)
1	3.6×10^{-6}	1.1×10^{4}
2	4.9×10^{-6}	1.5×10^{4}
3	6.3×10^{-6}	1.9×10^{4}
4	7.9×10^{-6}	2.4×10^{4}
5	9.8×10^{-6}	3.0×10^{4}

TABLE 2.5 Physical Parameters Affect the Test Run.

Symbol	Parameter, n	Dimensions, m	Units
Δh	Pressure head loss	L	m
V	Velocity of flow	LT^{-1}	m/s
D	Pipe diameter	L	m
L	Pipe length	L	m
μ	Blood viscosity	FTL^{-2}	Ns/m²
ρ	Blood density	$FL^{-4}T^{2}$	kg/m³

- According to Buckingham's Pi-Theorem from Table 2.5, n = number of parameters = 6; m = number of dimensions = 3. Therefore, number of π terms are $n - m = 6 - 3 = 3$. This means there are three dimensionless groups.
- Next, we formulate π groups using Power method as

$$[F]^{o}[L]^{o}[T]^{o} = h^{a}L^{b}D^{c}V^{d}\mu^{e}\rho^{f} \tag{2.10}$$

or, $$F^0L^0T^0 = L^a L^b L^c (LT^{-1})^d (FL^{-2}T)^e (FL^{-4}T^2)^f$$

or, $$F^0L^0T^0 = L^a L^b L^c L^d L^{-2e} L^{-4f} F^e F^f T^{-d} T^e T^{2f} \tag{2.11}$$

- Next, equating exponents of F, L, T on both sides, we get:

$$0 = e + f,\ 0 = 2f + e - d, \text{ and } 0 = a + b + c + d - 2e - 4f$$

- Next, we solve these equations for three variables and get values for e, d, and a:

$$e = -f,\ d = f, \text{ and } a = -b - c + f$$

- Substituting values in eq 2.11, we get

$$F^0L^0T^0 = h^{-b-c+f}L^bD^cV^f\mu^{-f}\rho^f \text{ or } 1 = h^{-b-c+f}L^bD^cV^f\mu^{-f}\rho^f$$

and by rearranging the terms lead us to a form as under

$$1 = \left[\frac{L}{h}\right]^b \left[\frac{D}{h}\right]^c \left[\frac{\rho V h}{\mu}\right]^f$$

- Next, regrouping let us to form new π groups as

$$\pi_{1,new} = \frac{h}{L},\quad \pi_{2,new} = \frac{D}{h}\bullet\frac{h}{L} = \frac{D}{L} \text{ and } \pi_{3,new} = \left[\frac{\rho V h}{\mu}\right]\left[\frac{D}{h}\right] = \left[\frac{\rho V D}{\mu}\right]$$

- Next step is to check whether these π groups are non-dimensional number or not. In the present case, all the π groups are non-dimensional quantities as shown below

$$\pi_1 = \frac{h}{L} = \frac{[L]}{[L]} = [L]^0$$

$$\pi_2 = \left[\frac{\rho V D}{\mu}\right] = \frac{\left[ML^{-3}\right]\left[LT^{-1}\right][L]}{\left[MLT^{-2}\right]\left[L^{-2}T\right]} = M^0L^0T^0$$

$$\pi_3 = \frac{L}{D} = [L]^0$$

- Finally, a generalized prediction equation can be written as

$$\pi_1 = C_1\left[\pi_2\right]^{a_3}\left[\pi_3\right]^{a5} \tag{2.12}$$

$$\frac{\Delta h}{L} = C_1\left[\frac{\rho V D}{\mu}\right]^{a_3}\left[\frac{L}{D}\right]^{a5} \tag{2.13}$$

- The final goal is to determine the unknown parameters C_1, a_3, a_5 in eqs 2.12 and 2.13 using the given data in Table 2.6.

TABLE 2.6 Determination of C_1, a_3, a_5.

Test #	Q (m³/s)	Δh = Δp/ρg	V (m/s) = Q/A	$\text{Re} = \frac{\rho V D}{\mu} = \pi_2$	L/D	$\frac{\Delta h}{L} = \pi_1$	ln π_1	ln π_2
1	3.60E–06	1.06	1.1459	600.616	150	3.54	1.2641	6.4062
2	4.90E–06	1.45	1.5597	824.311	150	4.82	1.5027	6.7145
3	6.30E–06	1.83	2.0054	1059.829	150	6.11	1.8099	6.9658
4	7.90E–06	2.31	2.5146	1328.992	150	7.72	1.9768	7.1921
5	9.80E–06	2.89	3.1194	1648.622	150	9.64	2.2659	7.4076

so that

$$\frac{\Delta h}{L} = C_1\left[\frac{\rho VD}{\mu}\right]^{a3}\left[\frac{L}{D}\right]^{a5} \quad \text{and} \quad \begin{array}{l} \left.\frac{h}{L}\right|_3 = C_1\left.\left(\frac{\rho VD}{\mu}\right)^{a_3}\right|_3 \left.\left(\frac{L}{D}\right)^{a5}\right|_3 \\ \left.\frac{h}{L}\right|_5 = C_1\left.\left(\frac{\rho VD}{\mu}\right)^{a_3}\right|_5 \left.\left(\frac{L}{D}\right)^{a5}\right|_5 \end{array}$$

- In the above set of equations, dividing second by first:

$$\frac{\left.\frac{\Delta h}{L}\right|_5}{\left.\frac{\Delta h}{L}\right|_3} = \left[\frac{\left.\left(\frac{\rho VD}{\mu}\right)\right|_5}{\left.\left(\frac{\rho VD}{\mu}\right)\right|_3}\right]^{a3}$$

$$a_3 = \left[\frac{\ln\left.\left(\frac{\Delta h}{L}\right)\right|_1 \Big/ \left.\left(\frac{\Delta h}{L}\right)\right|_2}{\ln\left.\left(\frac{\rho VD}{\mu}\right)\right|_1 \Big/ \left.\left(\frac{\rho VD}{\mu}\right)\right|_2}\right] \quad \text{and}$$

$$a_3 = \left[\frac{\ln\left(3.54\right)\big|_1 / \left(4.82\right)\big|_2}{\ln\left(605.61\right)\big|_1 / \left(824.31\right)\big|_2}\right] = 1.00$$

- We can also find C_1, a_3, a_5 by regression analysis: for constant $L/D = 150$, we have $\pi_1 = C_1 [\pi_2]^{a3}$. Taking the natural logarithm of both sides allows us to form an equation of straight line: $\ln \pi_1 = \ln C_1 + a_3 \ln \pi_2$ equivalent to $y = c + mx$. Here, the slope of the equation is given by a_3 and C_1 is antilog of c (see Figure 2.2).
- Next, analyzing Figures 2.2 and 2.3, we get $a_3 = 1$; and $C_1 = \text{Antilog}\ [-5.13] = 7.4 \times 10^{-6}$.

Finally, we get the following component equation:

$$\frac{\Delta h}{L} = 7.4 \times 10^{-6}\left[\frac{\rho VD}{\mu}\right]^{1}\left[\frac{L}{D}\right]^{a5} \tag{2.14}$$

2.8.3 EXAMPLE 3

Laplace's law[4,5,17] states that pressure difference (Δp) across a curved membrane in a state of tension is equal to membrane tension force (T) divided by its radius of curvature (r); or $T = p \times r$. The ventricles are considered as a thin-walled or thick-walled sphere or ellipsoid. Use "dimensional analysis" to predict the tension in the muscle, when the ventricle is considered as an ellipsoid, sphere, and cylinder.

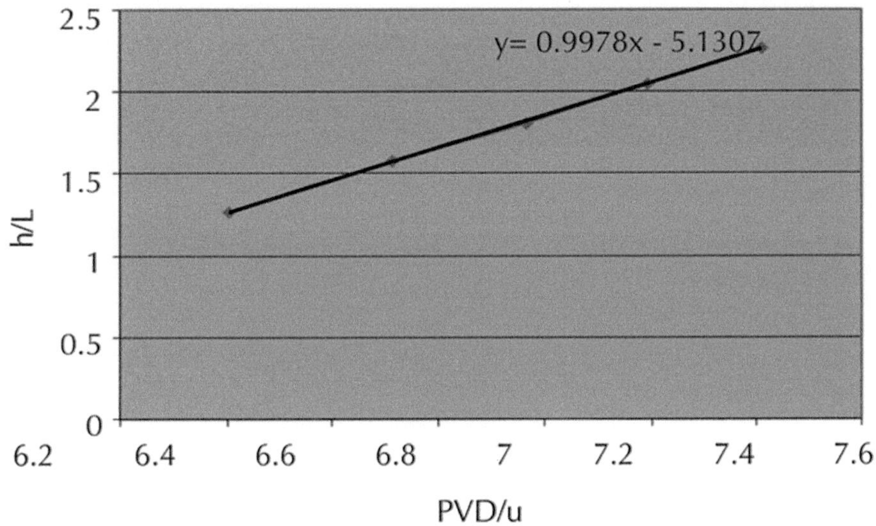

FIGURE 2.2 [h/L] versus [$\rho VD/\mu$] for example #2 of Table 2.6.
Source: Adapted from Ref. [1].

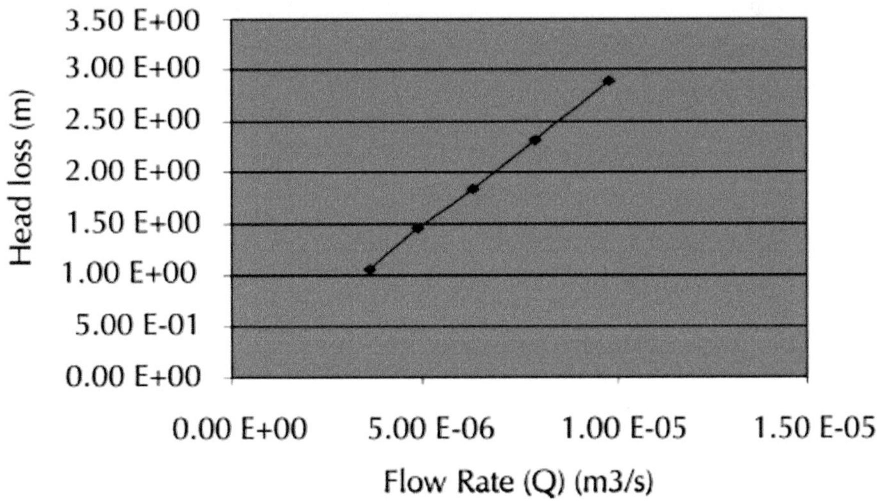

FIGURE 2.3 Head loss versus flow rate for example #2 of Table 2.6.
Source: Adapted from Ref. [1].

- For an ellipsoid[4,5]

$$p = T\left[\frac{1}{r_1} + \frac{1}{r_2}\right] \tag{2.15}$$

where p = transmural pressure; and r_1 and r_2 are major and minor radius of an ellipse, respectively.

- For a sphere,[4,5,10,17]

$$p = 2\left(\frac{T}{r}\right) \tag{2.16}$$

where sphere radius is r (such that both r_1 and r_2 of eq 2.15 are same for a sphere).

- For a cylinder,[4,16,17]

$$p = \frac{T}{r} \tag{2.17}$$

where one of the radii is infinite.

The physical parameters are identified in Table 2.7. Next, using FLT system dimensional matrix can be obtained as shown in Table 2.8.

TABLE 2.7 Physical Parameters.

Symbol	Parameter, n	Dimensions, m	Units
T	Membrane tension	FL^{-1}	dynes/cm
P	Transmural pressure	FL^{-2}	dynes/cm^2
R	Ventricle radius	L	cm
h	Membrane thickness	L	cm

TABLE 2.8 Dimensional Matrix.

	T	***P***	***r***	***h***
F	1	1	0	0
L	–1	–2	1	1

Rank of this matrix is 2.

- From BPT, the number of parameters, n = 4, number of dimensions = rank of matrix = m = 2, and number of dimensionless groups = $n - m$ = 2. Therefore, we have two π-groups. These two groups will completely define the geometric and mechanical behavior of the mammalian heart.[4,5,10,17] The π-groups are written as $\pi_1 = \frac{T}{ph}$ and $\pi_2 = \frac{r}{h}$. Next, dividing π_1 by π_2, we obtain new π as below

$$\pi_{new} = \frac{\pi_1}{\pi_2} = \frac{T}{pr} = \text{Constant} \tag{2.18}$$

- Applying analytical principles and Laplace's law, we get

$$T = pr \quad \text{or} \quad 1 = \left[\frac{T}{pr}\right] \tag{2.19}$$

Note, eqs 2.18 and 2.19 are identical. The value of constant in eq 2.18 must be determined experimentally. If we compare eqs 2.18 and 2.19, then the value of the constant in eq 2.18 is equal to one. Therefore, we do not have to conduct experiments to find this value. It indicates that the analytical solution further helps to save time and money. Also, eq 2.19 is a dimensionless number and the value of this dimensionless number must always be one.

2.8.4 EXAMPLE 4

Determine the dimensions of viscosity using Reynolds number, Newton's Law of viscosity, and Stokes' law.

- Reynolds number, Re: $T = pr \quad \text{or} \quad 1 = \left[\frac{T}{pr}\right]$ (2.20)
- Newton's law of viscosity: $\tau = \mu \frac{du}{dy} \quad \text{or} \quad \mu = \left[\frac{\tau}{du/dy}\right]$ (2.21)
- Stokes' law: $F_d = 3\pi\mu DV \quad \text{or} \quad \mu = \left[\frac{F_d}{3\pi DV}\right]$ (2.22)

Referring to Table 2.2, viscosity μ can be obtained from eq 2.20 as

$$\mu = \left[\frac{\rho VD}{\text{Re}}\right] = \frac{\left[ML^{-3}\right]\left[LT^{-1}\right]\left[L\right]}{\left[F^0L^0T^0\right]} = ML^{-1}T^{-1} = \text{kg}/(\text{ms})$$

Similarly, from eq 2.20 as

$$\mu = \left[\frac{\tau}{du/dy}\right] = \frac{\left[ML^{-1}T-2\right]}{\frac{\left[LT^{-1}\right]}{L}} = ML^{-1}T^{-1} = \text{kg}/(\text{ms})$$

and, from eq 2.21 as

$$\mu = \left[\frac{F_d}{3\pi DV}\right] = \frac{\left[MLT^{-2}\right]}{[L]\left[LT^{-1}\right]} = ML^{-1}T^{-1} = \text{kg}/(\text{ms})$$

This suggests us the dimensions for viscosity is same (i.e., kg/m.s) for all three laws using the MLT system of dimensions.

2.8.5 *EXAMPLE 5*

Determine the terminal velocity and the time to reach steady state for a sphere of density = 1.05 g cm^{-3} and radius of 10.35 μm that is settling in the water (ρ = 1 g/cm^3 and μ = 0.01 g/cm·s).

According to Stokes' law (eq 2.21)

$$F_d = 3\pi\mu DV$$

The gravitational force on the sphere is $\frac{2}{9}\left[\frac{\rho_s - \rho_w}{\mu}\right] g \left[\frac{4}{3}\pi R^3\right]$.

The settling velocity in water is $v = \frac{2}{9}\left[\frac{\rho_s - \rho_w}{\mu}\right] g \left[R^2\right]$

or, v = (2/9) [{(1.05–1.00) g cm^{-3}}/ (0.01 g cm^{-1} s^{-1})]

[(980 cm s^{-2})] [(0.001035 cm)^2] = 0.00117 cm s^{-1}.

The time constant for settling, T_c = (2/9)(1.05 g cm^{-3})

(0.01 g cm^{-1} s^{-1}) (0.001035 cm)^2 = 2.5 × 10^{-5} s

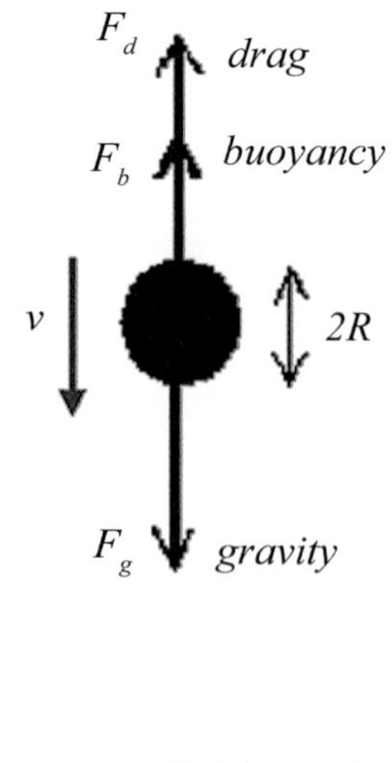

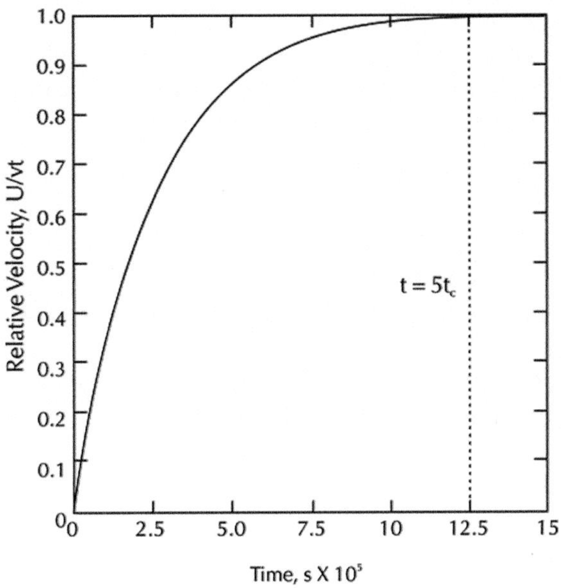

FIGURE 2.4 The terminal velocity of a sphere of radius 10.35 μm settling under the influence of gravity.
Source: Adapted from Ref. [1].

After five-time constants, the velocity is within 1% of the terminal velocity. The Reynolds number for the particle at its terminal velocity is 2.54 10^{-4}, i.e., the inertial force is negligible. Let us now calculate the distance moved during this time so that the particle will reach within 1% of its terminal velocity and then to compare this distance (x, cm) with the size of the particle. For given values of all variables for the fluid and particle, the particle moves a distance of 1.17×10^{-7} cm during the time it takes to reach within 1% of the terminal velocity. The distance is about 0.01% of the radius of the particle. Hence, because the distance is small relative to the particle size, we can safely neglect fluid acceleration and assume that the particle instantaneously adapts to a change in velocity. Figure 2.4 shows terminal velocity.

2.8.6 *EXAMPLE 6*[1]

A laboratory model of steady flow in the coronary artery is being developed. All dimensions are three times larger than the original dimensions. The working fluid is a glycerol–water mixture whose physical properties are assumed to be similar to blood, with a density of 1.1 g/cm^3, and viscosity of 0.03 cm^{-1}s^{-1}. Determine the velocity of the scale model relative to that of the original model. Assume the flow is steady; therefore, the effects of Strouhal number (or the Womersley parameter) can be neglected.

- To maintain dynamic similarity, we must have Re = Re_s, or $\rho_0 V_0 D_0 = \rho_s V_s D_s$ and $\mu_0 = \mu_s$ Since $\rho_0 = \rho_s$; $\mu_0 = \mu_s$; $3D_0 = D_s$, therefore, we will have: Vs = 3 [$V_{0/}$]. Now consider unsteady oscillatory flow in a coronary artery. In addition to requiring that Reynolds number should be the same, we also require that Wormersley number should also be the same or [$3R_0 (\omega_s/V_s)^{0.5} = R_0 (\omega_0/V_0)^{0.5}$]
- This yields a ratio of kinematics viscosities as: $(V_s/V_0) = 9$. Therefore, the given fluid does not have the same properties as blood. By substitution, we can find the following relationship for the velocities: $V_s = 3 (V_0)$

2.9 SUMMARY

Dimensional analysis is a formation of dimensionless groups to establish the relationships between dimensionless groups. It is applicable in all systems of units. It reduces the number of variables in the systems. With dimensional analysis, we can save time, money, and experimentation. The equations are universally applicable and are called general prediction equations. The technique is not as precise as analytical solutions. Results are empirical and include experimental errors. There is a need of professional expertise to identify the physical parameters that affect the system. It is necessary to identify dimensions of physical parameters by intuition or from library research.

KEYWORDS

- **Biot number**
- **Buckingham Pi theorem**
- **Cauchy number**
- **dimensional analysis**
- **dimensionless group**
- **dimensionless number**
- **Euler number**

REFERENCES

1. Goyal, M. R. Dimensional Analysis of Human Body Systems. In *Biofluid Mechanics of Human Body Systems*, 1st ed.; CRC and Apple Academic Press, 2013; pp 65–90.
2. Pardee, W. D. *Microsoft® Encarta® Encyclopedia.* Microsoft Corporation, 2002. (accessed 03/08/2006).
3. Murphy, G. *Similitude in Engineering*; Ronald Press Co., 1950.
4. Bronzino, J. D.; Peterson, D. R.; Eds. *The Biomedical Engineering Handbook,* Vol. 1–4; CRC Press LLC (Taylor & Francis Group): Boca Raton, FL, 2015; p 5422.
5. Cooney, D. *Biomedical Engineering Principles*; Marcel Dekker, 1976.
6. Fox, R. W.; McDonald, A. T.; Mitchell, J. W. *Fox and McDonald's Introduction to Fluid Mechanics*; John Wiley & Sons, 2020.
7. Gerhart, P. M.; Gerhart, A. L.; Hochstein, J. I. *Munson, Young and Okiishi's Fundamentals of Fluid Mechanics*; John Wiley & Sons, 2016.
8. Halliday, D.; Resnick, R.; Walker, J. *Fundamentals of Physics—Extended*; John Wiley and Sons, 2021.
9. *Microsoft® Encarta® Encyclopedia 2002. © 1993-2001* (accessed Mar 31, 2023).
10. Worstell, D. J. *Dimensional Analysis: Practical Guides in Chemical Engineering*; Butterworth-Heinemann, 2014.
11. Dawson, T. H. *Allometric Relations and Scaling Laws for the Cardiovascular System of Mammals*; United States Naval Academy: Systems, 2014. DOI: 10.3390/systems2020168.
12. Apsley, D. *Class Handout: Hydraulics—Topic 3 (Dimensional Analysis*; Department of Mechanical, Aerospace and Civil Engineering at the University of Manchester: Manchester, UK, 2013.
13. Delaplace, G.; Loubière, K.; Jeantet, R. *Dimensional Analysis of Food Processes*; ISTE Press: Elsevier, 2015; p 356.
14. Gibbings, J. C. *Dimensional Analysis*; Springer-Verlag, 2001.
15. Lindstedt, S. L.; Schaeffer, P. J. Use of Allometry in Predicting Anatomical and Physiological Parameters of Mammals. *Lab. Anim.* **2002**, *36*, 1–19.
16. Stuart, I. F.; Rompolski, K. *ISE Human Physiology (ISE HED Applied Biology)*; McGraw Hill, 2021.
17. Truskey, G. A.; Yuan, F.; Katz, D. F. *Transport Phenomena in Biological Systems*, 2nd ed.; Pearson Prentice Hall, 2009; p 888.
18. Chirinos, J. A.; Rietzschel, E. R.; De Buyzere, M. L.; De Bacquer, D. Arterial Load and Ventricular-Arterial Coupling Physiologic Relations with Body Size and Effect of Obesity. *Hypertension* **2009**, *54*, 558–566.

CHAPTER 3

Navier–Stokes Equation for Modeling Biofluid Dynamics

ARKA BHOWMIK, MEGH R. GOYAL, and RAMJEE REPAKA

3.1 INTRODUCTION

Biofluids are fluids with a heterogeneous mixture of biomaterials (e.g., cells, proteins, plasma, water, macromolecules, minerals, etc.), which make their properties distinct from normal fluids (e.g., water, air, oil, etc.). These properties vary with temperature, density, viscosity, viscoelasticity, compressibility, specific gravity, specific weight, surface tension, capillarity, etc. In fact, viscosity is one such property that only appears during fluid motion, and in particular exhibits a unique characteristic in case of biofluid dynamics. Biofluids are complex fluids secreted or absorbed by human body for proper functioning of living body. The movement or dynamics of biofluids through organs ensures living cells to function properly; in the absence of such a movement cells stop functioning properly. Therefore, developing an understanding of the dynamics of biofluids in different organs is a topic of interest to many.

Furthermore, the physical or experimental understanding of biofluid dynamics in different organs is not possible because it involves physical measurement inside the body. The alternate approach involves medical signal gathering or mathematical modeling to understand the behavior of biofluids in motion within the human body. On one hand, the medical signals could be in the form of image or electrical or sound/pressure values that allow clinicians to visually interpret the changes in the signal due to flow interruption. On the other hand, mathematical models allow one to mimic the behavior of biofluid dynamics in a particular organ and also allow the

Biofluid Dynamics of Human Body Systems: Expanded and Revised Edition.
Megh R. Goyal, Arka Bhowmik, & Anamika Chauhan (Eds.)

possibility of predicting the inherent properties of biofluids that cannot be gathered from image, electrical, sound, and pressure measurements.

In fact, mathematical modeling of biofluid dynamics is entirely based on general principles of fluid mechanics, such that most of the concepts are borrowed from transport phenomenon in channels and porous medium.

This chapter only focuses on mathematical models (Navier–Stokes and Poiseuille equations) for bulk motion of biofluids through channels like structures (such as arteries and veins) instead of models for fluid transport in a capillary or porous medium. The equations discussed in this chapter are not applicable for mass transport and particulate exchange across the porous tissue or cells.

The authors of this chapter emphasize that readers should look into the following excellent references[1–4] for more mathematical models of biofluid dynamics, especially advanced modeling techniques apart from Navier–Stokes model, such as turbulent flow, mixed phase flow, pulsatile flow, and lumped parameter modeling using electrical analogy. For other engineering applications of Navier–Stokes equation, one can refer the following books for reference.[5,6]

3.2 CONSTITUTIVE RELATIONS FOR NEWTONIAN AND NON-NEWTONIAN FLUIDS

The key differences between fluids and solids are their molecular force and deformation to applied force. Fluids are loosely bound molecules (i.e., weaker intermolecular force of attraction) and therefore can deform easily as compared to that of the solid. The externally applied force needed to deform any fluid is much less than that of a solid. The resisting force developed by fluid or solid against the externally applied force is known as *stress* which is primarily a bulk effect of intermolecular forces. On the other hand, deformation (or changes in shape) of solid is relatively linear after certain yield stress till it reaches a state of plasticity. That means if we increase *stress* within a solid, the *strain* (i.e., change in shape) also increases in a linear fashion.

In fact, the observed deformation is independent of time for constant force. However, in case of fluid (since loosely bound molecules) its *strain* varies with time for given fixed external force. It fails to reach a state of equilibrium immediately because of other invisible forces, unlike solid. Furthermore, another interesting phenomenon observed among fluids (especially of liquid nature) is that *stress* and *rate of strain* (or deformation rate) varies due to its constituents. In fact, some fluids display a linear relation between *stress* and *rate of strain* (e.g., water), whereas other fluid's *stress* v/s *rate-of-strain* (e.g., whole blood) shows a relation which is nonlinear. Again, the inherent reason for such an observation is the modification of intermolecular forces due to the presence of cells, long-chain proteins, sugar molecules, etc., in the blood. The constitutive relation in fluid mechanics, that is, in case of fluid exhibiting Newtonian behavior by *shear stress* v/s *shear strain rate* varying linearly is[3]

$$\tau = f(\dot{\gamma})$$
$$or, \tau = \mu\dot{\gamma} \tag{3.1}$$

where the shear stress is τ, shear strain rate $\dot{\gamma}$ is (= *du*/*dy*), *du* is change in velocity in the direction of flow, *dy* is the variation with depth, and μ is the property of fluid also referred to as viscosity.

It is important to note that we introduced a word "*shear*" before stress and strain rate for the theoretical reason of fluids being assumed to be lamina (or thin sheet) of molecules that moves over one another. The variation of tangential velocity changes (*du*) across the thickness of fluid layers (*dy*) arises due to the stress developed between two adjacent layers of fluids or at the interface of solid–fluid. Albeit, a constitutive relation of other fluids (non-Newtonian) which does not obey linear behavior *shear stress* v/s *rate-of-shear-strain* instead exhibits nonlinear relation[3]

$$\tau = f(\dot{\gamma}^n) \tag{3.2}$$

where shear stress and rate-of-shear-strain are τ and $\dot{\gamma}$, *n* is power which can vary from (1 > *n* > 1). If *n* = 0, eq 3.2 changes to Newtonian fluid. One should also note that the viscosity in case of non-Newtonian fluid cannot be constant.

The following equations are some of the constitutive relations for non-Newtonian fluid:

Power-Law fluids: $$\tau = \mu_{app}\dot{\gamma} = \left(\mu\dot{\gamma}^{n-1}\right)\dot{\gamma} \tag{3.3}$$

where μ_{app} is an instantaneous viscosity (also referred to as apparent viscosity).

For *n* > 1, the fluid acts like shear thickened (or dilatant), whereas for *n* < 1, the fluid behaves as a shear thinning (or pseudoplastic) fluid.

Herschel–Bulkley fluid: $$\tau = \tau_0 + \mu_{app}\dot{\gamma} = \tau_0 + \left(\mu\dot{\gamma}^{n-1}\right)\dot{\gamma} \tag{3.4}$$

for $\tau \le \tau_0$: $\dot{\gamma}$ is assumed 0 in eq 3.4

for $\tau > \tau_0$: eq 3.4 is used as it is

where τ_0 is yielding stress at zero strain rate (a limit must be crossed to have deformation).

Casson fluid: $\sqrt{\tau} = \sqrt{\tau_0} + \sqrt{\mu\dot{\gamma}}$ (3.5)

for $\tau \le \tau_0$: $\dot{\gamma}$ is assumed 0 in eq 3.5

for $\tau > \tau_0$: eq 3.5 is used as it is

Carreau model[7]: $\tau = \mu_{app}\dot{\gamma} = \left[\mu_\infty + (\mu_0 - \mu_\infty)\left(1 + \lambda^2\dot{\gamma}^2\right)^{0.5(n-1)}\right]\dot{\gamma}$ (3.6)

where λ denotes a parameter in time, viscosity μ_0 for fluid at zero strain rate, μ_∞ is the viscosity of fluid at infinite strain rate.

Herschel–Bulkley–Papanastasiou model[7]:

$$\tau = \mu_{app}\dot{\gamma} = \left[\mu\dot{\gamma}^{n-1} + \frac{\tau_0}{\dot{\gamma}}\left(1 - e^{-m_p\dot{\gamma}}\right)\right]\dot{\gamma} \quad (3.7)$$

where τ_0 is the yield stress, m_p is a fitting/smoothing parameter that control exponential growth of stress.

Casson–Papanastasiou model[7]:

$$\sqrt{\tau} = \sqrt{\tau_0}\left(1 - e^{-m_p\dot{\gamma}}\right) + \sqrt{\mu_p\dot{\gamma}} \quad (3.8)$$

where τ_0 is the yield stress, m_p is a fitting/smoothing parameter that controls exponential growth of stress, and μ_p is the viscosity due to plasticity.

It is important to consider that many other empirical models for non-Newtonian fluid exist in vast literature. One can refer literature[3,8,9] for various other non-Newtonian fluids and their empirical or semiempirical equations.

3.3 VISCOELASTIC FLUID

In true sense all biofluids are viscoelastic fluids since cells and structural proteins in biological fluids are elastic in nature. Elasticity is a property of material which has the tendency to return to its original shape after deformation. An ideal elastic material stores all its energy during deformation to regain its original shape after removal of applied force. Though often neglected, a small amount of energy in biofluids is stored due to its elasticity during deformation instead of only viscous dissipation in purely viscous fluid. In particular, a viscoelastic fluid has deformation component or strain component due to elasticity of its constituents and viscosity. The constitutive equation for viscoelastic fluid is given as[3]

$$\dot{\gamma} = f(\tau, \dot{\tau}, k) \quad (3.9)$$

where $\dot{\tau}$ is the rate of stress and k is the effective modulus for elasticity or rigidity of the fluid's elastic materials.

The constitutive relation of viscoelastic fluid could be a linear or nonlinear expression, and time-varying elasticity parameters could be frequency dependent.

3.4 FUNDAMENTAL EQUATIONS FOR MODELING BIOFLUID DYNAMICS

The bulk flow of biofluid in human body can be modeled using various methods such as 1D simplified network of blood vessel wherein Hagen–Poiseuille plug flow can be simulated or implementing full 3D flow domain with complex architecture wherein low Reynolds number or Laminar Navier–Stokes equation can be solved. In particular, the blood vessels in the human body differ in their radius (e.g., radius of aorta ~ 1.25 cm, vena cava ~ 1.5 cm, and capillary ~ 3 μm) and can have multiple branching. Due to this difference in diameter of individual vessels, the pressure gradient in different vessel sizes also changes which is entirely regulated by two pumps in heart (i.e., left and right sides of heart). Apart from blood flow, bulk flow of other fluids such as urine and lymphatic flow are modeled using equations deduced by Navier–Stoke. Exchange between capillary and tissue is ideally by diffusion and perfusion which is often modeled by porous media model or perfusion model. These models do not represent the bulk flow and are not discussed here.

3.4.1 CONCEPT OF CONTINUUM

Continuum is an assumption in fluid dynamic studies wherein fluid is assumed as a continuous medium instead of discrete particles. This assumption allows us to model bulk flow of fluid without calculating the net effect of particulate interaction in the fluid. Instead, using the concept of continuum all macroscopic properties were defined (such as density, viscosity, pressure) over space and time. However, one must realize at microscopic level only random interaction of particulates in the fluid medium exists.

The microscopic interaction is much more complex and the properties of interests are statistical in nature and cannot be fixed over space and time. In fact, this difference makes bottom-up approach computationally less viable for a large macroscopic domain. Therefore, continuum approach allows us a way to define fluid properties over space and time. However, continuum approach is only valid when the simulation domain or physical phenomenon is occurring at a much greater length scale compared to the distance between individual particles (see Figure 3.1).

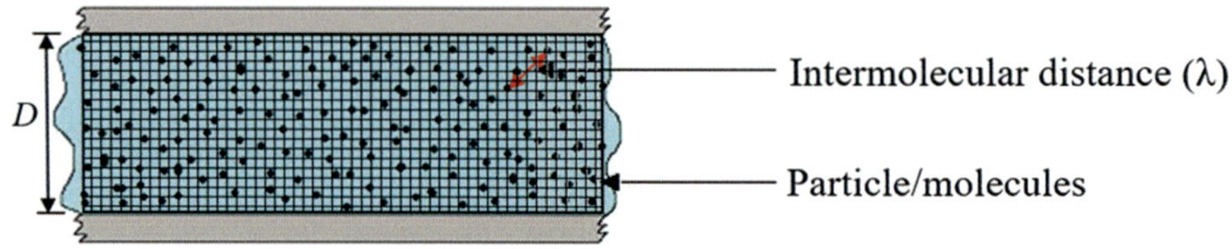

FIGURE 3.1 An illustrative figure of flow in a cylindrical channel to define concept of continuum.

The parameter that defines whether continuum approach should be used or not is referred to as Knudsen number (Kn). A fluid can be simulated using continuum approach only when Kn << 0.001.

$$Kn = \frac{\text{mean free path length}}{\text{physical length}} = \frac{\lambda}{D} \tag{3.10}$$

where the intermolecular distance and reference physical length are and *D*, respectively.

It can be realized from Figure 3.1 that each grid in a macroscopic domain where the macroscopic properties were defined is nothing but time-averaged interaction of a large number of molecules. This molecular interaction varies with both space and time. In other words, defining a fluid property at the microscopic level is practically impossible due to constant momentum of the particles. However, this movement is necessary for fluid to achieve all important functions. Especially, in case of biofluid this movement makes the human body animate. With the assumption of continuum, in the remaining section the mathematical models for bulk flow will be defined. The models discussed below are not applicable for microscopic interaction.

3.4.2 LAGRANGE AND EULER FLOW FIELD REPRESENTATION

To develop a mathematical model in fluid dynamics, an important consideration is the flow field representation. This flow field representation can be either control volume fixed in space (i.e., Eulerian representation) or control volume moving in space (i.e., Lagrange representation). It is important to know that a fluid dynamics model can be formulated in either of the two-flow field representation or sometimes in a mixed mode.

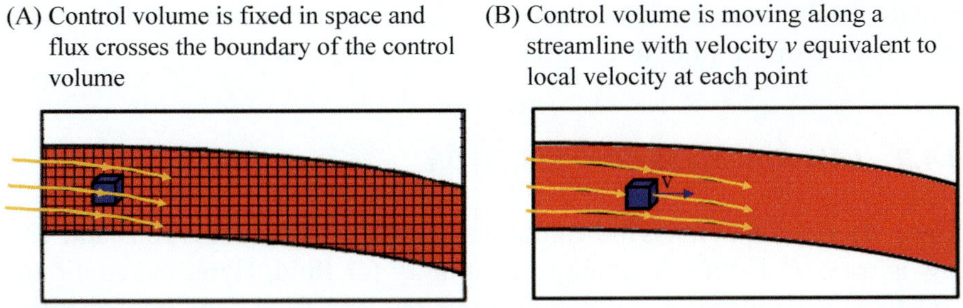

FIGURE 3.2 Illustration of flow field representation. (A) Eulerian representation and (B) Lagrange representation.

Figure 3.2 depicts a simplified version of two different flow field representations. In Eulerian representation (see Figure 3.2A), the control volume (in other words the discretized meshed domain) remains fixed in space and the fluid flux crosses the boundary. But, in Lagrange representation (see Figure 3.2B), reference volume moving (i.e., a meshfree simulation) means the same fluid packet moving from one location to another.

In the subsequent sections, we illustrated the derivation of Navier–Stokes (NS) equation in the Cartesian coordinate system and provided transformation to other coordinate systems, that is, cylindrical and spherical coordinate systems. Under continuum approximation, the starting point for formulating a mathematical model is to choose the fundamental laws of physics that governs the fluid dynamics. In a fluid dynamic problem, the applicable relations are mass, momentum, and energy conservation. We often neglect law of electrostatic and magnetism for a simple fluid flow problem. To accurately predict the dynamics of fluid mathematically, all these fundamental laws are needed to be solved in a coupled mode. The reason is that the macroscopic or the microscopic state properties of the fluid are dependent on its systems energy, which could vary over time and space because of energies such as thermal, magnetism, electrical, and gravitational. However, in special situations (i.e., negligible impact of other force fields on fluid properties) the energy equation can be completely decoupled from mass and momentum equation.

Therefore, for incompressible fluid, low Reynolds number, and single-phase fluid problem the flow dynamics can be represented by a simplified NS equation involving mass and momentum equations only. These equations were initially deduced by Navier, but the importance of viscous term was properly addressed in the work of Stokes. Hence, the equation was named as Navier–Stokes equation. The next step for mathematical formulation is to choose the reference flow field representation (either Euler or Lagrange).

Finally, apply these fundamental principles in a suitable reference flow field and obtain a differential or integral form of NS equation.

3.4.3 LAW FOR MASS CONSERVATION

The conservation of mass indicates that the total mass remains constant. This law is used to deduce equation of continuity for fluid. Here, the continuity equation is deduced in differential and integral forms by implementing conservation of mass.

We initially consider a control volume stationary in space (i.e., Euler reference flow approach) for the derivation of continuity equation (i.e., eqs 3.11 and 3.12) and subsequently using control volume moving in space (i.e., Lagrange approach) to deduce the continuity equation (i.e., eqs 3.13 and 3.14). It is worth noting that throughout the below derivation ρ is referred to as density, $\vec{v}$ is a velocity vector, u, v, and w are scalar velocities in the x, y, and z directions, V is used for volume, and S is used for representing surface.

1. *Euler approach*

Integral form

Step I: Let us consider a finite control volume (CV) fixed in space (*see* Figure 3.3).

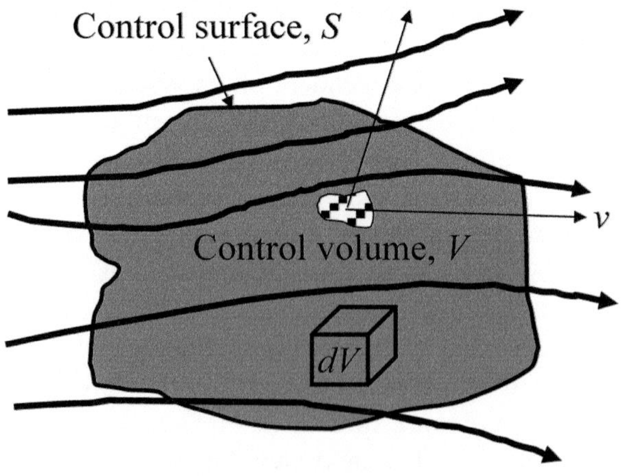

FIGURE 3.3 Representative finite CV fixed in space.

where v is flow velocity at a point on the surface, $d\text{V}$ is the elemental volume inside control volume V, dS is the vector element surface area, and ρ is the density.

Step II: Apply the fundamental principle "conservation of mass" to the assumed flow model.

$$\begin{bmatrix}\text{Mass flow out of CV}\\ -\text{Mass flow in CV}\end{bmatrix}=\begin{bmatrix}\text{Time rate of decrease}\\ \text{of mass inside the CV}\end{bmatrix}$$

$$\text{or,}\ \begin{bmatrix}\text{Net mass flow out}\\ \text{of CV through}\\ \text{surface S}\end{bmatrix}=\begin{bmatrix}\text{Time rate of decrease}\\ \text{of mass inside the CV}\end{bmatrix}$$

$$\text{or,}\ \iint_S \rho\vec{\text{v}}\bullet dS=-\frac{\partial}{\partial t}[\text{total mass}]$$

$$\text{or,}\ \iint_S \rho\vec{\text{v}}\bullet dS=-\frac{\partial}{\partial t}\left[\iiint_V \rho dV\right]$$

$$\text{or,}\ \frac{\partial}{\partial t}\left[\iiint_V \rho dV\right]+\iint_S \rho\vec{\text{v}}\bullet dS=0 \tag{3.11}$$

Differential form

Step I: Let us consider an infinitesimal fluid volume fixed in space (*see* Figure 3.4).

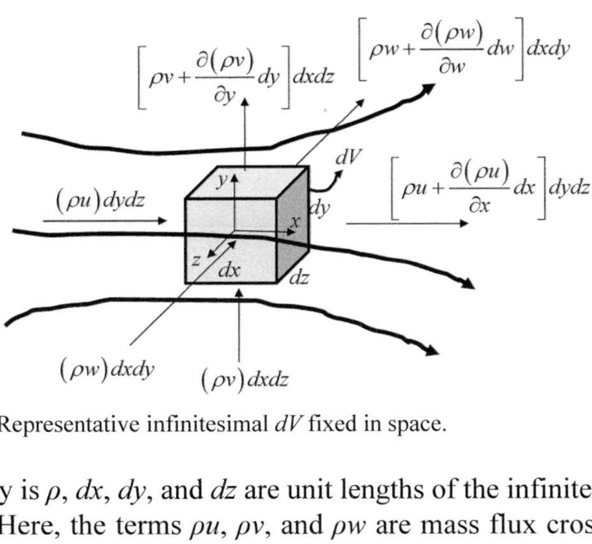

FIGURE 3.4 Representative infinitesimal dV fixed in space.

where density is ρ, dx, dy, and dz are unit lengths of the infinitesimal control volume dV. Here, the terms ρu, ρv, and ρw are mass flux cross the surface area ($dydz$), ($dxdz$), and ($dxdy$).

Step II: Apply the fundamental principle "conservation of mass" to the assumed flow model.

$$\begin{bmatrix}\text{Net mass flow out}\\ \text{of } dV\end{bmatrix}=\begin{bmatrix}\text{Time rate of decrease}\\ \text{of mass inside } dV\end{bmatrix}$$

$$\text{or,}\quad\begin{bmatrix}\text{mass flow out}\\ \text{of } dV \text{ in x-direc}\\ +\ \text{outward flow of}\\ \text{mass from } dV \text{ in y}\\ +\ \text{mass flow out}\\ \text{of } dV \text{ in z-direc}\end{bmatrix}=\begin{bmatrix}\text{Time rate of decrease}\\ \text{of mass inside } dV\end{bmatrix}$$

$$\text{or,}\quad\begin{bmatrix}\dfrac{\partial(\rho u)}{\partial x}\\ +\dfrac{\partial(\rho v)}{\partial y}\\ +\dfrac{\partial(\rho w)}{\partial z}\end{bmatrix}dxdydz=\begin{bmatrix}\text{Time rate of decrease}\\ \text{of mass inside } dV\end{bmatrix}$$

$$\text{or,}\quad\begin{bmatrix}\dfrac{\partial(\rho u)}{\partial x}\\ +\dfrac{\partial(\rho v)}{\partial y}\\ +\dfrac{\partial(\rho w)}{\partial z}\end{bmatrix}dxdydz=-\frac{\partial}{\partial t}(\text{Total mass within } dV)$$

$$\text{or,}\quad\left[\frac{\partial(\rho u)}{\partial x}+\frac{\partial(\rho v)}{\partial y}+\frac{\partial(\rho w)}{\partial z}\right]dV=-\frac{\partial}{\partial t}(\rho dV)$$

$$\text{or,}\quad\frac{\partial\rho}{\partial t}+\left[\frac{\partial(\rho u)}{\partial x}+\frac{\partial(\rho v)}{\partial y}+\frac{\partial(\rho w)}{\partial z}\right]=0 \qquad (3.12)$$

2. *Lagrange approach*

Integral form

Step I: Consider a finite CV moving in space (see Figure 3.5).

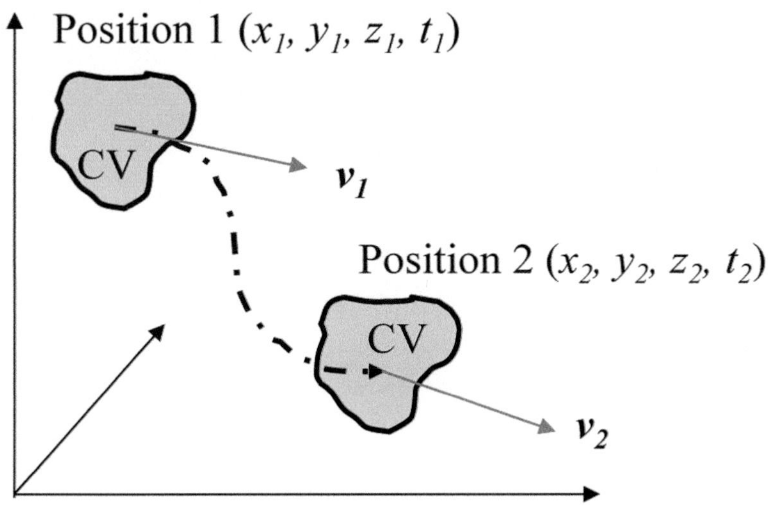

FIGURE 3.5 Representation finite CV moving in space.

In the situation of finite control volume (CV) moving in space and time (i.e., in the Lagrange approach), our eye is tracking the movement of the control volume from positions 1 to 2. But as it moves from positions 1 to 2, it has a local change inside the control volume ($\partial/\partial t$) due to exchange across the boundary of the control volume and a convective change ($u\partial/x + v\partial/\partial y + w\partial/\partial z$) due to preexisting gradient of velocity at that position compared to the control volume which is moving from one location to another location. In other words, there exists an instantaneous change in the content of the control volume also referred to as substantial derivative given as

$$\frac{D}{Dt} = \frac{\partial}{\partial t} + u\frac{\partial}{\partial x} + v\frac{\partial}{\partial y} + w\frac{\partial}{\partial z}$$

$$\text{or, } \frac{D}{Dt} = \frac{\partial}{\partial t} + \vec{v}\cdot\nabla$$

$$\text{where, } \vec{v} = u\hat{i} + v\hat{j} + w\hat{k}$$

$$\nabla = \frac{\partial}{\partial x}\hat{i} + \frac{\partial}{\partial y}\hat{j} + \frac{\partial}{\partial z}\hat{k}$$

Step II: Apply the fundamental principle "conservation of mass" to the assumed flow model. We know that although there are local + convective changes within the control volume while moving through position 1 to 2 but the mass flux going in and going out of the control volume such that total mass of the entire control volume remains constant over time. The total mass of the finite CV at any given time is given as $= \iiint_{CV} \rho(dxdydz)$. The time rate of change in instantaneous mass with CV while moving from one position to another is conserved as is given as

$$\begin{bmatrix} \text{Time rate of change} \\ \text{of mass inside } dV \end{bmatrix} = 0$$

$$\text{or, } \frac{D}{Dt}\iiint_{CV} \rho(dxdydz) = 0$$

$$\text{or, } \frac{\partial}{\partial t}\left[\iiint_{CV} \rho(dxdydz)\right] + \vec{v}\cdot\nabla\left[\iiint_{CV} \rho(dxdydz)\right] = 0 \quad (3.13)$$

Here, the first term on the left-hand side is time rate of change of mass in the control volume and the second term is time rate of change of mass due to mass flux from x, y, and z directions. Adding both instantaneous terms, that is, the local and convective terms leads to no change in the mass of the CV.

Differential form

Step I: Consider an infinitesimal fluid volume moving in space (*see* Figure 3.6).

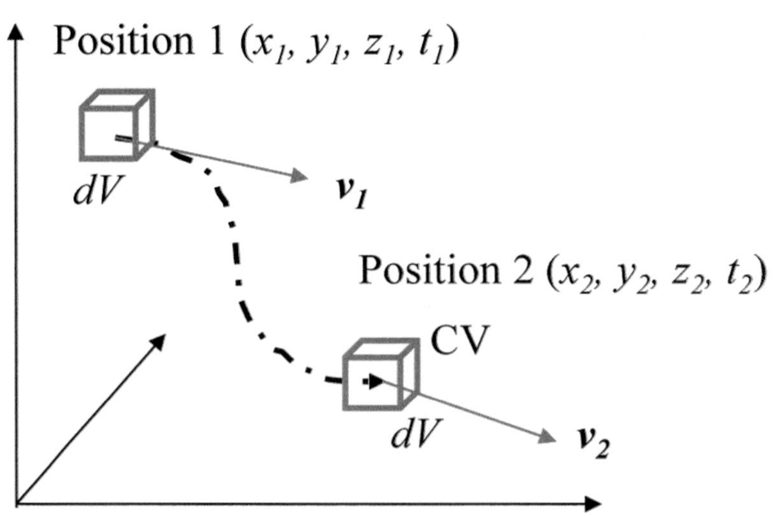

FIGURE 3.6 Representative infinite control volume (dV) moving in space.

As per earlier discussion, here the infinite small volume of fluid element (dV) is having constant mass (dm) while traversing. In fact, while moving from positions 1 to 2 the infinite small volume changes in such a way that the net mass of the fluid element remains constant. The infinitesimal fluid mass is given by $dm = \rho dV$.

Step II: Apply the fundamental principle "conservation of mass" to the assumed flow model; we can say rate at which mass within CV changes equals zero as the element traverses from positions 1 to 2. Therefore, implementing the concept of instantaneous change in infinitesimal fluid element over time as discussed earlier, the substantial derivative of the small fluid element can be written as $\frac{D}{Dt}(dm) = 0$.

$$\text{or, } \frac{D}{Dt}(\rho dV) = 0$$

$$\text{or, } dV\frac{D\rho}{Dt} + \rho\frac{D(dV)}{Dt} = 0$$

$$\text{or, } \frac{D\rho}{Dt} + \rho\left[\frac{1}{dV}\frac{D(dV)}{Dt}\right] = 0$$

Here, the second term $\frac{1}{dV}\frac{D(dV)}{Dt}$ is instantaneous change in volume of moving fluid per unit volume of fluid and is equivalent to divergence of the velocity $\nabla\vec{v}$ or surface flux for infinitesimally small fluid element in the limit when dV approaches zero or to a surface. Therefore, the equation of continuity for infinitesimally small element traversing with flow is

$$\frac{D\rho}{Dt} + \rho\nabla\vec{v} = 0 \tag{3.14}$$

3.4.4 LAW FOR MOMENTUM CONSERVATION

The law for momentum conservation defines the rate of momentum changes of the infinitesimal fluid element remains constant. This originates from the fact that the rate of momentum changes of two pushing isolated elements is equivalent to the combined rate of change of momentum of two individual fluid elements after collision.

This means that momentum cannot be created or destroyed like mass. Its magnitude can be changed by an applied force. Therefore, applying the law for motion, that is, mass × acceleration on infinitesimal fluid element assuming that the forces acting between two fluid elements boundary are equal and opposite (i.e., momentum is conserved) the governing force equation for fluid flow can be deduced. For convenience, we will adopt infinitesimal fluid element moving in space as a reference flow model (i.e., Lagrange approach) for the derivation (see Figure 3.7). Indeed, it is possible to deduce force equation using the Euler approach too but the presented Lagrange differential form is simpler than other methods.

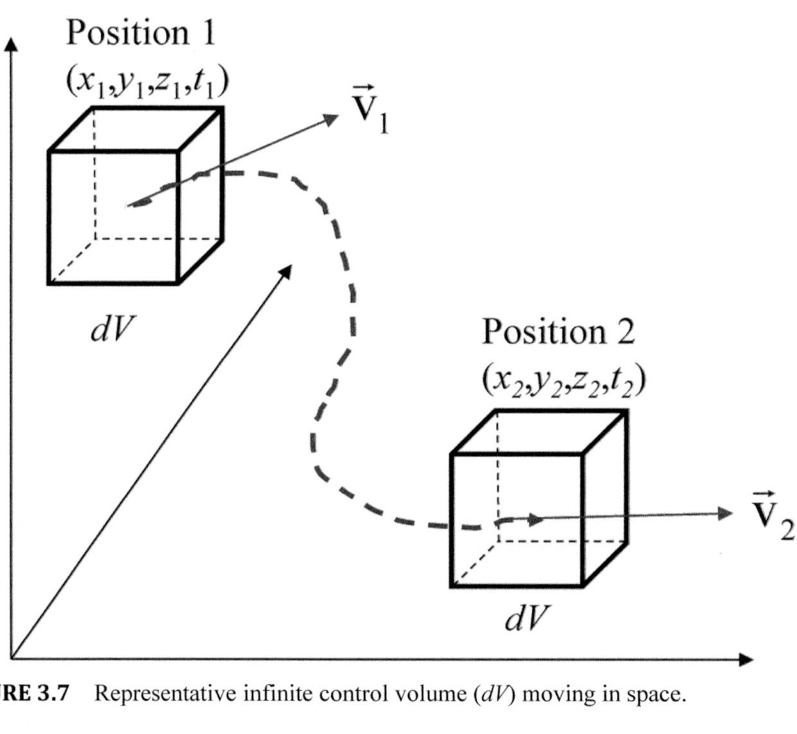

FIGURE 3.7 Representative infinite control volume (dV) moving in space.

Next, apply the equation of force for the infinitesimal fluid element dV in all directions as

$$\vec{F_i} = m\vec{a}_i = \frac{m\vec{v}_i}{t} \text{ where } i = x, y, z \tag{3.15}$$

where the force, fluid mass, acceleration, and velocity of element are denoted by F, m, $\vec{a}$, and $\vec{v}$, respectively. Next, if we only evaluate the left side of eq 3.15, then total force acting in any direction on dV is given by body force and surface force. Thus

$$[\text{Total force acting on a fluid element } dV]_i$$
$$= [\text{Body force} + \text{surface force}]_i \text{ where } i = x, y, z \quad (3.16)$$

The body forces are the forces that act on the entire volume of infinitesimal fluid element *dV,* whereas the surface forces are the forces that act only at the surface. The common forces acting on the body are weight and electromagnetic forces. The forces that act on the surfaces are originated due to static pressure and viscous effect. Next, let us deduce the net force acting in each direction as

a) [Net body force in x-direction] $= mf_{b,x}$

$$\text{or, } F_{body,x} = \rho dV\, f_{b,x} \quad (3.17)$$

where, $f_{b,x}$ is per unit mass body force.

b) $$[\text{Net surface force in x-direction}] = \begin{bmatrix} \text{Pressure} & + & \text{Viscous} \\ \text{Force} & & \text{Force} \end{bmatrix}$$

$$\text{or, } F_{surf,x} = F_{\text{Pres},x} + F_{\text{normal stress},x} + F_{\text{shear stress},x} \quad (3.18)$$

The pressure force on element *dV* at the surface is due force exerted by the surrounding fluids.

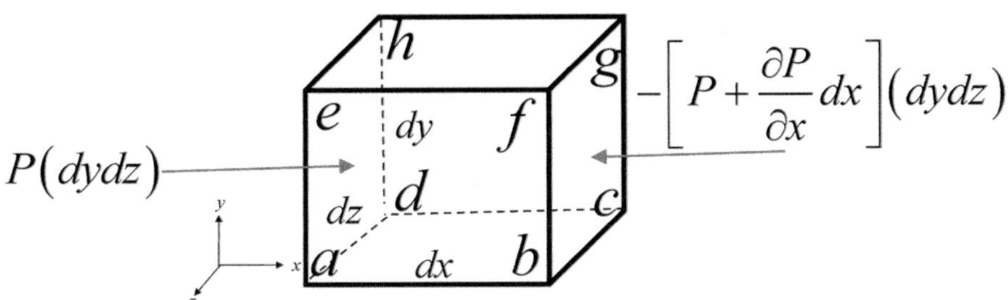

$$\begin{bmatrix} \text{Net pressure force} \\ \text{in x-direction} \end{bmatrix} = \begin{bmatrix} \text{Pressure force} & + & \text{Pressure force} \\ \text{on face } adhe & & \text{on face } bcgf \end{bmatrix}$$

$$\text{or, } F_{\text{Pres},x} = \left[-\frac{\partial P}{\partial x}\right] dxdydz \quad (3.18a)$$

where *P* is the static pressure.

On the other hand, viscous force on the element at the surface is due to the movement of surrounding fluid layers. It can be shearing force and normal force.

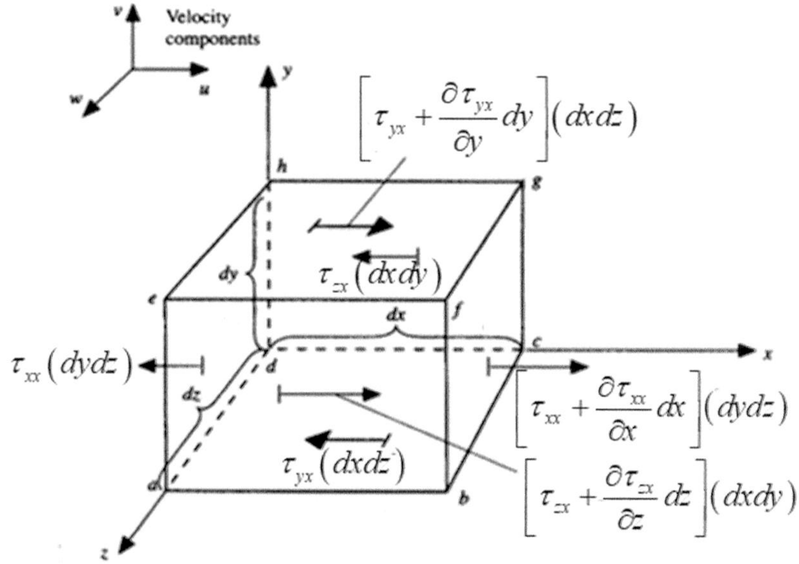

$$[\text{Net viscous force in x-direction}]=\begin{bmatrix}\text{Normal} & + & \text{Shearing}\\ \text{force} & & \text{Force}\end{bmatrix}$$

Therefore, normal shear force in the x-direction acts normal to surface (adhe) and (bfgc) is the components of normal shear stress (τ_{xx})

$$F_{\text{normal stress},x}=\left[\left(\tau_{xx}+\frac{\partial\tau_{xx}}{\partial x}dx\right)-\tau_{xx}\right](dydz) \tag{3.18b}$$

whereas force due to shear stress acting on faces "*abcd*," "*efgh*," "*abfe*," and "*dcgh*" in the x-direction are the components of shear stress (τ_{yx}) and (τ_{zx}).

$$F_{\text{shear stress},x}=\left[\left(\tau_{yx}+\frac{\partial\tau_{yx}}{\partial y}dy\right)-\tau_{yx}\right](dxdz)+\left[\left(\tau_{zx}+\frac{\partial\tau_{zx}}{\partial z}dz\right)-\tau_{zx}\right](dxdy)$$

$$\text{or, } F_{\text{shear stress},x}=\left[\frac{\partial\tau_{yx}}{\partial y}+\frac{\partial\tau_{zx}}{\partial z}\right](dxdydz) \tag{3.18c}$$

Therefore, the net force in the x-direction can be obtained by adding the pressure force (eq 3.18a), normal viscous force (eq 3.18b), shear viscous force (eq 3.18c), and body force (eq 3.17) in eq 3.16 as

$$F_x=\left[-\frac{\partial P}{\partial x}+\frac{\partial\tau_{xx}}{\partial x}+\frac{\partial\tau_{yx}}{\partial y}+\frac{\partial\tau_{zx}}{\partial z}+\rho f_{b,x}\right](dxdydz) \tag{3.19}$$

Now, let us formulate the right side of eq 3.15 as

$$ma_x = (\rho dV)\left[\frac{Du}{Dt}\right] = \rho\left[\frac{\partial u}{\partial t} + \vec{v}\nabla u\right]dV \tag{3.20}$$

where *D*/*Dt* is the substantial derivative used for the Lagrange model, $\vec{v}$ is the velocity vector given as $\left(= u\hat{i} + v\hat{j} + w\hat{k}\right)$, and ∇ is the gradient given as $\frac{\partial}{\partial x}\hat{i} + \frac{\partial}{\partial y}\hat{j} + \frac{\partial}{\partial z}\hat{k}$.

The reason for choosing a substantial derivative of velocity for an infinitesimal fluid element while moving in space has been discussed earlier during the derivation of continuity equation. The substantial derivative suggests that *u* in the direction of x changes while moving from positions 1 to 2 in flow field (see Figure 3.7). At position 2 the velocity itself fluctuates with time, that is, local derivative ($\partial u/\partial t$), and the velocity also changes over space coordinate while sweeping from position 1 to 2, that is, convective derivative ($\vec{v}\nabla u$).

Next, equating the left and right sides, that is, eqs 3.19 and 3.20 we obtain force equation in the x-direction acting on infinitesimal fluid element *dV* as

$$\rho\left[\frac{\partial u}{\partial t} + \vec{v}\bullet\nabla u\right] = \left[-\frac{\partial P}{\partial x} + \frac{\partial \tau_{xx}}{\partial x} + \frac{\partial \tau_{yx}}{\partial y} + \frac{\partial \tau_{zx}}{\partial z} + \rho f_{b,x}\right] \tag{3.21a}$$

Similarly, one can also implement a similar approach of balancing the net force in y and z directions to deduce the force equation for infinitesimal fluid element as

$$\rho\left[\frac{\partial v}{\partial t} + \vec{v}\bullet\nabla v\right] = \left[-\frac{\partial P}{\partial y} + \frac{\partial \tau_{xy}}{\partial x} + \frac{\partial \tau_{yy}}{\partial y} + \frac{\partial \tau_{zy}}{\partial z} + \rho f_{b,y}\right] \tag{3.21b}$$

$$\rho\left[\frac{\partial w}{\partial t} + \vec{v}\bullet\nabla w\right] = \left[-\frac{\partial P}{\partial z} + \frac{\partial \tau_{xz}}{\partial x} + \frac{\partial \tau_{yz}}{\partial y} + \frac{\partial \tau_{zz}}{\partial z} + \rho f_{b,z}\right] \tag{3.21c}$$

In eqs 3.21a–3.21c, the normal stress and shear stress can be determined by employing constitutive relations for Newtonian and non-Newtonian fluids discussed earlier in Section §3.2 and by deducing the stress–strain relationship for 3D fluid element similar to the treatment in solid mechanics. It is important to note, the total stress due to viscous forces adds new unknowns to the system of equations also referred to as "closure problem" and the derivation of stress–strain relation for Newtonian and non-Newtonian fluids requires some hypothesis and approximations. In fact, for incompressible laminar Newtonian fluid flow Stokes[10] suggested a linear relation for the viscous stress term in eqs 3.21a–3.21c as follows:

Normal stresses in eqs 3.21a–3.21c	Shear stresses in eqs 3.21a–3.21c
$\tau_{xx} = \lambda(\nabla \bullet \vec{v}) + 2\mu\frac{\partial u}{\partial x}$	$\tau_{xy} = \tau_{yx} = \mu\left[\frac{\partial v}{\partial x} + \frac{\partial u}{\partial y}\right]$
$\tau_{yy} = \lambda(\nabla \bullet \vec{v}) + 2\mu\frac{\partial v}{\partial y}$	$\tau_{xz} = \tau_{zx} = \mu\left[\frac{\partial w}{\partial x} + \frac{\partial u}{\partial z}\right]$ (3.22)
$\tau_{zz} = \lambda(\nabla \bullet \vec{v}) + 2\mu\frac{\partial w}{\partial z}$	$\tau_{yz} = \tau_{zy} = \mu\left[\frac{\partial w}{\partial y} + \frac{\partial v}{\partial z}\right]$

where λ denotes second viscosity and μ is the dynamic viscosity.

Now, according to Stokes hypothesis: $\lambda = -(2/3)\mu$ which is frequently used till date but still not proved experimentally. Therefore, the stress terms can be written as

Normal stresses in eq 3.22	Shear stresses in eq 3.22
$\tau_{xx} = \lambda(\nabla \bullet \vec{v}) + 2\mu\frac{\partial u}{\partial x} = \lambda\left[\frac{\partial u}{\partial x} + \frac{\partial v}{\partial y} + \frac{\partial w}{\partial z}\right] + 2\mu\frac{\partial u}{\partial x} = -\frac{2}{3}\mu\left[\frac{\partial u}{\partial x} + \frac{\partial v}{\partial y} + \frac{\partial w}{\partial z}\right] + 2\mu\frac{\partial u}{\partial x} = 2\mu\left[\frac{\partial u}{\partial x} - \frac{1}{3}(\nabla\vec{v})\right]$	$\tau_{xy} = \tau_{yx} = \mu\left[\frac{\partial v}{\partial x} + \frac{\partial u}{\partial y}\right]$
$\tau_{yy} = \lambda\left[\frac{\partial u}{\partial x} + \frac{\partial v}{\partial y} + \frac{\partial w}{\partial z}\right] + 2\mu\frac{\partial v}{\partial y} = -\frac{2}{3}\mu\left[\frac{\partial u}{\partial x} + \frac{\partial v}{\partial y} + \frac{\partial w}{\partial z}\right] + 2\mu\frac{\partial v}{\partial y} = 2\mu\left[\frac{\partial v}{\partial y} - \frac{1}{3}(\nabla\vec{v})\right]$	$\tau_{xz} = \tau_{zx} = \mu\left[\frac{\partial w}{\partial x} + \frac{\partial u}{\partial z}\right]$ (3.23)

Normal stresses in eq 3.22	Shear stresses in eq 3.22
$\tau_{zz} = \lambda\left[\frac{\partial u}{\partial x}+\frac{\partial v}{\partial y}+\frac{\partial w}{\partial z}\right]+2\mu\frac{\partial w}{\partial z}$ $= -\frac{2}{3}\mu\left[\frac{\partial u}{\partial x}+\frac{\partial v}{\partial y}+\frac{\partial w}{\partial z}\right]+2\mu\frac{\partial w}{\partial z}$ $= 2\mu\left[\frac{\partial w}{\partial z}-\frac{1}{3}(\nabla\vec{v})\right]$	$\tau_{yz} = \tau_{zy} = \mu\left[\frac{\partial w}{\partial y}+\frac{\partial v}{\partial z}\right]$

However, in most literature second viscosity λ is assumed to be –μ instead of (2/3)μ to obtain a simplified form. Therefore, the normal stresses in eq 3.23 can be expressed as

Normal stresses in eq 3.22	Shear stresses in eq 3.22	
$\tau_{xx} = \lambda(\nabla\bullet\vec{v})+2\mu\frac{\partial u}{\partial x}$ $= -\mu\left[\frac{\partial u}{\partial x}+\frac{\partial v}{\partial y}+\frac{\partial w}{\partial z}\right]+2\mu\frac{\partial u}{\partial x}$ $= \mu\left[\frac{\partial u}{\partial x}-\frac{\partial v}{\partial y}-\frac{\partial w}{\partial z}\right]$	$\tau_{xy} = \tau_{yx} = \mu\left[\frac{\partial v}{\partial x}+\frac{\partial u}{\partial y}\right]$	
$\tau_{yy} = \lambda\left[\frac{\partial u}{\partial x}+\frac{\partial v}{\partial y}+\frac{\partial w}{\partial z}\right]+2\mu\frac{\partial v}{\partial y}$ $= -\mu\left[\frac{\partial u}{\partial x}+\frac{\partial v}{\partial y}+\frac{\partial w}{\partial z}\right]+2\mu\frac{\partial v}{\partial y}$ $= \mu\left[\frac{\partial v}{\partial y}-\frac{\partial u}{\partial x}-\frac{\partial w}{\partial z}\right]$	$\tau_{xz} = \tau_{zx} = \mu\left[\frac{\partial w}{\partial x}+\frac{\partial u}{\partial z}\right]$	(3.24)
$\tau_{zz} = \lambda\left[\frac{\partial u}{\partial x}+\frac{\partial v}{\partial y}+\frac{\partial w}{\partial z}\right]+2\mu\frac{\partial w}{\partial z}$ $= -\mu\left[\frac{\partial u}{\partial x}+\frac{\partial v}{\partial y}+\frac{\partial w}{\partial z}\right]+2\mu\frac{\partial w}{\partial z}$ $= \mu\left[\frac{\partial w}{\partial z}-\frac{\partial u}{\partial x}-\frac{\partial v}{\partial y}\right]$	$\tau_{yz} = \tau_{zy} = \mu\left[\frac{\partial w}{\partial y}+\frac{\partial v}{\partial z}\right]$	

Now substituting the simplified form of normal and shear stress (i.e., eq 3.24) into the governing force equation (eqs 3.21a–3.21c) one can deduce a simplified expression of force equation for infinitely small fluid element moving in space in each direction asx-direction:

$$\rho\left[\frac{\partial u}{\partial t}+\left(u\frac{\partial u}{\partial x}+v\frac{\partial u}{\partial y}+w\frac{\partial u}{\partial z}\right)\right]=\left[-\frac{\partial P}{\partial x}+\mu\left(\frac{\partial^2 u}{\partial x^2}+\frac{\partial^2 u}{\partial y^2}+\frac{\partial^2 u}{\partial z^2}\right)+\rho f_{b,x}\right] \quad (3.25a)$$

y-direction:

$$\rho\left[\frac{\partial v}{\partial t}+\left(u\frac{\partial v}{\partial x}+v\frac{\partial v}{\partial y}+w\frac{\partial v}{\partial z}\right)\right]=\left[-\frac{\partial P}{\partial y}+\mu\left(\frac{\partial^2 v}{\partial x^2}+\frac{\partial^2 v}{\partial y^2}+\frac{\partial^2 v}{\partial z^2}\right)+\rho f_{b,y}\right] \quad (3.25b)$$

z-direction:

$$\rho\left[\frac{\partial w}{\partial t}+\left(u\frac{\partial w}{\partial x}+v\frac{\partial w}{\partial y}+w\frac{\partial w}{\partial z}\right)\right]=\left[-\frac{\partial P}{\partial z}+\mu\left(\frac{\partial^2 w}{\partial x^2}+\frac{\partial^2 w}{\partial y^2}+\frac{\partial^2 w}{\partial z^2}\right)+\rho f_{b,z}\right] \quad (3.25c)$$

Similarly, one can also deduce the generalized force equation using eq 3.23 by substituting into the governing force equation (eqs 3.21a–3.21c). We leave this for reader's practice. We are restricting ourselves to the form shown in eq 3.25 since we later want to use this simplified form to deduce the nondimensional Navier–Stokes equation.

Note, we need to develop a different set of equations or relations using the nonlinear constitutive equations discussed in section §3.2 for fluids that are non-Newtonian (e.g., blood) by substituting a different set of stress terms into eqs 3.21a–3.21c.

3.4.5 NAVIER–STOKES EQUATION IN CARTESIAN COORDINATES

In special situations (i.e., negligible impact of other force fields on fluid properties), the energy equation can be completely decoupled from mass and momentum equation. Therefore, for incompressible fluid, low Reynolds number, and single-phase fluid problem the flow dynamics can be represented by a simplified Navier–Stokes (NS) equation involving mass and momentum equations only as[5]

Continuity equation:

$$\frac{\partial \rho}{\partial t}+\left[\frac{\partial(\rho u)}{\partial x}+\frac{\partial(\rho v)}{\partial y}+\frac{\partial(\rho w)}{\partial z}\right]=0$$

$$\frac{\partial \rho}{\partial t}+\left[\nabla\cdot(\rho\vec{v})\right]=0 \quad (3.26a)$$

Momentum equation:

X-direction:

$$\rho\left[\frac{\partial u}{\partial t}+\left(u\frac{\partial u}{\partial x}+v\frac{\partial u}{\partial y}+w\frac{\partial u}{\partial z}\right)\right]=\left[-\frac{\partial P}{\partial x}+\mu\left(\frac{\partial^2 u}{\partial x^2}+\frac{\partial^2 u}{\partial y^2}+\frac{\partial^2 u}{\partial z^2}\right)+\rho f_{b,x}\right] \quad (3.26b)$$

Y-direction:

$$\rho\left[\frac{\partial v}{\partial t}+\left(u\frac{\partial v}{\partial x}+v\frac{\partial v}{\partial y}+w\frac{\partial v}{\partial z}\right)\right]=\left[-\frac{\partial P}{\partial y}+\mu\left(\frac{\partial^2 v}{\partial x^2}+\frac{\partial^2 v}{\partial y^2}+\frac{\partial^2 v}{\partial z^2}\right)+\rho f_{b,y}\right] \quad (3.26c)$$

Z-direction:

$$\rho\left[\frac{\partial w}{\partial t}+\left(u\frac{\partial w}{\partial x}+v\frac{\partial w}{\partial y}+w\frac{\partial w}{\partial z}\right)\right]=\left[-\frac{\partial P}{\partial z}+\mu\left(\frac{\partial^2 w}{\partial x^2}+\frac{\partial^2 w}{\partial y^2}+\frac{\partial^2 w}{\partial z^2}\right)+\rho f_{b,z}\right] \quad (3.26d)$$

However, a complex flow behavior wherein the effect of viscous dissipation due to flow would require coupling of energy equation (see Ref. [5] for a detailed derivation of energy equation) with the above momentum and continuity equations for the solution NS equation. The reason is that the fluid properties also vary with thermal energy. Then, the NS eqs 3.26a–3.26d take the form as follows neglecting kinetic energy per unit mass term from the energy equation

Continuity equation: $$\frac{\partial \rho}{\partial t}+\left[\frac{\partial(\rho u)}{\partial x}+\frac{\partial(\rho v)}{\partial y}+\frac{\partial(\rho w)}{\partial z}\right]=0 \quad (3.27a)$$

Momentum equation:

X-direction:

$$\rho\left[\frac{\partial u}{\partial t}+\left(u\frac{\partial u}{\partial x}+v\frac{\partial u}{\partial y}+w\frac{\partial u}{\partial z}\right)\right]=\left[-\frac{\partial P}{\partial x}+\mu\left(\frac{\partial^2 u}{\partial x^2}+\frac{\partial^2 u}{\partial y^2}+\frac{\partial^2 u}{\partial z^2}\right)+\rho f_{b,x}\right] \quad (3.27b)$$

Y-direction:

$$\rho\left[\frac{\partial v}{\partial t}+\left(u\frac{\partial v}{\partial x}+v\frac{\partial v}{\partial y}+w\frac{\partial v}{\partial z}\right)\right]=\left[-\frac{\partial P}{\partial y}+\mu\left(\frac{\partial^2 v}{\partial x^2}+\frac{\partial^2 v}{\partial y^2}+\frac{\partial^2 v}{\partial z^2}\right)+\rho f_{b,y}\right] \quad (3.27c)$$

Z-direction:

$$\rho\left[\frac{\partial w}{\partial t}+\left(u\frac{\partial w}{\partial x}+v\frac{\partial w}{\partial y}+w\frac{\partial w}{\partial z}\right)\right]=\left[-\frac{\partial P}{\partial z}+\mu\left(\frac{\partial^2 w}{\partial x^2}+\frac{\partial^2 w}{\partial y^2}+\frac{\partial^2 w}{\partial z^2}\right)+\rho f_{b,z}\right] \quad (3.27d)$$

Energy equation:

$$\rho c_p\left[\left(\frac{\partial T}{\partial t}+\vec{v}\cdot\nabla T\right)\right]=\left[\begin{array}{l}\rho\dot{q}+\nabla\cdot(k\nabla T)-\nabla P\cdot\vec{v}+\mu\left\{u\left(\frac{\partial^2 u}{\partial x^2}+\frac{\partial^2 u}{\partial y^2}+\frac{\partial^2 u}{\partial z^2}\right)\right\}\\ +\mu\left\{v\left(\frac{\partial^2 v}{\partial x^2}+\frac{\partial^2 v}{\partial y^2}+\frac{\partial^2 v}{\partial z^2}\right)\right\}+\mu\left\{w\left(\frac{\partial^2 w}{\partial x^2}+\frac{\partial^2 w}{\partial y^2}+\frac{\partial^2 w}{\partial z^2}\right)\right\}+\left(\rho\vec{f}_b\cdot\vec{v}\right)\end{array}\right] \tag{3.27e}$$

where ρ, k, c_p, P, and $\vec{v}$ are density, thermal conductivity, specific heat, static pressure, and velocity vector $\left(=u\hat{i}+v\hat{j}+w\hat{k}\right)$, respectively. Here, scalar velocities in x-, y-, and z- directions are u, v, and w. $\nabla$is the gradient given as $\frac{\partial}{\partial x}\hat{i}+\frac{\partial}{\partial y}\hat{j}+\frac{\partial}{\partial z}\hat{k}$, T is temperature, and μ is the dynamic viscosity. The per unit mass body force is f_b, volume is V, and volumetric heat addition per unit mass is $\dot{q}$.

3.4.6 NAVIER–STOKES (NS) EQUATION IN CYLINDRICAL COORDINATES

To deduce the NS equation for the cylindrical coordinate system, we often use the transformation method to the already-deduced equations for the Cartesian coordinate system (such as eqs 3.26a–3.26d). Figure 3.8 depicts the flow in the cylindrical domain and establishes the fact that in the cylindrical domain $\vec{v}$ is a function of (r, θ, z).

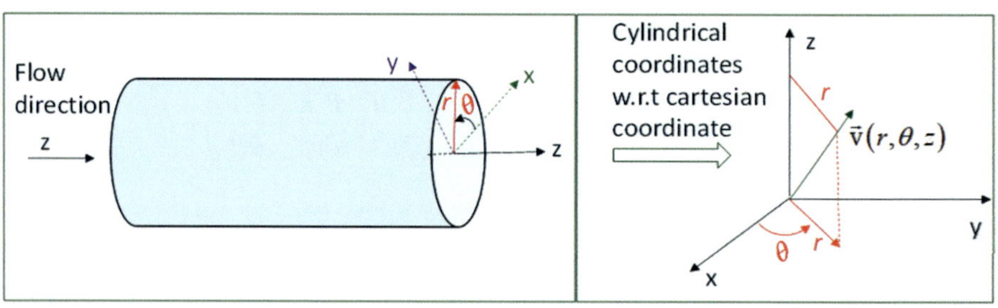

FIGURE 3.8 Illustrations for transformation from Cartesian coordinate to cylindrical coordinate.

We have seen earlier that in Cartesian coordinate, $\vec{v}$ is $\left(=u\hat{i}+v\hat{j}+w\hat{k}\right)$ such that velocities u, v, and w are in the x-, y-, and z-direction, and the unit vectors are *i*, *j*, and *k* in the x-, y-, and z-direction respectively, and the gradient term is given as $\left(\nabla \cdot \vec{v} = \frac{\partial u}{\partial x} + \frac{\partial v}{\partial y} + \frac{\partial w}{\partial z}\right)$.

However, in the cylindrical domain, $\vec{v}$ is $(= v_r \hat{e}_r + v_\theta \hat{e}_\theta + v_z \hat{e}_z)$ such that velocities in the *r*-, *θ*-, and z-direction are v_r, v_θ, and v_z, and the unit vectors are e_r, e_θ, and e_z in *r*-, *θ*-, and z- direction, respectively, and the gradient term takes a form $\left(\nabla \cdot \vec{v} = \frac{1}{r}\frac{\partial}{\partial r}(r v_r) + \frac{1}{r}\frac{\partial}{\partial \theta}(v_\theta) + \frac{\partial}{\partial z}(v_z)\right)$.

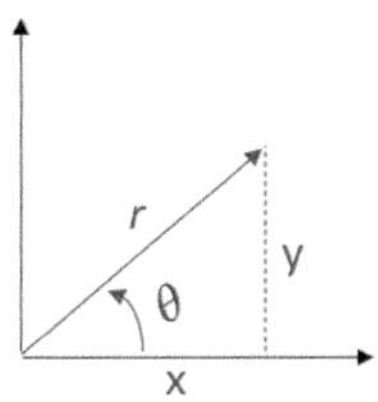

$$x = r\cos\theta, \quad y = r\sin\theta, \quad z = z$$

$$r = \sqrt{x^2 + y^2}, \quad \theta = \tan^{-1}\left(\frac{y}{x}\right), \quad z = z \tag{3.28}$$

To deduce the transformation, we use transformation equation between *x*, *y*, *z* and *r*, *θ*, z, and the derivatives in eqs 3.26a–3.26d and obtain a form shown below[11]:

Continuity equation: $$\frac{\partial \rho}{\partial t} + \frac{1}{r}\frac{\partial(\rho r v_r)}{\partial r} + \frac{1}{r}\frac{\partial(\rho v_\theta)}{\partial \theta} + \frac{\partial(\rho v_z)}{\partial z} = 0 \tag{3.29a}$$

Momentum equation:

r-direction:

$$\rho\left[\frac{\partial v_r}{\partial t} + \left(v_r \frac{\partial v_r}{\partial r} + \frac{v_\theta}{r}\frac{\partial v_r}{\partial \theta} - \frac{v_\theta^2}{r} + v_z \frac{\partial v_r}{\partial z}\right)\right]$$
$$= -\frac{\partial P}{\partial r} + \mu\left[\frac{\partial}{\partial r}\left(\frac{1}{r}\frac{\partial}{\partial r}(r v_r)\right) + \frac{1}{r^2}\frac{\partial^2 v_r}{\partial \theta^2} - \frac{2}{r^2}\frac{\partial v_\theta}{\partial \theta} + \frac{\partial^2 v_r}{\partial z^2}\right] + \rho f_{b,r} \tag{3.29b}$$

θ-direction:

$$\rho\left[\frac{\partial v_\theta}{\partial t} + \left(v_r \frac{\partial v_\theta}{\partial r} + \frac{v_\theta}{r}\frac{\partial v_\theta}{\partial \theta} + \frac{v_r v_\theta}{r} + v_z \frac{\partial v_\theta}{\partial z}\right)\right]$$
$$= -\frac{1}{r}\frac{\partial P}{\partial \theta} + \mu\left[\frac{\partial}{\partial r}\left(\frac{1}{r}\frac{\partial}{\partial r}(r v_\theta)\right) + \frac{1}{r^2}\frac{\partial^2 v_\theta}{\partial \theta^2} + \frac{2}{r^2}\frac{\partial v_r}{\partial \theta} + \frac{\partial^2 v_\theta}{\partial z^2}\right] + \rho f_{b,\theta} \tag{3.29c}$$

z-direction:

$$\rho\left[\frac{\partial v_z}{\partial t}+\left(v_r\frac{\partial v_z}{\partial r}+\frac{v_\theta}{r}\frac{\partial v_z}{\partial \theta}+v_z\frac{\partial v_z}{\partial z}\right)\right]$$

$$=-\frac{\partial P}{\partial z}+\mu\left[\frac{1}{r}\frac{\partial}{\partial r}\left(r\frac{\partial v_z}{\partial r}\right)+\frac{1}{r^2}\frac{\partial^2 v_z}{\partial \theta^2}+\frac{\partial^2 v_z}{\partial z^2}\right]+\rho f_{b,z} \quad (3.29d)$$

The full derivation of the NS equation in cylindrical coordinate is tedious. Readers can refer to Chapter 5 of the book titled, "*Fox and McDonald's Introduction of fluid mechanics*"[12] which provides a detailed derivation of the equation of continuity for the cylindrical domain in the similar manner as shown for the Cartesian coordinate system (in §3.4.3) or Chapter 6 of the book titled, "Fluid Mechanics."[13] The detailed derivation of the momentum equation in the cylindrical coordinate system can be found in Chapter 6 of the book titled, "*Introduction to Chemical Engineering Fluid Mechanics*"[14] and Chapter 9 of the book titled, "Fluid Mechanics Fundamentals and Applications."[11] It is important to note that each of the above books provides a different form of governing momentum equations with or without Stokes assumption for second viscosity discussed in §3.4.4. Eqs 3.29b–3.29d assume the situation when second viscosity is $-\mu$ instead of $-(2/3)\mu$.

3.4.7 NAVIER–STOKES (NS) EQUATION IN SPHERICAL COORDINATES

A similar manner of transformation from Cartesian to spherical can be employed for deducing NS equation for the spherical coordinate system. Figure 3.9 depicts the flow in the spherical domain and establishes the fact that in the spherical domain the control volume dV and $\vec{v}$ is a function of (r, θ, φ).

We have seen earlier that in the Cartesian coordinate system, $\vec{v}$ $\left(=u\hat{i}+v\hat{j}+w\hat{k}\right)$, and $\left(\nabla\cdot\vec{v}=\frac{\partial u}{\partial x}+\frac{\partial v}{\partial y}+\frac{\partial w}{\partial z}\right)$. However, in the spherical domain, $\vec{v}$ $\left(=v_r\hat{e}_r+v_\theta\hat{e}_\theta+v_\varphi\hat{e}_\varphi\right)$, where v_r, v_θ, and v_φ are radial component, polar component, and azimuth angle component, and e_r, e_θ, and e_φ are unit vectors of radial component, polar component, and azimuth angle component, respectively, and the gradient value takes a form $\left(\nabla\cdot\vec{v}=\frac{1}{r^2}\frac{\partial}{\partial r}\left(r^2 v_r\right)+\frac{1}{r\sin\theta}\frac{\partial}{\partial \theta}\left(v_\theta\sin\theta\right)+\frac{1}{r\sin\theta}\frac{\partial v_\varphi}{\partial \varphi}\right)$.

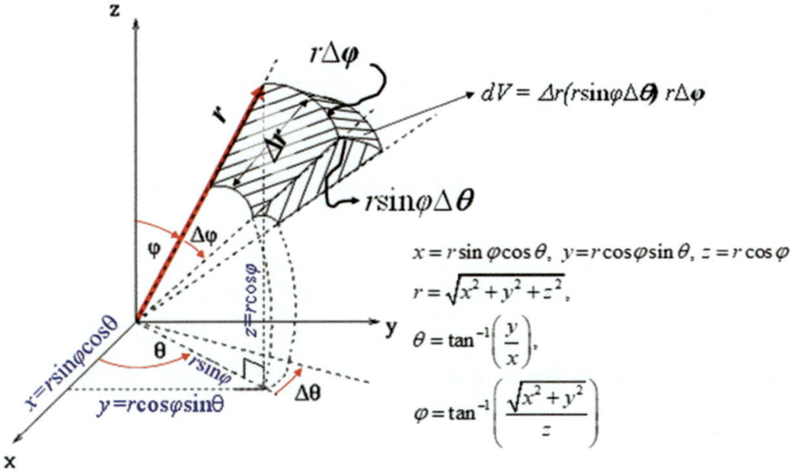

FIGURE 3.9 Illustrations for transformation from Cartesian coordinate to spherical coordinate.

To deduce the transformation, we use the transformation equation between x, y, z and r, θ, φ shown in Figure 3.9, and obtain the derivatives in eqs. 3.26a–3.26d to deduce a form shown below[13,15]:

Continuity equation:

$$\frac{\partial \rho}{\partial t}+\frac{1}{r^2}\frac{\partial\left(\rho r^2 v_r\right)}{\partial r}+\frac{1}{r\sin\varphi}\frac{\partial\left(\rho v_\theta\right)}{\partial \theta}+\frac{1}{r\sin\varphi}\frac{\partial\left(\rho v_\varphi \sin\varphi\right)}{\partial \varphi}=0 \quad (3.30a)$$

Momentum equation:

r-direction:

$$\rho\left[\frac{\partial v_r}{\partial t}+\left(v_r\frac{\partial v_r}{\partial r}+\frac{v_\theta}{r}\frac{\partial v_r}{\partial \theta}+\frac{v_\varphi}{r\sin\theta}\frac{\partial v_r}{\partial \varphi}-\frac{v_\theta^2+v_\varphi^2}{r}\right)\right]$$
$$=-\frac{\partial P}{\partial r}+\mu\left[\nabla^2 v_r-\frac{2}{r^2}v_r-\frac{2}{r^2}\frac{\partial v_\theta}{\partial \theta}-\frac{2}{r^2}v_\theta\cot\theta-\frac{2}{r^2\sin\theta}\frac{\partial v_\varphi}{\partial \varphi}\right]+\rho f_{b,r} \quad (3.30b)$$

θ-direction:

$$\rho\left[\frac{\partial v_\theta}{\partial t}+\left(v_r\frac{\partial v_\theta}{\partial r}+\frac{v_\theta}{r}\frac{\partial v_\theta}{\partial \theta}+\frac{v_\varphi}{r\sin\theta}\frac{\partial v_\theta}{\partial \varphi}+\frac{v_r v_\theta}{r}-\frac{v_\varphi^2\cot\theta}{r}\right)\right]$$
$$=-\frac{1}{r}\frac{\partial P}{\partial \theta}+\mu\left[\nabla^2 v_\theta+\frac{2}{r^2}\frac{\partial v_r}{\partial \theta}-\frac{v_\theta}{r^2\sin^2\theta}-\frac{2\cos\theta}{r^2\sin^2\theta}\frac{\partial v_\varphi}{\partial \varphi}\right]+\rho f_{b,\theta} \quad (3.30c)$$

φ-direction:

$$\rho\left[\frac{\partial v_\varphi}{\partial t}+\left(v_r\frac{\partial v_\varphi}{\partial r}+\frac{v_\theta}{r}\frac{\partial v_\varphi}{\partial \theta}+\frac{v_\varphi}{r\sin\theta}\frac{\partial v_\varphi}{\partial \varphi}+\frac{v_\varphi v_r}{r}+\frac{v_\theta v_\varphi}{r}\cot\theta\right)\right]$$

$$=-\frac{1}{r\sin\theta}\left(\frac{\partial P}{\partial \varphi}\right)+\mu\left[\nabla^2 v_\varphi-\frac{v_\varphi}{r^2\sin^2\theta}+\frac{2}{r^2\sin\theta}\left(\frac{\partial v_r}{\partial \varphi}\right)+\frac{2\cos\theta}{r^2\sin^2\theta}\left(\frac{\partial v_\theta}{\partial \varphi}\right)\right]+\rho f_{b,\varphi} \tag{3.30d}$$

$$\textit{where, } \nabla^2=\frac{1}{r^2}\frac{\partial}{\partial r}\left(r^2\frac{\partial}{\partial r}\right)+\frac{1}{r^2\sin\theta}\frac{\partial}{\partial \theta}\left(\sin\theta\frac{\partial}{\partial \theta}\right)+\frac{1}{r^2\sin^2\theta}\left(\frac{\partial^2}{\partial \varphi^2}\right).$$

Readers can refer Chapter 6 of the book titled, "*Fluid Mechanics*"[13] for a complete derivation of the continuity equation in the spherical coordinate system. The detailed derivation of momentum equation in the spherical coordinate system with Stokes assumption for second viscosity can be found in Chapter 6 of the book titled, "*Introduction to Chemical Engineering Fluid Mechanics*."[14] It is important to note that each of the above books provides a different form of governing momentum equations with or without Stokes assumption for second viscosity discussed in §3.4.4. Eqs 3.30b–3.30d[15] assume the situation when second viscosity is $-\mu$ instead of $(2/3)\mu$.

3.4.8 NONDIMENSIONAL NAVIER–STOKES EQUATIONS IN CARTESIAN COORDINATES

In many fluids flow problems, the nondimensional NS equation is used instead of a dimensional form because of geometric similarity and to deduce the nondimensional parameters, which govern or determine the nature or behavior of flow. Here, we consider a reference geometric dimension (such as diameter D for pipe or length L for plate) and reference scalar velocity or free stream velocity u_{ref} to deduce a nondimensional form of NS equation. We establish few dimensionless parameters that can be implemented to the earlier deduced NS equation (eqs 3.26a–3.26d).

$$\text{Geometric}: X^*=\frac{x}{D} \text{ and } Y^*=\frac{y}{D} \text{ and } Z^*=\frac{z}{D}$$

$$\text{Velocity: } U^*=\frac{u}{u_{ref}} \text{ and } V^*=\frac{v}{u_{ref}} \text{ and } W^*=\frac{w}{u_{ref}}$$

$$\text{Pressure: } P^*=\frac{P}{\rho u_{ref}^2}$$

First, we consider a steady state and incompressible continuity equation from eq 3.26a as follows:[16]

$$\frac{\partial u}{\partial x}+\frac{\partial v}{\partial y}+\frac{\partial w}{\partial z}=0 \qquad [\because \rho=\text{constant}] \tag{3.31}$$

Thus, a nondimensional form can be expressed as

$$\frac{u_{ref}}{D}\left[\frac{\partial\left(u/u_{ref}\right)}{\partial(x/D)}+\frac{\partial\left(v/u_{ref}\right)}{\partial(y/D)}+\frac{\partial\left(w/u_{ref}\right)}{\partial(z/D)}\right]=0$$

$$\text{or, } \frac{\partial U^*}{\partial X^*}+\frac{\partial V^*}{\partial Y^*}+\frac{\partial W^*}{\partial Z^*}=0 \tag{3.32}$$

Similarly, the steady state and incompressible dimensionless momentum equation in the x direction can be written from eq 3.26b by neglecting the body force as[16]

$$\text{x-direction: } u\frac{\partial u}{\partial x}+v\frac{\partial u}{\partial y}+w\frac{\partial u}{\partial z}=-\frac{1}{\rho}\frac{\partial P}{\partial x}+\frac{\mu}{\rho}\left(\frac{\partial^2 u}{\partial x^2}+\frac{\partial^2 u}{\partial y^2}+\frac{\partial^2 u}{\partial z^2}\right) \tag{3.33}$$

$$\text{or, } \frac{u_{ref}^2}{D}\left[\left(\frac{u}{u_{ref}}\right)\frac{\partial\left(u/u_{ref}\right)}{\partial\left(x/D\right)}+\left(\frac{v}{u_{ref}}\right)\frac{\partial\left(u/U_\infty\right)}{\partial\left(y/D\right)}+\left(\frac{w}{u_{ref}}\right)\frac{\partial\left(u/U_\infty\right)}{\partial\left(z/D\right)}\right]$$

$$=-\frac{\rho u_{ref}^2}{D}\left(\frac{1}{\rho}\frac{\partial\left(P/\rho u_{ref}^2\right)}{\partial\left(x/D\right)}\right)+\frac{\mu}{\rho}\times\frac{u_{ref}}{D^2}\left(\frac{\partial^2\left(u/u_{ref}\right)}{\partial\left(x/D\right)^2}+\frac{\partial^2\left(u/u_{ref}\right)}{\partial\left(y/D\right)^2}+\frac{\partial^2\left(u/u_{ref}\right)}{\partial\left(z/D\right)^2}\right)$$

$$\text{or, } U^*\frac{\partial U^*}{\partial X^*}+V^*\frac{\partial U^*}{\partial Y^*}+W^*\frac{\partial U^*}{\partial Z^*}=-\frac{\partial P^*}{\partial X^*}+\frac{1}{\mathrm{Re}_D}\left(\frac{\partial^2 U^*}{\partial X^{*2}}+\frac{\partial^2 U^*}{\partial Y^{*2}}+\frac{\partial^2 U^*}{\partial Z^{*2}}\right) \tag{3.34a}$$

Similarly, one can deduce in the *y* and *z* directions from eqs 3.26c–3.26d as follows:[16]

y-direction:

$$U^*\frac{\partial V^*}{\partial X^*}+V^*\frac{\partial V^*}{\partial Y^*}+W^*\frac{\partial V^*}{\partial Z^*}=-\frac{\partial P^*}{\partial Y^*}+\frac{1}{\mathrm{Re}_D}\left(\frac{\partial^2 V^*}{\partial X^{*2}}+\frac{\partial^2 V^*}{\partial Y^{*2}}+\frac{\partial^2 V^*}{\partial Z^{*2}}\right) \tag{3.34b}$$

z-direction:

$$U^*\frac{\partial W^*}{\partial X^*}+V^*\frac{\partial W^*}{\partial Y^*}+W^*\frac{\partial W^*}{\partial Z^*}=-\frac{\partial P^*}{\partial Z^*}+\frac{1}{\mathrm{Re}_D}\left(\frac{\partial^2 W^*}{\partial X^{*2}}+\frac{\partial^2 W^*}{\partial Y^{*2}}+\frac{\partial^2 W^*}{\partial Z^{*2}}\right) \tag{3.34c}$$

where **Re** is the nondimensional number called Reynolds number ($=\rho u_{ref} D/\mu$). It can be noticed from the above relations [eqs 3.34a–3.34c] that the flow is a function of dimensionless number, that is, Reynolds number alone.

Similarly, the steady state and incompressible dimensionless energy equation can be written as (neglecting the pressure work, external source, and generation term) in eq 3.27e as

$$u\frac{\partial T}{\partial x}+v\frac{\partial T}{\partial y}+w\frac{\partial T}{\partial z}=\frac{k}{\rho c_p}\left[\frac{\partial^2 T}{\partial x^2}+\frac{\partial^2 T}{\partial y^2}+\frac{\partial^2 T}{\partial z^2}\right]+\frac{\mu}{\rho c_p}\Phi \tag{3.35}$$

where, Φ = viscous dissipation term =

$$\left[2\left\{\left(\frac{\partial u}{\partial x}\right)^2+\left(\frac{\partial v}{\partial y}\right)^2+\left(\frac{\partial w}{\partial z}\right)^2\right\}+\left(\frac{\partial v}{\partial x}+\frac{\partial u}{\partial y}\right)^2+\left(\frac{\partial w}{\partial y}+\frac{\partial v}{\partial z}\right)^2\right]$$

The above energy equation is normalized by using additional dimensionless temperature parameter $T^*=\dfrac{T-T_{ref}}{T_s-T_{ref}}$ such that T_{ref} is the reference or free stream temperature and T_s is the measured surface temperature or center temperature of the pipe.

$$\frac{u_{ref}\left(T_s-T_{ref}\right)}{D}\left[\left(\frac{u}{u_{ref}}\right)\frac{\partial\left({}^{T-T_{ref}}/_{T_s-T_{ref}}\right)}{\partial\left({}^{x}/_{D}\right)}+\left(\frac{v}{u_{ref}}\right)\frac{\partial\left({}^{T-T_{ref}}/_{T_s-T_{ref}}\right)}{\partial\left({}^{y}/_{D}\right)}+\left(\frac{w}{u_{ref}}\right)\frac{\partial\left({}^{T-T_{\infty}}/_{T_s-T_{ref}}\right)}{\partial\left({}^{z}/_{D}\right)}\right]$$

$$=\frac{k}{\rho c_p}\times\frac{\left(T_s-T_{ref}\right)}{D^2}\left[\frac{\partial^2\left({}^{T-T_{ref}}/_{T_s-T_{ref}}\right)}{\partial\left({}^{x}/_{D}\right)^2}+\frac{\partial^2\left({}^{T-T_{ref}}/_{T_s-T_{ref}}\right)}{\partial\left({}^{y}/_{D}\right)^2}+\frac{\partial^2\left({}^{T-T_{ref}}/_{T_s-T_{ref}}\right)}{\partial\left({}^{z}/_{D}\right)^2}\right]+\frac{\mu}{\rho c_p}\times\frac{u_{ref}^2}{D^2}\Phi^*$$

or

$$U^*\frac{\partial T^*}{\partial X^*}+V^*\frac{\partial T^*}{\partial Y^*}+W^*\frac{\partial T^*}{\partial Z^*}=\frac{1}{\mathrm{Re}_D\,\mathrm{Pr}}\left(\frac{\partial^2 T^*}{\partial X^{*2}}+\frac{\partial^2 T^*}{\partial Y^{*2}}+\frac{\partial^2 T^*}{\partial Z^{*2}}\right)+\frac{\mathrm{Ec}}{\mathrm{Re}_D}\Phi^* \tag{3.36}$$

where, $\Phi^*=\left[2\left\{\left(\frac{\partial U^*}{\partial X^*}\right)^2+\left(\frac{\partial V^*}{\partial Y^*}\right)^2+\left(\frac{\partial W^*}{\partial Z^*}\right)^2\right\}+\left(\frac{\partial V^*}{\partial X^*}+\frac{\partial U^*}{\partial Y^*}\right)^2+\left(\frac{\partial W^*}{\partial Y^*}+\frac{\partial V^*}{\partial Z^*}\right)^2\right]$, **Pr** is another dimensionless number called Prandtl number $\left(=\mu c_p/k\right)$, and **Ec** is Eckert number $\left[=u_{ref}^2/c_p\left(T_s-T_{ref}\right)\right]$.

From eq 3.36, one can observe that the nondimensional temperature is a function of nondimensional numbers, that is, Reynolds number, Prandtl number, and Eckert number. It should be noted that the Eckert number originates from the viscous dissipation term and only becomes a significant factor at high Mach number. However, in case of low Mach number, it is often acceptable to neglect the viscous dissipation terms. Then, the governing nondimensional energy equation is only dependent on Reynolds and Prandtl numbers as follows:[16]

$$U^* \frac{\partial T^*}{\partial X^*} + V^* \frac{\partial T^*}{\partial Y^*} + W^* \frac{\partial T^*}{\partial Z^*} = \frac{1}{\mathrm{Re}_D \, \mathrm{Pr}} \left(\frac{\partial^2 T^*}{\partial X^{*2}} + \frac{\partial^2 T^*}{\partial Y^{*2}} + \frac{\partial^2 T^*}{\partial Z^{*2}} \right) \tag{3.37}$$

3.4.9 NEWTONIAN AND NON-NEWTONIAN HAGEN–POISEUILLE FLOW

The Hagen–Poiseuille flow model is for one-dimensional flow in the cylindrical domain. It is one of the simplest flow models for blood in vessel also referred to as a plug flow. One can assume plug flow using Lagrange flow representation as a one-directional flow of blood mimicked by cylindrical domain moving one position to another position wherein the centerline of the cylindrical domain is the same as the blood vessel axis as shown in Figure 3.10A.

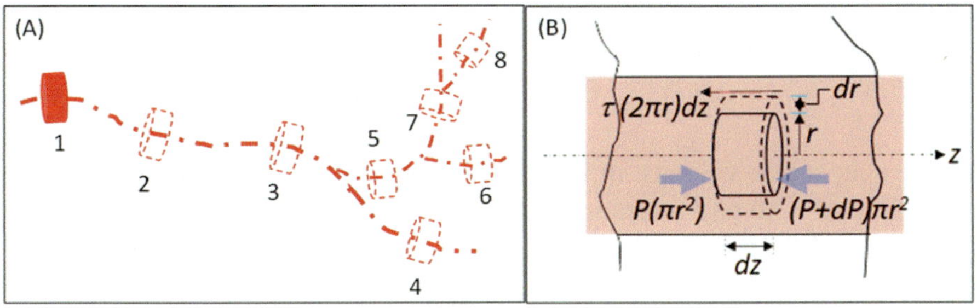

FIGURE 3.10 (A) Illustration of one-dimensional plug flow about the centerline of the vascular architecture, and (B) the cylindrical section represents forces acting on a small element by upstream and downstream fluid elements and force at circumference of the same element due to viscosity.

The Hagen–Poiseuille flow model is developed for steady, laminar, incompressible, viscous flow in a cylindrical tube of constant cross-section as shown in Figure 3.10B.[2] It assumes a region of fluid flow which is fully developed in a constant cross-section cylinder wherein the pressure gradient (dP/dz) is constant and viscous dissipation is negligible. If a force balance (or momentum conservation) in the direction of flow (i.e., z) is made, then we get[2]

$$P\left(\pi r^2\right)-\left(P+dP\right)\pi r^2-\tau\left(2\pi r\right)dz=0$$

$$\text{or } -dP\left(\pi r^2\right)=\tau\left(2\pi r\right)dz$$

$$\text{or } \tau=-\frac{r}{2}\frac{dP}{dz} \tag{3.38}$$

From eq 3.38, one can see the shear stress or resistance offered due to viscosity between two fluid layers varies with radius and gradient of pressure. However, at $r = 0$, the magnitude of stress becomes nearly zero and at $r = R_{vessel}$ the value is maximum. In fact, the stress observed the wall of vessel (i.e., R_{vessel}) is also referred to as wall shear stress. The wall shear stress can be written as

$$\tau_{wall}=-\frac{R_{vessel}}{2}\frac{dP}{dz} \tag{3.39}$$

3.4.9.1 *NEWTONIAN FLUID*

Now, in case of Newtonian fluid, stress is acting opposite to the direction of flow and can be given as

$$\tau=-\mu\dot{\gamma}=-\mu\left(\frac{dv_z}{dr}\right) \tag{3.40}$$

where v_z is scalar velocity in the z-direction or direction of flow. Equating eqs 3.38 and 3.40 and rearranging terms we can write change in velocity of Newtonian fluid as

$$dv_z=\frac{1}{2\mu}\frac{dP}{dz}\left(rdr\right)$$

Integrating both sides of the above equation leads to

$$v_z=\frac{1}{2\mu}\frac{dP}{dz}\frac{r^2}{2}+C_1 \tag{3.41}$$

Next, substituting a boundary equation at the vessel wall, we can determine the constant of integration as follows:

$$v_z\big|_{r=R_{vessel}}=0$$

$$or,\ \frac{1}{4\mu}\frac{dP}{dz}R_{vessel}^2+C_1=0$$

$$or,\ C_1=-\frac{1}{4\mu}\frac{dP}{dz}R_{vessel}^2$$

Therefore, the velocity v_z from eq 3.41 can be written as

$$v_z = \frac{1}{4\mu}\frac{dP}{dz}\left[r^2 - R_{vessel}^2\right] \tag{3.42}$$

In eq 3.42, v_z is always positive as we track velocity radially from 0 to R_{vessel}, since the dP/dz is a negative value of pressure drop to attain positive velocity (v_z) in the direction of flow. Again, it should be noted that maximum velocity ($v_{z,max}$) would exist at $r = 0$ of the vessel and minimum value would exist at the vessel wall (i.e., $r = R_{vessel}$). Therefore, substituting $r = 0$ into equation (eq 3.21) provides us with the maximum velocity

$$v_{z,\max} = -\frac{1}{4\mu}\frac{dP}{dz}R_{vessel}^2$$

$$or,\ \frac{1}{4\mu}\frac{dP}{dz} = -\frac{v_{z,\max}}{R_{vessel}^2}$$

Substituting the above expression back into eq 3.42, provides us with another form of velocity v_z in terms of vessel centerline velocity

$$v_z = v_{z,\max}\left[1 - \left(\frac{r}{R_{vessel}}\right)^2\right] \tag{3.43}$$

To determine the flux Q that crosses the cylindrical cross-section, we can determine the total flow passing through the vessel cross-section as

$$Q = \int_{r=0}^{r=R_{vessel}} (2\pi r) v_z(r)\, dr$$

$$\text{or, } Q = \frac{\pi}{2\mu}\left(\frac{dP}{dz}\right)\int_0^{R_{vessel}} \left(r^3 - rR_{vessel}^2\right) dr$$

$$\text{or, } Q = -\frac{\pi R_{vessel}^4}{8\mu}\left(\frac{dP}{dz}\right)$$

3.4.9.2 NON-NEWTONIAN FLUID

Similar to Newtonian fluid, in case of non-Newtonian, fluid stress is acting opposite to the direction of flow and can be given as follows:

$$\tau = -\mu\dot{\gamma}^n = -\mu\left(\frac{dv_z}{dr}\right)^n \tag{3.44}$$

Here, for illustration purposes the case of power-law fluids (eq 3.3) is considered. Now, equating eqs 3.38 and 3.44 and rearranging terms we can write change in velocity of non-Newtonian fluid as

$$dv_z = \left(\frac{1}{2\mu}\frac{dP}{dz}\right)^{1/n}\left(r^{1/n}dr\right)$$

Integrating both sides of the above equation leads to

$$v_z = \left(\frac{1}{2\mu}\frac{dP}{dz}\right)^{1/n}\left(\frac{n}{n+1}\right)r^{\left(\frac{n}{n+1}\right)} + C_1 \tag{3.45}$$

Next, substituting a boundary equation at the vessel wall, we can determine the constant of integration as follows:

$$v_z\Big|_{r=R_{vessel}} = 0$$

$$or,\ \left(\frac{1}{2\mu}\frac{dP}{dz}\right)^{1/n}\left(\frac{n}{n+1}\right)R_{vessel}^{\left(\frac{n}{n+1}\right)} + C_1$$

$$or,\ C_1 = -\left(\frac{1}{2\mu}\frac{dP}{dz}\right)^{1/n}\left(\frac{n}{n+1}\right)R_{vessel}^{\left(\frac{n}{n+1}\right)}$$

Therefore, the velocity v_z from eq 3.45 can be written as

$$v_z = \left(\frac{1}{2\mu}\frac{dP}{dz}\right)^{1/n}\left(\frac{n}{n+1}\right)\left[r^{\frac{n}{n+1}} - R_{vessel}^{\frac{n}{n+1}}\right] \tag{3.46}$$

To determine the flux Q that crosses the cylindrical cross-section, we can determine the total flow passing through the vessel cross-section as follows:

$$Q = \int_0^{R_{vessel}} (2\pi r)v_z dr$$

$$\text{or, } Q = 2\pi\left(\frac{1}{2\mu}\frac{dP}{dz}\right)^{1/n}\left(\frac{n}{n+1}\right)\int_0^{R_{vessel}}\left[r^{\frac{2n+1}{n+1}} - rR_{vessel}^{\frac{n}{n+1}}\right]dr$$

3.5 BOUNDARY CONDITIONS

In order to numerically solve the Navier–Stokes equation for a fluid flow domain, conditions at the domain's boundary are defined at inlet, outlet, and domain wall. We also need to define an initial condition for time-dependent problem which could be velocity or pressure field at time $t = 0$. The commonly known boundary conditions are[7]

- *Pressure boundary condition*: In this condition, the boundary either inlet or outlet was defined to have a constant or periodic function of pressure acting normal to the boundary, that is, $P_{boundary} = P_0$
- *No-slip boundary condition*: This condition is typically applied at the walls of flow domain wherein fluid is assumed to be at rest in the normal and tangential directions, that is, $u_{wall} = 0$.
- *Sliding wall boundary condition*: In this condition, the wall can be assumed to be moving or sliding in the tangential direction but no-slip condition in the normal direction, that is, $u_{wall} = 0$ in the normal direction but $u_{wall} = U_w$ in the tangential direction.
- *Mass flux boundary condition*: In this condition, a pointwise mass flux can be specified normal to the boundary which indirectly computes the normal velocity field from the mass flux, that is, $u = m_f/\rho$ such that m_f is the normal mass flux defined at the boundary.
- *Volumetric mass flow rate boundary condition*: In this condition, volumetric mass flow rate can be specified for the boundary which indirectly computes the normal velocity field at the boundary using an expression for mass flow rate, that is, $-\int_{\Omega} t_{boundary} \frac{\rho}{\rho_{ref}} u dS = \dot{Q}$, so that t is the thickness of the boundary, Q is the volumetric flow rate in the normal direction, Ω represents the hypothetical domain at the boundary having thickness t and surface S, ρ_{ref} is the reference density $\left(= P_{ref} M / RT_{ref}\right)$, where P_{ref} is standard pressure, M is the mean molar mass of fluid, constant for gas R, and T_{ref} indicates standard value. Note that the velocity field in the tangential direction of the boundary is set to zero.
- *Fully developed boundary condition*: In case of fully developed boundary condition, the flow domain extends a virtual boundary away from the true inlet or outlet boundary where constant pressure is assumed normal to the virtual boundary which gives sufficient distance to develop a fully developed velocity field at the actual boundaries. In general, the distance of virtual boundary is chosen in such a way that the flow becomes fully developed at the true boundary. In this virtual domain, a weak form of momentum equation is solved to get the velocity profile at the true boundary.

3.6 APPLICATIONS OF NAVIER–STOKES IN SIMULATION OF BIOFLUID DYNAMICS

In true sense all biofluid transport phenomena cannot be modeled using the Navier–Stokes flow model. Navier–Stokes are only useful for computer simulation of bulk flow of body fluid in the human body such as urine in bladder or blood flow in large blood vessels. Especially, blood flow is pulsatile in nature wherein the pressure waveform induces time-dependent pressure gradient for the positive downstream velocity waveform for blood. In such modeling a periodic function could be incorporated for gradient of pressure in the NS model or an explicit periodic pressure boundary condition can be applied at the boundary. Modeling flow of body fluids using the NS equation is widely employed (a) to estimate the vessel wall shear stress which is often considered one of the causes of arthrosclerosis, (b) to estimate flow of blood in brain and cardiovascular system with or without a diseased condition, and (c) to understand the flow behavior in case of aneurysm.

3.7 NUMERICAL EXAMPLES

1. Derive the average blood flow velocity (v_{avg}) in a circular cross-section blood vessel (*see* Figure 3.11) with (assume Newtonian fluid)

   a) parabolic velocity profile $v = v_{\max}\left[1-\left(\frac{r}{R}\right)^2\right]$ [ans: $v_{avg} = 0.5v_{max}$]

   b) arbitrary velocity profile $v = v_{\max}\left[1-\frac{r}{R}\right]^{1/7}$ [ans: $v_{avg} = 0.817v_{max}$]

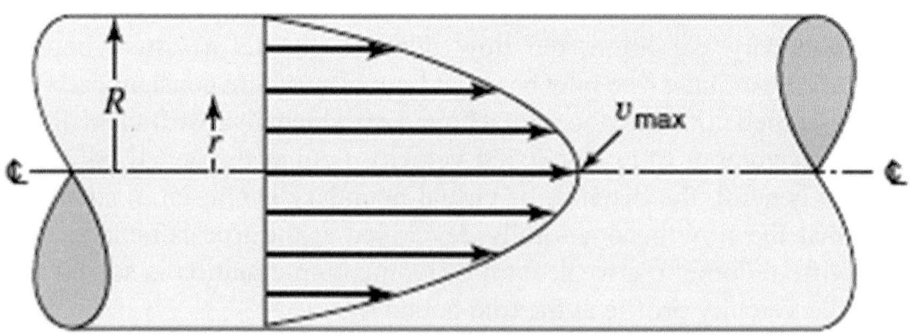

FIGURE 3.11 Parabolic velocity profile in circular vessel.

2. In the Eulerian fixed flow representation, the flow of fluid is approximated as steady 1-D velocity $v(z) = v_0 + \Delta v\left(1 - e^{-az}\right)$, which implies that at $z = 0$, fluid moves at v_0 and then accelerates downstream to $v(z \to \infty) = v_0 + \Delta v$. Derive an expression for $v = v(v_0, t)$ in Lagrange representative flow.[1]

Solution:

We know in 1-D flow in the z-direction, $v_z = dz/dt$, therefore

$$\frac{dz}{dt} = v_0 + \Delta v\left(1 - e^{-az}\right)$$

Arranging the parameters and integrating both sides leads to

$$\int_0^z \frac{dz}{v_0 + \Delta v\left(1 - e^{-az}\right)} = \int_0^t dt$$

$$\text{or, } z + \frac{1}{a}\ln\left[1 + \frac{\Delta v}{v_0}\left(1 - e^{-az}\right)\right] = \left(v_0 + \Delta v\right)t$$

Now, replacing the two z-terms in the above expression using the original v(z) expression, that is, $z = -\frac{1}{a}\ln\left[1 - \frac{v - v_0}{\Delta v}\right]$ and $e^{-az} = 1 - \frac{v - v_0}{\Delta v}$, the Lagrange form could be shown as

$$v(t) = \frac{v_0\left(v_0 + \Delta v\right)}{v_0 + \Delta v \exp\left[-a\left(v_0 + \Delta v\right)t\right]}$$

Prove that the hydraulic conductivity (κ) of a porous solid structure (a hypothetical tissue domain) with n parallel tubes represents pores as below

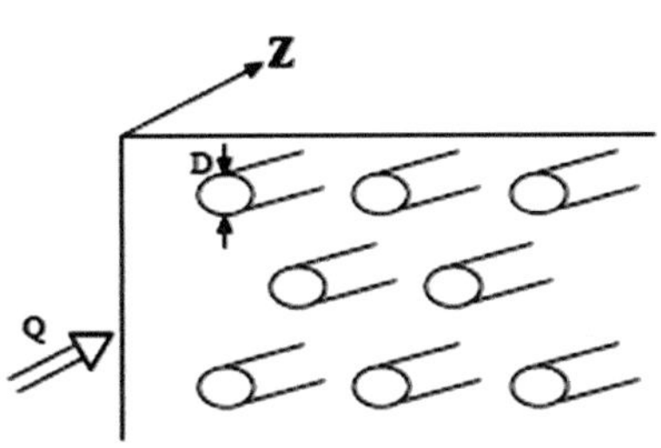

$$\kappa = \frac{D^2}{32\mu}$$

Assume 1-D Hagen–Poiseulle in the z direction, Darcy Law, Newtonian fluid, and laminar fully developed flow through the n pores to obtain the κ. The 1-D Darcy law can be defined as

$\frac{Q}{A} = v_0 = -\kappa \frac{dP}{dz}$, where the total volumetric flow rate via all n pores is denoted by Q and A is the total cross-sectional area of n parallel pores (i.e., $A = nD^2\pi/4$).[1]

3.8 SUMMARY

Blood flow through cardiovascular system and various branching vessels is one of the key aspects of living system particularly in a human body. It is often not possible to properly identify the behavior of the biofluids in the living organs based on noninvasive imaging devices mainly due to low resolution and associated noise created by machines. In fact, in many clinical settings preclinical models or patient/diseases specific models have been created to understand the behavior of biofluid flow in the organs. This chapter briefly discusses the models for bulk flow of biofluids in human body, such as simplified one-dimensional plug flow (also referred to as Hagen–Poiseuille flow model) in blood vessels to complete viscous flow of fluid in the living system (also referred to as Navier–Stokes model). The chapter also included some of the commonly viscous stress models for non-Newtonian fluid. Various boundary conditions necessary to obtain a solution of fluid flow are also discussed in this chapter. This chapter comprises the important mathematical models for biofluid dynamic modeling, and introduces some of the applications where these fluid flow models are used for practical purposes.

KEYWORDS

- **Carreau model**
- **Casson fluid**
- **Casson–Papanastasiou model**
- **Eckert number**
- **Euler flow field**
- **Hagen–Poiseuille flow**
- **Herschel–Bulkley fluid**

REFERENCES

1. Kleinstreuer, C. In *Biofluid Dynamics: Principles and Selected Applications*; CRC Press, 2006.
2. Waite, L. In *Biofluid Mechanics in Cardiovascular Systems*; McGraw-Hill, 2005.
3. Mazumdar, J. In *Biofluid Mechanics*; World Scientific, 2015.
4. Tu, J.; Inthavong, K.; Ahmadi, G. In *Computational Fluid and Particle Dynamics in the Human Respiratory System*; Springer Science & Business Media, 2012.
5. Anderson, J. D.; Wendt, J. In *Computational Fluid Dynamics*; Springer, 1995.
6. Younsi, R. In *Navier-Stokes Equations: Properties, Description and Applications*; Nova Science Publishers, 2012.
7. AB, C. *CFD Module User's Guide*. COMSOL 5.4, [Online] 2023. https://doc.comsol.com/5.4/doc/com.comsol.help.cfd/CFDModuleUsersGuide.pdf (accessed May 05, 2023).
8. Chhabra, R. P.; Richardson, J. F. In *Non-Newtonian Flow and Applied Rheology: Engineering Applications*; Butterworth-Heinemann, 2011.
9. Rubenstein, D.; Yin, W.; Frame, M. D. In *Biofluid Mechanics: An Introduction to Fluid Mechanics, Macrocirculation, and Microcirculation*; Academic Press, 2015.
10. Stokes, G. G. On the Theories of the Internal Friction of Fluids in Motion, and of the Equilibrium and Motion of Elastic Solids. *Trans. Camb. Phil. Soc.* **1845**, *8*, 287–305.
11. Cengel, Y.; Cimbala, J. In *Fluid Mechanics Fundamentals and Applications*; McGraw-Hill Education, 2013.
12. Fox, R. W.; McDonald, A. T.; Mitchell, J. W. In *Fox and McDonald's Introduction to Fluid Mechanics*; John Wiley & Sons, 2020.
13. Purohit, K.; Harsha, S. P.; Purohit, R. K. In *Fluid Mechanics*; Scientific Publishers, 2012.
14. Deen, W. M. In *Introduction to Chemical Engineering Fluid Mechanics*; Cambridge University Press, 2016.
15. Datta, A. K. In *Biological and Bioenvironmental Heat and Mass Transfer*; Marcel Dekker, Inc., 2002.
16. Oosthuizen, P. H.; Naylor, D. In *An Introduction to Convective Heat Transfer Analysis*; McGraw-Hill Science, Engineering & Mathematics, 1999.

CHAPTER 4

Bioheat Transfer in the Human Body*

MEGH R GOYAL, NOEL FARÍA GONZÁLEZ, JUAN FLORES LOZADA, and ÁNGEL E. RODRÍGUEZ BURGOS

4.1 INTRODUCTION

Biological organisms are comprised primarily of fluids, which are defined as any substance that deforms under applied shear stress. The human body lacks exemption from this generalization, with more than 70% of fluids in our body. A major purpose of these fluids is to help in the thermoregulatory processes that the body undergoes during daily activities.[1]

This chapter presents the salient features of how the human body maintains a constant temperature. In addition, it reviews the laws and principles that govern the heat transfer processes in our body.

4.2 HEAT TRANSFER[2–9]

Heat transfer processes in biological systems are regulated by three of the four thermodynamic laws.[10–12] According to the zeroth law, if two systems are in equilibrium, that is, thermal equilibrium with each other, and yet one system in equilibrium with a third one, then both systems are in thermal equilibrium with the third system (Figure 4.1). This allows a proxy evaluation of heat transfer between multiple systems or subsystems in an organism.

*This chapter has been modified from the review article prepared by students [Noel Faría González, Juan Flores Lozada and Angel E. Rodríguez Burgos] for the course on Fluid Mechanics, INGE 4015. Course Instructor: Megh R. Goyal, PhD, PE, Retired Professor in Biomedical Engineering, General Engineering Department, University of Puerto Rico–Mayaguez Campus. For details contact at goyalmegh@gmail.com.

Biofluid Dynamics of Human Body Systems: Expanded and Revised Edition.
Megh R. Goyal, Arka Bhowmik, & Anamika Chauhan (Eds.)

- The thermodynamics first law indicates changes to internal energy (U) are given by net heat added or removed (Q) minus work (W) done on the system or done by the system (eq 4.1).

$$\Delta U = Q - W \tag{4.1}$$

- On the other hand, the second law describes the property of entropy within a system and limits the direction of heat transfer within and between systems to transfer heat from states with entropy ranging from high to low.
- Whereas, the third law[2] states the system's entropy reaches a stagnant value called residual entropy as system progresses to an absolute zero temperature (0 K).

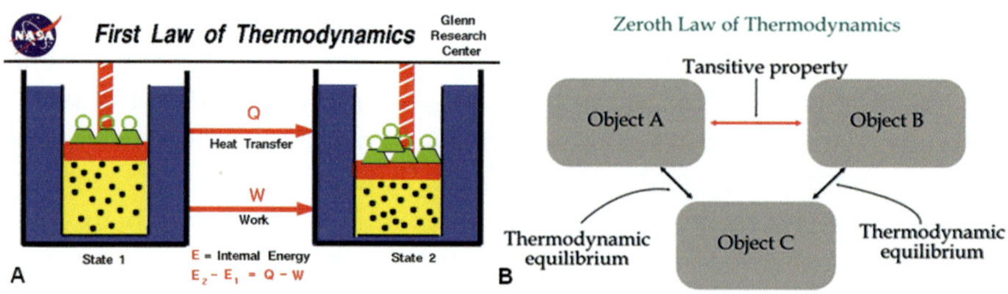

FIGURE 4.1 (A) First law of thermodynamics. (B) Thermal equilibrium among objects A, B, and C.

Source: (A) Reprinted from Ref. [13]; (B) Reprinted from Ref. [14].

Solutions to bioheat transfer equations using computational fluid mechanics and the finite element method are elaborated in numerous literature works.[11,15–18] There are the following three modes of heat transfer[2,15]: (1) **Conduction** is the process of heat energy transport by physical contact between two objects; (2) **Convection** is the transport of heat energy due to the bulk motion of fluid that carries heat from one location to another from the source of heat; and (3) **Radiation** is the transport of heat energy via electromagnetic waves, wherein the emitting object dissipates the heat energy.

4.2.1 DO BIOLOGICAL SYSTEMS OBEY THE LAWS OF THERMODYNAMICS?

In the first place, biological systems are not isolated systems that exchange energy and matter with the exterior world: they feed, they process the food, they breathe, and they excrete. A biological system dies soon after it becomes isolated. This demonstrates that the second law in the case of biological systems is not the one for isolated systems, but rather for nonisolated systems. For example, the entropy of the system may decrease if the entropy of its surroundings increases. Accordingly, biological behavior is compatible with the second law of thermodynamics since it involves the transfer of energy from higher to lower entropy systems.

4.3 FEVER

In most clinical situations, fever does no major harm and may benefit the human body by helping it fight off infection. Nevertheless, a feverish state can be a sign of severe sickness for adults and older individuals who lack strong immunity. Adults tend to run lower fevers than children and are usually less frequent or even absent in elderly patients. Fever itself can cause muscle aches, nausea, lightheadedness, weakness, and headache.

4.3.1 HOW DO WE GET FEVER?

Fever may be caused by many external or internal factors to which the human body is exposed. Causal stimuli include bacteria and their endotoxins, viruses, yeasts, spirochetes, protozoa, immune reactions, progestational hormones, as well as drugs and synthetic polynucleotides. All these substances that may cause fever are collectively referred to as exogenous pyrogens. The endogenous pyrogens act centrally on thermosensitive neurons in the preoptic hypothalamus to trigger increased heat production and minimize heat loss. Internal and external factors that cause fever are also related to other diseases. These factors may cause irritation and an acute inflammatory response. When the body notices the presence of a foreign organism, it responds with a rise in body temperature. Thus, fever is the first defense mechanism that our body uses when it detects something abnormal.

4.3.2 CLINICAL IMPORTANCE OF A FEVER

Body temperature is very helpful in identifying the condition of a patient. It is a simple, objective, and precise indicator of the physiologic state of a person and is less affected by external and psychogenic stimuli than other vital signs (such as pulse, respiratory rate, and blood pressure). Therefore, the determination of body temperature may help to identify the severity of an illness, its course and duration, and the effect of a therapy.[19]

4.3.3 BENEFITS OF FEVER

Fever therapy has been used to diminish the manifestations of some forms of chronic arthritis. Fever is a natural defense mechanism that our body uses to eliminate any strange factor that comes into our system and reduces the probability of having to take any medicine.

4.3.4 DETRIMENTAL ASPECTS OF A FEVER

The pathogenesis varies from activation of the inflammatory response to alterations in thermogenesis and heat regulation. Inflammatory diseases and intra-abdominal inflammatory processes (pancreatitis and hepatitis) may have fever associated with them. Diseases characterized by vascular inflammation or thrombosis, tissue infarction, or traumas (phlebitis; giant-cell arteritis; myocardial, pulmonary, and cerebral infarctions; and crushing mechanical trauma) are often accompanied by fever. In all these cases, complicating infections must be excluded.

Since fever is not a regular part of the daily metabolic activity of the human body, high fevers or prolonged fevers will have some negative consequences. In febrile patients, basal metabolic rates are increased, as are cardiac output and work. Sweating may aggravate the loss of salt and water. There may be general discomfort due to malaise, myalgias, headaches, photophobia, or an unpleasant sensation of warmth. In fact, the rigors and profuse sweats of hectic fevers are particularly unpleasant for many patients. In elderly patients with cardiac or cerebrovascular disease, fever and any illness associated with fever may be mortal. High or sustained fevers can activate defense mechanisms of the biological system that may lead to death.

4.3.5 MANAGEMENT OF A FEVER

The management of fever should be directed at its cause. Since fever ordinarily does little or no harm, imposes no great discomfort, and may benefit host defense mechanisms, antipyretic drugs are rarely essential to patient welfare and may obfuscate the effect of a specific therapeutic agent or of true reason. However, it is imperative where situation demands lowering of the body temperature, like heat stroke, hyperthermia, epileptic seizures, or when cardiac or central nervous system functions are impaired. Under these circumstances, we should proceed to lower the body temperature. The following techniques may be used to lower the body temperature:

- Drugs such as aspirin and acetaminophen are often used for reducing body temperature and relieving other manifestations of febrile illness.
- External cooling should be combined with massage of the skin in order to promote surface vasodilatation.
- Cooling blankets set at hypothermia temperature can be a highly effective means for external cooling. However, these effects may be negated if they induce further cutaneous vasoconstriction and shivering.
- Immediate immersion in a tub of ice water or packing with ice could be assumed to be an appropriate approach that can be lifesaving in patients with heat stroke if the internal body temperature is in excess of 42.2°C (108°F). In most cases, cooling blankets are safer and more comfortable.
- Sponging the body surface with cool saline or water or applying cool compresses to the skin and forehead may be helpful. There is no advantage to sponging with alcohol, which, because of its pungent odor, may add to a patient's discomfort.

4.4 MALIGNANT HYPERTHERMIA OR HYPERPYREXIA

Malignant hyperthermia (MH) or hyperpyrexia[20] happens during or after general anesthesia. The two main causes of MH are:

- **Triggering agents:** These are composed of instable anesthetics such as ether, halothane, chloroform, enflurane, desflurane, isoflurane, sevoflurane, and relaxant for muscle such as succinylcholine.
- **Genetic predisposition:** Some MH families have shown a defect on chromosome-19, on the ryanodine receptor gene, a calcium channel in the sarcoplasmic reticulum.

MH results from "over-excitable" muscle, which shrinks irrepressibly due to its rigidity, abnormal heart rhythms, severe fever, and kidney failure. An initial sign could be tachycardia and high fever, two of the late signs of MH. During these episodes, muscle tissue is destroyed, and breakdown products may damage the kidneys and cause acute renal failure. The urine may turn dark due to the release of muscle pigments (myoglobin). MH could be a dominant trait inherited by a child (it requires only one affected parent). Muscular dystrophy and central core disease could also be related to MH. The incidence of a genetic MH predisposition is 1:10,000; clinically identified cases are 1:30,000. The clinical symptoms of MH are not uniform, and the onset is quite variable in different patients. Symptoms of MH are given below:

- Dark brown coloration of urine
- Family history of MH or unexplained death during anesthesia
- Muscle ache without obvious exercise to explain sore muscles
- Muscle rigidity and stiffness following administration of anesthesia
- Rapid rise in temperature following administration of anesthesia

4.4.1 TREATMENT FOR MALIGNANT HYPERTHERMIA

The best medicine for MH is prevention. A family history of anesthesia-induced problems makes it important for the patient to alert the surgeon and anesthesiologist. The treatments consist of discontinuing the triggering agent and continuing the supportive and definitive therapy. Some supportive measures include 100% oxygen, administering sodium bicarbonate (1–2 mEq/kg) for acidosis, and beginning invasive cooling measures. It is also important to maintain urine output of at least 2 ml/kg/h by using the IV fluids with mannitol or furosemide. The alternate treatment is dantrolene sodium, a skeletal muscle relaxant that counteracts the high levels of intracellular calcium. Episodes of the disorder can often be prevented. Untreated episodes can be fatal; however, fortunately, the use of the drug dantrolene sodium has made fatal cases much less common.

Patient information may be obtained from the family and the patient's medical history. Episodes of MH may be prevented by giving dantrolene sodium before anesthesia, which is often a debated procedure. In susceptible individuals, it is important to avoid the administration of cocaine and amphetamine derivatives, as these may produce MH-like problems. Genetic counseling is essential for such patients or children.

4.5 HYPOTHERMIA

Hypothermia is related to a drop in temperature of the core body (i.e., <= 35°C/95°F). The central (core) temperature is always regulated by heat generation, dissipation, and exchange at the periphery.[20,21] During low-temperature environment, human physiology restricts blood from reaching the skin and the extremities to preserve, protect, and maintain core temperatures. Causes of hypothermia are:

I. Excessive heat loss
  - Environmental exposure: accidental and latrogenic.
  - Increased cutaneous blood flow: burns, psoriasis, and toxic epidermal necrolysis.

II. Inadequate heat production
  - Decreased metabolism: malnutrition, hypothyroidism, adrenal insufficiency, hepatic hypoglycemia, and diabetic ketoacidosis.
  - Altered thermoregulation: sepsis, uremia, hypothalamic dysfunction, spinal cord injury (T1 or above), and episodic spontaneous hypothermia with hyperhidrosis.
  - Drug-induced: phenothiazines, barbiturates, ethanol, opiates, clonidine, lithium, and benzodiazepines.

4.5.1 PHASES AND TREATMENT OF HYPOTHERMIA

- **Afterdrop:** The removal of cold environment victim can induce additional afterdrop of core temperature, a prominent reason for collapse after rescue. To prevent such an afterdrop, it is important to stabilize the body temperature by heating blankets or other equipment before transportation to a healthcare facility. This is primarily important since an afterdrop (~ 30°C/80°F) can cause heart ventricular fibrillation, and the patient could fail to respond to defibrillation. In other words, the rewarming of the core temperature by heating the blanket is the only solution for moderate or severe hypothermia.
- **Inhalation rewarming:** For rewarming core, one of the approaches is rewarming neck, head, and core by inhaling warm air that is water saturated at 43–45°C (107–122°F). It is also useful to rewarm the cardiac, respiratory, and hypothalamus systems. In fact, the method is equally useful for warming the central nervous system, improving consciousness and reviving the respiratory system. This also revives respiratory heat loss, typically 10–30% heat loss. In other words, this method is considered as basic approach to stabilize the thermal state of the core body and brain stem. This method of rewarming is useful for severe hypothermia patients, along with rewarming by blanket.

4.5.2 IMMERSION HYPOTHERMIA

Immersion hypothermia is a consequence of exposure to cold water, ordinary swimming, immersion, water accidents, shipwrecks, long-distance swimming, etc. Humans are at risk in the water if the water temperature is low enough to cause rapid loss of body heat. Many drowning deaths are due to exposure to cold water producing profound hypothermia. There is considerable literature on the subject of hypothermia after exposure to water during World War II. In very cold water, hypothermia can occur rapidly and cause death unless the individual is wearing a protective wet or dry suit, providing insulation against loss of body heat. Death in cold water may also be caused by difficulties related to impaired respiratory reflexes. Always consider that hypothermia may be present when a person is rescued from near-drowning and takes a rectal temperature.

4.5.3 ACCIDENTAL HYPOTHERMIA RELATED TO ALCOHOLISM

Alcoholism is well-known to predispose an individual to hypothermia.[20] Properties of alcohol (increasing loss of body heat) and the ability of alcohol to impair normal sensations, reducing awareness of dangerous exposure to cold, all contribute to making the alcoholic particularly vulnerable to accidental hypothermia. Based on animal experiments, alcohol appears to offer cardiac protection during profound hypothermia. As a result, the mortality in accidental hypothermia due to alcoholism alone is not as grave as when the individual also has some other serious systemic disorder.

4.5.4 INADVERTENT HYPOTHERMIA DURING SURGERY AND ANESTHESIA

The patient's body heat may be excessively reduced during surgical operations in operating rooms cooled by air conditioning. Anesthetics can contribute by causing cutaneous dilatation, while relaxants abolish shivering. Sweating may be altered, and the temperature-regulating center can be depressed. These and other factors may produce hypothermia. The body temperature can be monitored during the major surgery as a means of detecting body temperature changes.

4.5.5 HYPOTHERMIA DUE TO CERTAIN MISCELLANEOUS DISEASES

An important observation has been the occurrence of hypothermia in steatorrhea, which may cause unexpected death from hypothermia. Heat loss is related to a drop in body weight. Somewhat analogous is the observation that hypothermia may occur in Kwashiorkor, even in geographic areas near the equator in Africa. Hypothermia occurs in protein-calorie malnutrition, mostly in children under the age of three. Also to be noted is the frequent presence of hypothermia in terminal cirrhosis. It appears to be relatively unique to persons with cirrhosis dying of renal failure. Renal failure may be caused by hypothermia.

4.5.6 HYPOTHERMIA IN THE TROPICS

This is detailed in the literature on hypothermia in the tropics. It is reemphasized in this chapter that unlike the cold climates, malnutrition, loss of subcutaneous fat, and hypoglycemia following prolonged starvation are the main factors predisposing to hypothermia in the tropics.

4.6 OUR BODY TEMPERATURE

For a healthy human, the core body temperature[22,23] is approximately 98.6°F or 37.0°C, which is fairly constant. Though the body temperature of an individual can vary, this variation depends on the metabolic rate of each patient. A boost of the body's core temperature is associated with a higher rate of metabolism, and a sluggish metabolic rate leads to a lowering of the body temperature. In addition, body temperature may vary based on measurement site and time of measurement in a day, such as being high after food consumption or at the end of muscular activity at night and low in the morning due to a prolonged resting state. The common and standard site for measuring core temperature is oral temperature, typically at 37°C.

Measurement under armpits or skin folds could lead to inaccurate measurements and may lead to a value of 97.6°F/36.4°C. Whereas, internal measurement sites such as rectal can lead to an accurate value close to 99.6°F/37.6°C. But internal measurement is not a comfortable approach. The human core temperature differs due to age (Table 4.1).

TABLE 4.1 Body Temperatures by Age.[23]

Age	Temperature (°F)	Temperature (°C)
0–3 months	99.4	37.4
3–6 month	99.5	37.5
0.5–1 year	99.7	37.6
1–3 years	99.0	37.2
3–5 years	98.6	37.0
5–9 year	98.3	36.8
9–13 years	98.0	36.6
> 13 years	97.8–99.1	36.5–37.2

TABLE 4.2 Symptoms Due to Changes in Temperature.[7,23,24]

Temperature (°C)	Symptoms
28	Muscle failure
30	Loss of body temp. control
33	Loss of consciousness
37	Normal
42	Central nervous system breakdown
44	Death[a]

[a]By irreversible protein "denaturation," or unfolding; once their shape changes, they cease to function properly.

4.6.1 CIRCADIAN RHYTHM

A rhythm that operates synchronously in a 24-hour cycle, such as human functions both physiological and activity during sleep, body temperature, alertness, neurotransmitter levels, etc., is referred to as a circadian rhythm. Even if the individual is not aware of time, the body still maintains the cycle through 24-hour activities. However, it is not yet clear how the brain tracks time, which involves coordination of different brain areas.[22,25] The part of the brain involving the **hypothalamus** is referred as suprachiasmatic nucleus, which acts like a natural "pacemaker."

4.6.2 TEMPERATURE REGULATION[1,21,26–32]

The body has an inherent regulation system to maintain the core temperature at a constant level (Table 4.2). In fact, a healthy individual's core body temperature remains constant at approximately 98.6°F/37°C, irrespective of the changes in environmental conditions. The brain communicates with the central nervous system to regulate the core body temperature changes, such as activating evaporative cooling via perspiration at the skin surface in excessive core temperature, restricting the flow rate of blood to extremities in cold weather, or wearing warm clothes to restrict the heat dissipation in the colder environment. The different ways the body's core temperature is maintained are (Figure 4.2):

- Heat is produced through the metabolism of food. More heat is produced when metabolism is increased, and vice versa. For example, metabolism is increased when the diet is high in protein and when body temperature is elevated.
- Heat production in our body is increased by secretions of epinephrine, norepinephrine, and thyroxine.
- Exercise produces heat by stimulating metabolism.
- The body gains heat physically from its environment, but this is of lesser significance compared to the heat produced chemically. For example, clothing, the sun, and the ingestion of hot food may increase body temperature.
- Heat is dissipated from the body primarily through physical processes. As much as 95% of heat is lost through radiation, convection, and the evaporation of water from the lungs and skin. For example, the body is protected from overheating when it is cooled by the evaporation of perspiration. A negligible amount of heat is normally lost through conduction. The human body can lose heat by conduction when it is in contact with a cold surface, such as when a cooling blanket is placed over the body.
- Heat is lost in small amounts through the urine and feces and in raising the temperature of the inspired air to body temperature.
- Changes in the vascularity of the skin modify body temperature. When blood is directed to the skin through dilated vessels, heat loss is increased, and when the skin vessels contract, heat is conserved.
- The contraction of smooth muscles occurs when gooseflesh occurs. The involuntary movement of skeletal muscles occurs when shivering is present. Heat is produced by stimulating metabolism, and heat is conserved by constricting superficial vessels in the skin.

The body's thermostat is the hypothalamus, which maintains a constant body temperature. The anterior hypothalamus controls heat dissipation, and the posterior hypothalamus governs heat conservation efforts.

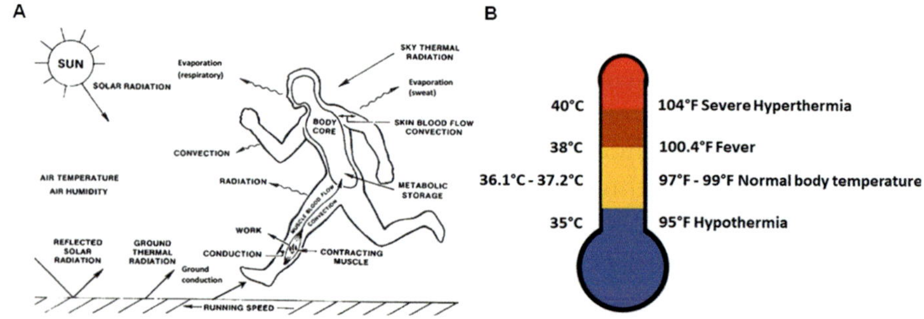

FIGURE 4.2 (A) Body heat production within active muscle, its transfer from the core to the skin, and the subsequent exchange with the environment. (B) Normal body temperature, hyperthermia, and hypothermia temperatures.

Source: (A) Adapted from Ref. [22]; (B) Adapted from Ref. [33].

4.6.2.1 NORMAL BODY TEMPERATURE

Variations of 0.3–0.6°C of the average normal temperature can occur in each person, and they are within normal limits (Figure 4.2B). However, wider variations from the average temperature have been found to be normal for certain persons. Factors such as endocrine secretions, environmental conditions, exercise, drugs, food, emotions, and age can have some effects on variations in body temperature. Newborns and young children normally have a higher body temperature. Oral temperature with an ordinary thermometer is most commonly used to measure our normal body temperature.

4.6.2.2 ELEVATED BODY TEMPERATURE

In general, the temperature of the body for adults changes little. Especially during menstrual cycle, the temperature of the body in women can increase by a few degrees. The temperature of the body can be elevated during the evening. Ideally, physical assertion, emotional changes, clothing, food consumption, room temperature, and humidity can contribute to the elevation of body temperature.[1,3,15,21,24] In the case of children, the internal temperature (i.e., rectal measurement) of 100.4°F/38°C could be assumed as no fever, but a temperature approach to or above 100.5°F shall be assumed as a fever. The fever in children can be caused by the body's immune system fighting the viral infection. The reason being that bacteria and viruses in the body find favorable condition at 98.6°F; the elevation in the body's temperature destabilizes the viral infections.[19] Additionally, fever is a triggering mechanism for the immune system to generate white blood cells, antibodies, and other defense mechanisms.

Fever-induced brain damage can only be possible at temperature >107.6°F/42°C. In general, untreated fever would not cross 105°F. It is necessary not to overdress the child in a hot environment, which could cause an increase in temperature. Ideally, the brain being the central system regulates the fever below 106°F. High fever can induce febrile seizures for a moment, which have fewer lasting consequences. Fever could also originate due to toxins, autoimmune diseases, and cancer, apart from viral infections.

Brain stroke induced by heat is a common cause of death at elevated environmental temperatures since the body lacks the ability to regulate the temperature in high-temperature environments.[22,25] It is important for parents not to leave a child in hot weather, or cars, and for players to avoid exercise in elevated temperature weather. Often, hidden infections or injuries can lead to unexplained origin of fevers, which are referred to by physicians as fevers of undetermined origin (FUO).

4.6.2.3 SUBNORMAL BODY TEMPERATURE

The human body has a thermal regulating or feedback system that maintains the body temperature at a subnormal level in cold and warm weather.[1,21] Additionally, various cultural techniques are adopted by humans to adjust the body's temperature in extreme temperatures and humidity.[23,24]

Especially in dire cold conditions, the human body is subjected to hypothermia, during which the body's temperature may drop to 94°F/34.4°C or 85°F/29.4°C since the body's temperature-regulating mechanism fails.[20] The decline in the body's temperature can lead to death, but in a few situations, with proper approach, individuals can be revived back after severe hypothermia, such as dropping to 57–60°F/13.9–15.6°C.

4.6.3 MEASUREMENT OF BODY TEMPERATURE[11,12,34,35]

4.6.3.1 GLASS THERMOMETER

Glass thermometers are accurate and economical. The most common types are filled with mercury. These also use spirits instead of mercury.

This is less dangerous because it is nontoxic in the case of glass breakage. Some of the applications for these thermometers include laboratories, dishwashers (used as a maximum registering), and fever. The accuracy is typically $\pm 1°$ scale division. Calibration is accomplished by returning the thermometer to a standard laboratory for verification against a NIST thermometer.

4.6.3.2 ELECTRONIC THERMOMETER

The electronic thermometer mostly uses thermistors for temperature measurement. It uses the principle of change in resistance due to temperature, which leads to the computation of temperature based on measured changes in the resistance of the thermometer.

4.6.3.3 DISPOSABLE SINGLE-USE THERMOMETER

A disposable single-use thermometer registers body temperature in seconds. These are nonbreakable because they are used only once and eliminate the danger of cross infection.

4.6.3.4 TEMPERATURE-SENSITIVE PATCH OR TAPE

A patch or tape is mostly applied to the abdomen or forehead. The temperature-sensitive tape or patch changes color in different temperature ranges. This type of equipment is often used with newborn infants. A thermometer may be used to check the temperature if the color on the tape or patch indicates that the temperature is either above or below the normal average range.

4.6.3.5 MONITORING DEVICE

Monitoring devices used in hospitals make it possible to have a constant measurement of the patient's vital signs, including body temperature.

4.7 RESPONSE OF OUR BODY TO HEAT AND COLD

4.7.1 ILLNESSES DUE TO LONG-TERM (CHRONIC) HEAT EXPOSURE

Chronic and long-term heat exposure can lead to liver, central nervous system, skin, heart, kidney disorders, and sleep disturbances.[36] High radiation or infrared radiation is damaging to the eye because the human eye is not equipped to filter radiation beyond the visible range. For example, thermal radiation can cause damage to human eye structures. In fact, many years of exposure to infrared radiation can lead to cataracts in human eyes, such as for workers in foundries, blacksmiths, and during high-heat operations in ovens. Many laboratory or animal studies have revealed adverse effects of heat on reproduction function, for instance, a lower conception rate in males and a vulnerable reproductive cycle in females. The effects of chemical toxicity and heat may simultaneously increase the reactivity to certain conditions. The rise in testicular temperature by 3–5°C reduces the count of sperm. Similarly, reduced fertility is a commonly observed phenomenon in women. However, there is a lack of evidence that indicates occupational heat exposure affects fertility cell production in men and women. Yet, many laboratory and animal studies have revealed that the chances of embryo deaths and malformation of critical parts such as the brain and central nervous system may occur due to prolonged hyperthermia.

4.7.2 THE EFFECTS OF HEAT ON THE CIRCULATORY SYSTEM, BLOOD, AND LYMPH

The caliber of the blood vessels, under hypothalamic influence, can be noted quickly when heat or cold receptors are stimulated.[19,22,25,26,36] The vasodilatation caused by heat and the vasoconstriction caused by cold are the bases for most therapeutic uses of heat and cold applications.

The peripheral blood vessels dilate when heat is applied to dissipate heat from the body to the environment.[11,23,25] This phenomenon can be observed in the skin, which becomes flushed and warm in the presence of heat. The caliber of vessels influences blood pressure. It falls in the presence of heat owing to vasodilatation, although this effect may vary somewhat among people. The heart's output of blood decreases with each stroke, and the pulse rate increases. The number and motility of leukocytes and lymph flow increase. Heat lowers the viscosity of blood, thereby helping it to flow more easily and discouraging blood clotting.

4.7.3 THE EFFECTS OF COLD ON THE CIRCULATORY SYSTEM, BLOOD, AND LYMPH

When cold is applied, vasoconstriction occurs to help conserve heat by not allowing heat dissipation from the skin, instead restricting it to the internal body. This phenomenon can be observed in the skin, which becomes pale and cool in the presence of cold. In the presence of cold and vasoconstriction, there is a rise in blood pressure and a decrease in lymph flow. The heart's output of blood increases with each stroke. Authorities indicate that the pulse rate under cold conditions may decrease or increase. Viscosity tends to increase, and the blood flows slowly in the presence of a cold. These reactions to the cold may facilitate blood clotting.

4.7.4 THE EFFECTS OF HEAT AND COLD ON THE INTEGUMENTARY AND MUSCULAR SYSTEMS

The skin's heat and cold receptors adjust readily.[22] For example, if the arm is placed in warm water, the sensation of warmth soon diminishes because of the adaptability of the heat receptors. The same phenomenon occurs if cool water is used. It is important to observe the ability of receptors to adapt to heat and cold. A person may become unaware of temperature extremes after tissue damage occurs.

- **Heat:** The density and location of heat receptors are located deep in the skin. When a person is exposed to warm surroundings, the sweat glands secrete perspiration. The body cools when the perspiration changes from liquid to vapor. Evaporation requires heat, which is given by the body's heat as a dissipation mechanism. Warmth tends to have an atonic, that is, relaxing effect on the body's muscles, while cold has an opposite effect. This can be observed in daily living. A warm bath is relaxing, while cold water tends to contract and tense muscles.
- **Cold:** Cold receptors are particularly numerous in the skin on the thorax and upper extremities; hence, these areas respond quickly to cold. The cold receptors are estimated to be approximately 8–10 times more numerous than the heat receptors. Also, receptors for cold lie superficially. When a person is exposed to a cold, perspiration is lacking. In addition, the pilomotor muscles contract and make the hair stand on end to cause gooseflesh to conserve heat. Shivering, also under hypothalamic control, generates considerable heat by agitating

muscles in the body. It also increases metabolism, oxygen usage, and blood circulation. The cold has an anesthetic effect on the area to which it is applied due to a rise in the threshold of pain receptors (cold therapy). Because of the anesthetic effect of cold, a person may become unaware of impulses from the skin that normally serve to warn that tissue damage is occurring.

4.7.5 THE EFFECTS OF HEAT AND COLD ON REMOTE AREAS OF THE BODY

A change in the location of blood in the body in response to vasodilatation and vasoconstriction affects tissues far removed from the area where heat and cold are being used. Heat applications decrease blood congestion in internal organs when blood is shifted to peripheral vessels. For example, if the feet are placed in a basin of warm water, congestion of blood in the abdominal organs is relieved as blood is diverted to the legs and feet. Blood in the body's internal organs increases when cold is applied.

4.7.6 THE EFFECTS OF HEAT AND COLD ON TISSUE METABOLISM

Tissue metabolism increases in the presence of heat due to an increased blood supply to tissues that allows more oxygen and carries wastes away more rapidly. There is a decrease in tissue metabolism in the presence of cold, as less oxygen is used and fewer wastes accumulate.[36]

4.7.7 THE EFFECTS OF HEAT AND COLD ON PROLONGED EXPOSURE

Because of the effects of prolonged exposure to heat and cold, it is best to remove applications of either heat or cold for a short period and then reapply again. This helps to bring on the intended initial effects. In some instances, heat and cold applications are applied alternately to prevent the effects of prolonged exposure.

- **Heat:** Blood flow decreases when the application of heat to a local area continues beyond 1 h. It is possible that excessively long applications of heat increase the local tissue metabolism more than transport via blood to different body organs. It is probably reflex vasoconstriction that results from the body's efforts at homeostasis through the autonomic nervous system. Prolonged applications of heat have other effects, too. A weakened patient becomes even weaker, especially if the area to which heat is being applied is large. Also, prolonged heat applications weaken cutaneous cells, and as a result, the skin becomes sensitive to injury.

- **Cold:** The immediate effect of cold application is vasoconstriction; the prolonged effect is vasodilatation, which is a defense measure of the body due to reflex action. The prolonged cold may cause damage to nerves in vessel walls, and following such damage, vasodilatation may result. Prolonged applications of cold, even with eventual vasodilatation, are contraindicated because the activity of cells is so decreased that their survival is threatened. The first danger signs include blueness and mottling of the skin and numbness and stiffness in the body part.

4.7.8 THE EFFECTS OF EXPOSURE TO INTENSE HEAT AND COLD

The almost immediate effect of the application of **intense heat** is body vasoconstriction. Muscles grow tense owing to a decrease in blood supply to the area. Providing nourishment and removing wastes from the cells cannot therefore be accommodated effectively. Cell destruction then occurs, and blisters form on the skin. It is generally believed that pain and heat receptors are stimulated when intense heat is placed on the skin, and the sensation of burning is the result of this double stimulation of receptors. A less widely accepted theory is that a second type of heat receptor in the skin with a very high threshold is stimulated when very hot objects touch the skin.

Intense cold causes frostbite, and eventually, freezing as ice crystals are formed in tissue fluids. This causes damage and death to cells that are being inadequately nourished in the presence of an intense cold.

4.8 ROLE OF BLOOD CIRCULATION IN INTERNAL HEAT TRANSFER

All body tissues are poor conductors of heat. In fact, unperfused (not permeated with flowing blood) muscle or fat tissues have essentially the same thermal conductivity as cork.[16,23,25,37] Hence, if the heat transfer by the body depended on conduction, then the ability to adapt to varying environmental conditions would be poor. Therefore, the convective flow of blood throughout the body is very important for internal heat transfer. The blood circulation affects heat transport in three major processes:

- Reducing the difference in temperature spatially and temporally. High metabolic rates (liver tissue) in tissue are associated with highly perfused tissues and are thus kept at nearly the same temperature in low active tissues. Cooler tissues are warmed by blood coming from active organs.
- Induces mechanisms to restrict or enhance heat dissipation from the skin. Figure 4.3 indicates how warm blood flow to the neighborhood of the skin is increased by "vasodilation" when the body wishes to reject heat and how the blood is bypassed from arteries to veins via deeper channels through "vasoconstriction," when conservation of body heat is vital.[1,21] These automatic mechanisms lower the temperature gradient for heat transfer by conduction in the subskin layers.
- Induces heat exchange in tissue due to countercurrent blood flow.[10] If heat conservation is necessary, arterial blood flowing to the extremities is precooked by the loss of heat to adjacent venous streams. This reduces the temperature of the limbs and lowers heat loss. Since most arteries lie deep, while veins appear both in superficial and deep regions, the extent of arteriovenous heat exchange depends on the route taken back to the body trunk by the venous blood (Figure 4.3). This is automatically regulated by the vasodilation–vasoconstriction mechanism.

4.9 CARDIOVASCUALR SYSTEM

The competition between thermal demands for skin blood flow and blood flow to other organs is possibly larger in humans due to the regulation of heat. First, total skin blood flow during sweating can represent a sizeable fraction of cardiac output.[10,25] The venous section of the skin vasculature is highly compliant, and the constrictor tone of the veins decreases with increasing skin and core temperatures. Thus, small increases in venous pressure following arteriolar dilation are accompanied by large increases in skin blood volume. Second, a majority of the blood vessels in upright humans are additionally subject to the distending effects of orthostatic pressure. The result is a pooling of blood in the lower part of the body and a decline in central blood volume, with adverse consequences on cardiac filling pressure, stroke volume, and arterial pressure.

During supine rest, the skin's requirements for large blood flow are often sufficed by higher cardiac output. However, there may be a redistribution of blood flow that can become pivotal in situations in which cardiac output approaches its upper limit during exercise. Compared to rest in a thermoneutral environment, splanchnic and renal blood flows are reduced during heat exposure and exercise. Thus, those vascular regions whose functions are not seriously compromised by a temporary shortage "lend" part of their blood flow to the skin. This option is used in humans and possibly all other species. During rest in the heat, the thermoregulatory mechanism in the skin ensures adequate skin blood circulation even though cardiac output remains constant.

The regional increase in sympathetic vasoconstrictor activity that contrasts with decreasing activity in the skin leads to reduced blood flow in the splanchnic. The input signal is purely thermal. Local warming of peripheral, spinal, or hypothalamic temperature sensors can induce not only increases in skin blood flow but also a decrease in splanchnic blood flow. In human subjects, resulting in a thermoneutral environment, the hepatic and splanchnic vessels contain approximately 25% of total blood volume, and most of it is on the compliant venous pressure and volume, which helps to counteract the drain on central blood volume caused by high skin blood flow. Therefore, there is a controlled redistribution of blood flow and volume.

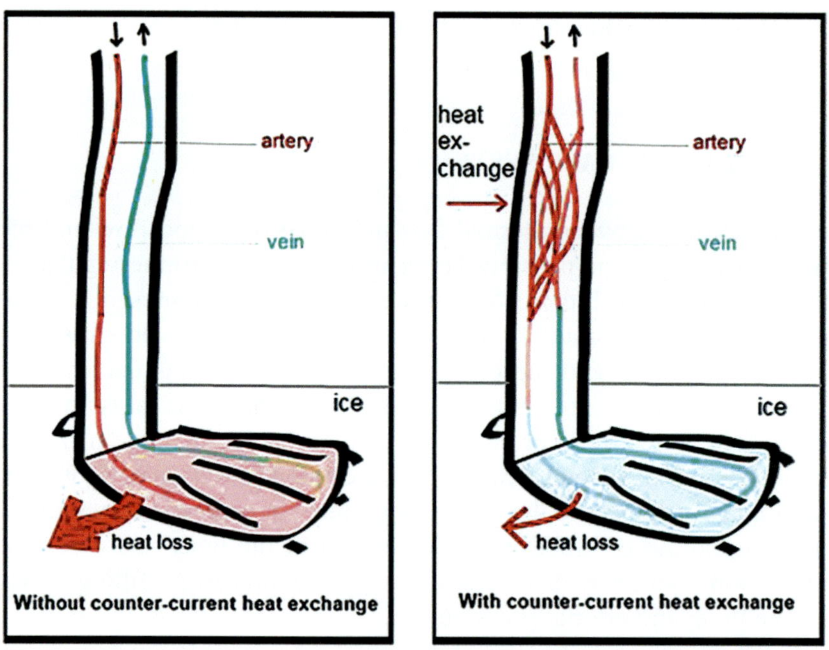

FIGURE 4.3 Countercurrent heat exchanges in the extremities when valve is open.
Source: Reprinted from Ref. [38].

During rest in an upright position or exercise, the splanchnic vasoconstriction is complemented by a relative vasoconstriction of the skin, so that the redistribution of blood flow to the skin is lower than in supine rest. The response is mediated by baroreceptor reflexes and a higher background level of general sympathetic activity, so a higher T_{core} is required for a certain level of skin blood flow. The compromise is tolerable provided T_{core} remains in the zone of proportional control (e.g., an adequate heat transfer from the core to the skin is accomplished by a relatively lower skin blood flow times a larger gradient between T_{core} and T_{skin}). Therefore, a larger fraction of cardiac output is available to other organs, such as exercising muscles, and the detrimental side effects of skin blood flow on skin blood volume are partially and temporarily mitigated.

4.10 HEAT EXCHANGE WITH THE ENVIRONMENT[2,3,6,9–12,29,34,39,40]

The thermal energy exchange from our body to the terrestrial environment takes place through radiation, conduction, convection, and evaporation (Figure 4.2). Of these, evaporation and convection (the forced convection of blood flow and to radiation a lesser extent the natural convection of airflow over the skin surface) can be altered by physiological processes. The other mechanisms of heat flow (radiation and conduction) are only indirectly under physiological control through variations in skin temperature, though these can be varied by behavioral responses.

The rate at which heat is transferred is given by units of energy, that is, Watt, which is dependent on the area of the tissue.[11,12] For instance, the transport of heat by convection, conduction, and radiation is dependent on the difference between the temperatures of the skin and its surroundings, whereas the differences in water vapor pressure on the skin and surroundings lead to evaporation phenomena.

4.10.1 HEAT RADIATION

Heat transfer by radiation is caused by the emission of electromagnetic waves that carry energy away from the emitting object. For short-wave radiation, organisms are always at the receiving end, whereas for long-wave radiation, they may lose or gain heat, depending on whether the surrounding surfaces are cooler or warmer than themselves. The Stefan–Boltzmann law defines the thermal radiation from the surface[2,10–12]:

$$Q = \varepsilon\sigma A\,(\theta^4 - \theta_c^4) \qquad (4.2)$$

where Q = heat transfer due to radiation, ε = emissivity = 1 for black body, A = radiating area, θ = temperature, σ = Stefan–Boltzmann constant[2] = 5.6703 × 10^{-8} watt/$m^2 k^4$, and θ_c = temperature of surroundings.

4.10.2 HEAT CONDUCTION

Heat energy transfer purely by vibration of molecules or lattices in materials without any bulk motion is referred to as conduction.[2,10–12] Conduction is energy transfer and involves no mass transfer. Thus, in the case of conduction between a solid and a fluid, the definition requires that the fluid should be still. Heat flow is due to the temperature gradient and follows the direction from hot to cold surfaces. Heat transfer by conduction is given by the following equation:

$$\frac{Q}{t} = \frac{kA}{d}\left(\theta_{hot} - \theta_{cold}\right) \qquad (4.3)$$

where Q, t, k, A, d, and θ are heat transfer due to conduction, time, conductivity, area, thickness, and temperature, respectively.

4.10.3 HEAT CONVECTION

The transport of heat energy due to the bulk motion of fluid that takes away heat from the source is referred to as convection. Heat is convected by air by the process of expansion or lowering the density of hot air. The hot air is less dense than the cold air, which sets up a density gradient and a convection current. Natural convection occurs when a warmer body is placed in cooler still air or water. Forced convection takes place when the medium around the body is kept moving by external forces (such as in currents of wind or water) or when the body moves itself through a still medium.

4.10.4 PERSPIRATION COOLING OF THE HUMAN BODY

In the situation of ambient temperature being higher than the skin or body temperature, heat is transported inside the body from the surroundings via all modes of heat transfer.[11,12] In other words, the only way heat is dissipated by the body is via perspiration, sweating, or exhaling moisture.[5,37] In a general day, there is about 600 g per day of "insensate loss"[9,34,41,42] of moisture from the skin (Figure 4.4). The cooling induced by perspiration is significant because it removes a significant amount of heat from water vaporization. The amount of vaporized heat is around 540 calories/g at the boiling point, but it is even higher (580 cal/g) at the normal skin temperature.

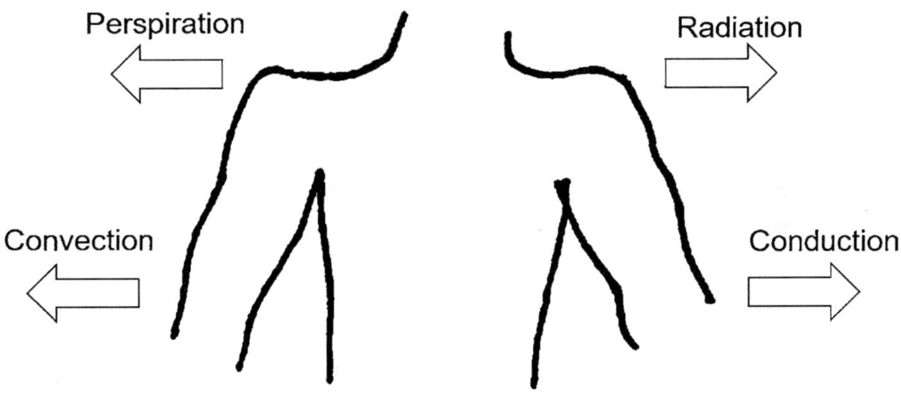

FIGURE 4.4 Perspiration (sweat) cooling of the body.

4.10.5 EVAPORATION

Evaporation is a heat transfer due to a state change from liquid into gaseous form. Evaporating sweat, which exchanges skin surface heat with the ambient atmosphere, is an example.[9,34,42] During sporting events, the individual body is at a high temperature, and the heat dissipation from the body takes place through four modes of heat transfer (Figures 4.2 and 4.4), with sweating being the dominant process (Figure 4.4). In this situation, when the body slows down, heat dissipation ultimately leads to a rise in temperature. The cooling by evaporation of the sweat depends on the humidity of the air, that is, the percentage of the vapor content of the air. In cases of high humidity (~ 100%), the water vapor content reaches a saturated condition with no more evaporation from the skin. This will lead to a rise in the body's temperature or more sweating.

The rate of sweating of about 2 quarts per hour can be observed during exercise, and it is highly recommended to consume excessive fluid to avoid dehydration. In geographic locations of high humidity and high temperatures, cooling of body by sweating is not possible, which may lead to severe health conditions if sufficient fluid is not replenished by the body.

4.10.6 HEAT OF VAPORIZATION[5,11,12,37]

The heat of vaporization is the energy required to change one gram liquid into vapor at boiling temperature. This energy breaks down molecular interaction in fluid and adds the requisite energy to expand to gas, typically energy of the amount of *PdV*. In the absence of potential energy linked to intermolecular attraction for ideal gas, the change is primarily due to the kinetic energy of molecules that contributes to internal energy.

4.10.6.1 WHY IS THE HEAT OF VAPORIZATION MORE AT BODY TEMPERATURE?

At normal core body temperature (37°C), the heat of vaporization of sweat is equivalent to 580 calories/gm, which is higher than the nominal 540 calories/gm at the boiling point.[41] It is larger at normal body temperature because at this temperature, the energy required to bind the molecules of water is higher, and at normal core temperature, it demands a larger amount of energy to convert water molecules into vapor.

4.11 DRY HEAT EXCHANGE

It is heat transport between skin and surroundings simply by convection and radiation, which means no heat is dissipated via evaporation.

4.11.1 EFFECT OF CLOTHING INSULATION

- **Thermal insulation of clothing:** Clothing plays a vital role in inducing an artificial insulation or resistance to heat transfer between skin and surroundings. The resistance due to clothing is represented by the ensemble resistance "Clo" to heat exchange, and the typical indoor combination has a Clo value of 1 (= 0.155 m^2 K/w).
- **Evaporative resistance of clothing:** The resistance offered by clothing to evaporative dissipation from the skin to its surroundings depends on the moisture permeability of the cloth and latent heat transfer. Since the metabolic rate is 58.2 W/m^2, there is neither body cooling nor body heating at an operating temperature of about 25.5°C for light-clothed people and 31°C for nude person. There is negative heat storage when

there is a drop in the temperature of the skin and superficial or deep tissues. When the operative temperature exceeds 29°C, the rate of evaporative heat loss is significantly increased, counterbalancing the reduction of dry heat exchange to maintain thermal equilibrium. The body temperature tends to rise only when the body is entirely wet, the evaporative heat loss is inadequate, and there exists a positive rate of heat storage. A body temperature >43°C may cause death.

4.12 THERMOREGULATION IN NEWBORNS

Thermal regulation plays a vital role in the acclimatization of newborns. The thermoregulation capability of newborns can efficiently work in a particular range of surrounding temperatures. The common disadvantages a newborn faces are (1) poor insulation, (2) low mass for heat sink, (3) a lesser propensity to restrict heat dissipation by changing posture, and (4) the lack of the possibility to adjust clothing requirements, etc. The ability of newborn thermoregulation can also be affected due to illness or hypoxia condition.

The exchange of heat energy between infants and their surroundings takes place through all four modes of heat transport (Figures 4.2 and 4.6). Heat loss from a newborn's body via conduction is relatively less than in other modes. Heat convection primarily depends on the convection current of air and the surrounding temperature.

On the other hand, the dissipation of heat by evaporation depends on the humidity of the surroundings and convection currents. In the case of a healthy infant, a clothed baby nursed in an incubator exhibits a small loss of heat due to evaporation. Whereas, for underdeveloped newborns nursed under an inadequate surrounding condition, evaporative loss plays a major role in heat dissipation.

A newborn generates sweat from the glands in the skin that produce sweat. The increase in skin metabolic rate led to an elevation of temperature, and an adverse internal rise in temperature could cause heat-related stroke or damage due to hyperthermia-induced dehydration in cerebral parts. In fact, newborns prior to standard 8 weeks lack the ability to sweat. For instance, a 3-week early birth can limit the sweating ability of the head and face. The ability to sweat matures over time.

An infant may shiver or perform other muscular activities to generate heat in cold surroundings, especially with the use of an automated surrounding-controlled incubator that uses a temperature sensor measuring the child's abdomen temperature to maintain the incubator temperature.

Prior to 31 weeks of gestation, heat loss is predominantly governed by evaporation, which is a result of incomplete skin keratinization, allowing larger water permeation to the skin surface. In general, the uncontrolled permeation lowers gradually in 7–10 days following birth. In fact, incubators are not always ideal for premature birth, since in situations when a premature baby undergoes respiratory problems, ideally the baby should be laid in a cradle with an overhead radiant heat source. The overhead radiant heat puts the heat back into the body while the baby is losing heat energy in some ways. During the online measurement, the baby's temperature is monitored using a thermistor. The physician also needs to overcome the issue of dehydration for premature newborns.

4.13 CONVECTIVE HEAT TRANSFER AROUND AN INFANT HEAD

Hyperthermia is thought to be the main factor associated with sudden infant death syndrome (SIDS). Normally, the mechanisms of heat and mass transfer (evaporation, convection, conduction, and radiation) are more than enough to transfer the metabolic heat produced by the body. However, when the body insulation increases (clothes and blankets), overheating may appear and should be transferred through the head, whose heat clearance efficiency can greatly vary with different environmental combinations.

4.14 HUMAN THERMOREGULATION[26,27]

Thermoregulation is any physiological or behavioral mechanism that ensures a controlled temperature suitable for proper human function in an elevated temperature environment and a lower temperature environment.[27]

There are two divisions in species according to temperature of body, such as homeotherms and poikilotherms. The species whose body temperature changes with environmental conditions are referred to as poikilotherms, whereas species that maintain a very narrow body temperature irrespective of changes in environment are known as homeotherms. For instance, humans are considered homeothermic since they maintain a stable core body temperature irrespective of environmental changes. In fact, humans exhibit different functions of the thermoregulatory system compared to other *Homo sapiens*, which originated from evolutionary challenges faced by humans over time.[27]

The key role of the thermoregulatory system in humans is to maintain a uniform and constant core temperature for all environmental conditions, such as adaptations made by the human body to limit the dissipation of heat in a cold environment. The core body temperature and mechanisms and adaptations can prevent excessive heat gain or can remove heat in hot environments to maintain the core body temperature. These mechanisms and adaptations are made by the thermoregulatory system to restrict heat input or loss.[27]

Thermal input to the human body can be realized through the transport of heat energy from air and water in situations where the body temperature is below the source temperature, via solar radiation or reflected radiation, due to shivering and human metabolism, etc., whereas, the loss of heat energy originates from the loss of heat by convection or dissipation by sweating and radiation.

The input of heat energy by the convection of hot air over the skin surface can be viewed as an important stress to the thermoregulatory system in situations of excessive heat transfer to the body. Similarly, excess radiative heat from surrounding inside the body via direct (solar) or indirect methods can be equally viewed as a challenge for the regulatory system. Furthermore, long-wavelength radiation has lower absorption by the skin, but skin pigments offer a strong absorbing capability of short-wavelength radiation. In fact, direct solar radiation can have other health-related consequences and should be avoided.[27]

Yet another input source is metabolic heat generation, which can rise to an exceedingly high value due to the heat load posed by environmental conditions. In fact, the metabolic heat generation load is an important thermoregulatory requirement to sustain balance. The commonly known sources of input are as follows:

- **Metabolic thermogenesis** takes place in the cycle of citric acid, wherein lipids and glucose are metabolized and heat is dissipated during the pathway. The value of metabolic heat generation is typically higher since it is performed in all tissue mass. The actions like potential, synaptic, and membrane transport are part of the brain's requirement for metabolic activity and lead to the dissipation of heat from brain tissue.
- **Contractile thermogenesis** refers to the generation of heat due to the movement of muscles and tendons due to friction. The movement of muscle or contraction and release needed nearly 22% of metabolic energy for work and rest is dissipated as heat.
- **Lipolysis** can be referred to as fatty tissue disintegration to produce metabolic heat. This is a dominant phenomenon in cold environment as considered important for human evolution under cold stress.

Ideally, heat dissipation by convection from humans to their surroundings contributes to stress on the thermoregulatory system in cold weather. Apart from convective loss, evaporative and radiation heat loss are also contributors to human heat dissipation mechanisms. For instance, evaporation can be activated in the chambers of the lungs or nasal compartments, in the oral mucosa and mucosa membranes of other organs, due to behavioral changes, etc. However, sweating plays a bigger role in evaporative cooling in humans. In humans, sweating actuates higher loss of heat through evaporative cooling.

4.14.1 THERMOREGULATORY ADAPTATIONS AND BEHAVIORS

Human thermoregulation is influenced by various factors, which include structural changes to the body, changes in metabolism, functional changes in vascular flow, changes to fat content, changes in human behavior, geographic locations, etc.[27]

The structural changes of the human body are a part of evolutionary modifications that could lead to variations in the average weight of similar homeotherm species populations than those living in lower-temperature environments. Also, populations of the same homeothermic species will have a shorter limb length in colder climates compared with those in high-temperature environments. The observed structural changes are important to reduce or increase heat loss because of their dependence on the mass of the body, exposed surface area, and surrounding temperature.

According to Fourier's mathematical law for thermal conduction, one can compute that the total heat loss rate from the body is directly dependent on surface area and inversely related to the thickness of the conducting tissue. A typical component is the variation in surface area to volume ratio, which tells us heat generation and dissipation ratio.

In other words, an increase in mass or fat thickness in the body would decrease the dissipation of heat from the skin to the surroundings.[26] This poses a challenge for the thermoregulatory system since an increase in body fat or mass restricts heat loss, and hence turns to a stressful situation to maintain the standard core temperature in a hot environment. Especially larger mammals implement other methods to maintain the core temperature.

Metabolism plays a vital role in heat generation and is regulated by the body depending on the environmental demand. For example, during cold weather, the rate of metabolism in the body is increased, but in hot weather, the rate of metabolism is restricted to a low value. The regulation of metabolism in the body is carried out by hormonal and neuronal activations by the brain. The thyroxine release by the thyroid gland can induce an increase or decrease in metabolic heat generation, and it can be a short- or long-term process. Additionally, adrenaline or noradrenaline released by the adrenal medulla could induce an increase in metabolic heat generation. On the other hand, neuronal activation could be the induction of shivering in muscles.

Human skin, on the other hand, has specialized thermosensitive receptors for sensing skin temperature and transporting the signals to the nervous system. These systems act as autonomous control units that restrict the blood flow to peripheral skin in case of signal revealing an increase in heat dissipation from the skin to cold surroundings, also referred to as vasoconstriction. In general, the process of vasoconstriction would drop close to 1/6–1/3 of heat loss for humans subjected to a low-temperature environment. On the other hand, the control unit ensures an increase in blood flow to peripheral skin in case of signal-revealing need for elevated heat dissipation, often referred to as vasodilation. In this process, the body ensures more heat is transferred through blood from the core to the skin and is equally responsible for sweating. Yet, vasoconstriction can increase arterial pressure, and during vasodilation, it can increase the restriction of internal blood flow caused by postural hypotension.

The subcutaneous fat content plays a vital role in the human body, that is, it either acts as an energy storage unit or as an insulator. It can also undergo microchemical steps to generate metabolic heat. Especially for mammals living in cold environments, the fat layer can form an insulation to restrict the dissipation of heat to the surroundings. On the other hand, for humans, it offers a method to store energy and can be used later for metabolic heat generation via lipolysis, wherein a significant amount of heat is generated by the oxidation of fatty acids. The evolutionary changes to thick fat layer development in certain parts of the body can be associated with hair loss and long-term exposure to cold surroundings.

Other approaches to acclimatizing the human body to surrounding temperature changes could be behavioral and cultural practices, such as indoor work during high solar activity, using shades or umbrellas in hot weather, creating nests, creating fires, building shelters, etc.[26] These approaches were essentially adopted by many ancestors in hot and cold environments.

The hair on the body, on the other hand, acts as an insulator to prevent heat transfer from the body to its surrounding or vice versa. It also acts as an insulator for radiation exposure for many mammals. A layer of hair on the body can restrict the flow of air over the body, creating a dead pocket to reduce the convection effect.

Human thermoregulation is dominated by another phenomenon called sweating, which ensures high heat removal from the skin surface due to the large value of specific heat capacity of water. Due to the large heat capacity of sweat, a considerable amount of heat is first absorbed by sweat, which is transported to the air by evaporation on the skin surface. The process of sweating is quite adaptive in nature since it varies by the production of eccrine glands, density of body hair, and flow of blood to the periphery.

- Eccrine glands are often referred to as glands that produce sweat.
- Body hair loss enhances the possibility of enhancing heat dissipation by evaporation.
- An increase in heat dissipation by evaporation is also possible by flow of blood from the core to the peripheral skin by vasodilation.

4.14.2 THE SWEATING MECHANISM IN HUMANS

The dissipation of heat from the body via evaporation is performed via sweating, a thermoregulatory mechanism. Sweating or evaporative cooling is better performed in the absence of body hair. The sweat glands in humans have some similarities with African apes in terms of structure and functioning. The human integumentary system is composed of three important glands: apocrine, sebaceous, and eccrine.

The gland name sebaceous is well known for producing oil or fluids for nourishing the hair and is attached to the hair follicle. It is also responsible for generating odor and is less significant in sweat production. Whereas, the gland name apocrine is responsible for producing cell components and breaking the apices of cells, which are also connected to the follicle of hair. These components of cells are often consumed by bacteria in the axillary areas of the body, producing body odors. The gland eccrine, on the other hand, is responsible for producing sweat, which is composed of water and mineral salts. Sweat production is primarily linked with thermoregulation.

All glands associated with the integumentary system display a standard distribution pattern throughout the body. The sebaceous gland exhibits no relation to thermoregulation but is found to be connected with all hair follicles. The gland apocrine is located in the axillary region for humans but is found in the entire body for primates. These glands are developed in the early development phase of the human fetus, along with hair follicles that are absorbed by the body during fetus development. The gland eccrine develops on all body surfaces, irrespective of the presence of hair. Especially for humans, this gland develops on the surface of the hands and feet during the 3 and ½ months of fetal development. However, it develops in the remaining body separately, close to 5 and ½ months of fetal development.

The gland eccrine can remain dormant for a long time and produce sweat only in situations of thermoregulatory need, whereas the other two glands can secrete continuously. It can follow two different secretions in a controlled manner, such as by adrenergic and cholinergic stimulation. The apocrine and eccrine glands of the friction surface are moderated through the activation of adrenergic sympathetic nerves. These nerves are activated by emotional changes, including the production of adrenaline hormones. The gland eccrine of other parts of the body is moderated by cholinergic nerves that are activated due to stress, such as heat. The gland eccrine is responsible for sweat secretion, which in the process of transport from the gland to the human surface, mixes with other glands' secretion and distributes in the axillary region. The content of the sweat is further broken down by bacteria on the surface of the skin, resulting in an odor. Essentially, the eccrine gland's true function in humans is to enhance evaporative heat loss via the secretion of sweat, whereas the eccrine glands are less reactive to heat stress in case of primates.

One of the effective dissipations of heat via the skin surface of the human body is attained by the secretion of sweat, which absorbs a large amount of body heat due to the high thermal capacity of water in sweat that finally undergoes evaporative cooling to the surrounding air. In fact, nonevaporative heat loss simply by conduction and convection is less effective than elevated evaporative cooling posed by a sweating body and a larger surface area of skin for a large body in warm climates of Africa.

Although dense hair can impede heat loss through sweating, it is important in terms of insulation from solar radiation and maintaining body heat at night or in cold surroundings. For instance, dense hair covers act as heat insulation for small animals. Larger animals, on the other hand, have less surface area compared to body mass and eventually have a smaller influence on thermoregulation due to their hair cover. In other words, a larger body increases metabolic heat generation compared to the amount of heat dissipated by the skin surface. These changes induce a restriction on the ability of a large body or animal to dissipate heat to its surroundings, and therefore the need for dense hair cover is not reasonable for the evolution of large body animals or humans. Therefore, with the increase in body size and the metabolic heat load, an elevated demand for heat dissipation originates for the body, and the demand for dense hair cover becomes less necessary, especially in hot weather. Many studies indicate dense body hair protects from solar radiation.[41]

4.15 HEAT LOSS BY SECRETION OF SWEAT (PERSPIRATION)

When an activity level rises above the basal state, additional heat is produced in the body. One of the automatic mechanisms for increasing heat loss is the sweating response, which provides secretions of a dilute electrolyte solution from numerous glands to the skin's surface. Evaporation from the wetted surface then occurs, removing about 570 kcal/g of water evaporated since.[3]

Table 4.3 gives typical amounts of sweat generated under different weather and exercise conditions. One can observe that the maximum amounts are quite large, up to 5 or more L/day. When one sweats copiously, not all of the sweat is effective in removing heat from the body since much of it may run off as liquid (this depends strongly on the ambient conditions).

The evaporation process can be controlled either by the rate at which sweat is supplied to the skin or by the rate at which evaporation occurs from the wetted surface. For example, when the surrounding air is dry and windy, all sweat reaching the surface will evaporate as fast as it arrives. Then the evaporative heat loss in kilocalories per hour is defined in eq 4.4. For stagnant and moist air, eq 4.4 reduces to eq 4.5.

$$Q_e = 570 m_w \tag{4.4}$$

$$Q_e = K_e A_w (P_s - P_a) \tag{4.5}$$

where m_w is the rate of water excretion by the sweat glands in kg/h, A_w is the wetted surface area, P_s is the water partial pressure corresponding to the surface conditions, P_a is the water partial pressure corresponding to the ambient conditions, and K_e is an evaporation transfer coefficient.

TABLE 4.3 Insensible Loss: Daily Loss of Water (mm).[22]

Body part	Normal temperature	Hot weather	Prolonged heavy exercise
Feces	200	200	200
Lungs	350	250	650
Skin	350	350	350
Sweat	200	1400	5000
Urine	1400	1200	500
Total	2500	3400	6700

TABLE 4.4 Coefficients for Forced Convection Evaporation: Heat Transfer from Nude Person in Air.[2]

Coefficient	Conditions
$12.7\ v^{0.634}$	v > 0.58 m/s, standing, cross flow
$9.66\ v^{0.25}$	v < 0.51 m/s
$19.1\ v^{0.66}$	–
$13.2\ v^{0.6}$	–
$18.4\ v^{0.37}$	–
$10.17\ v^{0.37}$	0.15 < v < 3.05 m/s
$11.6\ v^{0.4}$	–

Table 4.4 presents empirical correlations for K_e, which have been experimentally determined by several investigators. It may be noted that the dependence of K_e on air velocity is essentially the same as that of the convective coefficient h_c for the convection mode of heat transfer.[11]

4.15.1 FACTORS AFFECTING HEAT PRODUCTION

Body size can affect human heat production because heat is generated throughout all the mass of actively metabolizing tissue. The rate of heat production (watts per kg of body weight) is very closely related to the rate of heat loss, and this depends on the area of skin that allows dissipation of heat. For this reason, heat transfer rate is also expressed in Watts/m^2 of surface area. This measurement can be calculated using clinical studies for analytical purposes.[41] The body surface area (BSA) of a person is a height–weight monogram[41] and is calculated using eq 4.6. The metabolic rate is related to the body weight (eq 4.7).

$$BSA = 0.1\,W^{0.67} \tag{4.6}$$

$$M = aW^{b} \tag{4.7}$$

where M is the metabolic rate; a and b are constants; W is the body weight in kg; and BSA is the body surface area in m^2. Equation 4.7 determines the amount of food an animal in captivity will need. Wilmore et al.[23] have discussed in detail the effects of ambient conditions on heat production.

4.16 VENTILATION

Air circulation through a confined compartment can be referred to as ventilation.[41] It ensures the refreshment of new air to replace overheated and stale air in the room. The fresh air via ventilation improves the efficiency of the workers in the room. It uses an air conditioning system or fans to recirculate fresh air inside the room and is useful for the following reasons:

- Can operate in all weather conditions or temperature differences, even in humid, winter, and summer conditions.
- Can have precise control.
- Can be used without significant leakage of heat and often requires a small opening for refreshing the air.
- A typical air conditioning unit can provide filters for reducing dust and can also be turned into a heating system in the winter.

The ventilation rate is defined in liters per second and often needs to pass the below requirements:

- Adequate methods to refresh the fresh air and recirculation to avoid trapped stale air.
- Proper use of room temperature control by removing excess heat generated by appliances and heat from the sun.

The simplest method of determining the ventilation rate required is to make use of the accumulated experience of the industry. The volume in cubic meters (m^3) of the space to be ventilated is calculated and multiplied by the number of air changes per hour to give the ventilation rate in m^3 per hour. Finally, diving by a normalizing/conversion factor gives the rate in liters per second (*lps*) as shown in eq 4.8.

$$\text{Ventilation}\left(\text{in } lps\right) = \frac{1}{3.6}\left(\frac{\text{Volume}}{\text{Number of air changes/hr}}\right) \tag{4.8}$$

TABLE 4.5 Recommended Minimum Fresh Air Supply.[10]

Air space per person, m^3	Air supply per person, liters/s	Air changes per hour, ACH
3	17	20
6	11	6.5
9	8	3.2
12	6	1.8

Table 4.5 gives a guide to the number of air changes generally recommended. Provided air inlets and outlets are properly sited, the tabulated

values will be sufficient for air movement requirements. They will also meet requirements for pollution and moisture removal, in typical examples of the accommodation, and heat removal requirements in temperate climates without heat sources. However, installations should always be checked against the following questions:

- Are there specific bylaws or other legal requirements applicable in the locality?
- Is the space expected to become so crowded that a higher ventilation rate might be obtained by calculating on a per-person basis?
- Are there abnormal sources of heat or fumes?
- Is it desirable to provide a higher ventilation rate for summer cooling?

4.16.1 VENTILATION RATES PER PERSON

Individual human being produces heat, sweat, odor, and carbon dioxide that contribute to room condition. A ventilation challenge also involves the average fresh air it can maintain if a large group of humans gathers and odor concentrations elevate. In fact, the demand requires an increase in the rate of change of fresh air per person with the increase in the number of humans.

The fresh air ventilation rate needs to be further increased in situations of house smoking and lowered air quality. Often, a tolerance is set to the limited or permitted minimum of legal requisites. For instance, 1 *lps* of ventilation air is considered for one-person physiological responses such as oxygen requirement and carbon dioxide burden. Generally, consideration of odor concentration and dust removal rate is not considered when designing a ventilation system for an office.

The typical rate of ventilation for human comfort is around 2 ACH; any value above 2 ACH is considered a higher rate requirement. For instance, in a public or crowded space, the minimum air volume requirement for each human should be 28 m^3/h, which will ensure maintaining body odor concentration, whereas the content of carbon dioxide in the air should be maintained below 10 parts/10,000 humans.

There are exceptions in the case of humidity and dust, but they usually occur in more specialized applications where the requirements are either for full air conditioning with dehumidification by refrigeration for humidity control or for specific dust removal.

4.16.2 HEAT REMOVAL

Ventilation based on heat removal from rooms (where some considerable temperature drop is a main factor) needs special consideration. The amount of heat produced in the room must be estimated based on factors such as the body heat of occupants, electrical apparatus, other heat-producing processes, and solar heat gain through the structure. The volume of air required for ventilation can then be assessed by eq 4.9:

$$\frac{m^3}{hr}\text{ of air} = \frac{1}{3.6}\left(\frac{\text{Heat gain in kW} \times 3000}{\Delta\text{T in }^{\circ}\text{C}}\right) \tag{4.9}$$

The temperature difference in °C is the difference between the outside shade temperature and the required inside temperature. For a total heat gain of 11.75 kW per hour, an outside shade temperature of 21°C, and a maximum inside temperature of 27°C, one can calculate the volume of air required (m^3/h) as follows: m^3/h of air required = [11.75 × 3000]/[27 − 21] = 5875 m^3/h. Note that a small reduction in the temperature difference makes a considerable increase in the volume of air required. For instance, if the temperature difference is reduced by say 2°C, that is, required inside temperature 25°C, then the volume of air required would be increased to: m^3/h of air required = (11.75 × 3000)/(25 − 21) = 8812 m^3/h. It shows a 50% increase in the air volume. Always extract from as high as possible to prevent the collection of heated air under the roof or ceiling from extending downward to near breathing level. Fans at or near the ridge in such cases are ideal, with air replacement from side windows above the head level. This method assists the natural tendency for warmer air to rise, and the quicker this "fug" is removed from high levels, the better will be the conditions at breathing level.

Note that with ventilation alone, the temperature inside a room cannot be reduced to the outside shade temperature because it will always be a few degrees above. However, it is worth considering pulling the replacement air from the "cold" side of the building (the north-facing: north of the equator).

4.16.3 RADIANT HEAT

Radiant heat from hot surfaces (such as cooking vessels, steam presses, pipes, etc.) does not warm the surrounding air, but only solid bodies in the path of the radiation. The extraction of air, therefore, will not reduce or remove the radiant heat, but it will create air movement that would give relief by removing convicted heat.

4.16.4 HEAT INPUT

There are occasions when a ventilation scheme must include some form of heating for the replacement air to prevent the temperature of the air in the room from being lowered to an uncomfortable level. Unless the air replacement openings can be located behind existing radiators or the incoming air can enter from a room already warmed, the most convenient method is to use an intake fan in conjunction with a heater battery.

The required heat output is calculated by a similar formula as given for heat removal, using the known factors of air volume supplied and the temperature rise required, as shown by the following example:

A ventilating unit supplying 1530 m^3/h of air at 0°C is to be warmed to 19°C before being discharged into the room. What will be the required output of the heater battery? Equation 4.9 is rewritten as follows:

$$KW = [m^3/h \times °C]\,[1/3000] = [1500 \times 19]/(3000) = 9.7 \qquad (4.10)$$

4.17 NUMERICAL EXERCISES

1. How much does the human body radiate? When body temperature = θ = 37°C = 310 K; estimated surface of the body = 1.5 m^2; e = 0.70; $\sigma = 5.67 \times 10^{-8}$ watt/m^2K^4.

   $$Q = \varepsilon\sigma A\theta^4 = (0.70)(5.67 \times 10^{-8})(1.5)(310)^4 = 550 \text{ watt}$$

   The sun provides about 1000 watts per square meter at the earth's surface. Human skin reflects 30%. About 700 watts are absorbed per square meter.

2. If a human at temperature θ is surrounded by an environment at temperature θ_0, the net heat flow is [Hint: Use $Q_{net} = \varepsilon\sigma A(\theta^4 - \theta_0^4)$]:

   a. Standing outdoors on a hot day in August: Body temperature = θ = 37°C = 310 K; air temperature = θ_0 = 37C = 310 K. Then, $Q_{net} = \varepsilon\sigma A(310^4 - 310^4) = 0$.

   b. Standing outdoors on a cold morning in February: Body temperature = θ = 37°C = 310 K; air temperature = θ_0 = 0°C = 273 K; Estimated surface of the body = 1.5 m^2; ε = 0.70; $\sigma = 5.67 \times 10^{-8}$ watt/m^2K^4. Then: $Q_{net} = (0.70)(5.67 \times 10^{-8})(1.5)(310^4 - 273^4) = 219$ watt.

3. In problem number 2, the net heat loss is 219 Joules per second. How many food calories of energy will be lost in 1 hour? Given: 4186 Joules = 1 calorie
   $Q = 219 \times 3600 = 7.88 \times 10^5$ J = $(7.88 \times 10^5$ J)/(4186 J/calories) = 188 calories.
4. Calculate the cooling rate (Joules/s) of an adult who may lose 600 g/day of "insensate loss" of moisture from the skin surface.

$$\frac{Q}{t} = \left(600\frac{gm}{day}\right)\left(580\frac{cal}{gm}\right)\left(4.186\frac{J}{cal}\right)\left(\frac{1day}{24hr}\right)\left(\frac{1hr}{3600}\right) = 17W$$

5. The heat of vaporization change can be computed from the value of the specific heat of water as 1 calorie/g°C. It takes 37 calories to heat a gram of water from 0°C to 37°C. Find the change in kinetic energy at 37°C.

   We know,
   - Absolute temperature for zero °C = 0°C + 273 = 273 K
   - Absolute temperature for zero 37°C = 37°C + 273 = 310 K
   - K.E./mole at 0°C or at 273 K = 3400 J/mole = 45 cal/g
   - K.E./mole at 37°C or at 310 K = [45 cal/g] [310/273] = 51.1 cal/g
   - K.E./mole at 100°C or at 373 K = [45 cal/g] [373/273] = 61.7 cal/g at the boiling point.
   - The increase in K.E. for heating the water from 0°C to 100°C = 61.7 – 45 = 16.7 cal/g
   - The energy contributed to weakening the bonds = [37 – 6.1] = 30.9 calories

Therefore, the contribution to weakening the water bonds = 100 – 16.7 = 83.3 cal/g.

Conclusion: Using the results for water at 37°C, it is evident that 52.4 calories of additional energy must be supplied at 37°C to vaporize the water.

Using problem 5 above, comment on the work required to expand the water into its gaseous state. The "PdV" is the work required to expand the water into its gaseous form, which is slightly less at 37°C (our body temperature). By analogy with the work calculation above, work is found to be 34.2 calories/g, 6.8 calories/g less than that at 100°C. This model then suggests a heat of vaporization at 37°C as follows: Body temperature heat of vaporization = 539 cal/g + 52.4 cal/g – 6.8 cal/g = 585 cal/g. Therefore, this simple model agrees fairly well with the quoted value of 580 cal/g.[41]

4.18 SUMMARY

Heat transfer is a major consideration in the adaptation of the human body to our environment. The process affects the behavior and functions of the human body systems. Heat transfer is related to all vital processes in humans and many different diseases. Problems with heat transfer rates may cause some disorders in the most important systems of the body. A normal healthy person maintains a constant body temperature of 35.8°C, despite the temperature variations in the environment.

The human body is a regulatory system that balances input and output. Without proper expense of the input or improper metabolism, the body unavoidably stores the excess input. Thermal regulation is accomplished by body heat energy transport via conduction, evaporation, convection, and radiation mechanisms. The amount of heat dissipated by the body depends on the surroundings and the mode by which it interacts with air and water. In this chapter, we have learned the different ways the body responds to changes in temperature, like fever, hyperthermia, and hypothermia. The study of thermoregulatory processes is also important for recent investigations. Such investigations are important for engineers who need to develop thermoregulation systems, to make people feel comfortable in any given situation. An important part of mechanical engineering is to design systems for controlling the environment as a function of temperature, density, and velocity. We have learned that heat transfer is an important part of all systems especially the vital organs of the human body. Organs such as the kidney, liver, and pancreas, for example, need to maintain a certain temperature and are not allowed a high heat transfer rate. If this happens, the entire body can be put in a stressful situation where other organs can be affected.

ACKNOWLEDGMENT

We acknowledge the permission to use the copyright material at <http://www.modernhumanorigins.com/anth501.html> was granted and this chapter in Edition 2013 was reviewed by Dr. C David Kreger; Such reprint permission has been filed with Edition 2013 of this book by the Apple Academic Press; Authors dedicate this chapter to Prof. C David Kreger for his inspiration & dedication in teaching/ research on human anthropology.

KEYWORDS

- **acclimatization**
- **anesthesia**
- **convection**
- **evaporation**
- **entropy**
- **fever**
- **heat cramps**

REFERENCES

1. Blatteis, C. M.; Schneider, S. M. Thermal Physiology in the USA: A 100-Year History of the Science and its Scientists (1880–1980). In *Thermal Physiology: Perspectives in Physiology*; Blatteis, C. M., Taylor, N., Mitchell, D., Eds.; Springer, 2022; pp 239–255.
2. Alvarado, J.; Garcia, R.; Pena, H.; Chato, J. C. Eds.; In *Heat Transfer in Bioengineering*; Hemisphere Publication Corp, 1983.
3. Chato, J. C. Fundamentals of Bioheat Transfer. In *Thermal Dosimetry and Treatment Planning*; Gautherie, M., Ed.; Springer-Verlag: NY, 1990; pp 51–51.
4. Chato, J. C. Ed.; In *Advanced Heat Transfer*; University of Illinois: Champaign, 1969.
5. Jou, D.; Llebot, J. E. Eds.; In *Introduction to the Thermodynamics of Biological Processes*; Prentice Hall, 1990.
6. Roselli, R. J.; Diller, K. R. Bioheat Transport. In *Biotransport: Principles & Applications*, 1st ed.; Springer, 2011; pp 487–785.
7. Shitzer, A.; Eberhart, R. C. Ed.; In *Heat Transfer in Medicine & Biology: Analysis and Applications*; Plenum, 1985; vol 1.
8. Shrivastava, D. Ed.; In *Theory and Applications of Heat Transfer in Humans*; John & Wiley & Sons Ltd, 2018; vol 1, p 831.
9. Zolfoghari, A.; Maerefat, M. Bioheat Transfer. In *Developments in Heat Transfer*; Bernardes, M. A. d. S., Ed.; 2011; pp 155–170.
10. Cooney, D. In *Biomedical Engineering Principles*; Marcel Dekker, 1976.
11. Bronzino, J. D.; Peterson, D. R. Eds.; In *The Biomedical Engineering Handbook*; CRC Press LLC (Taylor & Francis Group): Boca Raton, FL, 2015; vol 1–4, pp 5422.
12. Goyal, M. R. Bioheat Transfer in the Human Body. In *Biofluid Mechanics of Human Body Systems*, First ed.; CRC and Apple Academic Press, 2013; pp 91–135.
13. Hall, N. In *First Law of Thermodynamics* [Online]. https://www.grc.nasa.gov/www/k-12/airplane/thermo1.html, 2021. https://www.grc.nasa.gov/www/k-12/airplane/thermo1.html (accessed Mar 10, 2023).

14. Eigenplus. *Laws of Thermodynamics:Definition & Examples* [Online]. https://www.eigenplus.com/laws-of-thermodynamics-definition-examples/, 2023 (accessed Mar 22, 2023).
15. Diller, K. R. Fundamentals of Bioheat Transfer. In *Physics of Thermal Therapy: Fundamentals of Clinical Applications*, CRC Press, 2013; pp 1–20.
16. Nyborg, W. L. Solutions of the Bio-Heat Transfer Equation. *Phys. Med. Biol.* **1988**, *33* (7), 785.
17. Reddy, J. N.; Gartling, D. K. In *The Finite Element Method in Heat Transfer and Fluid Dynamics*; CRC Press, 2010.
18. Pletcher, R. H.; Tannehill, J. C.; Anderson, D. In *Computational Fluid Mechanics and Heat Transfer*; CRC Press, 2012.
19. Glasser, O. In *Medical Physics*; Yearbook Publisher, 1944.
20. Foster, J.; Hodder, S. G.; Lloyd, A. B.; Havenith, G. Individual Responses to Heat Stress: Implications for Hyperthermia and Physical Work Capacity. *Front. Physiol.* **2020**, *11*, 541483.
21. Charkoudian, N. Skin Blood Flow in Adult Human Thermoregulation: How it Works, When it Does not, and why. *Mayo Clin. Proc.* **2003**, *78* (5), 603–612. DOI: 10.4065/78.5.603.
22. Gisolfi, C. V. Work-Heat Tolerance Derived from Interval Training. *J. Appl. Physiol.* **1973**, *35* (3), 349–354.
23. Wilmore, D. W.; Mason Jr, A. D.; Johnson, D. W.; Pruitt Jr, B. Effect of Ambient Temperature on Heat Production and Heat Loss in Burn Patients. *J. Appl. Physiol.* **1975**, *38* (4), 593–597.
24. Raymond, C.; Matthews, T.; Horton, R. M. The Emergence of Heat and Humidity too Severe for Human Tolerance. *Sci. Adv.* **2020**, *6* (19), eaaw1838.
25. Greenfield, A. D. The Circulation Through the Skin. In *Handbook of Physiology: Circulation*; American Physiology Society, 1963; vol 2, pp 1325–1351.
26. Kanosue, K.; Crawshaw, L. I.; Nagashima, K.; Yoda, T. Concepts to Utilize in Describing Thermoregulation and Neurophysiological Evidence for how the System Works. *Eur. J. Appl. Physiol.* **2010**, *109*, 5–11.
27. Kreger, C. D. *Human Thermoregulation and Hair Loss: Classroom Notes* [Online]. http://www.modernhumanorigins.net/anth501.html 2001 (accessed Mar 31, 2011).
28. McAllen, R. M.; Tanaka, M.; Ootsuka, Y.; McKinley, M. J. Multiple Thermoregulatory Effectors with Independent Central Controls. *Eur. J. Appl. Physiol.* **2010**, *109*, 27–33.
29. Today, M. N. *What is Thermoregulation, and How Does it Work?* [Online]. https://www.medicalnewstoday.com/articles/312628, 2022 (accessed Mar 31, 2023).
30. Osilla, E.; Marsidi, J.; Sharma, S. In *Physiology, Temperature Regulation. StatPearls*; StatPearls Publishing: Treasure Island, FL, 2021.
31. Parsons, K. In *Human Thermal Environments*; CRC Press, 2014. DOI: 10.1201/b16750.
32. Tansey, E. A.; Johnson, C. D. Recent Advances in Thermoregulation. *Adv. Physiol. Educ.* **2015**, *39*, 139–148.
33. Slutz, S. *What Is Normal Body Temperature for Humans?* [Online]. https://www.sciencebuddies.org/stem-activities/human-normal-body-temperature, 2023 (accessed Mar 22, 2023).
34. Benedict, R. P. In *Fundamentals of Temperature, Pressure, and Flow Measurements*; John Wiley & Sons, 1984.
35. Iso. *Liquid-in-glass Laboratory Thermometers — Principles of Design, Construction and Use*, 1977 [Online]. https://www.iso.org/standard/4370.html.

36. González-Alonso, J. Human Thermoregulation and the Cardiovascular System. *Exp. Physiol.* **2012**, *97* (3), 340–346.
37. Jessen, C. In *Temperature Regulation in Humans and Other Mammals*; Springer Science & Business Media, 2012.
38. ByBio. *Countercurrent Exchange-figure* [Online]. https://bybio.wordpress.com/2013/01/03/cold-feet/countercurrent-exchange-figure/, 2013 (accessed Mar 22, 2023).
39. Weiss, M. L.; Mann, A. E. In *Human Biology and Behavior: An Anthropological Perspective*, 4th Ed.; Little Brown, 1985.
40. Wilmore, J. H.; Costill, D. L.; Kenney, W. L. In *Physiology of Sport and Exercise*; Human Kinetics Champaign, IL, 1999.
41. Hall, J. E.; Hall, M. E. Overview of the Circulation; Biophysics of Pressure, Flow, and Resistance. In *Guyton and Hall Textbook of Medical Physiology*, 14th Ed.; Elsevier, 2020; pp 171–173.
42. Sherwood, L. In *Human Physiology: From Cells to Systems*; Cengage Learning, 2010.

APPENDIX – I

Metabolic Rate for Human Activities.[10,19,22]

Activity	Metabolic rate[a] (met)	Activity	Metabolic rate[a] (met)
Resting	0.7	Squash, singles	1.4–1.8
Sleeping	0.8	Basketball, half court, and intramural	1.4–2.0
Reclining	1.0	Miscellaneous occupations	1.2–2.4
Seated, quiet	1.2	Bakery	1.8–2.2
Standing, relaxed	2.0–3.8	Brewery	4.0–4.8
Walking	2.0–3.4	Carpentry	5.6–6.4
Domestic work	1.6–2.0	Machine sawing table	3.0–3.4
House cleaning	2.0–3.6	Sawing by hand	5.0–7.0
Cooking	1.4–1.8	Planning by hand	2.2–3.0
Washing by hand and ironing	1.2–1.4	Foundry work	1.4–1.8
Shopping	1.1–1.3	Using a pneumatic hammer	2.0–2.4
Office work	1.1–1.3	Tending furnaces	3.5–4.5
Typing	1.2–2.0	Garage work	2.0
Miscellaneous office work	3.0–4.0	General laboratory work	1.6
Drafting	2.4–4.4	Machine work	1.1
Leisure activities	3.6–4.6	Light	1.5
Stream fishing	5.0–7.2	Heavy	2.0
Calisthenics exercise	5.0–7.2	Shop assistant	3.2
Dancing, social	7.0–8.7	Teacher	1.4
Tennis, singles	1.4–2.6	Watch repairer seated	1.8

Note: [a]The unit of the metabolic rate is known as "met," which is equivalent to 58.2 W/m^2.

CHAPTER 5

Bioheat Transfer Models: Theory and Applications in Bioengineering

ARKA BHOWMIK, MANPREET SINGH, and RAMJEE REPAKA

5.1 INTRODUCTION

Bioheat is a specialized field of thermal and biomedical engineering that involves investigating the process of heat energy generation, heat transfer, heat absorption, thermal management within the human body, and prediction of thermally induced end damage to tissue. In particular, we study bioheat transfer since it enables us to quantify (1) the inherent thermoregulation processes in our body, (2) the diagnostic markers associated with abnormal (elevated and below normal) thermal response of the body, and (3) the effectiveness of thermally motivated treatment and management in case of health disorders, preservation and vitrification of stem cells/tissue/organ, and ecological endurance in space and ocean. In nature, bioheat is a form of energy crucial for proper functioning and development of the human body. In fact, it is the maximum amount of energy which is liberated by a human body during bio-chemical and physical processes in the body.[1] For instance, the energy associated with the nutrient content of food cannot be fully converted to biological work, majority of energy is liberated as heat. In fact, a detailed budget of energy shows only 50% of energy of food nutrient is transferred to adenosine triphosphate (ATP) molecules (energy capsules for cell), and another 50% is liberated as heat. Another 25% of the energy is liberated as heat during ATP expenditure by cells to perform work. In reality, only 25% of the energy derived from nutrient by the body is converted to biological work and rest 75% energy is liberated from the body as heat.[2]

The purpose of this chapter is to illustrate the physics of heat transfer and provide essential tools to investigate as well as manage the thermal

Biofluid Dynamics of Human Body Systems: Expanded and Revised Edition.
Megh R. Goyal, Arka Bhowmik, & Anamika Chauhan (Eds.)

response in the body during various applications of bioheat. A methodical elaboration involving application in normal, below normal, and supra normal temperature ranges is presented. The chapter also presents key mathematical models involving thermal transport at physical scales "*meso*" and "*macro*" also identified by tissue-organ level. Though intriguing, the bio-thermal interaction at other physical scale, that is, atomic (atoms), nano (bio-molecules), and micron (cell) level is beyond the scope of this book. In a couple of sections, the chapter also presents various instruments used for conducting bioheat transfer research. Finally, the chapter concluded with an analytical and numerical solution of bioheat equation, as well as some numerical examples in bioheat transfer.

5.2 HISTORICAL BACKGROUND

Heat is a form of energy in transit that cannot be measured directly without noticing the changes in the state property of matter. Ideally, the quantification of heat energy is a process of measuring the change in state properties (such as temperature, pressure, density, thermal capacity, conductivity, emissivity, absorptivity, transmissivity) and numeric computation of empirical relations (Stefan's-Boltzmann law for thermal radiation, Newton's law of cooling and Fourier or non-Fourier laws of heat conduction) using measurable or quantifiable state properties. For example, a physicist can compute the heat generated by our sun based on the color temperature. Similarly, a physician can identify abnormal function of the body based on measured temperature and blood pressure. In nature, heat is the only form of energy like gravity that can penetrate matter (via., conduction and radiation) and space (via., radiation) as indicated by famous mathematician and physicist Fourier in his book.[3] Though, we are aware of another mode of heat transfer due to motion of matter (in liquid, gas, and particulates), but in reality, convection is an effective transport quantity owing to combined effect of conduction and radiation when realized at a level of individual particles. However, this quantity provided us a useful way to quantify heat transport due to motion at a *meso* and *macro* scale.

From a historical viewpoint, the field of bioheat originated roughly around 20th century with the development of Pennes equation by Harry H. Pennes.[4] Though many equations down the line have been developed, yet nonetheless Pennes equation is the only model developed till date that enables easy analysis of local transport of bioheat in a human body with a certain degree of accuracy. It is worth noting that other models[1] were profoundly important to the development of this field, which overcomes the shortcomings in the Pennes equation as discussed later in this chapter.

5.3 THEORY AND MATHEMATICAL MODELS FOR BIOHEAT TRANSFER

5.3.1 BIO-PHYSICS OF HEAT TRANSFER IN TISSUE

In the field of bioheat transfer, within the body or organs the modes of transfer of heat energy take place in primarily three ways. These ways are convection due to the flow of blood in large vessels, heat transfer due to blood microcirculation during an exchange of nutrient and gases (O_2 and CO_2) between small capillaries and tissue heat conduction to some extent. All tissue and bio-fluids are poorly conducting (conductivity ≤ 0.5 W/m/K); thus, the majority of heat transfer within the organ takes place by the flow of blood in the vessels and exchange of nutrients in capillary beds. Especially, the body works mechanistically to maintain or regulates the heat balance within the body, for instance the excess temperature as a consequence of unbalanced heat within the body is liberated to surrounding by the flow of blood from the core to the extremities (skin) and by exhaled air. Skin covers the entire body and itself exchanges unbalanced heat to surrounding via evaporation accompanied by perspiration, convection, and surface radiation.

A demonstration of heat exchange within the body is shown in Figure 5.1 (inset). Here, blood flow in the vessel leads to convective heat exchange in the direction of flow, between the blood and vessel walls. As a consequence of temperature difference, the vessel walls conduct heat in the direction or from the surrounding tissues. Further, the arborescent distribution of permeable capillaries in the tissue bed led to the exchange of heat between tissue and blood, when fluid squeezes out of capillaries due to blood pressure and simultaneous reentering of fluid into the capillaries by osmotic action. This process of heat exchange is also realized as perfusion in the physics of bioheat. In fact, conduction in the tissue is identified as an effective phenomenon which is dominated by perfusion.

Apart from these above phenomena, the body has the ability to keep a trade-off between heat generation and heat transfer in order to maintain the core temperature which is centrally regulated by the brain. Metabolic heat is generated due to various biochemical reactions occurring within the body during the conversion of glucose to ATP molecules and its expenditure in biological work by individual cells/tissue. Heat is also produced by the effect of shivering, which is caused by successive muscle contraction and release as body temperature falls (be it skin temperature or core temperature).

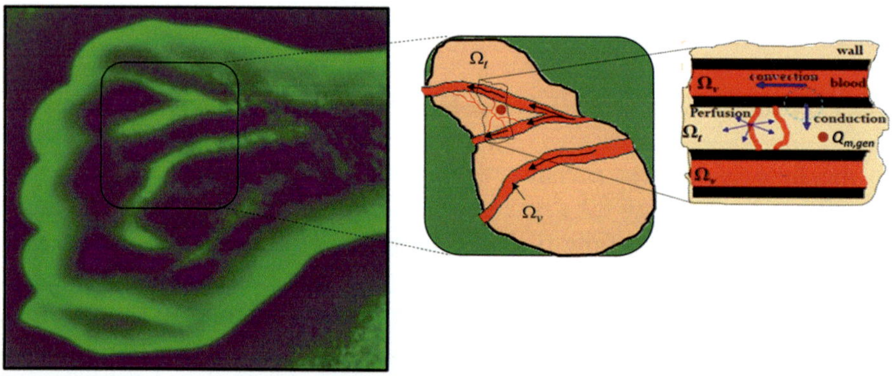

FIGURE 5.1 An optical vein viewer image of dorsum of human hand with demonstration of bioheat transfer. Nominal parameters: Ω_t (tissue domain), Ω_v (vascular domain), and $\Omega_{m,gen}$ (metabolic heat generation). [Image courtesy: Arka Bhowmik].

There is another mode of heat transfer, that is, volumetric radiation, which is often ignored by bioheat scientist since its calculation is complex. However, its contribution is significant in extremities especially in still ambient condition. Moreover, volumetric radiation is an important field for optical physicists and biophysicists since it involves understanding interaction of photon in visible, IR and UV wavelength range with human tissue, and for the interrogation of bioluminescence emitted by biological samples, and for the diagnosis of disease condition with the aid of many optical devices.

At present, the theoretical models are used to obtain a rational prediction of heat balance within the body. These models can be broadly classified into two groups. One group of models focuses on the loco-regional heat transfer for any body parts also referred as localized approach. Another group of models focuses on modeling complete body heat transfer by compartment models of individual sections of the body also referred as a whole-body approach. The key physics of heat transfer for whole-body approach remains the same as that of localized model. However, the consideration of complete body parts enables a thermal scientist to evaluate the entire thermal energy budget of the body in applications involving designing a room thermal comfort, high altitude clothing design, and HVAC design, etc.

5.3.2 ESSENTIAL CONTINUUM AND VASCULAR MODELS: LOCALIZED APPROACH

Here, some of the essential bioheat models were illustrated, which were used for making rational predictions of regional heat transfer within the tissue. These models were broadly classified either as continuum or vascular models.[5] A continuum model refers to the tissue configuration wherein the changes due to circulation of blood are assumed volume averaged or effective property. However, a vascular model involves a tissue configuration with vascular structures that encompasses the blood flow and heat transfer across the vessel and tissue.

5.3.2.1 CONTINUUM MODELS

5.3.2.1.1 Pennes Bioheat Equation

Pennes equation is a primitive and popular bioheat equation proposed by Harry H. Pennes in the year 1948 in his classic paper on the measurement and analysis of temperature in the interior of resting human forearm.[4] The paper presented methods and instruments for measuring skin surface and deep muscle temperature of the anesthetized human forearm, and proposed a heat transfer equation to illustrate the observation in the experimental data. In his analysis, Pennes modeled the forearm as an axis-symmetrical cylindrical domain with homogeneous thermophysical properties, and validated the analytical solution of the steady-state equation with that of the measured steady state temperature of the human forearm. Pennes illustrated a relation between tissue temperature and blood microcirculation and introduced a generation term related to metabolic reaction in the tissue, which was believed to be the basis of many models developed by other researchers. The general expression of Pennes equation is formulated as

$$\rho_t c_t \frac{\partial T_t}{\partial t} = \nabla \cdot (k_t \nabla T_t) + (1-\kappa)\rho_b c_b \omega_b (T_a - T_t) + Q_m''' \quad (5.1)$$

where the density is denoted by ρ, the specific heat at constant pressure is denoted by c, the thermal conductivity is denoted by k, the volumetric blood perfusion is denoted by ω, the arterial temperature (same as core temperature) is denoted by T_a, and Q_m''' is the metabolic generation of heat which is volumetric in nature. The thermal equilibration parameter whose value can vary from 0 to 1 is denoted by κ in eq 5.1. In case of $\kappa = 0$ (often considered), the heat transfer between blood in the capillaries and tissue is maximum since no thermal equilibrium exists, whereas for $\kappa = 1$ the blood and tissue heat exchange is zero since thermal equilibrium has already been achieved. Note, the subscripts t and b in eq 5.1 are referred to tissue and blood. The perfusion term in the Pennes equation could act as a heat sink or source. Especially, for tissue temperature above blood temperature as in case of thermotherapy the perfusion term acts as a sink. However, for tissue temperature below blood temperature as in case of low temperature application the perfusion term acts as a source.

5.3.2.1.2 Modified Pennes Bioheat Equation (m-BHT)

Recently, Bhowmik et al.[6] suggested an improvised version of the bioheat equation which defines new correlations to elaborate the balance of local energy of vascularized tissue by combining effective thermal properties of perfused tissue being dependent on the blood perfusion. The equation includes the local thermal diffusivity α_{eff}, thermal capacity qc_{eff} and generation of heat $Q_{m,eff}$ to be dependent on local site blood perfusion. The exchange of heat between capillaries and tissue, heat diffusion, and generation across the tissue is represented by[6]

$$\frac{\partial T_t}{\partial t} = \nabla \cdot \left(\frac{\{1-\phi_b\} k_t + \phi_b k_b}{\{1-\phi_b\} \rho_t c_t + \phi_b \rho_b c_b} \nabla T_t \right)$$

$$- \frac{\rho_b c_b \omega_b}{\{1-\phi_b\} \rho_t c_t + \phi_b \rho_b c_b} (T_t - T_b) + \frac{Q'''_{m,eff}}{\{1-\phi_b\} \rho_t c_t + \phi_b \rho_b c_b} \tag{5.2}$$

where the volume fraction of blood is denoted by ϕ_b in the tissue, expressed as product of blood perfusion and residence time of blood is the tissue ($\simeq \omega_b \Delta t$). The residence time Δt of blood in the tissue can be determined by the time interval between two consecutive frames of time series data. Ideally, the residence time should be small enough to capture the changes in local flux. In fact, the volume fraction computation would be precise for high temporal resolution measuring devices. The term T_b here is referred to as mean blood temperature depending on surface and core temperature [$= (1-\kappa)T_c + \kappa T_{i,j}$] such that κ is the weighting factor, T_c is the core body temperature, and $T_{i,j}$ represents the measured surface temperature of pixel number I, j and is equivalent to T_t. The un-perfused tissue properties, that is, ρ_t, c_t, and k_t are related to percentage of water, fat, and proteins,[6] whereas the properties of blood ρ_b, c_b, and k_b are considered constant at sea level. The heat of metabolism in living tissue is defined depending on oxygen concentration in the blood and blood perfusion, such that $Q_{m,\ eff}$ in eq 5.2 is nothing but the energy generated due to amount of oxygen consumed during aerobic reaction of metabolites (viz., glucose, fat, and proteins) as

$$Q_m = E_{O_2} \times \phi_{O_2} \times \omega_b \tag{5.3}$$

where the energy liberated by oxygen/volume of O_2 is denoted by E_{O2} and the volume fraction of oxygen in the blood is denoted by ϕ_{O2}. The E_{O2} can be expressed in J/mL of O_2

$$E_{O_2} = 5.88 \times 3600 \times 10^{-3} \times (0.23RQ + 0.77) \tag{5.4}$$

where the volume of CO_2 given off to that of the volume of O_2, is given by the respiratory quotient denoted as RQ. The value of RQ in the resting state is equal to 0.85.

Similarly, the volume fraction of blood O_2, ϕ_{O_2} can be expressed by an empirical correlation deduced from Pittman[7]

$$\begin{aligned}\phi_{O_2} &= \begin{bmatrix}\text{Volume of } O_2\\ \text{dissolved in blood}\end{bmatrix} + \begin{bmatrix}\text{Volume of } O_2\\ \text{bounded with hemoglobin in blood}\end{bmatrix}\\ \phi_{O_2} &= \left[\left(\frac{0.00165 \text{ ml } O_2}{100 \text{ ml blood}\bullet\text{mm Hg}}\right) \times P_{O_2}\right] + \left[\left(\frac{20.1 \text{ ml } O_2}{100 \text{ ml blood}}\right) \times S_{O_2}\right]\end{aligned} \tag{5.5}$$

where P_{O2} is the sea level partial pressure of O_2 in normal systemic arterial blood (~ 100 mm Hg) and the oxygen saturation value is denoted by S_{O2}.

The oxygen saturation value can be measured using a pulse oximeter or using Hill's correlation $S_{O_2} = \left(\frac{P_{O_2}}{P_{50}}\right)^n \Big/ \left[1 + \left(\frac{P_{O_2}}{P_{50}}\right)^n\right]$. The exponent n is the Hill's correlation that can be equated to 2.7 for human adult hemoglobin. But, the partial pressure of oxygen (in mm of Hg) at 50% oxygen saturation (P_{50}) can be obtained from oxygen-dissociation curve.[7] A detail account of the above model parameters is available in Ref. [6].

5.3.2.1.3 Chen and Holmes Equation

Chen and Holmes proposed an equation[8] in year 1980 for microvascular heat transfer, which was another improvised version of Pennes equation by consideration of effective thermal properties due to contribution of blood perfusion, an advection term due to local velocity of perfused blood in the small vessels, and a precise description of perfusion term.

$$(\rho c)_{eff} \frac{\partial T_t}{\partial t} = \nabla \cdot \left(k_{eff} \nabla T_t\right) + \omega_b^j \rho_b c_b \left[T_a^{in} - T_t\right] - \rho_b c_b u_p \cdot \nabla T_t + Q_m''' \tag{5.6}$$

where ρc_{eff} is the effective thermal capacity $\left[= (1-\phi_b)\rho_t c_t + \phi_b \rho_b c_b\right]$, ϕ_b is the volume fraction of blood given as $(= V_b/V << 1)$, k_{eff} is the effective thermal conductivity $(= k_t + k_p)$ so that k_p is the thermal conductivity change induced by blood perfusion, ω_b^j is the rate of blood perfusion in small vessels j within the control volume, T_a^{in} is the inlet temperature of jth blood vessel, which is not the same as the core body temperature assumed in Pennes equation.

The advection term $\rho_b c_b u_b \cdot \nabla T_t$ is applicable to blood flow in the vessels when the blood temperature is equal to tissue temperature throughout the vessel length. Here, u_p is the mean blood velocity in vessels in the direction of flow, and T_t is temperature of tissue as a consequence of thermal equilibrium with tissue and blood volume in the vascular bed.

They also suggested that the heat transfer across larger vessels should be treated separately. Based on convective analysis of individual vessels and deduced vascular geometrical parameter denoted by thermal equilibration length (which is a factual length of a particular vessel needed for maximum heat transfer across blood vessel and tissue to achieve thermal equilibrium), the study revealed that equilibration lengths for thermal balance between tissue and blood in larger arteries and veins are of the order of meters due to which blood is not fully equilibrated in these vessels. But, the equilibration length for precapillary arterioles, capillaries, and venues are few micrometers indicating blood in these vessels is already in equilibrium with tissue. The analysis revealed vessels with equilibration length comparable to actual length as in case of vessels between the terminal arterial branches and precapillary are thermally significant (i.e., 300 μm ≥ radius >> 10 μm).

5.3.2.1.4 Weinbaum–Jiji Bioheat Model

Weinbaum and Jiji[9] proposed a continuum bioheat equation in year 1985, with a detailed consideration of the counter-current effect of blood flow in parallel vascular geometry. The model was developed by assuming that heat exchange between counter-current vessels is prominent due to the temperature difference between artery and veins. However, the change in tissue temperature due to blood perfusion in the existing bioheat models is negligible. The resulting bioheat equation was given as

$$(\rho c)_{avg}\frac{\partial T_m}{\partial t}=\frac{\partial}{\partial x_i}\cdot\left(k_{ij,eff}\frac{\partial T_m}{\partial x_i}\right)-\frac{\pi^2 r_a k_b^2 \mathrm{Pe}}{4k\xi}\left[\frac{\partial(nr_a l_i \mathrm{Pe})}{\partial x_i}+\frac{2ng\mathrm{Pe}}{v_{avg}}\right]l_i\frac{\partial T_m}{\partial x_j}+Q_m''' \quad (5.7)$$

where T_m is the average tissue temperature $\left[=\frac{T_a+T_v}{2}\right]$, T_a and T_v are temperatures of blood in artery and vein, respectively, $(\rho c)_{avg}$ is the average heat capacity due to blood and tissue, $k_{ij,eff}$ is the effective conductivity in anisotropic tissue, which is equivalent to $k\left[\delta_{ij}+\left(1/4\xi k^2\right)n\pi^2 r_a^2 k_b^2 l_i l_j \mathrm{Pe}^2\right]$, k is tissue conductivity, r_a is radius of artery, v_{avg} is the velocity of vein and artery pairs (but averaged), n is the vessel pair number density crossing the control surface in the control volume per unit area, the Peclet number is denoted by Pe $(=\mathrm{Pr\,Re}=2\rho_b c_b r_a v/k_b)$, the volumetric bleed-off divided by vessel surface area is denoted by g, $l_{i/j}$ indicates direction cosines, δ_{ij} is the Kronecker delta, x denotes position, and ξ is the shape factor for conduction. Here, subscript i represents coordinate (i = 1, 2, 3) and j is the index of summation. Overall, they introduced many new parameters, which demand prior measurements of various parameters and details of vascular anatomy for its implementation.

5.3.2.1.5 Thermal-Wave (or Non-Fourier) Bioheat Model

The thermal-wave model is a wave-like energy equation for macroscopic thermal conduction[10] which was originally hypothesized dated back by Maxwell,[11] Morse and Feshbach,[12] Cattaneo,[13] and Vernotte.[14] It describes the heat flux's $q(t+\tau_q)$ lagging behavior than the temperature gradient $\nabla T(t)$ responses at time t in a nonequilibrium state due to high rate of heat transfer as in case of laser with short-pulse fluence to heat. The macroscopic thermal conduction relation for such a nonequilibrium state is described[10] as follows:

$$q\left(t+\tau_q\right)=-k\nabla T(t) \quad (5.8)$$

where the temporal phase-lag in heat flux response is denoted by τ_q.

Equation 5.8 indicates that in any nonequilibrium state at time t (represented by temperature gradient), the response (or heat flux) will take a finite time $(t+\tau_q)$ to transfer heat by conduction.[15] This statement is contradicting to the heat diffusion law by Fourier that explains instantaneous response or speed of heat transfer infinite for any nonequilibrium state at time t, and hence eq 5.8 is also referred to as non-Fourier heat diffusion equation. Taking Taylor series expansion of eq 5.8 and dropping terms of higher order due to the small value of τ_q ($<< t$) as $q(t)+\tau_q \frac{\partial q(t)}{\partial t} = -k\nabla T(t)$ followed by divergence on both sides of the resulting equation, we get $\nabla q(t)+\tau_q \frac{\partial}{\partial t}\left[\nabla q(t)\right] = -\nabla\cdot k\nabla T(t)$.

Next, putting the value of $\nabla q_c = -\rho_t c_t \frac{\partial T_t}{\partial t} + (1-\kappa)\rho_b c_b \omega_b (T_a - T_t) + Q'''_m$ from bioheat eq 5.1 in for eliminating ∇q_c, we obtain the non-Fourier bioheat model as

$$\begin{aligned}\tau_q \rho_t c_t \frac{\partial^2 T_t}{\partial t^2} &= \nabla\cdot(k\nabla T_t) + (1-\kappa)\omega_b \rho_b c_b (T_a - T_t) \\ &\quad - \left[\rho_t c_t + (1-\kappa)\tau_q \omega_b \rho_b c_b\right]\frac{\partial T_t}{\partial t} + \left[Q'''_m + \frac{\partial Q'''_m}{\partial t}\right]\end{aligned} \tag{5.9}$$

Here, all the symbols have their usual meaning as discussed earlier. However, subscript c was used for heat conduction. The experimental evidence of non-Fourier heat conduction in biological tissue was shown by Mitra et al.[16]

5.3.2.1.6 *Dual-Phase-Lag (DPL) Model*

Dual-phase-lag model is a generalization of the thermal wave theory, which introduces time lag for both flux of heat front (response) and gradient in temperature (effect) for all nonequilibrium states at time t. The macroscopic thermal conduction relation to such a nonequilibrium state is described by Ref. [17] as

$$q(t+\tau_q) = -k\nabla T(t+\tau_\theta) \tag{5.10}$$

where the temperature gradient-related phase-lag due to microstructural inhomogeneity is denoted by τ_θ.

Taking Taylor series expansion of eq 5.10 and dropping terms of higher order due to the small value of τ_q and τ_θ ($<< t$) as $q(t)+\tau_q \frac{\partial q(t)}{\partial t} = -k \left[\nabla T(t)+\tau_\theta \frac{\partial}{\partial t}\nabla T(t)\right]$ followed by divergence on both sides of the resulting equation, we get

$$\nabla q(t)+\tau_q \frac{\partial}{\partial t}\nabla q(t) = -\nabla\cdot\left[k\nabla T(t)+\tau_\theta \frac{\partial}{\partial t}k\nabla T(t)\right].$$

Next, putting $\nabla q_c = -\rho_t c_t \frac{\partial T_t}{\partial t} + (1-\kappa)\rho_b c_b \omega_b (T_a - T_t) + Q'''_m$ from bioheat eq 5.1 in $\nabla q_c + \tau_q \frac{\partial}{\partial t}\nabla q_c = -\nabla \cdot \left[k\nabla T_t + \tau_\theta \frac{\partial}{\partial t} k\nabla T_t \right]$ for eliminating ∇q_c, we obtain the DPL bioheat model as follows:

$$\tau_q \rho_t c_t \frac{\partial^2 T_t}{\partial t^2} = \nabla \cdot (k\nabla T_t) + \tau_\theta \frac{\partial}{\partial t}[\nabla \cdot (k\nabla T_t)] + (1-\kappa)\omega_b \rho_b c_b (T_a - T_t) - \left[\rho_t c_t + (1-\kappa)\tau_q \omega_b \rho_b c_b\right]\frac{\partial T_t}{\partial t} + \left[Q'''_m + \frac{\partial Q'''_m}{\partial t}\right] \quad (5.11)$$

Here, all the symbols have their usual meaning as discussed earlier.

5.3.2.1.7 Porous Media Bioheat Model

The porous media-based bioheat model was initially proposed by Roeztal and Xuan,[18] wherein two separate energy equations were suggested to compute heat transfer in vascular and extra-vascular/tissue regions, respectively. The model incorporates the thermal exchange between the vascular and extra-vascular regions by introducing effective thermal conductivities and interfacial convective heat transfer terms.[1] Later, Shih et al.[19] proposed a simplifying description of interfacial heat transfer between vascular and extra-vascular regions by retaining the blood perfusion term in the tissue domain and convection term due to seepage of fluid in vascular domain to volume-averaged energy equations for tissue and blood subdomain as follows:

$$(1-\varphi)\rho_t c_t \frac{\partial T_t}{\partial t} = (1-\varphi)\nabla \cdot (k_{eff}\nabla T_t) + (1-\varphi)\rho_b c_b \omega_b (T_a - T_t) + (1-\varphi)q'''_t \quad (5.12)$$

$$\varphi \rho_b c_b \left(\frac{\partial T_b}{\partial t} + \mathrm{v} \cdot \nabla T_b\right) = \varphi \nabla \cdot (k_b \nabla T_b) + \varphi q'''_b \quad (5.13)$$

where the porosity φ is given as $(= V_b/V)$, $(1 - \varphi)$ is $(= V_t/V)$, V is the total volume of the domain $(=V_b + V_t)$, $\mathbf{v}$ is the seepage velocity vector, q'''_t and q'''_b are source terms for tissue domain and vascular domain, respectively. The subscripts b and t are for blood and tissue, respectively.

Here, the interfacial exchange of blood from vascular domain to the tissue gives rise to enhanced thermal conductivity $k_{eff} = k_t(1 + \beta\omega_b)$, where $\beta = 0.02$ kg/m^{3}s. Assuming local thermal equilibrium (i.e., $T = T_t = T_b$) and adding eqs 5.12 and 5.13 enabled an overall bioheat equation for porous microvascular bed as follows:

$$\left[\varphi\rho_b c_b + (1-\varphi)\rho_t c_t\right]\frac{\partial T}{\partial t} + \varphi\,\rho_b\,c_b\left(\mathrm{v}\cdot\nabla T\right) = \left[\varphi k_b + (1-\varphi)k_{eff}\right]\nabla^2 T$$
$$+(1-\varphi)\rho_b c_b \omega_b\left(T_a - T\right) + (1-\varphi)q_t''' + \varphi\,q_b''' \tag{5.14}$$

5.3.2.1.8 Shrivastava and Vaughan Model

Shrivastava and Vaughan[20] proposed two separate energy equations for blood and tissue volumes for the prediction of tissue and blood volume temperature as an averaged quantity in the vascularized tissue. The model assumes source/sink terms in tissue and blood subvolume due to interfacial exchange between two subvolumes, and introduces new constants to account the change in thermophysical properties. The equation for tissue is given by

$$\rho_t c_t \frac{\partial \overline{T_t}}{\partial t} = C_{t1}\left[\nabla\cdot\left(k_t \nabla \overline{T_t}\right)\right] + \frac{\left(UA''\right)_{b-t}}{V_t}\left(\overline{T_b} - \overline{T_t}\right) + \overline{Q_t'''} \tag{5.15}$$

where $\overline{T_t}$ is the tissue temperature averaged over the volume $\left(=1/V_t \int T_t \cdot dV\right)$, C_{t1} is a constant originated due to volume averaging, U is the heat transfer coefficient which is also a volume averaged term and not same as standard coefficient, and $\overline{Q_t'''}$ is the volume averaged source term $\left(=1/V_t \int Q_t'''\cdot dV\right)$. Similarly, the equation for the volume of blood is given by

$$\rho_b c_b \frac{\partial \overline{T_b}}{\partial t} = C_{b2}\nabla\cdot\left[\left(PC\right)_b \overline{T_b}\right] + \frac{\left(UA''\right)_{b-t}}{V_b}\left(\overline{T_t} - \overline{T_b}\right) + \overline{Q_b'''} \tag{5.16}$$

where $\overline{T_b}$ is the volume averaged tissue temperature $\left(=1/V_b \int T_b \cdot dV\right)$, C_{b2} is the new constant term appeared due to simplification of volume averaging, P is perfusion mass flux (kg/m^{2}s) equivalent to $\left(=1/V_b \int \rho\,\mathrm{v}_b\, dV\right)$, $\mathbf{v}_b$ is blood velocity vector, $\overline{Q_b'''}$ is the volume averaged source term $\left(=1/V_b \int Q_b'''\cdot dV\right)$, and all other terms have their usual meaning.

Overall, the model introduced some new parameters that needed to be verified and evaluated based on detailed knowledge of the organ and the geometrical shape of the vascular structures.

5.3.2.1.9 Phase Change Bioheat Model

The bioheat equations incorporating the variability in heat energy due to change in phase were often implemented for thermal ablation process in the tissue, wherein high temperature of tissue causes a change in tissue phase due to coagulation ($T_t > 45$°C) and volatization ($T_t > 100$°C). A simplified energy equation involving a phase change process of vaporization of water and melting of fat in tissue was proposed by Dau and Chakraborty[21] as follows:

$$\rho_t c_t \frac{\partial T_t}{\partial t} = \nabla \cdot (k_t \nabla T_t) + (1-\kappa)\rho_b c_b \omega_b (T_a - T_t) + Q_m''' - \frac{\partial(\rho \Delta H)}{\partial t} \quad (5.17)$$

where ΔH is the latent enthalpy or heat released by tissue during the vaporization of water and melting of fat, so that ΔH at any location in the tissue is equivalent to latent heat, L if $T_t > T_p$ (phase transition temperature). However, if $T_t < T_p$ then $\Delta H = 0$.

Later, another modified bioheat equation involving liquid–vapor phase change process was proposed by Abraham and Sparrow[22] as follows:

$$\frac{\partial h_t}{\partial t} = \nabla \cdot (k_t \nabla T_t) + (1-\kappa)\rho_b c_b \omega_b (T_a - T_t) + Q_m''' \quad (5.18)$$

where h_t is referred to enthalpy per unit volume and enthalpy rate $\partial h_t / \partial t$ associated with change due to temperature and phase. The enthalpy rate can be associated with (1) no phase change for liquid-containing tissue (i.e., 37°C $\leq T_t \leq$ 99°C), (2) phase change due to latent heat of vaporization (i.e., 99°C $< T_t \leq$ 100°C), and (3) temperature variation post-phase change in the tissue (i.e., $T_t >$ 100°C). Hence, h_t was evaluated as

$$h_t = \begin{cases} (\rho_t c_t)^{\text{pre}} [T_t - 37\,°\text{C}] & 37 \leq T_t \leq 99\,°\text{C} \\ h_t(99) + h_{fg} \cdot C_{\text{liq}} \cdot \dfrac{(T_t - 99\,°\text{C})}{(100 - 99\,°\text{C})} & 99 < T_t \leq 100\,°\text{C} \\ h_t(100) + (\rho_t c_t)^{\text{post}} [T_t - 100\,°\text{C}] & T_t > 100\,°\text{C} \end{cases} \quad (5.19)$$

where h_{fg} is the latent heat of vaporization, C_{liq} is the liquid concentration in the tissue (~ 75%), h_t (99) and h_t (100) are referred value of h_t at 99°C obtained by the first relation and value at 100°C obtained by the second relation, respectively. The superscripts *pre-* and *post-* refer to the property value before phase change and after the phase change process, respectively.

5.3.2.1.10 Combined Continuum and Volumetric Radiation Model

The effect of volumetric radiation in biological tissue and its response to tissue temperature is given by Ref. [23] as follows:

$$\rho_t c_t \frac{\partial T_t}{\partial t} = \nabla \cdot (k_t \nabla T_t) - \nabla q_r + (1-\kappa)\rho_b c_b \omega_b (T_a - T_t) + Q'''_m \tag{5.20}$$

where q_r is the radiative flux of volumetric radiation and is written as

$$q_r = \int_{4\pi} I_j \cos\theta_i d\Omega_i \tag{5.21}$$

where I_j stands for intensity, θ_i is for incident angle, Ω_i is the solid angle for incident intensity, and subscript j stands for diffuse radiation (due to emission by tissue or external intensity source) as well as for collimated radiation (due to external intensity source that travels in one direction).

The component of climate and diffuse intensity can be determined based on equations for diffuse and collimated radiative transfer in a participating medium which are detailed in the Refs. [1, 23]. In the absence of any external intensity source, the I_j in eq 5.21 is primarily due to emission and a diffuse component of emitted radiation.

5.3.2.2 VASCULAR BIOHEAT MODELS

5.3.2.2.1 Keller and Seiler Counter-Current Model

The counter-current model proposed by Keller and Seiler[24] is composed of three steady-state energy equations of the vascularized human skin, wherein terminal arteries and veins run normal to the skin surface along the depth and capillaries joining the artery and vein run parallel to the skin surface. The flow in terminal artery takes place from datum ($z = 0$ to δ) and that of the vein is in the opposite direction, where δ is the thickness of the skin layer. The energy equation for tissue domain was established by balancing heat addition by thermal conduction and metabolic heat generation, heat convection by terminal artery to tissue and from tissue to a vein, and blood perfusion by transverse capillaries to the tissue. If blood in terminal artery and vein is at a higher temperature, then the arterial and venous convection acts as a source term for the tissue domain, and the perfusion of blood is introduced as a source term in the tissue domain.

$$k_t \frac{d^2 T_t}{dz^2} + (h_{ar} A'''_{ar} + c_b \omega_b)[T_{ar} - T_t] + h_{ve} A'''_{ve}[T_{ve} - T_t] + Q'''_m = 0 \tag{5.22}$$

where h_{ar} and h_{ve} are the convection coefficients for artery–tissue interface and tissue–vein interface, respectively, c_b is the specific heat of blood, A'''_{ar} and A'''_{ve} are total surface areas of arteries per unit volume of tissue and total surface area of veins per unit volume of tissue, respectively, ω_b is the perfusion of blood (in kg/s/m^3 of tissue), subscripts *t*, *ar*, and *ve* represent tissue, artery, and vein, respectively. The energy equation for terminal artery is due to the energy liberated by arterial volume to the tissue by blood perfusion and heat convection at artery–tissue interface as

$$\left[m_{ar,in} - \int_{z=0}^{z=\delta} \omega_b dz \right] c_b \frac{dT_{ar}}{dz} + h_{ar} A'''_{ar} \left[T_{ar} - T_t \right] = 0 \tag{5.23}$$

where $m_{ar,in}$ is the blood flow rate at the inlet of terminal artery in the considered skin layer (in kg/s/m^2).

Similarly, the energy equation for the vein domain is due to heat addition from tissue via blood perfusion from tissue and heat liberation due to convection and blood perfusion from vein to tissue as

$$-\left[m_{ve,in} + \int_{z=0}^{z=\delta} \omega_b dz \right] c_b \frac{dT_{ve}}{dz} + \left(h_{ve} A'''_{ve} + c_b \omega_b \right) \left[T_t - T_{ve} \right] = 0 \tag{5.24}$$

where $m_{ve,in}$ is the inlet blood flow rate of the terminal vein in the considered skin layer (in kg/s/m^2) which is often considered equivalent to $(= -m_{ar,in})$ due to its counter-current nature.

5.3.2.2.2 Discrete Vascular Model (DIVA Model)

The DIVA model was coined by Lagendijk and group[25,26] that incorporates the effect of counter-current flow and realistic vessel tree segmented from medical images. The model incorporates a three-dimensional domain with a realistic vessel tree in the tissue domain, and evaluates the energy conservation equations for tissue domain and vascular domain by applying appropriate boundary conditions. They defined tissue domain by three-dimensional voxels with or without cylindrical shapes also referred to as vessel segments. The tissue and vessel segments were connected via common boundary elements referred to vessel walls through which heat energy was balanced. The energy equation of tissue domain is the same as the Pennes equation considering the effect of smaller vessels and metabolic heat generation as

$$\rho_t c_t \frac{\partial T_t}{\partial t} = \nabla \cdot \left(k_t \nabla T_t \right) + \rho_b c_b \omega_b \left(T_a - T_t \right) + Q'''_m \tag{5.25}$$

On the other hand, the radially outward convective flux due to the flow of blood in vessel segments (or vascular domain) to the inner surface of vessel wall elements is given as

$$q_{r_{ves}} = -\frac{\text{Nu}k_b}{2r_{ves}}\left[T_w - T_{mix}(x)\right] \tag{5.26}$$

The energy equation for radial vessel wall elements was subjected to boundary condition eq 5.26 at its inner wall, whereas the outer wall of boundary elements is subject to continuous boundary condition across vessel wall and tissue domain. The wall temperature change can be evaluated as follows:

$$(\rho c)_{wall}\frac{\partial T_w}{\partial t} = \frac{1}{r}\frac{\partial}{\partial r}\left(k_w \frac{\partial T_w}{\partial r}\right) \tag{5.27}$$

where r_{ves} is the radius of vessel segment, k_b is the blood thermal conductivity, T_w is the wall temperature of the vessel, x is the axial direction within the vessel segment, Nu is the Nusselt number, T_{mix} is the mixing cup temperature of blood within the vessel segment equivalent to $1/v_m\pi r_{ves}^2 \int_0^{r_{ves}} 2\pi r v(r) T(r)\,dr$, and v_m is mean term for blood velocity $\left(= \int_0^{r_{ves}} 2\pi r v(r)\,dr\right)$. Note both vi_m and T_{mix} are functions of x due to blood flow within the vessel segment so that blood is simulated by a unique method of shifting and resampling along the vessel axis. This was achieved by shifting a small distance δ ($= v_m\Delta t$) in the direction of the vessel axis with fully developed flow assumption and without thermal variability in the time interval Δt, wherein the shifting to a new position δ leads to new wall temperature due to exchange of heat between the vessel wall and blood.

Overall, the DIVA model enables a detailed modeling of heat exchange across tissue and embedded vascular structures. This model was found to increase in the computation time due to tracking and re-estimating blood and tissue temperature.

5.3.2.2.3 Non-Newtonian Vascular Model

The realistic model of blood flow as a non-Newtonian fluid in counter-current vessels displayed an influence on skin temperature in Ref. [27]. Their model encompasses the full-scale Navier–Stokes simulation of the flow field of blood in counter-current vascular architecture with non-Newtonian behavior of blood along with blood velocity fluctuation due to heart rate. The acquired flow field of blood in the vessel is combined with the energy equation of a vascular domain for the prediction of local distribution of temperature within the vascular architecture which was also coupled with the energy equation for the perfused tissue domain by common boundary elements between vascular and tissue domain by continuous boundary condition. The model was already described earlier in chapter 3 and is not illustrated here to avoid repeatability.

5.3.3 ESSENTIAL COMPARTMENT MODELS: WHOLE-BODY APPROACH

Physics informed computational simulations mimicking sophisticated human thermoregulation mechanisms are the most efficient way to determine the effect of environments on the physiological responses in the human body. The aim is to predict the central thermal response by monitoring core-body temperature, average skin temperature, and regional skin temperature. In the last one decade, dedicated efforts were made to address heat- and cold-related injuries. The existing thermoregulation models in use may be classified as one-node (homogeneous-cylinder approach),[4] two-node or core-shell or one-segment two-layer lumped model,[28] multi-node (i.e., nodes for core, skin, arterial, vein, and superficial vein),[29–31] multi-layer models dividing body tissues into four different components (skin, fat, muscle, and core or bone) and multi-segments for forearms, hands, thighs, calves, feet, digits of hand, head, upper and lower chest, upper arms, and feet. The multi-segmented models are designed as annular concentric layers of cylinders for each part of human body tissue. Other approaches[32] employed a detailed representation of the human arterial vasculature considering geometrical, physiological, and anatomical considerations, while recent developments involve inclusion of detailed blood flow associated vasoconstriction in skin (closing of blood-supplying vessels) and vasodilation mechanisms (opening of blood supplying vessels). More recently, accurate three-dimensional virtual thermoregulatory models were developed to include heat transfer in capillaries, counter-current, and passing blood flow heat transfer.[33] The other approach was to develop model for each body segment such as hand, feet, finger, etc., thereby adding more anatomical details,[34–36] clothing details,[37] and physics of arterio-venous anastomoses (AVAs) more pronounced in outer-extremities of human bodies were incorporated into the model.[38–40]

5.3.3.1 THE STOLWIJK AND HARDY MODEL

In 1966, Stolwijk and Hardy suggested a heat transfer model for whole body subjected to surrounding environmental changes,[29] wherein simulations were performed to capture transport of heat to skin from body core, effect of central blood circulation, and to mimic controlling systems of the body such as vasodilation, vasoconstriction, shivering, etc. Later, a full-scale model was developed to study the thermal response of astronauts for NASA.[41,42]

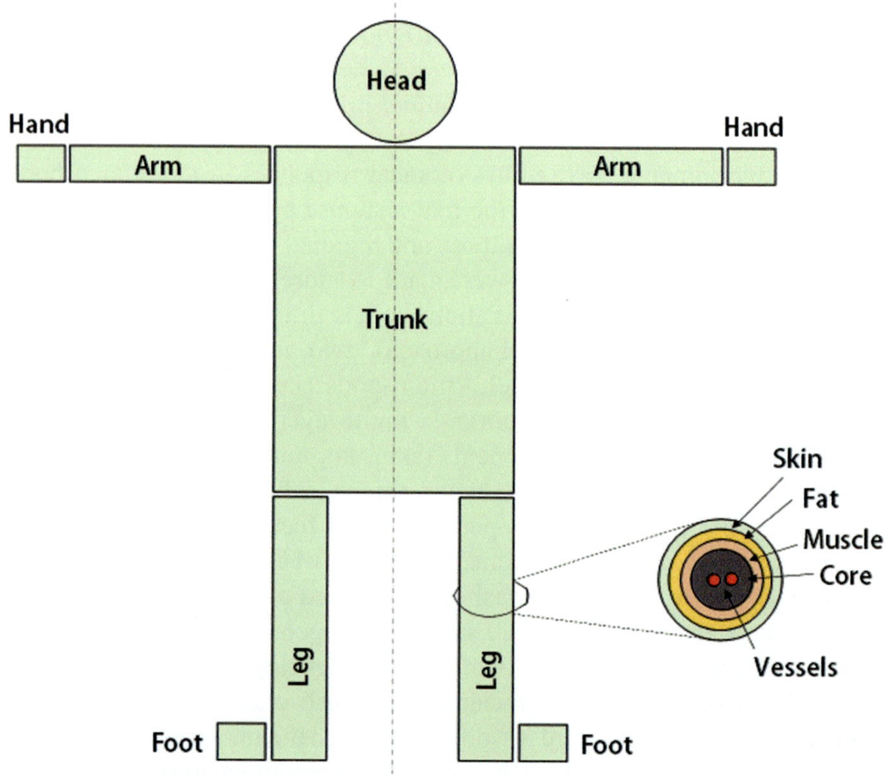

FIGURE 5.2 Representative passive system of the Stolwijk and Hardy model consisting of five cylindrical parts and one spherical part; head, trunk, arm, leg, feet, and central blood vessels. Each tissue part is composed of four compartments for skin, fat, muscle, and core.

The model is composed of two parts, a controlling and passive system. In passive part the model consists of six representative body parts by five cylinders for hands, legs, arms, trunk, and feet, and for head one sphere. These parts consist of four tissue types represented by concentric cylinders or spheres for the skin, fat, muscle, and inner core as shown in Figure 5.2. These parts also contain central blood vessels which supply blood to the parts, and exchange heat to the compartments by convection.

The thermal model of the passive system includes heat energy balance equation across a symmetry drawn for the entire geometry (*see* Figure 5.2), wherein the heat transfer takes place from bodies core to the surface of the skin via conduction, heat liberated by the skin due to convection, evaporation, and radiation to the surrounding, the heat liberated due to respiration, heat convected by blood to each compartment, and source terms due to shivering. Later, the total heat transfer was computed by multiplying the thermal response of each part twice to overcome the symmetry assumption. A generalized heat energy balance equation (in W) for passive system tissue parts and compartments is given as

$$C_{ij}\frac{\partial T_{ij}}{\partial t} = \pm K_{ij}\Delta T + \alpha_{ij}\rho c \times BF_{ij}\left(T_{CB} - T_{ij}\right) + M_{0,ij} + \Delta M + E - E_{r,ij} - E_{v,ij} - A_{ij}h_{ov}\left(T_{ij} - T_{amb}\right) \tag{5.28}$$

where the thermal capacitance is denoted by C (in J/°C), the thermal conductance between adjacent tissue layers/compartments is denoted by K (in W/°C), ΔT is the temperature difference between two adjacent tissue layers, α is the counter-current factor of blood flow, ρc is the thermal capacitance of blood (in J/°C), BF is the fraction of blood flow in tissue layer per second (in 1/s), M_0 is the basal metabolic heat (in W), ΔM is the fraction of heat generation due to shivering (in W), E is the fraction of heat addition due to exercise (in W), E_r is loss of heat during respiration (in W), E_v is the evaporation heat loss from skin surface (in W), A is the area of skin (m^2) exposed to ambient, h_{ov} is the overall heat transfer coefficient as a result of both convection and radiation (in W/m^2/°C), T_{amb} is the temperature of ambient (in °C). The subscripts ***i*** and ***j*** refer to the part and tissue layer/compartment, respectively. The subscript CB refers to central blood temperature.

Note that the heat lost/sink due to sweating/evaporation, convection, and radiation to environment is applicable to outermost tissue layer (i.e., skin). Similarly, the heat lost/sink by respiration (E_r) is applicable only to inner core tissue. On the other hand, heat source due to shivering and exercise is applicable to core and muscle compartments of certain parts. The sign associated with heat conduction differs due to changes in heat flow direction. A detailed model for each compartment is given in Ref. [29]. On the other hand, the generalized heat energy balance equation (in W) for central blood (CB) compartments is given as

$$C_{CB}\frac{\partial T_{CB}}{\partial t} = -\sum_{j=1}^{j=24} \alpha_{ij}\rho c \times BF_{ij}\left(T_{CB} - T_{ij}\right) \tag{5.29}$$

where $j = 1$ to 24 refers to the heat lost/exchanged by central blood compartment with tissue compartments (i.e., skin, fat, muscle, and core) of each six body parts of passive system. Note that the central blood compartment has no heat source or sink, except the amount of heat lost due to exchange of convective heat from central blood to each compartment of the body parts.

The passive system discussed above is controlled by a controlling system[42] which is composed of mimicking temperature sensors and actuators that are integrated via feedback loops to reference set points to sense the changes in surrounding environment and simultaneously actuate signals to incorporate body's functions like vasodilation, vasoconstriction, sweating, and shivering in the heat transfer model. The controlling system is a computer platform to simulate or compute human thermoregulation based on measured and set temperature via iteration and adjustment in the strength of blood flow, shivering, and sweating. These adjustment parameters were defined in the passive model as coefficients and the strength of the coefficients was defined by the control system. The details of the control system are presented by Stolwijk and Hardy[29,42] as analog circuit diagrams for each body part with symbols and scaling values.

5.3.3.2 THE FIALA MODEL

Fiala developed a full-scale thermal model of human body subjected to surrounding changes, implementing the effect of thermoregulatory functions, and effect of clothing in his thesis[43] in the year 1998. The model has its inspiration from Stolwijk and Hardy's model discussed above that it has two components: a passive component and an active (controlling) component. However, the model elaborated the thermoregulatory function with great details and provided various empirical correlations to compute the dynamic change in body temperature as a consequence of thermoregulatory response. In comparison to Stolwijk and Hardy's model, the passive component of the Fiala's model has 15 body parts which include head, face, neck, shoulders, arms, hands, thorax, abdomen, legs, and feet.

These parts were modeled using appropriate anatomical size and thermophysical properties for sphere and cylinders as shown in Figure 5.3, with annular tissue compartments. Particularly, the model considered multilayer compartments for each part with different tissue materials (brain, lung, bone, muscle, viscera, fat, inner skin, and outer skin) according to the composition of the body parts, and applied numerical discretization with more than one node for each compartment. The outer skin is of 1 mm thickness which is devoid of any thermally significant vessels and no heat source. Further, the thermal effect on blood circulation was mechanistically simulated by integrating three categories of blood temperature involving central blood pool, counter-current heat exchange in juxtaposed artery-vein pair, and perfusion of blood to individual tissue compartments.

The entire body was subjected to boundary heat transfer from outer skin surface to surrounding by convection, radiation, and evaporation, and exposed to external solar irradiation. The effect of clothing was introduced by incorporating overall heat transfer coefficient to the boundary flux. The generalized heat energy equation (in W/m^3) for the passive system of individual compartments of the body part was modeled by the Pennes bioheat equation as follows:

$$\left(\rho_t c_t\right)_{ij}\frac{\partial T_{ij}}{\partial t}=k_{ij}\left(\frac{\partial^2 T_{ij}}{\partial r^2}+\frac{\xi}{r}\frac{\partial T_{ij}}{\partial r}\right)+\rho_b c_b\times\omega_{b,ij}\left(T_a-T_{ij}\right)+Q_m'''-Q_{res}''' \quad (5.30)$$

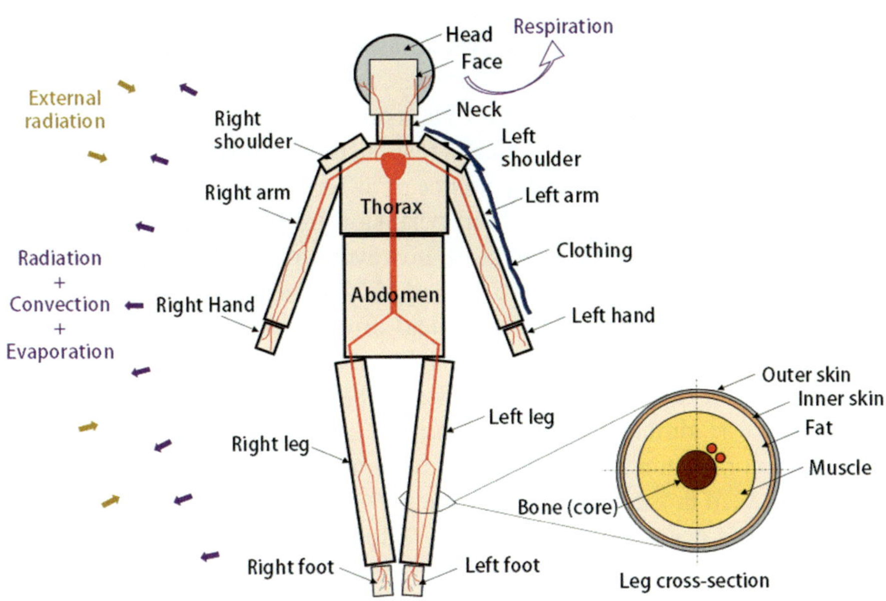

FIGURE 5.3 Representative Fiala's model passive system composed of 14 cylindrical parts and one spherical part (indicating feet, leg, abdomen, thorax, hands, arms, shoulders, neck, face, and head). Each part is a multi-layered tissue with bone/core, muscle, fat, inner skin and outer skin, and brain in case of head.

Source: Modified from Ref. [43].

The model assumed heat transfer in the polar and radial directions of the cylindrical and spherical domains, respectively. The heat transfer in other angular direction or axial direction of cylindrical domain is assumed to be negligible. Here, T_{ij} is the tissue temperature, ξ is a geometry factor (= 1 for cylinder and 2 for sphere), r is the radius, Q'''_{res} is the heat lost by respiration, subscripts t and b are for tissue and blood, respectively. Also, the subscripts $\boldsymbol{i}$ and $\boldsymbol{j}$ refer to the part and tissue compartments, respectively. All other parameters have their usual meaning discussed earlier.

Note that the strength of blood perfusion ω_b and metabolic heat generation Q'''_m in eq 5.30 will be influenced by the active component of the body. For instance, changes in vasomotion and involuntary heat are generated in the body as a part of the thermoregulation. On the other hand, the value of the heat lost through breathing depends on the strength of the metabolic activity of the entire body. Also, the heat lost by respiration is applicable to eq 5.30 for selected body parts as discussed in Table 5.1. The blood temperature of arterial T_a in eq 5.30 is a resultant blood simulation in the circulatory system.

The simulation of circulatory system assumes the warm blood T_{bp} from central pool passes through major arteries, small arteries, and capillaries in the extremities and exchanges heat two instances before returning to central pool via venous architectures, which is responsible for changes in new central blood pool temperature as a consequence of different surrounding conditions. The heat exchange is done by blood perfusion in the extremities and other parts of the body according to eq 5.30; however, the counter-current streams of blood in an artery–vein pair are only applicable to shoulders, arms, hands, leg, and feet in accordance with anatomical structure. The arterial temperature T_a in eq 5.30 is not fixed for the compartments and depends on the counter-current heat exchange with veins and core blood temperature of the body part which was simulated by assuming the decrease in temperature in the artery of a compartment $T_{bp} - T_a$ is equivalent to an increase in temperature of adjacent vein in the same compartment $T_{v,x} - T_v$. The energy conservation equation for an open system is given as follows:

$$\rho_b c_b \int \omega_b dV \left(T_{bp} - T_a\right) = \rho_b c_b \int \omega_b dV \left(T_{vc} - T_v\right) \tag{5.31}$$

where $\int \omega_b dV$ is the blood flow in the veins as a volumetric integral of capillary blood perfusion for the considered tissue compartment, T_{bp} is central blood pool temperature, T_{vc} is the venous blood temperature after counter-current heat exchange, T_a is arterial blood temperature in perfusing tissue after counter-current exchange, and T_v is venous blood temperature before counter-current exchange.

The net heat transfer (in W) from blood in artery to blood in the vein during a counter-current exchange can be written as $= h_{cc}(T_a - T_v)$, where h_{cc} is the heat transfer coefficient in W/°C due to the counter-current arrangement. Therefore, eq 5.31 and the value of T_a used in the bioheat equation can be deduced as

$$\rho_b c_b \int \omega_b dV \left(T_{bp} - T_a\right) = h_{cc}\left(T_a - T_v\right)$$

$$\text{or, } T_a = \frac{\rho_b c_b \int \omega_b dV \cdot T_{bp} + h_{cc} T_v}{\rho_b c_b \int \omega_b dV + h_{cc}} \tag{5.32}$$

Here, T_v is equivalent to $\left(= \int \omega_b T_{ij} dV \middle/ \int \omega_b dV\right)$ since capillary blood is often in thermal equilibrium with tissue T_{ij} which is an inherent assumption for Pennes bioheat equation. Once T_a and T_v are evaluated in individual tissue compartment then venous blood temperature after counter-current exchange T_{vc} can be determined by equating $h_{cc}\left(T_a - T_v\right) = \rho_b c_b \int \omega_b dV \left(T_{vc} - T_v\right)$ as

$$T_{vc} = \frac{\left[\rho_b c_b \int \omega_b dV - h_{cc}\right] T_v + h_{cc} T_a}{\rho_b c_b \int \omega_b dV} \tag{5.33}$$

Note that the core temperature for each body part is not the same which depends on the temperature of blood vessels branched from major artery and vein. In fact, the unknown central blood pool temperature T_{bp} was assumed to be numeric summation of venous blood temperature after counter-current exchange T_{vc} of all tissue compartments and body parts, which was approximated as

$$T_{bp} \approx \left[\sum_{i=1}^{N} \rho_b c_b \int_j \omega_b dV \times \rho_b c_b \int_j \omega_b \cdot T_{ij} dV\right] \Bigg/ \sum_{i=1}^{N} \left[\rho_b c_b \int_j \omega_b dV\right]^2 \tag{5.34}$$

where N refers to the total number of body parts. The values of h_{cc} for parts are given by Fiala[28] and the body parts (such as head, thorax, face, neck, and abdomen) where counter-current heat transfer is absent (h_{cc} was considered 0). On the outer surface of skin, a combined boundary condition encompasses the effect of heat convection, radiation, and evaporation to surrounding, the effect of insulation due to clothing, and effect of irradiation from external source.

$$q_{sk}\big|_i = q_c + q_r + q_e - q_{so} \tag{5.35}$$

where q_c is convection heat flux with ambient air, q_r is radiative heat flux with ambient, q_e is evaporation of moisture, and q_{so} is source radiation received by skin from external source.

The parts with clothing induce insulation to net heat flux, which was imposed in boundary condition with appropriate definition of overall heat transfer coefficient. Table 5.1 depicts the empirical relations suggested by Fiala for blood perfusion, metabolic heat generation, respiratory heat loss, and boundary heat flux. But, the other end/boundary of each body part is assumed to be having an adiabatic core condition as $dT_{i,0}/dr = 0$ at $r = 0$.

5.4 QUANTITATIVE MODELS FOR THERMAL DAMAGE

In the field of bioheat transfer, the application of heat energy for the treatment of cancer has significant implications in clinical management by raising the temperature of local cancer volume or tissue above normal temperature of body (also defined as hyperthermia: See chapter 4 in this book) or lowering the temperature of the unwanted volume sufficiently below bodies' normal range (also identify use in cryosurgery). Raising the temperature of tissue or freezing the tissue results in thermal denaturation of cells or various components in the cells (such as coagulation and vaporization, dehydration, or icing formation) which enable medical doctors to culminate the growth of unwanted cancer volume in the tissue. In such applications, bioheat models provide us with a tool to estimate the end effects prior to its application. Any thermal treatment involves a balance between temperature change, duration, and induced damage to the tissue volume during treatment. For instance, treatment time is longer for methods involving constant temperature, treatment time could be shorter for methods involving varying temperatures, and tissue damage could vary with rate of energy absorbed or released for a given treatment time. The thermal damage models are also found useful to predict the spread of damage useful for defining clinical margins during localized hyperthermia treatment.[44] Besides an estimate of temperature and time for treatment of bioheat models, another end effect (or damage induced) by any thermal treatment method can be modeled using quantitative assessment for thermal damage parameters discussed below. All thermal injury models can be briefly grouped into statistical models, kinetic models, modified kinetic models, and the enzyme-binding model.

TABLE 5.1 Empirical Correlation for the Physiological Parameters of Fiala's Model.

Physiological parameter	Regression correlations	Comments
Blood perfusion, ω_b	The coefficient of perfusion term in eq 5.30 is written as $\beta = \begin{cases} \rho_b c_b \omega_{b,0} & \text{for thermal neutrality} \\ \rho_b c_b \omega_{b,0} + \Delta\beta & \text{for non-neutrality} \end{cases}$	Here, $\omega_{b,0}$ is the perfusion rate in thermal neutrality, and $\Delta\beta$ is the gain factor due to increase or decrease in blood perfusion ($=\rho_b c_b \Delta\omega_b$) and can be written as a function of metabolism. The gain factor in the cutaneous plexus depends on the thermoregulation (as response to increase oxygen supply during metabolism by shivering and that of vasoconstriction and vasodilation). A detailed relation for gain factor due to shivering, vasoconstriction, and vasodilation was given by Fiala.[43] The values of $\omega_{b,o}$ for different tissue compartments are depicted in Ref. [43].
Convective heat flux, q_c	$q_c = h_{c,mix}(T_s - T_{amb})$ where, $h_{c,mix} = \sqrt{a_{nat}\sqrt{T_s - T_{amb}} + a_{frc} v_{eff} + a_{mix}}$	Here, $h_{c,mix}$ is the combined natural and forced convection coefficient deduced from experiment on a full-scale manikin. The coefficients a_{nat}, a_{frc}, a_{mix} were obtained by regression analysis given by Ref. [43]. The term v_{eff} is the effective velocity of air due to natural and forced convection.
Effect heat flux due to clothing insulation, q_{cl}	$q_{cl} = U_{cl}(T_s - T_{amb})$ where, $U_{cl,eff} = \dfrac{1}{1/f_{cl}(h_{c,mix} + h_r) + \sum_{k=1}^{L}(R_{cl,k})}$ $U_{cl,eff}$ is the local effective heat transfer coefficient (*in* W/m²/°C) that includes the effect of local radiation and mixed convection, and effect of resistance/insulation coefficient $R_{cl,k}$, f_{cl} is a nondimensional clothing area factor of the clothing layer in contact with ambient. The subscript k reflects the number of clothing layers for the body part.	The body parts subjected to clothing insulation were simulated by a modified localized effective heat transfer coefficient $U_{cl,eff}$ for the boundary heat flux. Note, in case of body parts with clothing $q_c + q_r$ in eq 5.35 will be replaced by q_{cl}. Here, the clothing effect includes insulation provided by layers of clothes and air trapped between the skin and clothes. A database of clothing-related parameters for items (trousers, shirt, underwear, socks, etc.) is given in Ref. [43].

TABLE 5.1 *(Continued)*

Physiological parameter	Regression correlations	Comments
Evaporative heat flux, q_e	$q_e = U_{e,cl}(P_s - P_a)$ $= [Evaportion\ by\ skin] + [Diffusion\ by\ moist\ skin]$ $= \left[\frac{\lambda_{H_2O}}{A_s}\frac{d}{dt}(m_{sw})\right] + \left[\frac{P_{os,sat} - P_s}{R_{e,s}}\right]$ where, $P_{os,sat} = 100 \cdot \exp\left(18.956 - \frac{4030}{T_s + 235}\right)$ $P_s = \frac{\frac{\lambda_{H_2O}}{A_s}\frac{d}{dt}(m_{sw}) + \frac{P_{os,sat}}{R_{e,s}} + U_{e,cl}P_a}{U_{e,cl} + 1/R_{e,s}}$ $Moisture\ permability = \frac{1}{R_{e,s}} = 0.003\ Wm^{-2}Pa^{-1}$ $U_{e,cl} = \frac{L_a}{[1/(f_{cl} \cdot h_{c,mix})] + \sum_{k=1}^{l}\left(\frac{R_{cl,k}}{\varphi_{cl,k}}\right)}$ $\frac{d}{dt}(m_{sw}) = a_{sw}\frac{A_s}{A_{s,j}}\mathbf{Sw} \times 2^{\frac{T_s - T_{s,0}}{10}}$ $\mathbf{Sw} = [0.65\tanh(0.82\Delta T_{s,m} - 0.47) + 1.15]\Delta T_{s,m} + [5.6\tanh(3.14\Delta T_{hy} - 1.83) + 6.4]\Delta T_{hy}$	Here, A_s is the area of skin sector, $U_{e,cl}$ is the effective heat transfer coefficient due to evaporation of sweat in the presence of clothing, (P_s–P_a) is referred to the evaporative potential such that P_s is water vapor pressure at skin surface and P_a is the vapor pressure of the ambient air, λ_{H2O} = 2256 KJ/kg is the heat of vaporization, dm_{sw}/dt is the rate of sweat production (in kg/s) by area A_s (depends on bodies thermoregulation function), $P_{os,sat}$ is the saturated vapor pressure within the outer cutaneous layer of the body sector which has sweat glands, T_s is the temperature of outer skin layer, L_a is Lewis constant for air (= 0.0165 K/Pa), $R_{cl.k}$ incorporates the effect of resistance/insulation coefficient due to clothing, φ_{cl} is the moisture permeability index, f_{cl} is a nondimensional clothing area factor of the clothing layer in contact with ambient, subscript *k* reflects number of clothing layers for the body part, a_{sw} is the local distribution coefficient of the skin sector associated with sweating, $A_s/A_{s,j}$ is skin area ratio of the skin sweating sector to area of skin of *j*th part, **Sw** is the sweat liquid production (in g/min) which depends on the afferent signals due to change in mean skin temperature $T_{s,m}$ and hypothalamus temperature T_{hy} below set references. Note, the active component of the model sets rate of sweat production based on afferent signal originated due to error Δ associated with T_{hy} and $T_{s,m}$ from its set references.

TABLE 5.1 *(Continued)*

Physiological parameter	Regression correlations	Comments
Heat lost by respiration, Q'''_{res}	$Q'''_{res} = \partial\left[a_{res}\left(E_{res} + C_{res}\right)\right]/\partial V$ where, $E_{res} = 4.373\int_{V_t} Q'''_m dV \begin{pmatrix} 0.028 - 6.5\times10^{-5}T_{amb} \\ -4.9\times10^{-6}P_a \end{pmatrix}$ $C_{res} = 1.948\times10^{-3}\int_{V_t} Q'''_m dV \begin{pmatrix} 32.6 - 0.066T_{amb} \\ -1.96\times10^{-4}P_a \end{pmatrix}$	Note that the total heat lost by respiration is only applicable in eq 5.30 involving body parts connected to pulmonary tract (i.e., lung, muscle layers of face and neck). The distribution of total respiratory loss is incorporated by a_{res} as a distribution coefficient, which varies in lung, muscle of the neck, inner and outer faces by 30%, 25%, 20%, and 25%, respectively. The term $\int_{V_t} Q'''_{m,ij}\cdot dV$ is whole-body metabolism (in W), V_t is volume of total body, T_{amb} is the ambient temperature, P_a is water vapor pressure in air, E_{res} and C_{res} are the heat loss by evaporation and convection, respectively, from inspired/expired air.
Metabolic heat due to muscle work, $Q'''_{m,work}$	$Q'''_{m,work} = \partial\left(a_{w,m}H\right)/\partial V_{mus}$	Here, $a_{w,m}$ is the coefficient associated with an extra source of metabolism during exercise in muscular layer, V_{mus} corresponds to a volume of the muscular layer of the body part, and H is the workload of the whole internal body. The value of $a_{w,m}$ and regression relation for H is given by Fiala [43].
Metabolic heat due to shivering, $Q'''_{m,shiv}$	$Q'''_{m,shiv} = a_{sh}\dfrac{\mathbf{Sh}}{V_{mus}}$ where, $\mathbf{Sh} = 10\left[\tanh\left(0.51\Delta T_{s,m} + 4.19\right) - 1\right]\Delta T_{s,m}$ $- 27.5\Delta T_{hy} - 28.2 + 1.9\Delta T_{s,m}\dfrac{d}{dt}\left(T_{s,m}\right)$	Note, $Q'''_{m,shiv}$ is the response of the body due to changes in body temperature sufficiently below thermal neutral (as a consequence of afferent signals due to change in mean skin temperature $T_{s,m}$ and hypothalamus temperature T_{hy} below set references). The value of a_{sh}, T_{hy}, and $T_{s,m}$ is given by Fiala.[43] The active component of the model sets shivering effect based on afferent signal originated due to error Δ associated with T_{hy} and $T_{s,m}$ from its set value.

TABLE 5.1 *(Continued)*

Physiological parameter	Regression correlations	Comments
Metabolic heat generation, Q'''_m	$Q'''_m = Q'''_{m,bas,0} + \Delta Q'''_{m,bas} + Q'''_{m,shiv} + Q'''_{m,work}$ **or,** $\Delta Q'''_{m,bas} = Q'''_{m,bas,0}\left[2^{(T_{ij}-T_0)/10} - 1\right]$ The value of $Q'''_{m,bas,0}$ is given in Ref. [43].	Here, $Q'''_{m,bas,0}$ is basal metabolic heat generation in case of thermal neutrality, $Q'''_{m,bas}$ is the change in basal metabolism in body, not in thermal neutrality, $Q'''_{m,shiv}$ and $Q'''_{m,work}$ are additional metabolism due to shivering and work. Note, the metabolism due to shivering and work were only applicable in muscle tissue. The temperature of thermal neutrality ($T_o = 30°$ C).
Radiative heat flux irradiated from external source, q_{so}	$q_{so} = \alpha_s \psi_{s-so} \cdot I_{so}$	Here, α_s is the absorption coefficient of skin, ψ_{s-so} is the view factor for curved skin surface and imaginary surrounding source envelop, and I_{so} is the intensity of external source.
Radiative heat flux, q_r	$q_r = h_r\left(T_s - T_{amb,m}\right)$ where, $h_r = \sigma \varepsilon_s \varepsilon_{amb} \psi_{s-amb}\left(T_s^2 + T_{amb,m}^2\right)\left(T_s + T_{amb,m}\right)$	Here, h_r is the radiative coefficient, T_s is absolute temperature skin surface, $T_{amb,m}$ is absolute mean temperature of imaginary curved surface/envelop of ambient, ε_s is the emissivity of skin, ε_{amb} is the emissivity of imaginary surface for ambient, ψ_{s-amb} is the view factor considering curved surface of skin and imaginary surface in surrounding, and σ is the Stefan's-Boltzmann constant.

5.4.1 ARRHENIUS THERMAL-DAMAGE MODEL OF CELL-DEATH

The irreversible kinetic model of first-order assumes that cell injury due to heating follows a rate process in which cells can be realized either in viable state V or in dead state D as

$$V \xrightarrow{k_r} D$$

where the reaction rate k_r can be evaluated as $k_r(t) = A\exp\left[-E_a/R\cdot\{T(t)+273\}\right]$.

For hyperthermia treatment in the range 37–95°C, the thermal damage parameter Ω for tissue was defined as a temperature and time-dependent function similar to the Arrhenius rate equation for the chemical reaction as proposed by Moritz and Henriques in 1947[45] and a convenient relation between undamaged tissue concentration with the damage parameter was given by[46]

$$\begin{aligned} \Omega(\tau) &= \int_0^{\tau} k_r\,dt \\ \ln\left[\frac{C_0}{C(\tau)}\right] &= \Omega(\tau) = \int_0^{\tau} A\exp\left[-E_a/RT(t)\right]dt \end{aligned} \tag{4.36}$$

where Ω is a nondimensional quantity, C_0 is the initial undamaged concentration of the tissue, $C(\tau)$ is the final undamaged concentration of the tissue over time $t = \tau$ in seconds, τ is the final time, A is the frequency factor (1/s) that defines the effective collision frequency of reacting biomolecules and is expressed as $\left[= RT/Nh\exp(\Delta S/R+1)\right]$.

Though the value of A is a function of temperature, yet it is often considered constant for numerical studies of tissue damage in the range of 37–95°C, E_a is the activation energy or energy barrier (J/mole) for transformation to damaged state, R is universal gas constant (8.32 J/mole/K), N is Avogadro's number (6.023×10^{23}), h is Planck's constant (6.627×10^{-34} Js), ΔS is the entropy of activation (J/mole/K), T is the absolute tissue temperature (K). Similarly, an expression of damage concentration can be established as follows:

$$\ln\left[\frac{C_0}{C_0 - C_D(\tau)}\right] = \Omega(\tau) = \int_0^{\tau} A\exp\left[-E_a/RT(t)\right]dt \tag{5.37}$$

The thermal damage index may be classified as

$$\Omega(\tau) = \begin{cases} 0; & \textit{no burn} \\ 0.5; & \textit{first degree burn occurs} \\ 1.0; & \textit{second degree burn occurs} \\ 10^4; & \textit{third degree burn occurs} \end{cases}$$

where $C_D(\tau)$ is the final damaged concentration, since undamaged concentration $C(\tau)$ is equivalent to $C_0 - C_D(\tau)$. In any chemical reaction, there is a potential barrier between reactants and final product formation. Likewise, during thermal treatment of tissue (or denaturation of protein) there is a potential barrier (known as activation energy) beyond which the original state of the tissue/protein molecules cannot be recovered. However, situations may also exist during which the premature treatment termination (before activation energy is reached) could reverse the state to its original state due to synthesis of enzyme or proteins, cells, and membranes as a part of biological repair mechanism. This is largely due to nonuniform spatial heating of tissue due to changes in property of the tissue and heat-induced vasomotion in the region of interest. Thus, it is often advisable to stop the thermal treatment for $\Omega \geq 1$, since $C_D \geq 0.63C_0$ which means that 63% of undamaged tissue is damaged. Further, an estimate of critical temperature beyond which thermal damage is evident can be computed as $T_{cr} = E_a/R\ln(A)$ for $d\Omega/dt$.

5.4.2 HEAT SHOCK PROTEIN (HSP) EXPRESSION MODEL

To capture the thermally induced heat shock protein response, a mathematical model was proposed by Rylander et al.[47] using the least-squares approach. The empirical model for the expression of Heat Shock Protein (HSP) is defined to be dependent on temperature and duration of heating. In fact, the HSP expression is defined directly by relating to its concentration $H = H(T, t)$ as follows:

$$\frac{\partial H(t,T)}{\partial t} = f(t,T)\cdot H(t,T) \tag{5.38}$$

where $f(t,T) = \alpha - \beta_1 t^{\gamma-1}$ is a general rate function which illustrates the nature of HSP induction. Here, α, β, and γ are temperature dependent and time independent with $\gamma>1$. $H(t,T)$ may be expressed as $H(t,T) = A\, exp(\alpha t - \beta t^{\gamma})$ where A is a temperature-dependent constant.

HSP 27 and HSP 70 expressions were used to assess the tumor cell viability ($HSP_{27,70} \leq 1.0$ mg/mL). To prevent the recurrence of tumor, HSP expression must be eliminated. This formulation presents an additional metric to assess the state of tissue along with damage metrics calculated by the Arrhenius model.

5.4.3 TEMPERATURE-DEPENDENT TIME-DELAY MODEL

Recently, experimental studies have been performed to further improve the accuracy of the traditional Arrhenius model of thermal cell-death. In the literature, the frequency factor, A, varies between 1×10^{40} and 1×10^{105} s^{-1}. But, activation energy, E_a, typically varies from 1×10^5 to 1×10^6 $Jmol^{-1}$.[46] It is observed that the kinetic rate coefficients should be both temperature and cell line/tissue dependent.[48] Also, it was evident from many related studies that such an assumption for tissue coagulation by limiting two damage parameters needed further consideration of the process of regeneration and effect of other protein functions.[49] To capture the thermal damage other research defined mathematical function with cut-off temperature for the initiation of denaturation as below.

$$\frac{\partial\Omega(x,y,z,t)}{\partial t} = \begin{cases} 0 & \text{for T}<44^{\circ}\text{C} \\ A\exp\left(\dfrac{-E_a}{R_u T(x,y,z,t)}\right) & \text{for T}\geq 44^{\circ}\text{C} \end{cases} \tag{5.39}$$

The constant-rate single-step irreversible unimolecular reactions based on traditional Arrhenius kinetic model of cell-death are essentially accurate and reasonable for structural proteins (collagen, elastin, and actin) at higher temperature range (≥ 55°C) in several cell lines (e.g., Chinese Hamster Ovary cells (CHO), PC3 human prostate cancer cells, and SN12 human renal cell carcinoma), but these are significantly insufficient with much larger error margin (15–20%) to accurately address slowly developing "shoulder region" with an acceptable accuracy.[50–52] The existence of cell survival curves within hyperthermic temperature range of 43–54°C temperatures is mainly attributed to the existence of intrinsic cell death processes (programmed cell-death) namely "*Apoptosis,*" "*Necroptosis,*" "*Autophagy,*" and "*Pyroptosis*" that could be treated mathematically by higher-order kinetic processes.

In the Arrhenius formulation, the nondimensional thermal damage parameter Ω is proportional to the duration of exposure and exponential variation with temperature. This nondimensional parameter Ω is noninformative regarding the state of severity of damage which is not ideal for biological feedback systems. In fact, it is proven by histopathology of thermally damage cells or tissue based on vital stain uptake, that is, propidium iodide (PI) staining and fluorescent staining, gene expression, loss of clonogenicity (in cancer cells), calcein leakage measurements, thermally induced birefringence loss suggests that the Arrhenius model compromises the accuracy by overestimating the death of cell in initial phase of heating and underestimates for late phase stages which thereby affects the heating effectiveness of tumor treatment.[50,51,53]

The higher-order chemical kinetics introduces a lag in temperature-dependent survival curve of cells which is a slowly developing process. Therefore, a lag must be incorporated into the rate equation, since the Arrhenius damage model only computes damage from starting time t = 0 which is not true in all situations. The temperature-dependent time delay can be incorporated to introduce lag in the Arrhenius damage model for cell-death. This indicates that a longer response time of hyperthermia treatment is needed to achieve damage parameter to reach a value of 4.6, a value that defines the success of treatment. Based on experimental observations for specific tissue/cell-line, the original Arrhenius kinetic rate coefficients were revised/modified with a new set of kinetic coefficients by Dr. John Pearce[51] and recently implemented by Liang Zhu and coworkers[54] numerically to include temperature-dependent time-delay for the initial cellular in activities as the following:

$$\Omega = A\int_0^{\tau} \exp\left(\frac{-E_a}{R_u T_{tumor}(x,y,z,t)}\right)dt - A\int_0^{\tau_d} \exp\left(\frac{-E_a}{R_u T_{tumor}(x,y,z,t)}\right)dt \quad (5.40)$$

where applying $\tau_d = 0$ simply the model to traditional Arrhenius equation and $\tau_d \neq 0$ is a full-scale Pearce thermal damage model.

This time-delay in eq 5.40 is specific to tumor cell and is also a function of local temperature of tissue. The required time delay τ_d was given as an empirical expression to describe its dependence on local temperature as follows:

$$\tau_d = b - mT \quad (5.41)$$

$$\frac{\partial \tau_d}{\partial T} = -m \quad (5.42)$$

$$\tau_d\big|_{eff} = b - \int m \frac{\partial T}{\partial t} dt \quad (5.43)$$

where both parameters *b* and *m* are tumor cell line dependent. *b* is a time constant-intercept (s) and *m* is a slope or gradient of the line (how steep the line is) (s°C^{-1}).

Note that this equation is valid for a narrow temperature range. In fact, studies revealed the introduction lag to damage parameter has a better fit to experimental data.

In principle, the modified thermal damage model would change the heating time when used in heating protocol design. This concept of modeling thermal damage was often implemented in a theoretical study focusing on accurate design of heating protocol.[54] It has been reported by the group that the inclusion of temperature-dependent time-delay would result in a 24% shorter heating time.[54] These models can also be mathematically extended for drug delivery applications.[55] Another possible implementation is to adopt three kinetic coefficient ranges ($T < 43°C$, $43°C \leq T \leq 54°C$, and $T > 54°C$) for attaining a well-fitted damage curve for temperature-time changes.

According to the published literature,[56,57] there are at least two breakpoints (transitional temperatures) 43°C and 54°C in the temperature range of 39°C–60°C. One such piecewise implementation of kinetic coefficients was defined in the work of Rylander et al., who define 54°C as the breakpoint.[47]

5.4.4 NON-ARRHENIUS KINETICS OF THERMAL DAMAGE

The non-Arrhenius kinetic damage model recently proposed by many groups[58–60] utilized an additional regeneration term defined by oxygen supply perfused blood to accommodate the simultaneous regeneration of cells while hyperthermia damage is in progress. This equation was recently revisited in many studies.[61–64]

$$\frac{\partial \Omega(x,y,z,t)}{\partial t} = \underbrace{A\left(1-\Omega(x,y,z,t)\right)\exp\left(\frac{-E_a}{R_u T(x,y,z,t)}\right)}_{\text{Arrhenius rate kinetics}} - \underbrace{B\omega_{c,0}\Omega(x,y,z,t)}_{\text{Regeneration term}} \tag{5.44}$$

where Ω is the parameter to quantify damage, A is the frequency factor, E_a is the activation energy, and $B = 9 \times 10^{-3}$ (dimensionless constant).

The parameter of thermal damage, $\Omega = 1$ indicates 63.21% protein denaturation which is enough to begin an irreversible death of cells. Including the term regeneration being perfusion dependent appears to suppress, prevent, and limit thermal damage front to spread into the healthy tissues from tumor region. Therefore, it restricts further thermal damages to accumulate within $\Omega \leq 1$ (only tumor region).

5.4.5 MODEL BASED ON ENZYME-CATALYZED REACTION IN TISSUE

A model for the thermal injury process was proposed by Xu and Qian[65] and recently implemented by Alzahrani and Abbas[63] which is based on enzyme-catalyzed reactions that are happening in living tissue in which the damage was quantified based on change in concentration of denatured proteins. This model takes into account temperature, pH, irradiation, chemical agents that affect the enzyme activity.

$$\frac{\partial \Omega}{\partial t} = \frac{A_o \exp\left[\alpha\left(1-\frac{T_o}{T}\right)\right]}{1+B_o \exp\left[-\beta\left(1-\frac{T_o}{T}\right)\right]} \tag{5.45}$$

where $\alpha = 100$ is a coefficient relating enzyme denaturation, $\beta = 195$ is another coefficient relating decomposition and complex formation, $B_0 = 8 \times 10^4$ and $A_0 = 1 \times 10^{-3}\ s^{-1}$. Using the finite difference scheme, the solution to the formulation may be solved at the targeted node and written as follows:

$$\Omega_{i+1} = \Omega_i + \frac{A_o \exp\left[\alpha\left(1-\frac{T_o}{T}\right)\right]}{1+B_o \exp\left[-\beta\left(1-\frac{T_o}{T}\right)\right]} \cdot \Delta t \tag{5.46}$$

This equation has an initial condition $\frac{\partial \Omega}{\partial t} = 0.$.

5.4.6 STATISTICAL THERMAL DAMAGE MODELS

5.4.6.1 TWO-STATE CELL DAMAGE MODELS

Feng et al.[52]developed a unified model using fundamental principles of thermodynamics and proposed following mathematical expression for cell viability relating heat temperature (T) and time (τ):

$$A(T,\tau) = \frac{e^{-(\gamma/T+\alpha\tau+\beta)}}{1+e^{-(\gamma/T+\alpha\tau+\beta)}} \tag{5.47}$$

where A is the viability of cells, and γ, α, and β are experimentally determined constants.

This model includes the slowly developing "shoulder region" in the initial states of heating ~ up to 56°C and heating duration of 30 minutes. The only limitation of this model is that it is nondynamic and may not be implemented during dynamic temperature simulation. The two-state model only has 3–5% error margin in comparison to the traditional Arrhenius model which has an error margin of 15–20%.

5.4.6.2 THREE-STATE CELL MODEL OF CELL-DEATH

The Arrhenius model considers only two biological states, that is, either alive or dead state with a single irreversible reaction. However, O'neill et al.[66] proposed a three-state model with an inclusion of an intermediate vulnerable state within alive and dead state of a cell having tendency to recover to an alive state. The model summarizes three states, such as native (alive) (N), unfolded (U), and denatured (D) where the unfolded state can recover or return to the native state with the constraint that the sum of all three states is equivalent to one (i.e., $N + U + D = 1$). The ordinary differential equations for the three-state thermal cell-death model can be mathematically expressed as follows:

$$\frac{dN}{dt} = -k_1 N + k_3 U$$

$$\frac{dU}{dt} = k_1 - k_3 U - k_2 U \tag{5.48}$$

$$\frac{dD}{dt} = k_2 U$$

The tissue viability parameter can be computed from the information on native and unfolded states (i.e., $G = N + U$). If the tissue viability parameter is less than the threshold value of 0.8, the tissue is considered to be completely dead.

5.4.7 ISO-EFFECTIVE DOSE MODEL FOR CUMULATIVE EQUIVALENT MINUTES

On the other hand, an equivalent time for a thermal dose can be computed by cumulative equivalent minutes (CEM)[67] as follows:

$$CEM = \int_0^{\tau} R_{CEM}^{(43-T)} dt \tag{5.49}$$

The parameter R is used to assess how the actual time step is equivalent to a time duration at 43°C, and its value was determined from experiments and cell culture in *in vitro* conditions. The R values below provide its dependence on temperature ranges as

$$R_{CEM} = \begin{cases} 0.25; & for\, 37°C \le T \le 43°C \\ 0.50; & for\, 43°C < T \le 48°C \\ 0.72; & for\, T > 48°C \end{cases} \quad (5.50)$$

where τ is total time for thermal treatment ($= t - t_0$), t_0 is the time at which number of viable cells is C_0, whereas $D_0(43)$ is cumulative equivalent minutes till time t_0 at which damage parameter is $\Omega = 1$.

In fact, the damage parameter (Ω) relates to the CEM calculation as $\Omega = CEM/D_0(43)$, and the final damage concentration in terms of CEM can be given as[68]

$$C_D(\tau) = C_0\left[1 - \exp\left(-CEM/D_0(43)\right)\right] \quad (5.51)$$

5.4.8 POLYNOMIAL CORRELATION OF THERMAL DOSE

Shih et al.[69] used the curve-fitting approach to establish a polynomial expression for the iso-survival curve data where the factor R is having functional dependence on the heating temperature instead of the two different constants suggested by Sapareto and Dewey. The modified equation presents a means to relate temperatures and heating duration and can be given as follows:

$$R = \left(\frac{t_2}{t_1}\right)^{\frac{1}{t_2 - t_1}} \quad (5.52)$$

where t_1 is the time of exposure, t_2 is the equivalent time at 43°C, T_1 is hyperthermia treatment temperature in tissue, and T_2 is the reference temperature of 43°C, and $R = 57151.44 - 5050.5203T + 167.36283T^2 - 2.464839T^3 + 0.013609776T^4$.

The thermal dose can be expressed as follows:

$$CEM_{43} = \int_{t_s}^{t_f} \left[57151.44 - 5050.5203T + 167.36283T^2 - 2.464839T^3 + 0.013609776T^4\right]^{T-43} dt$$

5.4.9 INTEGRATED MODEL BASED ON CEM_{43} AND ARRHENIUS EQUATION

An integral thermal dose model comprising two expressions (Arrhenius model and CEM_{43}) was proposed by Babbar et al.[70,71] to predict the biological tissue thermal injury and cellular necrosis.

$$\Omega(\tau) = k_r .\ \mathrm{CEM}_{43} \tag{5.53}$$

where $k_r = Ae^{\left(-\frac{E_a}{R.T(\tau)}\right)}$, $\mathrm{CEM}_{43} = \int_0^{\tau} R_{CEM}^{T_{break}-T}\, d\tau$, $R_{CEM} = e^{\left(-\frac{E_a}{RT_1(T_1+1)}\right)}$, and $T_1 = T_{\mathrm{break}} = 43°C = 43 + 273.15 = 316.15\mathrm{K}$.

5.4.10 CEM_{43} DOSE USING ARRHENIUS-TYPE VOGEL–TAMMANN–FULCHER EQUATION

Another Arrhenius-type piecewise VTF (Vogel–Tammann–Fulcher) equation was recently proposed by Assi et al.[72] In this formulation, the reaction velocity, k_r, may be defined as

$$k_r = \begin{cases} Ae^{-\left(\frac{a}{T-T_o}\right)}, & T > T_o \\ 0, & T \le T_o \end{cases} \tag{5.54}$$

where A [s^{-1}] is a frequency factor, a [K] is an activation energy that varies with temperature, T_o [°C] is the absolute temperature where the damage cannot be thermally activated, and T [°C] is the system temperature. The difference between Arrhenius and modified Arrhenius-VTF model is that system-based temperature replaces the absolute zero temperature (−273°C) in the traditional Arrhenius equation. The physical interpretation of this implementation is that the rate constant is equivalent to zero for any temperatures below T_o implying that there will be no dose accumulation below this temperature. Therefore, the traditional Arrhenius thermal dose model is modified to $\Omega(\tau) = A\int_0^{\tau} \exp\left(-\frac{a}{T(t)-T_o}\right)dt$.

5.4.11 UNIFIED THERMAL DOSE EQUATION

A mathematical unification of three models (CEM_{43}, Arrhenius, and Arrhenius-VTF) can be generalized into a thermal dose equation where the accumulative damage parameter Ω is depending on exponential coefficient β which is temperature dependent as

$$\Omega = \alpha \int_0^{\tau} e^{\beta(T)} dt \tag{5.55}$$

where α is the preintegral factor. The preintegral factor is $\alpha = 1$ for CEM_{43}, $\alpha = A$ for Arrhenius, $\alpha = A'$ for Arrhenius-VTF,

$$\beta(T) = \begin{cases} \ln(C)(T-43); \text{ CEM 43} \\ -\dfrac{E_a}{R(T+273)}; \quad \text{Arrhenius} \\ -\dfrac{a}{T-T_o}; \quad \text{Arrhenius - VTF} \end{cases} \tag{5.56}$$

The thermal dose at a constant temperature of 43°C for a duration t_{43} $\alpha e^{\beta(43°C)} t_{43} = \alpha \int_0^{\tau} e^{\beta(T)} dt$. The generalized CEM_{43} type equation $t_{43} = \int_0^{\tau} e^{\beta(T) - \beta(43°C) dt}$. The CEM_{43} for the Arrhenius–VTF model can be written as $t'_{43} = \int_0^{\tau} e^{\frac{a}{43°C - T_o}\left(1 - \frac{43°C - T_o}{T - T_o}\right) dt} \quad for\, T \geq T_o$.

The frequency factor, activation energy, and critical temperature for thermal damage estimation are tabulated in Table 5.2. In Table 5.2, the parameters were obtained by experiments performed on various tissues and by regression analysis.

5.4.12 THERMAL DAMAGE MODEL IN FREEZING TEMPERATURES

5.4.12.1 LOW-TEMPERATURE THERMAL INJURY

For cryotherapy/freezing temperature below 273.15 K, damage to the tissue is associated with ice formation that depends on the balance between cell dehydration rate and intracellular ice formation for a given cooling rate. Ideally, tissue damage during cryotherapy involves fast rate of cooling to induce intracellular ice formation of water and cellular compartments, since cells were unable to release intracellular water osmotically to eliminate intracellular freezing process at a faster rate of cooling.[93] The rate of cell dehydration process as a function of cell water permeability, potential barrier of the membrane, and freezing temperature was discussed in Ref. [94]. A more convenient approach for ice formation was proposed by Toner et al.,[95,96] suggesting that intracellular ice formation originates from a nucleation site

TABLE 5.2 Cell, Proteins, Tissue Damage Parameters Deduced from Experiment.

Material	A (s^{-1})	E_a (J/mole)	Critical temperature	Temperature range	Ref.
			Tissue		
Retina	0	-	-	$T \leq 42.85$ °C	[73]
	4.3×10^{64}	4.2×10^5	-	$42.85 < T \leq 49.85$°C	-do-
	9.3×10^{104}	6.7×10^5	-	$T > 49.85$°C	-do-
	3.1×10^{99}	6.28×10^5	57.6°C	-	[74]
	1.0×10^{44}	2.93×10^5	-	-	[75]
Skin	3.1×10^{98}	6.28×10^5	59.9°C	-	[76]
	2.2×10^{124}	7.83×10^5	55.4°C	$43.85 < T < 49.85$°C	[77]
	1.8×10^{51}	3.27×10^5	60.1°C	$T \geq 49.85$°C	-do-
	1.3×10^{95}	6.04×10^5	58.5°C	$T \leq 50$°C	[68]
	8.82×10^{94}	6.028×10^5	58.5°C	$T \leq 53$°C	-do-
	1.297×10^{31}	2.04×10^5	69.3°C	$T > 53$°C	-do-
	3.1×10^{98}	$[6.27 - 5.1(T\text{-}53)] \times 10^5$	69.3°C	$T > 53$°C	[78]
Kidney	1.48×10^{60}	3.996×10^5	73.7°C	-	[79]
	3.3×10^{38}	2.569×10^5	75.2°C	-	-do-
	5.73×10^{34}	2.404×10^5	88.0°C		[80]
	5.85×10^{28}	2.023×10^5	94.2°C		-do-
Prostate	2.08×10^{27}	1.866×10^5	83.6°C	-	[78]
	3.8×10^{57}	3.85×10^5	76.0°C	-	[79]
Muscle	2.94×10^{39}	2.596×10^5	70.4°C	-	[81]
Liver	5.51×10^{41}	2.769×10^5	73.4°C	-	[68]
	2.09×10^{33}	2.219×10^5	74.4°C		[82

TABLE 5.2 *(Continued)*

Material	A (s^{-1})	E_a (J/mole)	Critical temperature	Temperature range	Ref.
			Cells		
Erythrocytes	1×10^{31}	2.12×10^5	84°C	-	[83]
	6.8×10^{36}	2.49×10^5	80°C	-	[84]
	7.6×10^{66}	4.55×10^5	82.2°C	-	[85]
CHO cell	2.84×10^{99}	6.18×10^5	51.4°C	$T \geq 43$ °C	[68]
HeLa cell	9.09×10^{36}	2.49×10^5	78.8°C	-	[86]
HCC cell	2.04×10^{42}	2.814×10^5	73.8°C	-	[87]
AT1 cell	2.99×10^{37}	2.448×10^5	-	-	[88]
			Proteins		
HSP 70	6.90×10^{282}	1.74×10^6	48.2°C	-	[89]
Collagen (Aorta)	5.6×10^{63}	4.3×10^5	78.9°C	-	[90]
Collagen	1.77×10^{56}	3.676×10^5	68.2°C	-	[91]
	1.606×10^{45}	3.06×10^5	80.4°C	-	[92]]

(cell membrane in the presence of extracellular ice) which grows exponentially to form a cluster of ice. They suggested an appropriate heterogeneous intracellular ice formation (IIF) relation as

$$IIF = \Omega_{het} \exp\left(-k_{het} / \Delta T^2 T^3\right) \tag{5.57}$$

where Ω_{het} and k_{het} are the thermodynamic and kinetic parameters describing the rate of heterogeneous ice formation under isotonic solute, $\Delta T = T - T_f$ change in temperature of intracellular compartment and the temperature of the freezing intracellular compartment is denoted by T_f. The thermodynamic parameter is given as $\Omega_{het} = \Omega_o \sqrt{T/T_{fo}} \left(\eta_o N/\eta N_o\right)\left(f_\theta / f_{\theta o}\right)^{1/6}$ and the kinetic parameter is given as $k_{het} = k_o \left(T_f / T_{fo}\right)^4 \left(f_\theta / f_{\theta o}\right)$. Here, subscript "o" refers to parameters for isotonic condition, η is intracellular solution viscosity $\left(= 0.139 \times 10^{-3} \left[T/225 - 1\right]^{-1.64} \text{ kg/m.s}\right)$, T_{fo} is −0.52°C for an isotonic solution of 0.285 osm/kg.H_2O, f_θ is the factor for cluster formation as a function of nucleation contact angle θ given by Ref. [94]. Note that the damage to cell due to intracellular ice formation can be approximated by the formation of ice due to temperature change which is given as[95]

$$P_{IIF} = 1 - \exp\left[\frac{1}{B} \int_{T_f}^{T} \left(A_s \times IIF\right) dT\right] \tag{5.58}$$

where B is equivalent to the rate of temperature change (= – dT/dt), A_s is the surface area of the cell that involves surface-catalyzed nucleation. The above approach modified to mimic the probability of damage due to ice formation was integrated with the bioheat model for the freezing process.[96]

5.5 TABLES FOR INPUT PROPERTIES IN THE MODELS

Thermophysical properties of tissue and biofluid are the two key input parameters used in the bioheat equations discussed above to model the heat transfer phenomenon. These are thermal conductivity of perfused tissue, convection coefficients such as blood perfusion, metabolic heat energy generation, thermal capacity of perfused tissue and blood, and optical properties of tissue. The complexities associated with biological system such as perfusion in the tissue, blood flow variability due to cardiac function, oxygen and nutrient delivery by blood for enhanced metabolism, neovascularization for increased supply of blood in the diseased condition, neuronal regulation of nitric oxide, make them thermophysical properties function of blood perfusion, temperature, pressure, and time. For instance, the predication of thermal ablation often demands modeling the blood perfusion in the tissue as a function of temperature, damage integral, phase change parameters, since experimental observation reveals variability in blood perfusion during thermal treatment. However, the effect of blood perfusion, temperature, pressure, and time on tissue thermophysical properties is often neglected by assuming constant properties, except in case of strong functional dependence. In fact, assuming constant properties in bioheat equation permits prediction with reasonable accuracy. These responses or variation in properties is the resulting effect of normal and abnormal physiological conditions. The various input thermophysical properties and their correlations for different tissues (healthy/cancerous) and biofluids were depicted in 5A, whereas the optical properties are given in 5B.

5.6 APPLICATION OF BIOHEAT TRANSFER MODELING IN BIOENGINEERING

The broad area of bioheat transfer modeling and research in bioengineering involves numerical prediction, treatment planning and management of heat absorption and removal, rate of heat transfer, identification of disease-associated change in temperature or blood circulation. Ideally, these investigations are classified into (1) applications involved core temperature lower than the physiological range of the human body (< 30°C or subzero temperature), (2) temperature of physiological range (30–40°C), and (3) temperature above physiological range (40–120°C).

5.6.1 APPLICATIONS INVOLVE BELOW NORMAL TEMPERATURE RANGE

5.6.1.1 THERAPEUTIC HYPOTHERMIA

Therapeutic hypothermia is a clinical procedure of reducing and maintaining the bodies' core temperature or brain temperature few degrees below the normal physiological temperature (< 37°C) for 12, 24, 48, or 72 hours to prevent the damages to the brain. These damages include neuronal (such as physical, cognitive, and psychosocial dysfunction) or death due to neonatal hypoxia-ischemic encephalopathy after perinatal asphyxia,[97] and traumatic brain injury (TBI), ischemic strokes, hemorrhagic strokes, subarachnoid hemorrhage during cardiac arrest or surgery.[98] Figure 5.4A depicts MRI-generated apparent water diffusion coefficient map and diffusion-weighted map of brain function of neonates with perinatal asphyxia subjected to therapeutic hyperthermia.[99] In general, therapeutic hypothermia can be classified into mild hypothermia (34.5–36.5°C), moderate hypothermia (34.5–32°C), marked hypothermia (28–32°C), and severe hypothermia (< 28°C). Hypothermia can be induced externally using cooling blankets, ice packs, and special helmets designed for local cooling or internally by infusing Ringer's lactate via venous catheters. The exact mechanism of therapeutic hypothermia to arrest the brain injury can be related to decrease in metabolic rate at low temperature to maintain equilibrium in blood oxygen supply and demand, reducing neuronal damage, ATP expenditure, and free radical production, and increase in intracellular calcium to avoid cell death.[98]

Clinical application of whole body or local hypothermia often demands planning and monitoring the variation of core body temperature to induce cooling. To control the unwanted response of the body such as shivering and other associated damage to excess drop in core temperature, an ideal approach is to simulate the whole-body thermal response or local thermal responses to ascertain the set parameters for hypothermia. Extensive research on bioheat modeling of local brain hypothermia[100] and whole-body hypothermia[101,102] has been implemented in the past, which has led to the development of cooling helmet devices[103] for local hypothermia and cooling jacket with temperature monitored water flow feedback mechanism for whole-body hyperthermia. However, the scope of future investigation lies in the development of smart devices for simultaneous application, monitoring and control of hypothermia in a clinical setting. Especially, the dynamic integration of devices like electroencephalogram (EEG) and MRI for assessing the neurologic state and function of the brain due to the application of local or core hypothermia could lead to new findings.[104]

5.6.1.2 ORGAN PRESERVATION

Preservation of organ can be achieved by the suppression of organ metabolism while staring at hypothermia (0–8°C) by completely flushing blood by preservative fluid, or by continuous perfusion of oxygen and synthetic substrates (instead of blood) for metabolism by perfusion machine when the organ is kept in preservative solvent at low temperature. The key research in organ preservation science involves the development of compatible biochemical solution and innovative perfusion machine which would provide an environment to sustain the viability of tissue in hypothermia and enable longer preservation of organs ($\geq$ 24 or 36 hours) with low ischemic damage, limited hypothermia cell swelling, and relatively less blood reperfusion injury upon transplantation. At present many biochemical solutions exist[105,106] that has made transplant centers preserve kidney, liver, and pancreas successfully for nearly or greater than 24 hours either by flushing with a biochemical solution or by static storing in hypothermia. However, preservation of heart and lung has remained a great challenge with an effective preservation time of close to 8 hours. The primary difficulty faced in the transplantation of preserved heart is the revival of reduced myocardial electrical activity in cold storage.[107] New developments in organ preservation science are two artificially preserved organs on normothermia condition (around 32°C) by a continuous supply of oxygen and mimicking intracellular fluids, or by blood ex-vivo supply circuit for preserved organs at normothermia to ensure low injuries.[106] The research scope in the field of organ preservation involves bioheat model of heart, lung, kidney, liver, or pancreas to evaluate the effect of cold perfusion of preservative fluid, effect of perfusion rate in a perfusion machine, or thermal effect on tissue in static cold storage. A state-of-the-art perfusion machine capable of *ex vivo* preservation of lung and heart during transplantation is shown in Figures 4B and 4C, respectively.

5.6.1.3 CRYOPRESERVATION

Cryopreservation is the process of long-term conservation of the state of cells, tissue, embryos, and organs at ultra-low temperature (≤ -80 to -100°C) in the presence of cryo-protective chemicals. The idea of cryopreservation is popular for the fact that at an ultra-low temperature all biological activities such as protein functions, ion transfer, and metabolism can be put on hold for long duration. This will theoretically stop the damage to the organ in critical condition or enable long-term preservation of organ for future transplant. In fact, cryopreservation has potential applications in conservation of cadaver, genetic material, embryos, cells, and tissue for future assessment and/or breeding. However, the major challenge in cryopreservation is the associated injuries to biological samples during freezing and thawing. Thawing is the process of bringing the biological sample to its original state by appropriate warm up protocol. The injury associated with freezing could be minimized by imposing an optimal cooling rate for the biological sample or by using right cryo-protestant, though cooling-associated increasing in chemical concentration of cryo-protestant was also found to be toxic. Another method of cryopreservation involves cooling a biological sample in high concentration of cryo-protestant at a faster cooling rate to a temperature just before nucleation of ice, during which solidification was attained by increasing the viscosity of fluid instead of crystallization. This mode of cryopreservation is also known as vitrification. In fact, the increase in concentration of cryo-protectant during vitrification also increases the associated chemical damage to cells. A detailed discussion of injury caused by chemical, intra- and extra-cellular freezing of cells and tissue was given in Ref. [108–110]. Many advances were made in the last three decades in this field with some success attained in preservation of animal cells and embryos. However, the existing method of cryo-preservation is far from the clinical standard of viable preservation for human cell, tissue, and organs. The scope of future research in this field includes reducing the toxicity of cryo-protection to cells or tissue during cryo-preservation or vitrification, innovative method for freezing organs including directional/controlled ice crystallization during cooling, thermal modeling of cell, tissue, or organ to assess the effect of cooling rate, thawing rate, and associated intra- and extra-cellular freezing/crystallization processes during cryopreservation. A representative image of cryopreservation of biological cells in liquid nitrogen contained in the cryo-container is shown in Figure 4D.

5.6.1.4 CRYOSURGERY

Cryosurgery is a localized method of destroying the abnormal tissue or tumor volume by the process of fast freezing (≤ −20 to −50°C), slow rate of thawing, or by the process of subsequent freezing–thawing cycles using a specialized annular metallic probe which contains fluid at low temperature. The commonly used cryo-fluids are liquid nitrogen and carbon dioxide or its gaseous variants. The process of cell/tissue damage during cryosurgery depends on the cooling rate, thawing rate, and subsequent freeze–thaw cycles. The freezing of intra-cellular compartments and damage to extra-cellular proteins can be induced by rapid cooling rate, which causes physical and/or chemical damage to essential organelles during ice crystallization. On the other hand, slow rate of thawing after initial freezing causes formation of large volumes of fused ice at temperature > −40°C or swelling of the cell due to the large influx of water from an extracellular region which induces mechanical damage of cell membrane.[111] Overall, the tissue damage due to freezing and thawing process is heterogeneous in nature. Hence, in a practical scenario physicochemical damage to tissue volume is attained by repetitive freeze–thaw cycles in clinics to ascertain maximum damage. However, the optimal value of temperature, rate of cooling, and duration of cooling for freeze–thaw cycles are undefined parameters, which are often decided by surgeons based on experience.

Cryosurgery is one of the popular thermal treatment methods for treating lesions both at the periphery and within the body. Lesion at periphery is often treated by the process of freeze–thaw cycles (2–3 times) using cryo- gas or probe, which leads to scab formation after thawing. But, lesion at a depth within the body requires a surgical cut to guide cryo-probes to the desired site to apply freezing or subsequent freeze–thaw cycles to damage the tissue. Once the procedure ends, the lesion warms up in-situ and forms a scar which is eventually taken out by the body's immunity.[110] The key challenges encountered in cryosurgery (with or without image guidance) are to locate the precise margin of damaged tissue volume, since the freezing front travels nonuniformly outward from the probe surface. Therefore, in case of cryosurgery biopsy/complete removal of damage tissue volume with a small margin of the surrounding healthy region is a mandatory procedure to ascertain the nonviability of malignancy. Moreover, the freezing-associated damage is maximum near the probe and reduces away from the probe. Therefore, in practice maximum 4–5-inch effective diameter around the probe is assumed to be completely damaged or necrotized. However, to attain better treatment control over large lesion volume multiple cryo-probes were deployed. Figure 5.4E depicts cryocatheter used for cryoablation of multiple lesions in a canine model.

Numerical simulation of cryosurgery is often used to decide on the optimal exposure time, cooling rate, and temperature of cryo-probes to induce sufficient damage to tumor volume.[112] For this, bioheat models can be coupled with damage due to intracellular ice formation (eq 5.58) to evaluate the damage front within the tissue domain.

5.6.2 APPLICATION INVOLVES A NORMAL BODY TEMPERATURE RANGE

5.6.2.1 THERMAL DIAGNOSIS OF SUPERFICIAL CANCER

Thermal diagnosis of superficial cancer involves measuring the spatio-temporal change in surface temperature of lesion and peri-lesions due to change in microcirculation and metabolism using thermal cameras and/

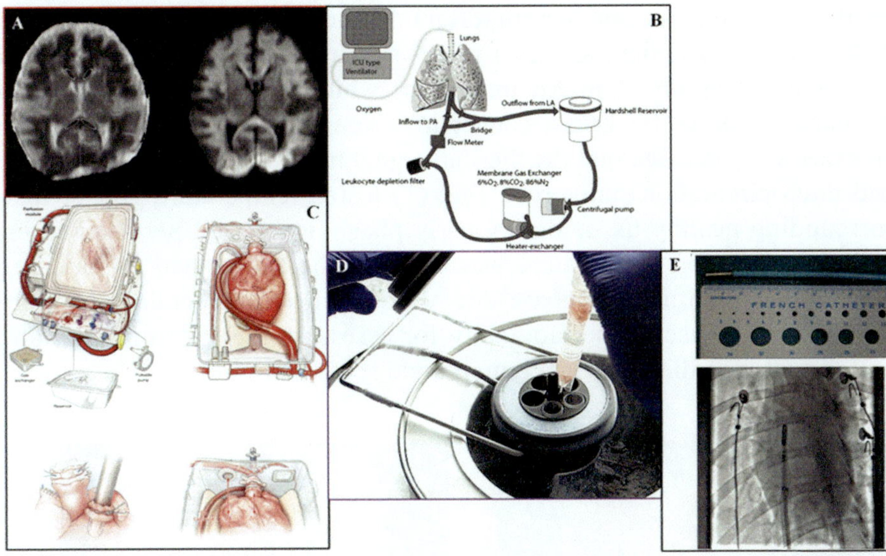

FIGURE 5.4 Clinical application of hypothermia and cryogenics in medical treatment and preservation. (A) Apparent water diffusion coefficient map (left) and diffusion-weighted MRI map (right) of the corpus callosum of neonates with perinatal asphyxia subjected to therapeutic hypothermia (B) Lung perfusion setup demonstrating protocol for ex vivo perfusion and ventilation of donor lung (C) A portable heart perfusion setup for ex vivo perfusion, maintain desired heart beating state, and dynamic assessment of heart state (D) Cryopreservation of biological sample in liquid nitrogen (E) Cryocatheter for cardiac tissue ablation (top) catheter placed in a canine model for cryotherapy of multiple spot and linear lesion (bottom) treatment

Source: (A) Reprinted from Ref. [99]. https://creativecommons.org/licenses/by/4.0/ (B) Reprinted from Ref. [113]. Image by Braithwaite and van der Kaaij, 2019. https://creativecommons.org/licenses/by/4.0/ (C) Reprinted from Ref. [113] https://creativecommons.org/licenses/by/4.0/ (D) Reprinted from Ref. [114]. https://creativecommons.org/licenses/by/4.0/ (E) Reprinted from Ref. [115]. https://creativecommons.org/licenses/by/4.0/)

or contact-based thermal sensors. The traditional approach for thermal diagnosis of cancerous lesion was performed using three methods[116] (1) illuminating the target lesion with light in infrared region and observing the transmitted or reflected light from the tissue, (2) cooling the lesion and healthy region below normal temperature and observing the spatio-temporal variation in surface temperature during thermal recovery, (3) passive measurement of spatiotemporal variation in surface temperature. In all these approaches, the fundamental principle of diagnosis is the fact that cancer exhibits increased metabolism and microcirculation as compared to healthy region which is inflicted in emanating thermal signature captured by thermogram. Yet another important feature observed in thermogram of breast cancer is the mosaic of hypo- and hyper-thermic region related to spatial variation of microcirculation, which was often considered a diagnostic marker for breast cancer involvement. However, quantitative analysis also revealed that the thermal signature of cancer varies with size and microcirculation status revealing (1) lesion temperature greater than surrounding healthy tissue during early phase or for well-perfused cases, and (2) the lesion temperature was found to be lower than surrounding healthy tissue during late phase due to poorly perfused condition. Figure 5.5 depicts the use of thermography for assessing the state of metastatic oral cancer revealing cervical lymph node spread.[117]

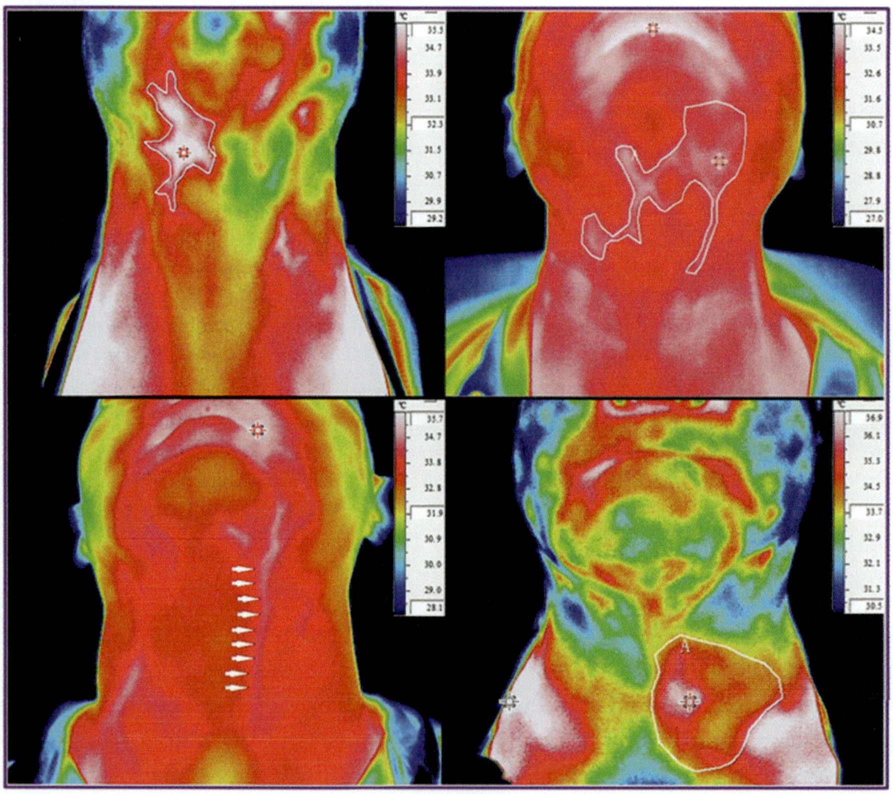

FIGURE 5.5 Clinical application of thermography revealed an increase in skin temperature and blood perfusion associated with cervical lymph node metastasis of oral cancer.

Source: Reprinted from Ref. [117]. https://creativecommons.org/licenses/by/4.0/.

The bioheat model of peripheral tissue with or without cancerous lesion subjected to appropriate boundary conditions was often implemented to measure the thermal response of lesion as compared to healthy tissue[27] and for prediction of effective properties of the lesion based on measured surface temperature.[118] Apart from a cancer diagnosis, thermal diagnosis was also found popular for non-contact detection of abnormal body temperature,[119] irregular breathing rate,[120] and cardiovascular diseases.[121]

More recently, an approach employing principles of inverse models has been developed to predict blood perfusion from thermal or infrared imaging maps on animal studies.[64] Both metabolic heat generation and blood perfusion in the mouse and the tumor were adjusted so that the theoretical predictions of the surface temperatures agree well with average temperature measurement from the infrared images.[122]

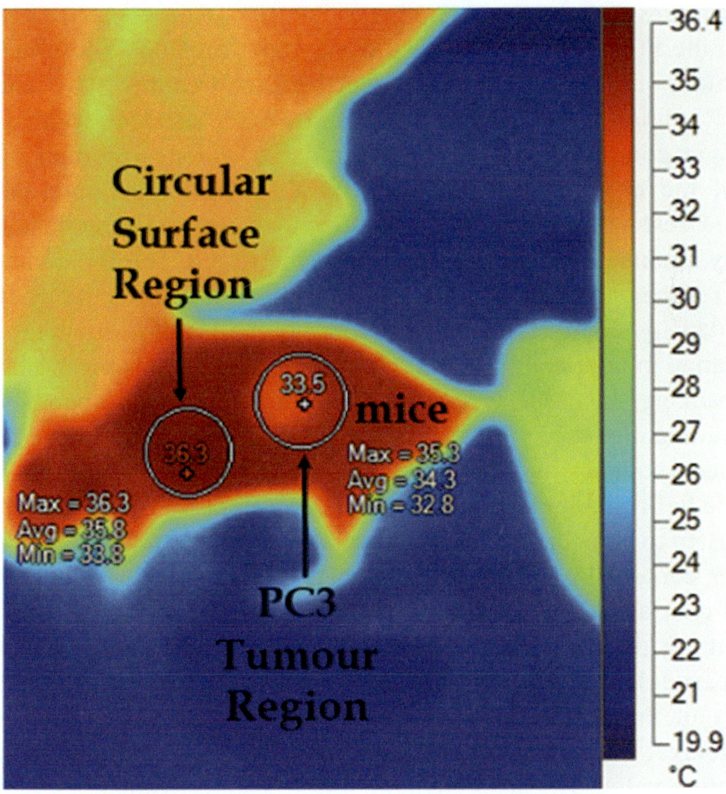

FIGURE 5.6 Preclinical application of thermography revealed temperature heterogeneity at mouse skin and tumor location. An infrared image of a Balb/c Nu/Nu male mouse placed in an ambient environment at 25°C. [Image courtesy: Manpreet Singh].

5.6.2.2 NONINVASIVE THERMOMETRY

Noninvasive thermometer is the method of extracting temperature data from tissue/organ at depth by various radiometric instruments, such as from magnetic resonance imaging (MRI), computed tomography (CT), microwave, and ultrasound scans. The method of CT,[123] microwave,[124] and ultrasound[125] thermometer is still at its developing stage, whereas 2D MR thermometry is an established method for predicting temperature of the deep tissue. In the clinical setting, it is essential to monitor the change in suspected tissue temperature either to control the treatment methods involving hyperthermia and cryosurgery, or to identify the normal or abnormal function of the brain since temperature variation in the brain is closely related to metabolism and cerebral blood flow (CBF). Here, we discuss an MR thermometry method in brief with its potential applications; readers may consider referring to the cited literature for deep understanding.

MR thermometry is one of the standard methods for noninvasive temperature measurement for hyperthermia treatment of cancerous lesion,[126] tumor ablation,[127] pyrexia-associated cerebral ischemic, and neurologic complications.[128] The method incorporates indirect measurement of the change in temperature or absolute temperature of proton magnetic resonance parameters of the measured radiofrequency signal intensity emanated by protons present in the tissue. These parameters are proton resonance frequency, T_1 and T_2 relaxation, diffusion coefficient, proton density, contrast agents, which are functions of tissue temperature. Any change in temperature led to variation in the signals picked up by the MRI scanners. In tissue water molecules are present in abundance, and hence tuning the receiving radiofrequency coil of MRI scanners to the resonance frequency of the proton (H atoms of water molecules) in a strong magnetic field enables acquisition of changes in radio frequency signals. It is important to note that most MRI scanners rely on the function of hydrogen of water content in the tissue. However, other nuclei or isotopes in the form of contrast agents can be administered to capture the specific signals upon tuning the radiofrequency coil to the proton resonance frequency of the particular type of isotope. In MR thermometry, measurement of a majority of the proton MR parameters (proton resonance frequency, T_1 and T_2 relaxation, diffusion coefficient, proton density) provides change in temperature value. The absolute temperature can be evaluated based on the known value of reference temperature, which is often difficult to measure *in-vivo*. However, methods involving contrast agents and spectroscopic imaging enable measurement of absolute temperature which is discussed in detail in the literature.[129,130] The potential applications of bioheat transfer models in the field of noninvasive thermometry involve development of hyperthermia treatment planning tools and prediction of contrast-free tissue microcirculation function.

5.6.3 APPLICATION INVOLVES ABOVE NORMAL BODY TEMPERATURE RANGE

5.6.3.1 HYPERTHERMIA TREATMENT

Hyperthermia is a treatment method utilizing temperature above normal body temperature in the range of 39–43°C to treat cancer for 1 hour using electromagnetic energy, along with the commonly used cancer treatment method such as radiation therapy and chemotherapy. In many clinical trials, hyperthermia along with chemotherapy and radiotherapy has shown better treatment outcome such as reduction in harmful radiation and chemo–therapy sessions, less treatment side-effects, and relatively greater survival in comparison to standard radiation and chemotherapy alone.[131–133] In fact, adjuvant hyperthermia treatment reveals a propensity of the damaging lipid membrane of cells, denaturation of proteins, and increase in blood perfusion and oxygen in radio resistant tumor. Hyperthermia treatment also suffers problems with nonuniform heating due to variation in effective properties of tissue and insufficient damage to tumor as a consequence of synthesis of heat shock proteins (HSP's) that appears to be one of the reasons for tumor thermotolerance. Therefore, it is essential to apply hyperthermia along with other traditional cancer therapy to ensure complete eradication of cancer and perform a biopsy to check the viability of cancer post treatment. Figure 5.7 depicts state-of-the-art setups for (A) noninvasive microwave hyperthermia with facility to assess target tissue temperature using MR thermometry, and (B) high-frequency ultrasound applicator used for treatment of superficial and deep tumors.

Apart from hyperthermia, two more high-temperature therapeutic methods coexist. One method found its application in physiotherapy by gentle induction of heat to body parts using electromagnetic energy or ultrasound, vibration, which works in the temperature range below hyperthermia range. The peripheral induction of electromagnetic and ultrasound energy causes an increase in the local temperature of the body section which improves

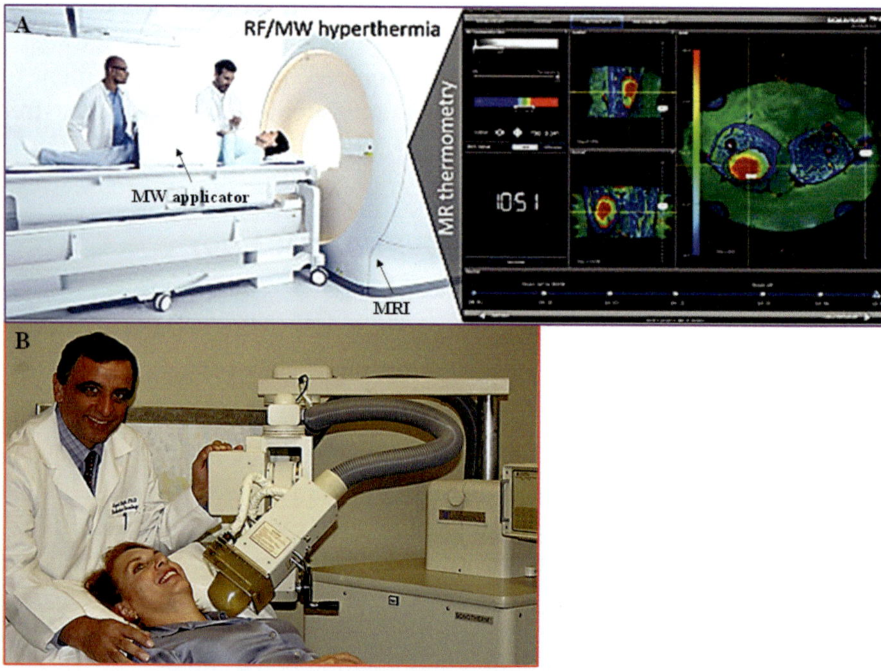

FIGURE 5.7 (A) BSD-2000 noninvasive microwave (MW) hyperthermia setup (Pyrexar Medical, USA) along with 3D MR thermometry facility (left) enabling better control during MW hyperthermia. MR thermometry data analysis console depicting the rise in target tumor temperature (right) (B) Application of thermal therapy to treat cancerous tumors by inducting high-frequency ultrasound to the target tissue.

Source: Reprinted from Ref. [134]. https://creativecommons.org/licenses/by/4.0/, (B) Reprinted from Ref. [135]. https://creativecommons.org/licenses/by/4.0/.

the blood flow, metabolism, ion-transport, relaxes the muscle, and relieves joint pains. Another method found its application in fast eradication of local cancerous tissue by coagulation necrosis, volatization, and charring, which works in the temperature range much higher than hyperthermia range. The tissue temperature range for irreversible damage caused by thermal therapy alone includes low-level coagulation necrosis (45–55°C), high-level coagulation necrosis (55–100°C), volatization, and carbonization (>100°C). In this method, abnormal tissue is either targeted by electromagnetic wave or pressure wave with minimal invasion percutaneously or by noninvasive focusing on the energy source from outside the body. The commonly employed devices used for this purpose are laser source, radio frequency, and microwave applicators (invasive or noninvasive), and high-frequency focused ultrasound. Methods involving photothermal therapy using plasmonic nanoparticle and magnetic heating using magnetic nanoparticle fluids are emerging fields. However, these methods are still limited to research laboratory or specialized centers, though they have the upper hand in selective damage of abnormal tissue. In clinical settings, different therapeutic methods and energy source can be implemented to induct prescribed tissue temperature of thermal therapy according to the requirement or clinical management protocol. However, the key factors needed to consider in clinical settings during high-temperature therapeutic procedures are precise control of temperature and time, invasive or noninvasive dynamic temperature monitoring, estimate of damage, and nanoparticle-associated toxicity. To address these issues, often clinical treatment was performed based on treatment planning tools/approximate numerical models, dynamic assessment using MRI thermometry and catheter temperature probes, and posttreatment biopsy to identify abnormal and normal tissue viability. The bioheat modeling plays a central part in the majority of thermal therapy for preclinical treatment planning, wherein a source term associated with an increase in internal energy of tissue is incorporated into the bioheat equation (eq 5.1) to evaluate the critical operating parameters. The modified bioheat equation can be written as

$$\rho_t c_t \frac{\partial T_t}{\partial t} = \nabla \cdot (k_t \nabla T_t) + (1-\kappa)\rho_b c_b \omega_b (T_a - T_t) + Q_m''' + Q_s \tag{5.59}$$

where Q_s is the volumetric source term that depends on the amount of energy received by tissue from a source and the fraction which is inducted in the tissue to raise its internal energy. All other parameters have the usual meaning as discussed earlier. To be more realistic blood perfusion ω_b should be considered dependent on the temperature or damage parameter, since perfusion varies during thermal therapy. The associated source and blood perfusion term involving various thermal therapy methods are detailed in Table 5.3.

Recently, studies have shown the feasibility to incorporate heterogeneous blood perfusion obtained from medical imaging such as contrast enhanced ultrasonography[158] and MRI[159] into Pennes bioheat equation. In such an implementation, blood perfusion would be varied over the whole biological tissue domain and monitored over all anisotropic locations extracted from imaging.

TABLE 5.3 Volumetric Source and Perfusion Term for Various Thermal Therapies.

Method	Relation	Comments	Ref
Noninvasive microwave hyperthermia	$Q_s = \rho SAR$ $SAR = \frac{\sigma}{2\rho}\lvert\vec{E}\rvert^2$	Here, ρ is tissue density, the specific absorption rate (SAR), $\vec{E}$ is electric field intensity, σ is the electrical conductivity. The value of SAR or $\vec{E}$ can be determined by solving Maxwell's EM equations[136] or its approximated equation for electric field (assuming tissue is nonmagnetic).	[137]
Minimally invasive radiofrequency ablation	$Q_s = \sigma E^2$	The invasive RF ablation method works on the principle of Joule heating due to resistance offered by tissue during electric passage in the tissue. Within the tissue, the distribution of electric field is determined by solving the potential distribution within the tissue as $\nabla \cdot (\sigma \nabla V) = 0$,such that $E = -\nabla V$, where V is the electric potential.	[138,139]
Minimally invasive microwave ablation	$Q_s = \rho SAR$ $SAR = \frac{\sigma}{2\rho}\lvert\vec{E}\rvert^2$	Here, the SAR value is determined by solving approximated model of electric field distribution deduced from Maxwell's electromagnetic equation discussed earlier.	[140,141]
High-frequency focused ultrasound (HIFU) therapy	$Q_s = 2\alpha \left\lvert \mathrm{Re}\left(\frac{1}{2}pv\right)\right\rvert$ where the attenuation coefficient (α) of the tissue, v is the velocity of acoustic wave and p is the acoustic pressure.	Here, attenuation coefficient is a function of the frequency of the wave and is given as ($\alpha_f = \alpha_{1\,\mathrm{MHz}} \times f^n$) such that f is the acoustic wave frequency and n is the exponent to raise acoustic wave above 1 MHz. The pressure wave distribution within the tissue is evaluated by governing equation of pressure waves in frequency domain.[142]	[143]
Cryotherapy/Cryo-ablation (enthalpy form)	For cryotherapy simulation, a generalized equation eq 5.59 is given in the enthalpy form as $\frac{\partial(\rho h_t)}{\partial t} = \nabla \cdot (k_t \nabla T_t) + (1-\kappa)\rho_b c_b \omega_b (T_a - T_t) + Q_m'''$	Note that during cryo-surgery the phase transition of water in tissue to ice/solid phase takes place from 0°C to –10°C. Here, β is defined using a linear relation for phase transition between liquid to solid as $\beta = \frac{T - T_{solid}}{T_{liquid} - T_{solid}},\ T_{solid} < T < T_{liquid}$	[112]

TABLE 5.3 *(Continued)*

Method	Relation	Comments	Ref
	where $h_t = \Delta H + h_s\left(T_{ref}\right) + \int_{T_{ref}}^{T} C_p\left(T\right)dT$ such that ΔH is the change in latent enthalpy ($= \beta L$) of water in the tissue during liquid–solid transition, β is liquid fraction, L is the latent heat, h_s is the sensible enthalpy, T_{ref} is the reference temperature/temperature at the onset of phase transition (equivalent to T_{liquid}). The above equation is subject to the boundary condition applied at the surface of annular cryoprobes that incorporates the convective heat removal from tissue due to cryo-fluid flow in the probe.	where T_{solid} and T_{liquid} is the range of phase transition. The blood perfusion value ceases as the freezing begins. Similarly, the metabolic heat generation term exponentially reduces in the unfrozen region and becomes zero in frozen and phase transition region. The tissue conductivity and specific heat also varies with unfrozen, frozen, and phase transition region, whereas the density can be approximated as constant or varying.	
	Photo-thermal therapy		
Continuous-wave (CW) laser ablation (based on Beer–Lambert law)	$Q_s = \kappa_a I_{cw}(z)$ where I_{cw} is the intensity of continuous-wave laser light within the tissue which is given as $I_{cw}\left(z\right) = I_{c,\max}\, e^{\left[-z(\kappa_a+\sigma_s)-\left(r^2/2\sigma(z)^2\right)\right]}$	Here, $I_{c,max}$ is the intensity at the irradiated surface of the tissue which is equivalent to $P/\pi R_0^2$ such that P is the laser power and R_0 is the radius of laser beam spot size, κ_a is the absorption coefficient, σ_s is the scattering coefficient, r is the radial coordinate as $\sqrt{x^2+y^2}$, z represents the depth, $\sigma(z)$ is the standard deviation for Gaussian beam radius. Note, for continuous wave laser σ does not vary with z (remains constant).	[144]
Continuous-wave (CW) laser ablation (based on Kubelka–Munk approximation)	$Q_s = \kappa_a I_{cw}(z)$ where I_{cw} is the intensity of continuous-wave laser light within the tissue which is given as $I_{cw}\left(z\right) = I_{c,\max}\, e^{\left[-\left(r^2/2\sigma(z)^2\right)\right]} e^{\left[-z\sqrt{2\kappa_a(2\kappa_a+4\sigma_s)}\right]}$	Here, all parameters have their usual meaning pointed out earlier. Note the Kulbeka–Munk approximation does not influence the Gaussian profile of laser light. It exponentially reduces the magnitude of intensity within the tissue.	[103]

TABLE 5.3 *(Continued)*

Method	Relation	Comments	Ref
Short-pulse (SP) laser ablation	$Q_s = \kappa_a I_{SP}(z,t)$ where the I_{SP} is the intensity of short-pulse laser light within the tissue which is given as $I_{SP}(z,t) = I_{c,\max} e^{\left[-(\kappa_a+\sigma_s)z-\left(r^2/2\sigma(z)^2\right)\right]} g(z,t)$ and $g(z,t) = e^{-4\ln 2\left(\frac{t-\frac{z}{c_m}-t_c-(N-1)T_p}{t_p}\right)^2}$	Here, $g(z, t)$ incorporates Gaussian temporal profile for laser pulses, the c_m is light speed in the medium ($= c_0/n$), c_0 is the speed of light in vacuum, n is the refractive index, to_p is the pulse width, N is the total number of pulses, T_p is the time period ($2t_c$), and T_c is the cutoff time ($3t_p$).	[145]
Focused laser ablation	Focusing of the laser beam is attained by using a lens between source and tissue. To obtain a relation for focused laser beam a depth-dependent change in the standard deviation $\sigma(z)$ of Gaussian beam radius can be incorporated in continuous-wave and short-pulse laser wave intensity distribution. Therefore, $\sigma(z)$ is given as $\sigma(z) = \begin{cases} \sigma(0)\left[1-\frac{z(R_0-R_{FD})}{R_0 FD}\right] & 0 \le z \le FD \\ \sigma(0)\left[\frac{z(R_0-R_{FD})}{R_0 FD}-\frac{R_0-2R_{FD}}{R_0}\right] & z > FD \end{cases}$	Here, during beam convergence the relation for $z \le FD$ is used and for beam divergence the relation for $z > FD$ is used, such that FD is the focal distance from irradiation surface, and R_{FD} is the radius of the beam at FD.	[146]
	Nanoparticle-assisted thermal therapy		
Nanoparticle-assisted laser therapy	$Q_s = \kappa_a I_{c,\max}\, e^{\left[-z(\kappa_a+\sigma_s)-\left(r^2/2\sigma(z)^2\right)\right]}$ where κ_a and σ_s are the absorption and scattering coefficients. The values of these coefficients change	Here, the volumetric source term Q_s due to laser therapy is more accurately deduced from the solution radiative transfer equation (RTE), wherein the heat flux/fluence due to collimated and diffuse light can be incorporated in the analysis. However, for simplicity the effect of diffuse/	[147,150]

TABLE 5.3 *(Continued)*

Method	Relation	Comments	Ref
	as the laser light traverses from only tissue region to tissue-nanoparticle region/tumor volume. Note, in the tissue region the optical properties of tissue are considered. However, for the tissue-nanoparticle region an effective optical property is considered. Here, we provide some relations for optical properties of tissue-nanoparticle region. **a) Gold-coated silica spherical nanoparticles:** $\kappa_a = \kappa_{\lambda,t} + \frac{3}{4} f_v \frac{Q_a}{a}$ and $\sigma_s = (1-g)\sigma_{\lambda,t} + \frac{3}{4} f_v \frac{Q_s}{a}$ where f_v is the volume fraction of nanoparticle $(=V_{np}/V_t+V_{np})$, g is the asymmetry factor, index λ is for wavelength and t for tissue, Q_a and Q_s are nondimensional absorption and scattering cross-section of single nanoparticles, and a is the radius for spherical geometry. All other parameters have their usual meaning discussed earlier. According to the Rayleigh theory[147] $Q_a = \frac{8\pi a n_t}{\lambda} \mathrm{Im}\left(\frac{\xi - \tilde{m}_g^2}{\tilde{m}_g^2 + 2\xi}\right)$ $Q_s = \frac{8}{3}\left(\frac{2\pi a}{\lambda} n_t\right)^4 \left\lvert \frac{\tilde{m}_g^2 - \xi}{\tilde{m}_g^2 + 2\xi} \right\rvert^2$ where, $\xi = \frac{1+\delta^3 M}{1-2\delta^3 M}$, $M = \frac{m_g^2 - n_{si}^2}{2m_g^2 + n_{si}^2}$, $\tilde{m}_g = \frac{m_g}{n_t}$, and $m_g = n_g - i\alpha_g$	scattered radiation is often neglected. Note that the optical properties of nanoparticles demonstrate strong dependence on the geometry, size, orientation, particle properties, and incident wavelength. The values of Q_a and Q_s for spherical nanoparticles can be evaluated based on the semi-analytical solution of Maxwell EM equations provided by the Mie theory (with or without effective complex index of refraction), or for complex geometry using full-scale numerical solutions provided by the T-matrix method or discrete dipole approximation (DDA). Considering the fact that the particle size is much smaller than the incident light wavelength, one can also use the Rayleigh theory with good accuracy for computation of Q_a and Q_s. Here, m_g is the complex index of refraction for gold nanoparticle, n_g is the index of refraction for gold, n_{si} is the index of refraction of silica, n_t is the refractive index of tissue, α_g is the index of absorption for gold, and δ is a parameter. The value of this parameter for gold-coated silica nanoparticle is given by Ref. [147]. In case of nanorod, a prolate ellipsoid shape does not have the same size and polarizability in all orientations in comparison to sphere. Therefore, it is important to consider an average absorption and scattering property of such out of shape particle with random orientations.[150] For the estimation of orientation-averaged properties, it is ideal to consider T-matrix or DDA method to estimate the incident field attenuation in all random orientations of the particle.	

TABLE 5.3 *(Continued)*

Method	Relation	Comments	Ref
	b) Gold spherical nanoparticle: $\kappa_a = \kappa_{\lambda,t} + \frac{3}{4} f_v \frac{Q_a}{a}$ and $\sigma_s = (1-g)\sigma_{\lambda,t} + \frac{3}{4} f_v \frac{Q_s}{a}$ According to the Rayleigh theory for spherical particle[107]		
	c) Gold nanorod (prolate ellipsoid shape): An equivalent analysis of absorption and scattering coefficient of gold nanorods can be obtained in the purview of the electrostatic approximation for small size particle ($2\pi a_{eff}/\lambda \ll 1$), where a_{eff} is the effective radius of prolate ellipsoid deduced by equating effective volume of ellipsoid to that of a sphere. $\kappa_a = \kappa_{\lambda,t} + \frac{3}{4} \frac{f_v Q_a}{a_{eff}}$ and $\sigma_s = (1-g)\sigma_{\lambda,t} + \frac{3}{4} \frac{f_v Q_s}{a_{eff}}$ The small size of particle compared to wavelength leads to an assumption of particle being subjected to uniform electric field E_0 of incident light ($I_{c,\max} = I_0 = c\lvert E_0\rvert^2/8\pi$) and an electric field experienced from surrounding particles due to dipole moments of the particles. In other words, the absorption and scattering efficiency factor depends on the external electric field, polarizability, and shape and size factor of the nanorods.[148,149]		

TABLE 5.3 *(Continued)*

Method	Relation	Comments	Ref
	$Q_{a,i} = C_{abs,i}/A_i$ and $Q_{s,i} = C_{sca,i}/A_i$ where, $i = 1,2,3;\ A_1 = 2\pi ah;\ A_2 = A_3 = \pi a^2$ $C_{abs,i} = k_h \operatorname{Im}(\alpha_i)$ and $C_{sca,i} = \frac{k_h^4}{6\pi}\lvert\alpha_i\rvert^2$ $\alpha_i = 4\pi(ha^2)\left[\frac{\varepsilon_d - \varepsilon_h}{3\varepsilon_d + 3L_i(\varepsilon_d - \varepsilon_h)}\right]$ $L_1 = \frac{1-e^2}{e^2}\left[-1 + \frac{1}{2e}\ln\left(\frac{1+e}{1-e}\right)\right],\ e^2 = 1 - \frac{a^2}{(h/2)^2}$ $L_2 = L_3 = \frac{1-L_1}{2}$, since $L_1 + L_2 + L_3 = 1$ Here, subscript I refers to three axes of ellipsoid such that 1 denotes height corresponding to major axis (= h), 2 and 3 are the lengths of another minor axis of ellipsoid which is equivalent to the radius (= a) for prolate ellipsoid. The cross-sectional area A_I denotes areas subjected to light, C_{abs} and C_{sca} denote the absorption and scattering cross-sections, k_h is the wavenumber for host medium (=2 π/λ), α_i is the polarizability, ε_d and ε_h are the permittivity of cylinder and host medium, L_i is the geometrical depolarization factor, and e is the eccentricity.		
Magnetic heating using magnetic nanoparticle fluids	**a) Rosenweig theory:** $Q_s = f_v \pi \mu_0 H_0 f \chi''$ where	Magnetic fluid thermal therapy (MFTT) is a noninvasive targeted therapy of deep-seated tumor, wherein fluid with magnetic particle attached with protein binding	[151,152]

TABLE 5.3 *(Continued)*

Method	Relation	Comments	Ref
	$\chi'' = \left[\frac{2\pi f \tau_{eff}}{1+(2\pi f \tau_{eff})^2}\right]\chi_0,\ \tau_{eff} = \frac{\tau_B \tau_N}{\tau_B + \tau_N},$ $\tau_B = \frac{3\mu V_H}{\kappa_B T},\ \tau_N = \frac{\sqrt{\pi}}{2}\tau_0 \frac{\exp(K_a V_M/\kappa_B T)}{\sqrt{K_a V_M/\kappa_B T}}$ $V_H = \frac{\pi D^3}{6}$, and $V_M = \frac{\pi(D+2\delta)^3}{6}$ Here, μ_0, f_v, H_0, f, and χ'' are the magnetic permeability in vacuum, volume fraction of magnetic particle in tissue, amplitude of alternating magnetic field, frequency, and magnetic susceptibility (imaginary quantity), respectively, τ is the relaxation parameter, subscripts $_{eff}$, $_B$, and $_N$ refer to effective parameter, Brownian, and Neel relaxation parameters, respectively, μ is the dynamic viscosity of fluid, τ_0 is the Larmor time constant, κ_B is the Boltzmann constant, T is local temperature, V_H is the hydrodynamic particle volume, V_M is the magnetic particle volume, δ is the ligand layer thickness, K_a is the magnetic anisotropy constant (=25 $\kappa_B T_B/V_M$), and T_B is the blocking temperature. Note, χ_0 static equilibrium magnetic susceptibility given as	moiety is delivered to the tumor via blood or directly by percutaneous incision, which was heated in an alternating magnetic field in an MRI machine. MFTT is advantageous since it can be administered along with other standard radiation therapy, chemotherapy, and immunotherapy. The heating of magnetic nanoparticle can be attained by three methods (1) hysteresis heating (due to magnetic reversal in the alternating field), (2) Neel relaxation heating (frictional heating due to rapid flipping of magnetic moments), and (3) Brownian relaxation heating (heating due to viscous drag caused by fluid during flipping). Note that the effect of eddy current heating due to induced electric field in the alternating magnetic field for nanoparticle is negligibly small as compared to bulk/conglomerated particles. The volumetric heat source term and various parameters of magnetic nanoparticle fluid are given by Ref. [151,152].	

TABLE 5.3 *(Continued)*

Method	Relation	Comments	Ref
	$\chi_0 = \chi_i \frac{3}{\xi}\left(\coth\xi - \frac{1}{\xi}\right)$ such that $\xi = \mu_0 M_d H V_M / \kappa_B T, \quad M_d = M_s / f_v$ $H = H_0 \cos(2\pi f t), \text{ and } \chi_i = f_v \mu_0 M_d^2 V_M / 3\kappa_B T$ Here, M_d and M_s are domain and saturation magnetization, respectively, and χ_I is the initial susceptibility.		
Magnetic heating using migration modeling of magnetic nanoparticle	1) Quantification of SAR under microCT image guidance[153] SAR= $\begin{cases} 0, & HU \le \mathrm{HU}_{th} \\ \mathrm{SAR}_{specimen} \frac{\mathrm{HU} - \mathrm{HU}_{tissue(th)}}{\mathrm{HU}_{ferrofluid\ at\ 5.8\%} - \mathrm{HU}_{tissue(th)}} & HU > \mathrm{HU}_{th} \end{cases}$ $HU = -1000 \times \frac{GS - GS_{water}}{GS_{air} - GS_{water}}$ where $\mathrm{HU}_{air} = -1000$; $\mathrm{HU}_{ferrofluid,\ 5.8\%} = 2750$; $\mathrm{HU}_{water} = 0$; $\mathrm{HU}_{th(threshold)}$ = without nanoparticles. By using this approach, heterogeneously distributed magnetic nanoparticles would determine the temperature distribution. The grayscale intensity with maximum value of tumor lacking nanoparticles is chosen as threshold or cutoff.[154]	Instead of using a contribution of single nanoparticles from Rosenweig theory, a microCT imaging-based experimental relationship can be used to establish correlations between pixel index or grayscale value and specific absorption rate of superparamagnetic iron-oxide nanoparticles. The pixel intensity varies in the range 0–255. Next, the grayscale value is converted to Hounsfield Unit (HU) which is a normalized attenuation coefficient of CT number. The calibrated SAR derived from a ferrofluid with 5.8% subject to $\mathrm{SAR}_{specimen} = 3.4 \times 10^6\ \mathrm{W/m^3}$.	[153,155]

TABLE 5.3 *(Continued)*

Method	Relation	Comments	Ref
	2) Singh Migration model This model connects the heat source SAR with nanoparticles volume-averaged concentration but the concentration C is varying with cell-necrosis over time. Overall concentration of magnetic nanoparticles varies due to mixing of intracellular fluids after cell-membranes ruptures. This causes concentration gradient of magnetic nanoparticles and transport from high concentration region to low concentration region. This mathematical model is coupled with interstitial space (porosity), thermal damage, temperature, diffusion coefficient. $Q_s = 2266.67 \times C(x,y,z,t)$ $\frac{\partial C(x,y,z,t)}{\partial t} = \nabla \cdot \left[\varphi D_n \nabla \left(\frac{C(x,y,z,t}{\varphi} \right) \right]$ $D_n = D_{n,f} \left[\frac{2\varphi}{3-\varphi} \right]$ $\varphi = \varphi_o + (X\% - \varphi_o)(1 - e^{-\grave{U}(x,y,z,t)})$ Later, it has been approximated 80% volume was occupied by interstitial fluid and 20% occupied by other constituents in total tumor volume. The tumor interstitial region accumulates more particles as cell-death rate increases. This is a robust model capable of capturing the nanoparticle redistribution following heating.[153,155]		

TABLE 5.3 *(Continued)*

Method	Relation	Comments	Ref
	Thermal therapy associated blood perfusion changes		
Temperature-dependent mass flow rate of blood in the tissue, $W_b(T_t)$	$$W_b(T_t)=\begin{cases} W_{b,0}\left(1+\omega_l \exp\left(-\frac{(T_t-T_{cr})^2}{s}\right)\right), & T_t \le T_{cr} \\ W_{b,0}(1+\omega_l), & T_t > T_{cr}\end{cases}$$ where $W_{b,0}$ is the baseline mass flow rate of blood per unit volume of tissue ($=\rho_b\omega_{b,0}$), ω_l and s are curve fitting parameters, and T_{cr} is a critical temperature that depends on the upper bound of hyperthermia treatment which is tissue-specific.	Note the mass flow rate of blood W_b in the tissue varies with increases in temperature during hyperthermia treatment. The unit of W_b and $W_{b,0}$ is kg/s/m³. The value of $W_{b,0}$, ω_l, and s is tissue specific.	[137]
Damage-dependent blood perfusion, $\omega_b(\Omega)$	A piecewise function[156] of blood perfusion variations within the tissue due to induced thermal damage is given as $$\omega_b(t)=\begin{cases} \omega_{b,0} & \text{for } \Omega(t)\le 0 \\ \omega_{b,0}\left[1+25\Omega(t)-260\Omega(t)^2\right] & \text{for } 0<\Omega(t)\le 0.1 \\ \omega_{b,0}\exp\left[-\Omega(t)\right] & \text{for } \Omega(t)>0.1\end{cases}$$ where $\omega_{b,0}$ is the base blood perfusion value prior to induced damage by heat. Here, $\Omega(t)$ is the thermal damage parameter which depends on tissue temperature given by eq 5.36.	Here, the blood perfusion has been approximated as constant before the onset of damage to tissue for a predefine threshold temperature. The perfusion increases briefly in the tissue due to heat-induced vasodilation, and finally decays exponentially. Note that the value of $\omega_{b,0}$ and $\Omega(t)$ is tissue specific.	[156]
Vascular-stasis-dependent blood perfusion, ω_b	A function of the blood perfusion is shown depending on the degree of vascular stasis. It is given as	Here, blood perfusion initial elevates due to vasodilation followed by a drop due to collapse of vascular structure as consequence damage induced by heating. Note that the	[64,157]

TABLE 5.3 *(Continued)*

Method	Relation	Comments	Ref
	$\omega_b(t)=\begin{cases}\omega_{b,o}(1+30\cdot\Theta); & for\, 0<\Theta\le 0.02\\ \omega_{b,o}(1.86-13\cdot\Theta); & for\, 0.02<\Theta\le 0.08\\ \omega_{b,o}(0.884-0.79\cdot\Theta); & for\, 0.08<\Theta\le 0.97\\ \omega_{b,o}(3.87-3.87\cdot\Theta); & for\, 0.97<\Theta\le 1.00\\ 0; & for <\Theta\, 1.00\end{cases}$ where $\omega_{b,0}$ represents base blood perfusion value before the onset of thermal damage, and $\omega_b(t)$ is the vascular-stasis parameter which is a function of tissue temperature $\Theta(x,y,z,t)=1-\exp(-\xi(x,y,z,t))$ $\xi(x,y,z,\tau)=\int_0^{\tau} A_{perf}\exp\left(\frac{-E_{a\,perf}}{R_u T(x,y,z,t)}\right)dt$ A first-order Arrhenius rate equation, the difference being the kinetic coefficients used to compute damage index, ξ. Here, the kinetic coefficients are $A_{perf} = 1.98 \times 10^{106}$ and $E_{aperf} = 6.67 \times 10^{5}$.	value of $\omega_{b,0}$ is tissue specific. Recently, inverse models were used to predict base perfusion value using thermal signatures from an infrared image.[122]	

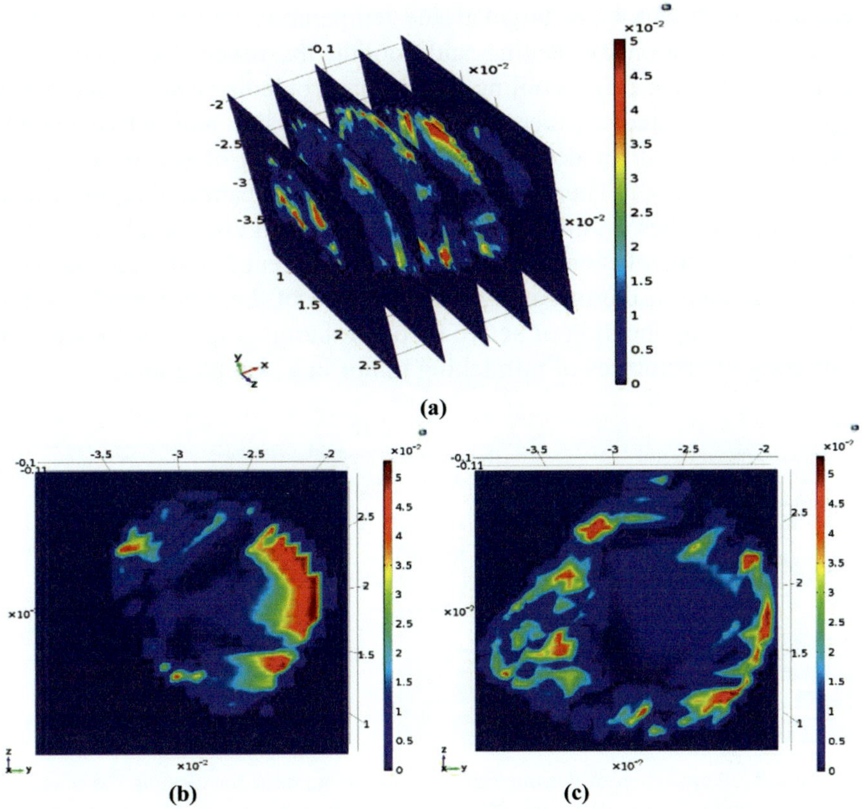

FIGURE 5.8 Three-dimensional spatial perfusion heterogeneity for a breast tumor in Sagittal Plane (Y–Z plane) with the clear illustration of presence of necrotic core. It possesses regions of no perfusion, low perfusion, moderate perfusion, high perfusion, respectively. Spatial coordinates can be mapped with the Euclidean distance approach. [Image courtesy: Manpreet Singh].

5.7 ESSENTIAL INSTRUMENTATION FOR BIOHEAT TRANSFER RESEARCH

5.7.1 THERMOGRAPHY

Thermography is widely used in the field of bioheat transfer research to measure surface/subsurface temperature of tissue for the estimation of the normotensive status of the human body and to monitor rise/fall in temperature during thermal therapy. Thermography is noninvasive and non-contact method which is classified into two categories: (1) passive thermography and (2) active thermography.[116] In passive thermography, thermal cameras were used to monitor the target tissue temperature variation without any external provocation by heating and cooling the tissue. But, active thermography employs either cooling or heating protocols prior to imaging the target tissue. In medical thermography, both passive and active methods have been explored for detection and differentiation of subsurface cancer of skin and breast.[116] The principle for subsurface cancer diagnosis using thermography relies on the fact that the metabolism and blood microcirculation of cancerous lesion is different from adjacent healthy region which can be indicated in thermal images in the form of thermal contrast. Figure 5.9 depicts the application of active mode of thermography for locating and monitoring temperatures of mimicking tumor in a skin phantom.

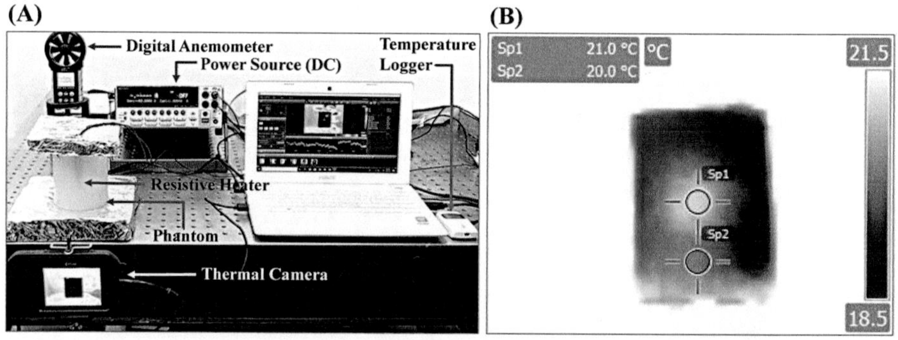

FIGURE 5.9 Application of thermography after cold stimulus for locating the mimicking tumor heat source in skin phantom. (A) Thermography setup for monitoring temperature mimicking tumor and (B) thermal image of a skin phantom with rise in tumor temperature compared to the surrounding healthy region. [Image courtesy: Arka Bhowmik].

All thermal cameras use radiometric sensor arrays which capture the spectral radiance emanated by a skin or breast in the near-to-far IR wavelength, that processes the signals to convert the captured radiometric to temperature values, and form a map of temperature corresponding to the sensor array. The conversion of radiometric value to the temperature value in the thermal camera is attained by employing equation based on Planck's law given as

$$E_{\lambda,cam} \approx \varepsilon(\lambda) \times \tau(\lambda) \times E_{\lambda,t}$$

$$\text{or, } E_{\lambda,cam} \approx \varepsilon(\lambda) \times 2\hbar c^2 \lambda^2 \left[\exp\left(\frac{\hbar c}{\kappa_B \lambda T_t} \right) - 1 \right]^{-1}$$

where E_λ is the spectral radiance, ε is the spectral emissivity of tissue, τ is the spectral transmissivity of air, c is the speed of light in air, $\hbar$ is the

Planck's constant, κ_B is the Boltzmann constant, λ is the wavelength, T_t is the tissue temperature, subscripts $_{cam}$ and $_t$ refer to spectral radiance received by camera and emanated by tissue, respectively. The transmission loss in air is assumed to be negligible, since IR radiation exhibits nearly 100% transmittance through the air (= 1) within the spectral range of 8–10 µm[160] which is an ideal range for thermal evaluation of the human body.

5.7.2 MAGNETIC RESONANCE (MR) THERMOMETRY

Magnetic resonance (MR) thermometer is one of the popular methods for noninvasive monitoring of the change in temperature of the brain and other organs which is in or not in normotensive status. The method is widely used for research and therapeutic purposes to evaluate the effective heating of the tumor or cancerous involvement during laser ablation, high frequency ultrasound (HIFU), microwave, and radiofrequency therapy. It encompasses various techniques to extract temperature data from tissue/organ based on measured change in radio frequency signals as a consequence of temperature-dependent proton magnetic responses. There are several proton resonance parameters that depend on the change in temperature of the tissue such as proton density, diffusion coefficient, proton resonance frequency shift, T_1 or T_2 relaxation, and temperature sensitive contrast agents. The detailed description of each method can be gathered from the literature.[128–130] Here, we will cover the frequently used method for MR thermometry, that is, based on spectroscopy method and phase mapping method associated with proton resonance frequency (PRF) shift, which had shown high accuracy in the indirect computation of temperature based on measured radiofrequency signals.

An MRI scanner typically measures the radio frequency signals emanated by protons of water and other macromolecules (lipids) from different tissue, and forms spatial map using an image construction method. These protons precess (i.e., rotate about its own axis due to electric field as well as rotates/wobbles the spin axis with a radius of gyration due to a magnetic field) at two different spin configurations at a frequency related to the strength of the magnetic field (also referred to as resonance frequency or Larmor frequency). Due to spatial variation in magnetic field strength, it exhibits a variation in resonance frequency of protons which serves as the basis of spatial contrast in MRI images. In fact, this resonance frequency differs for different tissue types which enables a clear distinction of healthy and cancerous tissue. Ideally, in an MRI scanner initially the proton was arranged in parallel and anti-parallel spin configuration in a continuous magnetic field. Next, the initial spin configuration was disturbed by short EM energy transmitted by a transmitting radio frequency (RF) coil. After a short period, the transmitting RF coil is stopped and a receiving RF coil close to the body receives radio frequency signals emanated by the protons while relaxing from the excited state to the initial state of Larmor frequency. The radio frequency signals thus received by MRI scanners depend on the parameters such as T1 and T2 relaxation time, proton density, and tissue temperature. MR thermometer based on PRF shift involves computation of absolute and relative temperature based on measured phase shift in the frequency of water proton or scanned images. The shift in the resonance frequency of the proton is mainly due to the temperature-dependent shielding effect of orbital electrons of hydrogen proton in water molecule to the applied magnetic field (B_0). In fact, the shielding effect increases with the rise in temperature since it weakens the hydrogen bond with an oxygen atom in water molecules, which in turn reduces the strength of the magnetic field and resonance frequency. The resonance frequency $\omega(T)$ due to shielding effect is given as

$$\omega(T) = \gamma B_0 \left[1 - \sigma(T)\right] \tag{5.60}$$

where γ is the gyromagnetic ratio (a constant parameter for protons), B_0 is incident magnetic flux, $\sigma(T)$ is the temperature-dependent electronic shielding/screening constant which is equivalent to [$= \beta T$], β is the change in shielding constant per unit temperature. The above description of the resonance frequency shift is used to compute absolute or relative temperature in two different scenarios: (1) proton resonance spectroscopy, and (2) phase mapping of PRF shift. During proton resonance spectroscopy, the absolute temperature of tissue is computed from a proton resonance frequency shift by comparing the resonance frequency of another reference hydrogen proton which are independent of temperature change such as hydrogen protons in N-Acetylaspartic acid (NAA) and choline, etc. The change in resonance frequency is related to absolute tissue temperature as

$$\Delta\omega = \omega_w(T) - \omega_{ref} = \gamma B_0 \left(1 - \beta_w T\right) - \omega_{ref} \tag{5.61}$$

where ω_w refers to the resonance frequency of protons in water, ω_{ref} is the resonance frequency of protons in reference compounds in the tissue that remains nearly constant with temperature $\left[= \gamma B_0 \left(1 - \beta_{ref} \times 37\,^{\circ}\mathrm{C}\right)\right]$, subscripts w and *ref* are for water and reference compound, respectively. In contrast, during the phase mapping method relative change of tissue temperature ΔT map can be computed based on measured phase change map $\Delta\,\varphi$ of the region-of-interest from two consecutive MR images over a short time interval as

$$\Delta\varphi = \varphi(T) - \varphi(T_{in}) = \gamma B_0 \beta_w \Delta T \times TE$$
$$\text{or,}\ \Delta T = \frac{\varphi(T) - \varphi(T_{in})}{\gamma B_0 \beta_w \times TE} \tag{5.62}$$

where ΔT is the relative change in temperature (= T-T_{in}), T_{in} is the initial tissue temperature before heating of tissue, $\varphi(T)$ and $\varphi(T_{in})$ are the phase changes due to the current temperature and initial temperature, respectively, and the *TE* is the short echo time. All other terms have been described earlier. Note that the phase change in $\varphi(T_{in})$ and $\varphi(T)$ is determined by taking difference in consecutive MR images taken before and after heating, respectively.

5.7.3 CONTACT AND CONTACTLESS DEVICES FOR MEASURING TISSUE PARAMETERS

In this section, various devices, sensors, and methods used in the field of bioheat transfer for measuring tissue temperature and blood perfusion, along with sensors for monitoring surrounding conditions such as temperature and humidity, are presented. In the purview of the bioheat transfer applications, both contact and contactless devices for tissue parameter estimation have been provided. The measurement of tissue temperature can be attained by both contact-based devices (such as thermometer, thermistors, thermocouple, and flexible resistive temperature detectors) and using contact-less devices (such as fiber optic sensor, thermography, and MR thermometry). Similarly, measurement of tissue blood perfusion can be attained by various contact and contact-free methods, whose working principle was also covered briefly in this section to maintain brevity and avoid repetition (discussed in other chapters).

5.7.3.1 THERMOMETER

The thermometer is also referred to as liquid-in-glass thermometer which uses mercury or other liquids that change volume with temperature. A typical thermometer is made of a cylindrical glass with a bulb at the bottom wherein the temperature sensitive liquid is stored. During temperature measurement the bulb portion is brought into contact with skin/tissue which resulted in volumetric expansion of the fluid (i.e., mercury). Due to expansion, the mercury raises along a smaller diameter capillary tube (partial vacuum) inside the glass and the height to which mercury was raised gives the absolute tissue temperature. The height marked on the thermometer is calibrated based on a standard reference temperature also known as ice point (at 0°C) which is detailed in Refs. [161,162]. Figure 5.10A depicts a general-purpose mercury thermometer for measuring body temperature. In fact, proper calibration could permit mercury thermometer to achieve an accuracy of ±0.10°C. Mercury thermometers are relatively fragile and use toxic substance, not immune to electromagnetic waves, and fail to capture dynamic change in the tissue temperature (ideal for steady-state measurement). However, mercury thermometers are handy and serve as the basis of calibration of many other devices. The expression of volumetric expansion of mercury in a thermometer is given as[162]

$$V = V_0\left(1+\alpha T+\beta T^2\right) \tag{5.63}$$

where V_0 is the volume of mercury at 0°C, α and β are coefficients for thermal expansion given as 1.8×10^{-4}°C^{-1} and 5.0×10^{-8}°C^{-2} for mercury, respectively.

5.7.3.2 RESISTIVE TEMPERATURE DETECTORS (RTD)

Resistive temperature detectors (RTDs) are one of the accurate methods for tissue temperature measurement with accuracy close to ±0.1°C to ±0.0001°C depending on the standard of calibration and material used.[163] It is a well-known fact that many conductors present linear or quasi-linear variation in resistance due to change in temperature. An RTD exploits this phenomenon to accurately compute the temperature of the tissue based on measured resistance (or voltage-current value). A typical RTD consists of a sensing unit and detector leads such that the sensing unit is represented by a coiled metallic conductor or thin conductive film, which experiences the change in temperature. On the other hand, the detector leads consist of two ends of the conducting metal which are connected to an electrical source. In RTDs, as the sensing unit temperature changes the resistance of the sensing unit also differs in a predictable manner (i.e., increase in temperature increases resistance and vice-versa), due to which small change in measured voltage is experienced when a constant current (around 0.8–1.0 mA) is set up in the detector leads. The measured voltage and known input current serves as the basis for the computation of bulk resistance of RTD which is correlated with tissue temperature or sensing unit temperature using a temperature-resistance calibration equation for the particular material. In general, one can construct a highly sensitive sensing unit using material such as platinum, gold, copper, or nickel. The calibration of RTDs was attained by measuring the change in temperature and resistance from a known reference temperature also referred to the ice point (at 0°C), and the calibration equation is known as Callender–Van Dusen equation that provides the temperature-resistance relation for platinum RTD as

$$R(T) = R_0\left(1 + \alpha T + \beta T^2\right) \tag{5.64}$$

where R_0 is the resistance of platinum RTD at 0°C, $R(T)$ is the resistance at temperature T (°C), α and β are calibration temperature coefficients for RTD in $\Omega/\Omega^{-1}/°C$ and $\Omega/\Omega^{-1}/°C^2$, respectively. To determine the temperature above 0°C from the change in resistance an iterative relation can be deduced from eq 5.64 as [163]

$$T = \left[\left\{R(T) - R_0\right\}/\alpha R_0\right] + \delta\left[(T/100 - 1)(T/100)\right] \tag{5.65}$$

where δ is the calibration coefficient. Here, the correct temperature is computed from the measured resistance at 0°C and resistance at temperature T (°C) by iteratively solving the equation based on guess value of temperature and new value of temperature such that the relative error $\leq 10^{-3}$. A state-of-the-art e-textile thin-film RTD is shown in Figure 5.10B. It is to be noted that RTDs are not immune to electromagnetic waves in thermotherapy, prone to measurement error due to impurities, deformation, and self-heating.

5.7.3.3 THERMISTORS

In comparison to RTDs, thermistors are semiconducting thermal resistors which also exhibit a large change in electrical resistance due to temperature. It is often used in silicon semiconducting devices and circuits for switching mechanisms and/or for cutouts by sensing the overheating temperature in the circuit or surrounding. It can be classified into two groups: (1) positive temperature coefficient (PTC) and (2) negative temperature coefficient (NTC) resistors, since some materials of thermistors reveal increase in resistance (positive) and decrease (negative) in resistance with temperature, respectively. It ideally displays a nonlinear characteristic curve (i.e., current–voltage and resistance–temperature curves) which makes them difficult to calibrate for temperature measurement. However, thermistors in combination with additional resistance network either in series or parallel could lead to good linear characteristic curves within limited temperature range. This approach is commonly used for most NTC thermistors for precise and accurate calibration for temperature measurement. Yet, the complexity in the calibration process arises due to the self-heating of thermistor as current flows through them, wherein the magnitude of power dissipated due to self-heating is equivalent to heat loss to the surrounding. In fact, the calibration process also needs to consider the relative strength of change in power dissipation due to self-heating to that of the resultant change in thermistor temperature indicated by the dissipation constant, δ. NTC thermistors such as ceramic bead-type thermistors or glass/epoxy coated chips are commonly used for temperature measurement due to its excellent thermal stability. The voltage–current characteristic of NTC thermistors reveals negligible self-heating at low current and displays nearly constant resistance revealing voltage and current to be proportional. This condition is also referred as zero-power resistance (negligible self-heating) which is the ideal range of temperature measurement or calibration. However, applications involving large current reveal increase in thermistor temperature due to self-heating which decreases the resistance indicating nonlinear voltage–current relation. In zero power resistance range, the resistance and temperature can be accurately related by establishing a calibration equation from resistance–temperature characteristic curve using third-order polynomial referred to as Seinhart and Hart equation as[164]

$$\frac{1}{T} = a + b \ln R(T) + c\left[\ln R(T)\right]^3 \tag{5.66}$$

This equation was found to be highly accurate within the temperature range of 0–70°C. Note that the temperature T is in absolute unit (K), the values of a, b, and c are constants of the fitting equation. It is important to note that thermistors can be used for temperature measurement, switching mechanism, and heating element depending on the temperature range and applications. Figures 5.10C and 5.10D depict thermistor units used for skin temperature measurement and switching of water supply to strawberry plant, respectively.

5.7.3.4 THERMOCOUPLE

Figures 5.10E and 5.10F depict commercial thermocouple units used for temperature measurement of any substance and that of tissue during hyperthermia treatment by placing a catheter tube, respectively. The thermocouple is one of the most common devices widely used for temperature measurement due to its simple construction and calibration. It uses two dissimilar metal thermos-elements joined at one end to sense the temperature and other ends of the thermos-elements for measuring the relative electromagnetic force (emf) which is observed due to Seeback effect. The Seeback effect can be referred to self-generated emf observed in each thermos-element due to the temperature difference between its ends, even if the thermos-elements are not joined at a common junction. However, the value of absolute emf is quite small to be observable. Hence, two dissimilar metallic elements are an ideal choice which originates relative emf. The relative emf can be evaluated as the difference in electromagnetic force between two thermos-elements A and B when one of their junctions is joined and in physical contact with measured temperature (T_m) and other ends of both elements are assumed to be at a reference temperature (T_r).[165]

$$E_{AB}(T_m,T_r) = \Delta E_A(T_m,T_r) - \Delta E_B(T_m,T_r) \tag{5.67}$$

where ΔE_A and ΔE_B are absolute emf given as

$$\begin{aligned} \Delta E_A(T_m,T_r) &= \int_{T_m}^{T_r} \sigma_A(T)\,dT = \left[E(T_m) - E(T_r)\right]_A \\ \Delta E_B(T_m,T_r) &= \int_{T_m}^{T_r} \sigma_B(T)\,dT = \left[E(T_m) - E(T_r)\right]_B \end{aligned} \tag{5.68}$$

Here, the corresponding Seeback coefficients σ_A and σ_B are material properties and functions of temperature. In thermometry, this relative voltage is directly observed in the termini. Once, the relative Seeback emf, E_{AB} is known then the temperature of measuring junction can be evaluated as follows:

$$\begin{aligned} &T_m - T_r \approx E_{AB}/\sigma_{AB} \\ \text{or,}\quad &T_m \approx T_r + E_{AB}/\sigma_{AB} \end{aligned} \tag{5.69}$$

where σ_{AB} is the relative Seeback coefficient given as $(={}_A - \sigma_B)$. Note that σ_A and σ_B are functions of temperature that can be assumed constant for small temperature difference. The above relation was obtained from the fact that the relative Seeback emf is proportional to the temperature difference between two junctions and relative Seeback coefficient of paired metals. For accurate measurement, the reference temperature T_r should be kept at 0°C but this value is arbitrary in the real thermocouple circuit. If T_r is not equivalent to 0°C, then a scaling function is added to correct the reference temperature. This scaling function or correction factor is available with the manufacturer which is equivalent to the difference between $E_{AB}(T_r)$ and $E_{AB}(T = 0°C)$. It is important to note that the thermocouples are not immune to strong magnetic field since the thermos-electric-magnetic effects can influence thermometry.

5.7.3.5 FIBER OPTIC TEMPERATURE SENSORS

Fiber optic temperature sensing is an indirect method of measuring temperature, wherein any change in reflected, transmitted, or emitted optical signals due to temperature variation of external probe and internal probe is used for temperature estimation. The external probe is attached at the end of the optical fiber and the internal probe is intrinsic within the fiber core. The fiber optic temperature sensors can perform measurement based on emitted thermal radiation, change in refractive index of the fiber core due to surrounding tissue temperature, change in fluorescence spectra emitted by an external probe with temperature, and changes in Raman or Brillouin scattering by an internal probe within the fiber itself. The fiber optic temperature sensor is ideal for a wide variety of applications since these sensors are small, immune to electromagnetic waves and are chemically inert. Figure 5.10G depicts a fiber optic temperature sensor with four intrinsic fiber Bragg grating (FBG). Here, the gratings were illuminated using broadband optical source, and any change in reflected and transmitted spectra due to temperature variation or variation in refractive index induces a shift in Bragg wavelength which acts as the basis for temperature measurement. The Bragg wavelength is the resonance wavelength originated in case of no perturbation in the grating refractive index, which means that the reflected waves and transmitted waves are in phase. The Bragg wavelength can be evaluated as $\lambda_B = 2n_{eff}P_g$, such that n_{eff} is the effective refractive index of fiber core and P_g is the period of gratings.[166]

5.7.3.6 LASER DOPPLER BLOOD PERFUSION IMAGING

Blood perfusion is volumetric flow of blood through the capillary network and extracellular spaces of the tissue. Measurements of blood perfusion provide functional information of various medical conditions such as lesion-induced changes in microcirculation, assessment of treatment protocol, diabetic-induced vasodilation, wound healing, etc. Laser Doppler method is one of the noninvasive measurement techniques to quantify the movement of red blood cells in superficial vessels and tissue, wherein the measured Doppler frequency shift between the coherent incident laser light and reflected light is used to compute the blood flow velocity in the superficial tissue. Doppler effect can be realized as the change in frequency of scattered light or spectral broadening of the incident light due to the movement of particles as compared to stationary particles. This change in frequency is

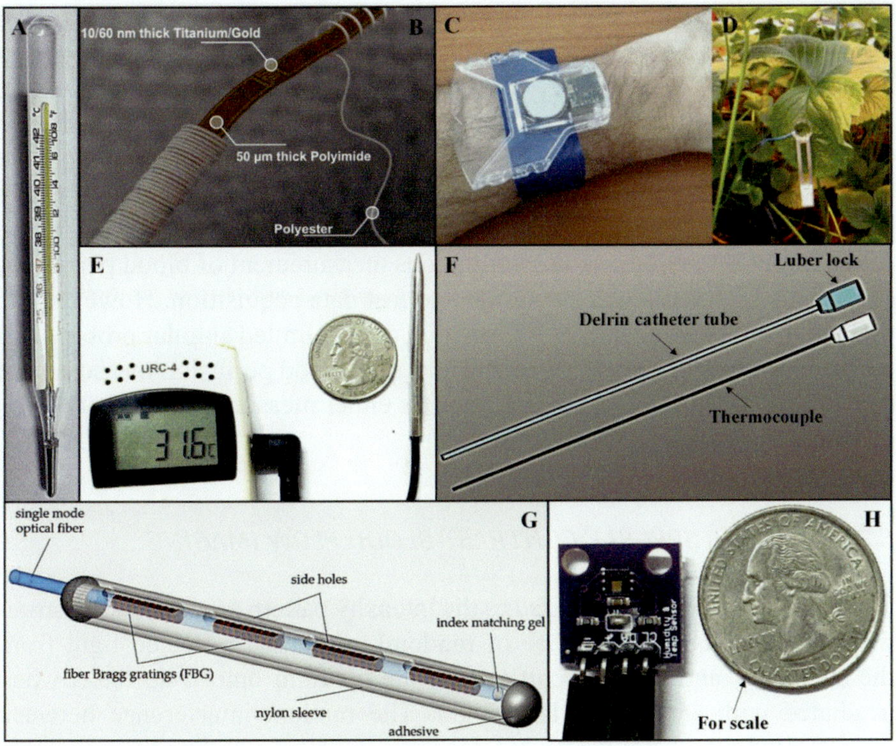

FIGURE 5.10 Portable contact and contact-less temperature sensors. (A) Mercury thermometer (B) Flexible thin-film resistive temperature detector (RTD) embedded in textile fabric (C) Thermistor setup for skin temperature measurement (D) Thermistor in strawberry leaf to switch on water treatment for plant (E) Thermocouple probe (F) CAD model of Delrin catheter tube and the thermocouple probe for hyperthermia application (G) Fiber optic temperature sensor and (H) HTU21D humidity and temperature sensor.

Source: (A) Image courtesy: Arka Bhowmik (B) Reprinted from Ref. [167]. https://creativecommons.org/licenses/by/4.0/ (C) Reprinted from Ref. [168] https://creativecommons.org/licenses/by/4.0/); (D) (Reprinted from Ref. [168]. https://creativecommons.org/licenses/by/4.0/ (E) Image courtesy: Arka Bhowmik (F) Image courtesy: Arka Bhowmik (G) (Reprinted from Ref. [169] https://creativecommons.org/licenses/by/4.0/ (H) Image courtesy: Arka Bhowmik.

related to incident light frequency and velocity of a single particle (relative to optical detector) explained by the equation[170]:

$$f_D = \left(2vf_0 \cos\theta\right)/c_m \tag{5.70}$$

where f_0 is the frequency of incident light, c_m is the speed of light in the medium, v is the velocity moving particle relative to the detector, θ is the angle between the moving particle direction and incident light, and f_D is the Doppler frequency (or shift in frequency). In practice, the objective is to compute the resultant velocity v of all moving particles (or red blood cells); hence, the net shift of reflected light frequencies is measured from the optical detector based on the knowledge of incident light frequency. This approach is adopted since the frequency shift of reflected light and the angle θ are not constant due to the random movement of a large number of particles. The Laser Doppler method enables continuous measurement of blood perfusion, featuring a high temporal resolution and fast data acquisition. However, the method is subjected to loss of information due to limited angular probing and stochastic nature of optical signals resulting in blood perfusion measurement in relative value of perfusion (i.e., flow is either measured in arbitrary units or volts).[171]

5.7.3.7 LASER SPECKLE CONTRAST BLOOD FLOW IMAGING

The term speckle can be realized as the intensity pattern aroused by constructive or destructive interference of randomly scattered reflected light from the stationary and moving particles of the medium onto a detector, when irradiated with a coherent laser light. The random interference between the reflected rays from scattering particles (with or without motion) in the medium leads to dark and white speckles in spatial images which are decorrelated (or blurred) in appearance.[170,171] The extent of decorrelation depends on the concentration of the scattering particles and movement of the scatterer. In other words, the spatial blurring effect varies for stationary particles as well as particles in motion. The spatial contrast of speckles is determined for each pixel by[172]

$$K_s = \frac{\sigma_s}{\text{mean}(I)} \tag{5.71}$$

where K_s is the speckle contrast value, σ_s is the spatial standard deviation of speckle intensity, and I is the reflected ray intensity. The raw speckle images are blurred in nature, but applying the speckle contrast relation eq 5.71 reveals the true spatial images. To capture the time-varying or flow effect of the speckle pattern due to the motion of particles in the region-of-interest, the variation in like pixel intensities in successive scans was used to compute the flow parameter or construct a time-varying 2D spatial image.[172] This method suffers from low signal-to-noise ratio and loss of information for thick tissue since the light scattering effect of stationary particles in the medium induces noise in the true signals coming from moving particles (blood cells). The laser speckle method is only capable of providing a relative flow parameter for microcirculation, which is ideal for flow measurement in superficial depth or surgically exposed vascularized tissue.[171]

5.8 SOLUTION OF TRANSIENT BIOHEAT EQUATION

5.8.1 ANALYTICAL SOLUTION

The bioheat equation, eq 5.1 is a nonhomogeneous partial differential equation (PDE) since the tissue temperature is a function of both time and space. The analytical solution of such a PDE is obtainable by many mathematical methods which include variable separable, Laplace transformation, Fourier transformation, and Green's function methods. These methods are well known for their ability to obtain analytical solutions of PDE subjected to complex boundary conditions which involve rigorous mathematical tricks to reach a final solution. Here, we will discuss one such mathematical method for solving the nonhomogeneous PDE eq 5.1, namely the variable separable method. We consider a one-dimensional transient bioheat equation in the Cartesian coordinate system subjected to a convective boundary condition at one boundary and an adiabatic condition on the other boundary. The first boundary refers to the exposed portion of skin of the surrounding and the second boundary is the nonexposed portion within the skin. Therefore, the governing bioheat equation, eq 5.1 for the skin subjected to a convective boundary condition can be written as

$$\frac{\partial T}{\partial t} = \alpha\left(\frac{\partial^2 T}{\partial z^2}\right) - \frac{(1-\kappa)\omega_b \rho_b c_b}{\rho c}\left(T - T_a\right) + \frac{Q_m'''}{\rho c} \tag{5.72}$$

where α is the thermal diffusivity (= $k/\rho c$). All other parameters have their usual meaning discussed earlier. Here, the thermophysical properties are assumed to be independent of space and time. The above equation is further simplified as

$$\frac{\partial T}{\partial t} = \alpha\left(\frac{\partial^2 T}{\partial z^2}\right) - m\left(T - T_a\right) + S \tag{5.73}$$

where $m = (1-\kappa)\omega_b \rho_b c_b / \rho c$ and $S = Q_m''' / \rho c$. Similarly, the boundary conditions can be written as

$$z = L_t : \quad -k\left.\frac{dT}{dz}\right|_{z=L_t} = \bar{h}_{conv}\left(T - T_{amb}\right)$$
$$z = 0 : \quad \left.\frac{dT}{dz}\right|_{z=0} = 0 \tag{5.74}$$

whereas the initial condition is given as

$$t = 0 : \ T\big|_{t=0} = T_o \tag{5.75}$$

where $\hbar_{conv}$ is the average heat transfer coefficient, T_{amb} is the ambient temperature, and T_o is the constant temperature. The transformed governing eq 5.73 and boundary conditions 5.70 are not homogeneous, except at z = 0. This implies that we cannot invoke the variable separable method. In order to obtain the solution, we first invoke the principle of superposition and decompose the problem into transient and steady-state part as

$$T(z,t) = T_t(z,t) + T_s(z) \tag{5.76}$$

Therefore, eq 5.73 is decomposed into a steady and transient part after substitution of eq 5.76 as

$$\alpha\frac{d^2T_s}{dz^2} - m\left(T_s - T_a\right) + S = 0 \tag{5.77a}$$

$$\frac{\partial T_t}{\partial t} = \alpha\frac{\partial^2 T_t}{\partial z^2} - mT_t \tag{5.77b}$$

Similarly, the boundary condition (eq 5.74) can be segregated into steady and transient part as

$$z = L_t : \quad \left.\frac{dT_s}{dz}\right|_{z=L_t} = \frac{\bar{h}_{conv}}{k}\left(T_{amb} - T_s\right) \qquad \left.\frac{dT_t}{dz}\right|_{z=L_t} = -\frac{\bar{h}_{conv}}{k}T_t \tag{5.78a}$$

$$z = 0 : \quad \left.\frac{dT_s}{dz}\right|_{z=0} = 0 \qquad \left.\frac{dT_t}{dz}\right|_{z=0} = 0 \tag{5.78b}$$

and initial condition eq 5.75 takes the form

$$T_t\big|_{t=0} = T_o - T_s(z) \tag{5.79}$$

It is worth noting, by segregating the steady and transient part we attempted to include as many nonhomogeneous terms with the easy steady-state problem. Clearly, adding up the corresponding governing equations and boundary conditions side-by-side by taking eq 5.76 into account, we recover the original eq 5.73 and boundary conditions eq 5.74. It is clear from eq 5.79

that the solution of transient part requires the solution of steady part. Next, we can further simplify eq 5.77a for our convenience by dividing both sides with α and assuming $\theta_s = T_s - T_a$ as

$$\frac{d^2\theta_s}{dz^2} - \psi^2\theta_s = -F \tag{5.80a}$$

where $\psi^2 = m/\alpha$, $\psi = \sqrt{m/\alpha} = \sqrt{(1-\kappa)\,\omega_b \rho_b c_b / k}$, and $F = S/\alpha = Q'''_m/k$. Similarly, the boundary conditions left to eqs 5.78a and 5.78b take a form

$$z = L_t: \quad \left.\frac{d\theta_s}{dz}\right|_{z=L_t} = \frac{\bar{h}_{conv}}{k}\left(T_{amb} - \theta_s - T_a\right) \tag{5.80b}$$

$$z = 0: \quad \left.\frac{d\theta_s}{dz}\right|_{z=0} = 0 \tag{5.80c}$$

Now, we first obtain the solution of nonhomogeneous steady eq 5.80a by its homogeneous and particular solution as

$$\theta_s(z) = \theta_h + \theta_{PI} \tag{5.81}$$

where θ_h and θ_{PI} are the homogeneous and particular solutions of eq 5.80a as

a) For homogeneous solution, θ_h

$$\frac{d^2\theta_s}{dz^2} - \psi^2\theta_s = 0$$

or, $\left[D_z^2 - \psi^2\right]\theta_s = 0 \qquad \left[\text{where, } D_z = \frac{d}{dz}\right]$

or, $\left[D_z^2 - \psi^2\right] = 0 \qquad [\because\ \theta_s \neq 0]$

or, $D_z = \pm\psi$

Hence, the roots are real and distinct. Therefore, complementary function can be given as

$$\theta_h = c_1 e^{\psi z} + c_2 e^{-\psi z}$$

Now,

b) For particular solution θ_{PI}

$\theta_{PI} = \dfrac{-F}{D_z^2 - \psi^2} \qquad \left[\text{where, } D_z = \frac{d}{dz}\right]$

or, $\theta_{PI} = -F\left(\dfrac{1}{D_z^2 - \psi^2} e^{0z}\right) \qquad$ [Here, the term is arranged to put the particular integral in one of its property]

Hence, the particular solution can be given as by substituting the coefficient of exponential in D_z^2

$\theta_{PI} = F/\psi^2$

Therefore, the solution of nonhomogeneous steady-state eq 5.80a according to eq 5.81 is given as

$$\begin{aligned} &\theta_s(z) = \theta_h + \theta_{PI} \\ \text{or,}\quad &\theta_s(z) = c_1 e^{\psi z} + c_2 e^{-\psi z} + F/\psi^2 \end{aligned} \tag{5.82}$$

Next, we apply the boundary conditions eqs 5.80b and 5.80c to determine the constants c_1 and c_2 such that

$$c_1 = c_2 = \frac{T_{amb} - T_a - F/\psi^2}{2\left[\left(\psi L_t/Bi\right)\sinh(\psi L_t) + \cosh(\psi L_t)\right]} \tag{5.83}$$

where Bi is a nondimensional number called Biot number ($= \bar{h}_{conv} L_t/k$). Next, substituting the values of c_1 and c_2 thus obtained after applying the boundary conditions back into eq 5.82, we get the solution of steady state (eq 5.80a) as

$$\theta_s(z) = F/\psi^2 + \left[\frac{T_{amb} - T_a - F/\psi^2}{\left(\psi L_t/Bi\right)\sinh(\psi L_t) + \cosh(\psi L_t)}\right]\cosh(\psi z)$$

$$\text{or,}\quad T_s(z) = \left[T_a + F/\psi^2\right] + \left[\frac{T_{amb} - T_a - F/\psi^2}{\left(\psi L_t/Bi\right)\sinh(\psi L_t) + \cosh(\psi L_t)}\right]\cosh(\psi z) \quad [\because\ \theta_s = T_s - T_a] \tag{5.84}$$

Next, we invoke the method of separation of variables applicable to eq 5.77b and separate the dependent variables by assuming that the solution of eq 5.77b is of product form as

$$T_t(z,t) = W(z)\xi(t) \tag{5.85}$$

Next, we substitute eq 5.85 into eq 5.77b and transform the partial differential eq 5.77b into two different ordinary differential equations as follows:

$$\begin{aligned} &\frac{1}{\alpha}\frac{\partial}{\partial t}\left[W(z)\xi(t)\right] = \frac{\partial^2}{\partial z^2}\left[W(z)\xi(t)\right] - \frac{m}{\alpha}\left[W(z)\xi(t)\right] \\ \text{or,}\quad &\frac{W(z)}{\alpha}\left[\frac{\partial \xi(t)}{\partial t}\right] = \xi(t)\frac{\partial^2 W(z)}{\partial z^2} - \frac{m}{\alpha}\left[W(z)\xi(t)\right] \\ \text{or,}\quad &\frac{1}{\alpha}\left[\frac{\xi'(t)}{\xi(t)}\right] = \frac{W''(z)}{W(z)} - \psi^2 = -\lambda^2 \text{(say)} \end{aligned} \tag{5.86}$$

where $\xi'(t) = \partial\xi/\partial t, W''(z) = \partial^2 W/\partial z^2$, and $-\lambda^2$ is a separation constant. Here, the minus sign is introduced for our convenience which enables us to reach solution smoothly. Now, we have two different ordinary differential equations such that t and z are well separated

$$\xi'(t) + \alpha\lambda^2\xi(t) = 0 \tag{5.87}$$

and

$$W''(z) + (\lambda^2 - \psi^2)W(z) = 0 \tag{5.88}$$

The solution of eqs 5.87 and 5.88 can be deduced using their operator as

$\xi'(t) + \alpha\lambda^2\xi(t) = 0$	$W''(z) + (\lambda^2 - \psi^2)W(z) = 0$
or, $(D_t + \alpha\lambda^2)\xi(t) = 0$	or, $\left[D_z^2 + (\lambda^2 - \psi^2)\right]W(z) = 0$
or, $(D_t + \alpha\lambda^2) = 0 \quad [\because \xi(t) \neq 0]$	or, $D_z^2 + (\lambda^2 - \psi^2) = 0 \quad [\because W(z) \neq 0]$
or, $D_t = -\alpha\lambda^2$	or, $D_t = \pm i\sqrt{(\lambda^2 - \psi^2)}$ (5.89)
Here, the root is real. Hence, using the complemetary function	Here, the roots are complex. Hence, using the complemetary function
$\xi(t) = c_1 e^{-\alpha\lambda^2 t}$	$W(z) = c_2\cos\left(z\sqrt{\lambda^2 - \psi^2}\right) + c_3\sin\left(z\sqrt{\lambda^2 - \psi^2}\right)$

According to product solution (eq 5.85), the solution of transient part is given as

$$\begin{aligned} T_t(z,t) &= W(z)\xi(t) \\ T_t(z,t) &= \left[c_2\cos\left(z\sqrt{\lambda^2 - \psi^2}\right) + c_3\sin\left(z\sqrt{\lambda^2 - \psi^2}\right)\right]c_1 e^{-\alpha\lambda^2 t} \\ T_t(z,t) &= \left[A\cos\left(z\sqrt{\lambda^2 - \psi^2}\right) + B\sin\left(z\sqrt{\lambda^2 - \psi^2}\right)\right]c_1 e^{-\alpha\lambda^2 t} \end{aligned} \tag{5.90}$$

where $A = c_1c_2$ and $B = c_1c_3$. Next, we apply the boundary conditions on the right side of eq 5.78b on eq 5.90 which led to constant $B = 0$. Substituting $B = 0$ into eq 5.90 would give the form

$$T_t(z,t) = \left[A\cos\left(z\sqrt{\lambda^2 - \psi^2}\right)\right]\cdot e^{-\alpha\lambda^2 t} \tag{5.91}$$

Applying the boundary condition on the right side of eq 5.78a on eq 5.91, we obtain a transcendental equation as a function of eigenvalue λ as

$$\cot\left(L_t\sqrt{\lambda_n^2 - \psi^2}\right) - \frac{\sqrt{\lambda_n^2 - \psi^2}}{Bi} = 0 \tag{5.92}$$

Note, the transcendental equation has many roots and the roots can be determined numerically (using the Newtons–Raphson method). Next, we apply the principle of superposition and express the solution eq 5.91 in terms of infinite series due to infinite number of eigenvalues λ_n as

$$T_t(z,t) = \sum_{n=1}^{\infty} A_n \cos\left(z\sqrt{\lambda_n^2 - \psi^2}\right) \cdot e^{-\alpha\lambda_n^2 t} \tag{5.93}$$

To determine the constant A_n, we further implement the initial condition (eq 5.79)

$$T_t(z,t)\Big|_{t=0} \Rightarrow \sum_{n=1}^{\infty} A_n \cos\left(z\sqrt{\lambda_n^2 - \psi^2}\right) = T_0 - T_s(z) \tag{5.94}$$

We now use the orthogonality property of cosine functions by multiplying $\cos\left(z\sqrt{\lambda_m^2 - \psi^2}\right)$ and integrating both sides of eq 5.94 from 0 to L_t as

$$\begin{aligned} A_n \int_0^{L_t} \cos\left(z\sqrt{\lambda_n^2 - \psi^2}\right)\cos\left(z\sqrt{\lambda_m^2 - \psi^2}\right)dz &= T_0 \int_0^{L_t} \cos\left(z\sqrt{\lambda_m^2 - \psi^2}\right)dz \\ &\quad - \int_0^{L_t} T_s(z)\cos\left(z\sqrt{\lambda_m^2 - \psi^2}\right)dz \end{aligned} \tag{5.95}$$

In fact, we can obtain a form based on orthogonality property such that eq 5.95 can be written as

$$A_n\left(\frac{L_t}{2} + \frac{\sin\left(2L_t\sqrt{\lambda_n^2 - \psi^2}\right)}{4\sqrt{\lambda_n^2 - \psi^2}}\right) = \frac{T_0 \sin\left(L_t\sqrt{\lambda_n^2 - \psi^2}\right)}{\sqrt{\lambda_n^2 - \psi^2}} - \int_0^{L_t} T_s(z)\cos\left(z\sqrt{\lambda_n^2 - \psi^2}\right)dz$$

$$\text{or,}\quad A_n = \frac{\dfrac{T_0 \sin\left(L_t\sqrt{\lambda_n^2 - \psi^2}\right)}{\sqrt{\lambda_n^2 - \psi^2}} - \displaystyle\int_0^{L_t} T_s(z)\cos\left(z\sqrt{\lambda_n^2 - \psi^2}\right)dz}{\left(\dfrac{L_t}{2} + \dfrac{\sin\left(2L_t\sqrt{\lambda_n^2 - \psi^2}\right)}{4\sqrt{\lambda_n^2 - \psi^2}}\right)} \tag{5.96}$$

The above eq 5.96 was deduced by assuming $n = m$ of the orthogonality property since

$$\int_0^{L_t} \cos\left(z\sqrt{\lambda_n^2 - \psi^2}\right)\cos\left(z\sqrt{\lambda_m^2 - \psi^2}\right)dz = \begin{cases} 0 & n \neq m \\ \dfrac{L_t}{2} + \dfrac{\sin\left(2L_t\sqrt{\lambda_n^2 - \psi^2}\right)}{4\sqrt{\lambda_n^2 - \psi^2}} & n = m \end{cases} \tag{5.97}$$

We can see that the constant A_n from eq 5.95 is undetermined if $n \neq m$ but the constant A_n has finite values for $n = m$ which are given by eq 5.96. Now, substituting the value of $T_s(z)$ from eq 5.84 within the integrand into eq 5.96 and performing integration by parts enables us to obtain the final expression of A_n as

$$A_n = \left(T_0 - T_b - \frac{F}{\psi^2}\right) \cdot \frac{\frac{\sin\left(L_t\sqrt{\lambda_n^2-\psi^2}\right)}{\sqrt{\lambda_n^2-\psi^2}}}{\left[\frac{L_t}{2} + \frac{\sin\left(2L_t\sqrt{\lambda_n^2-\psi^2}\right)}{4\sqrt{\lambda_n^2-\psi^2}}\right]}$$

$$- \frac{T_{amb}-T_a-F/\psi^2}{\frac{\psi L_t}{Bi}\sinh(\psi L_t)+\cosh(\psi L_t)} \cdot \frac{\left[\frac{\sqrt{\lambda_n^2-\psi^2}}{\lambda_n^2}\cosh(\psi L_t)\sin\left(L_t\sqrt{\lambda_n^2-\psi^2}\right) + \frac{\psi}{\lambda_n^2}\sinh(\psi L_t)\cos\left(L_t\sqrt{\lambda_n^2-\psi^2}\right)\right]}{\left[\frac{L_t}{2} + \frac{\sin\left(2L_t\sqrt{\lambda_n^2-\psi^2}\right)}{4\sqrt{\lambda_n^2-\psi^2}}\right]} \tag{5.98}$$

Hence, using eq 5.76, the final solution of the bioheat equation can be written as

$$T(z,t) = T_s(z) + T_t(z,t)$$

$$\text{or, } T(z,t) = T_a + \frac{F}{\psi^2} + \left[\frac{T_{amb}-T_a-F/\psi^2}{\frac{\psi L_t}{Bi}\sinh(\psi L_t)+\cosh(\psi L_t)}\right]\cosh(\psi z) + \sum_{n=1}^{\infty} A_n \cos\left(z\sqrt{\lambda_n^2-\psi^2}\right) \cdot e^{-\alpha\lambda_n^2 t} \tag{5.99}$$

where A_n is given by eq 5.98, $F = Q_m'''/k$, $\psi = \sqrt{(1-\kappa)\omega_b\rho_b c_b/k}$, $\alpha = k/\rho c$, $Bi = \bar{h}_{conv}L_t/k$, and λ_n is determined from the solution of eq 5.92.

5.8.2 FINITE DIFFERENCE METHOD

The finite difference method, also known by acronym FDM, is a method of numerical discretization to transform the partial differential equation (PDE) into algebraic finite difference equations. Using this method, the continuous physical domain is divided into discrete difference grids, such that the PDE corresponding to each grid is approximated by finite difference formulas. The finite difference approximation originates from the truncated series formula and hence it bears inherent errors known as truncation error.

Here, we implement the FDM on one-dimensional transient bioheat equation, eq 5.1

$$\rho c\frac{\partial T}{\partial t} = k\left(\frac{\partial^2 T}{\partial z^2}\right) - (1-\kappa)\omega_b\rho_b c_b\left(T - T_a\right) + Q_m''' \tag{5.100}$$

which is subjected to boundary conditions

$$z = L_t : \quad -k\left.\frac{dT}{dz}\right|_{i=N} = \bar{h}_{conv}\left(T - T_{amb}\right)$$
$$z = 0 : \quad \left.\frac{dT}{dz}\right|_{i=1} = 0 \tag{5.101}$$

We assumed that the domain is discretized into $N - 1$ number of grids such that N is the number of nodes joining the grids. The index for node at $z = 0$ is $i = 1$ and that of $z = L_t$ is $i = N$. Clearly, the finite difference formula for a boundary node at $i = 1$ and N can be determined by finite difference approximation of boundary conditions (eq 5.101), whereas the finite difference formula for internal nodes $i = 2$ to $N - 1$ can be determined by finite difference approximation of eq 5.100. Let us first develop the difference formula for eq 5.100, such that the transient term of the equation was discretized using the backward differencing scheme in time and the spatial term of the equation was discretized using the central differencing scheme. Therefore, the finite difference formula for eq 5.100 node $i = 2$ to $N - 1$ can be written as

$$\rho c\left[\frac{T_i^k - T_i^{k-1}}{\Delta t}\right] = k\left[\frac{T_{i-1}^k - 2T_i^k + T_{i+1}^k}{\Delta z^2}\right] - m\left(T_i^k - T_a\right) + Q_m''' \tag{5.102}$$

where m is a dummy variable equivalent to $(1 - \kappa)\ \omega_b \rho_b c_b$, Δt and Δz are referred to time step and grid size. The subscript i indicates the current node that varies from $i = 2$ to $N - 1$ and the superscript k refers to the current time. The above equation can be further simplified as

$$\rho c\Delta z^2\left(T_i^k - T_i^{k-1}\right) = k\Delta t\left(T_{i-1}^k - 2T_i^k + T_{i+1}^k\right) - \left(m\Delta z^2\Delta t\right)T_i^k$$
$$+\left(m\Delta z^2\Delta t\right)T_a + Q_m'''\left(\Delta z^2\Delta t\right) \tag{5.103a}$$

$$\text{or,}\quad \left(k\Delta t\right)T_{i-1}^k + I \times T_i^k + \left(k\Delta t\right)T_{i+1}^k = -\left(\rho c\Delta z^2\right)T_i^{k-1} + F \tag{5.103b}$$

where $\boldsymbol{I}$ and $\boldsymbol{F}$ are dummy variables given as

$$I = \left(-2 - m\Delta z^2\Delta t - \rho c\Delta z^2\right)$$
$$\text{and} \tag{5.104}$$
$$F = -\left(mT_a + Q_m'''\right)\Delta z^2\Delta t$$

Here, eq 5.99b gives the finite difference formula for internal nodes. Next, we obtain the finite difference formula for boundary nodes (i.e., $i = 1$) such that the spatial term of the equations was discretized again using the central differencing scheme as

At $z = 0$ ($i = 1$):

$$\left(\frac{T_{i+1}^k - T_{i-1}^k}{2\Delta z}\right) = 0 \tag{5.105}$$

$$\text{or,}\ T_{i+1}^k = T_{i-1}^k$$

Note, in the above equation $i-1$ is referring to fictitious node (i.e., node zero for boundary node $i = 1$). Hence, we have to substitute the value of T_{i-1}^k from eq 5.103b to make the boundary eq 5.105 depending only on T_{i+1}^k and T_i^k. Therefore,

$$\begin{aligned} T_{i+1}^k &= \tfrac{1}{(k\Delta t)}\left[-I \times T_i^k - (k\Delta t)T_{i+1}^k - \left(\rho c\Delta z^2\right)T_i^{k-1} + F\right] \\ \text{or,}\ I \times T_i^k &+ (2k\Delta t)T_{i+1}^k = -\left(\rho c\Delta z^2\right)T_i^{k-1} + F \\ \text{or,}\ 0T_{i-1}^k &+ I \times T_i^k + (2k\Delta t)T_{i+1}^k = -\left(\rho c\Delta z^2\right)T_i^{k-1} + F \end{aligned} \tag{5.106}$$

substituting $i = 1$ into eq 5.106, we obtain the finite difference eqution for node 1 as

$$I \times T_1^k + (2k\Delta t)T_2^k = -\left(\rho c\Delta z^2\right)T_1^{k-1} + F \tag{5.107}$$

Similarly, we obtain the finite difference formula for boundary nodes (i.e., $i = N$) as

At $z = L_t$ ($i = N$):

$$-k\left(\frac{T_{i+1}^k - T_{i-1}^k}{2\Delta z}\right) = \bar{h}_{conv}\left(T_i^k - T_{amb}\right) \tag{5.108}$$

Note, in the above equation $i + 1$ node is again a fictitious node (i.e., node $n + 1$ outside the boundary node n). Hence, we have to substitute the value of T_{i+1}^k from eq 5.103b to make the boundary eq 5.108 independent of T_{i+1}^k. Therefore,

$$\begin{aligned} T_{i-1}^k &= \left[\frac{1}{k\Delta t}\left\{\left(\rho c\Delta z^2\right)T_i^{k-1} + F - (k\Delta t)T_{i-1}^k - I \times T_i^k\right\}\right] - \left(\frac{2\bar{h}_{conv}\Delta z}{k}\right) T_{amb} + \left(\frac{2\bar{h}_{conv}\Delta z}{k}\right)T_i^k \\ \text{or,}\ (k\Delta t)T_{i-1}^k &= \left\{\left(\rho c\Delta z^2\right)T_i^{k-1} + F - (k\Delta t)T_{i-1}^k - I \times T_i^k\right\} - \left(2\bar{h}_{conv}\Delta z\Delta t\right) T_{amb} + \left(2\bar{h}_{conv}\Delta z\Delta t\right)T_i^k \\ \text{or,}\ (2k\Delta t)T_{i-1}^k &+ \left(I - 2\bar{h}_{conv}\Delta z\Delta t\right)T_i^k = \left(F - 2\bar{h}_{conv}\Delta z\Delta t \times T_{amb}\right) - \left(\rho c\Delta z^2\right)T_i^{k-1} \end{aligned} \tag{5.109}$$

Note, the boundary node is denoted by n and eq 5.109 can be written as

$$\left(2k\Delta t\right)T_{n-1}^{k}+\left(I-2\overline{h}_{conv}\Delta z\Delta t\right)T_{n}^{k}=\left(F-2\overline{h}_{conv}\Delta z\Delta t\times T_{amb}\right)-\left(\rho c\Delta z^{2}\right)T_{n}^{k-1}\quad(5.110)$$

Now, we have a system of equations, that is, eq 5.107 for node 1, eq 5.103b for nodes 2 to $n-1$, and eq 5.110 for node n. Here, a standard numerical procedure can be adopted to invert the coefficient matrix with the aid of Gauss–Jordan elimination method and determine nodal temperature using the expression $[T] = [A]^{1} \times [B]$. Here, [A] represents the coefficient matrix for temperature corresponding to nodes, [B] represents a matrix corresponding to constant parameters on the right side of system of equations, and [T] represents the matrix corresponding to an unknown temperature of each node.

5.9 NONDIMENSIONAL BIOHEAT EQUATION

Here, we obtain a nondimensional equation for bioheat equation, eq 5.1. This equation is dimensionally consistent and can be used directly without its nondimensional form. However, the idea of dimensional consistency can be used in another way, by grouping the variables into a dimensionless number which makes the solution depend on less variables than the original equation. Further, the combination of variables into fewer dimensionless numbers assists experimental analysis and correlation prediction to be easy. A nondimensional equation also plays a role in scaling problem, wherein for analysis the larger prototype can be scaled down to smaller geometry or vice versa. However, it is essential to note the scaling only works if it obeys the geometric similarity, kinematic similarity, and dynamic similarity between the original prototype and scaled geometry. Let us consider a three-dimensional bioheat equation as

$$\rho c\frac{\partial T}{\partial t}=k\left(\frac{\partial^{2}T}{\partial x^{2}}+\frac{\partial^{2}T}{\partial y^{2}}+\frac{\partial^{2}T}{\partial z^{2}}\right)-(1-\kappa)\omega_{b}\rho_{b}c_{b}\left(T-T_{a}\right)+Q_{m}'''\quad(5.111)$$

Next, we define nondimensional quantities for the variables in the above equation in terms of the reference length (L_{ref}) and reference temperature (T_{ref}) as

$$\frac{1}{\alpha}\frac{\partial T}{\partial t}=\frac{T_{ref}}{L_{ref}^{2}}\left[\frac{\partial^{2}\left(T/T_{ref}\right)}{\partial\left(x/L_{ref}\right)^{2}}+\frac{\partial^{2}\left(T/T_{ref}\right)}{\partial\left(y/L_{ref}\right)^{2}}+\frac{\partial^{2}\left(T/T_{ref}\right)}{\partial\left(z/L_{ref}\right)^{2}}\right]$$

$$-\frac{(1-\kappa)\omega_{b}\rho_{b}c_{b}T_{ref}}{k}\left(\frac{T}{T_{ref}}-\frac{T_{a}}{T_{ref}}\right)+\frac{Q_{m}'''}{k}$$

$$\text{or,}\quad \frac{\partial\left(T/T_{ref}\right)}{\partial\left(\alpha t/L_{ref}^2\right)} = \left[\frac{\partial^2\left(T/T_{ref}\right)}{\partial\left(x/L_{ref}\right)^2} + \frac{\partial^2\left(T/T_{ref}\right)}{\partial\left(y/L_{ref}\right)^2} + \frac{\partial^2\left(T/T_{ref}\right)}{\partial\left(z/L_{ref}\right)^2}\right] - \frac{(1-\kappa)\omega_b\rho_b c_b L_{ref}^2}{k}\left(\frac{T}{T_{ref}} - \frac{T_a}{T_{ref}}\right) + \frac{Q_m''' L_{ref}^2}{kT_{ref}} \tag{5.112}$$

Therefore, the nondimensional bioheat equation can be written as

$$\frac{\partial\theta}{\partial Fo} = \left[\frac{\partial^2\theta}{\partial x^{*2}} + \frac{\partial^2\theta}{\partial y^{*2}} + \frac{\partial^2\theta}{\partial z^{*2}}\right] - W_b^*\left(\theta - \theta_a^*\right) + Q^* \tag{5.113}$$

where the nondimensional variables are

Temperature ratios : $\theta = \dfrac{T}{T_{ref}}$ and $\theta_a^* = \dfrac{T_a}{T_{ref}}$

Fourier number : $Fo = \dfrac{\alpha t}{L_{ref}^2}$

Length ratios : $x^* = \dfrac{x}{L_{ref}}$, $y^* = \dfrac{y}{L_{ref}}$ and $z^* = \dfrac{z}{L_{ref}}$

Dimensionless blood perfusion term : $W_b^* = \dfrac{(1-\kappa)\omega_b\rho_b c_b T_{ref}}{k}$

Nondimensional source term : $Q^* = \dfrac{Q_m''' L_{ref}^2}{kT_{ref}}$

5.10 NUMERICAL EXERCISE

5.10.1 PROBLEM 1

A biological heat transfer problem of radiation/laser diathermy of living tissue is a well-known research problem in the field of bioheat transfer, wherein the temperature variation within the tissue depends on blood perfusion, metabolic heat generation, and external radiation absorption due to laser irradiation. The blood perfusion effect can be modeled as a cooling method during laser diathermy as $-(1-\kappa)\omega_b\rho_b c_b\left[T-T_a\right]$ (in W/m^3). The metabolic heat generation is denoted by Q_m''' (in W/m^3) and radiation absorption in the tissue is denoted by $Q_r'''=\kappa_a I_{cw}(z)$ (in W/m^3), where κ_a is the tissue absorption coefficient (in m^{-1}) and I_{cw} is the irradiated continuous-wave laser intensity as a function of tissue depth. All other parameters have the usual meaning as discussed earlier. Assuming that the thermophysical properties of the tissue are constant, obtain a finite differential equation for the one-dimensional bioheat equation subjected to radiative source, wherein the central differencing scheme is adopted for space derivative and backward differencing is applied to time derivative.

5.10.2 PROBLEM 2

In case of brain injury, the brain hypothermia procedure is often adopted to maintain core temperature of patient below the normal body temperature in order to reduce the metabolic activity of the brain. Reducing the metabolic activity of the brain helps doctor to treat brain injury efficiently since oxygen-deprived regions of injured site no longer demand an excess amount of oxygen supply. Using the finite difference method applied to the steady-state bioheat equation for the human brain given as

$$\alpha\left(\frac{\partial^2 T}{\partial x^2}\right)-\frac{W}{\rho c}\left(T-T_a\right)+\frac{Q_m'''}{\rho c}=0 \tag{5.114}$$

where T is the unknown temperature of the brain, α is the thermal diffusivity (= $k/\rho c$), T_a is the core temperature, W is the blood perfusion parameter, Q_m''' is the metabolic heat generation, k is the thermal conductivity, ρ is density, c is the specific heat, obtain the skin temperature distribution with depth (say x) post hypothermia during which the brain returns to its original temperature after certain time. The head is subjected to surrounding air at its exposed boundary and the interior boundary of the brain is at a constant core temperature of 37°C.

a) Formulate the set of algebraic equation in terms of unknown temperature at nodes such that the entire domain (Figure 5.11) is divided into 10 equidistant spaces, that is, number of nodes are 11.
b) Consider the following values for the constants in eq 5.114 and for boundary values, as depicted in Table 5.4. Write the final form of a set of algebraic equations in matrix representation.

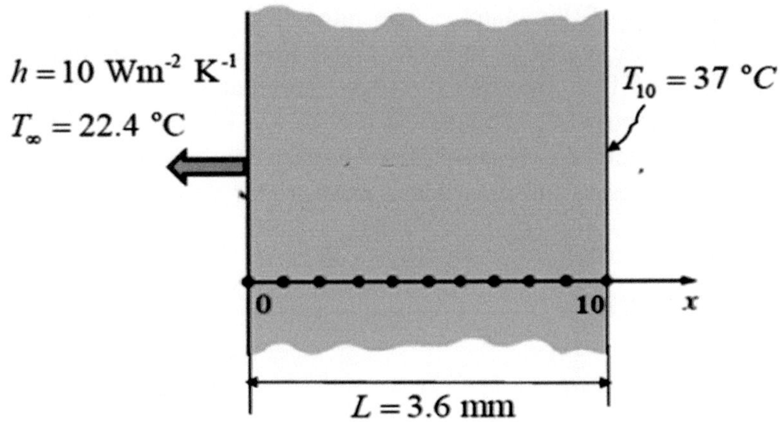

FIGURE 5.11 1D tissue slab.

TABLE 5.4 Thermal Properties of the Physical Domain.

Properties	Values
ρ (kg/m^3)	1088.88
k (W/m/K)	0.2947
c (J/kg/K)	2960.25
α (m^2/s)	9.1433×10^{-8}
W (W/m^3/K)	1454.6
Q_m''' (W/m^3)	357.875
T_a (K)	308.15

Compute the temperature distribution, T (in K) and show its variation with x (in mm) as shown in Figure 5.12 assuming an input parameter same as Table 5.4.

5.11 SUMMARY

Heat and mass transfer are one of the key aspects in human body at organ, tissue, and cellular levels that enable body to maintain homeostasis. This chapter in an orderly fashion elaborated the role of heat transfer in biological systems primarily focused on its application at tissue and organ level. The role of heat energy balance or transfer at cellular and molecular level is more fundamental in nature that can be understood by energy transfer mechanism in biomolecules involving the concept of bio-thermodynamics which is beyond the scope of this chapter. The interested students or readers may perform a literature search on these advanced topics to develop fundamental understanding of heat transfer at micro- and nano-scales. This chapter comprises the important mathematical models for bioheat transfer at tissue and organ level, introduces some of the applications and instruments used for research purposes, and provides a database for various thermophysical properties.

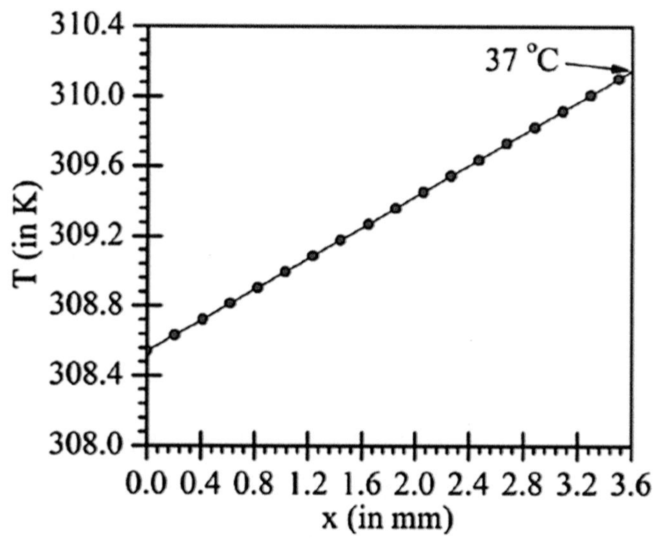

FIGURE 5.12 Temperature distribution in 1d tissue slab.

KEYWORDS

- **biofluids**
- **bioheat**
- **continuum model**
- **non-Fourier equation**
- **phase change equation**
- **vascular model**
- **whole body model**

REFERENCES

1. Bhowmik, A.; Singh, R.; Repaka, R.; Mishra, S. C. Conventional and Newly Developed Bioheat Transport Models in Vascularized Tissues: A Review. *J. Thermal Biol.* **2013**, *38* (3), 107–125.
2. Sherwood, L. In *Human Physiology: From Cells to Systems*; Cengage Learning, 2010.
3. Fourier, J. B. J. In *The Analytical Theory of Heat*; Cambridge University Press, 1878.
4. Pennes, H. H. Analysis of Tissue and Arterial Blood Temperatures in the Resting Human Forearm. *J. Appl. Physiol.* **1948**, *1* (2), 93–122. DOI: 10.1152/jappl.1948.1.2.93.
5. Baish, J. W. Microvascular Heat Transfer. In *The Biomedical Engineering Handbook*; Bronzino, J. D., Peterson, D. R., Eds.; CRC Press LLC (Taylor & Francis Group), 2015; vol 2, pp 1711–1724.
6. Bhowmik, A.; Ghosh, B.; Pal, M.; Paul, R. R.; Chatterjee, J.; Chakraborty, S. Portable, Handheld, and Affordable Blood Perfusion Imager for Screening of Subsurface Cancer in Resource-Limited Settings. *Proc. Natl. Acad. Sci. U. S. A.* **2022**, *119* (2). DOI: 10.1073/pnas.2026201119.
7. Pittman, R. N. In *Regulation of Tissue Oxygenation. Colloqoquium Series on Integrated Systems Physiolophy: From Molecule to Function*, 2011; vol *3*, pp 1–100.
8. Chen, M. M.; Holmes, K. R. Microvascular Contributions in Tissue Heat Transfer. *Ann. N. Y. Acad. Sci.* **1980**, *335* (1), 137–150. DOI: 10.1111/j.1749-6632.1980.tb50742.x.
9. Weinbaum, S.; Jiji, L. M. A New Simplified Bioheat Equation for the Effect of Blood Flow on Local Average Tissue Temperature. *J. Biomech. Eng.* **1985**, *107* (2), 131–139. DOI: 10.1115/1.3138533.
10. Tzou, D. Y. Thermal Shock Phenomena Under High Rate Response in Solids. *Annu. Rev. Heat Transf.* **1992**, *4* (4), 111–185. DOI: 10.1615/AnnualRevHeatTransfer.v4.50.
11. Maxwell, J. C. On the Dynamical Theory of Gases. *Philos. Trans. R. Soc. London* **1867**, *157*, 49–88. DOI: 10.1098/rstl.1867.0004.
12. Morse, P. M.; Feshbach, H. In *Methods of Theoretical Physics*; McGraw-Hill: New York, 1953.
13. Cattaneo, C. A Form of Heat Conduction Equation Which Eliminates the Paradox of Instantaneous Propagation. *Comptes Rendus* **1958**, *247*, 431–433.
14. Vernotte, P. Les Paradoxes de la Theorie Continue de l'Equation de La Chaleur. *Compute Rendus* **1958**, *246*, 3154–3155.
15. Özis¸ik, M. N.; Tzou, D. Y. On the Wave Theory in Heat Conduction. *J. Heat Transf.* **1994**, *116* (3), 526–535. DOI: 10.1115/1.2910903.
16. Mitra, K.; Kumar, S.; Vedevarz, A.; Moallemi, M. K. Experimental Evidence of Hyperbolic Heat Conduction in Processed Meat. *J. Heat Transf.* **1995**, *117* (3), 568–573. DOI: 10.1115/1.2822615.
17. Tzou, D. Y. In *Macro- to Microscale Heat Transfer: The Lagging Behavior*, 1997.
18. Roetzel, W.; Xuan, Y. Transient Response of the Human Limb to an External Stimulus. *Int. J. Heat Mass Transf.* **1998**, *41* (1), 229–239.
19. Shih, T. C.; Kou, H. S.; Lin, W. L. Effect of Effective Tissue Conductivity on Thermal Dose distributions of Living Tissue with directional Blood Flow during Thermal Therapy. *Int. Commun. Heat Mass Transf.* **2002**, *29* (1), 115–126. DOI: 10.1016/S0735-1933(01)00330-X.
20. Shrivastava, D.; Vaughan, J. T. A Generic Bioheat Transfer Thermal Model for a Perfused Tissue. *J. Biomech. Eng.* **2009**, *131* (7). DOI: 10.1115/1.3127260.

21. Dua, R.; Chakraborty, S. A Novel Modeling and Simulation Technique of Photo–Thermal Interactions Between Lasers and Living Biological Tissues Undergoing Multiple Changes in Phase. *Comput. Biol. Med.* **2005**, *35* (5), 447–462. DOI: 10.1016/j.compbiomed.2004.02.005.
22. Abraham, J. P.; Sparrow, E. M. A Thermal-Ablation Bioheat Model Including Liquid-to-Vapor Phase Change, Pressure- and Necrosis-Dependent Perfusion, and Moisture-Dependent Properties. *Int. J. Heat Mass Transf.* **2007**, *50* (13-14), 2537–2544. DOI: 10.1016/j.ijheatmasstransfer.2006.11.045.
23. Kim, K.; Guo, Z. Multi-Time-Scale Heat Transfer Modeling of Turbid Tissues Exposed to Short-Pulsed Irradiations. *Comput. Methods Programs Biomed.* **2007**, *86* (2), 112–123. DOI: 10.1016/j.cmpb.2007.01.009.
24. Keller, K. H.; Seiler, L. An Analysis of Peripheral Heat Transfer in Man. *J. Appl. Physiol.* **1971**, *30* (5), 779–786. DOI: 10.1152/jappl.1971.30.5.779.
25. Lagendijk, J. J. W. The Influence of Bloodflow in Large Vessels on the Temperature Distribution in Hyperthermia. *Phys. Med. Biol.* **1982**, *27* (1), 17–23. DOI: 10.1088/0031-9155/27/1/002.
26. Kotte, A. N. T. J.; Leeuwen, G. M. J. v.; Lagendijk, J. J. W. Modelling the Thermal Impact of a Discrete Vessel Tree. *Phys. Med. Biol.* **1999**, *44* (1), 57–74. DOI: 10.1088/0031-9155/44/1/006.
27. Bhowmik, A.; Repaka, R.; Mishra, S. C. Thermographic Evaluation of Early Melanoma Within the Vascularized Skin Using Combined Non-Newtonian Blood Flow and Bioheat Models. *Comput. Biol. Med.* **2014**, *53*. DOI: 10.1016/j.compbiomed.2014.08.002.
28. Gagge, A.; Stolwijk, J.; Nishi, Y. An Effective Temperature Scale Based on a Simple Model of Human Physiological Regulatiry Response. *Mem. Fac. Eng. Hokkaido Univ.* **1972**, *13* (Suppl), 21–36.
29. Stolwijk, J.; Hardy, J. Temperature Regulation in Man—A Theoretical Study. *Pflüger's Archiv für die gesamte Physiologie des Menschen und der Tiere* **1966**, *291* (2), 129–162.
30. Fiala, D.; Lomas, K. J.; Stohrer, M. A Computer Model of Human Thermoregulation for a Wide Range of Environmental Conditions: The Passive System. *J. Appl. Physiol.* **1999**, *87* (5), 1957–1972.
31. Fiala, D.; Lomas, K. J.; Stohrer, M. Computer Prediction of Human Thermoregulatory and Temperature Responses to a Wide Range of Environmental Conditions. *Int. J. Biometeorol.* **2001**, *45* (3), 143–159.
32. Avolio, A. Multi-Branched Model of the Human Arterial System. *Med. Biol. Eng. Comput.* **1980**, *18* (6), 709–718.
33. Gulati, T.; Hatwar, R.; Unnikrishnan, G.; Rubio, J. E.; Reifman, J. A 3-D Virtual Human Model for Simulating Heat and Cold Stress. *J. Appl. Physiol.* **2022**, *133* (2), 288–310.
34. Zhang, M.; Li, R.; Li, J.; Wang, F.; Subramaniam, S.; Lang, J.; Passalacqua, A.; Song, G. A 3D Multi-Segment Thermoregulation Model of the Hand With Realistic Anatomy: Development, Validation, and Parametric Analysis. *Build. Environ.* **2021**, *201*, 107964.
35. Yang, J.; Wang, F.; White, M. D.; Li, R.; Song, G.; Etter, C. V.; Gnatiuk, E. A.; Perrotta, A. S. A 7-Segment Numerical Hand-Glove/Mitten Model for Predicting Thermophysiological Responses of the Human Hand in Extremely Cold Conditions. *Comput. Biol. Med.* **2022**, 106351.
36. Fallahi, A.; Salimpour, M. R.; Shirani, E. A 3D Thermal Model to Analyze the Temperature Changes of Digits During Cold Stress and Predict the Danger of Frostbite in Human Fingers. *J. Therm. Biol.* **2017**, *65*, 153–160.

37. Joshi, A.; Wang, F.; Kang, Z.; Yang, B.; Zhao, D. A Three-Dimensional Thermoregulatory Model for Predicting Human Thermophysiological Responses in Various Thermal Environments. *Build. Environ.* **2022**, *207*, 108506.
38. Salloum, M.; Ghaddar, N.; Ghali, K. A New Transient Bioheat Model of the Human Body and its Integration to Clothing Models. *Int. J Therm. Sci.* **2007**, *46* (4), 371–384.
39. Karaki, W.; Ghaddar, N.; Ghali, K.; Kuklane, K.; Holmér, I.; Vanggaard, L. Human Thermal Response with Improved AVA Modeling of the Digits. *Int. J. Therm. Sci.* **2013**, *67*, 41–52.
40. Rida, M.; Karaki, W.; Ghaddar, N.; Ghali, K.; Hoballah, J. A New Mathematical Model to Simulate AVA Cold-Induced Vasodilation Reaction to Local Cooling. *Int. J. Biometeorol.* **2014**, *58* (9), 1905–1918.
41. Stolwijk, J. A. J. In *A Mathematical Model of Physiological Temperature Regulation in Man (NASA CR-1855)*; Washington, DC, 1971.
42. Stolwijk, J. A. J.; Hardy, J. D. Control of Body Temperature. In *Comprehensive Physiology*; John Wiley & Sons, Inc., 2011; pp 45–68.
43. Fiala, D. Dynamic Simulation of Human Heat Transfer and Thermal Comfort, 1998; pp 1–293.
44. Singh, M.; Singh, T.; Soni, S. Pre-operative Assessment of Ablation Margins for Variable Blood Perfusion Metrics in a Magnetic Resonance Imaging Based Complex Breast Tumour Anatomy: Simulation Paradigms in Thermal Therapies. *Comput. Methods Programs Biomed.* **2021**, *198*, 105781.
45. Moritz, A. R.; Henriques, F. C. Studies of Thermal Injury: II. The Relative Importance of Time and Surface Temperature in the Causation of Cutaneous Burns. *Am. J. Pathol.* **1947**, *23* (5), 695–720.
46. Pearce, J.; Thomsen, S. Rate Process Analysis of Thermal Damage. In *Optical-Thermal Response of Laser-Irradiated Tissue*; Springer: US, 1995; pp 561–606.
47. Rylander, M.; Feng, Y.; Zhang, Y.; Bass, J.; Stafford, R.; Volgin, A.; Hazle, J.; Diller, K. Optimizing Heat Shock Protein Expression Induced by Prostate Cancer Laser Therapy Through Predictive Computational Models. *J. Biomed. Opt.* **2006**, *11* (4), 041113.
48. Singh, M.; Gu, Q.; Ma, R.; Zhu, L. In *Temperature Distribution and Thermal Dosage Affected by Nanoparticle Distribution in Tumours During Magnetic Nanoparticle Hyperthermia*, ASME 2019 6th International Conference on Micro/Nanoscale Heat and Mass Transfer, 2019; vol V001T13A002. DOI: 10.1115/mnhmt2019-4233.
49. Fasano, A.; Hömberg, D.; Naumov, D. On a Mathematical Model for Laser-Induced Thermotherapy. *Appl. Math. Model.* **2010**, *34* (12), 3831–3840.
50. Pearce, J. A. Comparative Analysis of Mathematical Models of Cell Death and Thermal Damage Processes. *Int. J. Hyperthermia* **2013**, *29* (4), 262–280.
51. Pearce, J. A. Improving Accuracy in Arrhenius Models of Cell Death: Adding a Temperature-Dependent Time Delay. *J. Biomech. Eng.* **2015**, *137* (12), 121006.
52. Feng, Y.; Tinsley Oden, J.; Rylander, M. N. A Two-State Cell Damage Model Under Hyperthermic Conditions: Theory and in Vitro Experiments. *J. Biomech. Eng.* **2008**, *130* (4), 041016.
53. Pearce, J. Irreversible Tissue Thermal Alterations: Skin Burns, Thermal Damage and Cell Death. In *Theory and Applications of Heat Transfer in Humans*, 2018; vol *2*, pp 553–590.
54. Singh, M.; Gu, Q.; Ma, R.; Zhu, L. Heating Protocol Design Affected by Nanoparticle Redistribution and Thermal Damage Model in Magnetic Nanoparticle Hyperthermia for Cancer Treatment. *J. Heat Transf.* **2020**, *142* (7).

55. Singh, M.; Ma, R.; Zhu, L. Theoretical Evaluation of Enhanced Gold Nanoparticle Delivery to PC3 Tumors Due to Increased Hydraulic Conductivity or Recovered Lymphatic Function After Mild Whole Body Hyperthermia. *Med. Biol. Eng. Comput.* **2021**, *59* (2), 301–313.
56. Dewey, W.; Hopwood, L.; Sapareto, S.; Gerweck, L. Cellular Responses to Combinations of Hyperthermia and Radiation. *Radiology* **1977**, *123* (2), 463–474.
57. Rylander, M. N.; Diller, K. R.; Wang, S.; Aggarwal, S. J. Correlation of HSP70 Expression and Cell Viability Following Thermal Stimulation of Bovine Aortic Endothelial Cells. *J. Biomech. Eng.* **2005**, *127* (5), 751–757.
58. Dombrovsky, L. A.; Timchenko, V.; Pathak, C.; Piazena, H.; Müller, W.; Jackson, M. Radiative Heating of Superficial Human Tissues with the use of Water-Filtered Infrared-A Radiation: A Computational Modeling. *Int. J. Heat Mass Transf.* **2015**, *85*, 311–320.
59. Dombrovsky, L. A. Scattering of Radiation and Simple Approaches to Radiative Transfer in Thermal Engineering and Biomedical Applications. In *Springer Series in Light Scattering*; Springer, 2019; pp 71–127.
60. Dombrovsky, L. A. Laser-Induced Thermal Treatment of Superficial Human Tumors: An Advanced Heating Strategy and Non-Arrhenius Law for Living Tissues. *Front. Therm. Eng.* **2022**, 1.
61. Kumar, D.; Rai, K. A Study on Thermal Damage During Hyperthermia Treatment Based on DPL Model for Multilayer Tissues Using Finite Element Legendre Wavelet Galerkin Approach. *J. Therm. Biol.* **2016**, *62*, 170–180.
62. Liu, K. C.; Chen, T. M. Comparative Study of Heat Transfer and Thermal Damage Assessment Models for Hyperthermia Treatment. *J. Therm. Biol.* **2021**, *98*, 102907.
63. Alzahrani, F. S.; Abbas, I. A. Analytical Solutions of Thermal Damage in Living Tissues due to Laser Irradiation. *Waves Random Complex Media* **2021**, *31* (6), 1443–1456.
64. Singh, M. Incorporating Vascular-Stasis Based Blood Perfusion to Evaluate the Thermal Signatures of Cell-Death Using Modified Arrhenius Equation with Regeneration of Living Tissues During Nanoparticle-Assisted Thermal Therapy. *Int. Commun. Heat Mass Transf.* **2022**, *135*, 106046.
65. Xu, Y.; Qian, R. Analysis of Thermal Injury Process Based on Enzyme Deactivation Mechanisms. *J. Biomech. Eng.* **1995**, *117* (4), 462–465.
66. O'Neill, D. P.; Peng, T.; Stiegler, P.; Mayrhauser, U.; Koestenbauer, S.; Tscheliessnigg, K.; Payne, S. J. A Three-State Mathematical Model of Hyperthermic Cell Death. *Ann. Biomed. Eng.* **2011**, *39* (1), 570–579.
67. Sapareto, S. A.; Dewey, W. C. Thermal Dose Determination in Cancer Therapy. *Int. J. Radiat. Oncol. Biol. Phys.* **1984**, *10* (6), 787–800. DOI: 10.1016/0360-3016(84)90379-1.
68. Pearce, J. A. In *Relationship Between Arrhenius Models of Thermal Damage and the CEM 43 Thermal Dose*, Proceedings of SPIE; Ryan, T. P., Ed.; 2009; vol 2, pp 718104–718104. DOI: 10.1117/12.807999.
69. Shih, T. C.; Kou, H. S.; Liauh, C. T.; Lin, W. L. Thermal Models of Bioheat Transfer Equations in Living Tissue and Thermal Dose Equivalence due to Hyperthermia. *Biomed. Eng. Appl. Basis Commun.* **2002**, *14* (02), 86–96.
70. Babbar, A.; Jain, V.; Gupta, D. Thermogenesis Mitigation Using Ultrasonic Actuation During Bone Grinding: A Hybrid Approach Using CEM43° C and Arrhenius Model. *J. Brazilian Soc. Mech. Sci. Eng.* **2019**, *41* (10), 1–14.
71. Babbar, A.; Jain, V.; Gupta, D.; Agrawal, D. Finite Element Simulation and Integration of CEM43° C and Arrhenius Models for Ultrasonic-Assisted Skull Bone Grinding: A Thermal Dose Model. *Med. Eng. Phys.* **2021**, *90*, 9–22.

72. Assi, H.; Arsenault, M.; Whelan, W.; Kumaradas, J. A New Thermal Dose Model Based on Vogel-Tammann-Fulcher Behaviour in Thermal Damage Processes. *Int. J. Hyperthermia* **2022,** *39* (1), 697–705.
73. Tankata, A. N.; Goldfinch, L.; Hinds, J. K.; Kuan, L. P.; Thomopoulis, N. In *Thermal Model of Laser Induced Eye Damage, Final Rep. USAF School of Aerospace Medicine, Brooks AFB TX, Contract F41609-74-C-0005*; Chicago, IL, 1974.
74. Welch, A. J.; Polhamus, G. D. Measurement and Prediction of Thermal Injury in the Retina of the Rhesus Monkey. *IEEE Trans. Biomed. Eng.* **1984,** *BME-31* (10), 633–644. DOI: 10.1109/TBME.1984.325313.
75. Birngruber, R.; Hillenkamp, F.; Gabel, V. P. Theoretical Investigations of Laser Thermal Retinal Injury. *Health Phys.* **1985,** *48* (6), 781–796. DOI: 10.1097/00004032-198506000-00006.
76. Henriques, F. C. Studies of Thermal Injury V. The Predictability and Significance of Thermally Induced Rate Processes Leading to Irreversible Epidermal Injury. *Arch. Pathol.* **1947,** *43*, 489–502.
77. Weaver, J. A.; Stoll, A. M. Mathematical Model of Skin Exposed to Thermal Radiation. *Aerosp. Med.* **1969,** *40* (1), 24–30.
78. Thomsen, S.; Pearce, J. A. Thermal Damage and Rate Processes in Biologic Tissues. In *Optical-Thermal Response of Laser-Irradiated Tissue*; Springer: Netherlands, 2010; pp 487–549.
79. He, X.; McGee, S.; Coad, J. E.; Schmidlin, F.; Iaizzo, P. A.; Swanlund, D. J.; Kluge, S.; Rudie, E.; Bischof, J. C. Investigation of the Thermal and Tissue Injury Behaviour in Microwave Thermal Therapy Using a Porcine Kidney Model. *Int. J. Hyperthermia* **2004,** *20* (6), 567–593. DOI: 10.1080/0265673042000209770.
80. Pop, M.; Molckovsky, A.; Chin, L.; Kolios, M. C.; Jewett, M. A. S.; Sherar, M. D. Changes in Dielectric Properties at 460 kHz of Kidney and Fat During Heating: Importance for Radio-Frequency Thermal Therapy. *Phys. Med. Biol.* **2003,** *48* (15), 2509–2525. DOI: 10.1088/0031-9155/48/15/317.
81. Jacques, S. L.; Gaeeni, M. O. In Thermically Induced Changes in Optical Properties of Heart, Images of the Twenty-First Century. Proceedings of the Annual International Engineering in Medicine and Biology Society; IEEE, 1989; pp 1199–1200. DOI: 10.1109/IEMBS.1989.96158.
82. Matthewson, K.; Coleridge-Smith, P.; O'Sullivan, J. P.; Northfield, T. C.; Bown, S. G. Biological Effects of Intrahepatic Neodymium: Yttrium-Aluminum-Garnet Laser Photocoagulation in Rats. *Gastroenterology* **1987,** *93* (3), 550–557. DOI: 10.1016/0016-5085(87)90918-8.
83. Flock, S. T.; Smith, L.; Waner, M. Quantifying the Effects on Blood of Irradiation with Four Different Vascular-Lesion Lasers. In *Proceedings of SPIE*, 1993/7//; Jacques, S. L., Katzir, A., Eds.; 1993; pp 237–237. DOI: 10.1117/12.147664.
84. Moussa, N. A.; Tell, E. N.; Cravalho, E. G. Time Progression of Hemolysis of Erythrocyte Populations Exposed to Supraphysiological Temperatures. *J. Biomech. Eng.* **1979,** *101* (3), 213–217. DOI: 10.1115/1.3426247.
85. Lepock, J. R.; Frey, H. E.; Bayne, H.; Markus, J. Relationship of Hyperthermia-Induced Hemolysis of Human Erythrocytes to the Thermal Denaturation of Membrane Proteins. *Biochimica et Biophysica Acta (BBA)-Biomembranes* **1989,** *980* (2), 191–201. DOI: 10.1016/0005-2736(89)90399-4.
86. Moussa, N. A.; McGrath, J. J.; Cravalho, E. G.; Asimacopoulos, P. J. Kinetics of Thermal Injury in Cells. *J. Biomech. Eng.* **1977,** *99* (3), 155–159. DOI: 10.1115/1.3426283.

87. Reddy, G.; Dreher, M. R.; Rossmann, C.; Wood, B. J.; Haemmerich, D. Cytotoxicity of Hepatocellular Carcinoma Cells to Hyperthermic and Ablative Temperature Exposures: In Vitro Studies and Mathematical Modelling. *Int. J. Hyperthermia* **2013**, *29* (4), 318–323. DOI: 10.3109/02656736.2013.792125.
88. Bhowmick, S.; Swanlund, D. J.; Bischof, J. C. Supraphysiological Thermal Injury in Dunning AT-1 Prostate Tumor Cells. *J. Biomech. Eng.* **2000**, *122* (1), 51–59. DOI: 10.1115/1.429627.
89. Beckham, J. T.; Mackanos, M. A.; Crooke, C.; Takahashl, T.; O'Connell-Rodwell, C.; Contag, C. H.; Jansen, E. D. Assessment of Cellular Response to Thermal Laser Injury Through Bioluminescence Imaging of Heat Shock Protein 70. *Photochem. Photobiol.* **2004**, *79* (1), 76–85. DOI: 10.1111/j.1751-1097.2004.tb09860.x.
90. Agah, R. Quantitative Characterization of Arterial Tissue Thermal Damage, 1988.
91. Maitland, D. J.; Walsh, J. T. Quantitative Measurements of Linear Birefringence During Heating of Native Collagen. *Lasers Surg. Med.* **1997**, *20* (3), 310–318. DOI: 10.1002/(SICI)1096-9101(1997)20:3<310::AID-LSM10>3.0.CO;2-H.
92. Pearce, J. A.; Thomsen, S. L.; Vijverberg, H.; McMurray, T. J. In *Kinetics for Birefringence Changes in Thermally Coagulated Rat Skin Collagen*, *Proceedings of SPIE*, 1993/7//; Anderson, R. R., Bass, L. S., Shapshay, S. M., White, J. V., White, R. A., Eds.; 1993; pp 180–180. DOI: 10.1117/12.147029.
93. Mazur, P. Freezing of Living Cells: Mechanisms and Implications. *Am. J. Physiol. Cell Physiol.* **1984**, *247* (3), C125–C142. DOI: 10.1152/ajpcell.1984.247.3.C125.
94. Toner, M.; Cravalho, E. G.; Karel, M. Thermodynamics and Kinetics of Intracellular Ice Formation During Freezing of Biological Cells. *J. Appl. Phys.* **1990**, *67* (3), 1582–1593. DOI: 10.1063/1.345670.
95. Toner, M.; Tompkins, R. G.; Cravalho, E. G.; Yarmush, M. L. Transport Phenomena During Freezing of Isolated Hepatocytes. *AIChE J.* **1992**, *38* (10), 1512–1522. DOI: 10.1002/aic.690381004.
96. Devireddy, R. V.; Smith, D. J.; Bischof, J. C. Effect of Microscale Mass Transport and Phase Change on Numerical Prediction of Freezing in Biological Tissues. *J. Heat Transf.* **2002**, *124* (2), 365–374. DOI: 10.1115/1.1445134.
97. Rutherford, M.; Ramenghi, L. A.; Edwards, A. D.; Brocklehurst, P.; Halliday, H.; Levene, M.; Strohm, B.; Thoresen, M.; Whitelaw, A.; Azzopardi, D. Assessment of Brain Tissue Injury After Moderate Hypothermia in Neonates with Hypoxic–Ischaemic Encephalopathy: A Nested Substudy of a Randomised Controlled Trial. *Lancet Neurol.* **2010**, *9* (1), 39–45. DOI: 10.1016/S1474-4422(09)70295-9.
98. Andresen, M.; Gazmuri, J. T.; Marín, A.; Regueira, T.; Rovegno, M. Therapeutic Hypothermia for Acute Brain Injuries. *Scand. J. Trauma Resusc. Emerg. Med.* **2015**, *23* (1), 42–42. DOI: 10.1186/s13049-015-0121-3.
99. Alderliesten, T.; de Vries, L. S.; Khalil, Y.; van Haastert, I. C.; Benders, M. J. N. L.; Koopman-Esseboom, C.; Groenendaal, F. Therapeutic Hypothermia Modifies Perinatal Asphyxia-Induced Changes of the Corpus Callosum and Outcome in Neonates. *PLOS One* **2015**, *10* (4), e0123230–e0123230. DOI: 10.1371/journal.pone.0123230.
100. Zhu, L.; Diao, C. Theoretical Simulation of Temperature Distribution in the Brain During mild Hypothermia Treatment for Brain Injury. *Med. Biol. Eng. Comput.* **2001**, *39* (6), 681–687. DOI: 10.1007/BF02345442.
101. Silva, A. B. C. G.; Wrobel, L. C.; Ribeiro, F. L. B. A Thermoregulation Model for Whole Body Cooling Hypothermia. *J. Therm. Biol.* **2018**, *78*, 122–130. DOI: 10.1016/j.jtherbio.2018.08.019.

102. Silva, A. B. C. G.; Laszczyk, J.; Wrobel, L. C.; Ribeiro, F. L. B.; Nowak, A. J. A Thermoregulation Model for Hypothermic Treatment of Neonates. *Med. Eng. Phys.* **2016**, *38* (9), 988–998. DOI: 10.1016/j.medengphy.2016.06.018.
103. Wang, H.; Olivero, W.; Lanzino, G.; Elkins, W.; Rose, J.; Honings, D.; Rodde, M.; Burnham, J.; Wang, D. Rapid and Selective Cerebral Hypothermia Achieved Using a Cooling Helmet. *J. Neurosurg.* **2004**, *100* (2), 272–277. DOI:10.3171/jns.2004.100.2.0272.
104. Crepeau, A. Z.; Rabinstein, A. A.; Fugate, J. E.; Mandrekar, J.; Wijdicks, E. F.; White, R. D.; Britton, J. W. Continuous EEG in Therapeutic Hypothermia After Cardiac Arrest: Prognostic and Clinical Value. *Neurology* **2013**, *80* (4), 339–344. DOI: 10.1212/WNL.0b013e31827f089d.
105. Southard, M. D. J. H.; Belzer, M. D. F. O. Organ Preservation. *Ann. Rev. Med.* **1995**, *46* (1), 235–247. DOI: 10.1146/annurev.med.46.1.235.
106. Guibert, E. E.; Petrenko, A. Y.; Balaban, C. L.; Somov, A. Y.; Rodriguez, J. V.; Fuller, B. J. Organ Preservation: Current Concepts and New Strategies for the Next Decade. *Transfus. Med. Hemotherapy* **2011**, *38* (2), 125–142. DOI: 10.1159/000327033.
107. Jahania, M. S.; Sanchez, J. A.; Narayan, P.; Lasley, R. D.; Mentzer, R. M. Heart Preservation for Transplantation: Principles and Strategies. *Ann. Thorac. Surg.* **1999**, *68* (5), 1983–1987. DOI: 10.1016/S0003-4975(99)01028-0.
108. Mazur, P. Cryobiology: The Freezing of Biological Systems. *Science* **1970**, *168* (3934), 939–949. DOI: 10.1126/science.168.3934.939.
109. Fahy, G. M.; MacFarlane, D. R.; Angell, C. A.; Meryman, H. T. Vitrification as an Approach to Cryopreservation. *Cryobiology* **1984**, *21* (4), 407–426. DOI: 10.1016/0011-2240 (84)90079-8.
110. Rubinsky, B. Microscale Heat Transfer in Biological Systems at Low Temperatures. *Exp. Heat Transf.* **1997**, *10* (1), 1–29. DOI: 10.1080/08916159708946531.
111. Gage, A. A.; Baust, J. Mechanisms of Tissue Injury in Cryosurgery. *Cryobiology* **1998**, *37* (3), 171–186. DOI: 10.1006/cryo.1998.2115.
112. Zhang, J.; Sandison, G. A.; Murthy, J. Y.; Xu, L. X. Numerical Simulation for Heat Transfer in Prostate Cancer Cryosurgery. *J. Biomech. Eng.* **2005**, *127* (2), 279–294. DOI: 10.1115/1.1865193.
113. Braithwaite, S. A.; van der Kaaij, N. P. New Techniques for Optimization of Donor Lungs/Hearts. *Anesthesiol. Clin.* **2019**, *37* (4), 639–660. DOI: 10.1016/j.anclin.2019.08.010.
114. Wzsuzsanna3. Cryopreservation. In *Wikipedia ORG [Online] [Available: https://en.wikipedia.org/wiki/Cryopreservation],* 2020
115. Baust, J. M.; Robilotto, A.; Guerra, P.; Snyder, K. K.; Buskirk, R. G. V.; Dubuc, M.; Baust, J. G. Assessment of a novel cryoablation device for the endovascular treatment of cardiac tachyarrhythmias. SAGE Open Medicine **2018**, *6,* 205031211876979-205031211876979. DOI: 10.1177/2050312118769797.
116. Bhowmik, A.; Repaka, R. Passive and Active Thermography for the Detection of Subsurface Malignant Lesions. In *Thermal Imaging: Types, Advancements and Applications*; Strickland, C., Ed.; Nova Science Publishers, Inc., 2017; pp 25–60.
117. Dong, F.; Tao, C.; Wu, J.; Su, Y.; Wang, Y.; Wang, Y.; Guo, C.; Lyu, P. Detection of Cervical Lymph Node Metastasis from Oral Cavity Cancer Using a Non-radiating, Noninvasive Digital Infrared Thermal Imaging System. *Sci. Rep.* **2018**, *8* (1), 7219–7219. DOI: 10.1038/s41598-018-24195-4.
118. Bhowmik, A.; Repaka, R. Estimation of Growth Features and Thermophysical Properties of Melanoma Within 3-D Human Skin Using Genetic Algorithm and Simulated Annealing. *Int. J. Heat Mass Transf.* **2016**, *98*. DOI: 10.1016/j.ijheatmasstransfer.2016.03.020.

119. Liu, C. C.; Chang, R. E.; Chang, W. C. Limitations of Forehead Infrared Body Temperature Detection for Fever Screening for Severe Acute Respiratory Syndrome. *Infect. Control Hosp. Epidemiol.* **2004**, *25* (12), 1109–1111. DOI: 10.1086/502351.
120. Procházka, A.; Charvátová, H.; Vyšata, O.; Kopal, J.; Chambers, J. Breathing Analysis Using Thermal and Depth Imaging Camera Video Records. *Sensors* **2017**, *17* (6), 1408–1408. DOI: 10.3390/s17061408.
121. Ahmadi, N.; Nabavi, V.; Nuguri, V.; Hajsadeghi, F.; Flores, F.; Akhtar, M.; Kleis, S.; Hecht, H.; Naghavi, M.; Budoff, M. Low Fingertip Temperature Rebound Measured by Digital Thermal Monitoring Strongly Correlates with the Presence and Extent of Coronary Artery Disease Diagnosed by 64-slice Multi-Detector Computed Tomography. *Int. J. Cardiovasc. Imag.* **2009**, *25* (7), 725–738. DOI: 10.1007/s10554-009-9476-8.
122. Singh, M.; Flores, H.; Ma, R.; Zhu, L. In *Extraction Of Baseline Blood Perfusion Rates in Mouse Body and Implanted PC3 Tumor Using Infrared Images And Theoretical Simulation*, Summer Biomechanics, Bioengineering, and Biotransport Conference, Virtual Meeting, May 17, 2022; p 2.
123. Fani, F.; Schena, E.; Saccomandi, P.; Silvestri, S. CT-Based Thermometry: An Overview. *Int. J. Hyperthermia* **2014**, *30* (4), 219–227. DOI: 10.3109/02656736.2014.922221.
124. Rodrigues, D. B.; Maccarini, P. F.; Salahi, S.; Oliveira, T. R.; Pereira, P. J. S.; Limao-Vieira, P.; Snow, B. W.; Reudink, D.; Stauffer, P. R. Design and Optimization of an Ultra Wideband and Compact Microwave Antenna for Radiometric Monitoring of Brain Temperature. *IEEE Trans. Biomed. Eng.* **2014**, *61* (7), 2154–2160. DOI: 10.1109/TBME.2014.2317484.
125. Ebbini, E. S. Ultrasound Thermography: Principles, Methods, and Experimental Results. In *Theory and Applications of Heat Transfer in Humans*; John Wiley & Sons Ltd, 2018; pp 301–332.
126. Carter, D. L.; MacFall, J. R.; Clegg, S. T.; Wan, X.; Prescott, D. M.; Charles, H. C.; Samulski, T. V. Magnetic Resonance Thermometry During Hyperthermia for Human High-Grade Sarcoma. *Int. J. Radiat. Oncol. Biol. Phys.* **1998**, *40* (4), 815–822. DOI: 10.1016/S0360-3016(97)00855-9.
127. Puccini, S.; Bär, N.-K.; Bublat, M.; Kahn, T.; Busse, H. Simulations of Thermal Tissue Coagulation and Their Value for the Planning and Monitoring of Laser-Induced Interstitial Thermotherapy (LITT). *Magn. Reson. Med.* **2003**, *49* (2), 351–362. DOI: 10.1002/mrm.10357.
128. Dehkharghani, S.; Qiu, D. MR Thermometry in Cerebrovascular Disease: Physiologic Basis, Hemodynamic Dependence, and a New Frontier in Stroke Imaging. *Am. J. Neuroradiol.* **2020**, *41* (4), 555–565. DOI: 10.3174/ajnr.A6455.
129. Rieke, V.; Butts Pauly, K. MR Thermometry. *J. Magn. Reson. Imaging* **2008**, *27* (2), 376–390. DOI: 10.1002/jmri.21265.
130. Odéen, H.; Parker, D. L. Non-Invasive Thermometry with Magnetic Resonance Imaging. In *Theory and Applications of Heat Transfer in Humans*; John Wiley & Sons Ltd, 2018; pp 267–299.
131. Datta, N. R.; Ordóñez, S. G.; Gaipl, U. S.; Paulides, M. M.; Crezee, H.; Gellermann, J.; Marder, D.; Puric, E.; Bodis, S. Local Hyperthermia Combined with Radiotherapy and-/or Chemotherapy: Recent Advances and Promises for the Future. *Cancer Treat. Rev.* **2015**, *41* (9), 742–753.
132. van der Zee, J.; González, D.; van Rhoon, G. C.; van Dijk, J. D. P.; van Putten, W. L. J.; Hart, A. A. M.; et al. Comparison of Radiotherapy Alone with Radiotherapy Plus

Hyperthermia in Locally Advanced Pelvic Tumours: A Prospective, Randomised, Multicentre Trial. *Lancet* **2000**, *355* (9210), 1119–1125.

133. Barsukov, Y. A.; Gordeyev, S. S.; Tkachev, S. I.; Fedyanin, M. Y.; Perevoshikov, A. G. Phase II Study of Concomitant Chemoradiotherapy with Local Hyperthermia and Metronidazole for Locally Advanced Fixed Rectal Cancer. *Colorectal Dis.* **2013**, 15 (9), 1107–1114. DOI: 10.1111/codi.12281.
134. Paulides, M. M.; Dobsicek Trefna, H.; Curto, S.; Rodrigues, D. B. Recent Technological Advancements in Radiofrequency- and Microwave-Mediated Hyperthermia for Enhancing Drug Delivery. *Adv. Drug Delivery Rev.* **2020**, *163–164*, 3–18. DOI: 10.1016/j.addr.2020.03.004.
135. Gracefairchildfisher. In *Ultrasound Therapy*, [Online] 2023. https://fr.wikipedia.org/wiki/Ultrasonothérapie (accessed Mar 10, 2023).
136. Boyse, W. E.; Lynch, D. R.; Paulsen, K. D.; Minerbo, G. N. Nodal-Based Finite-Element Modeling of Maxwell's Equations. *IEEE Trans. Antennas Propag.* **1992**, *40* (6), 642–651.
137. Rodrigues, D. B.; Hurwitz, M. D.; Maccarini, P. F.; Stauffer, P. R. Optimization of Chest Wall Hyperthermia Treatment Using a Virtual Human Chest Model. In *2015 9th European Conference on Antennas and Propagation (EuCAP)*; IEEE, 2015; pp 1–5.
138. Ekstrand, V.; Wiksell, H.; Schultz, I.; Sandstedt, B.; Rotstein, S.; Eriksson, A. Influence of Electrical and Thermal Properties on RF Ablation of Breast Cancer: Is the Tumour Preferentially Heated? *Biomed. Eng. Online* **2005**, *4*, 41–41. DOI: 10.1186/ 1475-925X-4-41.
139. Singh, S.; Repaka, R. Temperature-Controlled Radiofrequency Ablation of Different Tissues Using Two-compartment Models. *Int. J. Hyperthermia* **2017**, *33* (2), 122–134. DOI: 10.1080/02656736.2016.1223890.
140. Cavagnaro, M.; Amabile, C.; Bernardi, P.; Pisa, S.; Tosoratti, N. A Minimally Invasive Antenna for Microwave Ablation Therapies: Design, Performances, and Experimental Assessment. *IEEE Trans. Biomed. Eng.* **2011**, *58* (4), 949–959. DOI: 10.1109/TBME. 2010.2099657.
141. Wu, T.; Li, P.; Shao, Q.; Hong, J.; Yang, L.; Wu, S. A Simulation-Experiment Method to Characterize the Heat Transfer in Ex-Vivo Porcine Hepatic Tissue with a Realistic Microwave Ablation System. *Numer. Heat Transf. A Appl.* **2013**, *64* (9), 729–743. DOI: 10.1080/10407782.2013.798202.
142. Huang, J.; Holt, R. G.; Cleveland, R. O.; Roy, R. A. Experimental Validation of a Tractable Numerical Model for Focused Ultrasound Heating in Flow-Through Tissue Phantoms. *J. Acoust. Soc. Am.* **2004**, *116* (4), 2451–2458. DOI: 10.1121/1.1787124.
143. Bhowmik, A.; Repaka, R.; Mishra, S. C.; Mitra, K. Thermal Assessment of Ablation Limit of Subsurface Tumor During Focused Ultrasound and Laser Heating. *J. Therm. Sci. Eng. Appl.* **2016**, *8* (1). DOI: 10.1115/1.4030731.
144. Welch, A. The Thermal Response of Laser Irradiated Tissue. *IEEE J. Quantum Electron.* **1984**, *20* (12), 1471–1481. DOI: 10.1109/JQE.1984.1072339.
145. Bhowmik, A.; Repaka, R.; Mishra, S. C.; Mitra, K. Analysis of Radiative Signals from Normal and Malignant Human Skins Subjected to a Short-Pulse Laser. *Int. J. Heat Mass Transf.* **2014**, *68*. DOI: 10.1016/j.ijheatmasstransfer.2013.09.032.
146. Jaunich, M.; Raje, S.; Kim, K.; Mitra, K.; Guo, Z. Bio-Heat Transfer Analysis During Short Pulse Laser Irradiation of Tissues. *Int. J. Heat Mass Transf.* **2008**, *51* (23-24), 5511–5521. DOI: 10.1016/j.ijheatmasstransfer.2008.04.033.
147. Dombrovsky, L. A.; Timchenko, V.; Jackson, M.; Yeoh, G. H. A Combined Transient Thermal Model for Laser Hyperthermia of Tumors with Embedded Gold

Nanoshells. *Int. J. Heat Mass Transf.* **2011**, *54* (25-26), 5459–5469. DOI: 10.1016/j.ijheatmasstransfer.2011.07.045.
148. Bohren, C. F.; Huffman, D. R. In *Absorption and Scattering of Light by Small Particles*; John Wiley & Sons, 2008.
149. Hanson, G. W.; Patch, S. K. Optimum Electromagnetic Heating of Nanoparticle Thermal Contrast Agents at rf Frequencies. *J. Appl. Phys.* **2009**, *106* (5), 54309–54309.
150. de Hulst, H. C. V. In *Light Scattering by Small Particles*; Dover Publications, Inc., 1981.
151. Maenosono, S.; Saita, S. Theoretical Assessment of FePt Nanoparticles as Heating Elements for Magnetic Hyperthermia. *IEEE Trans. Magn.* **2006**, *42* (6), 1638–1642. DOI: 10.1109/TMAG.2006.872198.
152. Kaddi, C. D.; Phan, J. H.; Wang, M. D. Computational Nanomedicine: Modeling of Nanoparticle-Mediated Hyperthermal Cancer Therapy. *Nanomedicine* **2013**, *8* (8), 1323–1333. DOI: 10.2217/nnm.13.117.
153. Singh, M. Biological Heat and Mass Transport Mechanisms Behind Nanoparticles Migration Revealed Under MicroCT Image Guidance. *Int. J. Therm. Sci.* **2023**, *184*, 107996.
154. LeBrun, A.; Joglekar, T.; Bieberich, C.; Ma, R.; Zhu, L. Identification of Infusion Strategy for Achieving Repeatable Nanoparticle Distribution and Quantification of Thermal Dosage Using Micro-CT Hounsfield Unit in Magnetic Nanoparticle Hyperthermia. *Int. J. Hyperthermia* **2016**, *32* (2), 132–143.
155. Singh, M.; Ma, R.; Zhu, L. Quantitative Evaluation of Effects of Coupled Temperature Elevation, Thermal Damage, and Enlarged Porosity on Nanoparticle Migration in Tumors During Magnetic Nanoparticle Hyperthermia. *Int. Commun. Heat Mass Transf.* **2021**, *126*, 105393.
156. Singh, S.; Bhowmik, A.; Repaka, R. Thermal Analysis of Induced Damage to the Healthy Cell During RFA of Breast Tumor. *J. Therm. Biol.* **2016**, *58*, 80–90.
157. Schutt, D. J.; Haemmerich, D. Effects of Variation in Perfusion Rates and of Perfusion Models in Computational Models of Radio Frequency Tumor Ablation. *Med. Phys.* **2008**, *35* (8), 3462–3470.
158. El Kaffas, A.; Hoogi, A.; Zhou, J.; Durot, I.; Wang, H.; Rosenberg, J.; Tseng, A.; Sagreiya, H.; Akhbardeh, A.; Rubin, D. L. Spatial Characterization of Tumor Perfusion Properties from 3D DCE-US Perfusion Maps are Early Predictors of Cancer Treatment Response. *Sci. Rep.* **2020**, *10* (1), 1–11.
159. Singh, M. Heterogeneous Blood Perfusion Mapping in Cancerous Lesions. In *Summer Biomechanics, Bioengineering, and Biotransport Conference*, Virtual Meeting, June 14, 2021; p 2.
160. Smith, R. A.; Jones, F. E.; Chasmar, R. P. *The Detection and Measurement of Infra-Red Radiation*; Oxford University Press, 1968.
161. Iso. *Liquid-in-glass Laboratory Thermometers—Principles of Design, Construction and Use*, [Online] 1977. https://www.iso.org/standard/4370.html.
162. Nicholas, J. V. Liquid-In-Glass Thermometers. In *Measurement Instrumentation and Sensors Handbook*; Webster, J. G., Ed.; CRC Press LLC, 1999; pp 32–117.
163. Burns, J. Resistive Thermometers. In *Measurement Instrumentation and Sensors Handbook;* Webster, J. G., Ed.; CRC Press, 1999; pp 32–32.
164. Sapoff, M. Thermistor Thermometers. In *Measurement Instrumentation and Sensors Handbook*; Webster, J. G., Ed.; CRC Press, 1999; pp 32–32.
165. Reed, R. P. Thermocouple Thermometers. In *Measurement Instrumentation and Sensors Handbook*; Webster, J. G., Ed.; CRC Press, 1999; pp 32–32.

166. Culshaw, B. Fiber-Optic Thermometers. In *Measurement Instrumentation and Sensors Handbook*; Webster, J. G., Ed.; CRC Press, 1999; pp 32–32.
167. Lugoda, P.; Costa, J. C.; Oliveira, C.; Garcia-Garcia, L. A.; Wickramasinghe, S. D.; Pouryazdan, A.; Roggen, D.; Dias, T.; Münzenrieder, N. Flexible Temperature Sensor Integration into E-Textiles Using Different Industrial Yarn Fabrication Processes. *Sensors* **2019**, *20* (1), 73–73. DOI: 10.3390/s20010073.
168. Goumopoulos, C. A High Precision, Wireless Temperature Measurement System for Pervasive Computing Applications. *Sensors* **2018**, *18* (10), 3445–3445. DOI: 10.3390/s18103445.
169. Roriz, P.; Silva, S.; Frazão, O.; Novais, S. Optical Fiber Temperature Sensors and Their Biomedical Applications. *Sensors* **2020**, *20* (7), 2113–2113. DOI: 10.3390/s20072113.
170. Swain, I. D.; Grant, L. J. Methods of Measuring Skin Blood Flow. *Phys. Med. Biol.* **1989**, *34* (2), 151–175. DOI: 10.1088/0031-9155/34/2/001.
171. Daly, S. M.; Leahy, M. J. 'Go With the Flow': A Review of Methods and Advancements in Blood Flow Imaging. *J. Biophotonics* **2013**, *6* (3), 217–255. DOI: 10.1002/jbio.201200071.
172. Boas, D. A.; Dunn, A. K. Laser Speckle Contrast Imaging in Biomedical Optics. *J. Biomed. Opt.* **2010**, *15* (1), 011109–011109. DOI: 10.1117/1.3285504.
173. Hasgall, P. A.; Di Gennaro, F.; Baumgartner, C.; Neufeld, E.; Lloyd, B.; Gosselin, M. C.; Payne, D.; Klingenböck, A.; Kuster, N. IT'IS Database for Thermal and Electromagnetic Parameters of Biological Tissues, 2018.
174. Cooper, T. E.; Trezek, G. J. A Probe Technique for Determining the Thermal Conductivity of Tissue. *J. Heat Transf.* **1972**, *94* (2), 133–140. DOI: 10.1115/1.3449883.
175. Bowman, H. F. Heat Transfer and Thermal Dosimetry. *J. Microw. Power* **1981**, *16* (2), 121–133. DOI: 10.1080/16070658.1981.11689231.
176. Chato, J. C. Fundamentals of Bioheat Transfer. In *Thermal Dosimetry and Treatment Planning*; Gautherie, M., Ed.; Springer-Verlag: NY, 1990; pp 51–51.
177. Lang, J.; Erdmann, B.; Seebass, M. Impact of Nonlinear Heat Transfer on Temperature Control in Regional Hyperthermia. *IEEE Trans. Biomed. Eng.* **1999**, *46* (9), 1129–1138. DOI: 10.1109/10.784145.
178. Gautherie, M.; Qenneville, Y.; Gros, C. H. Thermogenesis of Mammary Epitheliomas. III. Study, by Means of Fluvography, of the Termal Conductivity of Mammary Tissue and of the Influence of Tumor Vascularization. *Biomedicine* **1975**, *22* (3), 237–245.
179. Cohen, M. L. Measurement of the Thermal Properties of Human Skin. A Review. *J. Invest. Dermatol.* **1977**, *69* (3), 333–338. DOI: 10.1111/1523-1747.ep12507965.
180. Dumas, A.; Barozzi, G. S. Laminar Heat Transfer to Blood Flowing in a Circular Duct. *Int. J. Heat Mass Transf.* **1984**, *27* (3), 391–398. DOI: 10.1016/0017-9310(84)90286-2.
181. Balasubramaniam, T. A.; Bowman, H. F. Thermal Conductivity and Thermal Diffusivity of Biomaterials: A Simultaneous Measurement Technique. *J. Biomech. Eng.* **1977**, *99* (3), 148–148. DOI: 10.1115/1.3426282.
182. Valvano, J. W.; Cochran, J. R.; Diller, K. R. Thermal Conductivity and Diffusivity of Biomaterials Measured with Self-Heated Thermistors. *Int. J. Thermophys.* **1985**, *6* (3), 301–311. DOI: 10.1007/BF00522151.
183. Valvano, J. W. Tissue Thermal Properties and Perfusion. In *Optical-Thermal Response of Laser-Irradiated Tissue*; Springer: US, 1995; pp 445–488.
184. Thurau, K.; Levine, D. Z. The Renal Circulation. In *The Kidney: Morphology, Biochemistry, Physiology*; Rouiller, C., Mulle, r. A. F., Eds.; Academic Press, Inc., 1971; pp 1–70.

185. Sekins, K. M.; Dundore, D.; Emery, A. F.; Lehmann, J. F.; McGrath, P. W.; Nelp, W. B. Muscle Blood Flow Changes in Response to 915 MHz Diathermy with Surface Cooling as Measured by Xe133 Clearance. *Arch. Phys. Med. Rehab.* **1980**, *61* (3), 105–113.
186. Lassen, N. A.; Lindbjerg, J.; Munck, O. Measurement of Blood-Flow Through Skeletal Muscle by Intramuscular Injection of Xenon-133. *Lancet* **1964**, *283* (7335), 686–689. DOI: 10.1016/S0140-6736(64)91518-1.
187. Keele, C. A.; Neil, E. In *Samson Wright's Applied Physiology*; Oxford Press, 1971.
188. Johnson, J. M.; Brengelmann, G. L.; Hales, J. R.; Vanhoutte, P. M.; Wenger, C. B. Regulation of the Cutaneous Circulation. *Fed. Proc.* **1986**, *45* (13), 2841–2850.
189. Yoon, G. Absorption and Scattering of Laser Light in Biological Media-Mathematical Modeling and Methods for Determining Optical Properties, 1988.
190. Wilson, B. C. Measurement of Tissue Optical Properties: Methods and Theories. In *Optical-Thermal Response of Laser-Irradiated Tissue*; Springer: US, 1995; pp 233–303.
191. Keijzer, M.; Richards-Kortum, R. R.; Jacques, S. L.; Feld, M. S. Fluorescence Spectroscopy of Turbid Media: Autofluorescence of the Human Aorta. *Appl. Opt.* **1989**, *28* (20), 4286–4286. DOI: 10.1364/AO.28.004286.
192. Splinter, R.; Cheong, W. F.; van Gemert, M. J. C.; Welch, A. J. In Vitro Optical Properties of Human and Canine Brain and Urinary Bladder Tissues at 633 nm. *Lasers Surg. Med.* **1989**, *9* (1), 37–41. DOI: 10.1002/lsm.1900090108.
193. Flock, S. T.; Wilson, B. C.; Patterson, M. S. Total Attenuation Coefficients and Scattering Phase Functions of Tissues and Phantom Materials at 633 nm. *Med. Phys.* **1987**, *14* (5), 835–841. DOI: 10.1118/1.596010.
194. Reynolds, L.; Johnson, C.; Ishimaru, A. Diffuse Reflectance from a Finite Blood Medium: Applications to the Modeling of Fiber Optic Catheters. *Appl. Opt.* **1976**, *15* (9), 2059–2059. DOI: 10.1364/AO.15.002059.
195. Pedersen, G. D.; McCormick, N. J.; Reynolds, L. O. Transport Calculations for Light Scattering in Blood. *Biophys. J.* **1976**, *16* (3), 199–207. DOI: 10.1016/S0006-3495(76)85681-0.
196. Wilson, B. C.; Muller, P. J.; Yanch, J. C. Instrumentation and Light Dosimetry for Intra-Operative Photodynamic Therapy (PDT) of Malignant Brain Tumours. *Phys. Med. Biol.* **1986**, *31* (2), 125–133. DOI: 10.1088/0031-9155/31/2/002.
197. Svaasand, L. O.; Ellingsen, R. Optical Properties of Human Brain. *Photochem. Photobiol.* **1983**, *38* (3), 293–299. DOI: 10.1111/j.1751-1097.1983.tb02674.x.
198. Sterenborg, H. J. C. M.; Gemert, M. J. C. V.; Kamphorst, W.; Wolbers, J. G.; Hogervorst, W. The Spectral Dependence of the Optical Properties of Human Brain. *Lasers Med. Sci.* **1989**, *4* (4), 221–227. DOI: 10.1007/BF02032451.
199. Key, H.; Jackson, P. C.; Wells, P. N. T. Light Scattering and Propagation in Tissue. In *World Congress on Medical Physics and Biomedical Engineering*; San Antonio: TX, 1988.
200. Marchesini, R.; Bertoni, A.; Andreola, S.; Melloni, E.; Sichirollo, A. E. Extinction and Absorption Coefficients and Scattering Phase Functions of Human Tissues In Vitro. *Appl. Opt.* **1989**, *28* (12), 2318–2318. DOI: 10.1364/AO.28.002318.
201. Eichler, J.; Knof, J.; Lenz, H. Measurements on the Depth of Penetration of Light (0.35–1.0 μm) in Tissue. *Radiat. Environ. Biophys.* **1977**, *14* (3), 239–242. DOI: 10.1007/BF01323942.

202. Doiron, D. R.; Svaasand, L. O.; Profio, A. E. Wavelength And Dosimetry Considerations In Photoradiation Therapy (PRT). In *Laser in Medicine and Surgery*, 1982/12//; Goldman, L., Ed.; 1982; pp 48–52. DOI: 10.1117/12.976072.
203. Qu, J.; MacAulay, C.; Lam, S.; Palcic, B. Optical Properties of Normal and Carcinomatous Bronchial Tissue. *Appl. Opt.* **1994**, *33* (31), 7397–7397. DOI: 10.1364/AO.33.007397.

APPENDIX 5A

Thermophysical Properties of Human Tissue and Biofluids

TABLE 5.5 Detailed Database of Thermophysical Properties (Mean Value) of Biofluid and Tissue Given by IT'IS Foundation.

Tissue	ρ (kg/m³)	c (J/kg/°C)	k (W/m/°C)	ω_b (mL/min/kg)	Q_m (W/kg)
Adrenal Gland	1028	3513	0.44	1458	22.58
Air	1	1004	0.03	-	-
Bile	928	4037	0.58	-	-
Blood	1050	3617	0.52	10,000	-
Blood Plasma	1020	3930	0.58	-	-
Blood Serum	1024	-	-	-	-
Blood Vessel Wall	1102	3306	0.46	150	2.32
Bone (Cancellous)	1178	2274	0.31	30	0.46
Bone (Cortical)	1908	1313	0.32	10	0.15
Bone Marrow (Red)	1029	2666	0.28	135	2.09
Bone Marrow (Yellow)	980	2065	0.19	30	0.46
Brain	1046	3630	0.51	559	11.37
Brain (Grey Matter)	1045	3696	0.55	764	15.54
Brain (White Matter)	1041	3583	0.48	212	4.32
Breast Fat	911	2348	0.21	47	0.73
Breast Gland	1041	2960	0.33	150	2.32
Bronchi	1102	3306	0.46	238	3.69
Bronchi lumen	1	1004	0.03	-	-
Cartilage	1100	3568	0.49	35	0.54
Cerebellum	1045	3653	0.51	770	15.67
Cerebrospinal Fluid	1007	4096	0.57	-	-
Cervix	1105	3676	0.53	700	10.84
Commissura Anterior	1041	3583	0.48	212	4.32
Commissura Posterior	1041	3583	0.48	212	4.32
Connective Tissue	1027	2372	0.39	37	0.58
Diaphragm	1090	3421	0.49	99	2.44
Ductus Deferens	1102	3306	0.46	188	2.91
Dura	1174	3364	0.44	380	5.89
Epididymis	1082	3778	0.52	200	3.09
Esophagus	1040	3500	0.53	190	2.94

APPENDIX 5A *(Continued)*

Tissue	ρ (kg/m³)	c (J/kg/°C)	k (W/m/°C)	ω_b (mL/min/kg)	Q_m (W/kg)
Esophagus Lumen	1	1004	0.03	-	-
Eye (Aqueous Humor)	994	3997	0.58	-	-
Eye (Choroid)	1050	3617	0.52	5950	92.15
Eye (Ciliary Body)	1090	3421	0.49	3800	93.72
Eye (Cornea)	1062	3615	0.54	-	-
Eye (Iris)	1090	3421	0.49	1632	40.26
Eye (Lens)	1076	3133	0.43	-	-
Eye (Retina)	1036	3696	0.55	240	4.89
Eye (Sclera)	1032	4200	0.58	380	5.89
Eye (Vitreous Humor)	1005	4047	0.59	-	-
Eye Lens (Cortex)	1076	3133	0.43	-	-
Eye Lens (Nucleus)	1076	3133	0.43	-	-
Fat	911	2348	0.21	33	0.51
Fat (Average Infiltrated)	911	2348	0.21	33	0.51
Fat (Not Infiltrated)	911	2348	0.21	-	-
Gallbladder	1071	3716	0.52	30	0.46
Heart Lumen	1050	3617	0.52	10,000	0.00
Heart Muscle	1081	3686	0.56	1026	39.45
Hippocampus	1045	3696	0.55	764	15.54
Hypophysis	1053	3687	0.51	885	13.71
Hypothalamus	1045	3696	0.55	764	15.54
Intervertebral Disc	1100	3568	0.49	35	0.54
Kidney	1066	3763	0.53	3795	18.05
Kidney (Cortex)	1049	3587	0.53	3874	18.43
Kidney (Medulla)	1044	3745	0.54	599	2.85
Large Intestine	1088	3655	0.54	765	11.85
Large Intestine Lumen	1045	3801	0.56	-	-
Larynx	1100	3568	0.49	35	0.54
Liver	1079	3540	0.52	860	9.93
Lung	394	3886	0.39	401	6.21
Lung (Deflated)	1050	3886	0.39	401	6.21
Lung (Inflated)	394	3886	0.39	401	6.21
Lymph	1019	-	-	-	-
Lymphnode	1035	3739	0.46	461	7.15
Mandible	1908	1313	0.32	10	0.15

APPENDIX 5A *(Continued)*

Tissue	ρ (kg/m³)	c (J/kg/°C)	k (W/m/°C)	ω_b (mL/min/kg)	Q_m (W/kg)
Medulla Oblongata	1046	3630	0.51	559	11.37
Meniscus	1100	3568	0.49	35	0.00
Midbrain	1046	3630	0.51	559	11.37
Mucous Membrane	1102	3150	0.34	594	9.19
Muscle	1090	3421	0.49	37	0.91
Nerve	1075	3613	0.49	160	2.48
Ovary	1048	3778	0.52	236	3.65
Pancreas	1087	3164	0.51	767	11.89
Penis	1102	3306	0.46	12	0.19
Pharynx	1	1004	0.03	-	-
Pineal Body	1053	3687	0.51	885	13.71
Placenta	1018	3807	0.52	1700	26.33
Pons	1046	3630	0.51	559	11.37
Prostate	1045	3760	0.51	394	6.10
Subcutaneous fat	911	2348	0.21	33	0.51
Salivary Gland	1048	3760	0.51	383	5.93
Seminal vesicle	1045	3760	0.51	394	6.10
Skin	1109	3391	0.37	106	1.65
Skull Cancellous	1178	2274	0.31	30	0.46
Skull Cortical	1908	1313	0.32	10	0.15
Small Intestine	1030	3595	0.49	1026	15.89
Small Intestine Lumen	1045	3801	0.56	-	-
Spinal Cord	1075	3630	0.51	160	2.48
Spleen	1089	3596	0.53	1557	24.11
Stomach	1088	3690	0.53	460	7.13
Stomach Lumen	1045	3801	0.56	-	-
Tendon\Ligament	1142	3432	0.47	29	0.45
Testis	1082	3778	0.52	200	3.09
Thalamus	1045	3696	0.55	685	13.93
Thymus	1023	3043	0.34	247	3.83
Thyroid Gland	1050	3609	0.52	5624	87.10
Tongue	1090	3421	0.49	78	1.21
Tooth	2180	1255	0.59	-	-
Tooth (Dentine)	2063	1338	0.52	-	-
Tooth (Enamel)	2958	711	0.79	-	-

APPENDIX 5A *(Continued)*

Tissue	ρ (kg/m³)	c (J/kg/°C)	k (W/m/°C)	ω_b (mL/min/kg)	Q_m (W/kg)
Trachea	1080	3568	0.49	35	0.54
Trachea Lumen	1	1004	0.03	-	-
Ureter\Urethra	1102	3306	0.46	188	2.91
Urinary Bladder Wall	1086	3581	0.52	78	1.21
Urine	1024	4178	0.56	-	-
Uterus	1105	3676	0.53	458	7.09
Vagina	1088	3655	0.54	98	1.52
Vertebrae	1908	1313	0.32	10	0.15
Water	994	4178	0.60	-	-

The statistical information (i.e., standard deviation and spread of individual tissue properties) can be obtained from the database.

Source: Adapted from Ref. [173].

TABLE 5.6 Thermal Conductivity, Thermal Diffusivity, and Specific Heat of Normal Un-perfused Human Brain Tissue.

Tissue	T (°C)	k (W/m/K)	ρ (kg/m³)	α (= $k/\rho c$) (10^{-3} cm²/s)	c (J/kg/K)	ω_b (kg/s/m³)	Refs
Whole brain	5–20	0.528	–	1.38 ± 0.11	–	–	[174, 175]
	37	0.503–0.576	–	–		–	
Cerebral cortex	5–20	0.565	–	1.43 ± 0.09	3680	–	[174]
White matter	5–20	0.565	–	1.34 ± 0.10	3600	–	[174]
Bone	–	0.410–0.630	–	–	–	–	[176]
Fat	–	0.210	900	–	3500	ω_{fat}^*	[177]
Benign tumor	–	0.642	1000	–	3500	ω_{tum}^*	[177]
Malignant	–	0.280	1000	–	3500	ω_{tum}^*	[177,178]
Bladder	–	0.600	1000	–	3500	5.0×10^{-6}	–do–
Kidney	–	0.577	1000	–	3500	66.67×10^{-6}	–do–
Liver	–	0.640	1000	–	3500	6.67×10^{-6}	–do–
Muscle	–	0.642	1000	–	3500	ω_{mus}^*	–do–
Intestine	–	0.550	1000	–	3500	3.333×10^{-6}	–do–
Aorta	–	0.506	1000	–	3500	83.33×10^{-6}	–do–
Dermis (skin)	–	0.293–0.322	–	–	–	–	[179]
Epidermis (skin)	–	0.209	–	–	–	–	[179]
Whole blood	–	0.505	–	–	–	–	[180,
		0.492 ± 0.009	–	–	–	–	181]

TABLE 5.6 *(Continued)*

Tissue	T (°C)	k (W/m/K)	ρ (kg/m³)	α (= $k/\rho c$) (10^{-3} cm²/s)	c (J/kg/K)	ω_b (kg/s/m³)	Refs
Plasma	–	0.582	–	–	–	–	[180,
		0.570 ± 0.010	–	–	–	–	181]

$$^{*}\omega_{fat} = \begin{cases} 0.36\left[1+e^{-\frac{(T-45)^2}{12}}\right] & T \le 45\,^{\circ}C \\ 0.72 & T > 45\,^{\circ}C \end{cases}, \quad \omega_{mus} = \begin{cases} 0.45\left[1+7.88e^{-\frac{(T-45)^2}{12}}\right] & T \le 45\,^{\circ}C \\ 4.0 & T > 45\,^{\circ}C \end{cases}$$

$$\text{And: } \omega_{tum} = \begin{cases} 0.833 & T < 37\,^{\circ}\text{C} \\ 0.833 - \dfrac{(T-37)^{4.8}}{5.483 \times 10^3} & 42 \le T < 37\,^{\circ}\text{C} \\ 0.416 & T > 42\,^{\circ}\text{C} \end{cases}$$

TABLE 5.7 Temperature-dependent thermal conductivity and thermal diffusivity of normal and cancerous human tissue was given by Ref. [182,183] based on *in vitro* measurements of tissue properties taken at 3, 10, 17, 23, 30, 37, and 45°C.

Species	Tissue	k_0 (W/m/K)	k_1 (W/m/°C)	α_0 (mm²/s)	α_1 (mm²/s/°C)
	$k_t = k_0 + k_1 T$ (in°C) and $\alpha_t = k_t/\rho_t c_t = \alpha_0 + \alpha_1 T$ (in°C)				
Human	Renal pelvis	0.4795	0.001923	0.1329	0.00011
Human	Renal medulla	0.4994	0.001102	0.1278	0.00055
Human	Renal cortex	0.4989	0.001288	0.1266	0.00055
Human	Myocardium	0.4925	0.001195	0.1289	0.00050
Human	Pancreas	0.4365	0.002844	0.1391	0.00084
Human	Lung	0.4071	0.001176	0.1192	0.00031
Human	Liver	0.4692	0.001161	0.1279	0.00036
Human	Spleen	0.4913	0.001300	0.1270	0.00047
Human	Fat of Spleen	0.3431	−0.000254	0.1321	−0.00002
Human	Cerebral cortex	0.5043	0.000296	0.1283	0.00050
Human	Adenocarcinoma (Breast)	0.4194	0.003911	0.1617	−0.00049
Human	Colon cancer	0.5450	(at 19°C)	0.1340	-

Thermal conductivity and diffusivity of human arterial tissue

Species	Tissue	k_t (W/m/°C)[a]				α_t (10^{-3} cm²/s)[a]			
		35°C	55°C	75°C	90°C	35°C	55°C	75°C	90°C
Human	Normal aorta	0.476 (0.041)	0.503 (0.060)	0.559 (0.037)	0.612 (0.012)	1.27 (0.07)	1.33 (0.11)	1.44 (0.10)	1.56 (0.05)
Human	Fatty plaque aorta	0.484 (0.041)	0.497 (0.049)	0.546 (0.054)	0.081 (0.577)	1.28 (0.05)	1.32 (0.06)	1.41 (0.11)	1.46 (0.15)
Human	Fibrous plaque aorta	0.485 (0.010)	0.507 (0.003)	0.538 (0.007)	0.577 (0.056)	1.29 (0.22)	1.36 (0.30)	1.411 (0.38)	1.52 (0.20)
Human	Calcified plaque aorta	0.502 (0.059)	0.526 (0.073)	0.581 (0.082)	0.619 (0.085)	1.32 (0.07)	1.37 (0.12)	1.53 (0.17)	1.66 (0.20)

[a]These values represent mean (standard deviation).

Similarly, properties for human arterial tissue were measured by Ref. [183] at 35, 55, 75, and 90°C.

TABLE 5.8 Measured Value of Blood Perfusion of Human Tissue (mL/min/100 gm).[183]

Tissue	ω_b (mL/min/100 gm)	Standard deviation	Refs
Cortex (Kidney)	400–500	–	[184]
Outer medulla (Kidney)	120	–	-do-
Inner medulla (Kidney)	25	–	-do-
Resting thigh muscle (1.5 cm depth)	1.8	1.1	[185]
Resting thigh muscle (3.0 cm depth)	2.6	1.3	[185]
Resting thigh muscle	2.0–2.2	–	[186]
Skeletal muscle, average	2.7	–	[187]
Forearm (in cold)	2.0		[188]
Forearm (in thermoneutral)	4–5		[188]
Forearm (in hyperthermic)	> 20		[188]

APPENDIX 5B

Optical Properties of Human Tissue

TABLE 5.9 Optical Properties of Normal and Abnormal Human Tissues for *In Vitro*.

Tissue	λ(nm)	μ_t (cm^{-1})	μ_a (cm^{-1})	μ_s (cm^{-1})	μ_s (1−g) (cm^{-1})	G	μ_{eff}(cm^{-1})	Refs.
Aorta								
	632.8	316	0.52	316	41	0.87	–	[189]
	1060	–	2.0	–	–		–	[190]
Intima (Aorta)								
	476	252	14.8	237	45	0.81	–	[191]
	580	191	8.9	183	34.8	0.81	–	-do-
	600	182	4.0	178	33.8	0.81	–	-do-
	633	175	3.6	171	25.7	0.81	–	-do-
Media (Aorta)								
	476	252	7.3	410	45.1	0.89	–	-do-
	580	191	4.8	331	33.1	0.90	–	-do-
	600	182	2.5	323	35.5	0.89	–	-do-
	633	312	2.3	310	31.0	0.90	–	-do-
Adventitia (Aorta)								
	476	252	18.1	267	69.4	0.74	–	-do-
	580	191	11.3	217	49.9	0.77	–	-do-
	600	182	6.1	211	46.4	0.78	–	-do-
	633	201	5.8	195	37.1	0.81	–	-do-

APPENDIX 5B *(Continued)*

Tissue	λ(nm)	μ_t (cm^{-1})	μ_a (cm^{-1})	μ_s (cm^{-1})	μ_s (1− g) (cm^{-1})	G	μ_{eff}(cm^{-1})	Refs.
Bladder								
	633	30.7	1.40	29.3	2.64	0.91	–	[192]
Blood								
Whole	633	29.0	–	–	–	0.974	–	[193]
HbO_2, Hct = 0.41	665	1247	1.30	1246	6.11	0.995	–	[194]
	685	1416	2.65	1413	–	0.99	–	[195]
	960	508	2.84	505	3.84	0.992	–	[194]
Normal brain (Whole)								
	630	–	–		–	–	2.2–3.7	[196]
	630	–	–		–	–	4.8–10.0	-do-
Adult brain (Whole)								
	488	–	–		–	–	14–25	[197]
	514	–	–		–	–	14–16.7	-do-
	630	–	0.3–1.0	–	30–40	–	8.3	[198]
	660	–	–		–	–	7.0–12.5	-do-
	1060	–	–		–	–	2.3–3.4	-do-
Neonates brain (Whole)								
	488	–	–		–	–	5.9–7.9	[197]
	514	–	–		–	–	5.8–9.0	-do-
	660	–	–		–	–	2.5–3.3	-do-
	1060	–	–		–	–	1.1–1.4	-do-

APPENDIX 5B *(Continued)*

Tissue	λ(nm)	μ_t (cm^{-1})	μ_a (cm^{-1})	μ_s (cm^{-1})	μ_s (1− *g*) (cm^{-1})	*G*	μ_{eff}(cm^{-1})	Refs.
Brain white matter	633	52.6	1.58	51.0	2.04	0.96	–	[192]
Brain grey matter	633	62.8	2.68	60.2	7.22	0.88	–	-do-
Brain tumor	630	–	–	–	–	–	2.4	[196]
	630	–	–		–	–	2.2–6.6	-do-
Brain glioma	630	–	5.0	–	7.0	–	–	[198]
Melanoma	630		5.0	–	7.0	–	–	-do-
Breast fibrous tissue								
	514	202	–	–	–	–		[199]
	633	189		–	–	–	–	-do-
	1060	165		–	–	–	–	-do-
Breast fats								
	514	775	–	–	–	–		-do-
	633	676		–	–	–	–	-do-
	1060	524		–	–	–	–	-do-
Breast tissue	635	–	≤ 0.2	395 ± 35	–	–	–	[200]
Heart								
Endocardium	1060	–	0.07	136		0.973	–	[192]
Epicardium	1060	–	0.35	167	–	0.983	–	-do-
Kidney	630		–	–	–	–	4.0	[201]
Liver								
	630	–	–	–	–	–	11.0	[201]

APPENDIX 5B *(Continued)*

Tissue	λ(nm)	μ_t (cm^{-1})	μ_a (cm^{-1})	μ_s (cm^{-1})	μ_s (1− g) (cm^{-1})	G	μ_{eff}(cm^{-1})	Refs.
	635	315	2.3	313	–	0.68	26.6	[200]
	515	304	18.9	285	–	–	–	-do-
Normal lung								
Deflated	633		–	–	–	–	11.0	[202]
	635	332	8.1 ± 2.8	324 ± 46	–	0.75	–	[200]
Epithelium	633		1.1 ± 0.5	204 ± 32	–	0.95 ± 0.01	–	[203]
	515		1.6 ± 0.7	273 ± 42	–	0.94 ± 0.01	–	-do-
Submucosa	633		1.8 ± 0.2	207 ± 13	–	0.94 ± 0.01	–	-do-
	515	–	3.7 ± 0.5	236 ± 18	–	0.92 ± 0.01	–	-do-
Cartilage	633		1.1 ± 0.1	246 ± 16	–	0.95 ± 0.01	–	-do-
	515	–	2.1 ± 0.2	270 ± 18	–	0.93 ± 0.01	–	-do-

CHAPTER 6

Biomass Transfer in the Human Body

MEGH R. GOYAL, JORGE ALVARADO COLON,
REYINALD GARCIA MONTAÑEZ, and HECTOR PEÑA SANTIAGO

6.1 INTRODUCTION

The human body experiences a lot of variation in the body temperature. The human body is 70% water. Fluids help in the thermoregulatory process, through which the body goes through during each daily activity. Human body deals with thermoregulatory processes to maintain a constant temperature between 35.8 and 37.2 °C. As water absorbs heat, human body sweats to eliminate heat from water absorption. Water helps maintain a constant temperature in the human body. The budget of heat generation by the body of human being is related to amount of activities performed by them and the effect of changes in surrounding temperature. When body temperature rises, radiant energy or heat is lost by human skin due to flow of blood from core to periphery and dilation of vasculature. When the body temperature is low, vessels constrict to reduce blood flow and heat loss to the skin.

This chapter presents how biomass transfer in our body is affected by body heat loss or gain, heat production, sweating, wearing clothes as insulated material; and how skin reacts to the change in temperature and other factors relating to biomass transfer.

[*] This chapter has been modified from the review article prepared by the students (Jorge Alvarado Colon, Reyinald Garcia Montañez, and Hector Peña Santiago) for the course on Fluid Mechanics, INGE 4015. Course instructor: Megh R. Goyal, PhD, P.E., Retired Professor in Biomedical Engineering, General Engineering Department, University of Puerto Rico—Mayaguez Campus; For details contact author at goyalmegh@gmail.com.

Biofluid Dynamics of Human Body Systems: Expanded and Revised Edition.
Megh R. Goyal, Arka Bhowmik, & Anamika Chauhan (Eds.)

6.2 PRINCIPLES OF BIOMASS TRANSFER

6.2.1 HEAT PRODUCTION AND HEAT LOSS[1]

Heat is continuously produced by human body due to biochemical process also known as metabolism, and body heat is also continuously being lost to the surroundings. A thermal balance in the body can only exist if rate of production of heat energy is equal to rate of dissipation of heat energy from the body. However, when the two rates are out of equilibrium, body heat or temperature will either increase or decrease. Heat is lost from the body by radiation, conduction, convection, and evaporation. Also, convection of the air plays a major role in heat loss by both conduction and evaporation. The percentage of dissipated heat energy associated with each of these different mechanisms obviously varies with atmospheric conditions. For instance, about 2.257 KJ of heat is dissipated by per gram of water evaporating from the skin surface. The evaporation of water from lungs and skin is about 600 mL/day. This causes continuous heat loss at a rate of 50.23–66.98 J/h. Unfortunately, this insensible evaporation of water is directly through the skin, due to continuous diffusion of water molecules. In addition, it is important to note that evaporation is one of the dominant methods of heat dissipation of heat from body to surrounding when surrounding temperature is less than body temperature. In contrast, evaporative loss gets restricted when temperature of surrounding is greater than that of the body; this ensures increase in core temperature (for more details, the reader is referred to Chapters 4 and 5 in this book).

6.2.2 BODY TEMPERATURE[1]

The temperature of core body remains almost constant, within ± 0.56 °C throughout the day with average value close to 37 °C. When a nude person is subjected to lower (12.78 °C) and higher (60 °C) temperatures, the bodies' thermoregulatory system maintains the core temperature. Therefore, the mechanism of thermoregulation in body is a perfectly designed control system. For measurements, normal temperatures may vary, from approximately 36.11 to 37.4 °C. The rectum temperature is approximately 0.6 °C greater than the oral temperature. However, due to increased metabolic activity during strenuous exercise, the rectal temperature can rise to 38.33 °C.

6.2.3 THERMAL SENSATIONS[1]

The humans can perceive thermal gradations of cold and heat. The gradient is perceived by three receptors for sensing in human body. These receptors are called warm, cold, and pain receptors. However, the receptor involving pain can be activated by the body along with warm and cold receptors during sense of extreme "freezing" and "burning" sensing.

The warm and cold receptors are embedded below the skin at different locations having shape of approximately 1 mm diameter. Throughout the body, the cold and warm receptors are present is large numbers. The quantity of these receptors varies from around 15–25 receptors/cm^2 in the lips and 3–5 receptors/cm^2 in the fingers; and to lower as 1 receptor/cm^2 in entire surface of the trunk. In fact, the cold receptors are larger in amount than warm receptors. Especially, a cold receptor can be viewed as a small receptor with Aδ myelinated nerve that has numerous arms and whose tip extends to epidermal cells of basal layer.

6.3 REGULATION OF BLOOD FLOW IN THE SKIN

6.3.1 NERVOUS CONTROL OF CUTANEOUS BLOOD FLOW[2]

To control temperature of body, the flow of blood in skin is regulated with the aid of nervous system of the body. In other words, the skin blood flow is principally regulated by the nervous system as compared to local regulation.

6.3.2 TEMPERATURE RECEPTORS[2]

The preoptic vicinity of hypothalamus (anterior) controls the temperature of body. Rise in heat of this area causes vasodilation of essentially all vessels found in skin and enhances the process of sweating. Thus, the hypothalamus controls blood flow in the skin to balance the body temperature or changes to temperature via vasoconstrictor and vasodilator mechanisms.

- **The vasoconstriction mechanism:** The skin throughout the body is supplied by sympathetic vasoconstrictor fibers that secrete norepinephrine at their endings. This constrictor system is extremely active across foot, hand, ear, nose, and lips that are exposed to severe cold and which are also the areas with large number of arteriovenous anastomoses. At a normal body temperature, the sympathetic vasoconstrictor nerves keep these anastomoses almost totally closed. When the body becomes overly heated, the number of sympathetic impulses is greatly reduced so that the anastomoses relaxes to entertain flow of blood to venous plexuses, thus promoting heat loss from the body. For the rest of the body (such as, arms, legs, and body trunk) surfaces, almost no arteriovenous anastomoses are present. Nevertheless, vasoconstrictor control of the nutritive vessels can still affect the major decrease in blood flow by constricting the local arterioles. On the contrary, when the body becomes overheated, the sympathetic vasoconstrictor impulses cease, and the blood flow to the skin increases about two folds.

- **The vasodilator mechanism:** The body temperature becomes excessive and overwhelming amount of sweating takes place which is due to increase in flow of blood through arm and trunk as a consequence of vasodilation. The excessive vasodilation is absent if the sweating is absent. Hence, it can be hypothesized that the sympathetic fibers that secrete acetylcholine to activate the sweat glands cause a secondary vasodilation as follows—the increased activity of sweat gland activates the expression of kallikrein, which in turn splits the polypeptide bradykinin from globulin in the interstitial fluids. Bradykinin in turn is a powerful vasodilator substance that accounts for contrary to increased blood flow when sweating begins to occur. In this theory, restraining the mechanism involving bradykinin fails to completely block the increased blood flow associated with sweating. Therefore, other vasodilator peptides have also been suggested as the cause of vasodilation.

6.3.3 BLOOD TRANSFER FROM THE SKIN TO THE REMAINING PARTS OF CIRCULATORY SYSTEM

Skin of body parts that are connected to surrounding has an overwhelming capability to increase or decrease the flow of blood. For instance, exercising the body is subject to hemorrhage and human anxiety can trigger sympathetic stimulation of the venous plexuses that can deliver blood at the rate 5–10% of the total volume of internal vessels. Especially, a blood reservoir can be formed in veins of subcutaneous region, which provides blood to serve other circulatory functions when needed.[3]

6.3.4 BLOOD FLOW TO THE SKIN AND HEAT TRANSFER FROM THE BODY CORE

The blood vessels penetrate the subcutaneous insulator tissues and are distributed profusely in the subpapillary portions of the skin wherein the blood flow and heat transfer rate are high. Beneath the skin over the entire body, there is an extensive venous plexus that functions to control the body heat. In areas of hands, feet, and ears, the blood is provided via arteriovenous anastomoses beginning in arterioles of venous plexus. The rate of blood flow is monitored through activation of sympathetic nervous system as a means to control heat loss from the body. This flow can vary from barely above zero in cold weather to as much as 30% of the total cardiac output in hot weather proving that the skin being a perfect radiator, wherein skin blood flow plays an important role dissipation of heat energy.

6.4 SWEATING AND ITS REGULATION BY THE AUTONOMIC NERVOUS SYSTEM

The body demands a rapid evaporative cooling by enhancing the sweating process in situations of unexplained overheating of the body. The sweating is activated by anterior part of the hypothalamus stimulation which leads to origination of impulses transmitted to autonomic pathways toward the cord and followed by outflow of signals to sweat glands in the skin everywhere in the body. Table 6.1 shows physical and chemical properties of sweat.

- **Rate of sweating:** The production of sweat in cold surrounding is close to zero; however, in very hot weather, the maximum rate of sweat production is from 0.7 liter per hour (lph) in an unacclimatized person to 1.5–2 lph in a person maximally acclimatized to heat. Due to excessive sweating, about 1.36 kg of body weight/h can be lost by an individual.[4]
- **Mechanism of sweating and secretions:** The sweat gland is a tubular structure consisting[4]: (a) A deep spiral coil that allows sweat creation; and (b) A duct portion passing outward to the surface of the skin. The salivary glands and production structure of sweat glands are also responsible for precursor secretion. Consequently, certain constituents of this fluid are altered as it flows through the duct. Large amounts of sodium chloride are lost in the sweat. When the rate of sweat secretion is very small, concentration of NaCl is also very low, because most of

it is absorbed by the secretion of precursor nature before reaching to the surface of the body. NaCl concentration is sometimes as low as 5 mEq/L. On the other hand, when the rate of secretion becomes progressively greater, the rate of NaCl reabsorption does not increase so that the concentrations in the sweat of the normal unacclimatized person rise to maximum levels of about 60 mEq/L, or nearly half the levels in plasma. Table 6.1 shows physical and chemical properties of sweat.

- **Effect of aldosterone on sodium loss in the sweat-acclimation to heat:** Aldosterone increases the rate of active reabsorption of sodium by the ducts.[5] The reabsorption of sodium also carries chloride ions due to the electrical gradient that develops across the epithelium. This aldosterone minimizes loss of NaCl in the sweat when the blood NaCl concentrations are already low.
- **Extreme sweating:** Extreme sweating occurs in hot surroundings. It can deplete extracellular fluids of electrolytes, especially of sodium and chloride. A person with profuse sweat may lose about 15–20 g of NaCl per day until the person becomes acclimatized to the heat. For example, post 4–6 weeks stabilization, the loss of NaCl ranges from 3 to 5 g/day.
- **The insulator system of the human body:** The insulators for our body is the skin that covers the entire body.[1] The fat is particularly the major layer of skin that acts as an insulator as it only enables one-third heat transfer as compared to other tissues. Especially, in the absence of blood flow from core, the ability of male body is approximately equal to three quarters of the insulating properties of usual clothes. The insulation is much higher for women. The insulation below skin surface ensures the equilibrium between skin surface and surroundings.

6.5 HUMAN SKIN[5, 7–9]

Contracting and relaxing of the capillary bed of the skin regulates our body temperature. There are approximately 15 miles of capillaries for each square inch of skin, so any contraction of this capillary bed serves to hold a large volume of blood deep within the body, which otherwise would come close to the surface and lose its heat. This serves to retain heat within the body. When the weather is hot, or with vigorous and/or sustained exercise, the capillary bed dilates bringing larger volumes of blood near to the surface, thus losing heat. The sweat glands assist in this process by transporting body water to skin surface.[5,10–13]

TABLE 6.1 Physical and Chemical Properties of Sweat.[6]

Property or constituent	Units	Value
Physical properties		
pH	–	3.8–6.5
Rate of production	mL/min	17.7–38.2
Specific gravity	–	1.001–1.006
General chemical components		
Solids total	%	1.174–1.597
Water	%	99.0–99.5
Electrolytes		
Calcium	mg/100 mL	1–8
Chloride	mg/100 mL	36–468
Copper	mg/100 mL	0.006
Iodine	µg/100 mL	0.9
Iron	mg/100 mL	0.027
Magnesium	mg/100 mL	0.004–0.286
Manganese	mg/100 mL	0.006
Phosphorous	mg/100 mL	0.009–0.045
Potassium	mg/100 mL	21–126
Sodium	mg/100 mL	24–312
Nitrogenous substance		
Amino acid N	mg/100 mL	1.1–10.2
Amino acids Arginine	mg/100 mL	13.6
Creatine	mg/100 mL	0.1–1.3
Histidine	mg/100 mL	8
Isoleucine	mg/100 mL	2.27
Leucine	mg/100 mL	2.69
Lysine	mg/100 mL	2.26
Nitrogen	mg/100 mL	33.2
Non-protein *N*	mg/100 mL	27–64
Phenylalanine	mg/100 mL	2.19
Threonine	mg/100 mL	5.38
Tryptophan	mg/100 mL	1.12
Tyrosine	mg/100 mL	3.15
Urea	mg/100 mL	12–57
Urea N	mg/100 mL	5–36

TABLE 6.1 *(Continued)*

Property or constituent	Units	Value
Uric acid	mg/100 mL	0.16
Valine	mg/100 mL	2.96
Carbohydrates, miscellaneous organic acids, hormones		
Sugar	mg/100 mL	
Carbolic	mg/100 mL	2–8
Lacrid acid	mg/mL	285
Corticoids	μg/100 mL	4–8
Vitamins		
B1 (thiamine)	μg/100 mL	0.15 (0–0.6)
B2 (riboflamin)	μg/100 mL	0–0.5
B6 (pyridoxine)	μg/100 mL	0.04–0.17
Bc (folic acid)	μg/100 mL	0.53–0.88
Bx (paminobenzoic acid)	μg/100 mL	0.24 (0.08–1.70)
C (dehydroascorbic acid)	μg/100 mL	70.5
Choline	μg/100 mL	7.1
Inositol	μg/100 mL	21 (15–36)
Niacin (nicotinic acid)	μg/100 mL	1.7–8.7
Oantothenic acid	μg/100 mL	1.5–7.7

The evaporation of water is an endothermic phenomenon that needs heat for water to evaporate. Part of this heat is derived from the body (from the dilated capillary bed) and part from the environment. The part, which is derived from the body, is lost, thus leaving the body cooler. The central control of body temperature is located in the hypothalamus of the brain. Skin is useful for losing heat up to an ambient temperature of 46.11°C. Above this temperature, it does not work, as the skin absorbs as much heat to removal by sweating and vascular perfusion. Table 6.2 indicates chemical/physical attributes of sebum.

6.5.1 NEURAL SENSORS OF THE SKIN

The neural sensors on the skin include specialized organs to sense hot/cold, pressure, and light touch. Pain is the individual's interpretation of the sensation caused by the release of large numbers of neurotransmitters into the synapses between nerve cells in the skin. This happens in response to severe hot, cold, pressure, or tissue injury.

TABLE 6.2 The Chemical/Physical Attributes of Sebum.[6]

Property or constituent	Units	Value
Physical properties		
Melting point	°C	35.8
Specific gravity at 4°C	ratio	0.911
Surface tension	dynes/cm	24.89
Viscosity at 30°C	millipoises	859.7
Chemical constituent		
Fatty acid combined: Forearm	g/g	0.346
Scalp	g/g	0.28
1. Triglycerides: Forearm	g/g	0.325
Forehead	g/g	0.44
2. Waxes: Scalp	g/g	0.16
3. Free fatty acids: Forearm	g/g	0.28
Fore head	g/g	0.38
Usaponifiable matter: Forearm	g/g	0.30
Scalp	g/g	0.34
1. Aliphatic alcohols: Forearm	g/g	0.062
Scalp	g/g	0.09
2. Straight chain: Forearm	g/g	0.024
Scalp	g/g	0.045
3. Branched chain: Forearm	g/g	0.038
Scalp	g/g	0.009
4. Cholesterol: Forearm	g/g	0.041
Scalp	g/g	0.035
5. Dihydrocholesterol: Forearm	g/g	0.001
6. Hydrocarbons: Forearm	g/g	0.081
Scalp	g/g	0.09
7. Phosphatides: Fore head	g/g	0.009
8. Squalene: Forearm	g/g	0.055
Scalp	g/g	0.07

6.5.2 SKIN COLOR

Skin color affects heat and mass transfer rates. Skin color is determined by the concentration of melanocytes in the skin. These cells contain melanosomes, which produce the pigment melanin. Melanin serves the function of protection from ultraviolet rays of the sun and is gathered together in clumps within the cell. Tanning represents the production and dispersion of melanin throughout the melanocytes, and it happens as a result of measured exposure to sunlight.

6.5.3 STRUCTURE OF HAIR FOLLICLE[2,5,7,8,11,12,14]

Figure 6.1 shows different types of gland in the human body. The structures of human skin and hair follicle are shown in Figures 6.2 and 6.3, respectively. The hair follicle varies in histological appearance according to whether it is in the growing or resting stage. Hair grows in cycles—a period of growth during which new development occurs and consequent period of rest, when the hair is lost. About two-thirds of the distance from the bulb to the neck of the follicle, cells from the external root sheath grow outward to form a sebaceous gland. Every hair follicle is joined to a sebaceous gland and the gland secretes its product into the space around the hair and the secretion reaches the skin surface. The hair keratinization at root takes place during keratogenous zone briefly.

6.6 WHY DO WE PERSPIRE OR SWEAT?

We perspire to maintain our body temperature at 37 °C. The body is composed of various temperature activation region in brain which are composed of heating and cooling regions. The body uses about 10.47 kJ of our daily intake of calorie-laden food to fuel the body. This process of oxidation burns the calories, producing enough heat to bring 23.66 L of water to the boiling point. Therefore, body dissipates the heat of core to blood and activates the cooling regions of brain.[2]

The cooling region restrains the burning of calories and simultaneously activates the process of blood vessel dilation in skin to dissipate the heat from blood to surroundings. This process is accompanied by perspiration, wherein water from skin goes out through the pores. It leads to concentration of water droplets or particles that achieves a quick evaporation and assists in the cooling of body's temperature. In humid environment, the rate of perspiration slows down and often requires fans and air conditioners (ventilation) in order to drive the moist air. This returns our internal thermostat to its normal setting of 37 °C. Normal body temperature (37 °C) is an equilibrium achieved between heat dissipation and generation within the body. In a dry environment, most of the sweat produced evaporates into the air.

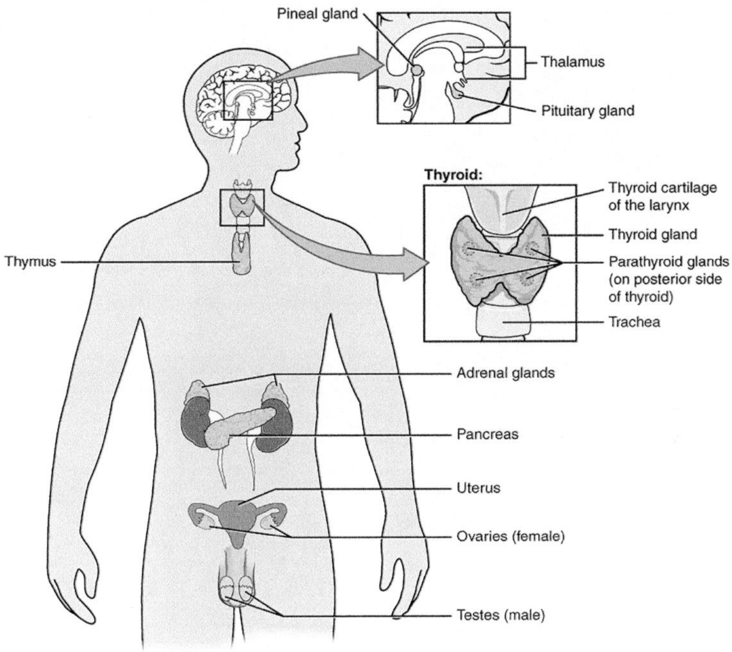

FIGURE 6.1 Endocrine system.

Source: Reprinted from Ref. [15] https://creativecommons.org/licenses/by-sa/4.0/

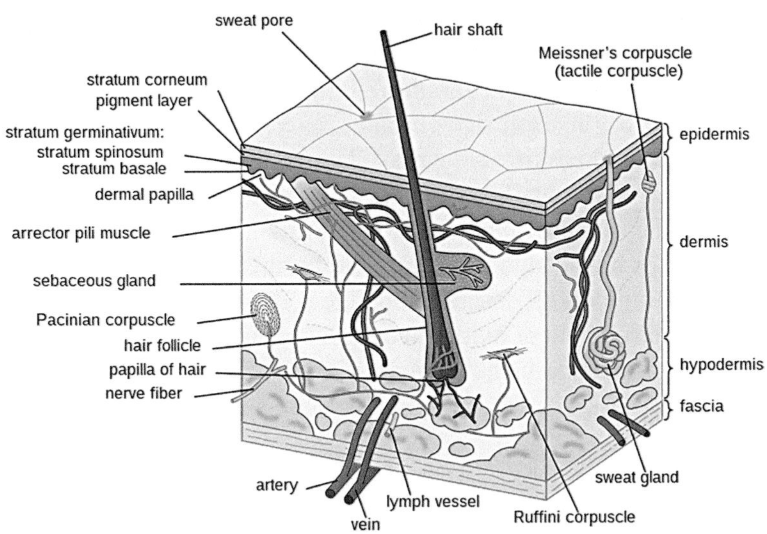

FIGURE 6.2 Schematic diagram of the integumentary system.

Source: Reprinted from Ref. [16]. https://creativecommons.org/licenses/by-sa/4.0/

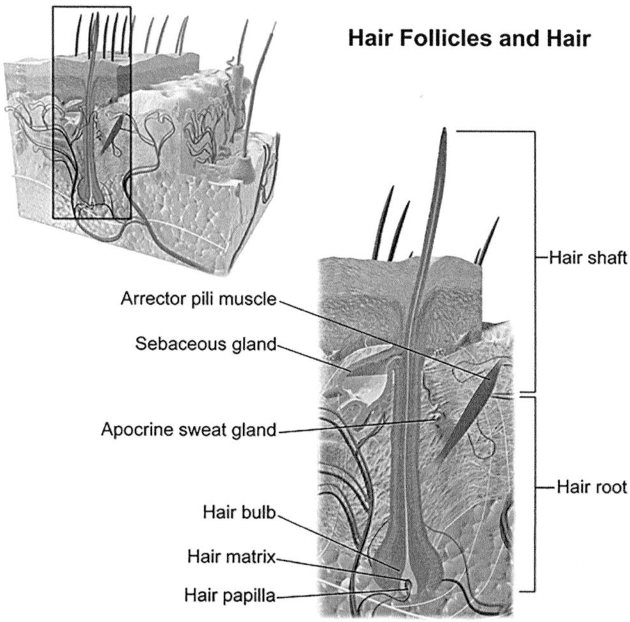

FIGURE 6.3 Diagram of a follicle.

Source: Reprinted from Ref. [17]. https://creativecommons.org/licenses/by-sa/4.0/

Body water is routinely lost via perspiration and by the humidification of the air that enters and leaves the lungs. This insensible water loss amounts to about 50 mL/h. Some water is lost in the voiding of wastes via urine and feces. The average adult loses about 0.7 L of sweat/day but sweat losses can be as much as 2.5 lph. Hot, dry climates, exercise, and fever can increase water losses from the skin and lungs. Humid environment decreases the degree to which sweat vaporizes. Heat is constantly produced in the body due to the exothermic chemical reactions during the metabolism.

Water must be replaced to maintain fluid balance. Water imbalance of only 1–2% can lead to illness. A general guideline is drinking eight glasses of water per day. Increased amount of fluid is needed for people who engage in strenuous exercise or work outside in hot climates; have fever, vomiting, diarrhea, or excessive urine losses; and those on high protein diets or taking diuretics.

6.6.1 SEBACEOUS (OIL) GLANDS

The sebaceous glands (Figure 6.3) are developed as an outgrowth of the external root sheath of the hair follicle.[7,8,14] These secrete an oil substance called sebum that coats not only the hair, but also the surface of the skin.

Properties of sebum are shown in Table 6.2. Sebum is produced as a holocrine secretion, that is, the entire cell becomes filled with the fatty product; the cell undergoes progressive disruption and then damage as it fills cell; and ultimately, both cellular junk and chemical products are released from the gland. As the cells become filled with numerous lipid droplets separated by thin partitions of cytoplasm, these moves away from the basal layer. The excessive accumulation of sebum in the glands gives rise to the condition called acne.

6.6.2 SUDORIFEROUS (SWEAT) GLANDS

Sweat glands are simple coiled glands found in the skin in two forms: eccrine and apocrine. Most sweat glands are endocrine glands that produce the secretion without destruction of the cells of the glands. Sweat glands is extensively present in palms of the hand and soles of the feet. All endocrine glands do not function all the time, but they respond quickly to heat stress. Endocrine sweat is an ultra-filtrate of plasma forming a colorless, hypotonic solution that is 99% water and 1% solutes consisting of sodium, chloride, potassium, calcium, urea, ammonia, uric acid, and phosphorus. Aldosterone increases the reabsorption of sodium from sweat sudoriferous glands (see Figures 6.1–6.3).

- **Eccrine sweat glands** are distributed over the entire surface of the body surface except at a relatively few locations (such as, lips and external genitalia). These lack any association with follicles of hair. The secretory portion of this gland is identified inside dermis-hypodermis junction. There are three types of cells identified within glandular portion of the gland, that is, myoepithilical cells, clear cells, and dark cells. **Myoepithilical cells** move secretions out of the gland by contraction. **Dark cells** contain abundant *rER* and granules with a glycoprotein content, and these are thought to secret mucopolysaccharide. The **clear cells** show no secretory granules and miniscule quantity of rER. The duct of the sweat glands begins as a coiled element in the same nest as the coiled secretory portion of the gland. The cells of the duct are smaller and looks less lightened cells in secretory regions. The sweat gland called eccrine produces a watery solution low in protein and contains varying amounts of NaCl, urea, uric acid, and ammonia. When urea in the body cannot be released by kidneys, concentration of urea increases in sweat and post evaporation of water crystals are seen on the skin surface particular in upper lip. The eccrine sweat glands strike an important contribution for temperature regulation through the cooling that results from the evaporation sweat (see Chapter 4 of this book).

- **Apocrine glands** are common in the armpit, axilla, areola, nipple of the mammary gland, labia majora, the circumanal region, and external genitallia. Apocrine secretions are those, where the apical region of the secreting cells is lost with the secretion. This type of sweat contains protein and carbohydrates to produce odor in the presence of bacteria on the skin. Apocrine sweat glands empty into hair follicles while endocrine glands open at the skin surface. Moll in the eyelid is also an apocrine gland. This gland contains only one type of secretory cell and its cytoplasm is stained with eosin. Myoepithelial cells are also in the secretory portion of the gland situated adjacent to the basal lamina. The duct portion of the gland resembles the duct of the eccrine gland. The duct opens into the hair follicle at the level above the duct of the sebaceous gland. The secretions of the glands vary with the anatomical location. In the axilla, the glands produce a proteinaceous milky fluid containing organic compounds, including steroid.

6.6.3 INNERVATION OF SWEAT GLANDS

Eccrine and apocrine sweat glands are innervated by the autonomic nervous system. The eccrine sweat glands are stimulated by cholinergic transmitters and apocrine sweat glands are stimulated by adrenergic transmitters. The gland eccrine stimulates to nervous and thermal strain. Apocrine glands respond to emotional and sensory stimuli. Sweat glands are innervated by cholinergic nerves found in sympathetic nerve system. Especially, in hot weather or rigorous exercise can lead to secretion of more than 1 lph sweat in an adult. On the other hand, sweat glands found in soles and palms also stimulate due to stress caused by emotional factors such as anxiety along with increase in temperature of body or surrounding. The secretion of waxy fluid by gland called ceruminous is found in the ear canal, which protects the skin from desiccation and irritation.

Regulatory heat exchange is accomplished by the action of the endocrine sweat glands and the cutaneous blood vessel network (see Chapter 4 of this book). As the liquid sweat evaporates and cools the skin, the cutaneous blood vessels dilate or constrict to remove or conserve heat. A thermostat in the hypothalamus regulates the function of the endocrine glands and the cutaneous vasculature. Perspiration is controlled by sympathetic nervous system (cholinergic) fibers that innervate the sweat glands. Heat is the primary stimulus to the endocrine glands but nervousness or eating spicy foods can cause sweating. Acclimation to heat in the environment is a requirement for the body to function properly. The time of acclimation required depends on temperature, humidity, and duration of exposure. Full acclimation to hot environments can take 2 weeks to 1 month. The properties of sweat are shown in Tables 6.3A and 6.3B.

6.6.4 HOW DOES ENDOCRINE SYSTEM WORK?

Hormones from the endocrine organs are secreted directly into the bloodstream, where special proteins usually bind to them to keep the hormones intact. The proteins also act as a reservoir, allowing miniscule portion of hormone circulating in the blood. Specialized proteins in the target tissue (called receptors) bind with the hormones in the bloodstream; and induce chemical changes in response to the need of our body. Too much or too little hormone can be harmful to the body; therefore, hormone levels are regulated by a feedback mechanism that works like a household thermostat. The change that a hormone produces also serves to regulate the secretion of hormone. Other mechanisms may also influence feedback relationships. If an individual becomes ill, the adrenal glands increase the secretions of certain hormones to help the sick body. The adrenal glands work along with the pituitary gland and the brain to increase the tolerance of these hormones in the blood, thus preventing normal feedback mechanism from decreasing secretion levels until the illness disappears. Long-term changes in hormone levels can influence the endocrine glands. For instance, if hormone secretion is chronically low, the increased stimulation by the feedback mechanism leads to a growth of the gland. For example, in the thyroid, if a person's diet has insufficient iodine.

6.7 DEHYDRATION OF A HUMAN BODY

6.7.1 WHAT IS A PROCESS OF DEHYDRATION?

When the body lacks enough fluids to function at an optimal level dehydration is noticed.[12,18–20] Water plays an important role in various functions of the body and severe reduction water content of body can be detrimental. Dehydration presents in a progression from thirst, fatigue, weakness, then delirium, coma, and finally death. Environmental factors (such as, temperature and humidity) will determine the rate of progression as well as hydration status when the dehydrating situation began. Normally, the water depletion is replenished by drinking water. However, if the water lost is not replaced adequately within the next few days, severe dehydration may occur. Depending on age and based on the percentage of body weight lost during the acute illness, dehydration is classified as:

- **Mild dehydration:** a loss of 3–5% of body weight.
- **Moderate dehydration:** a loss of 6–10% of body weight.
- **Severe dehydration:** a loss of more than 9–15% of body weight.

6.7.2 CONDITIONS THAT FAVOR DEHYDRATION

Dehydration can be caused by fluid loss, inadequate intake, or a combination of both. Other causes of dehydration could be due to infectious elements such as rotavirus, enterotoxigenic *Escherichia coli* (ETEC), epidemics, *Vibrio cholerae* O1 or O139, etc. In infants and children, acute gastroenteritis is a major cause of dehydration.

- **Causes of severe loss of fluid:** uncontrolled diarrhea and vomiting observed during cholera; discharge of uncontrolled urine referred to as polyuria; excessive sweating; and fever.
- **Inadequate intake leading** to dehydration may be due to: nausea; refusing to drink because of mouth sores or a sore throat; and acute illness with loss of appetite.
- **Dehydration within children** is a mixture of gastroenteritis with vomiting and diarrhea; and acute illness where the child refuses to intake fluids and fluid loss by sweating during fever.

6.7.3 SIGNS AND SYMPTOMS OF DEHYDRATION

The mouth drying and thirstiness is considered as early signs of dehydration.[2,4,5,7,8,11,14,19,20] Other symptoms of dehydration could be irritation in behavior, lacking rest, loss of moisture in skin, mouth dryness, sunken eyes, lack of tears, etc. Similarly, for mild and early signs of dehydration could be flushing of face skin, thirstiness, reduction in urine volume, vertigo, arm and leg camps, insomnia, headache and feeling of dryness, etc.

The elevation in level of dehydration can lead to shock called hypovolemic, losing consciousness, stop in urine discharge, reduced pulse, hypotension condition, cyanosis in periphery, and death, etc. Other symptoms may include stomach bloating, tachycardia, loss of elasticity of skin, habitual deep breathing, lack of capillary refilling, etc.

6.7.4 INTAKE OF FLUIDS REDUCES INCIDENTS OF DEHYDRATION

Water is vital to maintaining our normal bodily functions. It regulates body temperature at 37 °C, since the body produces heat throughout the day from several body processes.[18–21] Water helps regulate body temperature through perspiration by evaporating on our skin to cool it (see Chapter 4 in this book). The healthy kidneys can balance the amount of water lost through urination and perspiration (sweating) with the number of fluids that we drink. Through dehydration, vital minerals called electrolytes and salts are lost. Our body responds by absorbing fluids from our blood and other fluids. If one loses too much water, blood vessels may collapse.

TABLE 6.3A Effect of Heat, Exercise, and Chemical Substance on Composition of Sweat.[6]

Constituent	Condition or variable	Units	Value
Calcium	Normal	mM/L	1.12–2.5
	Hot & humid environment	mM/L	0.50 (0.1–1.1)
Chloride	**Acclimatization to heat:** Before: Arm	mM/L	29
	Face	mM/L	99
	Thigh	mM/L	57
	Torso	mM/L	43
	• After: Arm	mM/L	36
	Face	mM/L	57
	Thigh	mM/L	18
	Torso	mM/L	16
	• 2nd day of exposure, resisting	mM/L	89
	• 14th day of exposure, resisting	mM/L	59.4
	• 2nd day of exposure, walking	mM/L	104.7
	• 14th day of exposure, walking	mM/L	66.9
	Warm	mM/L	34
	Cool	mM/L	25
	Sweat rate: High	mM/L	31.9
	Sweat rate: Low	mM/L	13
	Sweat rate: Continuous	mM/L	108 (90–173)
	Sweat rate: Intermittent	mM/L	308 (194–463)

TABLE 6.3A *(Continued)*

Constituent	Condition or variable	Units	Value
	Work: Before	mM/L	56
	Work: After	mM/L	37.6
Uric acid	Normal	mM/L	0.048
	Acclimation to heat	mM/L	0.014–0.018
	Heat chamber	mM/L	0.018
	Sweat rate: profuse	mM/L	0
	Sweat rate: Intermittent	mM/L	0.048
Copper	Hot & humid environment	mM/L	0.0025
Iodine	Normal	μg/L	3.0 (1.5–6.3)
	Elevated iodine intake	μg/L	36.5
Iron	Normal	mM/L	0.005
	Hot & humid environment	mM/L	0.0049
	Cool & dry environment	mM/L	0.037
Magnesium	Normal	mM/L	0.058–1.8
	Hot & humid environment	mM/L	0.0018–0.17
Manganese	Hot & humid environment	mM/L	0.0011
Phosphorus	After administration of 15g of Na2HPO4	mM/L	0.97
	Low phosphorus diet	mM/L	0.58
	Hot & humid environment	mM/L	0.008
Potassium	Normal	mM/L	12.8
	Acclimatization to heat:		
	• 2nd day of exposure, resisting	mM/L	7.55
	• 14th day of exposure, resisting	mM/L	4.29
	• 2nd day of exposure, walking	mM/L	8.22
	• 14th day of exposure, walk	mM/L	4.44
	After administration of adrenocortical extract	mM/L	6.1
	Sweat rate: High	mM/L	9.2
	Sweat rate: Low	mM/L	2.7
	Before work	mM/L	10.5
	After work	mM/L	9.4
	Work and rest	mM/L	4.86

It is noted that metabolic-related heat production elevates with the increase in activity of the muscle during exercise which is subsequently released by production of sweat. A longer duration of exercise can increase the level of dehydration which can lead to lowering of athletic performance, muscle cramp, etc. Therefore, such symptoms for athletes should be carefully assessed.

6.7.5 HOW TO KNOW IF ONE IS ADEQUATELY HYDRATED?

If urine is dark and scant, one is likely dehydrated. If it is clear and plentiful, one is likely well hydrated. Certain vitamins and some foods can make urine darker. If one is urinating four or more times a day, one is usually well hydrated. Less often is a sign of dehydration.

6.7.6 WHY DOES AN ATHLETE NEED TO DRINK A LOT OF WATER?

Dehydration from inadequate fluid replacement is the most common contributor of decreased athletic performance in contributing to fatigue and increased risk of injuries.[18–21] Dehydration also leads to muscle cramps. Moderately active people need six to eight glasses of water daily. More active individuals need more than eight glasses. During a strenuous workout, weight loss from water loss can easily be ½ to 1 kg, an amount that inevitably triggers the thirst mechanism. The American College of Sports Medicine endorses 16 ounces of water 2 h before exercise and 5–12 ounces every 15 min during exercise. Athletes and exercisers should always drink water prior to, during, and following all training and events every day. The higher the intensity of exercise, the greater is heat production by the muscles. Overheating causes more sweat production.

6.7.7 HUMIDITY AS A NEGATIVE FACTOR

The body temperature may be elevated by surrounding heat and due to muscular activity. However, the dominant cooling is achieved by sweat evaporation. But, the hindrance to evaporative cooling is imposed by humid environment.[18,21] The rate of sweat formation can be elevated in humid condition because of continuous rise in body temperature since evaporative cooling is minimized as surrounding water vapor in already saturated. The uncontrolled sweat formation or water release by body as a consequence of elevated body temperature can lead to dehydration. This also effect the performance of individual and could set an injury. In fact, possibility of death is present especially for temperature of air less than 75 °F (24 °C) at relative humidity of >95%.

TABLE 6.3B Effect of Heat, Exercise, and Chemical Substance on Composition of Sweat.[2,6,22,23]

Constituent	Condition or variable	Units	Value
Nitrogen total	Acclimatization to heat:	mg/mL	66.4/100
	• 2nd day exposure		
	• 14 st day of exposure resting	mg/mL	51.5/100
	• 2nd day of exposure, walking	mg/mL	53.8/100
	• 14th day of exposure, walk	mg/mL	25.8/100
	Sweat rate: high	mg/mL	28/100
	Sweat rate: Low	mg/mL	32.2/100
	Carbohydrates		
Reducing sugars (as glucose)	Normal	mM/L	1.1
	Hot chamber	mM/L	0.83
	After phlorizin, intradermally	mM/L	1.5
	Sweat rate: profuse	mM/L	0–0.061
	Sweat rate: Intermittent	mM/L	0–0.89
	Sweat rate: After 20g glucose, intravenously	mM/L	0–1
	Salts		
Lactate	normal	mM/L	5–8.7
	Acclimatization to heat:	mM/L	22.3
	• Before: Skin		
	• After: Skin	mM/L	13.6
	• Before: Arm	mM/L	33.3
	• After: Arm	mM/L	11
	• Before: Body surface	mM/L	24.4
	• After: Body surface	mM/L	12.3
	• Before	mM/L	9.4
	• After	mM/L	8.1
	• Eccrine sweat apocrine sweat	mM/L	31.6–37.5
	• Apocrine sweat	mM/L	27.5–36.5
	Elevated blood lactate	mM/L	7.5
	Hot chamber	mM/L	8.2
	Alternate work and rest	mM/L	9.7
	Nitrogenous substances		

TABLE 6.3B *(Continued)*

Constituent	Condition or variable	Units	Value
Ammonia	Normal	mM/L	6.5
	• Eccrine sweat	mM/L	4
	• Apocrine sweat	mM/L	29
	Sweat rate: Profuse	mM/L	4.5
	Sweat rate: Intermittent	mM/L	17
Sodium	Normal	mM/L	34.3
	Acclimatization to heat:		
	• Before	mM/L	51
	• After	mM/L	17–5.1
	• 2nd day of exposure, resting	mM/L	89.3
	• 14 st day of exposure resting	mM/L	63.2
	• 2nd day of exposure, walking	mM/L	105
	• 14th day of exposure, walking	mM/L	75.1
	Administration of desoxycorticosterone acetate:		
	• Before	mM/L	90
	• After	mM/L	55
	Sweat rate:		
	• High	mM/L	63
	Sweat rate:		
	• Low	mM/L	18
	Nitrogenous substances		
Creatinine	Normal	mM/L	0.08
	Acclimatization to heat:	mM/L	0.022–0.065
	• 2nd day of exposure, resting	mM/L	0.048
	• 14 st day of exposure resting	mM/L	0.034
	• 2nd day of exposure, walking	mM/L	0.06
	• 14th day of exposure, walking	mM/L	0.029
	Hot	mM/L	0.05
	Work in heat	mM/L	0.028–0.069
	Alternate work and rest	mM/L	0.009–0.019
	After drinking 500 mL of 5% of creatine	mM/L	0.025
Urea	Normal	mM/L	8.62
	Acclimatization to heat:		
Before : Skin	mM/L	5	

TABLE 6.3B *(Continued)*

Constituent	Condition or variable	Units	Value
	After :Skin	mM/L	5
	Before: Arm	mM/L	15.5
	After: Arm	mM/L	9.2
	Before: Body surface	mM/L	5.4
	After: Body surface	mM/L	6
	Hot	mM/L	9.7
	Sweat rate: Profuse	mM/L	11.2
	Sweat rate: Intermittent	mM/L	46
	Sweat rate: After 3 g urea, orally	mM/L	11.5–24
	Rest	mM/L	4.7

6.7.8 HEAT INJURY

Heat-related injuries are inception cramps, feeling exhaustion, and inception of heat associated strokes. There is no clear boundary between these conditions. The generation cramps are linked with muscle and exercise often originates in arms, legs, or abdomen.

6.7.9 CHEMICAL REACTIONS DURING DEHYDRATION

The equilibrium-balanced state of chemical compositions varies with dehydration such as level of sodium (salt) and potassium found in the blood and other body fluids.[5,6,11] In order for cells to function adequate level, the concentration of analytes should be in the preferred range. The decline in required level of the analytes may be a serious issue for infants which suffers larger death rates because of gastroenteritis being cause by dehydration. Various tests can be performed to monitor patients with dehydration, such as:

- Elevation in concentration of creatinine;
- Elevation in nitrogen concentration in blood;
- Identifying the level of sodium, potassium, and bicarbonate concentration in blood.
- Level of blood sugar;
- Specific gravity test for urine;
- Total blood count changes.

The dehydration process makes the body to lack important minerals or discharges in the form of body waste. For instance, loss of water is associated with respiration or inception of dehydration, lacking desired quantity of salts, and variation in level of urea, etc.

6.8 MASS TRANSFER IN LIVING SYSTEMS

Mass transfer is transfer of constituents from concentrated region to less concentration region (called concentration gradient similar to temperature gradient or pressure gradient).[3] When the concentration of two or more species of chemical differs in spatial domain it can be assumed that process of mass transfer has been established to achieve equilibrium between the concentration of a chemical within the system and moving towards the system. The physical principles for the mass transfer are analogous to those in heat transfer (see Chapters 1 and 4 in this book). The situation is usually more complex in mass transfer, because instead of having one major dependent variable (temperature), there are as many dependent variables as there are species present. Only one mode of heat transfer is considered at one time within a phase; however, mass transfer always involves at least two fundamental mechanisms simultaneously[24].

The molecular mass transfer is analogous to the conduction heat transfer and convective mass transfer is analogous to convective heat transfer.[3,4,7,8,13,14,24–28] The two mechanisms often act simultaneously. The transfer of mass can take place from surface to fluid in motion which depends on the characteristics of moving fluid as well.

Living organisms must be supplied with nutrients and must dispose of waste products to sustain life; and are naturally systems in which mass transfer and diffusion are of critical importance. In many of the control mechanisms of the human body, a primary mechanism involves a variation of the permeability of membranes to selected solutes. These adjustments of the diffusion properties of materials are stimulated by chemical agents. In the design and operation of artificial organs, mass transfer quite often is a major problem and the key process that must be controlled. In addition to concentration driven diffusion, there are several other chemical and physical processes, which are due to concentration gradients, free energy, and hydrostatic pressure.

6.8.1 GENERAL PRINCIPLES[1,3,4,7,8,13,14,22–29]

The natural tendency of more than one gas in mixture is to reach the state of equilibrium especially in the absence of bulk flow is achieved by a microscopic transport phenomenon called diffusion. Especially, the diffusion of gases species take place from region of higher concentration to region of lower concentration with the sole purpose to attain a mixture with equal concentration of molecules. The diffusion of mass purely depends on the gradient of concentration between species. Table 6.4 depicts molar concentration of different gaseous species.

The species concentration in a mixture can be computed in numerous ways such as mass concentration per unit volume, density of a unit volume, mass fraction, molar concentration, etc. A species mass concentration can be identified as ratio of species mass per unit mixture volume. Similarly, overall mass concentration can be defined by term density which sum of all mass concentration of all the species (eq 6.1). Yet, mass fraction is computed by ratio of mass concentration of species and overall mass concentration of mixture (eq 2) In fact, adding all species mass fraction would lead to unity (eq 3). Again, species molar concentration is computed by ratio of quantity of moles of specific species and overall volume of mixture. A mole of species is mass or molecular weight (*M*) of species. Table 6.5 depicts mass fraction of different species.[6,10,22]

$$\rho = \sum_{i=1}^{n} \rho_i \tag{6.1}$$

$$\omega_a = \frac{\rho_a}{\sum_{i=1}^{n} \rho_i} = \frac{\rho_a}{\rho} \tag{6.2}$$

$$\sum_{i=1}^{n} \omega_i = 1 \tag{6.3}$$

$$C_A = \frac{\rho_a}{M_A} \tag{6.4}$$

$$c_A = \frac{C_A}{R\theta} \tag{6.5}$$

When dealing with a gas phase under conditions in which the Ideal Gas Law applies, the molar concentration is given by eq 6.5 in which P is a total pressure, p_A is a partial pressure of the species A in the mixture, T is an absolute temperature and R is the gas constant. The total molar concentration (c) is the total moles of mixture contained in a unit volume (eq 6.6).

For a gas mixture obeying the Ideal Gas Law, we have eq 6.7. The mole fraction for liquid or solid mixtures (x_A) and for gaseous mixtures (y_A) are the molar concentrations of species A divided by the total molar concentration in eqs 6.8 and 6.9. For a gaseous mixture obeying the ideal gas law, the mole fraction (y_A) can be written in terms of pressures (eq 6.10). The eq 6.10 is an algebraic representation of Dalton's Law for gas mixtures. The sum of the mole fractions must be equal to one (eq 6.11).

$$c = \sum_{i=1}^{n} c_i \tag{6.6}$$

$$c_a = \frac{P}{R\theta} \tag{6.7}$$

$$x_a = \frac{c_A}{c} \text{ (liquids and solids)} \tag{6.8}$$

$$y_A = \frac{c_A}{c} \text{ (gases)} \tag{6.9}$$

$$y_A = \frac{c_A}{c} = \frac{P_A}{P} \tag{6.10}$$

$$\sum_{i=1}^{n} y_i = \sum_{i=1}^{n} x_i = 1 \tag{6.11}$$

Based on Fick's Law, diffusion is quantified in an ideal condition such as in an isothermal and isobaric system which ensures species movement only due to concentration gradient in the mixture. In fact, this movement is computed by mass molar average velocity which is averaging of each species velocity in the mixture and can be written as eq 6.12. In eq 6.12, v_i represents species absolute velocity. Similarly, averaged molar velocity of the mixture is given by eq 6.13.

$$v = \frac{\sum_{i=1}^{n} \rho_i v_i}{\sum_{i=1}^{n} \rho_i} = \frac{\sum_{i=1}^{n} \rho_i v_i}{\rho} = \sum_{i=1}^{n} \omega_i v_i \tag{6.12}$$

$$v = \frac{\sum_{i=1}^{n} c_i v_i}{\sum_{i=1}^{n} c_i} = \frac{\sum_{i=1}^{n} c_i v_i}{c} = \sum_{i=1}^{n} x_i v_i \tag{6.13}$$

The rate of diffusion can be defined in flux. For instance, quantity (mass or molar description) of a species that crosses a normal unit area in the direction of motion (vector) per unit time can be referred to as a mass flux. In fact, we can use a reference to define flux. This reference can be coordinates fixed in space or moving with average mass velocity or moving with average molar velocity. For a species, *i* in a fixed coordinate system mass flux can be given by eq 6.14. By adding mass flux of all the species, we can obtain total flux as eq 6.15. Similarly, for a fixed reference molar flux is given by eqs 6.16, and 6.17 provides total molar flux.

$$n_i = \sum_{i=1}^{n} \rho_i v_i \tag{6.14}$$

$$n = \rho v \tag{6.15}$$

$$N_i = \sum_{i=1}^{n} c_i v_i \tag{6.16}$$

$$n = cv \tag{6.17}$$

The mass diffusion flux for any species in terms of average molar velocity as reference frame is provided by eq 6.18a. The molar diffusion flux for species *i* with respect to the molar-average velocity is shown in eq 6.18b. The mass flux (n_i) depends on the mass diffusion flux by eq 6.18. The molar flux (N_i) is related to the molar diffusion flux as given in eq 6.19. The proportionality factor (D_{12}) is a diffusion coefficient or diffusivity in the Fick's Law and its fundamental dimensions are given in eq 6.20. The units of mass diffusivity are in cm^2/s and m^2/s.

Mass diffusivity is identical to other transport properties: kinematic viscosity (ν) and thermal diffusivity (α). The coefficient of diffusion inherently dependent on state properties of the system such as temperature, pressure, and concentration. It also varies due to intermolecular forces, that is, value of diffusion coefficient higher in gases than liquids. In the absence of experimental data, semi-theoretical expressions have been developed to give approximations. The effect of intermolecular forces is quantified by Leonard-Jones approximation, and diffusion coefficient for non-reacting and nonpolar species pair of gas molecules can be approximated by Hirschfelder (eq 6.21). These approximations are made for nominal conditions of low and moderate pressure. In eq 6.21, D_{AB}, θ, P, σ_{AB}, Ω_D are coefficient of diffusion, temperature, pressure, collision diameter defined by Lennard Jones, and diffusion collision integral (dimensionless), respectively. Here, molecular weights of A and B is denoted by M_A and M_B.

$$j_i = \rho_i (v_i - v) \text{ and } \sum_{i=1}^{n} j_i = 0 \tag{6.18a}$$

$$J_i = c_i\left(v_i - V\right) \text{ and } \sum_{i=1}^{n} J_i = 0 \tag{6.18b}$$

$$n_i = j_i + \rho_i v = j_i + \omega_i n \tag{6.18c}$$

$$N_i = J_i + c_i V = J_i + y_i N \tag{6.19}$$

$$D_{12} = -\frac{J_1}{c\nabla x_1} = \left(\frac{M}{L^2 T}\right)\left(\frac{1}{M/L^3 \cdot 1/L}\right) = \frac{L^2}{T} \tag{6.20}$$

$$D_{AB} = \frac{0.002660\theta^{3/2}}{PM_{AB}^{1/2}\sigma_{AB}^2\Omega_D} \tag{6.21}$$

$$\text{where, } M_{AB} = 2\left[\frac{1}{M_A} + \frac{1}{M_B}\right]^{-1}$$

$$\Omega_D = \frac{a}{\left(\theta^*\right)^b} + \frac{c}{\exp\left(d\theta^*\right)} + \frac{e}{\exp\left(f\theta^*\right)} + \frac{g}{\exp\left(h\theta^*\right)} \tag{6.22}$$

The collision integral, Ω_D in eq 6.22, is dependent on intermolecular potential field and temperature of molecules A and B. Where, $\theta^* = \kappa\theta/\varepsilon_{AB}$ such that κ being Boltzmann constant (= 1.38×10^{-16} erg/K) and ε_{AB} (Lennard-Jones parameter in erg) defining an approximation of energy of molecular interaction for binary systems. Often the parameters of eq 6.22 are a = 1.06036; e = 1.03587; b = 0.15610; f = 1.52996; c = 0.19300; g = 1.76474(d) = 0.47635; and h = 3.89411. Additionally, the parameters of Lennard-Jones of a binary species consisting non-polar molecules can be approximated by eq 6.23a. Further, Hirschfelder provides an adaptive relation for polar-polar and polar-nonpolar molecular interaction. In general, parameters of Lennard-Jones are obtained from value of viscosity for pure species.

Due to lack of experiment evidences, an empirical equation is employed to evaluate the parameters for pure species (see eq 6.23b). Where, V_b and T_b are volume of moles of substance in in cm^3/g mole and at normal boiling point, respectively. The value of V_b for common compounds are tabulated Table 6.4. Tyn and Calus, on the other hand, used eq 6.23c for species not available in Table 6.4.

$$\sigma_{AB} = \frac{\sigma_A + \sigma_B}{2} \text{ and } \varepsilon_{AB} = \sqrt{\varepsilon_A \varepsilon_B} \tag{6.23a}$$

$$\sigma = 1.18V_b^{1/3} \quad \text{and} \quad \varepsilon_A/\kappa = 1.15\theta_b \tag{6.23b}$$

$$V_b = 0.285V_c^{1.048} \tag{6.23c}$$

Overall, employing the fundamental principal of continuity explains the change in concentration of a species over time within a control volume is fixed in space through bulk motion of mass is taking place (see eq 6.24). The divergence term is expanded in eq 6.25. Use eq 6.26 for each species in the mixture. In case of steady state diffusion without reaction eq 6.27a can be used. In situation of diffusion only in z-direction, we can use one-dimensional diffusion eq 6.28 that can be simplified to eq 6.29.

$$\nabla N_A + \frac{\partial c_A}{\partial t} - R_A = 0 \tag{6.24}$$

$$\nabla N_A = \frac{\partial N_{A,x}}{\partial x} + \frac{\partial N_{A,y}}{\partial y} + \frac{\partial N_{A,z}}{\partial z} \tag{6.25}$$

TABLE 6.4 Molar Volumes at Normal Boiling Point.[22]

Compound	Volume, V_b (cm^3/gmole)
Air	29.9
Ammonia, NH_3	25.8
Bromine, Br_2	53.2
Carbon dioxide, CO_2	34.0
Carbon Monoxide, CO	30.7
Carbonyl sulfide, COS	51.5
Chlorine, Cl_2	48.4
Hydrogen sulfide, H_2S	32.9
Hydrogen, H_2	14.3
Iodine, I_2	71.5
Nitric oxide, NO	23.6
Nitrogen, N_2	31.2
Nitrous oxide, N_2O	36.4
Oxygen, O_2	25.6
Sulfur dioxide, SO_2	44.8
Water, H_2O	18.9

$$\nabla N_i + \frac{\partial c_i}{\partial t} - R_i = 0 \quad \text{for } i = 1, \ldots, n \tag{6.26}$$

$$\nabla N_i = 0 \quad \text{for } i = 1, \ldots, n \tag{6.26a}$$

$$\frac{dN_{i,z}}{dz} = 0 \quad \text{for } i = 1, \ldots, n \tag{6.27a}$$

$$N_i = N_{i,z} = \text{constant} \quad \text{for } i = 1, \ldots, n \text{ and } \sum_{i=1}^{n} N_i = \text{constant} \tag{6.27b}$$

$$N_i = J_i + c_i V = J_i + y_i N \tag{6.28}$$

$$N_c = -cD_{AB}\frac{dy_A}{dz} + y_A\left(N_A + N_B\right) \tag{6.29}$$

$$\frac{-dy_A}{N_A - y_A\left(N_A + N_B\right)} = \frac{dz}{cD_{AB}} \tag{6.30}$$

TABLE 6.5 Symbols in Eqs 6.39–6.51.

Quantity	Symbol	Relation	Sum	Dimensions
Mass density of *I*	ρ_I	m_i/V	ρ	M/(L3)
Mass fraction of *I*	ω_I	m_i/m_{tot}	1	–
Mass of species *I*	m_i	–	m_{tot}	*M*
Mass ratio of *I*	Ω_1	$m_i/[m_{tot} - m_i]$	–	–
Molar density of *i*	C_i	n_i/V	C	M/(L3)
Mole fraction of *I*	Xy_i	n_i/n_{tot}	1	–
Mole ratio of *I*	X_i, Y_i	n_i/n_{tot}	–	–
Moles of species *I*	n_i	–	n_{tot}	Mole
Overall mass density	ρ	m_{tot}/V	–	M/(L3)
Overall molar density	C	n_{tot}/V	–	M/(L3)
Partial Volume of *i*	V_i	–	V	(L3)
Specific molar volume	V_i	V_i/n_i	–	(L3) per mole
Specific volume of *i*	V_i	V_i/m_i	–	(L3)/M

After separating the variables in eq 6.29, we have eq 6.30. For a gaseous mixture, we assume that:

a. The temperature and pressure are constant;
b. Parameters C and D_{AB} are independent of the position and composition, and both are constant;
c. All molar fluxes are constant; and
d. Boundary conditions are: at $Z = Z_1$, $Y_A = Y_{A1}$ and at $Z = Z_2$, $Y_A = Y_{A2}$. Subscript 1 is for starting of diffusion path, and subscript 2 is for diffusion path end. Also, $Z_2 - Z_1 = Z$.

The eq 6.30 is integrated and is multiplied both sides by N_A to give eq 6.31. The molar flux (Ψ_A) is defined as in eq 6.32, where N_A gives value in positive z-direction. The flux in opposite z-direction will give a negative sign. The molar flux fraction Ψ_A depends exclusively on the relation between N_A. Combining eqs 6.31 and 6.32 will yield eq 6.33. The eq 6.33 is a fundamental relation for computation of mass transfer. The most challenging problem is computation of diffusion during turbulent nature.

$$N_A = \frac{N_A}{N_A + N_B}\left(\frac{cD_{AB}}{z}\right)\ln\left[\frac{(N_A/N_A + N_B) - y_{A2}}{(N_A/N_A + N_B) - y_{A1}}\right] \quad (6.31)$$

$$\psi_A = \frac{N_A}{N_A + N_B} \quad (6.32)$$

$$N_A = \psi_A\left(\frac{cD_{AB}}{z}\right)\ln\left[\frac{\psi_A - y_{A2}}{\psi_A - y_{A1}}\right] \quad (6.33)$$

If the flow inside a fluid has past the boundary of phase, it would lead to a velocity gradient within the fluid that results in a momentum transfer. In fact, the process of mass, momentum, and heat transfer are all related and should be considered in all realistic computation process. In case of flow over a flat plate, velocity of fluid at the surface is zero which leads to creation of boundary layer where flow is laminar. In this boundary layer, fluid stress that is shearing in nature establishes a velocity gradient (see eq 6.34), where v is the kinematic viscosity (see eq 6.35). The kinematic viscosity has the same dimensions as mass diffusivity, length^2/time. The quantity *uρ* can be interpreted as a volumetric momentum concentration. The shearing stress (τ) may also be interpreted as a viscous momentum flux toward the solid surface. Therefore, the eq 6.35 is a rate equation and is analogous to Fick's First Law. The Schmidt number (S_c) is defined as the dimensionless ratio of the momentum and the mass diffusivity (eq 6.36).

$$\tau = -\mu\left(\frac{du}{dz}\right) \quad (6.34)$$

$$\tau = -\frac{\mu}{\rho}\frac{d(u\rho)}{dz} = -v\frac{d(u\rho)}{dz} \quad (6.35)$$

$$S_c = \frac{\mu}{\rho D_{AB}} \quad (6.36)$$

When a temperature gradient exists between the fluid and the plate, the rate of heat transfer in the laminar region is given by eq 6.37, which can be rewritten as eq 6.38, where K is a thermal conductivity of the fluid; C_p is a specific heat at constant pressure; the quantity ($\theta C_p \rho$) is a volumetric thermal concentration; and $\alpha = k/C_p\rho$ is a thermal diffusivity with dimensions of length^2/time. The eq 6.38 is a heat rate equation analogous to the corresponding equations for momentum and mass transfer.

$$q = -k\frac{d\theta}{dz} \tag{6.37}$$

$$q = -\frac{k}{\rho c_p}\frac{d\left(\rho c_p \theta\right)}{dz} = -\alpha\frac{d\left(\rho c_p \theta\right)}{dz} \tag{6.38}$$

6.9 GENERALIZED MASS BALANCE EQUATION

6.9.1 REYNOLD'S TRANSPORT THEOREM

The Law of Conservation for any species (i) in any given system is an accumulation in the system, which is equal to sum all inputs minus sum of all outputs. The eq 6.39 is a algebraic form of this law, where m_i = the total amount of i in the system; t = time variable; $\sum_{in} w_i$ = the sum of all mass flows of i into the system; $\sum_{out} w_i$ = the sum of all mass flows of i out of the system; $\sum_{out} w_i$ = the sum of all mass flows of i out of the system; and r_i = the net rate at which i is produced within the system. Note that the sum of the $\dot{r}_i$ terms must equal zero in eq 6.39, since no mass can be created or destroyed (Law of Conservation of Mass). All symbols in eqs 6.39–6.51 are defined in Table 6.5.

$$dm/dt = \sum_{in} w_i - \sum_{out} w_i + r_i \tag{6.39}$$

When the eq 6.39 is summed over all the species present, we get generalized mass balance eq 6.40. For molar equivalents, eq 6.39 reduces to eq 6.41, where n_i = the total number of moles of i in the system; N_i = rate of molar flow through the system of i. Unlike the mass equivalent, when the individual mole balances for each species are summed over all species, the generation term does not necessarily disappear, since moles are not necessarily conserved during chemical reactions. Therefore, we get eq 6.42

$$dm_{tot}/dt = \sum_{in} w - \sum_{out} w \tag{6.40}$$

$$dn_i/dt = \sum_{in} N_i - \sum N_i + R_i \tag{6.41}$$

$$dn_{tot}/dt = \sum_{in} N - \sum_{out} N + \sum_{i} R_i \tag{6.42}$$

6.9.2 SPECIAL CASES

For a **steady state** process, left-hand side is zero in eqs 6.40–6.42 that reduce to eqs 6.43 and 6.44 for mass flow; and eqs 6.45 and 6.46 for molal flow

$$\sum_{in} w\omega_i = \sum_{out} w\omega_i - r_i \qquad \text{for mass} \tag{6.43}$$

$$\sum_{in} Q\rho_i = \sum_{out} Q\rho_i - r_i \qquad \text{for mass} \tag{6.44}$$

$$\sum_{in} Nx_i = \sum_{out} Nx_i - R_i \qquad \text{for moles} \tag{6.45}$$

$$\sum_{in} Qc_i = \sum_{out} Qc_i - R_i \qquad \text{for moles} \tag{6.46}$$

The Ideal Gas Law is defined in eqs 6.47–6.51:

$$PV = nR\theta \tag{6.47}$$

$$PV = R \tag{6.48}$$

$$p_i V = n_i R\theta \tag{6.49}$$

$$PV_i = n_i R\theta \tag{6.50}$$

Since,

$$p_i/P = v_i/V = n_i/n \tag{6.51}$$

$$\sum_i p_i = P \tag{6.52}$$

$$\sum_i V_i = V \tag{6.53}$$

$$\sum_i n_i = n \tag{6.54}$$

Let us now introduce the concept of *mole ratios.* The mole ratio of a species i (Y_i) in a mixture is a ratio of moles of i to molar of mixture free of i as shown in eq 6.55a for gases and eq 6.55b for liquids.

$$Y_i = \frac{n_i}{n - n_i} = \frac{y_i}{1 - y_i} \tag{6.55a}$$

$$X_i = \frac{x_i}{1 - x_i} \tag{6.55b}$$

6.10 NUMERICAL EXERCISES

What is the mass ratio for 3 L of water with density = 1000 kg/m^3 and the blood density with 1057 kg/m^3?

Solution: Volume of water, $V = 3L\left(\frac{0.001m^3}{1L}\right) = 0.003m^3$

We know $\rho = m/V$ or $m = \rho V$

or, m_a = 3 kg = (1000)*(0.003), and m_b = 3.171 kg = (1057)*(0.003)

or, $\frac{m_b}{m_a} = \frac{3.17}{3} = 1.057$, Therefore, mass ratio = 1.057

Find blood velocity and Reynolds number at the exit of a vessel with a given data?

V_1 = 2 m/s, P_1 = 68.947 N-m, P_2 = 69.637 N-m, ρ = 1057 kg/m^3, h = 0.03 m, and the viscosity is 0.004 N-s/m^2.

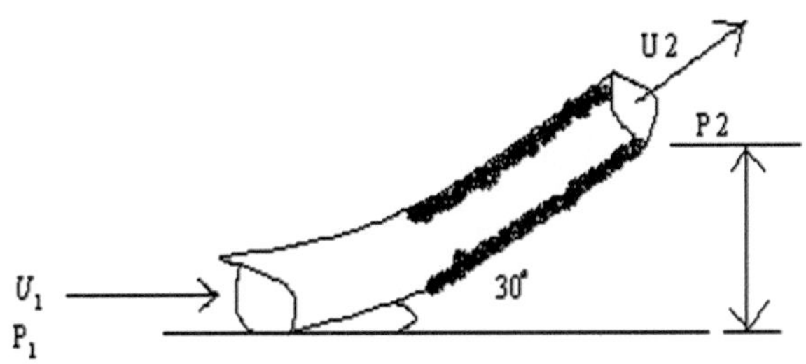

Solution: If

$$V_2 = \sqrt{2g\left(\frac{v_1^2}{2g} + \frac{P_1}{\rho g} - \frac{P_2}{\rho g} - z\right)}$$

$$= \sqrt{2\times 9.81\left(\frac{2^2}{2\times 9.81} + \frac{68.947}{1057\times 9.81} - \frac{69.637}{1057\times 9.81} - 0.03\right)} = 1.45\frac{m^3}{s}$$

Reynolds number, Re = $\rho VL/\mu$

$$\text{or, Re} = \frac{1057\times 1.45\times \frac{0.03}{\sin 30}}{0.004} = 23007$$

This means the flow is turbulent in nature.

The value of pressure P_1 and P_2 at the ends of cylindrical pipe oriented horizontally are (P_1) 0.3 and 0.6 kg_f/cm^2, that are having diameter D_1 and D_2 value of 7.5 and 15 cm, respectively. What would be the flow direction incase flow rate of water is 8.5 kg/s by considering the given parameters and mass flow rate as product of density and volumetric flow rate (AV)?

Solution:

Volumetric flow rate = 8.5/1000 = 8.5×10^{-3} m^3/s

$$V_1 = 8.5 \times 10^{-3}/(\pi D12/4) = 8.5 \times 10^{-3}/(\pi \times 0.0752/4)$$
$$= 8.5 \times 10^{-3}/0.00441 = 1.924 \text{ m/s}$$

$$V_2 = 8.5 \times 10^{-3}/(\pi D22/4) = 8.5 \times 10^{-3}/(\pi \times 0.152/4)$$
$$= 8.5 \times 10^{-3}/0.01767 = 0.481 \text{ m/s}$$

$$P_1 = 0.3 \text{ kgf/cm}^2 = 0.3 \times 9.812\ N/\text{cm}^2 = 2.9436 \times 104\ N/\text{m}^2$$

$$P_2 = 0.6 \text{ kgf/cm}^2 = 0.6 \times 9.812\ N/\text{cm}^2 = 5.8872 \times 104\ N/\text{m}^2$$

With the assumption that the direction of flow is from 1 to 2, and implementing Bernoulli's Equation in x-direction, we get:

$$(2.9436 \times 104/1000) + (1.9242/2) = (5.8872 \times 104/1000) + (0.4812/2)$$
$$+ (h + w - q)\ 29.436 + 1.851 = 58.872 + 0.116 + h + w - q$$

Therefore, here the work done in pumping the fluid, q and done by fluid, w is zero. This makes the pressure head, h a negative quantity which is not physically viable. Hence, the problem is indeed correct when flow takes place from 2→ 1

$$58.872 + 0.116 = 29.436 + 1.851 + h$$

The solution or the flow direction should be end 2 to 1.

A container includes air and CO_2 mixture of 4.5 and 2 kmol, respectively, at 25°C and at 1 bar. Whereas, air is a mixture of oxygen and nitrogen of 21 and 79%, respectively, by volume. Compute for the mixture: (1) mass of CO2, O_2, and N_2, and the total air mass; (2) % content of carbon by mass; (3) gas constant and molar mass; (4) specific volume of the mixture.

Solution:

a. Number of moles of O_2 = 0.21 × 4.5 = 0.945 kmol,
   Number of moles of N_2 = 0.79 × 4.5 = 3.55 kmol
   Mass of CO_2 = 2 × 44 = **88 kg;**
   Mass of O_2 = 0.945 × 32 = **30.24 kg**

Mass of N_2 = 3.55 × 28 = **99.54 kg**
The total mass = 88 + 30.24 + 99.54 = **217.48 kg**

b. Percentage of carbon in the mixture: (24/217.48) × 100 = **11.035% by mass.**
Molar mass = $[(n_1/n)\ m_1 + (n_2/n)\ m_2 + (n_3/n)\ m_3]$ = [(2/6.5) × 44 + (0.945/ 6.5) × 32 + (3.555/6.5) × 28] = **33.5 kg/kmol**

c. Gas constant of the mixture = [Universal gas constant / molar mass of mixture] = 8314/33.5 = **248.18 J/kgK.**

d. Specific volume of the mixture, $v = RT/p$ = 248.18 × 298/ (1 × 10^5) = **0.7395 m3/kg**

6.11 SUMMARY

This chapter has focused on the perspiration from the human body (biomass transfer). Perspiration is the method that our body uses to cool down and maintain mass equilibrium in our body. The endocrine system plays a big role in this perspiration process, to eliminate impurities throughout the pores meaning that our body breathes through our skin. The character of the skin differs over most of the body surface. There are two locations where the skin is obviously different at gross level and where there are no hairs for example soles and hands. Additionally, the epidermis in these two locations is much thicker than elsewhere and it is called thick skin. The epidermis is thinner elsewhere, and this skin is called thin skin. It contains hairs except in certain locations. The actual process of the perspiration takes place throughout different kinds of sweat glands that help to eliminate the impurities from our body. They assist in this process by placing water on the surface of the skin. The skin regulates body temperature by contracting and relaxing capillaries. It works by keeping blood closer to the surface to lower the temperature or vice versa, keeping it within the core of the skin to gain heat and get a higher temperature. The skin acts as a heat exchanger, exchanging heat to the environment (Chapter 4 in this book). This process is controlled by the hypothalamus of the brain that signals to cells and capillaries what to do (phenomenon of mass transfer).

There are several types of sweat glands, that is, sebaceous and sudoriferous that are for oil and sweat, respectively. This whole process of perspiration is made by designating body parts to have a certain overall control. Hormones send a chemical substance into the body fluids by cells, which are secreted by gland named as endocrine. The nervous system and the endocrine systems are main systems that help in the mass transfer in our body. We discussed only the endocrine system, because the nervous system relates to electric stimulus, but we emphasize that these two systems must work together to achieve the whole process of the human perspiration. The equations for mass transfer in the human body are presented.

KEYWORDS

- **active transport**
- **axilla**
- **biomass transfer**
- **blood**
- **density**
- **eccrine sweat glands**
- **endocrine glands**

REFERENCES

1. Caplan, S. R.; Essig, A. *Bioenergetics and Linear Nonequilibrium Thermodynamics: The Steady State*; Harvard University Press, 1983.
2. Brooks Cole, R. *Cell Biology and Genetics*; Campbells, 2001.
3. Mikulecky, D. C. Biological Aspects of Transport. In: *Transport Phenomena in Fluids*, Hanley, H. J. M., Ed.; Decker, 1969.
4. Solomon, E. P.; Berg, L. R.; Martin, D. W.; Villee, C. *Biology*; Saunders College Publishing, 1993.
5. Hall, J. E.; Hall, M. E. *Guyton and Hall Textbook of Medical Physiology*; Pearson & Prentice Hall, 2020.
6. Altman, P. *Blood and Other Body Fluids*; Federation of American Societies for Experimental Biology, 1961.
7. Ross, M. H.; Reitth, E. J.; Romrell, L. J. Integumentary System. In: *Histology: Text and Atlas*, 2nd ed.; Williams & Wilkins, 1989.
8. Sherwood, L. *Human physiology: from cells to systems*; Cengage Learning, 2010.
9. Weiss, M. L.; Mann, A. E. *Human Biology and Behavior: An Anthropological Perspective*, 4th ed.; Little Brown, 1985.
10. Alvarado, J.; Garcia, R.; Pena, H.; Chato, J. C., Eds. *Heat Transfer in Bioengineering*; Hemisphere Publication Corp, 1983.
11. Glasser, O. *Medical Physics*; Yearbook Publisher, 1944.

12. Greenfield, A. D. The Circulation Through the Skin. In: *Handbook of Physiology: Circulation*, Vol. 2; American Physiology Society, 1963; pp 1325–1351.
13. Roselli, R. J.; Diller, K. R. Biological Mass Transport. In: *Biotransport: Principles & Applications*, 1st ed.; Springer, 2011; pp 789–1215.
14. Skalak, R.; Chein, S. *Handbook of Bioengineering*; McGraw Hill Book Co, 1986.
15. OpenStaxCollege. The Endocrine System. https://commons.wikimedia.org/wiki/File:1801_The_Endocrine_System.jpg, (accessed 03/10/2023).
16. Kebert, T. *Human skin structure hand-redrawn into svg*. https://commons.wikimedia.org/wiki/File:Human_skin_structure.svg, (accessed 03/10/2023).
17. Blausen.com. *Hair Follicle and Hair.* https://commons.wikimedia.org/wiki/File:Blausen_0438_HairFollicleAnatomy_02.png, 2014. (accessed 03/10/2023).
18. Bronzino, J. D.; Peterson, D. R., Eds. *The Biomedical Engineering Handbook,* Vol. 1–4; CRC Press LLC (Taylor & Francis Group): Boca Raton, 2015; p 5422.
19. Leonard, E. F. *Transport and Rate Phenomena in Biological Systems;* Pearson & Prentice Hall, 2023.
20. Lih, M. M. *Transport Phenomena in Medicine and Biology;* Wiley-Interscience, 1975.
21. Goyal, M. R. Biomass Transfer in the Human Body Systems. In: *Biofluid Mechanics of Human Body Systems*, 1st ed.; CRC and Apple Academic Press, 2003, pp 137–174.
22. Benìtez, J. *Principles and Applications of Mass Transfer Operations*; John Wiley & Sons, Inc, 2002.
23. Fournier, R. L. *Basic Transport Phenomena in Biomedical Engineering*; CRC Press, 2017.
24. Reddy, J. N.; Gartling, D. K. *The Finite Element Method in Heat Transfer and Fluid Dynamics*; CRC Press, 2010.
25. Peattie, R. A.; Fisher, R. J.; Bronzino, J. D.; Peterson, D. R. *Transport Phenomena in Biomedical Engineering: Principles and Practices*, 1st ed.; CRC Press, 2012; p 197.
26. Seagrave, R. C. *Biomedical Applications of Heat and Mass Transfer*, 1st ed.; The Iowa State University Press, 1971.
27. Tosun, I. *Modeling in Transport Phenomena: A Conceptual Approach*; Elsevier, 2007.
28. Truskey, G. A.; Yuan, F.; Katz, D. F. *Transport Phenomena in Biological Systems*; Prentice Hall, 2009.
29. Geankoplis, C. J. *Mass Transport Phenomena*; Holt, Rinehart and Winston, 1972; p 495.

CHAPTER 7

Biofluid Dynamics of the Human Circulatory System*

MEGH R. GOYAL, HUMAN ARJOMANDI, STEPHANIE S. BARCELONA, SIOBHAIN L. GALLOCHER, and MARTHA VALLEJO

7.1 INTRODUCTION

The human circulatory system consists of cardiovascular and lymphatic systems that help to maintain the normal functioning of all cells within our body. The cardiovascular system functions in transportation, regulation, and protection; while the lymphatic system transports excess and the accumulated fluid from the interstitial spaces to the blood and also plays an immunological role.[1–5]

The objectives of this chapter are to discuss the properties of blood; and apply principles of fluid mechanics to the circulatory system. For further study, the reader is advised to consult the books *Biofluid Dynamics: Principles and Selected Applications* by C. Kleinstreuer; and *Biofluid Mechanics: The Human Circulation* by K. B. Chandran, Stanley E. Rittgers, and Ajit Prithiviraj Yoganathan.

7.2 DESCRIPTION OF THE SYSTEM

The cardiovascular system is comprised of the heart, blood, and blood vessels. Our heart acts as two pumps in series (see Figure 7.1). The right-hand side of the heart receives blood from the venous system and is responsible for pumping it through the pulmonary circuit. The oxygenated blood returns to

* This chapter has been modified from the review article that was prepared by my students (Human Arjomandi, Stephanie S. Barcelona, Siobhain L. Gallocher, and Martha Vallejo) for the course on Biofluid Mechanics, BME 4994. Course instructor: Megh R. Goyal, Ph.D., P.E., Visiting Professor in Biomedical Engineering, Biomedical Engineering Department, Florida International University (FIU), FL, USA. For details, contact at <goyalmegh@gmail.com>. We acknowledge the contribution by Dr James Moore Jr. and Dr Richard Schoephoerster and the faculty at FIU.

Biofluid Dynamics of Human Body Systems: Expanded and Revised Edition.
Megh R. Goyal, Arka Bhowmik, & Anamika Chauhan (Eds.)

the left-hand side of the heart, where it is pumped into the systemic circuit. Each different blood vessel type is responsible for performing a specific function. Under normal resting conditions, 70% of the blood are located in the venous system, and only 17% are located in the arterial system.[1,6,7] The cardiovascular and lymphatic systems are intertwined.

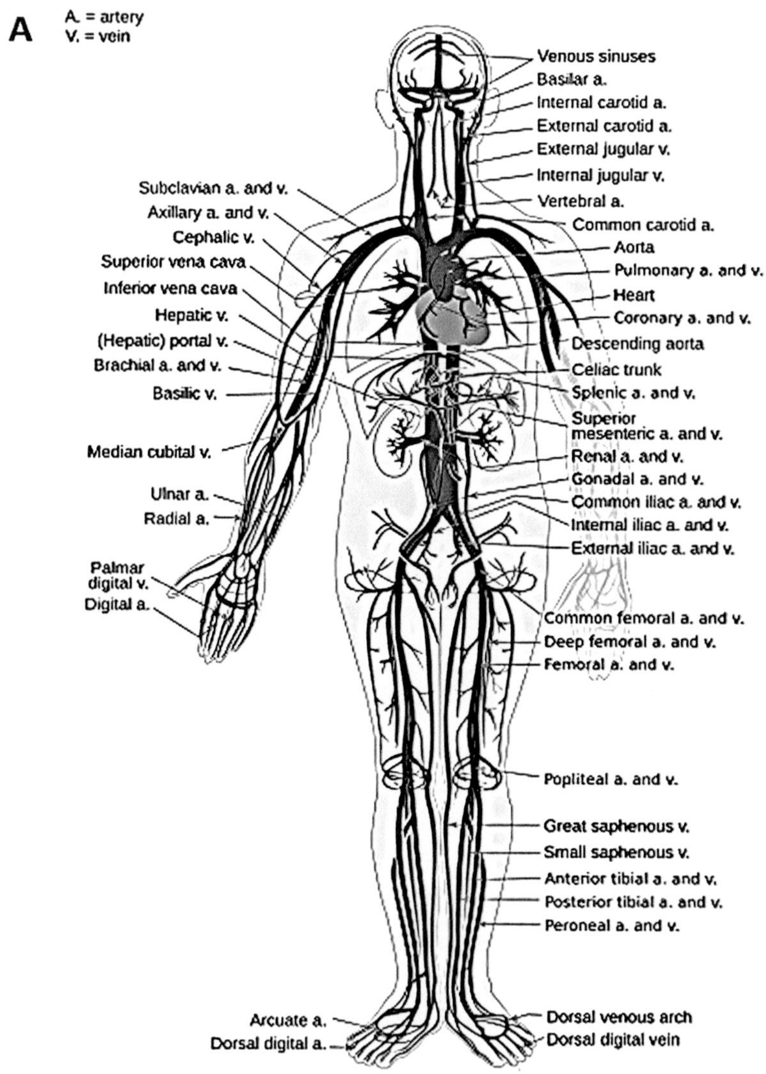

FIGURE 7.1 Schematic diagram of the circulatory system

Source: Reprinted from Ref. [10]. https://creativecommons.org/licenses/by-sa/4.0/

The lymphatic system transports capillary ultrafiltrate (combined with proteins, bacteria, fat micelles, and immune cells) from the interstitial spaces back to the cardiovascular circulation. The lymphatic system consists of lymph, a network of lymphatic vessels, lymphocytes, and lymphoid tissues and organs.[1,4,8,9]

7.3 FUNCTIONS OF THE BIOFLUID

The three types of body fluids are extracellular, intracellular, and transcellular fluids.[7,12–16] These fluids form about 40 L (60% of the body weight) for an average 70 kg male (see Table 7.1). The intracellular fluid makes up 36% of the body weight. Extracellular fluid accounts for 17% of the body weight, and the blood plasma comprises about 4% of the body weight.[10] Here, we shall consider only lymphatic fluid and natural blood. Without the blood, our body will cease to function. The constituents and functions of the blood are shown in Table 7.2. At the alveoli capillaries, the blood picks up oxygen and releases carbon dioxide. When the oxygenated blood reaches the capillaries within the tissues, the oxygen is released. The carbon dioxide is a by-product of cellular respiration and is picked up for transporting it back to the lungs.

TABLE 7.1 Human Body Fluids.[10]

Fluid	Fluid volume (L)	% of body weight
Intracellular	25	36
Extracellular	15	21
Interstitial	12	17
Plasma		4
Transcellular	-	-
Total	40	57

The blood collects and releases nutrients and hormones for distribution to target cells throughout the body, and waste products are transferred to the kidneys for excretion out of the body. The blood regulates pH and ion compositions of interstitial fluids. The diffusion gradient between the interstitial fluid and the blood results in the diffusion of water and ions down their concentration gradient until normal physiological conditions persist. Excess ions become excreted by the kidneys and through perspiration.

TABLE 7.2 Functions of the Constituents in Blood Cells.[3]

Constituents		Main functions
Formed elements (37–54%)		
Erythrocytes (RBC) (99.9%)		CO_2 and O_2 transport
Leukocytes (WBC) (<0.1%)	Neutrophil (50–70%)	Phagocytosis of bacteria
	Eosinophil (2–4%)	Defense against parasitic helminths; modulation of inflammatory processes
	Basophil (<1%)	Release of histamine and other inflammation mediators
	Monocyte (2–8%)	Generation of myonuclear-phagocyte system cells in tissues.
	Lymphocytes (20–30%)	T cells kill virus-infected cells; B cells generate antibody-producing terminal cells; Natural killer cells kill tumor and virus-infected cells
Thrombocytes (Platelets) (<0.1%)		Blood clotting
Plasma (46–63%)		
Plasma proteins (7%)	Albumins (60%)	Contributes to the osmotic pressure of plasma; transports fatty acids, thyroid hormones, and some steroid hormones.
	Globulins (35%)	Immunoglobulins attack foreign proteins and pathogens; transport globulins bind ions, hormones, and other compounds that might be lost at kidneys or have very low solubility in water.
	Fibrinogen (4%)	Converted to fibrin, which provide the basic framework for a blood clot.
	Regulatory proteins (<1%)	Catalyze specific biochemical reactions and regulate activities of other cells.
Other solutes (1%)	Electrolytes	Maintain normal extracellular fluid ion composition essential for cellular activities; contribute to osmotic pressure of body fluids.
	Organic nutrients	Needed for ATP production, growth, and cell maintenance.
	Organic wastes	Carried to sites of breakdown or excretion.
Water (92%)	Transports organic and inorganic molecules, formed elements, and heat.	

TABLE 7.3 Blood Flow to Different Organs and Tissues under Basal Conditions.[26]

Organ	%	mL/min	mL/min/100g
Adrenal glands	0.5	25	300
Bone	5	250	3
Brain	14	700	50
Bronchial	2	100	25
Heart	4	200	70
Kidneys	22	1100	360
Liver	27	1350	95
Muscle (inactive state)	15	750	4
Skin (cool weather)	6	300	3
Thyroid gland	1	50	160
Other tissues	3.5	175	1.3
Total	100.0	5000	–

TABLE 7.4 Blood Distribution by Volume and Flow Rate.[3]

Physiological structure	Typical volume		Typical flow rate, mL/min
	%	mL	
Aorta	4.5	156	5345
Arteries, larger distributing	5.4	187	695
Brain, cerebral circulation	2.3	78	–
Coronary circulation	2.3	78	267
Heart chambers	9.8	338	–
Lungs	15	520	5185
Lymphatic circulation	0.4	15	5
Muscle, inactive skeletal	21	728	802
Renal circulation	1.5	52	1149
Skin	7.5	260	428
Veins, large collecting	26.7	925	–
Vena cava	3.6	125	5185
Total	100	3462	5345

If the pH of blood drops, the blood accepts hydrogen ions and vice versa. In this manner, the blood is capable of regulating the pH. The blood prevents excessive fluid loss at sites of injury by inducing a cascade of events that leads to the formation of a clot to temporarily plug the injury site. The blood contains leukocytes and antibodies that defend against any toxins and pathogens that may enter the system. Finally, the blood, regulates overall body temperature by absorbing excess heat that is generated and redistributing it (see Chapter 4 in this book). If the body temperature rises, the blood is directed to the surface of the skin, where the heat can be lost to the environment. If the body temperature is lowered, the blood is redirected to the brain and other temperature-sensitive organs.[1,3,8,9,14,17–20]

The lymphatic system plays an essential part in maintaining a healthy body.[10,18,20] The lymph consists of fluid and solutes from the interstitial spaces. There is an imbalance in the amount of fluid the capillaries add to and remove from the tissues, so that the lymphatic system removes this fluid containing solutes as a lymph for a return to the venous system. Many substances from the tissues that are not capable of entering the bloodstream directly through the capillaries enter it through the lymphatic system. It is in this fashion that the lymph transports hormones, nutrients, waste products, and dietary lipids. The lymphatic system produces, maintains, and distributes lymphocytes within the lymph.

The lymphocytes are part of the immune response and react to the presence of invading pathogens (viruses and bacteria), abnormal body cells, and foreign proteins by eliminating them or adapting them.[1,7,21,22] Lymph is considered to be a part of the extracellular fluid system, and it is in many ways similar to plasma.

Lymph differs from plasma in that it contains a significantly lower percentage of suspended proteins. The difference in concentrations of ions between the lymph and plasma is not significant. Lymphocytes accounts for 83% of the cells in lymph. Neutrophils, eosinophils, and erythrocytes are also found in the lymph contributing to 0.9, 0.2, and 15% of the total number of cells, respectively.[23,24]

Table 7.3 summarizes the blood distribution to the different organs and tissues in the human body. Table 7.4 summarizes the blood distribution by volume and flow. From these tables, the greater the metabolism in an organ, the greater the blood flow. Table 7.5 lists the physical properties of an adult human blood. Table 7.6 indicates laboratory analysis of blood for a healthy 20-year-old female.[2] Table 7.7 presents lymph components of a healthy male.

7.4 PHYSICAL PROPERTIES OF THE BLOOD

The physical characteristics of blood remain the same for all the different sources of blood (venous blood, arterial blood, etc.). Blood temperature is roughly 38 °C (100.4 °F), pH is between 7.35 to 7.45, and viscosity is five times greater than that of water. The composition of the blood is affected by age, sex, genetic factors, environmental conditions, the physiological state of the individual, and disease.[25] Lymph is a clear, watery fluid found in the lymphatic vessels. The rate of lymph flow is determined by the product of interstitial fluid pressure times the activity of the lymphatic pump.[26]

TABLE 7.5 Physical Properties for an Adult Human Blood.[1,2]

Property	Value	
	Range	Normal
Colloid osmotic pressure: plasma (mmHg)	280–480	330
Density (g/cm^3)	1.050–1.064	1.057
pH	7.35–7.45	7.4
Refractive index	16.2–18.5	17.4
Specific gravity	1.0501–1.0619	1.057
Specific heat (g/cal)	–	0.92
Surface tension (dynes/cm)	55.5–61.2	–
Viscosity (cP)	2.18–3.59	3.0

TABLE 7.6 Laboratory Analysis of Blood for a Healthy 20-Year-Old Female.[2]

Component	Units	Average	Range
Complete metabolic profile (CMP)			
A/G ratio	–	1.3	1.0–1.9
Albumin	g/dl	4.6	3.4–5.1
Alkaline phosphatase	U/L	76	20–150
ALT	U/L	8	1–45
AST	U/L	14	1–45
Bilirubin: Direct	mg/dl	0.14	0.00–0.40
Bilirubin: Total	mg/dl	0.5	0.2–1.3
Bun/Creatinine ratio	—	11.1	7–23
Calcium	mg/dl	9.5	8.5–10.5
Carbon dioxide	mEq/L	23.3	23.0–32.0
Chloride	mEq/L	105	95–110
Cholesterol	mg/dl	141	120–200
Cholesterol/HDL ratio	–	2.7	2.9–4.3
Creatinine	mg/d	0.9	0.5–1.4

TABLE 7.6 *(Continued)*

Component	Units	Average	Range
GGT	U/L	11	1–70
Globulin, (Calculated)	g/dl	3.6	2.2–3.7
Glucose	mg/dl	91	65–115
HDL Cholesterol	mg/dl	53	35–95
Iron	mg/dl	83	40–180
LD	U/L	126	80–235
LDL Cholesterol (Calculated)	mg/dl	69	50–130
PBG	mg/gm	0.73	0.00–1.20
Phosphate	mg/dl	3.7	2.4–4.9
Potassium	mEq/l	4.1	3.5–5.5
Protein, total	g/dl	8.2	6.2–8.4
Sodium	mEq/l	142	134–147
TIBC	mg/dl	357	250–390
Trans. Sat. percent	%	23	15.0–50.0
Tryglycerides	mg/dl	94	40–200
Urea nitrogen	mg/dl	10	6.0–22.0
Uric Acid	mg/dl	4.2	1.7–7.4
Complete blood count (CBC) & diff.			
Basophils	%	0.8	0.0–2.0
Basophils, ABS	$10^3/mm^3$	0.05	0.00–0.13
Eosinophil, ABS	$10^3/mm^3$	0.06	0.03–0.55
Eosinophils	%	1.0	0.0–6.0
ESR, Westergreen	mm/h	17	0–20
Hematocrit, Hct	%	37.2	34.0–45.0
Hemoglobin, Hb	g/dl	12.5	11.6–15.1
Lymphocytes, ABS	$10^3/mm^3$	1.61	1.10–3.50
MCH	PG	30.4	26.8–33.2
MCHC	%	33.6	32.0–36.0
MCV	fl	90.5	80.0–96.0
Monocytes	%	6.3	0.0–10.0
Monocytes, ABS	$10^3/mm^3$	0.40	0.04–0.90
MPV	fl	7.3	7.5–11.5
Platelet count	$10^3/mm^3$	277	150–400
RBC	$10^6/mm^3$	4.11	3.80–5.10
RDW	%	11.9	11.0–15.0
Segmented Neutrophils	%	66.2	40.0–75.0
Segmented Neutrophils, ABS	$10^3/mm^3$	4.17	1.70–8.50
Total lymphocytes	%	25.6	15.0–50.0
WBC	$10^3/mm^3$	6.3	3.9–11.4

TABLE 7.7 Lymph Components for a Healthy Male Female.[2]

Constituent	Units	Thoracic duct lymph
Electrolytes		
Chloride	mEq/l	87–103
Phosphorous	mEq/l	2.0–3.6
Potassium	mEq/l	3.9–5.6
Sodium	mEq/l	118–132
Nitrogenous substances		
Albumin	g/100 mL	1.50–2.67
Bilirubin	g/100 mL	0.8
Creatinine	g/100 mL	0.8–8.9
Globulin	g/100 mL	1.50–4.80
Nitrogen, non-protein	g/100 mL	13.4–139.0
Protein, Total	g/100 mL	2.91–7.33
Uric Acid	g/100 mL	1.7–10.8
Lipids, carbohydrates		
Cholesterol, Total	mg/100 mL	34–106
Fat, Total	mg/100 mL	1.1
Free	mg/100 mL	15–51
Glucose	mg/100 mL	136
Enzymes		
Aldolase	units/mL	10
Alkaline	units/mL	1.9
Phosphatase	units/mL	0.7
Transaminase	units/mL	12

7.5 CHEMICAL AND BIOLOGICAL COMPOSITION OF THE HUMAN BLOOD

Blood is composed of a fluid matrix, plasma, with suspended formed elements. The formed elements with a percentage volume of 37–54% consist of three basic cell types: Erythrocytes (red blood cells), leukocytes (white blood cells), and thrombocytes (platelets). The erythrocytes with 95% of the formed elements are specialized in the transportation of oxygen in the blood. The leukocytes act as a defense mechanism for our body and are capable of identifying and removing foreign substances from the human body. The leukocytes account for less than 0.15% of the formed elements. The agranulocytes, the monocytes, and lymphocytes can identify foreign matter; and the granulocytes, the neutrophils, eosinophils, and basophils dispose-of the foreign matter. The thrombocytes account for approximately 5% of the formed elements and are involved in the process of clotting.[3,23]

Plasma makes up 46–63% by volume of whole blood. It is an aqueous, saline suspension medium that is 92% water by volume. The composition of the plasma closely resembles that of the interstitial fluid, with the major plasma ions being those similar to the interstitial fluid.

Plasma proteins make up approximately 6.5–8% by weight of the plasma. The three major plasma proteins are albumins, globulins, and fibrinogens: (a) albumins contribute to the osmotic pressure of plasma and transport lipids and steroid hormones; (b) globulins transport ions, hormones, and lipids, and participate in the immune response; and (c) fibrinogen is a vital component of the clotting system.

The remaining 2% of the plasma consists of mineral, trace elements, and electrolytes that are required by the cells for normal functioning and for the acid-base balance of the blood.[3,23] Table 7.6 shows a laboratory analysis of blood for a healthy 20-year-old female along with the normal ranges.[2]

7.6 RHEOLOGY OF THE BLOOD

Blood is a non-homogenous, ionic, and anisotropic fluid. The blood has a high molecular weight with long-chain polymers that behave in a complex way under shear stress.[3] It consists of a mixture of blood cells and plasma. Thus, blood exhibits a non-Newtonian and viscoelastic flow. In contrast, a Newtonian fluid exhibits a linear relationship between the shear stress and fluid velocity gradient[10] and the slope of this line is called a viscosity. For the non-Newtonian fluids, the slope of the line at a given value of the shear rate is the apparent viscosity. Blood can behave both as Newtonian and non-Newtonian fluid depending on the shear rate. At low shear rates, the apparent viscosity of the blood is quite high due to the presence of rouleaux and aggregates, thus exhibiting a non-Newtonian behavior. However, at shear rates above 100/s, only individual cells subsist and blood behaves as a Newtonian fluid.[10]

Blood flow in arteries, veins, and the smaller vessels is Newtonian near the vessel wall and non-Newtonian as the flow approaches the centerline of the vessel.[10] The anomalies in the viscous behavior of blood result partially from the presence and orientation of the cells suspended in plasma and partially from the accumulation of RBCs in the axial portion of the bloodstream.

Therefore, the apparent viscosity decreases with increasing velocity of flow and with decreasing vessel radius.[28] As blood leaves the aorta and flows into capillaries, the velocity of the blood decreases. The average velocity of blood in the aorta is approximately 0.3 m/s compared with 10^{-3} m/s in a capillary.[27] Blood velocity is inversely related to the total cross-sectional area of the vessels carrying the blood (Figure 2A-D). The velocity of blood at the center of the vessel is greater than the velocity near the walls (Figure 2E). This affects the distribution of RBCs in the circulatory system. It implies that the distribution of RBCs is not uniform so that there are more RBCs in the center than on the wall. The effect of this is two-folds: (a) as blood enters a small vessel branching from the side of a main vessel, the hematocrit (percentage of red blood cells in the blood) will be less than that of the main vessel; (b) extremities have a greater percentage of RBCs than blood from the heart.[27]

The rheology of blood can be significantly affected by the hematocrit and the plasma protein concentrations. Increasing the hematocrit increases the viscosity of blood (Figure 7.2F). The same effect is also seen when the fibrinogen concentration increases (eq 7.2). Fibrinogen is primarily responsible for the non-Newtonian fluid behavior of the blood. Other plasma proteins like albumin do not contribute to the non-Newtonian behavior of blood.[4,5] Viscoelastic deformation characteristics of blood can be modeled by the Power Law (eq 7.1).

$$\tau = k\left(\frac{du}{dy}\right)^{n} + \tau_{y} \tag{7.1}$$

Where τ is the time dependent stress, du/dy is the time dependent strain, K is a material dependent constitutive parameter, which reduces to the real dynamic coefficient of viscosity for a Newtonian fluid, n is the material-dependent non-linearity index that is > 0, and τ_y is the yield stress. Typical values for K and n are 0.1352×10^{-1} Pa-s and 0.7844, respectively.[3] $\tau_{y,}$ for blood (dynes per cm^2) can be calculated using eq 7.2

$$(r_y)^{0.5} = (H - 0.1)\,(C_F + 0.5) \tag{7.2}$$

where H is the hematocrit (fraction), and C_F is the plasma concentration of fibrinogen (g per 100 mL). It has been shown that the deformation properties of RBC suspensions are influenced by shear rate, hematocrit, degree of cellular aggregation, and the mechanical properties of each of the cells themselves.[3] Figure 7.2F reveals that an increase in shear rate leads to a decrease in viscosity, and an increase in hematocrit leads to an increase in viscosity. The effects of shear rate and temperature can

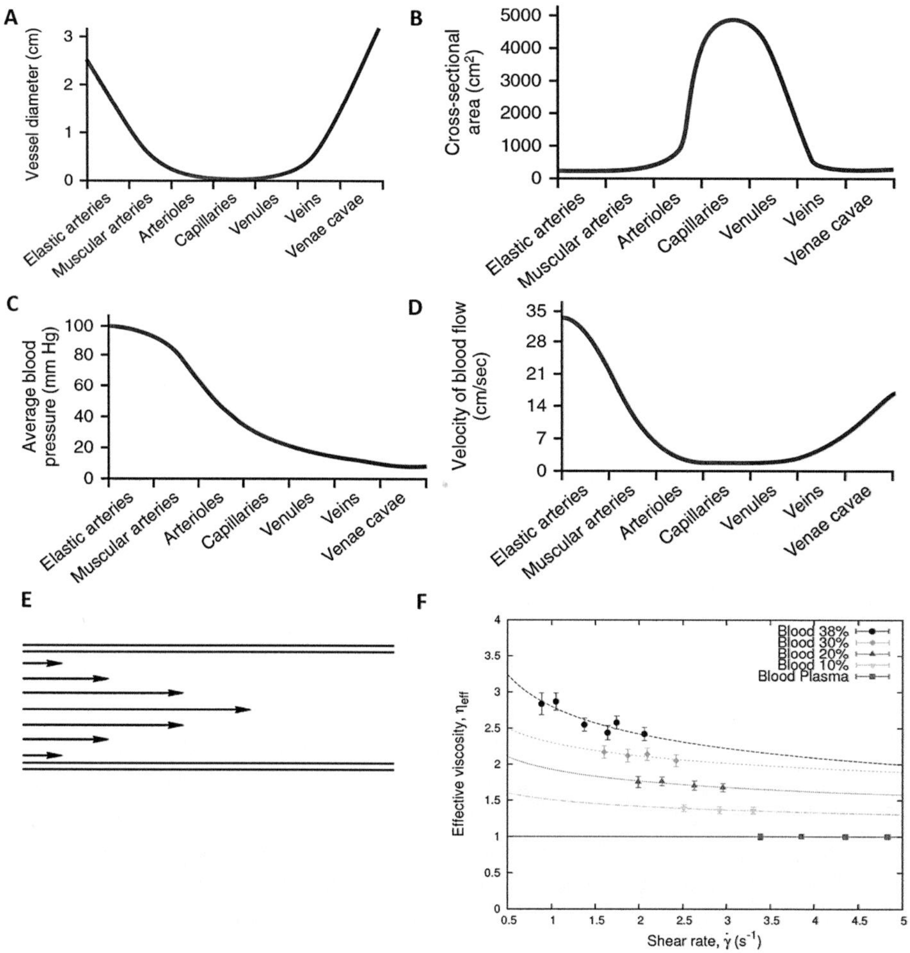

FIGURE 7.2 (A) The relationships among blood vessels that can be compared include (A) vessel diameter, (B) total cross-sectional area, (C) average blood pressure, and (D) velocity of blood flow. (E) Velocity profile of blood flow, and (F) The effective viscosity vs. shear rate relations for different percentage of hematocrit concentrations and plasma

Source A-D: Reprinted from Ref. [28]. https://creativecommons.org/licenses/by-sa/4.0/; E-F: Reprinted from Ref. [29]. https://creativecommons.org/licenses/by-sa/4.0/

be observed. Again, viscosity is shown to decrease with increasing shear rate, and a decrease in temperature led to an increase in viscosity.

The relative viscosity of blood at a shear rate >100 s^{-1} with that of suspensions with rigid spheres. At 50% concentration, the viscosity of a suspension of rigid spheres reaches almost infinity so that the suspension is not able to flow. On the contrary, normal blood remains fluid even at a hematocrit content of 98% due to the deformability of its RBCs.[30] Casson's equation is another method to describe the shear stress of whole blood

$$\sqrt{\tau} = \sqrt{\tau_y} + \eta\sqrt{\frac{du}{dy}} \tag{7.3}$$

Where the shear stress, yield stress, and shear rate is defined by eqs 7.1 and 7.2, and η is a constant. Experimental data of whole blood at 37°C and a hematocrit of 51.7%. The impact of hematocrit on the yield stress of whole blood is represented. As is expected, the yield stress increases with increasing hematocrit. The blood behaves like a non-Newtonian at low shear rates, Casson fluid; but as the shear rate increases, the blood begins to act more and more like a Newtonian fluid. Rouleaux formation increases the blood viscosity, whereas breaking up rouleaux decreases blood viscosity.[5,7]

For blood flowing through a cylindrical vessel (such as is found in the human circulatory system), the apparent viscosity decreases with decreasing vessel diameter when all other parameters are kept constant. While the blood is flowing through the vessels, the RBCs drift toward the centerline and equilibrate and align themselves at a particular radial location to decrease the amount of profile drag in the direction of flow. With this arrangement of RBCs, a cell-free layer persists at the walls of the vessel, so that fluid flowing along the walls of the vessels consists only of plasma. The cell-free layer is what leads to a decrease in the hematocrit with decreasing vessel diameter, as the size of the cell-free layer remains constant in all sized vessels. This implies that by decreasing vessel size, the cell-free layer occupies a greater volume of the vessel leading to a decreased apparent viscosity. A phenomenon called plasma skimming occurs in the body, as small branching vessels draw the majority of their fluid from the cell-free layer, thereby decreasing the hematocrit in the smaller vessels.[3,5] Lymph obeys Newton's Law of Viscosity.[3–5]

$$\tau = \mu\frac{du}{dy} \tag{7.4}$$

Where μ is the dynamic viscosity or absolute viscosity or viscosity.

7.7 CONCEPTUAL MODEL OF THE CIRCULATORY SYSTEM

An analog model of the human cardiovascular system provides a basic idea of how cardiovascular system functions. The Oxygen Plant represents the lungs, where oxygen is absorbed, and carbon dioxide is removed from the blood. The pipeline leading from the plant to pump 1 represents the pulmonary veins, pump 1 represents the left-hand side of the heart, and the pipeline leading from the pump 1 to the factory represents the systemic arterial circulation. The factory represents the capillary bed where oxygen and nutrients diffuse into the tissues. The pipeline leading from the factory to the inlet valve represents the systemic venous system, pump 2 represents the right-hand side of the heart, and the pipeline leading from the pump 2 to the plant represents the pulmonary arteries.[32]

7.8 BIOFLUID DYNAMICS

The mass blood flow problem requires the determination of equations of motion and pressure as functions of position and time. Blood is non-Newtonian fluid; flow is not steady but pulsatile; the vessels are elastic and multi-branched conduits of constantly changing diameter and shape. Therefore, a biomedical engineer must calculate blood flow in each segment of the cardiovascular system.[33] Blood flow through an organ or any vascular network is driven by a pressure gradient or perfusion pressure that is normally represented by the difference between the arterial and venous pressures across the organ. The actual blood flow at any given pressure gradient is determined by the resistance to blood flow.[33]

Blood flow can be either laminar or turbulent. Flow is laminar when the fluid travels smoothly or in regular paths. The velocity, pressure, and other flow properties at each point in the fluid remain constant. Turbulent flow occurs when the fluid undergoes irregular fluctuations or mixing. The speed of the fluid at a point is continuously undergoing changes in magnitude and direction, which results in a swirling and eddying as the bulk of the fluid moves in a specific direction. The Reynolds number is defined in eq 7.5, which can be used know if the flow is laminar or turbulent[21,27,29]:

$$\mathrm{Re} = \frac{u_m D}{\nu} \tag{7.5}$$

$$R = \frac{8 \cdot u \cdot l}{\pi a^4} \tag{7.6}$$

$$Q = \frac{\Delta P}{R} \tag{7.7}$$

$$R = R_A + R_a + R_c + R_v + R_V \tag{7.8}$$

$$\frac{1}{R} = \frac{1}{R_1} + \frac{1}{R_2} + \frac{1}{R_3} \tag{7.9}$$

The diameter (or radius), length, and viscosity of the blood are three primary factors that will determine the resistance to blood flow within a single vessel. The secondary factors are an organization of the vascular network (series and parallel arrangements), and extravascular mechanical forces acting upon the vasculature.[21,22,24] If only the primary factors are considered, Poiseuille's law can be used to find the resistance and flow in eq 7.6.

In Poiseuille's law, vessel resistance R is directly proportional to the length (l) of the vessel and the viscosity (μ) of the blood, and inversely proportional to the radius to the fourth power (a^4).[5,9,27] Volume flow rate (Q), then, can be related to the resistance (R) and pressure difference between two points (ΔP) as shown in eq 7.7. However, the eqs 7.5–7.7 are only valid for long and straight tubes, and a Newtonian fluid (e.g., water, and blood is a non-Newtonian fluid), steady and laminar flow conditions. Nevertheless, the relationship clearly shows the dominant influence of the vessel radius of resistance and flow and therefore serves as an important concept to understand how physiological and pathological changes in vessel radius affect flow.[2] The properties of the venous system are shown in Table 7.8.

TABLE 7.8 Some Properties of the Venous System.[23,28,34]

Vessel	Velocity (cm/s)	Diameter (cm)	Re
Venule	0.2	0.002	0.01
Vein	10	0.5	140
Vena Cava	38	3.0	3300

As the vessel radius decreases, there is a dramatic fall in the flow, because flow is directly related to the radius to the fourth power. Therefore, very small changes in vessel radius can have dramatic effects on flow.[33]

Within a given organ, the vascular arrangement is a combination of series and parallel elements.[4,5,33] For example, the vascular segments labeled A (large artery), a (arterioles), c (capillaries), v (venules), and V (large vein) are in-series with each other. All of the blood that flows through the large artery (A), also flows through each of the other segments. Within each of the series segments, there may be many parallel components (e.g., several parallel capillaries may arise from a single arteriole). Each vascular segment will have a resistance value (R_x) that is determined by the length and radius of each of the individual parallel vessels.[4–6] For the in-series resistance network, the total resistance (R_T) equals the sum of the individual resistances as shown in eq 7.8. For the parallel resistance network, with three parallel vessels, the total resistance is calculated using eq 7.9.

The total resistance of a network of parallel vessels is less than the resistance of the vessel having the lowest resistance. Therefore, a parallel arrangement of vessels greatly reduces resistance to blood flow. That is why capillaries, which have the highest resistance of individual vessels because of their small diameter, constitute only a small portion of the total vascular resistance of an organ or micro vascular network.

7.8.1 CORONARY CIRCULATION

Since the heart works continuously, cardiac muscle cells need reliable supplies of oxygen and nutrients. The coronary circulation supplies the heart with blood while the heart pumps the blood into the systemic circulation. During maximum exertion, oxygen demand rises and blood flow to the heart increases nine times that of resting levels. Reduced circulatory supply (or coronary ischemia) results from partial or complete blockage of the coronary arteries due to the formation of a fatty deposit, or plaque, in the wall of a coronary vessel. Blood flow is reduced as a result of the narrowing of the passageway.[32]

During maximum exertion, oxygen demand rises and blood flow to the heart increases nine times that of resting levels. Geometry and elasticity of vessel walls will determine the flow behavior. High velocity fluctuations should be avoided. Flow rate ratio, pressure and velocity gradients, flow behavior, velocity distribution, shear stress on the wall, and on blood cells are the dynamic factors that are responsible for the deposit of blood cells and lipids that are the leading cause of atherosclerosis. Interaction between blood cells and vessel wall leads to the formation of plaques and agglomeration.[9,21,22,33]

7.8.2 LAMINAR FLOW IN A TUBE—HAGEN-POISEUILLE LAW

The vessels of the arterial system are considered round; therefore, we can use equations for the flow of liquids (incompressible fluids) in a long straight tube and for steady state, laminar and Newtonian flow. The velocities in the x, y, and z directions are: $u = u(y, z)$; $v = 0$; $w = 0$. Using the Law of

Conservation of Mass,[10,18,19,21,22,36–38] we can use the continuity equation (eq 7.10).

Equation 7.11 is a Navier-Stokes equation for flow through a cylindrical tube, where *dp*/*dx* is a pressure gradient. The Navier-Stokes equation in cylindrical coordinates (x, r, θ) is shown in eq 7.12. For a symmetrical flow, the eq 7.12 reduces to eq 7.13. On integrating eq 7.13 for boundary conditions in eq 7.14a, we get eq 7.14, which is a parabolic velocity profile for the Hagen-Poiseuille flow.[14,15,39]

$$\frac{\partial u}{\partial x}+\frac{\partial v}{\partial y}+\frac{\partial w}{\partial z}=0 \tag{7.10}$$

$$\frac{\partial^2 u}{\partial y^2}+\frac{\partial^2 u}{\partial z^2}=\frac{1}{u}\frac{dp}{dx} \tag{7.11}$$

$$\frac{1}{r}\frac{\partial}{\partial r}\left(r\frac{\partial u}{\partial r}\right)+\frac{1}{r^2}\frac{\partial^2 u}{\partial \theta^2}=\frac{1}{u}\frac{dp}{dx} \tag{7.12}$$

$$\frac{1}{r}\frac{d}{dr}\left(r\frac{dr}{dr}\right)=\frac{1}{u}\frac{dp}{dx} \tag{7.13}$$

No-slip condition at the tube wall at $r = a$ and the symmetry at the center line give following boundary conditions:

$$u = 0 \text{ at } r = a \text{ and } \frac{du}{dr} = 0 \text{ at } r = 0 \tag{7.14a}$$

$$u=\frac{-1}{4\mu}\left(a^2-r^2\right)\frac{dp}{dx} \tag{7.14}$$

$$\dot{Q}=2\pi\int_0^a urdr \tag{7.15}$$

$$\dot{Q}=-\frac{\pi a^4}{8\mu}\frac{dp}{dx}=\frac{\pi a^4}{8\mu}\frac{\Delta p}{L}=\frac{\Delta p}{R} \tag{7.16}$$

$$\tau_w=-\frac{a}{2}\frac{dp}{dx}=-\frac{a}{2}\frac{\Delta p}{L}=-\left(\frac{4\mu}{\pi a^3}\right)\dot{Q} \tag{7.17}$$

Volume flow rate (= $A*V$ = continuity equation) is defined as in eq 7.15. Equations 7.14 and 7.15 will result in Hagen-Poiseuille formula as shown in eq 7.16. In eq 7.16, R = flow resistance; a = vessel radius; A = vessel area; V = average velocity; ΔP = pressure difference between two points; and μ is the dynamic viscosity or absolute viscosity or viscosity.

Wall shear stress is given by eq 7.17. As was shown in Figure 7.5B, the most important factor controlling blood flow through the artery is the radius of blood vessels. If the size of the vessel is halved, the flow rate decreases by a factor of 16, and if the vessel size is doubled, the flow rate increases by a factor of 16. A decrease in vessel size by a factor of 2 would require an increase in pressure by a factor of 16 to maintain the constant flow rate, and vice versa. Physiologically speaking, a successful way to adjust blood pressures is by changing vessel diameters. Hagen-Poiseuille's equation is not applicable to large arteries with the turbulent and pulsatile flow. Another requirement for eq 7.16 is that the tube should be rigid and long enough to avoid entrance effects. Therefore, many of the assumptions are not applicable under actual physiological conditions in a human body.[8,18,19,25] Therefore, Hagen-Poiseuille's formula can only be used for blood flow under very specific conditions.[9,11,13,18,21,22,29,31,40] The cross section of an artery with a variable vessel radius that will change the volume and volume flow rate due to the continuity equation (= AV).

7.8.3 PULSATILE FLOW IN A TUBE

Blood flow in the arteries is pulsatile, and the pulsatility decreases with distance from the heart. To investigate the flow of blood through the arteries, a simplified model is needed, and its complexity must then be increased to give an accurate representation of blood flow.[1–3,8–10,12,17,25,31] The Navier-Stokes equation for unsteady, laminar flow and an incompressible liquid is given as follows:

$$\frac{\partial^2 u}{\partial r^2}+\frac{1}{r}\frac{\partial u}{\partial r}+\frac{1}{\mu}\frac{\partial p}{\partial x}=\frac{1}{\nu}\frac{\partial u}{\partial t} \tag{7.18}$$

In the case of pulsatile flow, the pressure gradient is assumed to be a simple harmonic motion:

$$\frac{\partial p}{\partial x}=A^* e^{i\omega t} \tag{7.19}$$

Where A^* is the complex conjugate of A, and $\omega = 2\pi f$, the angular frequency in radians per second. Substituting eq 7.18 into eq 7.19, we get:

$$\frac{\partial^2 u}{\partial r^2}+\frac{1}{r}\frac{\partial u}{\partial r}=-\frac{A^*}{\mu}e^{i\omega t} \tag{7.20}$$

Equation 7.20 can be simplified by substituting $U = ue^{i\omega t}$, we get:

$$\frac{d^2 U}{dr^2}+\frac{1}{r}\frac{dU}{dr}-\frac{i\omega}{\nu}U=-\frac{A^*}{\mu} \tag{7.21}$$

Equation 7.21 is a form of the Bessel's equation. The solution with appropriate boundary conditions is as follows:

$$U = \frac{A^*}{i\omega\rho}\left\{1 - \frac{\Im_0\left[r\sqrt{(\omega/\nu)}i^{3/2}\right]}{\Im_0\left[R\sqrt{(\omega/\nu)}i^{3/2}\right]}\right\} \tag{7.22}$$

The eq 7.22 includes the Bessel function. The quantity, $R\sqrt{(\omega/\nu)}$, is a dimensionless parameter referred as Womersley number (α). It is a ratio of unsteady to viscous effects (see Chapter 2 in this book). When α becomes large, the velocity profile becomes blunt. The volume flow can be calculated in the same manner as Poiseuille flow rate. The Bessel function provides a solution to the Navier-Stokes equation for pulsatile, unsteady flow; and this more closely approximates what is happening in the arteries.[7,9,14,15,27,30,31,35]

7.8.4 MICROCIRCULATION

The forces acting on the capillary walls are shown in Figure 7.3. Capillary blood flow is often identified with the microcirculation. The microcirculation also includes blood flow to the arterioles, and venules. To distinguish between macrocirculation and microcirculation, two basic parameters are taken into account: Womersley number (α), and the Reynolds number (Re). If the Womersley number and the Reynolds number are both smaller than 1, then inertial force is not taken into account, and the flow is said to be microcirculation. If on the other hand, both numbers are larger than 1, fluid viscosity is ignored, and the flow is said to be macrocirculation. These numbers are dimensionless[9,23]:

$$\alpha = N_w = \frac{D}{2}\sqrt{\left(\frac{\omega}{\nu}\right)} \tag{7.23}$$

$$\mathrm{Re} = \frac{VD}{\nu} \tag{7.24}$$

In eqs 7.23 and 7.24, *D* is the blood vessel diameter, ν is the kinematic viscosity of the blood, and ω is the circular frequency of oscillation of the blood velocity fluctuations, V is the mean velocity of flow in the vessel. Some of the general features of the microcirculation are: hematocrit is lower in the blood than in the heart, apparent viscosity of blood decreases due to the declining concentration of red blood cells, and activated white blood cells have greater chance to stick to the wall vessel because of closer contact with the endothelium.[5]

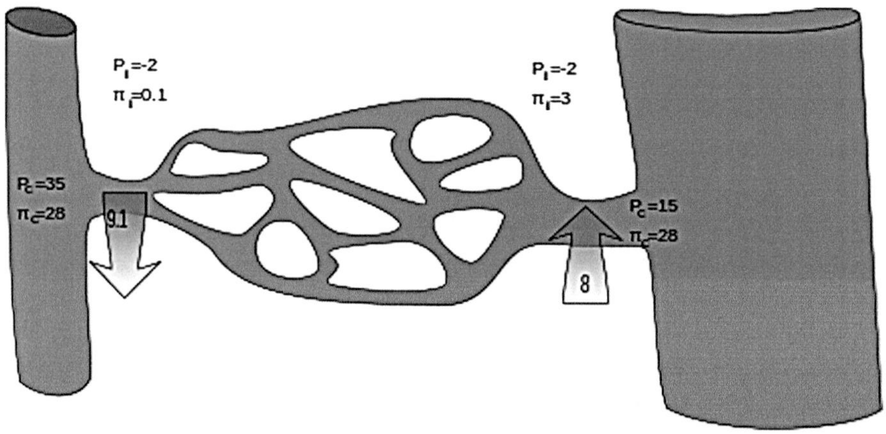

FIGURE 7.3 Forces acting across capillary walls and mechanisms of capillary exchange.
Source: Reprinted from Ref. [45]. https://creativecommons.org/licenses/by-sa/3.0/

The exchange of materials between blood and interstitial fluid across the capillary membrane is regulated by three specific mechanisms: (1) hydrostatic pressure on each side of the capillary wall; (2) osmotic pressure in the plasma and in the tissue fluid; and (3) the characteristics of the capillary wall (Figure 7.7). The pressure outside the capillary wall plays an important role in mass transport; and the flow of fluid across the capillary wall is driven by a difference in pressure across this capillary wall (see chapter 6 on biomass transfer in this book). This difference takes place due to the hydrodynamic effects, and the difference in osmotic pressure between the fluids separated by the membrane.[10] The following Starling equation represents the fluid flow across the capillary membrane, and it can also be used to describe the flow across any permeable membrane. The units of this equation are mm Hg.

$$\frac{J}{L_p S} = \left[\left(P_c - P_{if}\right) - \left(\pi_c - \pi_{if}\right)\right] = \overline{\Delta P} \tag{7.25}$$

In eq 7.25, L_p stands for hydraulic conductance, J represents the volumetric fluid transfer rate, S is the total circumferential surface area of the capillary wall, and ΔP is the effective pressure. The first term in brackets on the right side is a hydrodynamic pressure difference between the capillaries and the interstitial fluid. If this term is positive, fluid will leave the capillary; if it is negative, it fluid will flow from the interstitial fluid into the capillary. The second term is an osmotic pressure difference between the capillary and the interstitial fluid. If this term is positive, the osmotic flow of water into the capillary will occur; and if it is negative, the opposite will occur. If the first term in brackets is positive, this net osmotic pressure flows opposite to the flow induced by the hydrodynamic pressure difference.[10] The stokes eq 7.26 is for plasma, which gives pressure-flow relationship in a capillary blood vessel:

$$0 = \frac{1}{\rho}\frac{\partial p}{\partial x_i} + \upsilon \nabla^2 u_i \tag{7.26}$$

Since this is the flow in capillaries, it can be assumed that the Reynolds number and the Womersley number are less than 1, thus the acceleration term on the left side of the equation can be set to zero. Here, u_i is the velocity vector. To compute the flow in the vessel, the pressure distribution and other stresses at the entry and exit sections need to be specified.[5,9] Flow through a vessel is based on two factors: (1) pressure difference; and (2) vascular resistance.

The diffusion is another phenomenon that takes place in the capillaries (see Chapter 6 in this book). Diffusion is a mechanism for transferring nutrients and waste products between blood and tissues. Lipid-soluble substances (such as:\, oxygen and carbon dioxide) diffuse rapidly down their concentration gradients across the lipophilic membranes. Water and water-soluble substances (such as:\, ions and glucose) pass through the capillary wall at a slower rate, mainly through intercellular clefts and fenestrae.[26]

7.8.5 MECHANISMS OF CAPILLARY EXCHANGE

Fluid, electrolytes, gases, and other substances transverse the capillaries by different mechanisms, such as: (1) diffusion; (2) bulk flow; (3) vesicular transport; and (4) active transport. Diffusion is important for O_2, CO_2, and lipid-soluble substances. Fick's First Law of Diffusion for movement (or flux) across capillaries is shown in eq 7.27.

$$\frac{dn}{dt} = -DA\frac{\Delta C}{\Delta X} \tag{7.27}$$

Where $\frac{dn}{dt}$ = flux rate (moles/s); D = diffusion constant; A = surface area; ΔC = concentration difference; and ΔX = thickness of barrier to diffusion.

7.8.6 CAPILLARY FLUID EXCHANGE

Water, electrolytes, and small molecules are freely exchanged between the intravascular and extravascular compartments of our body. The primary site for this exchange is the capillary. Several mechanisms are involved in this exchange with bulk flow; and diffusion being the most important ones. The rate of exchange is determined by physical factors, such as, hydrostatic pressure and oncotic pressure.[9,26,27,31]

- Hydrostatic pressure is the principal force infiltration across capillary walls. It consists of two opposing forces: capillary hydrostatic pressure (P_C) and tissue (interstitial) hydrostatic pressure (P_T). Since $P_C >> P_T$, the net hydrostatic pressure gradient across the capillary is positive and fluid flows from the capillary into the interstitium.[9]
- Capillary hydrostatic pressure, P_c is dependent on the arterial pressure, the venous pressure, precapillary (artery and arteriolar) resistance, and postcapillary (venules and small veins) resistance. An increase in arterial or venous pressure results in an increase in the capillary hydrostatic pressure, and vice versa. Since the capillary hydrostatic pressure is higher at the arteriolar end of the capillary and lowest at the venular end, filtration is favored at the arteriolar end and reabsorption is favored at the venular end.[9,27,34] Capillary hydrostatic pressure[9] is
- determined by the equation 7.28:

$$P_c = \frac{\left[\left(\frac{R_v}{R_a}\right)P_a + P_v\right]}{\left[1+\left(\frac{R_v}{R_a}\right)\right]} \tag{7.28}$$

where P_c = capillary hydrostatic pressure; P_a = arterial pressure; P_v = venous pressure; R_a = precapillary resistance (on the arterial side); and R_v = postcapillary resistance (on the venous side).

- Tissue (interstitial) hydrostatic pressure, P_T, is dependent on the interstitial fluid volume and by tissue compliance. The normal value for P_T is nearly zero. However, during enhance filtration or lymphatic blockage, P_T dramatically increases as a result of increased interstitial fluid volume. It results in a decrease in the hydrostatic gradient across the capillary thereby limiting filtration.[9,40]
- Oncotic pressures are also two opposing oncotic pressures influencing fluid exchange: capillary plasma oncotic pressure (π_C) and tissue (interstitial) oncotic pressure (π_T).
- Capillary plasma oncotic pressure: Since the capillary barrier is permeable to ions, the osmotic pressure within the capillary is determined by plasma proteins that are impermeable. This pressure, which is generated by colloids, is referred to as capillary plasma

oncotic pressure. Albumin is responsible for about 70% of the oncotic pressure, which is typically at 25–30 mm Hg. Along the length of the capillary especially along capillaries with high net filtration, π_C increases along its length, because the filtering fluid leaves behind proteins thereby increasing the concentration of these proteins.

- Tissue (interstitial) oncotic pressure depends on the interstitial protein concentration and the reflection coefficient of the capillary wall so that increased permeability of the capillary barrier to proteins has a positive effect on π_T. This pressure is also dependent on the amount of fluid filtered into the interstitium so that an increase in capillary filtration decreases the interstitial protein concentration and it reduces π_T.[4,9,26,28] Oncotic pressure difference, $\pi = \pi_C - \pi_T$, is determined by the following equation:

$$\Delta\pi = (\sigma)\, RT\, (C_i - C_o) \tag{7.29}$$

where $\Delta\pi$ = oncotic pressure difference; σ = reflection coefficient; R = gas constant = universal gas constant/molecular weight; T = absolute temperature; C_i = solute concentration inside the capillary; and C_o = solute concentration outside the capillary. Since $\pi_C >> \pi_T$, then the oncotic pressure difference ($\pi = \pi_C - \pi_T$), if unopposed by hydrostatic forces, it would reabsorb fluid from the interstitial into the capillary.[9,28]

7.8.7 BALANCE OF HYDROSTATIC AND ONCOTIC PRESSURE

Starling Hypothesis

The Starling hypothesis is based on the relationship between hydrostatic pressure and oncotic pressure and their role in regulating fluid passage across the capillaries. This relationship can be expressed by the following equations:

$$J = k\,(\Delta P - \sigma\Delta\pi) \tag{7.30}$$

$$J = k\left\lfloor (P_c - P_i) - \sigma(\pi_p - \pi_i) \right\rfloor \tag{7.31}$$

$$J = k\,(P_c - P_o) \tag{7.32}$$

In eqs 7.38–7.40, J = rate of fluid movement; k = filtration constant for the capillary endothelium; P_c = capillary hydrostatic pressure; P_i = interstitial fluid hydrostatic pressure; σ = reflection coefficient; π_p = plasma protein oncotic pressure; π_i = intersititial fluid oncotic pressure; and $P_o = \sigma(\pi_p - \pi_i) + P_i$. Filtration occurs when the algebraic sum of the hydrostatic and oncotic pressures is positive; while absorption occurs when this sum is negative. In an idealized capillary, filtration occurs at the arterial end, while absorption occurs at the venous end. However, this is not the case in most capillaries (such as: in the renal glomerulus), where filtration can be observed over their entire lengths. In the intestinal mucosa, Starling forces change when intestinal epithelium actively pumps water from the intestinal lumen into the interstitium; thereby promoting absorption of fluid. The rate of fluid movement across the capillary (J) also depends on the filtration constant of the capillary endothelium (k):

$$k = Lp \times A \tag{7.33}$$

where k = capillary filtration constant; L_p = hydraulic conductivity of the endothelium; and A = area of the capillary wall available for filter.

7.9 ANALOG MODELS OF BLOOD FLOW

The circulatory system consists of a complex network of branching tubes, and it can adjust itself in response to certain stimuli from different systems within the body. Analog models provide a means to analyze single vessels, or even the entire cardiovascular system.

7.9.1 WINDKESSEL MODEL

The heart beats rhythmically and creates pressure and flow wave forms within the arteries. When the heart contracts, the pressure within the left ventricle increases, until it is greater than the pressure in the aorta, leading to the opening of the aortic valve. Blood flows out of the ventricle into the aorta; and through the contraction cycle, both flow and pressure decrease, until pressure within the aorta becomes higher than in the ventricle, leading to the closing of the aortic valve. The pressure in the ventricle decreases rapidly as the muscle relaxes; however, the decrease in pressure in the aorta occurs at a much slower rate as the elastic vessel acts as a reservoir. The Windkessel model is a linear mechanical model of the circulatory system.[3,6,9]

The aorta and large vessels are modeled as a compliant chamber, while the peripheral vessels are modeled as rigid tubes. According to this model:

$$V = C \cdot P \text{ and } Q_{out} = \frac{P}{R} \tag{7.34}$$

where V, P, and C are the volume, pressure, and compliance in the compliant chamber; Q_{out} is the outflow of the chamber; R is the linear resistance of the rigid tube; and P_{venous} is assumed to be zero. Law of Conservation of Mass requires:

$$Q_{in} = C\frac{dP}{dt} + \frac{P}{R} \tag{7.35}$$

7.9.2 ELECTRICAL ANALOG MODELS

Electrical analog models are used to estimate blood flow trends; and they result from the derivation of linear differential equations. In the electrical models, the vessel's resistance to blood flow is represented by a resistor, the tube compliance by a capacitor, and the blood inertia by inductance. An electrical current can represent the blood flow and the voltage can represent the blood pressure. Using Kirchoff's Law of currents, the governing differential equation for the system becomes:

$$\frac{d^2P}{dt^2} + RC\frac{dP}{dt} + P = L\frac{dQ}{dt} + RQ \tag{7.36}$$

When vessels are attached in series, a more complex model is required to explain the system that is composed of numerous compartments, $i = 1, 2, 3\ldots$ etc. Kirchoff's Law can be applied to each loop, resulting in:

$$P_i - P_{i-1} = R_iQ_i + L_i\frac{dQ_i}{dt} \tag{7.37}$$

$$P_i = \frac{V_i}{C_i} \tag{7.38}$$

$$Q_i - Q_{i-1} = \frac{dV_i}{dt} \tag{7.39}$$

It is possible to solve eq 7.38 analytically with specific boundary conditions. Therefore, this equation can be used to determine flow, pressure, compliance, or resistance in the cardiovascular system.[3,6,9,20,31] Again, the system can be solved with the appropriate boundary conditions. The RLC characteristics can be derived from the specific properties of the blood and vessels.[33]

7.9.3 FLOW IN THE ASCENDING AORTA

Flow in the ascending aorta is complex, pulsatile, and three-dimensional. The aortic flow patterns with a "Magnetic Resonance Velocity Mapping technique" has been visualized. They found that the right-handed helical flow through the ascending aorta to the aortic arch was unstable. They attributed the skewed pattern to flow in a curved tube. Both groups also noticed a retrograde flow in the inner curvature of the ascending aorta during late systole and diastole but could not speculate on the cause.[29,41] A combination of magnetic resonance imaging (MRI) and CFD has also been used to simulate flow patterns in the ascending aorta.[32,35,40] Their results revealed that: (1) in contrast to the previously reported retrograde flow in the inner curvature of the ascending aorta,[34,35] a relatively steady, large vortex during late systole and most of diastole has been observed; and (2) the skewing flow pattern seemed to be formed as a result of the acceleration movement of the ascending aorta rather than the geometric influence of curvature as previously thought.[21,27,29] Unsteady movement of the lumen had a strong effect on the flow.

7.10 RESEARCH ADVANCES

Biologists at the California Institute of Technology have uncovered an unsuspected but fundamental genetic rule governing the formation of the cardiovascular system. The cells destined to form arteries and veins are already genetically distinct at the earliest stages of blood vessel formation in the embryo. It is likely that arteries and veins will differ in their expression of many other genes that have yet to be discovered. Such genes may lead to the development of new artery- or vein-specific drugs, or may help to target known drugs specifically to either arteries or veins. Such advances could potentially enhance the efficacy or specificity of blood vessel-directed anticancer drugs (see Chapters 5 and 15 in this book).

7.11 SUMMARY

The human circulatory system, consisting of the cardiovascular and lymphatic systems, is a complex network of branching vessels, each with a different compliant nature and flow characteristics. The non-Newtonian behavior of blood makes it a unique fluid. The pressure and flow in the circulatory system is under the autonomic control, so certain vessels can dilate and constrict, and the properties of the fluid can change over time. It is almost impossible to accurately model the circulatory system as it is not possible to include all known variables in a single model. If all the known variables are included, the model would become complex and difficult to solve.

An analysis of the fluid dynamics of the arteries is not an easy task due to the many variables: fluid properties, vessel properties, and Poiseuille flow was adapted to take into account the pulsatile flow in the large vessels. Finally, wave propagation speeds were derived which took into account the compliant nature of the vessels. Analog models of blood flow, such as the Windkessel model and Kirchoff's electrical model are presented.

In the venous system, veins have less well-developed intima and media than comparable sized arteries. They represent a large reservoir for blood volume so that 50% of the circulating blood volume is in the systemic venous system at rest. In addition to viscosity of blood, the main determinants in finding the resistance and flow are the vessel diameter and length. The capillaries are most important part of the cardiovascular system because they are the only part of the system in which exchange between blood and interstitial fluid occurs. The rate of fluid exchange in the capillaries is determined by hydrostatic pressure and by oncotic pressure. The Starling hypothesis shows their relationship to each other. Hemodynamics and the interaction of blood cells with vessel wall help to understand blood flow behavior within the heart.

KEYWORDS

- **abdominal aorta**
- **alveoli**
- **aorta**
- **aortic valve**
- **arterioles**
- **artery**
- **Bessel functions**

REFERENCES

1. Aaronson, P. I. *The Cardiovascular System at a Glance*; Blackwell Science, 1999.
2. Altman, P. *Blood and Other Body Fluids*; Federation of American Societies for Experimental Biology, 1961.
3. Enderle, J.; Bronzino, J. *Introduction to Biomedical Engineering*; Academic Press, 2012.
4. Fung, Y.-C. *Biomechanics: Motion, Flow, Stress, and Growth*; Springer-Verlag, 2013.
5. Fung, Y.-C. *Biomechanics: Mechanical Properties of Living Tissues*; Springer-Verlag, 2013.
6. Kutz, M. *Standard Handbook of Biomedical Engineering Design*; McGraw-Hill, 2003.
7. Tortora, G. J.; Grabowski, R. R. *Principles of Anatomy and Physiology*; Harper Collins, 1996.
8. Cooney, D. *Biomedical Engineering Principles*; Marcel Dekker, 1976.
9. Fung, Y.-C. *Biomechanics: Circulation*; Springer-Verlag, 2013.
10. Ladyofhats, Villarreal, M. R. Simplified diagram of the human Circulatory system in anterior view. https://en.wikipedia.org/wiki/File:Circulatory_System_en.svg, (accessed 03/10/2023)
11. Hur, S.; Kim, J.; Ratnam, L.; Itkin, M. Lymphatic intervention, the frontline of modern lymphatic medicine: part I. history, anatomy, physiology, and diagnostic imaging of the lymphatic system. *Korean Journal of Radiology* **2023**, *24* (2), 95.
12. Professor, B. *Lesson on Blood Circulation—Biology 4th Grade Children's Biology*; Speedy Publishing Co, 2017.
13. Greenfield, A. D. The Circulation Through the Skin. In: *Handbook of Physiology: Circulation*, Vol. 2; American Physiology Society, 1963; pp 1325–1351.
14. Smith, D. L.; Fernhall, B. *Advanced Cardiovascular Exercise Physiology*; Human Kinetics, 2022.
15. To, A. Simple Ways to Improve Blood Circulation: The Benefits of Improving Blood Circulation for a Healthy Life. CreateSpace Independent Publishing Platform, 2017; p 35.
16. Truskey, G. A.; Yuan, F.; Katz, D. F. *Transport Phenomena in Biological Systems*; Prentice Hall, 2009.
17. Caplan, S. R.; Essig, A. *Bioenergetics and Linear Nonequilibrium Thermodynamics: The Steady State*; Harvard University Press, 1983.
18. Glasser, O. *Medical Physics*; Yearbook Publisher, 1944.
19. Huxley, T. H. *William Harvey and the Circulation of the Blood*; Amazon Books, 2022.
20. Jou, D.; Llebot, J. E., Eds. *Introduction to the Thermodynamics of Biological Processes*; Prentice Hall, 1990.
21. Geankoplis, C. J. *Transport Processes: Momentum, Heat and Mass*; Allyn and Bacon Series in Engineering, 1983; p 538.
22. Goyal, M. R. Biofluid Dynamics of Circulatory System. In: *Biofluid Mechanics of Human Body Systems*; CRC and Apple Academic Press, 2013; pp 175–214.
23. Martini, F. H. *Fundamentals of Anatomy and Physiology*; Prentice Hall, 2001.
24. Nieweg, O. E.; Essner, R.; Reintgen, D. S.; Thompaon, J. F. *Lymphatic Mapping and Probe Applications in Oncology*; Marcel Dekker Inc, 2000.
25. Bronzino, J. D.; Peterson, D. R., Eds. *The Biomedical Engineering Handbook*, Vol. 1–4; CRC Press LLC (Taylor & Francis Group): Boca Raton, 2015; p 5422.
26. Hall, J. E.; Hall, M. E. *Guyton and Hall Textbook of Medical Physiology*; Pearson & Prentice Hall, 2020.
27. Skofronich, J. G.; Cameron, J. R.; Grant, R. M. *Physics of the Body*, 2nd ed.; Medical Physics Publishing, 1999.

28. MountCastle, V. B. *Medical Physiology*; C. V. Mosby Company, 1974.
29. Geankoplis, C. J. *Mass Transport Phenomena*; Holt, Rinehart and Winston, 1972; p 495.
30. Stanely, A. B.; Stroud, J. S.; Rayz, V. Hemodynamic Simulations of Flow in Veins. In: *ASME International Mechanical Engineering Congress and Exposition*, 2001.
31. Vlachopoulos, C.; O'Rourke, M.; Nichols, W. W. *McDonald's Blood Flow in Arteries: Theoretical, Experimental and Clinical Principles*; CRC Press, 2011.
32. Dzwinel, W.; Boryczko, K.; Yuen, D. A. A Discrete-Particle Model of Blood Dynamics in Capillary Vessels. *J. Colloid Interface Sci.* **2003**, *258* (1), 163–173.
33. Lightfoot, E. N. *Transport Phenomena and Living Systems*; John Wiley & Sons Inc, 1974.
34. Moore Jr, J. E.; Weydahl, E. S.; Santamarina, A. Frequency Dependence of Dynamic Curvature Effects on Flow Through Coronary Arteries. *J. Biomech. Eng.* **2001**, *123* (2), 129–133.
35. Wood, N. Aspects of Fluid Dynamics Applied to the Larger Arteries. *J. Theoret. Biol.* **1999**, *199* (2), 137–161.
36. Reddy, J. N.; Gartling, D. K. *The Finite Element Method in Heat Transfer and Fluid Dynamics*; CRC Press, 2010.
37. Roselli, R. J.; Diller, K. R. Biofluid Transport. In *Biotransport: Principles & Applications*, 1st ed.; Springer, 2011; pp 105–485.
38. Seagrave, R. C. *Biomedical Applications of Heat and Mass Transfer*, 1st ed.; The Iowa State University Press, 1971.
39. Skalak, R.; Chein, S. *Handbook of Bioengineering*; McGraw Hill Book Co, 1986.
40. Van de Vosse, F. N.; Gijsen, F. J.; Wolters, B. J. Numerical Analysis of Coronary Artery Flow. In *Proceedings of the 2001 Bioengineering Conference*, Snowbird, Utah, June 27–July 1, 2001; American Society of Mechanical Engineers.
41. Jin, S.; Oshinski, J.; Giddens, D. Effects of Inflow Conditions and Wall Motion on Flow in the Ascending Aorta. *ASME-Publications-Bed* **2001**, *50*, 13–14.

CHAPTER 8

Biofluid Dynamics of the Human Respiratory System*

MEGH R. GOYAL, JULIAN BEDOYA, TEJAS CHOKSI, HONG GUO, and AMAURY PITA

8.1 INTRODUCTION

Respiration is an exchange of gases between the atmosphere, blood and cells; and it consists of three basic phases: pulmonary ventilation, external respiration, and internal respiration.[1,2] Pulmonary ventilation is the process of air flowing into the lungs during inspiration (inhalation) and out of the lungs during expiration (exhalation). The exchange of oxygen and carbon dioxide between the alveoli and the blood is called external respiration. Internal respiration is the exchange of oxygen from systemic blood capillaries into cells, and it relies on diffusion and partial pressure for exchange to take place. The proper functioning of the above three processes is affected by the pH of blood, partial pressure of CO_2 in the blood, temperature of the blood, and O_2 diffusion capacity. The lack of any of these factors may result in acute respiratory failure or distress. Such conditions can be regulated using mechanical ventilators that help in reproducing body's normal breathing mechanism using intrapulmonary pressure.[2–10] Recent advances have also been made in embryonic mouse cells as lung cells to treat various respiratory disorders.

This chapter focuses on the mechanics of human respiration system.

* This chapter has been modified from the review article that was prepared by students [Julian Bedoya, Tejas Choksi, Hong Guo and Amaury Pita] for the course on Biofluid Mechanics, BME 4994. Course Instructor: Megh R. Goyal, PhD, P.E., Visiting Professor in Biomedical Engineering (2003), Biomedical Engineering Department, Florida International University (FIU), 10555 West Flagler Street, Miami–FL 33174, USA. For details contact at <goyalmegh@gmail.com>. We acknowledge the contribution and suggestions by and the faculty at FIU.

Biofluid Dynamics of Human Body Systems: Expanded and Revised Edition.
Megh R. Goyal, Arka Bhowmik, & Anamika Chauhan (Eds.)

8.2 PHYSIOLOGY OF THE HUMAN RESPIRATION SYSTEM

Respiration is the process by which oxygen in the air is brought into the lungs and into close contact with the blood, which absorbs it and carries it to all parts of the body. At the same time, the blood gives up carbon dioxide, which is carried out of the lungs during exhalation. Respiration is achieved through the mouth, nose, trachea, lungs, and diaphragm (Figure 8.1) as described below:[4–7,9]

- The sinuses regulate the temperature and humidity of air breathed in, as well as to lighten the bone structure of the head and to give resonance to the voice.
- The nose cavity is the entrance of outside air into the respiratory system. The hairs that line the wall help in the air-cleaning system.
- Air also enters through the mouth (oral cavity), especially in people, who have a mouth-breathing habit or whose nasal passages may be temporarily obstructed.
- The tonsils are lymph nodes in the wall of the throat (pharynx), which are part of the germ-fighting system of the body.
- The throat (pharynx) collects incoming air from the nose and mouth; and passes it downward to the windpipe (trachea).
- The epiglottis guards the entrance to the windpipe (trachea), closing when anything is swallowed that should go into the esophagus and stomach.
- The voice box (larynx) is a place where moving air being breathed in and out creates voice sound.
- The ribs support and protect the chest cavity.
- The windpipe (trachea) divides into two main bronchial tubes, one for each lung, which subdivides into each lobe of the lungs. These, in turn, subdivide further.
- The right lung is divided into three lobes, or sections. Each lobe is like a balloon filled with sponge-like tissue. The left lung is divided into two lobes.
- The pleura is a membrane that separates the lungs from the chest wall.
- The diaphragm separates the chest cavity from the abdominal cavity.
- The smallest subdivisions of the bronchial tubes are called bronchioles, at the ends of which are the air sacs or alveoli (plural of alveolus).
- The alveoli are very small air sacs that are the destination of air breathed in.
- The capillaries are blood vessels that are imbedded in the walls of the alveoli.
- Blood is brought to capillaries by the pulmonary artery and is taken away by the pulmonary vein.

- The bronchial tubes are lined with cilia (like very small hairs) that have a wave-like motion.

This wave-like motion carries mucus (sticky phlegm or liquid) upward and out into the throat, where it is either coughed up or swallowed.

8.3 FUNCTIONS OF BIOFLUIDS[11–17]

Blood is a multifunctional fluid connective tissue; and it has five following functions:[6,11,12]

- **Transportation:** Dissolved gases (oxygen and carbon dioxide) are delivered to all tissues in the body; and the distribution of nourishment to cells through circulating nutrients is: (a) absorbed in the digestive tract or released by the liver in the post-absorptive state; (b) the distribution of hormones from appropriate endocrine glands to their target cells; and (c) the transport of metabolic wastes generated by different tissues. These wastes are filtered and expelled via the kidneys.
- **pH regulation:** Blood acts as a buffer by acting as a weak acid or weak base. The proteins in the blood are major contributors to this mechanism; and this supplies the proton donor and proton acceptor.
- **Ion composition** is also regulated by blood. Concentration imbalances are controlled by diffusion between interstitial fluids and blood.
- **Restriction of fluid losses at injury sites:** When there is a tissue injury, there are three pathways by which blood clots begin to form. The extrinsic pathway, the intrinsic pathway, and the common pathway combine to create a clot to plug the injury site. The clots are created by a series of enzymatic reactions.
- **Defense against toxins and pathogens:** White blood cells are transported within the blood and are directed to fight infections or any foreign substances or reactions that might have occurred.
- **Stabilization of body temperature:** When skeletal muscles are used, heat is generated. This heat can be used to maintain a constant body temperature of 37°C.

Blood is a non-Newtonian fluid (see Chapter 7 in this book) due to the formed elements contained in the liquid matrix.[18,19] This adds to the nonlinearity. Plasma is a non-cellular portion of the blood and is considered as a Newtonian fluid due to the nature of the absence of formed elements in the matrix.[18,19] It is part of the ECF and communicates continuously with the interstitial fluid through pores in the capillaries. Loss of plasma from the circulatory system through the capillary system and the pores is minimized by the colloid osmotic pressure exerted by the plasma proteins (see Chapter 7 in this book). The ECF and the ICF exist to regulate the electrolytic balance of the body that is crucial for maintaining homeostasis. Ions (such as potassium.) are very important for muscle contraction and therefore for regular heartbeat and heart rate. Calcium is also important in muscle contraction and control, and it is a blood-clotting factor in the intrinsic pathway. Bicarbonate ions are important for maintaining adequate pH values.

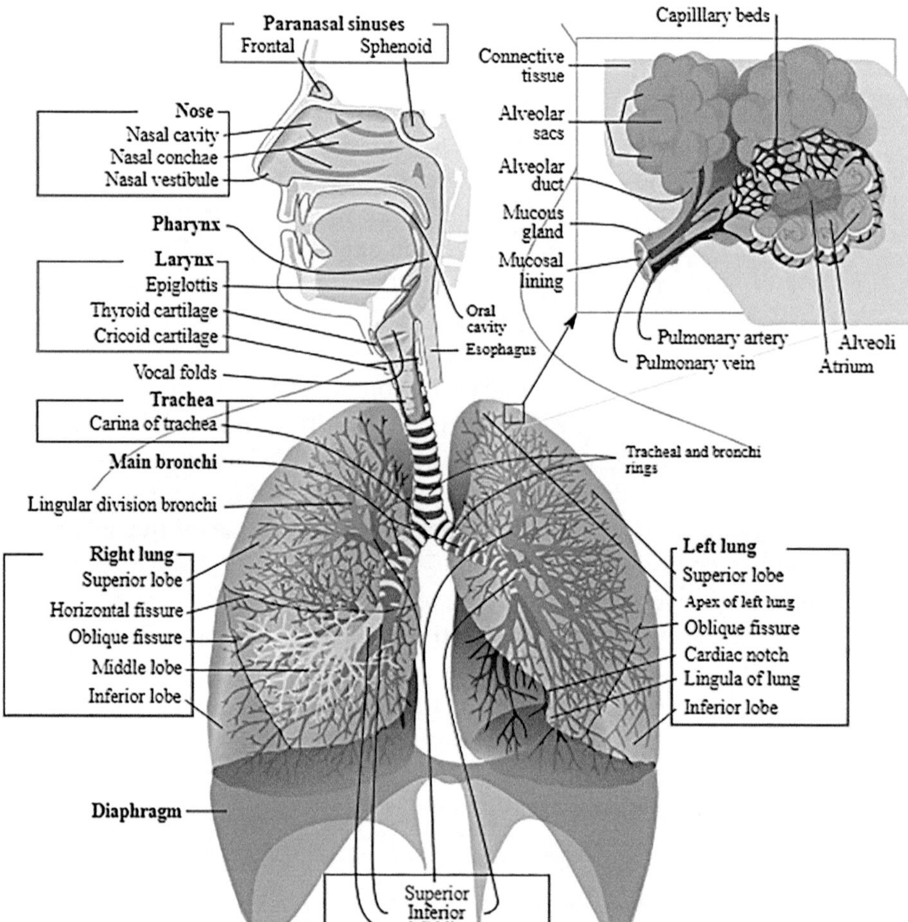

FIGURE 8.1 The human respiratory system.

Source: Reprinted from Ref. [18]. https://creativecommons.org/licenses/by-sa/4.0/

The daily secretion of saliva normally ranges between 1000 and 1500 mL. Saliva has relatively large amounts of potassium and bicarbonate ions. Saliva is mainly used to maintain oral hygiene. Since there are a large number of antibodies present in saliva, It is therefore capable of fighting off bacteria and viruses. Although saliva is mostly water, yet the organic compounds with mucous secretions are enough to render the fluid as non-Newtonian.

8.3.1 INTERSTITIAL FLUID AND LYMPH

In eq 8.1 for transcapillary water transport,[1,5,20–24] *P* and *COP* are hydrostatic pressure and colloid osmotic pressure; Subscripts *c*, *p*, and *i* denote capillary, plasma, and interstitial, respectively; σ is the capillary reflection coefficient for protein; L_p is the hydraulic conductivity of the capillary membrane; and *S* is the surface area for exchange.

In eq 8.2 for transcapillary transport of protein,[1,5,20–24] J_s and J_v are solute flux and fluid flux, respectively; C_p and C_i are protein concentrations in plasma and interstitial fluid, respectively; PS is the permeability-surface area product (the diffusive capacity of the capillary membrane).

In Starling eq 8.3 for fluid balance in microcirculation,[1,5,20–24] P_c and P_t denote capillary and tissue hydrostatic pressure; π_{pl} and π_t are plasma and tissue colloid osmotic pressures.

In Darcy's Law (eq 8.4), *k* is the permeability of fluid, *Q* is the volumetric flow rate, *A* is the average cross-sectional area of the specimen, *L* is the total length of the specimen in the flow direction, μ is the Newtonian fluid viscosity, ΔP is the hydrostatic pressure drop, and $(\Delta P/L)$ is a pressure gradient.[18]

$$J_v = L_p S\left[(P_c - P_i) - \sigma(COP_p - COP_i)\right] \tag{8.1}$$

$$J_s = J_v(1-\sigma)C_p + (C_p - C_i)PS \tag{8.2}$$

$$P_s - P_t = \pi_{pl} - \pi_t \tag{8.3}$$

$$k(\Delta P/L) = -\eta(Q/A) \tag{8.4}$$

8.4 PHYSICAL AND CHEMICAL COMPOSITION OF BIOFLUID

Table 8.1 indicates the physical properties of selected biofluids. Oxygen, carbon dioxide and nitrogen (extra cellular fluids) are dissolved gases in the blood. The oxygen allows the cells to live and get energy from the metabolism of

TABLE 8.1 Physical Properties of Biofluids in the Human Body.[6,25–28]

Biofluid	MW	Density	Specific volume	Specific gravity	pH	Surface tension	Relative viscosity (compared to water)	Gas constant	Specific heat	Colloid osmotic pressure	Average volume
Units	g/gmol	g/cm^3	cm^3/g	<——>	<——>	dynes/cm	<——>	J/kg.K	g.cal	mm Hg	ml × (10^3)
Blood	NA	1.06	0.94	1.06	7.35–7.45	56–61	4.71	NA	0.92	NA	5
Plasma	NA	1.03	0.97	1.03	7.30–7.40	70	1.32	NA	0.94	25	3
Extra-cellular fluid	NA	NA	NA	NA	NA	25–40	NA	NA	NA	0-5	16
Intra-cellular fluid	NA	NA	NA	NA	NA	25–40	NA	NA	NA	0-5	23
Lymph	NA	NA	NA	NA	NA	25–40	NA	NA	NA	NA	NA
Oxygen	32	0.0014	714	0.0014	NA	NA	NA	260	909 J/kg.K	NA	NA
Carbon dioxide	44	NA	NA	NA	NA	NA	NA	189	840 J/kg.K	NA	NA
Nitrogen	28	NA	NA	NA	NA	NA	NA	297	1039 J/kg.K	NA	NA
Air at STP	28	0.0012	869	0.0016	NA	NA	0.019	289	1004	NA	NA

this gas.[11–17] Carbon dioxide is a by-product of respiration. Nitrogen is present always, but it cannot be absorbed from the atmosphere, because it is an inert gas. Table 8.2 indicates the chemical composition of biofluids in the human body for an adult. Table 8.3 presents the body water content for different age groups of a person. The inhaled and exhaled gases have the same amount of inert and N_2 gases (Figure 8.1C). The inhaled gases are rich in O_2 and deficient in CO_2 while the exhaled gases are rich in CO_2 and deficient in O_2.

TABLE 8.2 Composition of Biofluids.[25]

Dissolved materials	Intracellular fluid (mEq/lmEq/l)	Extracellular fluid (mEq/lmEq/l)			Pleural fluid (mEq/lmEq/l)
		Interstitial fluid	Plasma	Lymph	
Cations					
Na^+	10	145	153	127	140
K^+	159	4	4.3	4.7	4.8
Ca^{2+}	1	3	5.4	4.2	4.3
Mg^{2+}	14	2	2.2	2.4	1.71
Anions					
HCO_3^-	7	28	27	23	23.8
Cl^-	3	117	111	98	100
Other (PO_4^{3-}, SO_4^{2-}, etc)	155	9	9	8	9
Protein	45	1	18	11	1.77

TABLE 8.3 Body Water Content for Various Age Groups.[25]

Subjects	Total body water (mL/kg)	Range (mL/kg)
Newborn	754	647–861
Newborn-6 months	NA	720–830
6 mo-11 yr	NA	530–630
1–12 mo	627	504–750
1–10 yr	569	491–647
18–46 yr	611	430–729
17–39 yr	548	NA
20–39 yr	444	NA
26–65 yr	497	463–531
40–59 yr	545	NA
>60 yr	516	NA

8.5 RHEOLOGY OF BIOFLUIDS[18]

Water, air, ethanol, benzene, glycerin, and gases are Newtonian fluids.[18] This means that when shear stress is plotted against the shear rate at a given temperature, the plot shows a straight line with a constant slope that is independent of shear rate. The slope of this line is called the viscosity of the fluid. The non-Newtonian fluids do not obey Newton's law of viscosity (for more details, refer to Chapters 1 and 7 in this book).

8.5.1 NON-NEWTONIAN FLOW (SEE CHAPTER 7 IN THIS BOOK)

In the constitutive eqs 8.5 and 8.6 for blood,[1,5,20–24] σ_{ij} is the stress tensor, V_{ij} is the strain rate tensor, u_i is the velocity component, δ_{ij} is the isotropic tensor, p is the hydrostatic pressure, and μ is the viscosity. In eq 8.7 for the shear rate and shear stress relationship of a Bingham fluid,[1,5,18,20–24] η_B is the Bingham viscosity and τ_B is the yield stress. In eq 8.8 for shear rate and shear stress relation for Casson fluid,[1,5,20–24] η_c is the Casson viscosity and τ_c is the Casson yield stress.

$$\sigma_{ij} = -p\,\delta_{ij} + 2\mu V_{ij} \tag{8.5}$$

$$V_{ij} = \frac{1}{2}\left(\frac{\partial u_i}{\partial u_j} + \frac{\partial u_j}{\partial u_i}\right) \tag{8.6}$$

$$\dot{\gamma} = \begin{Bmatrix} \frac{1}{\eta_B}(\tau - \tau_B), & for\ \tau \rangle\ \tau_B \\ 0, for\ \tau\ \langle \tau_B & \end{Bmatrix} \tag{8.7}$$

$$\sqrt{\dot{\gamma}} = \begin{Bmatrix} \frac{1}{\sqrt{\eta_c}}(\sqrt{\tau} - \tau_c), for\ \sqrt{\tau}\rangle\ \sqrt{\tau_c} \\ 0, for\ \sqrt{\tau}\ \langle\ \sqrt{\tau_c} \end{Bmatrix} \tag{8.8}$$

8.6 BIOFLUID MECHANICS[8,11–14,16,17,29]

The change in alveolar ventilation is inversely related to changes in arterial P_{CO2} and directly proportional to total body CO_2 production. The eq 8.9 shows that the arterial partial pressure of CO_2 is directly proportional to the amount of CO_2 produced by metabolism and delivered to the lungs (V_{CO2}); and is inversely proportional to the alveolar ventilation (V_A).

Eq 8.10 shows arterial oxygen content and CaO_2. In eqs 8.9 and 8.10, P_{CO2} = arterial partial pressure of CO_2, V_{CO2} = CO_2 production by the body, V_A = alveolar ventilation (L/min), CaO_2 = arterial oxygen content, PaO_2 = arterial P_{O2} (mmHg), Hb = hemoglobin content, and SaO_2 = saturation of hemoglobin with O_2 expressed as a percentage.

The effects of pH on the percent saturation of hemoglobin. Figure 8.3C shows the percent saturation of hemoglobin at various partial pressures of oxygen. At a partial pressure of O_2 of 100 mmHg, the hemoglobin is 100% saturated. This is important because blood entering the lungs picks up and binds to oxygen readily.

$$Pa_{CO_2} = 0.863 \times \left({}^{V_{CO_2}}\!/_{V_A} \right) \tag{8.9}$$

$$CaO_2 = (SaO_2 \times Hb \times 1.34) + 0.003\ (PaO_2) \tag{8.10}$$

8.6.1 *HENDERSON-HASSELBALCH EQUATION*

In eq 8.11 for the changes in arterial P_{CO2}, HCO_3^- is bicarbonate ions, and $PaCO_2$ = arterial P_{CO2} in mmHg. The Henderson–Hasselbalch equation shows changes in pH. As pH drops, the affinity of Hb for oxygen drops for the oxygen dissociation curve.

$$pH = pKa + \log \left\{ \frac{[HOC_3]}{[0.03 \times P_{CO_2}]} \right\} \tag{8.11}$$

8.7 TRANSPORT OF GASES[2,3,5,8,9,29,30]

The transport of O_2, CO_2, anesthetics, toxins or any other soluble gas occurs by convection, diffusion and their interplay. Deoxygenated blood is pumped from the right ventricle of the heart to the lungs.

The partial pressure of oxygen in the deoxygenated blood is lower than that of the oxygen in the air in the alveoli. Owing to this difference, there is a net diffusion of oxygen from the alveoli across the cell membranes and into the blood. At the same time, carbon dioxide present in the blood diffuses into the alveoli, because there is an imbalance in the partial pressures in the opposite direction. Three integrated steps are involved in external respiration:

- **Pulmonary ventilation:** Breathing involves the physical movement of air into and out of the lungs. When the person is relaxed and breathing

quietly, the pressure difference between atmospheric and intrapulmonary pressures is relatively small. On inhalation, the lungs expand, and the intrapulmonary pressure drops to about 759 mmHg. Because the intrapulmonary pressure is 1 mmHg below atmospheric pressure, it is generally reported as "–1 mmHg." On exhalation, the lungs recoil and intrapulmonary pressure rises to 761 mmHg, or ±1 mmHg.
- **Gas diffusion** occurs across the respiratory membrane between alveolar air spaces and alveolar capillaries.
- **Alveolar ventilation** is the amount of air reaching the alveoli each minute. At rest, the alveolar ventilation rate is approximately 4.2 liters per minute (Lpm).

8.7.1 OXYGEN TRANSPORT

Once the oxygen is in the blood stream, it is transported to the cells. Oxygen does not easily dissolve in water; therefore, only a small amount is carried in the blood plasma while the remainder is transported by combining with hemoglobin in RBCs. Approximately, each 100 mL of oxygenated blood contains about 20 mL of oxygen that consists of 19.7 mL of O_2 combined with the hemoglobin and 0.3 mL of O_2 dissolved in the plasma. Oxygen and hemoglobin combine in an easily reversible reaction to form oxyhemoglobin. The factors (Table 8.4) to determine how much oxygen combines with the hemoglobin are:

- The partial pressure of oxygen in arterial blood.
- pH of the blood: hemoglobin has less affinity for oxygen at higher levels of acidity.
- The partial pressure of carbon dioxide in the blood: at higher P_{CO2}, oxygen is released by hemoglobin more readily.
- At high temperatures, oxygen disassociates more readily from hemoglobin.

The hemoglobin structure and active sites for attachment with oxygen.[8] The method of cooperative binding and release between oxygen and hemoglobin can be seen in the dissociation curve for hemoglobin. The active tissues may cause an alteration in the structure of hemoglobin, which in turn, causes hemoglobin to give up its oxygen. In other words, the oxygen–hemoglobin dissociation curve "shifts to the right." This means that at a given partial pressure of oxygen, the percent saturation for hemoglobin is lower. For example, a value of P_{O2} = 40 gives hemoglobin saturation of about 75%. While for a "right-shifted" curve: a value of $P_{O2} = 40$ gives hemoglobin saturation of about 60%. Therefore, a "shift to the right" in the oxygen–hemoglobin dissociation or saturation curve means that more oxygen is being released by hemoglobin. O_2 saturation is given as the fraction or percentage of all the hemoglobin-binding sites that are currently occupied by oxygen (as shown in eqs 8.12 and 8.13).

$$Hb + O_2 \Leftrightarrow HbO_2 \tag{8.12}$$

$$\%Sat = \frac{[Total\,O_2\,Content - Dissovled\,O_2\,Content]}{[O_2\,Capacity\,of\,Hb]} \tag{8.13}$$

8.7.2 CARBON DIOXIDE TRANSPORT

Carbon dioxide, once produced in the tissues, needs to be transported to the lungs for elimination. There are three mechanisms involved in CO_2 transport: (a) about 7% of CO_2 crosses the red cell membrane and dissolves into the plasma; (b) about 23% of CO_2 binds to the multiple amino groups of hemoglobin (carboxyl hemoglobin); and (c) about 70% is carried as bicarbonate ions.

Carbon dioxide created by respiring cells diffuses into the blood plasma and then into the RBCs, where most of it is converted to bicarbonate ions. It first reacts with water forming carbonic acid, which then breaks down into H^+ and HCO_3^-, as shown in eqs 8.14 and 8.15. Sixty percent of CO_2 is converted to HCO_3^- and H^+ by carbonic anhydrase within RBCs. HCO_3^- then diffuses out of the RBCs, and Cl^- diffuses into the RBC to restore the electrical gradient. This is called the chloride shift or Hamburger shift. The chloride shift is rapid and is complete before the cells exit the capillary. The osmotic effect of the extra HCO_3^- and Cl^- in venous red cells causes the venous RBCs volume to increase slightly.

$$R - NH_2 + CO_2 \Leftrightarrow R - NH - COO^- + H^+ \tag{8.14}$$

$$CO_2 + H_2O \rightarrow H_2CO_3 \rightarrow HCO_3^- + H^+ \tag{8.15}$$

The Haldane effect is the removal of O_2 from hemoglobin increases the ability of hemoglobin to pick up CO_2 and CO_2 generated H^+. Increasing O_2 tension decreases the affinity of hemoglobin for CO_2. As a result, the CO_2 "dissociation curve" shifts downward. This effect shows that high P_{O2} promotes CO_2 unloading in the lungs. Increases in $[H^+]$ cause a reduction in oxygen affinity for hemoglobin, and this is called the Bohr Effect. Table 8.4 presents the effect of CO_2 on fluids with and without buffer.

8.7.3 RESPIRATORY REGULATION OF ACID–BASE BALANCE

Blood has a normal pH in ranging from 7.35 to 7.45. When the blood pH is < 7.35, the condition is called acidosis, and when the pH is >7.45, it is called alkalosis. Inadequate ventilation of the lungs causes respiratory acidosis (the rate at which carbon dioxide is eliminated from the body fluids through the lungs decreases). This increases the concentration of carbon dioxide in the body fluids. As carbon dioxide levels increase, excess carbon dioxide reacts with water to form carbonic acid. The carbonic acid dissociates to form hydrogen ions and bicarbonate ions. The increase in hydrogen ion concentration causes the pH of the body fluids to decrease. The major effect of acidosis is depression of the central nervous system.

Respiratory alkalosis results from hyperventilation of the lungs. This increases the rate at which carbon dioxide is eliminated from the body fluids and results in a decrease in the concentration of carbon dioxide in the body fluids. The resulting decrease in the concentration of hydrogen ions causes the pH of the body fluids to increase. If the pH of body fluids increases above 7.45, symptoms of respiratory alkalosis become apparent. A major effect of alkalosis is hyperexcitability of the nervous system resulting in spontaneous nervous stimulation of muscles.

The Henderson-Hasselbach eq 8.16 gives a relationship between P_{CO2} and $[H^+]$ concentration. The eq 8.16 shows that an increase in P_{CO2} results in respiratory acidosis while a decrease in P_{CO2} results in respiratory alkalosis. Effects of pH of arterial blood on the alveolar ventilation are shown in Figure 8.5E.

$$pH = 6.1 + \log \frac{[HCO_3^-]}{(0.03 mEq / L / mmHg)(P_{CO_2})} \tag{8.16}$$

8.7.4 PULMONARY VOLUMES AND CAPACITIES

A spirometer yields a graph that is called a spirogram. The graph describes the air movement in and out of the lungs. To describe the events of pulmonary ventilation, the air in the lungs is subdivided at different points on the diagram into four volumes and four capacities. The four different pulmonary volumes are:[4,5,7,9]

- **Tidal volume** is a volume of air inspired or expired with each normal breathes.
- **Inspiratory reserve volume** is a volume of air that can be inspired above the normal tidal volume.

- **Expiratory reserve volume** is a volume of air that can be expired by forced expiration after normal tidal expiration.
- **Residual volume** is a volume of air remaining in the lungs after forceful expiration.

The four different pulmonary capacities are:

- **Inspiration capacity** = [Tidal + Inspiratory Reserve]
- **Functional residual capacity** = [Expiratory reserve + Residual]
- **Vital Capacity** = [Inspiratory reserve + Tidal + Expiratory Reserve]
- **Total Lung Capacity** = [Vital Capacity + Residual]

8.8 CONCEPTUAL MODEL OF THE HUMAN RESPIRATORY SYSTEM

Researchers have developed a physical model of the human respiratory system to simulate both the mechanics of air transport and the electromechanical forces responsible for gas exchange. This model includes four distinct regions, each characterized by a particular set of functions (Table 8.1). These regions are connected. The dotted line is the control volume and it represents the thoracic cage. When air is inhaled, the pressure difference between pleural pressure (P_{PL}) and atmospheric pressure increases and air flows through the three different airways and into the alveolar region, where it interacts with blood. The upper resistance airway has a resistance R_{uaw}, which is a function of the flow rate Q. The resistance in the collapsible airway R_c and the recoil pressure P_{TM} are functions of the local volume, V_c. On exhalation, it follows the opposite path to the atmosphere. P_{EL} characterizes the compliance of the collapsible and lung compartments.

TABLE 8.4 Effects of CO_2 on Fluids With and Without Buffer Other Than Bicarbonates.[8,30]

Type of fluid	pH	$[HCO_3^-]$ (mEq/L)	$[H^+]$ (mEq/L)	P_{CO2} (mmHg)
Fluid with no non-bicarbonate buffers of importance (CSF)	7.6	24.2	25	25
	7.2	24.2	63	63
Fluid with buffers in addition to bicarbonate (blood)	7.6	20	28	23
	7.3	30	56	70

8.9 MECHANICS OF PULMONARY CELLS

The lung is a dynamic organ that is subjected to varying mechanical forces throughout life. During fetal development, the lung is subject to both tonic distention and fetal breathing movements. After birth, cyclical expansion and relaxation is brought about by changes in transpulmonary pressure produced by the muscles of the chest – wall and diaphragm. Changes in lung volume not only result in airflow but can also affect many aspects of lung metabolism, function and growth. Mechanical forces may be secondary to gradients in gravity, motion, osmotic forces and interactions between cells and/or matrix components. Cells within a complex three-dimensional structure (such as: the lung) are subjected to a variety of different physical forces, thus causing some degree of strain.

The main types of cells in the lung are: lung fibroblasts, pleural mesothelial cells (PMC), airway smooth muscle cells, cells of the pulmonary artery wall, airway epithelial cells, alveolar type II cells, and PMC. PMC are subjected to mechanical stimuli from distention of the lung during inflation and to shear forces exerted by a fluid film between the visceral and parietal pleura. It has been reported that the average diameter of visceral PMC was increased from 27 to 45 μm during lung expansion in rabbits. Shear force was as high as 20 dynes/cm^2 at a breathing frequency of 20 min^{-1}.[32,33]

There have been several different approaches to quantifying the extent to which cells undergo mechanical deformation with changes in the extent of lung inflation. Because changes in either collagen fibril length or basement membrane surface area may correlate with the extent of distention/contraction of the basal portion of the epithelial cell; therefore, these measurements may reflect how inflation and deflation affect the basal cell surface area. These studies suggest that, when lung volume increases from 40 to 100% of TLC, epithelial cell basal surface area increases from 16 to 34%.[13,32]

The response of cells to mechanical deformation may not be linear, may be due to alterations of cell–matrix adhesions. This underlines the importance of measuring the extent of cell deformation in different experimental models. Shear stress forces may play a role in cells other than vascular endothelium, such as the effects of pleural fluid on PMC and the fluid hypophase on airway and alveolar epithelial cells. Vascular endothelium might also be subjected to strain and hydrostatic pressure. Also, physical forces can be generated within cells. The strain can be generated by cytoskeletal rearrangements (such as actin contraction) that caused tension to be transmitted throughout the cell, including the nucleus.

The term "autobaric effects" has been proposed for these types of physical forces that are generated intracellularly in contrast to "parabaric effects" generated by intercellular or cell–matrix interactions. The authors have proposed a model in which the basal equilibrium stress state of cells can be altered by extra cellular perturbations. As a consequence of perturbations and to minimize intracellular strain, the interactions between cytoskeletal components and integrins may be strengthened, actin cables may polymerize, and tension on focal adhesions may increase, resulting in the stimulation of matrix protein synthesis and secretion and an increase in focal adhesion complexes. These complex interactions ultimately result in the realignment of cell structure. The opposite may occur, when strain is reduced.

8.10 BIOFLUID DYNAMICS OF THE HUMAN RESPIRATORY SYSTEM[11,12]

The functions of the upper human respiratory system are: (a) to filter incoming air from pathogens, foreign substances and debris; and (b) to humidify and temper the air to suit the delicate environment within the lower respiratory system. This is accomplished by the flow characteristics within the nasal cavity.

The nasal passage creates flow disturbances, such as turbulent flow and recirculation. The air is saturated with water vapor and its temperature increases closer to 37°C.[5,6] The flow field characteristics are of interest in pulmonary mechanics. The flow in the respiratory system is described by eqs 8.17 and 8.18, where is density, V is the general velocity vector, and the del operator ∇ is the gradient of a vector producing a spatial vector field.

The air density in the respiratory tract can be considered to be constant since there are no large-scale deformations, large pressures or extreme temperatures that would change this parameter.[18,34] In the Law of Conservation of Momentum Equation,[18] only the x-component is shown for simplicity (the u-component of velocity). *P* represents the pressure; *g* is the acceleration due to gravity and μ is the dynamic viscosity. The second-order derivatives in the right side of the equation yield a second-order partial differential equation. The spatial derivatives in the left side of the equation multiplied by the velocity components introduce nonlinearity to this equation. If these terms are not negligible, the equations will become difficult to solve analytically.

$$\nabla \rho \vec{v} + \frac{\partial \rho}{\partial t} = 0 \tag{8.17}$$

$$\rho\left(\frac{\partial u}{\partial t} + u\frac{\partial u}{\partial x} + v\frac{\partial u}{\partial y} + w\frac{\partial u}{\partial z}\right) = \rho g - \frac{\partial p}{\partial x} + \mu\left(\frac{\partial^2 u}{\partial x^2} + \frac{\partial^2 u}{\partial y^2} + \frac{\partial^2 u}{\partial z^2}\right) \tag{8.18}$$

In a study by Renotte, Bouffioux and Wilquem,[35] the laryngeal channel was modeled and analyzed numerically in three dimensions with an oscillatory flow to quantify shear stresses and pressure drops in the human larynx. It is an established fact that the upper respiratory system accounts for 30% of the total airway resistance.[35] The study included the compliance of the larynx and the variation in the cross-sectional area that occurs with each cycle of respiration.

The glottis become larger during inspiration, thus lowering the overall flow resistance from the mouth to the lungs. During expiration, the glottis reduce the cross-sectional area thus increasing flow resistance. On the whole, the effective flow area changes from 70 ± 7 to 126 ± 8 mm^2 during inspiration to expiration process.[35] It is known that changes in area will affect the pressure of the fluid and the velocity (Continuity equation).[18] Therefore, a strong jet stream is expected when the area of the glottis is small, and not when the area is larger.[18,35] Other characteristics of this study included a maximum flow rate of 0.75 L/s that corresponds to a quiet breathing in a healthy adult. The corresponding breathing frequency was 0.25 Hz (one full cycle of inspiration and expiration every 4 seconds). Another parameter is a Reynolds number that is a relative measure of inertial effects to viscous effects (See Chapter 2 in this book).

A laminar regime consists of a flow without circulations, secondary flows, separations or wake turbulence, and the velocity profile is parabolic in a two-dimensional plane. A turbulent flow has fluctuations, and the velocity profile has a blunter characteristic.[6,23] In many cases, the flow regime in the respiratory system is transitional. It contains characteristics of both flow regimes at different points in time. The bends, bifurcations, and high velocities generally lead to turbulent flow. The analysis is complicated due to different characteristic branches of the respiratory system. These branches begin at the trachea for generation zero and the main bronchus is of generation one. Successive generations follow until generation number 23 at the alveoli. Generations 0–16 are termed as conducting airways, generation number 16 is termed terminal bronchiole and generations 17–23 have alveoli and are known as the respiratory zone. It is in this zone where gas exchange occurs.[36] The speed decreases with the generation until it reaches the alveoli, where the flow is nearly stagnant. This is also confirmed by the decreasing magnitude of the Reynolds number. This is evidently a manifestation of how frictional losses become increasingly important as the diameter of the airway decreases. As a result, the pressure also decreases accordingly.

Liu, So, and Zhang used computational fluid dynamics (CFD) for a three-generation bifurcation that was modeled in three dimensions.[34] The motivation to use a complex three-dimensional model lies in the fact that backflow and vortices are common in respiratory flow fields. A two-dimensional model does not include these disturbances, which greatly hinders the validity of results.[34] These observations are best illustrated by examining the Dean number, which describes the ratio of curvature effects to viscous effects. At low Dean numbers, the flow field acquires circulating vortices symmetric along a longitudinal plane. As shown by Liu, these vortices were seen in their investigation. Figure 8.8C depicts these eddies.

In addition, the unsteady flow intensifies these vortices. With each cycle, the pressure gradient rises and falls, changing direction between inspiration and expiration. The inertia carried by the flow is enough so that when the pressure gradient suddenly changes direction then mixing occurs between forward flow and backward flow.

Womersley number is a ratio of unsteady to viscous effects (see Chapters 2 and 6 in this book); and it describes this irregularity in the flow field. Moreover, the velocity profiles of unsteady flow have a characteristic M-shaped profile. The intensity of the M-shaped profile changes as the Womersley number increases. In general, a Womersley parameter of 3 indicates a mostly viscous characteristic and the velocity profile is more parabolic.[32,37] As the Womersley parameter increases, the center of the parabola (the central streamlines), which have the highest velocity and inertia,[18,38] lag with respect to the streamlines that have lower velocities. The lower velocity streamlines are located near the wall of the pipe, tube or airway, where the no-slip condition applies. In eq 8.19 for Womersley number: ω = Circular frequency and ν = Kinematic viscosity.

Finally, bifurcations themselves induce changes in the flow behavior. At the point of bifurcation, there is a stagnation region where the fluid collides abruptly with the vessel. This sudden impact creates a region of high pressure at the bifurcation and a corresponding low-pressure region on the outer wall of the airway. This interface between high and low pressures further exacerbates the circulation region. One key feature in bifurcations of similar size parent and daughter vessels is the bifurcation angle (eq 8.19). The greater the angle, the more pronounced is the effect of induced eddies.[34]

$$\alpha = a\sqrt{\frac{\omega}{\nu}} = \sqrt{\frac{\rho\omega}{\mu}} \tag{8.19}$$

8.11 COMPUTATIONAL FLUID DYNAMICS (CFD)

In CFD for complex problems in biofluid mechanics, the parameters in the model can easily be modified with the least effort. Therefore, a series of cases can be investigated in a relatively short time. Furthermore, better spatial resolution can be achieved using CFD.

Experimental investigation can be expensive and time consuming. Another limitation is relatively scarce measurements that can be obtained. Measurements can only be made at discrete points under experimental situations. Finally, in numerical models detailed information of a flow field (such as velocity, pressure, vorticity, strain rate, shear stress, temperature, and density) can be extracted at all locations with high accuracy, once the computation is completed provided only when the model is spatially and temporally defined. There are three distinct steps for a typical numerical simulation.[28]

- The first step consists of discretizing a volume of interest or solution space into smaller control volumes. This is accomplished by generating a grid or mesh that can be structured or unstructured.
- Secondly, boundary conditions, initial conditions, and the equations that need to be solved at each control volume must be defined.
- Lastly, once the simulations have taken place, the information must be extracted from the data generated in the solution process.

Simulations may be done in one, two, or three dimensions plus a time dependence or independence option. The higher dimensional simulations are more accurate, because the nature of flow fields is often heterogeneous. This non-homogeneity of properties can be included when other dimensions are added. Furthermore, the mesh created dictates how these individual control volumes or cells interact with each other.

For illustration purposes, only the three-dimensional characteristic of the meshes and elements will be described here. Generally, more elements in the model give more accurate results. However, as the model grows in complexity, the addition of elements will increase the computing time. In many cases, simulations may run for weeks or even months.[28,37,39] Yet, there are a number of elements at which the model will run accurately enough to the point which if additional elements are included, accuracy will not be increased; only computing time.[40] In CFD, the Navier-Stokes equations are converted into finite difference equations[41] as follows:

$$\phi_1 = \phi_2 - \Delta x\left(\frac{d\phi}{dx}\right)_2 + \frac{1}{2}(\Delta x)^2\left(\frac{d^2\phi}{dx^2}\right)_2 \text{..........} \tag{8.20}$$

$$\phi_3 = \phi_2 - \Delta x\left(\frac{d\phi}{dx}\right)_2 + \frac{1}{2}(\Delta x)^2\left(\frac{d^2\phi}{dx^2}\right)_2 \text{..........} \tag{8.21}$$

$$\left(\frac{d\phi}{dx}\right)_2 = \frac{\phi_3 - \phi_1}{2\Delta x} \tag{8.22}$$

$$\left(\frac{d^2\phi}{dx^2}\right)_2 = \frac{\phi_1 + \phi_3 - 2\phi_2}{(\Delta x)^2} \tag{8.23}$$

Determining the shape of the elements is also of importance. In a three-dimensional simulation, the two most popular choices for element shapes[42] are tetrahedrons and hexahedrons. Furthermore, the meshes, where the elements are placed, may be structured or unstructured. In general, a structured mesh is used for simulations, where the geometry is simple. Examples would include rectangular channels, rectangular air ducts, and straight lines in general. Unstructured meshes would be applied to curved geometries and irregular shapes. Unstructured meshes are often times used in biomedical engineering problems since the morphology of the organs can be very complex.[28,43] A well-known commercial code for CFD is a FLUENT Europe Ltd (Sheffield, England).

8.12 VENTILATORS

Mechanical ventilation is used in patients with acute and chronic respiratory distress.[1,11,22] The goal of mechanical ventilation is to reproduce the body's normal breathing mechanism. To achieve this, mechanical ventilators utilize positive intrapulmonary pressure. Machines based on positive intrapulmonary pressure literally blow air into the lung, that is, they create a positive pressure within the alveoli.

8.12.1 TYPES OF VENTILATORS

- **Volume ventilators** are time-cycled, volume regulated, volume limited; and were first used in anesthesia and are being used in pediatric and adult intensive care units (NICU). Initially, these ventilators were not used in the NICU due to the difficulty of achieving consistent small volumes (5–7 mL/kg in infants). Current volume ventilators are able to deliver small volumes consistently.

- **High frequency ventilators:** The rate in "high frequency, Hz" ranges from 3 to 15 Hz (i.e., 180–900 breaths per minute). Gas exchange occurs due to enhanced diffusion. This ventilator uses a piston with a diaphragm unit to actively move gas in and out of the lung. Since the tidal volume generated by these ventilators approximates to the dead space; therefore, simple pulmonary mechanics, physics, and physiology are inadequate to explain their mode of operation.
- **Pressure ventilators** are on most common ventilators in the NICU. Traditional "pressure ventilators" are of constant flow, time-cycled, and pressure limited. Pressure limited means that once the pre-set pressure has been reached, it is maintained for the duration of the inspiratory cycle. Time-cycled implies that breaths are given at fixed intervals. Newer "pressure ventilators" can sense infant's breaths and synchronize with them.

8.12.2 OPERATING MODES OF A VENTILATOR

The ventilator basically has three types of operating modes: (a) In Assist Mode, the inspiration is triggered by patient and pressure sensor detects a negative pressure when the patient tries to inhale. Positive pressure is then applied; (b) In Control Mode, the respirator runs at a fixed rate, and for patients who are unable to initiate inspiration; and (c) In Assist-Control Mode, it is the combination of the above two modes.

8.13 OXYGENATORS

There are direct contact oxygenators and membrane oxygenators. The direct contact oxygenators (Figure 8.10B) have oxygen in direct contact with the blood. In these oxygenators, O_2 is bubbled up through the blood. In eq 8.24 for oxygen transport in direct contact oxygenators, a = solubility of O_2 in blood = 0.3×10^{-4}mL/mL-torr; dm = mass of O_2 transferred across a membrane element of area dS; P_i = partial pressure of O_2 at interface (700 torr); P_b = partial pressure of O_2 in venous blood (40 torr); D = diffusivity of O_2 in blood = $0.8*10^{-5}$ cm^2/s; L = characteristic length (cm); and N_u = local Nusselt number.

$$\frac{dm}{dS} = \frac{at}{L}(P_i - P_b)\, DN_u \tag{8.24}$$

In the membrane oxygenator, gas and blood layers contain a fine screen mesh. The major drawback of this oxygenator is that the CO_2 transfer is less efficient than O_2 transfer. In eq 8.25 for the membrane oxygenator, x = length of a parallel plate membrane oxygenator to reduce the partial pressure of CO_2 in blood from P_1 to P_2; K = 2.5*10^{-4} mol/L torr; Q = blood flow rate (mL/s); D = thickness of membrane (cm); W = width of the plates; and D = permeability of membrane to CO_2 = 8.33*10^{-11} mol-cm/cm^2-min-torr.

$$x = \frac{KQd}{2WD}\ln\left(\frac{P_1}{P_2}\right) \tag{8.25}$$

8.13.1 MATHEMATICAL RELATIONS FOR OXYGENATORS

With the help of the various mathematical relationships described below, the performance of an oxygenator can be evaluated.[29,30]

- **Gas phase arterial oxygen gradient:** Gradient between the partial pressure of oxygen in ventilating gas and resultant PaO_2 is calculated. This helps for estimating the oxygenator's ability to diffuse oxygen across the membrane. When the blood flow increases, the time that blood stays in contact with oxygen in the gas phase is reduced. These values can also be represented in a form that has been corrected for blood flow.

$$O_2\ gradient = (FIO_2 \times 760\ mmHg) - P_aO_2(corrected) \tag{8.26}$$

$$O_2\ gradient\,/\,1\ blood\ flow = \frac{mean\ oxygen\ gradient}{mean\ blood\ floow} \tag{8.27}$$

- **Diffusing capacity of oxygen** is the pressure difference that is needed to transfer 1 mL of oxygen per minute (eq 8.28). Diffusion through a membrane depends mainly on the area and the thickness of the membranes. When DmO_2 is divided by the membrane surface area, we get eq 8.29 for the diffusing capacity of oxygen per unit membrane surface area:

$$DmO_2 = [(O_2\,transferred,\ mL/min.) - (P_{gas\text{-}phase,}\ O_2 - P_v O_2)] \tag{8.28}$$

$$\frac{DmO_2}{m^2} = \frac{DmO_2}{membrane\ surface\ area} \tag{8.29}$$

- **Oxygen transfer slope** is an oxygen consumption (VO_2/min) from eq 8.30, where: CaO_2 = Arterial oxygen content; and CvO_2 = Venous oxygen content.

$$VO_2 = (CaO_2 - CvO_2) \times Q_b \tag{8.30}$$

- **Maximum oxygen transfer:** The oxygen transfer slope can be extrapolated to find the maximum oxygen transfer, when FIO_2 is 100% (eq 8.31). The maximum oxygen transfer per m^2 of membrane surface area can be calculated using eq 8.32.

$$Maximum\ oxygen\ transfer = \frac{1.0 - intercept}{slope} \tag{8.31}$$

$$\frac{maximum\ O_2\ transfer}{m^2} = \frac{maximum\ O_2\ transfer}{membrane\ surface\ area} \tag{8.32}$$

- **Shunt fraction** is the fraction of venous blood that flows through the oxygenator without being effectively oxygenated. In eq 8.33: CiO_2 = Ideal oxygen content; CaO_2 = arterial oxygen content; and CvO_2 = venous oxygen content.

$$Q_s / Q_t = \frac{C_iO_2 - C_aO_2}{C_iO_2 - C_vO_2} \tag{8.33}$$

- **Heat exchanger performance** is evaluated using eq 8.34, where: T = temperature.

$$Performance\ factor = \frac{T_{arterial} - T_{venous}}{T_{water} - T_{venous}} \tag{8.34}$$

8.14 RECENT ADVANCES

8.14.1 EMBRYONIC STEM CELLS CHANGED INTO LUNG CELLS

Scientists have used embryonic stem cells from mice to transform them into a type of lung cell to create artificially grown cells and tissues for humans. Stem cells are master cells that have the capability of developing into any type of human tissue, having the potential of regenerating tissue and organs to treat a range of diseases. The process basically involved placing the mouse embryonic stem cells in a specialized growth system and encourage these to change into cells that will line a part of the lung, where oxygen is absorbed and carbon dioxide is excreted. These findings using human embryonic stem cells are being eventually used in replacing damaged cells and tissues by implanting new ones generated from stem cells.

It can be seen that the transformed cells could help reline the lungs in patients, who had lung damage; or in premature infants, whose lungs were not fully developed. The cells could be developed in such a way that the body would not reject them. Currently, these cells are grown on natural and artificial scaffolds with an aim to construct lung tissue, which could be used in implantation in an entire range of diseases, such as: cystic fibrosis, and cancer.

8.14.2 INTRAVENOUS OR IMPLANTABLE MEMBRANE OXYGENATOR (IMO)

This device is really just a balloon surrounded by gas absorbing fibers. When the balloon inflates, it helps the fibers to switch the gases. It is intended for patients with acute respiratory distress syndrome, pneumonia and chronic lung disease, as well as organ transplant patients and patients in intensive care units. Through a small incision, the device is inserted into the patient's vena cava, a large vein that carries blood back to the heart. The device is several meters in length and consists of about 1000 hollow fiber membranes. Oxygen enters these membranes through an external tube, and flows through the fibers under vacuum pressure. Oxygen within each fiber diffuses through tiny pores in the fiber wall into the blood, and is exchanged for carbon dioxide, which diffuses into the fibers and exits the device through a second tube.

8.14.3 ARTIFICIAL IMPLANTABLE LUNG

An artificial implantable lung uses tiny hollow fibers and the heart's own pumping power to oxygenate blood.[24,44] This device shows promising results in pre-clinical studies, and may reach clinical trials for patients with lung failure awaiting a lung transplant. The artificial lung uses no mechanical pump; instead, it relies on the heart's own pumping force to send blood from the pulmonary artery into the chamber, passing the hollow fibers with tiny micropores that exchange oxygen from the air with carbon dioxide from the blood. The lack of a pump cuts costs and reduces damage to blood cells. The blood can then flow back to the pulmonary artery and circulate through the lungs or directly into the heart's left atrium before being sent to the body.

8.14.4 LIQUID VENTILATION

Liquid ventilation support technique in which the lungs are partially or totally filled with a neat perfluorocarbon liquid that acts as the oxygen and carbon dioxide carrier. This technique promises to improve the two major problems facing clinicians, who treat acute and neonatal respiratory distress syndrome: hypoxemia and low pulmonary compliance. The method also circumvents high-pressure-induced lung injury associated with conventional mechanical gas ventilation. It has been seen that liquid ventilation quickly and effectively returns arterial oxygen and carbon dioxide blood gas pressures to normal physiologic values in premature lambs suffering from respiratory distress syndrome.

Fluorocarbon solvents like perfluorooctyl bromide, Perflubron, n-$C_8F_{17}Br$, FX-80 (mainly (perfluorobutyl) tetrahydrofuran), FC-75 (a perfluorinated cyclic ether mixture composed mainly of FC-43 (perfluor-tributyl amine)) are used for liquid ventilation, as they have higher solubilities for oxygen and carbon dioxide and have lower surface tensions than water. These properties allow them to flow through extremely narrow airways, whereas water would obstruct the airways without filling the lung completely. Fluorocarbons are also biocompatible. A solution of a drug in perfluorooctyl bromide could be administered to the lung during conventional mechanical ventilation. Perfluorooctyl bromide that evaporates during the treatment would be replenished as needed. This process is called partial liquid ventilation, because the volume of the fluorocarbon solvent instilled into the lung does not exceed the lung functional residual capacity. The drug would be released from the fluorocarbon solvent and transported to the intended site of action by passive diffusion.

8.15 SUMMARY

The respiratory system is a complex process that maintains oxygen and blood circulation through the lungs and the rest of the body. It also helps in the removal of unwanted gases (such as: carbon dioxide), dust and other impurities by the means of secretions, for example, mucus. Its malfunctioning can result in respiratory diseases, such as: asthma, bronchitis, and cystic fibrosis. Recently, engineers have helped in the design of artificial lungs, which can be vital for persons with incurable diseases.

The transport of CO_2 and O_2 gases forms a vital part in describing the functional anatomy of the human respiratory system. The effects of higher levels of CO_2, pH, and temperature also play an important role in the regulation of the respiratory system. The oxygen saturation and the hemoglobin–oxygen relationship are some critical aspects of affecting the proper functioning of the respiratory system. To regulate the arrhythmia, mechanical ventilators are generally used. CFD allows for detailed analysis of parameters, such as velocity fields, shear stress gradients, shear rates, temperature, and density. CFD has become a standard in the analysis of the respiratory systems.

KEYWORDS

- **alveolus**
- **angina**
- **aorta**
- **apex**
- **artificial lung**
- **atrium**
- **biofluid**

REFERENCES

1. Goyal, M. R. Biofluid Dynamics of Respiratory System. In *Biofluid Mechanics of Human Body Systems*; CRC and Apple Academic Press, 2013; pp 215–252.
2. Lightfoot, E. N. *Transport Phenomena and Living Systems*; John Wiley and Sons, 1974.
3. Geankoplis, C. J. *Transport Processes: Momentum, Heat and Mass*; Allyn and Bacon Series in Engineering, 1983; p 538.
4. Glasser, O. *Medical Physics*; Yearbook Publisher, 1944.
5. Grotberg, J. B. Respiratory Fluid Mechanics and Transport Processes. *Annu. Rev. Biomed. Eng.* **2001**, *3* (1), 421–457.
6. Martini, F. H. *Fundamentals of Anatomy and Physiology*; Prentice Hall, 2001.
7. MountCastle, V. B. *Medical Physiology*; C. V. Mosby Company, 1974.
8. Roselli, R. J.; Diller, K. R. Biofluid Transport. In *Biotransport: Principles & Applications*, 1st ed.; Springer, 2011; pp 105–485.
9. Skofronich, J. G.; Cameron, J. R.; Grant, R. M. *Physics of the Body*, 2nd ed.; Medical Physics Publishing, 1999.

10. Hayashi, K.; Ed. Abstracts of the Congress Lectures and Symposia and Free Communication. In *The Third World Congress of Biomechanics*; Sapporo, Japan, 2–8 August, 1998; p 4.
11. Bronzino, J. D.; Peterson, D. R.; Eds. *The Biomedical Engineering Handbook*, Vol. 1–4; CRC Press LLC (Taylor & Francis Group): Boca Raton, FL, 2015; p 5422.
12. Cooney, D. *Biomedical Engineering Principles*; Marcel Dekker, 1976.
13. Enderle, J.; Bronzino, J. *Introduction to Biomedical Engineering*; Academic Press, 2012.
14. Kutz, M. *Biomedical Engineering Fundamentals*; McGraw-Hill Education, 2021.
15. Kutz, M. *Standard Handbook of Biomedical Engineering Design*; McGraw-Hill, 2003.
16. Seagrave, R. C. *Biomedical Applications of Heat and Mass Transfer*, 1st ed.; The Iowa State University Press, 1971.
17. Skalak, R.; Chein, S. *Handbook of Bioengineering*; McGraw Hill Book Co, 1986.
18. Fox, R. W.; McDonald, A. T.; Mitchell, J. W. *Introduction to Fluid Mechanics*; Wiley Publishers, 2020.
19. Moore, J. E. Personal Communication. Biomedical Engineering Institute, Florida International University: Miami, FL, 2003.
20. Aukland, K.; Reed, R. K. Interstitial-Lymphatic Mechanisms in the Control of Extracellular Fluid Volume. *Physiol.Rev.* **1993**, *73* (1), 1–78.
21. Hall, J. E.; Hall, M. E. *Guyton and Hall Textbook of Medical Physiology*; Pearson & Prentice Hall, 2020.
22. Hess, D. R.; Kacmarek, R. M. *Essentials of Mechanical Ventilation*; McGraw Hill Education, 2019.
23. Huang, Z.; Tarbell, J. Numerical Simulation of Mass Transfer in Porous Media of Blood Vessel Walls. *Am. J. Physiol.* **1997**, *273* (1), H464–H477.
24. Ionescu, C. M. *The Human Respiratory System: An Analysis of the Interplay Between Anatomy, Structure, Breathing and Fractal Dynamics*; Springer-Verlag, 2013.
25. Altman, P. *Blood and Other Body Fluids*; Federation of American Societies for Experimental Biology, 1961.
26. Crummett, W. P.; Western, A. B. *University Physics: Models and Applications*. C. Brown Communications: Iowa, 1994.
27. Drzewiecki, G. M.; Li, J. K. *Analysis and Assessment of Cardiovascular Function*; Springer-Verlag, 1998.
28. Duraiswamy, N. Personal Communication. Biomedical Engineering Institute, Florida International University: Miami, FL, 2003.
29. Fournier, R. L. *Basic Transport Phenomena in Biomedical Engineering*; CRC Press, 2017.
30. Truskey, G. A.; Yuan, F.; Katz, D. F. *Transport Phenomena in Biological Systems*; Prentice Hall, 2009.
31. Menon, A.; Weber, M.; Chang, H. Model Study of Flow Dynamics in Human Central Airways. Part III: Oscillatory Velocity Profiles. *Respir. Physiol.* **1984**, *55* (2), 255–275.
32. Wirtz, H. R.; Dobbs, L. G. The Effects of Mechanical Forces on Lung Functions. *Respir. Physiol.* **2000**, *119* (1), 1–17.
33. Chang, H.; El Masry, O. A. A Model Study of Flow Dynamics in Human Central Airways. Part I: Axial Velocity Profiles. *Respir. Physiol.* **1982**, *49* (1), 75–95.
34. Liu, Y.; So, R.; Zhang, C. Modeling the Bifurcating Flow in a Human Lung Airway. *J. Biomech.* **2002**, *35* (4), 465–473.
35. Renotte, C.; Bouffioux, V.; Wilquem, F. Numerical 3D Analysis of Oscillatory Flow in the Time-Varying Laryngeal Channel. *J. Biomech.* **2000**, *33* (12), 1637–1644.
36. Fung, Y.-C. *Biomechanics: Motion, Flow, Stress, and Growth*; Springer-Verlag, 2013.

37. Looga, R. Reflex Cardiovascular Responses to Lung Inflation: A Review. *Respir. Physiol.* **1997**, *109* (2), 95–106.
38. Caro, C. G.; Pedley, T. J.; Schroter, R.; Seed, W. *The Mechanics of the Circulation*; Oxford University Press, 1978.
39. Tu, J.; Inthavong, K.; Ahmadi, G. *Computational Fluid and Particle Dynamics in the Human Respiratory System*; Springer Science & Business Media, 2012.
40. Prakash, S.; Ethier, C. R. Requirements for Mesh Resolution in 3D Computational Hemodynamics. *J. Biomech. Eng.* **2001**, *123* (2), 134–144.
41. Patankar, S. *Numerical Heat Transfer and Fluid Flow*; Taylor & Francis, 2018.
42. Chandrupatla, T.; Belegundu, A. *Introduction to Finite Elements in Engineering*; Prentice Hall, 2002.
43. Fung, Y.-C. *Biomechanics: Mechanical Properties of Living Tissues*; Springer-Verlag, 2013.
44. Klaas, M.; Koch, E.; Schroder, W. Fundamental Medical and Engineering Investigations on Protective Artificial Respiration. In *A Collection of Papers from the DFG funded Research Program PAR*, 1st ed.; Springer-Verlag, 2011; p 196.

CHAPTER 9

Biofluid Dynamics of the Human Brain*

MEGH R. GOYAL, GUILLERMO COLÓN, CARLOS A. CRUZ, and ROBERTO LA TORRE

9.1 INTRODUCTION

By utilizing magnetic resonance imaging (MRI) to observe brain activity, researchers can examine the brain's active areas during specific tasks. For instance, when a person moves their right foot, a particular region in the left hemisphere's parietal area shows the corresponding activity. Brain mapping enables the identification of connections between localized brain damage and the consequent neural functioning.[1] The intricate human nervous system relies on a multitude of distinct yet interconnected functions working harmoniously. Facilitated by cerebrospinal fluid (CSF), communication between the brain and the body occurs. This fluid transports neurons responsible for relaying information throughout the body.[1,2]

This chapter focuses on the mechanics of biofluids that maintain the functioning of the brain.

9.2 THE PHYSIOLOGY OF HUMAN BRAIN AND THE CENTRAL NERVOUS SYSTEM[1,3–15]

The brain collaborates with the spinal cord to transmit various messages through specialized nerve cells called neurons. These messages enable the brain to execute diverse functions, including the coordination of 12 bodily

*This chapter has been modified from the review article prepared by students [Guillermo Colón, Carlos A. Cruz and Roberto La Torre] for the course on Fluid Mechanics, INGE 4015 - Course Instructor: Megh R. Goyal, PhD, P.E., Retired Professor in Biomedical Engineering, General Engineering Department, University of Puerto Rico – Mayaguez Campus; For details contact at <goyalmegh@gmail.com>.

Biofluid Dynamics of Human Body Systems: Expanded and Revised Edition.
Megh R. Goyal, Arka Bhowmik, & Anamika Chauhan (Eds.)

systems, regulation of breathing, heart rate, body temperature, metabolism, body movements, shaping personality and behavior, and processing our five senses. Each segment of the brain has distinct roles and interconnects with other segments to create more intricate functions.[11] The reticular activating system filters incoming sensory information for the brain, relaying only the important data to the conscious mind. Meanwhile, the Limbic system serves as the emotional hub within the human body, governing fundamental emotions. Situated in the lower brain region, this system encompasses the hypothalamus and a portion of the thalamus.[1]

To delve deeper into brain structure and the roles of its distinct components, interested individuals can refer to the book "Anatomy of a Human Body." The cerebrum, the brain's largest section, hosts processes, such as perception, imagination, thinking, judgment, and decision-making. This region forms the outermost layer of the telencephalon. Gray matter, positioned at the cerebrum's outer edges, rests atop the white matter. Due to this peripheral placement of gray matter, the number of neurons corresponds to the cerebrum's surface area.[1,10] The complete cerebrum is split into two halves known as cerebral hemispheres. The right hemisphere governs the left side of the body, whereas the left hemisphere governs the right side. The brain harbors four uniquely shaped cavities known as ventricles, which house CSF.[1]

Neurons, act as couriers for transmitting information throughout the body. These cells are enveloped by a cell membrane and contain a nucleus housing genetic material. Neurons also hold "cytoplasm, mitochondria, and other cellular constituents" known as "organelles." Dendrites and axons represent specialized projections of neurons, with dendrites conveying information towards the cell body, while axons transmit it away. Neurons communicate through an electrochemical procedure and possess distinctive elements such as synapses and neurotransmitters. Any disruption to neurons and nerve pathways can influence an individual's cognitive functions, behaviors, feelings, and physical motions.[1,7,17]

9.3 ANATOMY OF CEREBROSPINAL FLUID AND VENTRICULAR SYSTEM

CSF is primarily found within the ventricular system and the subarachnoid space. This fluid-filled system is derived from the neural tube and consists of the lateral ventricles, third ventricle, and the mesencephalic aqueduct.[1,2,17–20]

The fourth ventricle continues into the central canal of the spinal cord. The space within the cranial cavity is enclosed, requiring ongoing control of intracranial pressure (ICP). Within this confined space, the elements that play a role in regulating this pressure encompass the brain tissue, CSF, and blood.[1]

CSF is produced by the choroid plexus, which consists of choroidal epithelium, blood vessels, and connective tissue in between. This plexus is well-supplied with blood vessels, reflecting its heightened metabolic activity. It features intricate folds and microvilli that give it a structure resembling other epithelia responsible for fluid transport. Tight junctions (TJ) between adjacent choroidal epithelial cells form a barrier that prevents the passage of large molecules. Similar TJ are found in brain endothelial cells and arachnoid membrane cells. The resistance to electrical flow on the apical surface of the choroidal epithelium is a result of these TJ. This barrier effectively protects the integrity of capillaries against challenges such as osmotic changes or sudden increases in blood flow-related pressure. The cerebral ventricles are coated with ependymal cells, often accompanied by a subependymal layer consisting of glial fibers and cells. This location facilitates an exchange between the fluid in the ventricles and the adjacent extracellular brain fluid in the subependymal region.

Around 15–20% of the brain's volume is occupied by interstitial fluid encompassing its cells. This fluid-filled area is more expansive in gray matter than in white matter, primarily because gray matter contains a greater proportion of water compared with white matter.[1,21] A connection exists between the scala tympani and the subarachnoid space via the cochlear aqueduct. Fluctuations in pressure within the three compartments of the cochlea correspond to variations seen in CSF across different animal species. However, the exact rate at which fluid moves between CSF and perilymph through the cochlear aqueduct is not definitively known. The process of draining CSF is facilitated by a valve referred to as the arachnoid granulation.[1,10,17]

> "Artery blood serum –> choroid plexus –> CSF secreted by choroid plexus –> Lateral ventricles –> foramen of Monro third ventricle–> Aqueduct of Sylvius –> fourth ventricle –> Foramina of Magendie and Luschka –> Subarachnoid space over brain and spinal cord –> dural sinuses –> Reabsorption into venous sinus blood via arachnoid granulations"

9.3.1 CIRCULATION OF THE CEREBROSPINAL FLUID

CSF is generated in the lateral ventricles and exits via both the lateral apertures and the central aperture of the fourth ventricle. This fluid then enters the subarachnoid spaces and spreads throughout the brain and spinal cord, as shown in Figure 9.1.

Around 430–450 mL of CSF are produced every day, requiring replacement approximately every 6–7 hours. Changes in breathing and blood circulation are believed to impact pressure changes within the closed system, aiding the mixing and distribution of the fluid.[23] CSF traverses interconnected subarachnoid cisterns situated at the base of the brain. Most of the CSF moves in an upward direction, spreading across the cerebral hemispheres, while lesser amounts flow downward along the spinal cord.[16–18,23–25]

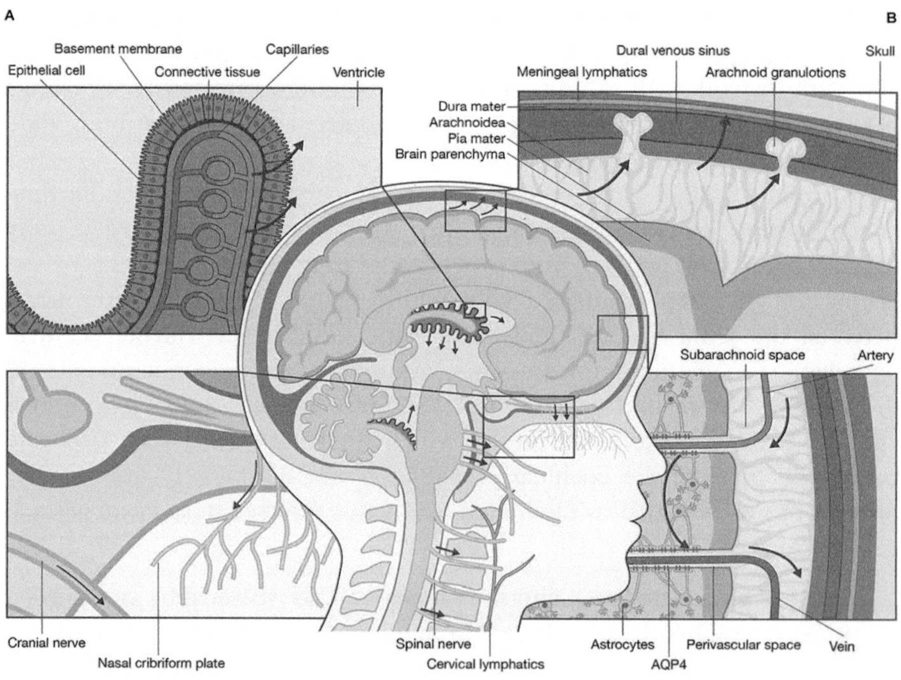

FIGURE 9.1 (A) Location of CSF production, (B) Anatomical sites are responsible for CSF absorption e.g., arachnoid granulations, meningeal lymphatics, (C) Cranial nerve sheaths and nasal cribriform plate to the cervical lymphatics, (D) One essential function of CSF is the delivery of nutrients and the removal of waste products.

Source: Reprinted from Ref. [23]. https://creativecommons.org/licenses/by/4.0/

9.3.2 INTRACRANIAL PRESSURE (ICP)

The bones surrounding the cranial cavity encase its contents, maintaining a fixed volume within. Three crucial factors contribute to the pressure within the cranial cavity: the brain, CSF, and blood. Changes in ICP can lead to brain displacement through herniation or pressure-related shrinkage, as observed in conditions, such as hydrocephalus. The quantities of blood and CSF within the cranial cavity are interconnected, working together to regulate normal ICP levels. This process involves the automatic control of cerebrospinal blood flow and CSF absorption, as well as adjustments in body positioning.[1,2,17–19]

When evaluating CSF opening pressure, the patient is required to lie on their side, ensuring that their legs and neck are positioned neutrally. During this process, alterations of the fluid's meniscus ranging from 2 to 5 mm will synchronize with the patient's heartbeat, while variations of 4–10 mm will align with their breathing. It is essential to instruct the patient to avoid exertion, as this could elevate the opening pressure, and to warn them against excessive breathing, which might cause a reduction in the opening pressure. Normal opening pressure ranges:

- "10–100 mm H_2O in young children,
- 60–200 mm H_2O after eight years of age, and
- up–250 mm H_2O in obese patients"

Intracranial hypotension is identified by an opening pressure measuring below 60 mm H_2O, whereas opening pressures exceeding 250 mm H_2O point to the existence of intracranial hypertension. Elevated pressure within the cranial cavity is noticeable in diverse pathological scenarios, such as meningitis, intracranial bleeding, and the presence of tumors. Idiopathic intracranial hypertension is commonly seen in women who are overweight and in their reproductive years.

When encountering an elevated opening pressure, it is advisable to gradually withdraw CSF while consistently monitoring the pressure throughout the procedure. It is important to halt the CSF removal once the pressure reaches 50% of the initial opening pressure. In most dogs under general anesthesia, their CSF pressure typically falls within the range of 5–12 mmHg, while slightly lower pressures are common in cats. Conditions like space-occupying lesions, such as tumors, which compress venous sinuses and impede CSF absorption in arachnoid villi, can lead to heightened CSF pressure. Similarly, elevated CSF pressure is observed in cases of cerebral edema, often associated with brain injuries, hydrocephalus, and inflammatory lesions.[1,16]

9.3.3 PHYSIOLOGY AND FUNCTIONS OF CSF[1,2,9,10,17–19]

9.3.3.1 SECRETION

CSF has multiple origins. In addition to its production by the choroid plexus, CSF is also generated directly by the brain through the ependymal lining of the ventricular system and the pia-glial membrane. It can also stem from blood vessels within the pia-arachnoid. The rate at which CSF is produced varies among different species, spanning from 0.2 to 0.5 mL per minute per gram of choroid plexus tissue. The specific generation rates per minute, contingent upon the size of the animal, stand at approximately 0.047 mL for a dog, 0.017 mL for a cat, 0.002 mL for a rat, and 0.35 mL for an adult human.

CSF is produced through a combination of ultrafiltration from blood plasma and active transport processes. The pressure exerted by the fluid within capillaries initiates the movement of water and ions into the interstitial space, and subsequently, these substances are transported to the choroidal epithelium.[23,26–29] The subsequent transfer occurs through tightly sealed apical junctions and across cellular structures. Both of these transfers across cell membranes likely rely on ion pumps and are closely associated with sodium transport. Sodium transport is dependent on the presence of sodium-potassium-activated ATPase, an enzyme bound to the membrane, which is situated on the apical surface and intercellular gaps. Despite the similarity in sodium content between plasma and CSF, CSF shows an excess of chloride and magnesium, and a deficiency of potassium and bicarbonate. While plasma is composed of 93% water, CSF consists of 99% water. The differences in concentration ratios of major ions between CSF and protein-free filtrate highlight the fact that CSF composition is influenced by secretory mechanisms. Furthermore, facilitated diffusion enables the transportation of vitamins, nucleosides, purines, glucose, and essential amino acids into the CSF, supporting brain development and metabolism. Conversely, toxic metabolites are eliminated from CSF into the plasma. An overview of the properties of CSF can be found in Table 9.1.[2,18]

9.3.3.2 ABSORPTION

CSF moves from the ventricular system into the subarachnoid space, where it undergoes circulation. In animals, there is a continuous flow along the spinal cord through the central canal. In humans, there is both a downward flow toward the spinal cord and an upward flow toward the head. To maintain balance, the rate at which CSF is absorbed must be equal to its rate of formation, in accordance with the Law of Conservation of Mass. The primary location for CSF absorption is the arachnoid villi. The process of bulk flow reabsorption into the venous system is influenced by the hydrostatic pressure within the subarachnoid space. Other sites for absorption are:[2,10,18] "the choroid plexus, diffusion in the brain and capillaries, veins and lymphatic placed around spinal nerve roots."

9.3.3.3 FUNCTIONS

CSF serves a variety of purposes, which encompass providing physical support, serving as an excretory mechanism, enabling intracerebral transport, and maintaining the chemical environment within the central nervous system (CNS). It carries neurons containing information for transmission to different parts of the body. Moreover, CSF plays a role in protecting the brain against sudden fluctuations in blood pressure, thus contributing to the regulation of ICP. Essentially, the brain is supported by CSF, acting as a cushion against shocks. Because of its function as a medium for intracerebral transport, CSF also holds significant value for clinical research endeavors.[1,2,17–19]

9.3.3.4 COMPOSITION

The CSF is a watery solution containing ions and different substances[2,24,29–31] to serve as an intracerebral transport medium for nutrients, neuroendocrine substances and neurotransmitters. Although CSF has a distinct composition from plasma, their osmolality remains consistent at 289 mOsm/L. In comparison to plasma, CSF contains slightly lower glucose levels (approximately 80%) and significantly fewer proteins, mainly albumin. Additionally, a small number of leukocytes can be observed in the CSF. In the CSF of healthy dogs, the count of lymphocytes typically ranges from 0 to 3 cells per microliter.

9.3.3.5 ACQUISITION (TECHNIQUES OF CSF TAP)

CSF is collected through a cisternal puncture, performed on an anesthetized animal positioned in lateral recumbency. After surgical preparation, a puncture is made in the cerebello-medullary cistern, found between the occipital bone and the atlas. This is achieved using 22-gauge, 1.5-inch spinal needles equipped with a stylet. Throughout the procedure, the head is maintained at a right angle to the vertebral column to prevent excessive head bending that could obstruct the airway. Continuous monitoring is crucial to ensure proper ventilation throughout the process. The reference point for this procedure is a triangular area formed by the occipital protuberance and the wings of the atlas, with the puncture centered within this triangle.

With careful precision, the needle is carefully inserted, and one can feel the puncture of both the dura mater and the atlantooccipital membrane. Occasionally, a slight muscular twitch might be observed, indicating the correct placement of the needle. After removing the stylet, the visible flow of CSF is the reliable indicator of a successful puncture. The collected CSF is directed into a sterile tube. While spinal manometers can measure pressure, their additional insights are somewhat restricted. The usual CSF volume in an animal is about 1 mL per 5 kg of body weight. Typically, around 1–2 mL of CSF is extracted, which is generally sufficient for most examinations. In cases where only a small amount can be obtained, a mere 100–200 μL of CSF is adequate for measuring protein content and cell count.[1,2]

An alternative approach to obtaining a CSF sample involves lumbar puncture, usually conducted around the L5–L6 region. While this method is suitable for most cats, its execution in dogs can pose greater difficulty. At times, this approach yields only a minimal quantity of CSF. Additionally, when compared with other techniques, the cell count obtained from lumbar puncture tends to be slightly higher in normal CSF samples.[2,20,32] The measurement of CSF pressure within the subarachnoid space can be conducted following a puncture. Changes in body positioning have an impact on the recorded pressure values for measurements taken from both the cervical and lumbar spine. Tilting the body at angles of 30, 60, and 90° leads to a decrease in pressure.

9.3.3.6 NORMAL AND PATHOLOGICAL FINDINGS (COLOR)

The CSF is a watery, clear, and colorless fluid.[2,17,18] Any departure from its typical transparent appearance resembling water is considered abnormal in CSF. Cloudiness in CSF occurs when there's an elevated cellular count (>500/μL). Increased protein levels can intensify this opacity, leading to a thicker consistency. The presence of a red hue indicates hemorrhage, often resulting from the puncture of radicular vessels. If the CSF becomes clear after centrifugation, the discoloration is attributed to a traumatic tap. However, the existence of red or yellow staining post-centrifugation signifies pathological bleeding, as red blood cells break down readily in CSF. Xanthochromia, characterized by a yellow color in CSF, is commonly observed in subarachnoid hemorrhage even without hyperbilirubinemia.[2]

9.3.3.7 CELL COUNTS

CSF and its cellular constituents offer valuable insights for both diagnostic and research purposes, providing a partial reflection of the immune response taking place within the CNS. Analyzing CSF cells is especially advantageous when there's suspicion of an ongoing inflammatory reaction, as seen in conditions like inflammatory or infectious CNS diseases and tumors. The overall cell count is determined using a cell counting chamber such as the Fuchs-Rosenthal chamber. Ideally, this counting process should occur within around 30 minutes of CSF collection due to the rapid deterioration of cells in a low-protein fluid. Notably, granulocytes are particularly delicate and can experience up to a 40% breakdown rate after 2 hours at room temperature. Cooling measures can help mitigate this degradation.

In the CSF of healthy cats and dogs, a cell count ranging from 0 to 3 cells per microliter is normal, primarily consisting of mononuclear cells like lymphocytes and monocytes. It is important to note that a normal cell count of three leukocytes per microliter of CSF corresponds to 3000 white blood cells per milliliter. Roughly, for every 500–700 red blood cells present, one white blood cell should be subtracted from the CSF leukocyte count.

For a more detailed examination of CSF, cell cytospin preparations are commonly used. A cytocentrifuge is employed to concentrate all cells within 0.5–1.0 mL of CSF. To prevent rapid cell deterioration during laboratory processes and ensure effective cytospin preparations, protein is added to the CSF sample. This involves mixing approximately one-third of a 10% bovine serum albumin solution with two-thirds of the CSF. CSF samples with elevated protein content are centrifuged without the addition of albumin. An intensely stained background can complicate cell evaluation. Following staining techniques, such as DiffQuick or Papanicolaou, the types of cells—such as lymphocytes, plasma cells, monocytes, macrophages, neutrophils, and eosinophils—are quantified. Furthermore, cells are assessed based on their size, appearance, and the presence of mitosis or tumor cells.[1,2,17,18]

9.3.3.8 GLUCOSE

CSF glucose concentration is impacted by both blood glucose levels and the metabolic rate of the CNS. Typically, CSF glucose levels range from 60 to 80% of blood levels. Glucose enters the CSF through glucose transporter proteins and simple diffusion. It takes a variable period (approximately 2–4 hours) for equilibrium to be established between CSF and blood glucose levels. CSF glucose concentration is generally influenced by blood glucose levels from the preceding 4 hours. It is essential to measure CSF and blood glucose levels simultaneously. An elevation in CSF glucose concentration does not hold significant diagnostic value; rather, it indicates hyperglycemia within the 4 hours leading up to the puncture. When CSF glucose levels are low in the absence of concurrent hypoglycemia, it might suggest the presence of a widespread meningeal disorder.[2,16,18]

9.3.3.9 PROTEIN

Most of the proteins that are typically found in CSF originate from the bloodstream. The entry of proteins into CSF is mainly reliant on processes such as pinocytosis across capillary endothelial cells and the isoelectric point. This mechanism is thought to involve more intricate processes beyond simple filtration. The rate of protein exit from CSF back into the bloodstream is about 200 times faster than the rate of entry. This exit usually occurs through the arachnoid villi into the venous blood, likely via macrovesicular transport. To qualitatively assess protein levels, the Pandy solution is used, which is a 10% carbolic acid solution that causes globulin to precipitate. In normal CSF samples, no cloudiness is observed after mixing with the Pandy solution. Depending on the protein concentration, turbidity is categorized as 1+ to 4+ positivity. For quantification, the biuret method is commonly used, but it lacks the sensitivity needed to accurately measure CSF protein levels. Various quantitative measurement techniques for CSF protein levels have been adapted for use in animals. These methods include turbidometric approaches using trichloroacetic acid, benzethonium chloride in an alkaline environment, or nephelometry.[2,10,17,18]

9.3.3.10 IGG INDEX

The IgG index is a calculated ratio that utilizes the concentrations of IgG and albumin in both CSF and serum to detect intrathecal IgG synthesis. This index is used to differentiate inflammatory or infectious disorders of the CNS from other conditions. The assessment of intrathecal IgG synthesis involves comparing the immunoglobulin levels in CSF and serum, using albumin as a reference protein. Although albumin can cross the blood–brain barrier (BBB), unlike IgG, it is not synthesized within the CNS itself. Various methods have been employed to quantify these proteins in animal CSF, including techniques like "rocket immune electrophoresis, single radial immunodiffusion, agarose electrophoresis, enzyme-linked immunosorbent assay (ELISA), and laser nephelometry. Immunoglobulin IgA" can also be measured in CSF using ELISA. However, the measurement of these proteins is mainly of theoretical interest.[2,10,19]

TABLE 9.1 Physical Properties and Chemical Composition of the Human Cerebrospinal Fluid.[2,18]

Properties	Units	Range
Aluminum	–	Trace
Barium	–	Trace
Bicarbonate	mg/ml	0.00483
Boron		Trace
Bromide	mg/ml	0.0014–0.0038
Calcium	mg/ml	0.039–0.051
Carbon dioxide	%Volume	57–62
Chloride	mg/ml	4.18–4.52
Cholesterol	mg/ml	0.0024–0.005
Citric acid	mg/ml	0.0004
Copper	mg/ml	0.00006–0.0002
Freezing point depression	°C	0.540–0.603
Iodine	mg/ml	0.00001
Iron	mg/ml	0.00035
Lymphocytes	Cells/mm cu	0–10"
pH	–	7.35–7.70
Phosphorus	mg/ml	0.0125–0.0210
Potassium	mg/ml	0.085–0.115
Pressure	mm H_2O	70–180
Protein cisternal	mg/ml	0.15
Protein Lumbar	mg/ml	0.2-0.4
Protein Total	mg/ml	0.12-0.43
Protein Ventricular	mg/ml	0.1
Refractive index		1.3351
Sodium	mg/ml	5.01–5.43
Solid	%	0.85–1.70
Specific gravity	–	1.0062–1.0082
Strontium	mg/ml	Trace
Sulfur	mg/ml	0.006
Vitamin C	mg/ml	0.018
Volume	ml^3	90–150
Water	%	99"

9.3.3.11 OTHER METABOLITES

CSF has been examined for various metabolites, often within studies focusing on disease pathogenesis. These assessments aim to improve diagnostic strategies and enhance the predictive value of CSF analyses. Alterations in levels of lactate and pyruvate may signal mitochondrial disorders, and the ratio of lactate to pyruvate offers insights into the brain's redox state. Although the measurement of these metabolites has been largely reported in case studies, comprehensive data across a range of small animals and diverse diseases are still limited.

An association has been established between concentrations of gamma-aminobutyric acid (GABA) in CSF and susceptibility to seizures. Additionally, CSF examinations and the quantification of phenobarbital within CSF have been utilized to monitor the treatment of seizures in dogs.[1,10] The flow path of CSF is shown in Figure 9.1C.

9.4 CEREBRAL BLOOD FLOW (CBF)

9.4.1 FUNCTIONS AND CHARACTERISTICS OF BLOOD

Blood plays vital functions including transport, regulation, and defense. It acts as a carrier for moving nutrients, gases, hormones, and waste products. Blood is also instrumental in maintaining fluid and electrolyte balance, regulating acid-base levels, and controlling body temperature. White blood cells provide protection against pathogens, and the blood clotting mechanism prevents excessive bleeding in case of injuries.[2,24,29–31]

Blood is a liquid connective tissue that measures about 5 liters in an adult human and accounts for 8% of the body weight. Its normal pH range is 7.35–7.45. The transportation activity of blood is divided into:[17] "Carries O_2 and nutrients to the cells; Transports CO_2 and wastes from the tissues to the lungs and the kidneys, where wastes can be removed from the body; Carries hormones to the endocrine glands to the target tissues."

The regulation activities of blood are:[24,31] "(a) Helps to regulate body temperature by removing heat from active areas (such as: skeletal muscles) and transporting it to other areas of the skin so that the heat can be dissipated (see Chapter 4 in this book); and (b) Plays a significant role in fluid and electrolyte balance because salt and plasma proteins contribute to the osmotic pressure by providing weight and bulk to our blood. Blood functions as a transport medium. It also has roles in temperature regulation, fluid and electrolyte balance, pH regulation, prevention of fluid loss, and disease prevention.[16,24]"

9.4.2 CHARACTERISTICS OF BLOOD (SEE CHAPTERS 1 AND 7 IN THIS BOOK)

- **Amount:** A person has 4–6 liters of blood, depending on the body size. Of the total blood volume in the human body, 38–48% is composed of formed elements. The remaining 52–62% of the blood volume is the plasma or the liquid portion of the blood.[24]
- **Color:** Arterial blood is bright red because it contains high levels of oxygen. Venous blood has given up much of its oxygen in tissues and has a darker, dull red color. This may be important in the assessment of the source of bleeding.[24]
- **pH:** The normal pH range of blood is 7.35–7.45, which is slightly alkaline.[24]
- **Viscosity** is a resistance to flow. Blood is thicker than water. The density of blood is 1057 kg/m^3. Viscosity is increased by the presence of blood cells and the plasma proteins, and this thickness contributes to normal blood pressure.[24,31]"

9.4.3 COMPOSITION OF BLOOD

When blood is subjected to centrifugation, the denser cellular components and fragments get separated from the liquid. Because of their higher density compared with the liquid matrix, these formed elements are propelled to the bottom of the tube due to the centrifugal force. The pale-yellow liquid that accumulates at the top is referred to as plasma. In a given volume of blood, approximately, 55% is plasma, while the remaining 45% comprises formed cells.

Plasma, constituting the fluid part of blood, is primarily composed of around 90% water. The rest of the composition includes more than 100 various organic and inorganic solutes that are dissolved within the water. Given that plasma acts as a transport medium, its solute makeup is in a continual state of change, with substances being added or removed by cells.[2,24,29–31]

Plasma mainly holds a significant concentration of proteins that stay within the blood and interstitial fluid without being utilized for energy. The primary plasma proteins can be grouped into three classes: albumins, globulins, and fibrinogen. These proteins in plasma serve distinct purposes, with many of them being synthesized in the liver. One of their roles involves facilitating the transport of nutrients, oxygen, and carbon dioxide, acting as solutes within the plasma.[2]

9.4.4 BLOOD PRESSURE

- **High blood pressure (hypertension):** During activity, the blood pressure rises to supply blood faster to the hard-working muscles. This is caused by the body producing adrenaline the so-called "fear, flight or fright" hormone. When relaxed, the pressure drops as the body's demands are normally much lower. A person has high blood pressure, called hypertension, if their resting blood pressure is higher than about 140/90 mmHg. If this happens in the brain, it leads to a stroke. A blood clot forms and deprives a part of the brain of blood. This damages nerve cells and can lead to paralysis.[2,24,31]
- **Low blood pressure (hypertension):** Low blood pressure can be a sign of heart failure and may be due to a dramatic loss of blood due to severe bleeding after something like a road accident. If it is not treated quickly, oxygen is not delivered to the brain and this is fatal.[2,24,31]"

9.4.5 COLLATERAL BLOOD SUPPLY OF THE BRAIN

The specific arrangement of collateral blood flow is contingent upon the location of significant vessel stenosis or occlusion, as well as the availability of unobstructed collateral channels. In general, the establishment of collateral pathways is more successful when major vessel occlusion takes place gradually over a period of weeks or months, rather than abruptly.[20] Collateral blood flow can emerge through mechanisms such as:

"Extra cranial connections (ECA): In the orbit, branches of ECA anastomose with branches of ophthalmic artery, vertebral artery and subclavian artery. Branches of vertebral arteries anastomose with branches of the subclavian artery.[1,10]

Intra cranial connections (ICC): Circle of Willis is a network of blood vessels present at the base of the brain. It is formed by the proximal parts of the two anterior cerebral arteries connected by the ACoA and the proximal parts of the two posterior cerebral arteries connected to the distal internal carotid arteries by the posterior communicating arteries. However, 50% of circles have absent segments and the potential for collateral flow is not always as good as it might first appear.[10] Leptomeningeal anastomoses lie on the surface of the brain. They develop between the cortical branches of the anterior, middle, and posterior cerebral arteries.[10] Dural anastomoses occur between meningeal branches of the ICA, ECA and the vertebral arteries, anterior choroidal artery is a branch of ICA can anastomose with the posterior choroidal artery and the posterior choroidal artery that is a branch of ECA.[10]"

9.4.6 VENOUS DRAINAGE

Blood from the outer regions is transported by superficial cerebral veins, whereas blood from the central areas is directed through deep cerebral veins. Both of these venous pathways contribute to the formation of venous sinuses, which ultimately empty into the internal jugular vein. These cerebral veins lack valves and possess thin walls. An intricate network of connections exists between cerebral veins, dural sinuses, and the venous systems of the meninges, skull, scalp, and nasal sinuses. These connections play a role in facilitating the potential movement of blood clots or the spread of infections between these vessels.[1,10]

9.4.7 REGULATION OF CEREBRAL BLOOD FLOW

CBF within the brain operates at an approximate rate of 50 mL per minute for every 100 grams of brain tissue. "CBF, cerebral blood volume (CBV), and cerebral energy metabolism (measured as cerebral metabolic rate of oxygen ($CMRO_2$) or glucose (CMRglu)" are interconnected, showing higher values in gray matter compared with white matter. This balance maintains a relatively consistent oxygen extraction fraction (OEF) of around 40% across the brain. In a normally resting human brain, CBF serves as a reliable indicator of CMRO2. CBF is influenced by "cerebral perfusion pressure (CPP)" and cerebrovascular resistance. CPP is determined by the difference between "systemic arterial pressure (SAP)" and venous pressure (approximated by ICP) at the exit of the subarachnoid space.[10]

9.4.8 AUTOREGULATION

One remarkable characteristic of the brain is its ability to autonomously control its own blood supply, allowing it to ensure an adequate oxygen delivery for its daily functions. In individuals without underlying health conditions, the CBF remains relatively consistent when the mean arterial pressure (MAP) is within the range of 60–160 mmHg. Generally, MAP aligns closely with CPP, especially when ICP is negligible. The precise mechanisms that drive this self-regulatory process, whether involving myogenic, metabolic, or neurogenic factors, are not yet fully understood. In areas of the brain that have experienced damage (such as from ischemia or trauma), this autoregulation might be compromised or absent. Consequently, CBF becomes responsive to pressure changes and becomes reliant on variations in CPP.[24,31]

9.5 BLOOD–BRAIN BARRIER[1,6,12,18,20–22,33–35]

The brain functions within a well-controlled environment separate from the milieu of the periphery.[6,21] The mechanisms that control the unique environment of the brain are collectively referred to as "blood–brain barrier (BBB)" (see Figure 9.2).

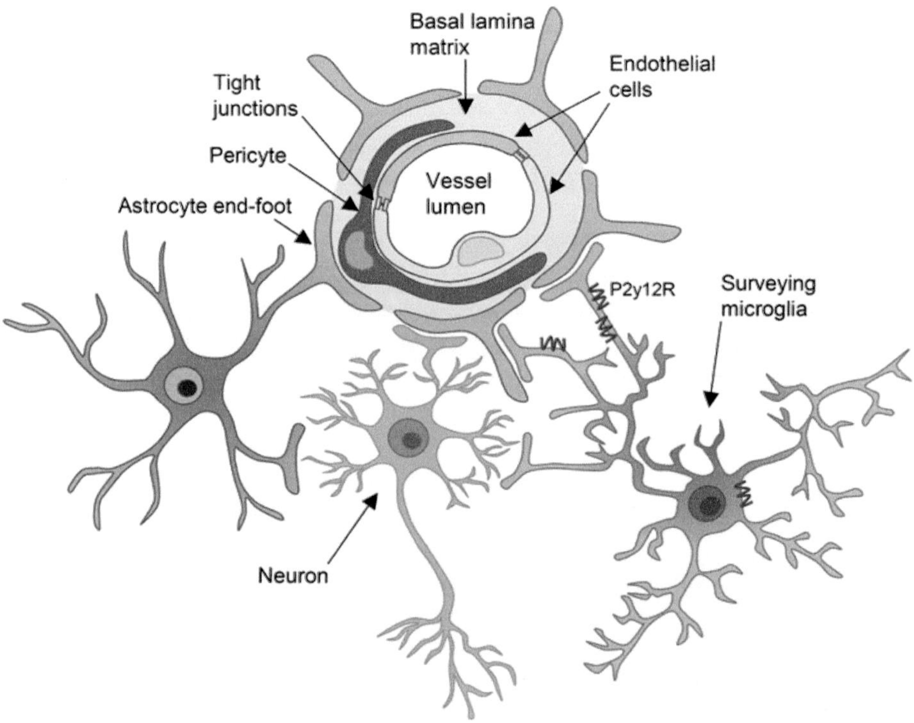

FIGURE 9.2 Schematic representation of the Neurovascular Unit structure and the reciprocal interaction among all components. The endothelial cell (EC), whose intracellular spaces are sealed by the TJs, are surrounded by mural cells (pericytes or smooth muscle cells). ECs and mural cells are enwrapped by the two basal laminas, which provide structural and functional support to the Blood brain barrier (BBB). The astrocytic end feet cover almost the whole blood vessel surface and make direct contact with ECs and mural cells. Neurons make contact with ECs and astrocytes, and their activity can regulate the vascular tone either directly on ECs or indirectly through astrocytes. Microglial processes make dynamic and transient contacts with the vasculature, astrocytes and neurons and contribute to the neurovascular coupling through P2Y12R.

Source: Reprinted from Ref. [39]. https://creativecommons.org/licenses/by/4.0/

9.5.1 PHYSIOLOGY OF THE CEREBRAL CAPILLARY ENDOTHELIA

The microvasculature of the brain exhibits a surface area of roughly 100 cm^2 for every gram of tissue. This arrangement is accompanied by a capillary volume constituting 1% and an endothelial cell volume accounting for 0.1% relative to the tissue volume. On average, there's a distance of approximately 40 micrometers between capillaries in the human brain. Such closely situated capillaries facilitate the swift equilibration of solutes within the brain's interstitial space for small molecules, once they have crossed the blood–brain barrier (BBB).[1,10]

Figure 9.2 depicts the cerebral capillary endothelial cells. These cells are equipped with TJ that effectively seal the connections between adjacent endothelial cells, forming a continuous blood vessel. These TJ play a vital role in establishing a strong barrier among blood–brain barrier (BBB) endothelial cells, resulting in remarkably high endothelial electrical resistance. In pial vessels, this resistance is within the range of 1500–2000 Ω cm^2, which is notably different from the 3–33 Ω cm^2 seen in other tissues. For non-pial cerebral microvessels in vivo, the estimated electrical resistance across endothelial cells is approximately 8000 Ω cm^2. This elevated resistance ultimately leads to minimal paracellular permeability across the BBB.[1,10]

The gaps between cells, known as intercellular clefts, have a width of approximately 200 Å, allowing for the convenient diffusion of tissue solutes. To restrict the passage of substances, these clefts are narrowed. The junctional complex creates an interconnected arrangement of intramembrane strands that form multiple barriers. A crucial component of the integral membrane structure of TJ is occludin, a 65 kDa protein responsible for bringing the opposing cell surfaces into contact. Zonula occludens (ZO-1, ZO-2, and ZO-3) are cytoplasmic proteins that act as recognition molecules, facilitating the proper positioning of TJ to offer structural support for signal transduction proteins.[1,10]

ZOs belong to the "MAGUK family of proteins (membrane-associated guanylate kinase-like proteins) and have a number of binding sites for cytoskeletal proteins, signal transduction molecules, and kinases. AF6 is a Ras effector molecule associated with ZO-1. The 7H6 antigen is a phosphoprotein found at TJ impermeable to ions and macromolecules."

"Junctional adhesion molecules (JAM)" are localized at the TJ and belong to the "immunoglobulin superfamily (IgSF)". Additionally, TJ microdomain on the cell membrane rich in cholesterol, which contain caveolin-1.[1,10]

An enzymatic barrier is present at the cerebral endothelial level, capable of metabolizing drugs and nutrients. These enzymes predominantly focus on the metabolism of neuroactive substances present in the bloodstream. Cerebral microvessels contain elevated concentrations of specific enzymes, including glutamyl transpeptidase (GTP), alkaline phosphatase, and aromatic acid decarboxylase. In contrast, these enzymes are generally present in low concentrations or are absent in capillaries unrelated to neurons.[1,10]

9.5.2 FUNCTIONS OF THE BBB

The blood–brain barrier (BBB) functions as the gatekeeper that controls which substances from the bloodstream are permitted to enter the brain and which are restricted. This specialized system of capillary endothelial cells has the primary role of shielding the brain from potentially harmful compounds present in the blood, while also ensuring the supply of essential nutrients for optimal brain function. In contrast to capillaries found in other body areas, the BBB rigorously manages the passage of substances into the brain through a combination of physical mechanisms (tight junctions) and metabolic processes (enzymes). As a result, the BBB often holds a key role in determining the degree to which therapeutic drugs can access the brain, serving as a limiting factor for their penetration. Additionally, the disruption of the BBB is hypothesized to play a significant role in various conditions associated with the CNS.[1,12,20,35]

Certain substances present in the bloodstream can effectively penetrate the brain, including glucose, specific ions, and oxygen. Moreover, small molecules that are fat-soluble, like ethanol, can cross the blood–brain barrier (BBB). However, certain water-soluble molecules, such as glucose, amino acids, L-dopa, choline, purine bases, and nucleosides, might access the brain by binding to specialized proteins on the plasma membrane of endothelial cells. Conversely, larger molecules, such as proteins, toxins, most antibiotics, and monoamines (such as neurotransmitters) are restricted from entry due to their potential to disrupt brain function. To prevent their passage, certain undesirable molecules are actively transported out of the endothelial cells.[1,10]

The investigation of the blood–brain barrier (BBB) has evolved and grown over time, involving a diverse group of experts, such as pharmacologists, neuroscientists, pathologists, physiologists, and clinical professionals.[20]

9.5.3 TRANSPORT AT THE BBB

The blood–brain barrier (BBB) is comprised of two sequential membranes: the inner membrane facing the blood (lumenal) and the outer membrane facing the brain (ablumenal). There is an approximately 300 nm space of endothelial cytoplasm between these membranes. Consequently, for the process of solute transcytosis from the blood to the brain to occur, transport mechanisms (as illustrated in Figures 9.3A and 9.3B) are required on both the lumenal and ablumenal aspects of the endothelial cell.[1,10]

The blood–brain barrier (BBB) employs specialized carrier transport systems to facilitate the entry of vital nutrients from the bloodstream into the brain. These carriers are responsible for the rapid movement of solutes, such as glucose, amino acids, choline, purine bases, and nucleosides across the BBB, occurring within milliseconds. In contrast, receptor-mediated transport across the BBB involves peptides and takes a longer duration, typically several minutes, to complete. The BBB features a variety of peptide-receptor systems that allow specific peptides to pass through it, for instance, insulin, which gains access to the brain through the BBB's insulin receptor.

There are four basic mechanisms by which solute molecules move across membranes:[23,24,26–29,32,36–41] "(a) simple diffusion, which proceeds from low to high concentrations; (b) facilitated diffusion, a form of carrier-mediated endocytosis, in which solute molecules bind to specific membrane protein carriers, also from low to high concentration; (c) Simple diffusion through an aqueous channel, formed within the membrane; and (d) active transport through a protein carrier with a specific binding site that undergoes a change in affinity. Active transport requires ATP hydrolysis and it conducts movement against the concentration gradient. Movement between cells is referred to as paracellular diffusion. The BBB has a number of highly selective mechanisms for transport of nutrients into the brain. Diffusion of substances into the brain can be divided into:[1,10]

- paracellular (i.e., between cells);
- transcellular (i.e., across cells) diffusion"

Processes involving non-saturable and non-competitive mechanisms are not prevalent at the blood–brain barrier (BBB) due to the presence of "tight junctions," which prevent paracellular diffusion. Transcellular diffusion is the more common process at the BBB, where substances with higher lipophilicity exhibit enhanced diffusion into the brain.

In the comparison of two substances with identical characteristics except for molecular weight, the smaller substance will penetrate at a faster rate. This means that smaller inorganic molecules like O_2, CO_2, NO, and H_2O have high permeability. Additionally, the reduction of hydrogen bonding in a compound improves its ability to permeate membranes. By removing or masking the hydrogen bonding donor group from a compound, the energy required for the compound's transfer from water into the cell membrane is decreased.[1,10]

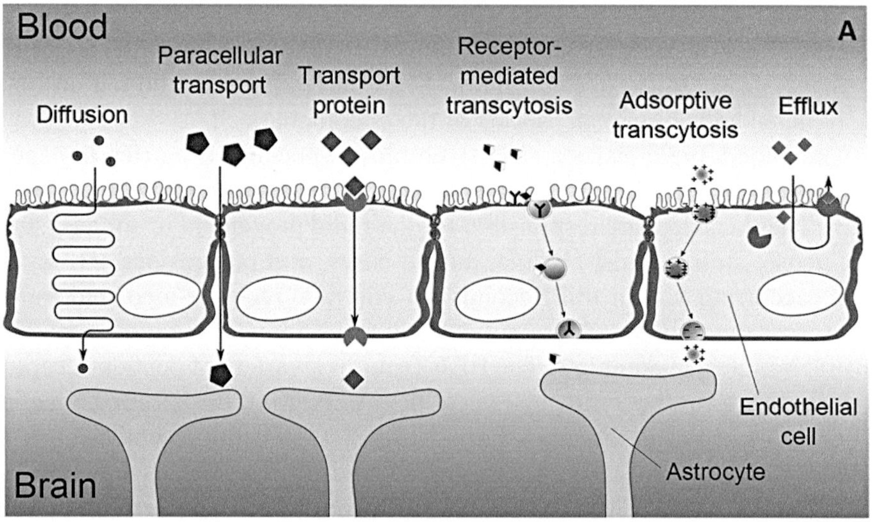

B

Factors Affecting Drug Transport Across BBB

• Concentration gradient of drug/polymer	• Molecular charge
• Molecular weight of the drug	• Affinity for receptors or carriers
• Lipophilicity of the drug	• Cerebral blood flow
• Sequestration by other cells	• Systemic enzymatic stability
• Affinity for efflux proteins (e.g. Pgp	• Metabolism by other tissues
• Pathological status	• Clearance rate of drug/polymer
• Flexibility, conformation of drug/polymer	• Cellular enzymatic stability

FIGURE 9.3 (A) Transport mechanisms at the BBB. (B) Potential factors can alter drug uptake into the brain.

Source: Reprinted from Ref. [46]. https://creativecommons.org/licenses/by/3.0/

9.5.3.1 SIMPLE DIFFUSION

This is an intrinsic mechanism guided by the stochastic movement of solutes. The change in free energy during the diffusion of a solute through a membrane is directly affected by the intensity of the concentration gradient. In the case of non-charged substances (nonelectrolytes), their traversal across the membrane conforms to the equation outlined below:[23]

$$G = (R) x (T) x [ln x (Ci)/(Co)], \quad (9.1)$$

where G is the Gibbs free-energy change (for example, the change during a process in the energy available to do work), R is a gas constant (1.987 cal/mol-K), T is an absolute temperature (Kelvin), and [Ci]/[Co] is the ratio of the concentration of solute on the inside and outside of the membrane.

If the solute is a charged species (i.e., electrolyte), the charge difference between compartments must be considered. The net movement of a solute with the same charge as the membrane is thermodynamically unfavorable due to mutual repulsion of ions. The greater is the potential difference or voltage (i.e., charge) between compartments, the greater is the difference in free energy. Therefore, for an electrolyte to diffuse between compartments, two gradients must be considered: (a) a chemical gradient, determined by differences in concentration; and (b) an electric potential gradient determined by difference in charge. Free-energy change for diffusion of an electrolyte across a membrane is:[23]

$$G = (R) \times (T) \times [ln \times (Ci)/(Co)] + [(z)\ (F) \times (Em)], \qquad (9.2)$$

where z is the charge of the solute, F is Faraday's constant (=23.06 kcal/V-equivalent), and E_m is the potential difference between compartments."

The time required for a molecule to diffuse between two points is the square of the distance separating the two points. "Einstein relationship for diffusion" relates the time of diffusion to the distance traveled (i.e., the average molecular displacement; X):[23]

$$(X)^{2} = 2 \times Distance \qquad (9.3)$$

9.5.3.2 FACILITATED DIFFUSION (A FORM OF CARRIER-MEDIATED ENDOCYTOSIS)

This mechanism involves the binding of a solute to a transporter positioned on a distinct side of the membrane. This interaction triggers a change in the protein's structure, facilitating the transfer of the substance across the membrane to the opposite side. This movement occurs from regions of higher concentration to lower concentration. Facilitated diffusion operates without the requirement for energy and is accountable for the transportation of a range of substances at the blood–brain barrier (BBB). These substances encompass monocarboxylates, hexoses, amines, amino acids, nucleosides, glutathione, and small peptides, among others.[23]

9.5.3.3 CARRIER-MEDIATED TRANSPORT

This process can be categorized into various mechanisms based on energy dependence and/or the simultaneous transport of another substance. This co-transport can occur in either the same direction (symport) or the opposite direction (antiport). The movement takes place from an area of higher concentration to an area of lower concentration.[1,10,23]

- **Endocytosis** can be segregated into bulk-phase (also known as fluid phase), endocytosis and mediated endocytosis (receptor and absorptive-mediated). Bulk-phase endocytosis (pinocytosis) is the nonspecific uptake of extracellular fluids and occurs at a constitutive level within the cell via mechanisms, which are independent of ligand binding. Bulk-phase endocytosis is temperature and energy dependent, non-competitive, and non-saturable. Bulk-phase endocytosis occurs to a very limited degree in the endothelial cells of the cerebral microvasculature.
- **Receptor-mediated endocytosis** (RME) provides a means for selective uptake of macromolecules. Cells have receptors for uptake of many different types of ligands, including hormones, growth factors, enzymes, and plasma proteins. RME occurs at the brain for substances, such as transferrin.

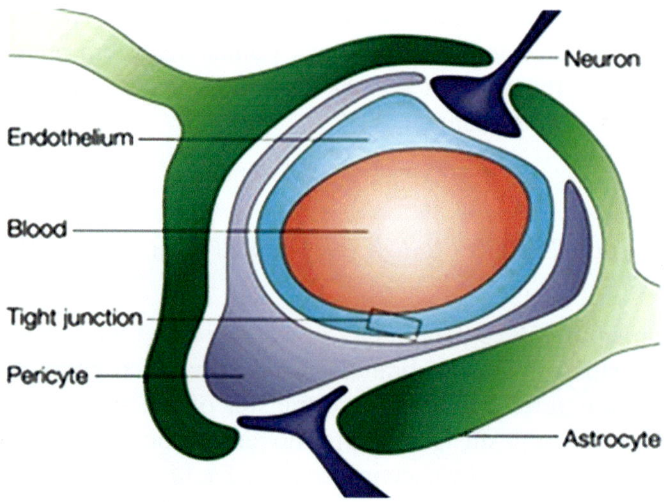

FIGURE 9.4 Four cells comprising the CNS micro-vasculature.

Source: Reprinted from Ref. [47]. https://creativecommons.org/licenses/by/4.0/.

9.5.3.4 *ABSORPTIVE-MEDIATED TRANSPORT (AME)*

It is triggered by an electrostatic interaction between a positively charged substance, usually a charge moiety of a peptide, and the negatively charge plasma membrane surface (i.e., glycocalyx). AME has a lower affinity and higher capacity than RME. The development of many new drug delivery technologies focuses on AME.[23]

9.5.3.5 CARRIER-MEDIATED EFFLUX

This mechanism is involved in extruding drugs from the brain and is a major obstacle for many pharmacological agents, with the ABC (ATP binding cassette) transporter P-glycoprotein being the principal efflux mechanism of these agents. There also exists efflux transporters for organic anions via multidrug resistance-associated protein (MRP), and anionic and cationic cyclic peptide. Additionally, the peptide transport system (PTS)-1 shows the efflux transport of synthetic opioid peptide Tyr-MIF-1.[23]"

9.5.4 TRANSPORT OF SUBSTANCES INTO THE BRAIN

The ability of a specific substance to cross the blood–brain barrier (BBB) and reach the brain relies on multiple factors associated with the BBB itself. These factors include aspects like the concentration difference between different compartments, the size of the molecule (measured by its molecular weight), its shape and flexibility, the amino acid composition, its lipophilicity, stability against cellular enzymes, containment within cells, susceptibility to efflux mechanisms (such as "P-glycoprotein"), ability for hydrogen bonding (including charge), affinity for carrier systems, and potential influences from underlying pathological conditions. External factors encompass systemic enzymatic stability, binding to plasma proteins, CBF, uptake by other tissues, clearance rate, and the presence of pre-existing pathological conditions. Figures 9.3A and 9.3B illustrate the potential elements that can alter the transport of drugs into the brain.[23]

9.5.5 FUNCIONS OF CELL MEMBRANES (Figure 9.4)

- Membranes compartmentalize, providing continuous, relatively unbroken sheets.
- Membranes prevent the unrestricted exchange of molecules, providing a selectively permeable barrier.
- Membranes contain the machinery for the physical transport of substances from one side of the membrane to the other.
- The membrane is involved in the response of the cell to external stimuli, via signal transduction.
- Membranes allow cells to recognize one another, to adhere, and to exchange materials in an intracellular interaction.

- Membranes provide a means to organize cellular biochemical activities, through an extensive framework or scaffolding within which components can be arranged for effective interaction.
- Membranes maintain cell polarity.
- Membranes are involved in the process of energy transduction (i.e., conversion of one type of energy to another), such as the transfer of chemical energy from carbohydrates and fats to ATP.[23]

9.6 ACTIVITIES THAT MAY DAMAGE THE BRAIN

- **Alcohol Abuse:** When a person drinks a product with alcohol, the stomach absorbs about 20% of the alcohol. The remaining 80% is absorbed in the small intestines. Once in the bloodstream, the alcohol is carried to all parts of the body, including the brain.[12] Alcohol causes a chemical wash on the transmitters and receptors, altering the speed of the brain's electrical activity. No part of the brain is immune to the effects of alcohol, but some parts of it are more sensitive to its effects. Various factors cause alcohol to be absorbed and processed differently, such as age, body mass, general health, and heredity.
- **Drug Abuse:** Heroin, marijuana, cocaine, inhalants, ecstasy, LSD, steroids, Ritalin.

9.7 BRAIN INJURIES

Traumatic brain injury typically occurs when the head forcefully collides with an immobile object, like a windshield or the ground. This impact leads to the brain being compressed, twisted, and contorted within the skull, resulting in both localized and widespread damage across the brain. Often, this trauma is accompanied by a period of altered consciousness or unconsciousness. In cases of severe brain injury, unconsciousness can last for hours, days, or even weeks. Memory gaps regarding the injury itself, the moments before, and the subsequent time are common. This memory loss is due to direct damage to brain regions responsible for memory and learning. Additional brain damage can occur after the initial impact due to factors like bleeding, breathing cessation, or brain tissue swelling. Various types of brain injuries exist.

The primary types of brain injuries are skull fractures, when the bony skull breaks and hits the brain directly, concussions, that are bruises close to the point of impact and hematomas that are blood clots in the brain.

Secondary internal insults and cerebrovascular accidents: The obstructive and hemorrhagic stroke; shaken baby syndrome (SBS).

9.8 BRAIN DISEASES[45]

- Alzheimer's and other dementias
- Amblyopia
- Blindness
- Depression
- Hydrocephalus: Communicating hydrocephalus and non-communicating hydrocephalus
- Multiple sclerosis
- Parkinson
- Schizophrenia
- Seizures (epilepsy)
- Stroke

9.9 SPECIFIC RESEARCH AND TECHNOLOGY

- **Computerized Tomography (CT):** CT scan enables a physician to visualize detailed and accurate cross-sectional anatomy of the brain. This method can be used to show several abnormalities in the human brain (See Chapter 14 and 15 in this book).
- **Magnetic Resonance Imaging (MRI):** The MRI images demonstrate excellent contrast between various organs, with high spatial resolution, thin slices and small pixels (See Chapters 14 and 15 in this book). It differs from the CT because the MR signal from various organs is characteristic not only of the physical density of the organ, but also of its biochemical nature, the procedure may permit both identification and characterization of diseases like the distinction between benign and malignant tumors. This type of examination can be used to identify the following brain diseases: Neurological disorders (such as stroke, hemorrhage, tumor, infection, multiple sclerosis, demyelinating disease);[17] Spinal disorders (i.e., bulging or herniated disk, spinal stenosis, and metastatic disease).[17]
- **MRA** (A-angiography) can image blood vessels and can aid in determining blood flow direction and speed without the need for contrast injection (See Chapters 14 and 15 in this book). MRA visualization

has a major role in the assessment of vascular disease, particularly the major vessels of the head, neck, heart, major organs, and peripheral limb vasculature.[17]

- **Position Emission Tomography (PET)** produces images of the body's biological functions. PET (See Chapters 14 and 15 in this book) is capable of detecting certain diseases before other imaging modalities because PET is able to capture chemical and physiological changes related to metabolism. PET images may therefore demonstrate pathological changes long before they would be evident on CT or MRI. The PET procedure can detect cancer, Parkinson's disease, Epilepsy, Huntington's disease, Alzheimer's disease, and others dementias that are known diseases associated with the brain.[45]
- **Drug Delivery Across the BBB:** The BBB protects the brain against toxic substances that circulate in the blood stream. While this is a life-supporting protection for the brain, the existence of the blood–brain barrier is a severe limitation for the delivery of most drugs to the brain because they do not cross in sufficient amounts. Therapeutic drugs targeted for the CNS must overcome the challenge of crossing the (BBB). The BBB typically allows only very small molecules or lipid soluble molecules to cross into the brain. A large number of potentially useful drugs (such as cytostatic) and CNS *in vitro* active agents do not cross the BBB at all or in insufficient quantities. A few common compounds are currently known to readily cross the BBB, including caffeine, alcohol, nicotine, and a few antidepressant drugs. Unfortunately, the most effective drugs are neither small nor lipid soluble. However, because of therapeutic reasons, it is essential to increase the bioavailability of drugs in the brain. Therefore, methods are needed to deliver drugs to the brain that are usually blocked from entering the brain by the BBB.[34] See Appendix – I[20] in this chapter for "*In Vitro Blood–Brain Barrier Model*" for Drug Delivery Studies by Anthony McGoron, Associate Dean, College of Engineering, Florida International University (FIU)."

9.10 SUMMARY

The brain governs conscious actions like walking and thinking through electrochemical signals, while also regulating involuntary behaviors and influencing internal organ functions. Notably, the brain itself lacks pain receptors. CSF and CBF play distinct roles in transporting substances to and protecting the brain. CSF, primarily composed of water, is produced by the choroid plexus from blood, with specialized cells filtering it. This selective process prevents larger cells, proteins, and most drugs from passing through, resulting in CSF being colorless and transparent. Regulatory mechanisms cause variations in ion and glucose concentrations between blood and CSF. CSF is integral to brain support, excretion, transportation, and maintaining the CNS's environment. The blood–brain barrier (BBB), formed by capillary endothelial cells, safeguards the brain by preventing harmful substances from the bloodstream while supplying vital nutrients. Unlike peripheral capillaries, the BBB restricts passage into the brain through TJ and enzymatic barriers. Recent neuroscience studies, employing brain imaging and graphical data, are expanding our knowledge of the mind–brain connection. To understand this dynamic interaction, integrating this knowledge with a broader understanding of brain processes is crucial. Recognizing the importance of brain fluids enhances our grasp of bodily functions, as they facilitate neural information transfer, enabling the brain to control the body through the nervous system. Among these fluids, cerebrospinal fluid stands out, enabling communication between neurons and coordinating the intricate operations of the human body.

KEYWORDS

- **aneurysm**
- **angiotensin artery**
- **anterior cerebral artery**
- **anterior choroidal artery**
- **external carotid artery**
- **internal carotid artery**

REFERENCES

1. Vanderah, T. W.; Gould, D. I. *Nolte's The Human Brain: An Introduction to its Functional Anatomy*, 8th ed.; Elsevier, 2021; p 672.
2. Altman, P. *Blood and Other Body Fluids*; Federation of American Societies for Experimental Biology, 1961.

3. Ashwell, K.; Restak, R. The Brain Book: Development, Function, Disorder, Health. In *The American Biology Teacher*, 2nd ed.; Vol. 83; BioOne, 2019; p 352.
4. Carter, R. *The Human Brain Book: An Illustrated Guide to Its Structure, Function, and Disorders*; DK Penguin Random House, 2019.
5. Britanica. *Cerebrospinal Fluid.* https://www.britannica.com/science/cerebrospinal-fluid, 2023. (accessed 03/30/2023).
6. Chudler, E. H.; Chudler, K. *Neuropedia: A Brief Compendium of Brain Phenomena*; Princeton University Press, 2022.
7. Danaila, L. *The Cells of the Brain—A Review Book*; Independently Published, 2018.
8. Glasser, O. *Medical Physics*; Yearbook Publisher, 1944.
9. Greenfield, A. D. The Circulation through the Skin. In *Handbook of Physiology: Circulation*, Vol. 2; American Physiology Society, 1963; pp 1325–1351.
10. Hall, J. E.; Hall, M. E. *Guyton and Hall Textbook of Medical Physiology*; Pearson & Prentice Hall, 2020.
11. Boeree, C. G. *The Cerebrum* [Online], 2023. https://webspace.ship.edu/cgboer/genpsy-cerebrum.html (accessed Mar 30, 2023).
12. Mancall, E. L.; Brock, D. G. *Gray's Clinical Neuroanatomy: The Anatomic Basis for Clinical Neuroscience*; Elsevier Health Sciences, 2011.
13. Martini, F. H. *Fundamentals of Anatomy and Physiology*; Prentice Hall, 2001.
14. MountCastle, V. B. *Medical Physiology*; C. V. Mosby Company, 1974.
15. Tortora, G. J.; Grabowski, R. R. *Principles of Anatomy and Physiology*; Harper Collins, 1996.
16. UTMB. *Cerebral Spinal Fluid (CSF) Bacterial Culture* [Online], 2023. https://www.utmb.edu/lsg2/Home/Details?id=1172 (accessed Mar 30, 2023).
17. Goyal, M. R. Biofluid Dynamics of Brain. In *Biofluid Mechanics of Human Body Systems*; CRC and Apple Academic Press, 2013; pp 253–271.
18. Davson, H.; Segal, M. B. *Physiology of the CSF and Blood-Brain Barriers*; CRC Press, 1996.
19. Telano, L. N.; Baker, S. *Physiology, Cerebral Spinal Fluid*; StatPearls Publishing LLC, 2022.
20. McGoron, A. In-Vitro Blood-Brain Barrier Model for Drug Delivery. In Personal Communication, Associate Dean (Academic Affairs), College of Engineering, Florida International University (FIU), 2013.
21. Abbott, N. J.; Rönnbäck, L.; Hansson, E. Astrocyte–Endothelial Interactions at the Blood–Brain Barrier. *Nat. Rev. Neurosci.* **2006**, *7* (1), 41–53.
22. Chudler, E. *Glossary of Neuroscience Words* [Online], 2023. http://faculty.washington.edu/chudler/gloss.html (accessed Mar 30, 2023).
23. Truskey, G. A.; Yuan, F.; Katz, D. F. *Transport Phenomena in Biological Systems*; Prentice Hall, 2009.
24. Bronzino, J. D.; Peterson, D. R.; Eds. *The Biomedical Engineering Handbook,* Vols. 1–4; CRC Press LLC (Taylor & Francis Group): Boca Raton, FL, 2015; p 5422.
25. BrainSource. *Emphasizing the Science in Neuroscience* [Online], 2023. http://brainsource.com/ (accessed Mar 30, 2023).
26. Reddy, J. N.; Gartling, D. K. *The Finite Element Method in Heat Transfer and Fluid Dynamics*; CRC press, 2010.
27. Roselli, R. J.; Diller, K. R. Biological Mass Transport. In *Biotransport: Principles & Applications*, 1st ed.; Springer, 2011; pp 789–1215.

28. Roselli, R. J.; Diller, K. R. Biofluid Transport. In *Biotransport: Principles & Applications*, 1st ed.; Springer, 2011; pp 105–485.
29. Seagrave, R. C. *Biomedical Applications of Heat and Mass Transfer*, 1st ed.; The Iowa State University Press, 1971.
30. Kutz, M. *Standard Handbook of Biomedical Engineering Design*; McGraw-Hill, 2003.
31. Enderle, J.; Bronzino, J. *Introduction to Biomedical Engineering*; Academic Press, 2012.
32. Datta, A. K. *Heat and Mass Transfer: A Biological Context*; CRC Press, 2017.
33. Daneman, R.; Prat, A. The Blood–Brain Barrier. *Cold Spring Harb. Perspect. Biol.* **2015,** *7* (1), a020412.
34. Wikipedia. *Blood–Brain Barrier* [Online], 2023. https://en.wikipedia.org/wiki/Blood%E2%80%93brain_barrier (accessed Mar 30, 2023).
35. Kadry, H.; Noorani, B.; Cucullo, L. A Blood–Brain Barrier Overview on Structure, Function, Impairment, and Biomarkers of Integrity. *Fluids Barriers CNS* **2020,** *17* (1), 1–24.
36. Cooney, D. *Biomedical Engineering Principles*; Marcel Dekker, 1976.
37. Fournier, R. L. *Basic Transport Phenomena in Biomedical Engineering*; CRC Press, 2017.
38. Fung, Y.-C. *Biomechanics: Circulation*; Springer-Verlag, 2013.
39. Geankoplis, C. J. *Transport Processes: Momentum, Heat and Mass. Allyn and Bacon Series in Engineering*, 1983; p 538.
40. Jou, D.; Llebot, J. E.; Eds. *Introduction to the Thermodynamics of Biological Processes*; Prentice Hall, 1990.
41. Lightfoot, E. N. *Transport Phenomena and Living Systems*; John Wiley & Sons Inc, 1974.
42. Liu, F.; Li, X.; Zhang, L.-Y.; Song, Q.-R.; Zhang, M.; Zhao, C.-X.; Wang, J.; Sun, G.-t.; Liu, Z.-H. Stimuli-Responsive Nanocarriers for Drug Delivery to the Central Nervous System. *Curr. Nanosci.* **2016,** *12* (1), 4–17.
43. Pathan, S. A.; Iqbal, Z.; Zaidi, S.; Talegaonkar, S.; Vohra, D.; Jain, G. K.; Azeem, A.; Jain, N.; Lalani, J. R.; Khar, R. K. CNS Drug Delivery Systems: Novel Approaches. *Recent Pat. Drug Deliv. Formul.* **2009,** *3* (1), 71–89.
44. Pardridge, W. M. Drug and Gene Targeting to the Brain with Molecular Trojan Horses. *Nat. Rev. Drug Discov.* **2002,** *1* (2), 131–139.
45. NLM. *Brain Diseases* [Online], 2023. https://medlineplus.gov/braindiseases.html (accessed Mar 30, 2023).

APPENDIX I

In Vitro Blood–Brain Barrier (BBB) Model for Drug Delivery Studies by Dr. Anthony McGoron, FIU.[20]

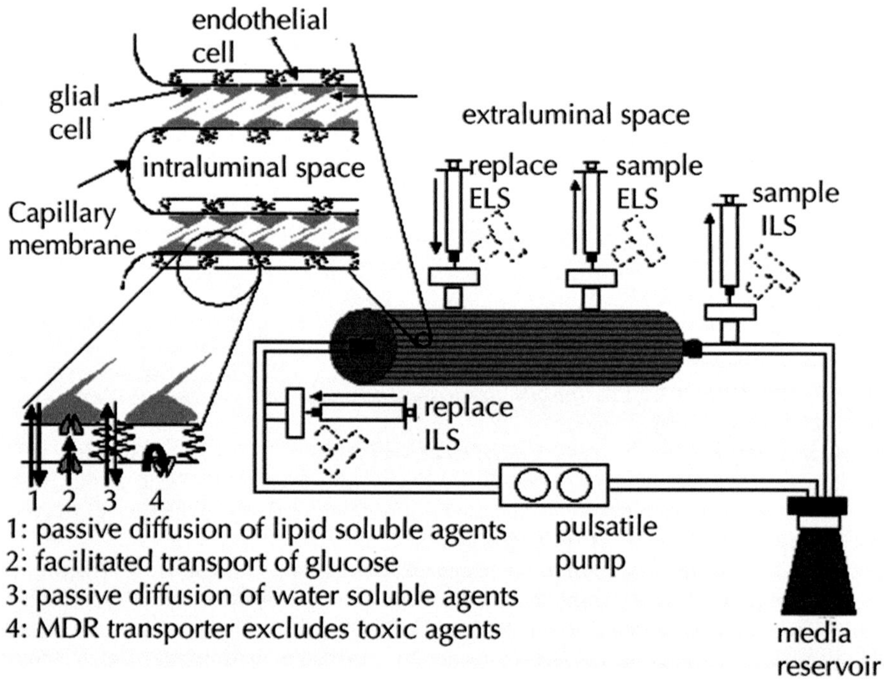

OBJECTIVES

- Develop anticancer drugs that transport across the BBB.
- Endothelial cells are co-cultured with glioma cells in a hollow fiber tube.
- Effective BBB with endothelial cells attached to the luminal side of the capillary membrane and glioma cells attached to the extraluminal side of the capillary membrane.

CHAPTER 10

Biofluid Dynamics of the Human Urinary System*

MEGH R. GOYAL, LUIS DE JESUS DIAZ, YARIMAR PADUA ROSADO, and MELISSA PÉREZ GONZÁLEZ

10.1 INTRODUCTION

Water circulates through our bloodstream and lymphatic system, delivering vital oxygen and nutrients to our body cells. It also aids in waste removal through sweat and urine. Maintaining the right balance of dissolved salts and water inside and outside our cells is crucial; and water provides cushioning to our joints and soft tissues. Additionally, water helps with digestion and waste elimination in the digestive system. The human body comprises approximately 65% water, and it is essential to replenish the water lost each day, which typically amounts to about 8–10 cups. The amount of water needed depends on factors like daily urine and sweat output. If someone experiences diarrhea, vomiting, or engages in intense physical exercise, their water needs increase. It is best to drink liquids devoid of caffeine and alcohol to rehydrate since these chemicals might increase urine production and cause dehydration. In addition to drinking liquids, many foods also contribute to our water intake, for example, fruits and vegetables containing 80–95% water, meats containing around 50% water, and grains (such as, oats and rice) containing up to 35% water.[1–7]

The human body's survival heavily relies on water, as a person can only last a few days without it, whereas the absence of food may take weeks or

* This chapter has been modified from the review article prepared by students (Luis de Jesus Diaz, Yarimar Padua Rosado, and Melissa Pérez González) for the course on Fluid Mechanics, INGE 4015. Course instructor: Megh R. Goyal, Ph.D., P.E., Retired Professor in Biomedical Engineering, General Engineering Department, University of Puerto Rico—Mayaguez Campus. For details, contact at goyalmegh@gmail.com.

Biofluid Dynamics of Human Body Systems: Expanded and Revised Edition.
Megh R. Goyal, Arka Bhowmik, & Anamika Chauhan (Eds.)

even months to cause death (for more details, refer to Chapters 1, 4, 5, and 6 in this book).

The urinary system (depicted in Figs. 10.1A–E) consists of the urinary bladder, urethra, kidneys, and ureters, and has the vital job of removing the majority of metabolic waste from the body.[8–12] The nephron, which includes Bowman's capsule, tubules, and a web of capillaries, is the kidney's functional unit. The glomerulus is responsible for filtering most of the blood's fluid into Bowman's capsule, forming a substance known as filtrate. As this filtrate moves through the tubule portion of the nephron, the body reabsorbs necessary materials, leaving behind the remaining filtrate to become urine.

This chapter specifically focuses on the physiology and biofluid dynamics of the human urinary system.

10.2 PHYSIOLOGY OF THE URINARY SYSTEM[9, 16–19, 21–24]

Urine formation involves three main processes, such as: filtration, reabsorption, and tubular excretion.[24,25]

- **Filtration:** In the filtration process, glomerular excretion is pushed through pores present in the walls of both the glomerular capillaries and the Bowman's capsule. The resulting filtrate contains dissolved substances in the same concentration as the blood, except for the formed elements and plasma proteins, which are large enough to pass via pores.
- **Reabsorption:** As the filtrate travels through the nephron's tubules, the blood reabsorbs water and numerous dissolved substances. Remarkably, 99% water within filtrate is reabsorbed during its journey through the tubules.
- **Tubular excretion:** Additionally, the tubules eliminate substances from the blood. This process, known as tubular excretion, complements the initial glomerular filtration.

Even though the kidneys only weigh about 0.5% of our total body weight, they process around 20% of blood, which is sent by heart by pumping. In fact, blood supply kidneys' abundant allows them to perform various vital functions:

- Regulate blood composition.
- Maintain stable concentrations of essential ions and other substances.
- Keep the body's water volume balanced.

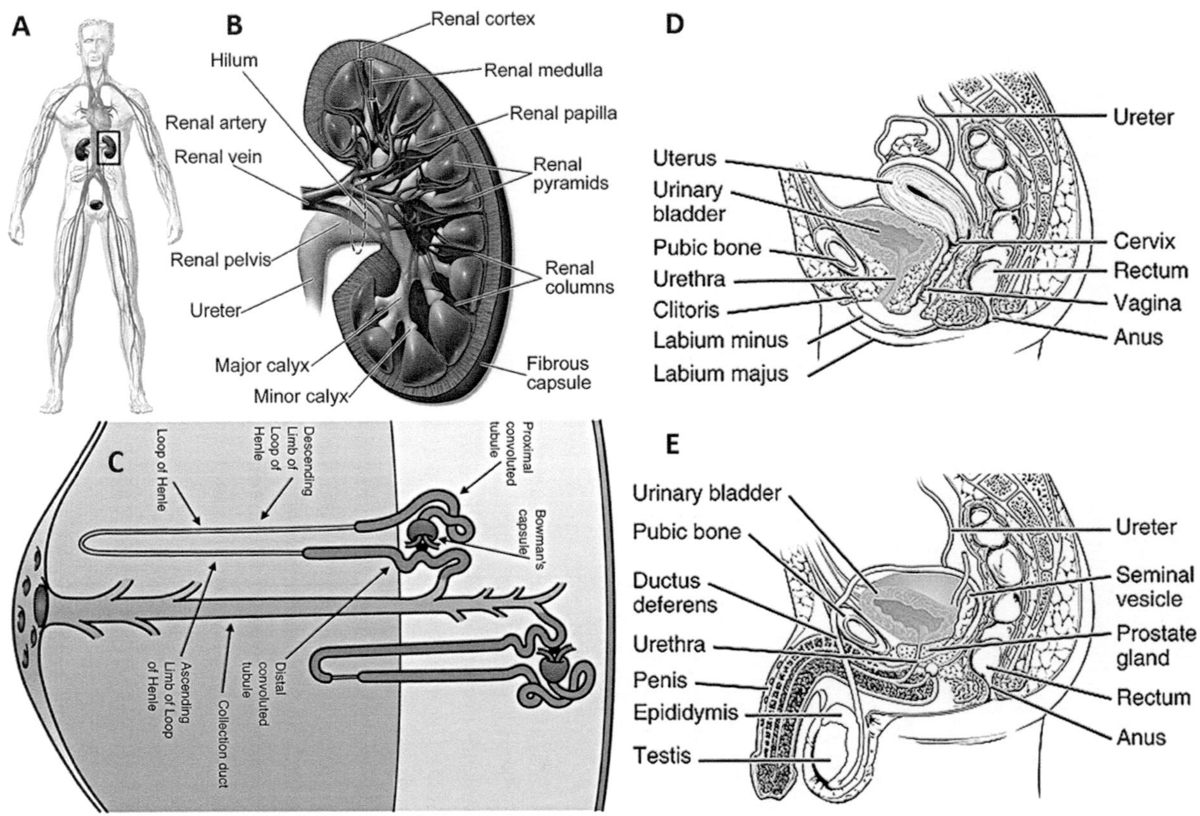

FIGURE 10.1 (A) Human urinary system. (B) Human kidney. (C) A nephron. (D) Median sagittal section of female pelvis. (E) The male urethra.

Source: (A-B) Reprinted from Ref. [8]. https://creativecommons.org/licenses/by/3.0/ (C) Reprinted from Ref. [9]. https://creativecommons.org/licenses/by/3.0/ (D-E) Reprinted from Ref. [10]. https://creativecommons.org/licenses/by/3.0/

- Eliminate waste products like urea, ammonia, drugs, and toxins.
- Stabilize the acid–base concentration of the blood.
- Aid in blood pressure regulation.
- Encourage red blood cell production.
- Maintain proper calcium levels.

Blood enters the kidneys through the renal artery, wherein it is cleaned before replenishing the body again. At the same time, they eliminate waste and other unwanted substances. The ureters carry the urine that the kidneys make to the bladder, where it is held until the urethra allows the body to expel it. If the kidney is cut, distinct components become visible:

- Renal capsule: A thin, protective outer membrane surrounding the kidney.
- Cortex: The outer region of the kidney with a lighter color.
- Medulla: The inside portion of kidney, which appears darker and reddish-brown.
- Renal pelvis: A flat, funnel-shaped cavity responsible for collecting urine and guiding it into the ureters.

The nephron, which plays a vital function in the kidney, possesses a distinctive blood supply unlike any other organ. It is the sole organ in the body where veins and arteries are joined by two series of capillary beds. This unique arrangement plays a crucial role in ensuring a consistent blood flow through and around the nephron, even when there are fluctuations in systemic blood pressure.[6,7,26–30] Through the process of peristalsis, these muscular tubes can propel urine forward. Unlike check valves, backflow prevention in the ureters is achieved through a combination of bladder filling pressure and the muscle tone in the bladder wall.[8]

Ureteral pressure can range from 10 to 40 mmHg when influenced by indwelling catheters containing fluids. The ureter is an adaptive organ, functioning as an adaptive system. It can modify one or more of its characteristics and dimensions to fulfill its functional needs.[2,6,7,24]

10.3 BIOMECHANICS OF URINATION[25,31–34]

10.3.1 WHEN TO URINATE?

The sensation of needing to urinate is caused by the micturition reflex, a process involving nerve signals between the bladder and spinal cord.

Interestingly, this reflex can operate independently of the brain's input. When the bladder fills and stretches, it activates stretch receptors in its walls. These receptors send signals via pelvic nerves to the lower spinal cord, which, in turn, triggers a reflex signal back to the bladder through parasympathetic nerve fibers. This signal prompts the contractions bladder muscles, leading to increased pressure inside the bladder causing the need for urinate. The strength of this reflex is related to the pressure in the bladder, determining the intensity of the urge to urinate.

The micturition reflex informs us when we need to urinate, but actual urination occurs when we consciously decide to relax the external sphincter, a ring-shaped muscle found close to bladder neck, which is under voluntary control by the brain. Interestingly, during physical activities like walking or running, the brain effectively suppresses the micturition reflex more than when we are at rest.[35]

10.3.2 CONTROL OF MICTURITION REFLEX BY THE BRAIN

The micturition reflex serves as a protective mechanism for the kidneys which is composed of two phases: the filling or storage phase and the emptying phase. At birth, this reflex is involuntary, but as we grow, we learn to control it. The ability to control urination depends on two crucial factors: receptors and chemicals that maintain muscle balance to ensure proper function. Additionally, there must be an intact neuro/sensory pathway connecting the brain, spinal cord, and bladder, allowing communication between the bladder's receptors and the brain to regulate the micturition reflex. These receptors[16–18,36] are:

- **Alpha receptors** are found in the neck region of bladder; chemically activated during the filling phase to contract and maintain closure of the bladder neck; chemically inhibited during the emptying phase to relax and open the bladder neck and urethra.
- **Beta receptors** are in the bladder; chemically stimulated during the filling phase to relax the muscles; chemically blocked during the emptying phase to contract and cause the bladder to "collapse."
- **Cholinergic receptors** are distributed throughout the bladder; chemically blocked during the filling phase to relax the muscles; chemically activated during the emptying phase to enhance the contraction of the bladder muscle.

10.4 URINE COMPOSITION

10.4.1 CHEMICAL AND PHYSICAL CHARACTERISTICS OF URINE

A standard urinalysis involves examining the urine for: pH, density, color, protein, glucose, ketones, blood, bilirubin, urobilinogen, nitrite, and bacteria. To conduct this analysis, a specially designed multi-test strip called Multistix is dipped into the urine sample, and the resulting colors are compared to values on a chart. The chart also provides corresponding numerical pH values, which are reported. Table 10.1 displays the physical and chemical properties of urine, while Table 10.2 presents the urinalysis results for an adult female.[34,37,38]

pH Measurement[40]

- **How the test is performed**: To obtain a "clean-catch" (midstream) urine sample, the procedure varies for males, females, and infants. For males or boys, the head of the penis is wiped clean before urination. For females or girls, the area between the vagina lips is washed with soapy water and rinsed thoroughly. Afterward, about 30–60 mL of urine is acquired in container, away from the urine stream, and is handed to the laboratory technician. For infants, the process involves cleaning the area around the urethra. A urine collection bag, which has adhesive paper on one end, is placed on the infant's penis for males, ensuring that the entire penis is inside the bag and securing it in place. For females, it is collected via a bag on the labia. This method allows for the collection of a urine sample in a hygienic manner.[11] The steps to be followed are: Place a diaper over the baby and a urine collection bag to obtain a urine sample from a newborn. Once the infant has urinated into the bag, carefully observe the child and remove the bag. In infants, multiple attempts may be necessary as their movements can displace the bag, making it challenging to obtain the sample. After collecting the urine, drain it into a container for transportation to the healthcare provider. Before the test, the healthcare provider may advise discontinuing certain medications which may hamper the results. It is advised to follow a typical diet couple of days before test.[40]
- **Why the test is performed?** The negative logarithm of the concentration of hydrogen ions (H+) is used to compute pH, a measurement of acidity or alkalinity. It is given as a pH value between 0 and 14, with 7.0 denoting a neutral pH. In the plasma, the pH is tightly regulated in

TABLE 10.1 Physical and Composition of Human Urine.[37]

Property or constituent		Method	Value average (range)
Nitrogenous substances			
Amino acids Argine	Total	Carsten-Cannon ion-exchange chromatography; Van Pilsmun calorimetric (Sakaguchi); microbiological assay	0.45 (0.34–0.50
	Free	Carsten-Cannon ion-exchange chromatography; Van Pilsmun calorimetric (Sakaguchi); microbiological assay	0.16 (0.07–0.30)
Asparatic acid	Total	Carsten-Cannon ion-exchange chromatography; microbiological assay	1.7 (1.2–2.7)
	Free	Carsten-Cannon ion-exchange chromatography; microbiological assay	0.04 (0.01–0.07)
Citrulline		Young et al., filter paper chromatography	0.9 (0–2.8)
Cystine	Total	Microbiological assay	1.7 (1.0–2.6)
	Free	Microbiological assay	1.3 (0.6–1.9)
Glutamic acid	Total	Carsten-Cannon ion-exchange chromatography; microbiological assay	(3.7–5.0)
	Free	Carsten-Cannon ion-exchange chromatography; microbiological assay	0.8 (0–1.5)
Glycine	Total	Microbiological assay	6.5
	Free	Carsten-Cannon ion-exchange chromatography; microbiological assay	2.2
Histidine	Total	Carsten-Cannon ion-exchange chromatography; microbiological assay	2.7 (1.0–5.0)
	Free	Carsten-Cannon ion-exchange chromatography; microbiological assay	2.0 (1.2–2.7)
Hydroxypoline,	Total	Carsten-Cannon ion-exchange chromatography; microbiological assay	0.02
Isoleucine	Total	Microbiological assay	0.20 (0.1–0.3)
	Free	Microbiological assay	0.08 (0.04–0.20)
Leucine	Total	Microbiological assay	0.30 (0.22–0.45)
	Free	Microbiological assay	0.13 (0.05–0.17)
Lysine	Total	Carsten-Cannon ion-exchange chromatography; microbiological assay	0.8 (0.48–1.70)
	Free	Carsten-Cannon ion-exchange chromatography; microbiological assay	0.4 (0.17–0.67)

TABLE 10.1 *(Continued)*

Property or constituent		Method	Value average (range)
Methionine	Total	Carsten-Cannon ion-exchange chromatography; microbiological assay	0.14 (0.10–0.17)
	Free	Carsten-Cannon ion-exchange chromatography; microbiological assay	0.05 (0.03–0.10)
Ornithine		Carsten-Cannon ion-exchange chromatography; microbiological assay	0.15
Phenylalanine	Total	Carsten-Cannon ion-exchange chromatography; microbiological assay	0.30 (0.21–0.54)
	Free	Carsten-Cannon ion-exchange chromatography; microbiological assay	0.17 (0.09–0.23)
Proline	Total	Microbiological assay	0.61 (0.3–0.9)
	Free	Microbiological assay	0.12 (0.03–0.20)
Serine	Total	Carsten-Cannon ion-exchange chromatography; microbiological assay	0.6 (0.5–0.7)
Amino acids (concluded) Serine (concluded)	Free	Carsten-Cannon ion-exchange chromatography; microbiological assay	0.3 (0.2–0.5)
Taurine		Filter paper chromatography	(0.11–0.20)
Threonine	Total	Carsten-Cannon ion-exchange chromatography; microbiological assay	0.50 (0.36–2.60)
	Free	Carsten-Cannon ion-exchange chromatography; microbiological assay	0.25 (0.11–0.35)
Tryptophan	Total	Microbiological assay	0.40 (0.23–0.70)
	Free	Microbiological assay	0.20 (0.11–0.36)
Tyrosine	Total	Microbiological assay	0.7 (0.44–0.82)
	Free	Microbiological assay	0.2 (0.15–0.30)
Valine	Total	Microbiological assay	0.3 (0.25–0.42)
	Free	Microbiological assay	0.09 (0.04–0.18)
Bilirubin		Naumann talc absorption	0.70
Coproporphyrin I and III		Schwarz et al., ultraviolet fluorescence	(0.00024–0.00400)
Creatine		Van Pilsum colorimetric (Sakaguchi); Folin picric acid; Petera picric acid; Ennor et al., diacetyl	0.8 (0–2.0)

TABLE 10.1 *(Continued)*

Property or constituent	Method	Value average (range)
Creatinine	Folin picric acid	23 (15–30)
Guanidinoacetic acid	Dubnoff-Borsook ion exchange; Van Pilsum et al., barium hydroxide, zinc sulfate precipitation	(0.2–0.5)
Guanine	Weissman et al., ion-exchange chromatography	0.006 (0.003–0.009)
8–Hydroxy–7–methyl-	Weissman et al., ion-exchange chromatography	0.02 (0.016–0.030)
7–Methyl	Weissman et al., ion-exchange chromatography	0.09 (0.08–0.11)
N^2–Methyl	Weissman et al., ion-exchange chromatography	0.007 (0.006–0.009)
Hippuric acid		1–10
Histamine	Barsoum-Gaddum guinea pig ileum	(0.0002–0.0010)
Hypoxanthine	Weissman et al., ion-exchange chromatography	0.14 (0.08–0.19)
1–Methyl	Weissman et al., ion-exchange chromatography	0.006 (0.003–0.010)
Imidazole derivatives	Koessler-Hanke p-phenyldiazonium sulfonate	(2–3)
Indoxylsulfuric acid (indican)	Sharlit alcoholi thymol	1.0 (0.5–2.0)
Methionine Sulfoxide	Carsten-Cannon ion-exchange chromatography; Young et al., filter paper chromatography	(0–0.31)
Purine bases	Krüger-Schmid copper oxide precipitation	(0.2–1.0)
6–Succinopurine	Weissman et al., ion-exchange chromatography	0.014
Theophylline	Dikstein et al., ion-exchange chromatography	Trace
Urea	Urease	(200–500)
Uric acid	Brown colorimetric (phosphotungstic acid)	2.0 (0.8–3.0)
1–3-Dimethyl	Dikstein et al., ion-exchange chromatography	Trace
Urobilin	Sparkman p-dimethylaminobenzaldehyde	(0.143–1.857)

TABLE 10.1 *(Continued)*

Property or constituent		Method	Value average (range)
Urobilinogen		Sparkman p-dimethylaminobenzaldehyde	(0.043–0.357)
Xanthine		Weissman et al., ion-exchange chromatography	0.09 (0.07–0.12)
Nitrogen Total *N*		Kjeldahl digestion and distillation	(130–300)
Amino acid N		–	(3–6)
Ammonia N		Folin aeration of alkaline solution	(3–13)
Lipids, carbohydrates, miscellaneous organic acids			
Cholesterol			(0–0.0714)
Reducing substances			(7–21)
Acetone bodies, total		Weichselbaum-Somogyl acetone precipitation; Nessler-Abels; Messiger iodine titration; Behre-Benedict salicylic aldehyde	0.2 (0.03–0.30)
Acetoacetic acid		Weichselbaum-Somogyl acetone precipitation	0.04 (0.03–0.06)
Carbolic acid (phenol)	Total	Schmidt ether extraction; Deichmann-Schafer spectrophotometric	(0.2–0.6)
	Free	Schmidt ether extraction; Deichmann-Schafer spectrophotometric	(0–0.05)
Carbonic acid			2.7 (2.1–3.3)
Citric acid		Taussky-Shorr pentabromoacetone, Pucker et al., pentabromoacetone; Thumberg enzymatic	(3–20)
Formic acid		Dakin et al., ether extraction, mercuric chloride, mercuric chloride	0.80 (0.4–2.0)
Lactic acid		Friedmann-Graesertitration	3.0 (2–5)
Oxalic acid		Powers-levatin	0.5 (0.3–0.7)
Vitamins, and metabolites			
B_1 (thiamine)		Hennessy-Cerecedo; Meckelsen et al., Schultz et al.; Jowett fluorometric	0.003 (0.0006–0.0060)
B_2 (riboflavin)		Microbiological assay; Ferrebee fluorometric; Conner-Straub absorption	0.0124 (0.002–0.024)

TABLE 10.1 *(Continued)*

Property or constituent	Method	Value average (range)
B_6 (pyridoxine)	Microbiological assay	(0.00008–0.00270)
B_{12} (cyanocobalamine)	Microbiological assay	0.44×10^{-6} (0.23×10^{-6} to 0.79×10^{-6})
B_c (folic acid)	Microbiological assay	0.000058 (0.00003–0.0003)
B_x (p-aminobenzoic acid)	Microbiological assay	(0.002–0.003)
C (ascorbic acid)	Roe et al., dinitrophenylhydrazine	(0.1–0.4)
H (biotin)	Microbiological assay	0.0005 (0.0002–0.0010)
Choline	Microbiological assay	0.079 (0.068–0.130)
Citrovorum factor	Microbiological assay	0.000037 (0.000023–0.000069)
Inositol	Microbiological assay	0.2
Niacin (nicotinic acid)	Microbiological assay; Ellinger-kader	0.0034 (0.002–0.020)
Niacinamide (nicotin-amide)	Microbiological assay; Ellinger-kader	0.02 (0.01–0.05)
Pantothenic acid	Microbiological assay	0.045 (0.016–0.100)
Dehydroascorbic acid	Roe et al., dinitrophenylhydrazine	(0.19–0.29)
Dehydroascorbic plus diketogulonic acids	Roe-Kuether, for dehydroascorbic and diketogulonic acids together	0.23 (0–1.28)
Diketogulonic acid	Roe et al., dinitrophenylhydrazine	(0.14–0.19)
N'- Methylnicotinamide	Najjar, and Huff-Perlzweig acetoe	(0.04–0.60)
Pyridoxal	Microbiological assay	0.001 (0.0007–0.0053)
Pyridoxamine	Microbiological assay	0.0016 (0.0004–0.0030)
4–pyridoxic acid	Microbiological assay; Huff-Perlzweig fluorometric	(0.009–0.16)
Trigonelline	Perlzweig; Kodicek-wang; Sarett modification of Kodicek	(0.030–0.300)

TABLE 10.1 *(Continued)*

Property or constituent		Method	Value average (range)
Hormones			
Aldosterone	♂	Paper chromatography	0.00005 (0.00001–0.00013)
	♀	Paper chromatography	0.00006 (0.00003–0.00010)
Androgens		Zimmermann reaction: ♂, 3–5 yr	0.21
		Zimmermann reaction: ♂, 20–40 yr	0.26 (0.20–0.33)
		Zimmermann reaction: ♂, 60 + yr	0.07 (0.03–0.13)
		Zimmermann reaction: ♀, 3–5 yr	0.05
		Zimmermann reaction: ♀, 20–40 yr	0.20 (0.18–0.21)
		Zimmermann reaction: ♀, 60 + yr	0.04 (0.015–0.130)
Androsterone	♂	Paper chromatography; Zimmermann reaction	0.05 (0.035–0.060)
	♀	Paper chromatography; Zimmermann reaction	0.06 (0.05–0.08)
Etiocholanolone	♂	Paper chromatography; Zimmermann reaction	0.06 (0.04–0.07)
	♀	Paper chromatography; Zimmermann reaction	0.05 (0.03–0.06)
Estradiol	♀	Follicular phase	0.00003 (0–0.00005)
	♀	Luteal phase	0.0001 (0.00007–0.00017)
	♀	Post-menopause	0.00001 (0–0.00009)
Estriol	♀	Follicular phase	0.0001 (0–0.0003)
	♀	Luteal phase	0.0004 (0.00013–0.00130)
	♀	Post-menopause	0.00005 (0–0.00018)
Estrone	♀	Follicular phase	0.00008 (0.00006–0.00012)
	♀	Luteal phase	0.0002 (0.00017–0.00040)
	♀	Post-menopause	0.00003 (0–0.00012)

TABLE 10.1 *(Continued)*

Property or constituent		Method	Value average (range)
17-Hydroxysteroids	♂	Porter-Silber reaction	0.08 (0.04–0.17)
	♀	Porter-Silber reaction	0.06 (0.02–0.14)
17-Ketogenic adrenocorticoide	♂	Norymberski reaction	0.21 (0.15–0.31)
	♀	Norymberski reaction	0.18 (0.12–0.30)
α–Ketol steroids		Madar-Buck B. T. Reaction	0.26 (0.13–0.47)
Pregnanediol	♂	Klopper	0.13 (0.005–0.020)
	♀	Klopper follicular phase	0.018 (0.013–0.025)
	♀	Klopper luteal phase	0.055 (0.03– 0.07)
	♀	Klopper post-menopause	0.01 (0.005–0.014)
Pregnanetriol	♀	Paper chromatography: follicular phase	0.025
	♀	Paper chromatography: luteal phase	0.032
	♀	Paper chromatography: post-menopause	0.011
Tetrahydrocortisol		Paper chromatography	0.024 (0.008 –0.050)
Tetrahydrocortisone		Paper chromatography	0.054 (0.02–0.12)

small range (7.35–7.45). The body achieves this precise pH balance by utilizing buffers, which are weak acids or bases capable of converting between two forms by either absorbing or releasing hydrogen ions. Most important among the body's buffer systems is the bicarbonate system,[6,7,30] as follows:

$$H^+ + HCO^-_3 \leftrightarrow CO_2 + H_2O \tag{10.1}$$

TABLE 10.2 Urinalysis of an Adult Woman.[37–39]

Property	Units	Range
Color	–	
Turbidity	–	
Specific gravity	–	1.005–1.030
Reaction pH	–	4.6–8.0
Occult Blood	–	Negative–Positive
Glucose	–	Negative–Positive
Bilirubin	–	Negative–Positive
Ketone	–	Negative–Positive
Protein	–	Neg–Trace
Urobilinogen	–	0.3–1.0
Nitrite	–	Negative–Positive
Leukocytes	–	Negative–Positive
RBC	HPF	0–2
WBC	HPF	0–5
Bacteria	HPF	None–Few
Mucous	LPF	None
Crystals	LPF	None
Amorphous urates	HPF	None
Yeast cells	HPF	None
Trichomonas	HPF	None
CAST	TPF	None
Epithelial cells	TPF	None
Squamous	HPF	None–Few
Transitional	HPF	None–Few
Renal	HPF	None
Oval fat bodies	HPF	None

The regulation of HCO3- (bicarbonate) is carried out by the kidneys, while carbon dioxide (CO_2) levels are controlled by the lungs. In case of any dysfunction of kidneys, the lungs will attempt to compensate, and vice versa. However, this balance can be disrupted, particularly when excessive organic acids neutralize HCO3-. Unlike blood pH, urine pH is not as tightly controlled and can be influenced by various factors after collection. If urine is left uncovered in a container, CO_2 can escape. Moreover, bacteria can increase urine pH by possessing enzymes to convert to ammonia, which then reacts with hydrogen ions.

Serum pH (blood pH) is generally a more reliable indicator of acid–base imbalances compared to urine pH. In cases of respiratory acidosis caused by chronic obstructive pulmonary disease, the kidneys (if functioning properly) will attempt to counteract the increased pH by reabsorbing and synthesizing more bicarbonate, resulting in the production of more acidic urine. The test for urine pH may be conducted in the presence of certain conditions, including alkalosis, interstitial nephritis, distal renal tubular acidosis, and sepsis. Many factors, including food choices and drug use, influences the urine pH. For example, certain foods and medications like chlorothiazide diuretics, methenamine mandelate, and ammonium chloride can affect urine pH. Additionally, drugs like acetazolamide, potassium citrate, and sodium bicarbonate can also have an impact on urine pH.[40]

- **Normal values of pH** range from 4.6 to 8.0.[40]
- **Meaning of abnormal results:** Elevated urine pH (alkaline urine) can indicate conditions such as gastric suction, renal failure, renal tubular acidosis, urinary tract infection, and vomiting. On the other hand, reduced urine pH (acidic urine) may suggest conditions like chronic obstructive pulmonary disease (e.g., emphysema), diabetic ketoacidosis, diarrhea, and starvation.

Protein Measurement[40]

- **How the test is performed:** A urine albumin test is performed to determine the available concentration of proteins in urine. To ensure accurate test results, the healthcare provider may advise the individual to stop using medications that could interfere with the test. For a quick "spot test," a chemically treated stick is dipped into a random urine sample to check for detectable protein. However, for a precise measurement of the protein excreted in the urine, a 24-h urine sample is necessary. A 24-h urine sample, including the first urine in the morning on day two, must be collected as part of the protocol. The container should be capped and refrigerated during the collection period, then labeled with the name, date, and time of completion before being returned to the laboratory. Cleaning the region around the urethra and applying a urine collection bag—a plastic bag with adhesive paper on one end—to an infant are both part of the operation. With the adhesive bonded to the skin, the bag can completely enclose the penis in males, while in females, it is placed over the labia. The diaper is worn on top of the bag with a lock. With active babies, it could take a few tries since they can shift the bag, allowing the samples sucked in diaper. The bag should be replaced once the baby has urinated in it, and the infant should be examined periodically. After that, the collected pee is emptied into a container and brought to the laboratory.[40]

- **Why the test is performed:** A urine protein test is useful for screening kidney disease when there are concerns about its presence. In typical situations, routine dipstick tests do not identify protein in urine. This is because the kidneys can only filter out tiny pollutants during the process of generating urine, as they are naturally made to retain large molecules in the blood, such as protein.[40]
- **Normal values:** The typical range for protein in urine, as determined by a dipstick spot test, is roughly 0 to 8 mg/dL. The typical value for a 24-h urine test is less than 150 mg every 24-h period. Please take note that mg/dL refers to milligrams per deciliter.[40]
- **Meaning of abnormal results[41–43]:** Abnormal results may indicate an increase in urine protein, and this may indicate: amyloidosis; bacterial pyelonephritis; bladder tumor; congestive heart failure (inadequately perfusing the kidneys); diabetic nephropathy; glomerulonephritis; goodpasture's syndrome; heavy metal poisoning; lupus erythematosus; malignant hypertension; multiple myeloma; nephrotic syndrome; nephrotoxic drug therapy; polycystic kidney disease; and preeclampsia.

Glucose Measurement ("Urine Sugar, Urine Glucose, or Glucosuria")

- **How the test is performed:** A dipstick with a color-sensitive pad is used to measure urine glucose. Specific chemicals in the pad interact with glucose to change the pad's color in a manner that is proportionate to the amount of glucose present. Cleaning the region around the urethra and using a urine collection bag—a plastic bag with glue on one end—are both steps in the procedure for babies. With the adhesive adhered to the skin, the bag covers the entire penis in males, and the labia in females. The diaper is then placed over the infant, including the bag. The bag must be checked frequently, and when the baby has urinated, it must be replaced. With active infants, multiple attempts may be necessary as their movements can displace the bag, making it challenging to obtain the sample. The collected urine is then drained into a container and analyzed in laboratory.[40]

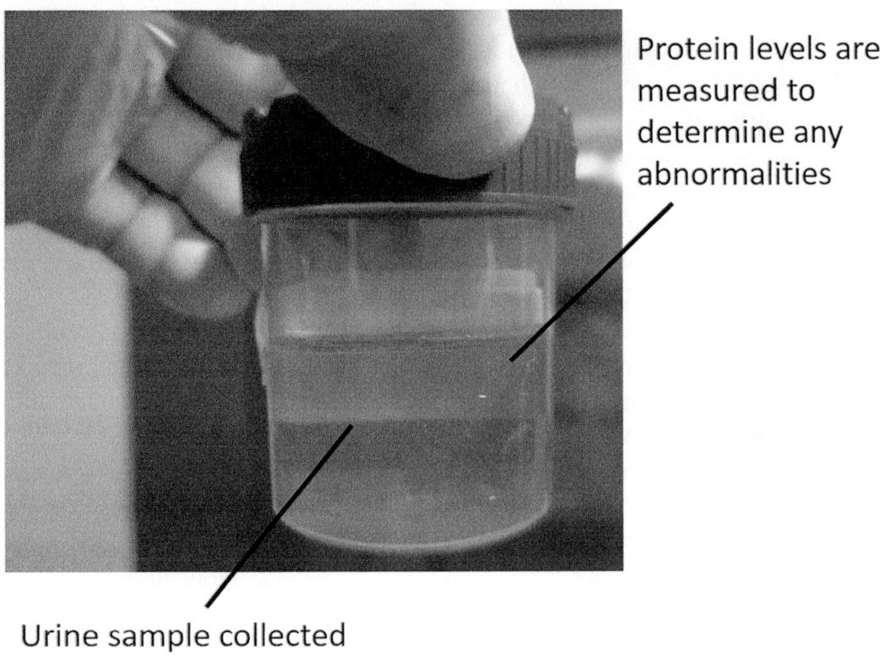

FIGURE 10.2 Protein test.

Source: Reprinted from Ref. [41]. https://creativecommons.org/licenses/by-sa/4.0/

- **How to prepare for the test:** Prior to performing the test, it is advised to discontinue any medications that may affect the results. In most cases, there is no specific preparation required for the test. However, for urine collection in infants, it may be beneficial to have additional urine collection bags ready in case multiple attempts are needed to obtain the specimen.[40]
- **Why the test is performed:** Most carbohydrates consumed in the diet are converted into glucose in the bloodstream. Glucose plays a

vital role as an energy source for most cells in the body, especially the brain and red blood cells, which heavily rely on blood glucose for their energy needs. Maintaining specific glucose concentrations is crucial for normal brain function. If glucose levels drop below around 30 mg/dL, it can lead to confusion or unconsciousness. Similar symptoms can also result from high glucose levels (over 300 mg/dL) when they are coupled with acidosis, illness, or dehydration. Although other hormones like glucagon, epinephrine, and cortisol also have an impact on glucose levels in the body, insulin is the main hormone in charge of controlling them. In order to diagnose diabetes or to keep track of diabetes management, glucose levels are frequently measured. Diabetes may be brought on by insufficient insulin production or diminished insulin sensitivity. In cases of renal glycosuria, where glucose escapes into the urine even when blood glucose levels are normal, abnormal findings in a urine glucose test may be observed.[40]

- **Normal values:** Under normal circumstances, glucose is not present in urine and should not be detectable. If glucose is detected in a urine sample, further testing is recommended.[40]
- **Meaning of abnormal results:** Elevated concentration of glucose found in urine may suggest conditions like diabetes mellitus or renal glycosuria. Renal glycosuria can either be a standalone abnormality or coexist with other kidney disorders, such as Fanconi's Syndrome.[40]
- **Special considerations:** (a) Amino salicylic acid, cephalosporins, chloral hydrate, chloramphenicol, dextrothyroxine, diazoxide, loop and thiazide diuretics, estrogens, isoniazid, levodopa, lithium, nafcillin, nalidixic acid, and nicotinic acid (high doses) are medications that may increase urine glucose; (b) drugs like 40: acetylsalicylic acid, amino salicylic acid, ascorbic acid, cephalothin, chloral hydrate, nitrofurantoin, streptomycin, and sulfonamides may cause false positive findings with Clinitest but not Clinistix or Tes-tape; and (c) ascorbic acid (using Clinistix, Tes-tape), levodopa (using Clinistix), and phenothiazines (using Clinistix, Tes-tape) are medicines that could provide erroneous negative results.

Urine Ketone Measurement ("Acetone, Acetoacetic Acid, and Beta-Hydroxybutyric Acid")[40]

Urine ketones are typically assessed using a "spot test" with a dipstick that has a color-sensitive pad saturated with specific chemicals. These substances change the color of the dipstick when they interact with the ketone bodies in the urine. A qualitative sign of the presence of ketones is the color shift.

- **How the test is performed:** The healthcare provider may suggest a special diet, and it is advisable to discontinue medications that could influence the test results. For urine collection from an infant, having a few additional collection bags on hand might be beneficial.[40] For adults and children, a "clean-catch" (midstream) urine sample is collected. For males or boys, cleaning the head of the penis is necessary, while females or girls should wash the area between the vaginal lips with soapy water and rinse well. A minute volume of urine is allowed to fall into the toilet bowl initially to clear the urethra of contaminants. Subsequently, around 30–60 mL of urine is collected in a clean container, away from the urine stream, and then handed to the technician. For infants, the procedure involves thoroughly washing the area around the urethra. To collect urine from an infant, a urine collection bag is used. The bag is a plastic bag with adhesive on one end. For males, the entire penis is placed inside the bag which is stuck to skin via adhesive. For females, the bag is positioned over labia. Once the bag is secured in place, the diaper is then placed over the infant, covering the bag. Frequent checks are necessary, and the bag should be removed after the infant has urinated into it. With active infants, multiple attempts might be needed as their movements can displace the bag, making it difficult to obtain the specimen. The collected urine is then drained into a container to be transported back to the laboratory.[40]
- **Why the test is performed:** Ketones, including beta-hydroxybutyric acid, acetoacetic acid, and acetone, are produced as the end products of rapid or excessive breakdown of fatty acids. Like glucose, ketones are present in urine when their levels in the blood surpass a certain threshold. Hormones like glucagon, epinephrine, and growth hormone stimulate the secretion of fatty acids within adipose tissue. These hormone levels are elevated in situations such as starvation (regardless of whether it is associated with excessive alcohol consumption or not), uncontrolled diabetes mellitus, and various other medical conditions.[40]
- **Normal values:** It's typical to have a negative test result. When acetone is detected in the urine, the results are often labeled as minor, moderate, or large with these associated numbers: small < 20 mg/dl; moderate 30–40 mg/dl; large, < 80 mg/dl.
- **Meaning of abnormal results:** A positive result could mean: metabolic problems, such as uncontrolled diabetes or a disease of the

glycogen storage organ; abnormal dietary practices, such as fasting, anorexia, and high-protein or low-carbohydrate diets; prolonged vomiting, such as hyperemesis during pregnancy; conditions that cause an increase in metabolism, such as hyperthyroidism, fever, acute or serious illnesses, burns, pregnancy, lactation, or after surgery.

- **Special considerations:** Test results can be affected by specific dietary patterns. For example, a diet low in carbohydrates but high in protein and fat can influence ketone levels. Moreover, certain medications, such as glucocorticoids, can result in false positive readings in the measurements.

Bilirubin Measurement ("Conjugated Bilirubin-Urine and Direct Bilirubin-Urine")

Bilirubin metabolism starts with reduction of red blood cells (RBCs) by phagocytic cells. Hemoglobin from RBCs is converted into heme and globin. Heme is then transformed into bilirubin, which is carried in the blood by albumin to the liver. Most of the bilirubin in the liver is converted to conjugated bilirubin, sometimes referred to as direct bilirubin (DB), by combining with glucuronic acid. The type of bilirubin that exists in the liver prior to conjugation is known as indirect bilirubin (IB).

$$Total\ bilirubin = DB + IB \tag{10.2}$$

The process of bilirubin metabolism commences with reduction of RBCs by phagocytic cells. Hemoglobin from RBCs is converted into heme and globin. Subsequently, heme is converted into bilirubin, which is transported in the bloodstream by albumin to the liver. The majority of bilirubin in the liver reacts with glucuronic acid to generate conjugated bilirubin, commonly referred to as "direct bilirubin (DB)." Unconjugated bilirubin, known as "indirect bilirubin (IB)," is the form that exists prior to conjugation in the liver.[40]

- **How the test is performed:** Follow the healthcare provider's instructions very carefully to obtain a 24-h urine sample, and stay away from any drugs that can affect the results. On the first day, start by urinating into the toilet in the morning and then collect all subsequent urine in a designated container for the next 24 h. Repeat the process on the second day by urinating into the container upon waking up, capping it, and keeping it refrigerated throughout the collection period. Clean the region around the urethra in babies completely. Affix an adhesive-sided urine collection bag to the infant's body. Place the bag close to labia in the case of girls, and the entire penis in the case of boys, with the adhesive connected to the skin. Put a diaper over the infant, including the bag. Check the baby regularly and change the bag after they have urinated into it. With active infants, it might be necessary to make multiple attempts as their movements can displace the bag, making it challenging to collect the specimen. Finally, drain the collected urine into the container for transportation to the laboratory.[40]

- **How to prepare for the test:** No additional precautions before test.[40]
- **Why the test is performed:** This test may be performed when problems with liver or gallbladder are suspected.[40]
- **Normal values:** Bilirubin often not located in urine.
- **Meaning of abnormal results:** When the bile ducts are blocked, the level of direct bilirubin in the blood can significantly increase, causing its release from the liver into the bloodstream. As a result, in cases of elevated direct bilirubin levels, some of it may be present in the urine. The presence of direct bilirubin in urine can indicate various conditions, including obstruction of the liver's biliary ducts, biliary strictures, cirrhosis, gallstones in the biliary tract, hepatitis with associated biliary obstruction, surgical trauma affecting the biliary tract, and liver or gallbladder tumors.[40]
- **Special considerations:** The bilirubin molecule can disintegrate when exposed to light. Several drugs, such as allopurinol, certain antibiotics, barbiturates, chlorpromazine, diuretics, ethoxazene, oral contraceptives, phenazopyridine, steroids, and sulfonamides, can result in false elevations in the test results. Conversely, indomethacin and ascorbic acid can cause false negative results.[40]

Blood in the Urine Measurement (Hematuria)

Blood in the urine can present as pale pink or bright red discoloration in the toilet water or as spots of blood after urination. When blood appears in the urine in women, it may be a sign of vaginal bleeding, while in men, it could be mistaken for urinary bleeding but might actually be a result of bloody ejaculation, often associated with prostate issues. Blood-like discoloration in the urine may be brought on by specific medications, beets, or other foods. It is important to remember that blood in the urine may not always be obvious to the human eye and cannot always be identified during normal testing but can be identified through microscopic examination. If visible blood is present, immediate and thorough evaluation is necessary.

In children, this condition often requires hospitalization for a comprehensive investigation. The physician will meticulously monitor the situation to ascertain the underlying reason and choose the best line of action.[40]

- **Common causes:** There are several possible reasons for blood in the urine, many of which are related to problems with the kidneys or other parts of tract. The doctor might look into the possibility of a bleeding disease if the prostate, genitals, kidneys, urinary tract, and urinary tract are all found to be in good health. The presence of kidney or bladder stones, chronic and recurrent bladder, kidney, or urethral infections, bladder or kidney cancer, kidney trauma, fractured pelvis, or recent treatments involving the urinary tract. Other probable causes include polycystic kidney disease, acute or chronic kidney failure, and post-streptococcal glomerulonephritis, which is a typical reason for children to pass blood in their urine after strep throat. Thorough evaluation is necessary to determine the specific underlying cause in each case.[42–45]
- **Home care:** The treatment for blood in the urine varies depending on the underlying cause, which is determined through a thorough evaluation by the doctor. It is crucial not to ignore the presence of blood in the urine. In particular if you are suffering unexplained weight loss, a burning sensation during urination, frequent urination, it is crucial that you let your doctor know about this symptom and undertake the necessary evaluation. Prompt medical attention is vital to identify the cause and determine the most suitable course of action.[40,42,43]
- **What to expect in the office of a physician[40]:** A medical history will be taken, and a physical examination will be done by the doctor. Some possible medical history inquiries are:
    - "When was the first noticed blood in the urine?
    - How the urine color is?
    - Does pain experience during urination?
    - Has volume of the urine increased or decreased?
    - Does the urine have an odor?
    - How frequently urinates?
    - Which drugs being taken, including over the counter drugs?
    - Were beets, berries, or rhubarb eaten?
    - Is there a past urinary issue or kidney issues?[42]
    - Are there any allergies?
    - Was there any recent injury?

- **Diagnostic tests[36,40] may include:**
  - Blood studies: a CBC, blood differential, C3, and creatinine.
  - Urinalysis.
  - Urine culture.
  - 24-h urine collection for creatinine, protein, and calcium.
  - Tests for strep.
  - Cystoscopy.
  - Kidney biopsy.[36]
  - X-rays of the kidneys.[36]
  - IVP.
  - Abdominal ultrasound.
  - CT scan of the abdomen.

Urine Volume Measurement (24-h Urine Collection)

- **How the test is performed[40]:** To collect a 24-h urine sample, follow these steps: For adults, on the first day, urinate into the toilet in the morning and then collect all subsequent urine in a designated container for the next 24 h. On the second day, repeat the process by urinating into the container upon waking up, cap it, and refrigerate storage. Make sure to label the container with your name, date, and time of completion before returning it to the laboratory. Clean the region around the urethra in babies thoroughly. Place the infant on a urine collection bag that has an adhesive end open. Place the bag over the labia for girls and the entire penis for boys inside the bag with the adhesive connected to the skin. Put a diaper over the infant, securing the bag in place. Change the bag after the infant has urinated in it and keep an eye on the baby at all times. As active babies can move the bag and cause the specimen to be sucked into the diaper, this procedure might need to be repeated multiple times. Drain the urine from the bag into the provided container for transportation to the laboratory. Once the sample reaches the laboratory, they will measure the volume of urine in the specimen to analyze it properly. It is essential to follow the healthcare provider's instructions carefully and ensure accurate collection for accurate test results.[40]
- **Why the test is performed:** Tests that measure the amount of certain compounds expelled each day, such as protein, aldosterone, sodium, potassium, or urea nitrogen, such as creatinine clearance or other assessments, frequently analyze urine volume. It is also tested in people

who have polyuria, a medical disorder known for the abnormally high production of urine, such as "diabetes insipidus."

- **Normal values:** With a typical daily fluid intake of roughly 2 L, the range is 800–2000 mL. Individuals may have slightly varying normal value ranges.
- **Meaning of abnormal results:** Dehydration, insufficient fluid intake, and renal insufficiency or failure are indication that result in decreased urine volume. Diabetes mellitus, diabetes insipidus, renal diabetes, central diabetes, end-stage renal disease, high fluid intake, and usage of diuretic drugs are a few of the illnesses that result in increased urine volume.

Urine Color Measurement

Urine often ranges in color from straw to light amber. Pee that is concentrated is often darker than pee that is dilute. The urine's hue can be described using the words below:

- No color.
- Straw: light or dark
- Amber: light or dark
- Redness.

When blood, medications, or diagnostic dyes are present in the urine, the color can alter.

It's possible for urine to seem clear, hazy, barely cloudy, cloudy, or extremely cloudy. Freshly discharged pee typically appears translucent or clear. However, it may appear foggy if there is blood, phosphate, crystal, pus, or germs present. The urine's transparency is only significant if the sample is still fresh. Due to bacterial activity, salts, and breakdown over time, all urine has the potential to become murky. When urine sits and cools, a fine cloud of "mucus, leukocytes, and epithelial cells" may form. This cloud condenses near the container's bottom and is typically not clinically significant.

Urine Odor Measurement[40]

- **How the test is performed:** Males should clean the penis, while females should use soapy water to clean the vaginal area between the lips in order to get a "clean-catch" or midstream urine sample. To clean the urethra of impurities, start urinating and let a small amount fall into the toilet. After that, collect 30–60 mL of urine in a clean container, and remove the container from the urine stream. When using an adhesive-tipped urine collection bag on a baby (for males, the entire penis can be placed in the bag; for females, the bag is placed on the labia), it is best to disinfect the region around the urethra first. A diaper is then placed over the infant, including the secured bag. Check the baby frequently and remove the bag after the infant has urinated into it. For active infants, multiple attempts may be needed as they can displace the bag. The urine is then drained into a container for transport back to the laboratory. Finally, the health care provider may smell the specimen to determine its odor.[40]

- **Normal values:** The normal odor of urine is aromatic.
- **Meaning of abnormal results:** Hepatic (liver) failure, ketonuria, maple sugar urine disease (extremely rare), phenylketonuria (rare), rectal fistula, and urinary tract infection are among the conditions that might cause abnormal urine odor.
- **Special considerations:** Some of the compounds in urine may break down when it is left to stand at room temperature, giving off an ammonia-like odor. Urine from asparagus causes a distinctive odor.[40]

Urine Concentration Measurement (Water Loading Test and Water Deprivation Test)[40]

- **How the test is performed:** The kidneys' capacity to conserve or eliminate water in the right amounts is assessed using a urine concentration test. Men or boys should clean the head of the penis, and women or girls should wash the area between the lips of the vagina with soapy water and rinse properly in order to get a "clean-catch" or midstream urine sample. Start urinating into the toilet to clear the urethra of contaminants, then place a clean container under the urine stream to capture 1 to 2 ounces. Cap the container, mark it with relevant information, and give it to the healthcare provider or assistant.

  For infants, wash the area around the urethra and place a urine collection bag (a plastic bag with adhesive on one end) on the infant. With the adhesive connected to the skin, the entire penis can be inserted within the bag for males, while the labia are covered by the bag for females. The regular diaper is then worn over the closed bag. This procedure could require a few tries because active babies can knock the bag around and let the sample get absorbed by the diaper. After the infant has urinated into the bag, it should be changed and the baby should be examined frequently. The urine is then drained into a container for transport to the laboratory. In this specific test, the specific gravity of urine is measured before and after specific procedures, such as water loading (consuming excessive water or receiving it through intravenous solutions), water deprivation (abstaining from fluids for a period), or administration of ADH (anti-diuretic hormone, thus ensures urine to become concentrated). The changes in specific gravity help assess the urine's concentration and dilution abilities under different conditions.[40]

- **How to prepare for the test:** For accurate test results, it's essential to follow these steps: maintain a normal, balanced diet for a few days before the test, and then provide a urine sample to the laboratory. If contamination is possible (due to vaginal discharge or bleeding), collect a clean-catch sample. Depending on the needs of the test, specific guidelines will be provided for either water loading or water deprivation. The healthcare provider may advise discontinuing medications that could influence the test outcome. If the collection is from an infant, it might be helpful to have extra urine collection bags on hand.[40]
- **Why the test is performed:** The urine concentration test is commonly used to investigate suspected diabetes insipidus and distinguish it from nephrogenic diabetes insipidus. It is also valuable in diagnosing and treating syndrome of inappropriate ADH (SIADH), a condition characterized by excessively high urine concentrations due to water conservation, despite the individual's hydration status. The kidneys normally maintain urine concentration by adjusting water excretion, which is partly regulated by ADH (anti-diuretic hormone).[40]
- **Normal values** are as follows:
  - 1.010 to 1.025 (normal specific gravity).
  - 1.001 after water loading.
  - 1.025 after water deprivation.
  - Concentrated (increased specific gravity) after ADH administration".
- **Meaning of abnormal results:** Elevated urine specific gravity can be a sign of a number of illnesses, including dehydration, diarrhea, excessive sweating, glucosuria, renal artery stenosis, syndrome of unfitting antidiuretic hormone secretion (SIADH), nausea, and water

restriction. Heart failure can also be caused by diminished blood supply to the kidneys. On the other side, a decreased urine specific gravity could be a sign of pyelonephritis, renal failure, diabetes insipidus, or excessive fluid intake.

- **Special considerations[40]:** The urine concentration test may also be conducted for conditions like complicated urinary tract infection (pyelonephritis), nephrogenic diabetes insipidus, and in cases where certain drugs like dextran and sucrose, or radiographic contrast medium have been used within the past 3 days, as these substances can elevate the values of urine specific gravity.

Specific Gravity Measurement for Urine

The specific gravity of a urine sample measures its density in comparison to a volume of distilled water of the same volume. It is influenced by quantity of dissolved solids present in the urine and typically falls within the range of 1.015 to 1.030 over a 24-h period.

10.5 URODYNAMICS[12,28,46–48]

Urodynamics is a diagnostic process that examines the working of peripheral urinary system, including the bladder and urethra, by assessing the patient's condition and evaluating physical factors such urine flow rate and pressure. The evaluation starts with a comprehensive medical history and examination, which may reveal any abnormalities in the lower abdomen or pelvis that could be related to urinary symptoms. To gather additional information, patients are asked to maintain a urination diary for three days, recording their fluid intake, output, and any episodes of incontinence.[49,50] The urination diary offers valuable insights into bladder capacity, frequency of urination, episodes of incontinence, and nighttime urination patterns. It can also identify issues like excessive fluid intake. Additionally, a midstream urine sample is collected and sent to the laboratory for analysis to rule out any potential infection.

A Pad test is conducted to assess urinary incontinence in females. During the test, the patient consumes 500 mL of water and carries out regular daily activities while wearing a pre-weighed pad. Afterward, the pad is weighed again, and if there is an increase of more than 1 g/h, it indicates urinary incontinence. The test helps objectively measure the severity of incontinence and provides a clear demonstration of the symptom.[49,50]

The detrusor muscle, responsible for bladder contraction, can sometimes become "unstable," leading to lower urinary tract symptoms like frequent urination, urgency, and nocturia. The unstable detrusor can contract even between voidings, causing high bladder pressures that result in urgency and urine leakage. In men, similar symptoms can be caused by an enlarged prostate gland (associated with BPH), and urodynamic studies can help distinguish between these two causes.[14]

According to the International Continence Society, a urodynamic pressure-flow study is currently considered the most objective approach to diagnose and evaluate the severity of obstruction. Additionally, there are different types of dissoluteness.[49,50] There are two types of incontinence: stress incontinence, which is brought on by a weak bladder closure mechanism, and urge incontinence, which is brought on by an overactive bladder.

During a urodynamic pressure-flow study, a narrow urethral catheter is used, allowing voiding to occur around it. The key measurement is the detrusor pressure at maximum flow, which helps distinguish between obstruction (high pressure, low flow) and a loss of strength in detrusor muscle (low pressure, low flow). The amount of leftover urine in the bladder is calculated after voiding. In more complicated situations, flow/pressure studies, and X-ray screening can be coupled (Video cystourethography) to understand anatomy of bladder and urethra. Ultrasound scanning is a useful method for detecting abnormal low bladder neck. While not all patients with symptoms of the lower urinary tract require a comprehensive urodynamic examination, these tests are especially helpful when other treatments have failed. They are essential before considering surgical interventions for incontinence or bladder prolapse.

10.5.1 CYSTOMETRY

Cystometry, also known as urodynamic studies, is conducted to assess the bladder's functioning. It is performed when there might be an issue with the muscles or nerves that affect the bladder's ability to hold or release urine properly. The process of urination is a complex includes communication between the bladder, spinal cord, and brain. As the bladder fills, nerves in its wall signal the spinal cord and brain, indicating the need to urinate. In response, the spinal cord triggers the bladder to contract (voiding reflex). When one consciously holds in the urine, the brain suppresses this reflex. Allowing the reflex to occur leads to urination.

Bladder dysfunction can develop from any issue affecting this neural system or the muscles of the bladder wall. Cystometry helps identify and evaluate such issues to determine appropriate treatment.

During cystometry, the bladder is filled with water or gas to assess its ability to hold and expel the fluid. Medications may be administered to observe the bladder's response to them, checking if it contracts or relaxes appropriately. To measure abdominal pressure during bladder filling, a small tube (catheter) can be inserted into the rectum. Muscle function near the anus may be assessed using a pad or needle. The catheter used in cystometry has two channels—one for filling and the other for recording pressure. Notably, around 85% of all incontinence cases are observed in women,[11] with three-quarters of those experiencing stress incontinence (leakage without over-activity). Cystometry helps to diagnose and to understand the underlying causes of these urinary issues.

Compliance refers to the elasticity of the detrusor muscles, which allows the bladder to stretch and reach its normal capacity while maintaining low pressures. It is an assessment of how well the bladder can expand without generating high pressures. When the bladder is being filled to its normal capacity, the detrusor muscles are watched to gauge stability. This evaluation helps determine whether there is detrusor overactivity or instability, indicating whether the bladder contracts or behaves abnormally during the filling phase.

Cystometry is a process used to measure the amount of pee still in the bladder after a person thinks they have completely emptied it and to evaluate the bladder's ability to store urine. Although a catheter is inserted into the bladder, there is a slight risk of developing a urinary tract infection despite the fact that the surgery is generally regarded as safe.. There is a small chance that a bladder infection will spread to the kidneys and bloodstream, which would be extremely dangerous. If an infection develops, antibiotics can be used to treat it. The results of cystometry are interpreted as either normal or abnormal based on the findings.

- **Normal results:** The speed at which someone's bladder empties during a typical urination. When there is 150–200 mL of liquid in the bladder, the time when one first feels the urge to urinate is within the typical range: (a) Tests of the function of the nerves that regulate the bladder are normal; (b) Tests of the maximum volume of liquid the bladder can hold are within the normal range, 400–500 mL; and (c) The stress test does not cause dribbling.
- **Abnormal results:** (a) The rate at which urine flows from the bladder when one urinates is slower than normal; (b) it is challenging to start

the flow of pee; (c) the need to urinate is felt sooner or later than usual, or not at all; (d) the bladder's maximal capacity is lower than usual, or one cannot feel it; and (e) the urge to urinate does not occur. When the nerves that control the bladder are examined, normal feelings and reactions are absent; also, the stress test causes dribbling.

10.5.2 UROFLOWMETRY

Uroflowmetry is a diagnostic technique used to evaluate how quickly urine is expelled during urination. The electronic measurement of urine flow determines the flow rate in milliliters of urine passed per second. With the use of this test, average and maximum flow rates can be calculated.

The patient is instructed to consume four cups of fluids around 4 h before to the test to make sure the bladder is sufficiently filled. The patient will be instructed to urinate into a funnel that is attached to an electronic uroflowmeter, which records the urine flow information on a flow chart, during the test. After the test, the patient can resume their regular daily activities immediately.

The peak flow rate, Q (max), is a significant indicator for assessing the severity of blockages or obstructions. A low flow rate may suggest conditions like urethral obstruction, enlarged prostate, or impaired bladder function. Flow rate measurement is a valuable diagnostic tool due to its non-invasive nature and cost-effectiveness.

Among the commonly measured urinary flow parameters, the peak flow rate (Qmax) shows a strong correlation with the degree of outflow obstruction (as shown in Table 103A and B). In cases of obstruction, the total voiding time is prolonged, and Qmax is reduced. Uroflowmetry alone cannot differentiate between poor detrusor contractility and bladder outflow obstruction. Therefore, additional urodynamic techniques like cystometry are necessary to make a precise diagnosis[12,28,46–48] should be performed. The abnormal patterns of uroflowmetry are:

TABLE 10.3A Interpretation of Uroflowmetry Results.[8,13,51]

Flow rate, Q_{max}	Interpretation
> 15 mL/s	Unlikely to be significant obstruction.
< 10 mL/s	Likely to be significant obstruction or weak detrusor activity.
10–15 mL/s	Equivocal.

TABLE 10.3B Normal Flow Rates (mL/sec).[8,13,51]

Flow rate, Q_{max}	Interpretation
> 15 mL/s	Unlikely to be significant obstruction.
< 10 mL/s	Likely to be significant obstruction or weak detrusor activity.
10–15 mL/s	Equivocal.

- The flowmetry pattern of a patient with bladder outlet obstruction. The maximum flow reached is around 10 mL/s and the flow rate is prolonged. A diagnosis cannot be made from this reading alone but is certainly characteristic of someone with obstructed voiding and having a normal detrusor muscle function.
- The intermittent flow pattern, usually representing the abdominal straining in an attempt to overcome outflow obstruction. The peak flow rate may be normal or high, especially if outlet resistance is reduced. The curve's shape provides more information[13]: In a healthy person, the flow rate quickly increases to a peak and then is maintained until the bladder is almost empty; in a person with outflow blockage, the curve is drawn out and the maximum flow rate is lower. Peak flow rate will be:

	Male	Female
Young adult (< 55 years)	0	8
Over 55 years	2	5

10.5.3 MEASUREMENTS FOR URINE FLOW RATE

- The flow time, the time over which measurable flow occurs.
- Time to maximize flow, time elapsed from the onset of flow to maximum.
- Maximum rate of flow.
- Voided volume, total volume expelled from the urethra.
- Flow pattern: continuous, interrupted, or specifically described.

Urine volumes of at least 150 mL are considered good. Smaller voided volumes need special attention. Volumes >500 mL are also accompanied by reduced flow rates, perhaps because of overstretching the detrusor fibers. Flow rates, however, are also influenced by the patient's age and sex. Table 10.3A shows the interpretation of uroflowmetry results. Table 10.3B indicates urine flow rate for different age groups. Studies demonstrate that females have significantly higher flow rates than males matched for age and voided volumes.

The urine flow rate measurement depends on several factors, such as: (a) patient's comfort with the environment in which the test takes place and the intimidation experimented; (b) the patient must have an adequately full bladder, as the urine flow rate is dependent on voided volume; (c) abnormal results can be obtained if—any circular muscles that normally regulate urine flow, becomes weak or dysfunctional resulting in an increase in urine flow and/or urinary incontinence; (d) a weak bladder muscle could lead to a reduction in urine flow. **Note:** Abnormal results should be confirmed with other urodynamic studies after the evaluation by a urologist.[50]

10.5.4 URINE FLOW CURVE

The uroflow test is a noninvasive method that is used to quantitatively and qualitatively assess urinary stream of a patient. During the test, the patient is asked to urinate into a special container connected to an electronic urology machine that continuously records the urine flow rate. The test should be performed with a normally full bladder and the natural urge to urinate, avoiding forced urination, or waiting until the bladder is overly full. The uroflow test provides valuable information to urologists about a patient's urination patterns, including maximum and average urine flow rates, total urine volume, and urination duration. It supplements the overall assessment of the patient's urinary condition and can be repeated over time to monitor progress or treatment response. The uroflow test is frequently used to examine the symptoms of benign prostatic hyperplasia (BPH), but it can also be used to assess other urinary disorders. However, it alone does not make a diagnosis and should be combined with a thorough history, physical examination, and possibly more advanced urodynamic studies or cystoscopy for accurate diagnosis in complex cases. The uroflow test is especially helpful for monitoring treatment effectiveness in BPH cases.

Several flow curves are: (i) The normal flow curve is bell-shaped, with a steep acceleration and deceleration; (ii) The undulating flow curve is unbroken, but characterized by a less steep acceleration and deceleration. Flow fluctuates during voiding without reaching zero before the end of voiding; (iii) A fractioned flow curve is the result of a micturition in several times. The flow falls to zero in between the flow episodes; and (iv) The fourth pattern is the long curve–low Q_{max} pattern.[51]

10.5.5 BALLOON CATHETER TO ELIMINATE THE OBSTRUCTION

The balloon catheter system (Figure 10.3) is designed to alleviate obstructive prostatism symptoms. It consists of an expandable dilatation catheter and a long sheath for insertion through the urethra. The sheath serves as a guide for the catheter and a cystoscope lens. The catheter features irrigation ports near its proximal end, allowing a saline solution to flush away blood and debris, providing a clear view for the urologist. Once positioned correctly, the dilation balloon is inflated, opening the affected prostatic urethra, and relieving the obstruction. This system ensures effective treatment while maintaining a clear visual field during the procedure.

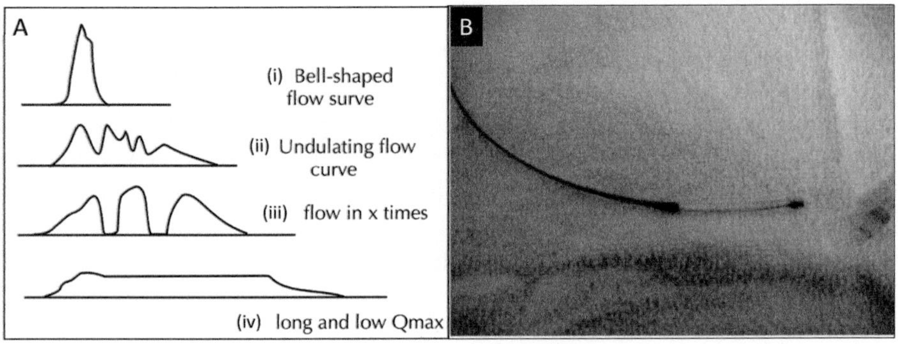

FIGURE 10.3 Foley catheter in urinary tract.

Source: Reprinted from Ref. [53]. https://creativecommons.org/licenses/by-sa/4.0/

10.5.6 CATHETER TIP TRANSDUCER

The catheter tip transducer technique involves measuring the urethral pressure profile (UPP). A catheter with pressure transducers in the bladder and urethra is used for this purpose. The bladder transducer assesses bladder spasms, while the urethral transducer evaluates urethral tone and the presence of intrinsic sphincter defect (ISD). By analyzing pressure differences between the bladder and urethra during coughing and bearing down, the test helps assess conditions such as ISD, obstruction, overflow, and detrusor hyperactivity with impaired contractility.

10.5.7 RENAL PLASMA FLOW

The amount of plasma that travels to the kidneys in one unit of time is known as renal plasma flow (RPF). Fick's Principle (Law of Conservation of Mass) provides renal plasma flow (eq 10.5) to balance the renal inputs (renal artery) and renal outputs (renal vein and ureter). The renal vein and the ureter are thought to be the two exit destinations for a non-metabolizable solute that enters the kidney through the renal artery. It is necessary for the mass arriving through the artery to equal the mass leaving by the vein and ureter in relation to time (eq 10.3, Continuity Equation for Mass Flow).

$$RPF_a \text{ x } P_a = [RPF_v \text{ x } P_v] + [V^*U_x] \quad (10.3)$$

$$RPF \text{ x } P_a = [RPF \text{ x } P_v] + [V^*U_x] \quad (10.4)$$

$$RPF = [V^*U_x]/[P_a - P_v] \quad (10.5)$$

$$RPF = [V^*U_x]/[P_a] \quad (10.6)$$

$$eRPF = [V^*U_{PAH}]/[P_{PAH}] \quad (10.7)$$

In eqs 10.3–10.5, the substance's venous plasma concentration is Pv, its urine concentration is Ux, and the flow rate of the urine is V. Pa is the concentration of the substance in the arterial blood. Mass per unit of time is the result of flow and concentration. There is little distinction between venous and arterial blood flow, therefore, RPF_a is approximated equivalent to RPF_v. Therefore, eq 10.3 reduces to eq 10.4. Solving eq 10.4 for RPF will give us eq 10.5 for Fick's Law.

Values of P_v are problematic to get in patients. Instead, *PAH* (*para-aminohippurate*) clearance is used to compute the effective renal plasma flow (*eRPF*). In the nephron, PAH is freely filtered and not reabsorbed. Even though they are easily filtered, not all PAH enters Bowman's capsule's main urine; yet, *PAH* remaining in the vasa recta is taken up actively by proximal convoluted tubule epithelial cells, who then discharge the substance into the tubular lumen. In this manner, tiny dosages of PAH are entirely eliminated from the blood during a single kidney transit. As a result, the concentration of PAH in venous plasma is close to zero. Setting P_v to zero in eq 10.5 for *RPF* yields the eq 10.6 for effective renal clearance (*eRPF*). For *PAH*, the eq 10.5 is commonly represented as eq 10.7.

As a result, eRPF typically underestimates RPF by around 10% because the concentration of PAH in venous plasma is non zero (in fact, it is typically 10% of the concentration of PAH in arterial plasma). This error margin is often acceptable because measuring eRPF with PAH injection is so straightforward.

Residual Urine

The volume of pee in the bladder that is post-void residual (PVR) refers to the amount of urine that is still present after urination. To measure PVR, a catheter is inserted into the urethra shortly after voiding. Using a catheter has the added benefit of allowing urine collection for further analysis.

- **Radiography:** excretion urography, micturition cystography.
- **Ultrasound** is useful in determining the volume of urine.
- **Radioisotopes** (clearance, gamma camera): The measurement of the volume of voids as well as the interval between voiding and residual urine estimation should be noted when calculating residual pee: If the patient is going through a diuretic phase, this is very crucial. Under vesicoureteric reflux, urine may re-enter the bladder after micturition and can falsely be interpreted as residual urine. The presence of urine in bladder diverticula following micturition presents special problems of interpretation, since a diverticulum may be regarded either as part of the bladder cavity or as outside the functioning bladder.

Each method for measuring residual urine has its limitations in terms of accuracy and suitability for different clinical scenarios. It is essential to select the appropriate method based on the specific clinical issues at hand. While the absence of residual urine can be clinically relevant, it does not definitively rule out intravesical obstruction or bladder dysfunction. If residual urine is detected as an isolated finding, it should be verified before being considered medically significant.

Measuring post void residual urine volume (PRUV) is an important measurement for urinary incontinence. This is the amount of urine left in the bladder after urination:

- Normal PRUV: about 50 mL or less of urine is left;
- Abnormal PRUV: >100 mL suggests an abnormality and requires further tests;
- Abnormality sign: >200 mL.

10.5.8 URETHRAL PRESSURE PROFILE (UPP)

Urethral pressure refers to the fluid pressure required to open a closed or collapsed urethra. Urethral pressure profilometry (UPP), a method, can be used to measure this pressure along the urethra's length. UPP is commonly used to assess urinary incontinence cases, where "high maximum urethral pressure (MUP) and maximum urethral closure pressure (MUCP)" indicate good sphincteric function, while low pressures suggest poor function.

During UPP, a catheter with side openings near its tip is inserted into the urethra and slowly withdrawn, while the pressure necessary to maintain a constant flow is recorded. This provides an evaluation of the urethral wall's activity. Urethral pressure measurements can be recorded during both the storage phase and voiding, allowing for the assessment of pressure distribution and locating any site of obstruction by identifying pressure gradients.[46]

10.5.9 STATIC URETHRAL PROFILE[11,20,45]

In order to identify whether there is detrusor overactivity or instability, stability is tested by watching the detrusor muscle as the bladder fills to its normal capacity.

Using a specific catheter made for pressure monitoring in the urinary tract, static urethral pressure is recorded within the bladder. It accounts for the force that the detrusor muscle applies to the fluid as well as the weight or pressure of the urine in the bladder and the pressure that the abdominal contents impose on the bladder. Resting pressure, which varies depending on the patient's position and health, is the pressure in an empty bladder. Normal bladder resting pressures typically range from 8 to 40 cm of water (pressure applied at the base of a water column 40 cm high). A specialized catheter is inserted either in the vagina or the rectum to measure abdominal pressure. Because the bladder is situated in the floor of the abdominal cavity, it's critical to identify pressures and activities that are unique to the bladder.[11]

10.5.10 STRESS URETHRAL PRESSURE PROFILE

Closure pressure, which is determined as the difference between urethral pressure and intravesical pressure, is the effective pressure that preserves continence. If intravesical pressure ever exceeds urethral pressure (resulting in a negative closure pressure), the possibility of leakage and stress incontinence arises. Patients who have had urethral injury may feel this due to factors such as childbirth, obesity, or previous surgery, causing the pressure transmitted to the urethra to be lower than in the bladder.

In static cases, negative closure pressure is rare, but the situation becomes more complex when the patient applies fast transient pressures to the bladder during activities like coughing or exercise. During these activities, a spike in pressure occurs in the bladder, but it should also be transmitted to the urethra, maintaining a closure pressure greater than zero to prevent leakage.[11,20,45]

It is difficult clinically to differentiate genuine stress incontinence from detrusor instability secondary to stress, and this is one of the main application of urodynamics.[48] An ambulatory trace over 2 min. There is leakage of about 20 mL in one episode. The bladder is stable could induce movement.

The patient is bending down and up again. Because the bladder is stable, the urologist can make the diagnosis for genuine stress incontinence.

There are three types of incontinence, based on urethral pressure measurements and x-ray visualization of the bladder neck under stress.[49,50]

- **Type 1 incontinence:** (a) Proximal urethral closure pressure >10 at 5 mm from vesical outlet; (b) Leakage associated with minimal rotational descent of urethra < 45°
- **Type 2 incontinence:** (a) Proximal urethral closure pressure > 10 at 5 mm from vesical outlet; (b) Leakage associated with gross rotational hypermobility > 45°
- **Type 3 incontinence:** (a) Proximal urethral closure pressure < 10 at 5 mm from the vesical outlet (non-functioning "open" internal sphincter); (b) Leakage not necessarily associated with rotational descent.

10.5.11 DYNAMIC PRESSURE PROFILE

The dynamic pressure profile (DPP) demonstrates fluctuations in sphincteric closure pressure during different physiological events, stresses, and commands. It is obtained using a membrane catheter or micro transducer. The dynamic pressure can be influenced by the patient's position, with maximum pressure typically observed in an upright position. Sphincter dyssynergia is diagnosed when there is a sudden drop in bladder pressure between the bladder neck and the bulbo-membranous urethral junction.[11,20,22,25,31–34,45]

10.5.12 PERFUSION URETHRAL PRESSURE PROFILE (PUPP)

The perfusion urethral pressure profile (PUPP) involves using a catheter with side holes to infuse gas or saline solution with a syringe pump. The pressure recorded by the transducer represents the resistance to flow from the catheter side hole. The initial pressure recorded is the intravesical pressure, followed by a positive deflection at the bladder neck and a gradual increase in urethral pressure toward the middle portion of the urethra in females and the membranous urethra in males. Subsequently, the pressure gradually decreases until the external meatus is reached.

These studies demonstrate the relationship between the emptying of the renal pelvis into the ureter and the process of bolus formation. The ureter extracts the bolus of urine from the pelvis, initiating a peristaltic wave to transport the bolus distally.[11]

10.5.13 CYSTITIS

Interestial Cystitis (IC)

Figure 10.6A and B show kidney stones and Interstitial cystitis, respectively. Interstitial cystitis (IC) is a disorder that causes persistent pain or discomfort in the area around the bladder and the pelvic floor (see Figure 10.6B). Interstitial cystitis (IC) patients may experience a variety of feelings, including slight discomfort, pressure, soreness, or severe pain in the bladder and pelvic area. Their signs may include the need to urinate immediately (urgency), a need to urinate frequently (frequency), or both. The intensity of pain might vary as the bladder fills with urine or empties. In women, symptoms tend to worsen during menstruation.

In interstitial cystitis (IC), the bladder wall may experience irritation, leading to scarring or stiffness. On the bladder wall, there exist glomerulations, which are pinpoint bleeding due to persistent irritation. Individuals with IC may find that their bladders have reduced capacity to hold urine, resulting in increased urination frequency. Additionally, those who have IC frequently feel pain during sexual activity.

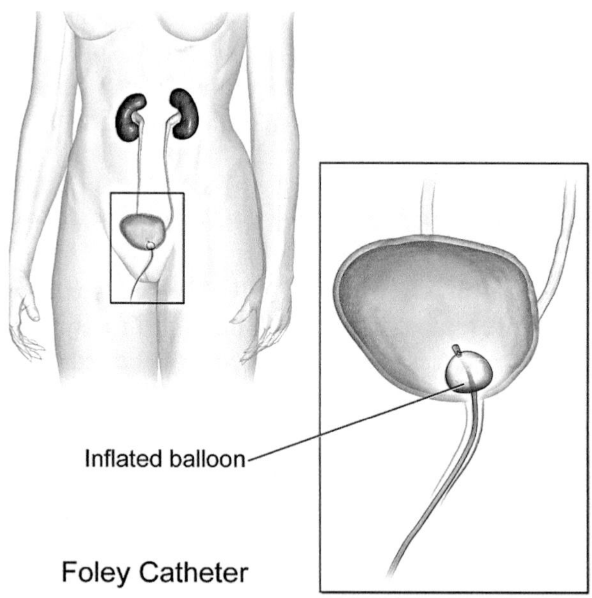

FIGURE 10.4 Kidney stones.

Source: Reprinted from Ref. [55]. https://creativecommons.org/licenses/by/3.0/deed.en)

IC can be diagnosed using a variety of diagnostic techniques, including urine analyses, urine cultures, cystoscopies, bladder wall biopsies, bladder distention under anesthesia, urine cytology, and, in men, laboratory study of prostate secretions. Urinalysis or culturing urine can identify common organisms causing urinary tract infections, but they may not detect certain infections like "Chlamydia." A negative culture does not rule out all types of infections.

A urine sample can be collected using catheterization or the "clean catch" method. With the clean catch method, the patient washes the genital area and collects midstream urine in a sterile container. A urinary tract infection, which is treatable with medications, may be indicated by the presence of bacteria, white and red blood cells, and bacteria in the urine. If symptoms persist, and the urine remains sterile for an extended period, IC may be considered as a possible diagnosis. In men, the doctor may examine prostatic fluid for signs of infection, which can also be treated with antibiotics.

During a cystoscopy, the bladder and urethra are inspected using a cystoscope, a hollow tube containing lenses and a light (Figure 10.5). To make the examination more comfortable for patients with IC who may experience pain, anesthesia is administered. This test can reveal bladder wall inflammation, a thick and stiff bladder wall, and Hunner's ulcers. Glomerulations, which are usually visible only after the bladder is fully stretched, can also be detected.

The physician might administer a test to determine the patient's maximal bladder capacity, or the most liquid or gas the bladder can store. Such test requires anesthesia since pain or a strong urge to urinate can limit the bladder's capacity. A small bladder capacity under anesthesia might indicate IC.

Researchers are currently investigating the presence of specific substances in urine samples from people with IC that are not found in normal urine. If a reliable IC marker is identified in urine, it could potentially serve as a diagnostic tool, eliminating the need for a cystoscopic examination or biopsy to diagnose IC.[40]

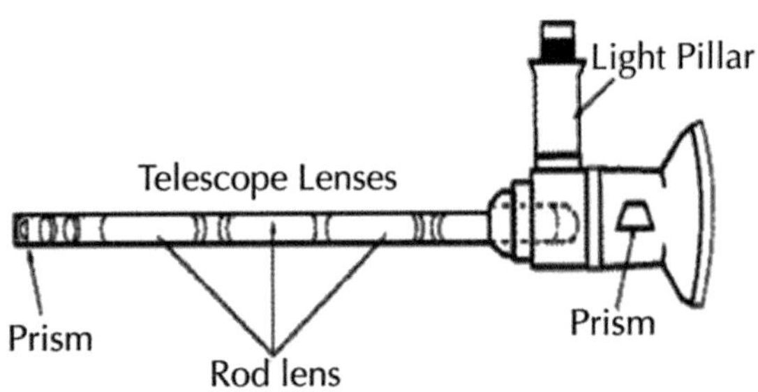

FIGURE 10.5 Cysctoscope.

Source: Cancer Research UK, 2014. https://creativecommons.org/licenses/by-sa/4.0/deed.en

Transcutaneous Electrical Nerve Stimulation (TENS)

Transcutaneous Electrical Nerve Stimulation (TENS) is a potential method to alleviate IC pain. This involves applying mild electric pulses to the body for some time, either once or multiple times a day.

It is an affordable treatment that enables patients to take an active role in their care. Patients can choose when, how long, and at what strength to use TENS by adhering to a few simple rules. In individuals with Hunner's ulcers, TENS has shown potential efficacy in reducing discomfort and frequency.

10.5.14 UTEREOCELE

A ureterocele is a congenital abnormality where the ureters may enlarge and form a pouch. This condition leads to blockage and abnormal urine flow (Table 10.4). Ureteroceles are more common in females and can affect one or both ureters. They can be associated with other congenital anomalies like ectopic ureter (misplacement of the ureter), obstruction between the kidney and ureter, or vesicoureteral reflux (urine flowing back into the ureter from the bladder) on the opposite side. Symptoms of a ureterocele include flank pain spreading to the groin, genital area, and thighs; back and abdominal pain, possibly on one side; urinary tract infection; fever; painful urination; foul-smelling urine; blood in the urine; excessive nighttime urination; urinary incontinence; and the presence of an abdominal mass.

10.6 NUMERICAL EXERCISE[30]

10.6.1 PROBLEM 1

Saadatmand et al.[30] used numerical orthogonal collocation method to compute concentration of urine (in mammals) under steady state situations. Here, the method is used to solve nonlinear equations for transport of electrolyte in nephron, central core, and renal medulla outer, which was also comparable with results obtained from finite difference method (FDM). Their model, however, was vulnerable to non-ideal countercurrent exchange between the vasa recta and the medulla interstitial. Distal tubule will influence the input to the collecting duct even if it is located in the cortical interstitial. It was demonstrated in their work that altering the distal tubule wall's membrane transport properties had a significant impact on the characteristics of the outflow from the outer medulla collecting duct.

TABLE 10.4 Human Urine Electrolyte Excretion Rate.[37]

"Electrolyte	Concentration MEq/L	Excretion rate	
		MEq/day	MEq/kg/day
Ammonia	27 (20–50)	38 (30–70)	0.7 (0.4–1.5)
Bicarbonate	1.1 (0–26)	1.5 (0–36)	0.02 (0–0.5)
Calcium	7.5 (1.4–13.0)	8 (2–18)	0.14 (0.08–0.23)
Chloride	110 (49–210)	118 (58–250)	1.6 (1.0–2.2)
Magnesium	8.2 (5.8–11.6)	9 (6–16)	0.15 (0.11–0.21)
Phosphate	28 (22–41)	36 (22–54)[4]	0.6 (0.5–0.9)
Potassium	50 (27–119)	64 (53–91)	1.1 (0.8–1.2)
Sodium	114 (35–167)	120 (40–186)	1.9 (1.3–2.4)
Sulfate	26 (13–56)	37 (11–56)	0.6 (0.2–0.9)

Note: Water excretion = 1400 (500–3000) mL/day, and 20 (7–40) mL/kg/day/day".

The cortex (C), the outer medulla (OM), and the inner medulla (IM) make up the majority of mammalian kidneys. Fluid inside the cortex, the kidney's outer layer, is virtually isosmotic to arterial blood. While a cortico-papillary gradient of osmolality is sufficient within the medulla to concentrate the urine. Urine is concentrated in the nephrons, which are the kidney's main functioning units.

Their model comprised a distal tubule (DT), a collecting duct (CD), and a short-looped nephron with a loop of Henle (descending Henle's loop (DHL) and ascending Henle's loop (AHL)). Figure 10.6 displays the configuration of the central core model. The medulla interstitial (MI) is a channel through which the Henle loop and CD can communicate with one another. All medulla interstitial components, such as interstitial cells, interstitial spaces, and vasculature, are included in the central core model.

The most significant solutes of the renal medulla, Na+, Cl-, and K+, as well as urea (subscripted by k = 1, 2, 3, and 4, respectively), were considered in their model. The steady-state fluid flow rate, solute concentrations, electric potential, and pressure profile are predicted by model formulations based on conservation of solute and water in the renal outer medulla as a function of tubule and OM lengths.

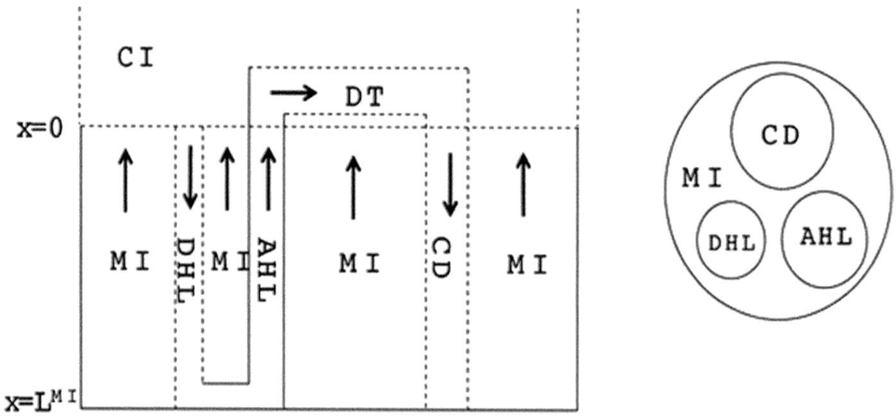

FIGURE 10.6 Schematic of central core model. Legends are (a) **Left**—Tubules (DHL, AHL, DT and CD) and surrounding interstitial (CI and MI); **Dashed lines**—water permeable boundaries; **Arrows**—steady-state flow directions. (b) **Right:** Medulla cross-section showing connectivity between MI and other tubules. (c) **Henle's loop and CD** are represented by a tubular compartment. (d) The DHL, AHL, CD, and MI (superscripted by 1, 2, 4 and MI, respectively) are represented by rigid tubules that are oriented along the corticomedullary axis, which extends from x = 0 at the corticomedullary boundary to $X = L^{MI}$ at the boundary of the inner and outer medulla. A stiff tubule in the cortical interstitial (CI) reveals the DT (indexed by 3) that connects AHL to CD.

Source: Reprinted with permission from Ref. [30].

Equations for the Models

For the *i*-th tubule of the nephron (i = 1, 2, 3, 4) in steady-state, the conservation equation for water (about equal to solution flow) is provided as follows:

$$\frac{dF_v^i}{dx^i} = -J_v^i \tag{10.8}$$

where x^i, F_v^i, J_v^i are axial length of tubule, flow rate of water and net transtubular water flux outflow per unit tubular length, respectively. They assumed that there are no chemical or physical processes in any tubule that would make or consume solutes, hence the conservation of the *k*th solute (k = 1, 2, 3, 4) for the *i*th tubule (i = 1, 2, 3, 4) in the steady state is represented as follows:

$$\frac{dF_k^i}{dx^i} = -J_k^i \tag{10.9}$$

where F_k^i is axial solute flow rate and J_k^i is net outward trans tubular solute flux per unit tubular length. Since they believed Poiseuille's Law could account for the pressure drop in each tubule, the pressure drop along tubule i (i = 1, 2, 3, 4, MI) is determined by:

$$\frac{dP^i}{dx^i} = -R_F^i F_v^i \tag{10.10}$$

where P^i is hydrostatic pressure; $R_F^i = 8\mu / \pi (r^i)^4$ is flow resistance; μ is fluid viscosity; and r^i is tubule radius. Single barrier model equations that contain a variety of forces between the tubule and the interstitial surrounding the tubule (medullary interstitial for DHL, AHL, and CD) were used to represent the trans-tubular fluxes. In general, the trans-tubular water flux across the tubular wall will be determined by both pressure gradient and osmotically induced forces. Passive and active transport work together to determine the trans-tubular solute flux. Therefore, trans tubular water flux, J_v^i, and solute flux, J_k^i, can be given as:

$$J_v^i = h_v^i \left[\sum_k RT\left(C_k^{MI} - C_k^i\right)\sigma_k^i + P^i - P^{MI} \right] \quad (i = 1,2,4) \tag{10.11}$$

$$J_v^i = h_v^i \left[\sum_k RT\left(C_k^{CI} - C_k^i\right)\sigma_k^i + P^i - P^{CI} \right] \quad (i = 3) \tag{10.12}$$

$$J_k^i = J_v^i(1-\sigma_k^i)\frac{(C_k^i + C_k^{MI})}{2} + h_k^i\left(\left(C_k^i - C_k^{MI}\right) + \frac{z_k F}{RT}\left(\Psi^i - \Psi^{MI}\right)\frac{C_k^i + C_k^{MI}}{2}\right)$$
$$+ \frac{V_{\max,k}^i C_k^i}{K_{M,k}^i + C_k^i} \quad (i = 1, 2, 4) \tag{10.13}$$

$$J_k^i = J_v^i(1-\sigma_k^i\frac{\left(C_k^i + C_k^{CI}\right)}{2} + h_k^i\left(\left(C_k^i - C_k^{CI}\right) + \frac{z_k F}{RT}\left(\Psi^i - \Psi^{CI}\right)\frac{\left(C_k^i + C_k^{CI}\right)}{2}\right)$$
$$+ \frac{V_{\max,k}^i C_k^i}{K_{M,k}^i + C_k^i} \quad (i = 3) \tag{10.14}$$

where h_v^i is hydraulic permeability, h_k^i is passive permeability for k^{th} solute, C_k^i is Staverman reflection coefficient of the wall for the kth solute, P^i is concentration of kth solute, P^i is pressure and ψ^i is electric potential relative to the cortical interstitial taken as the reference voltage and assumed to be zero. The last terms in eqs 10.13 and 10.14 are due to active transport for k^{th} solute in i^{th} tubule, $V_{\max,k}^i$ is maximum rate of transport; $K_{M,k}^i$ as Michaelis constant, R is the universal gas constant, T is the absolute temperature, F is the Faraday constant and Z_k is the valence of the kth solute. These water flow and solute flux values According to eqs 10.15 and 10.16, MI is caused by the trans-tubular water flux and the solute flux from other tubules in the medulla:

$$J_v^{MI} = \sum_i J_v^i \quad (10.15)$$

$$J_k^{MI} = \sum_i J_k^i \quad (10.16)$$

where i is summed over all tubules present in the medulla interstitial ($i = 1, 2, 4$). Two integrated mass balances for water flow rate and solute flow rate in the medulla were obtained by integrating the combination of eqs 10.8 and 10.9 with eqs 10.15 and 10.16. This suggests that the sum of medullary water flow rate and solute flow rate at any depth is equal to the final collecting duct (CD) water flow rate and solute flow rate, respectively. Solute flow rate and water flow rate at every level of the medullary interstitial can be calculated using these integrated equations as follows:

$$F_v^{MI} = F_v^{CD}(out) - \sum_i F_v^i \quad (10.17)$$

$$F_k^{MI} = F_k^{CD}(out) - \sum_i F_k^i \quad (10.18)$$

where i is the total number of tubules in the medullary interstitial space ($i = 1, 2, 4$), $F_v^{CD}(out)$ is the rate of water flow leaving the CD in eq (10.17); and $F_k^{CD}(out)$ is rate of solute flow leaving the CD eq (10.18); and for each tubule and medulla (i = 1, 2, 3, 4, MI), solute flow rate (F_k^i) is related to water flow rate (F_v^i) concentration (C_k^i); and potential (ψ^i) by the following Nernst-Planck eq 10.19:

$$F_k^i = F_v^i C_k^i - A^i D_k^i \frac{dC_k^i}{dx^i} - A^i u_k^i C_k^i \frac{d\psi^i}{dx^i} \quad (10.19)$$

where A^i is the cross sectional area of the ith tubule; D_k^i is the diffusion coefficient of the kth solute; and u_k^i is its mobility, which is related to D_k^i by the following eq (10.20):

$$u_k^i = \frac{F_k^i z_k F}{RT} \quad (10.20)$$

In all tubules except for the medullary interstitial, they believed that the diffuse term is insignificant in comparison to the convective flow term. Nernst-Planck eq (10.19) for tubules ($i = 1, 2, 3, 4$) hence becomes simpler to:

$$F_k^i = F_v^i C_k^i \quad (10.21)$$

Since Na^+, Cl^-, and K^+ are ions, the equations below must be followed to preserve electroneutrality (i = 1, 2, 3, 4, MI):

$$\sum_{k=1}^{3} z_k C_k^i = 0 \tag{10.22}$$

$$\sum_{k=1}^{3} z_k J_k^i = 0 \tag{10.23}$$

The cortical interstitial was thought to be a uniformly mixed bath. Therefore, it was assumed that all cortical interstitial solute concentrations, pressure, and potential remained constant. Therefore, the 35 differential equations (eqs 10.8–10.10) and algebraic equations (eqs 10.17, 10.18, 10.22, 10.23) that make up the governing equations for the steady-state electrolyte transport model of a single nephron of the outer medulla are coupled together.

Dimensionless variables were created by splitting dimensional variables by dimensional reference values, making it more convenient to work with them in equations (see Chapter 2 in this book).

For example: Dimension X^i was normalized through division by L^i, that is *i-th* tubule's length; C_k^i by $C_1^1(0)$ is the level of salt in the water entering the DHL; F_v^i by $F_v^1(0)$ is the water flow rate at the DHL's entrance; P^i by $P^4(1)$ which is exit flow pressure from CD, ψ^i by RT/F water flux, J_v^i by $F_v^1(0)/L^i$ is the total water flow rate's uniform rate of trans tubular water transport $F_v^1(0)$ to exit through the walls of the ith tubule; also J_k^i by $F_v^1(0)\ C_1^1(0)/L^i$, which means that all of solute flow rate, $F_v^1(0)\ C_1^1(0)$, exits through the ith tubule's wall at constant rate".

The normalized parameters were obtained as follows:

$$h_v^i \Big/ \left(\frac{F_v^i(0)}{L^i}\right),\quad h_k^i \Big/ \left(\frac{F_v^i(0)}{L^i}\right),\quad V_{\max,k}^i \Big/ \left(\frac{F_v^i(0)C_1^1(0)}{L^i}\right),$$

$$K_{M,k}^i / C_1^1(0), \text{ and } D_k^i \Big/ \left[\frac{F_v^1(0)L^1}{\pi\left(r^i\right)^2}\right]$$

Boundary Conditions

Flow rate and solute concentrations have been specified as the boundary conditions at the entrance to DHL. Up until the exit pressure for the CD has been established, the pressure at the entry of DHL must be adjusted. For the medullary interstitial at the cortico-medullary boundary, solute concentrations, pressure, and potential have also been provided. The model in this section was considered to be closed since the volume and solute flow rate to the outer medulla from the inner medulla had little to no impact on the profiles of the outer medulla at $X = L^{MI}$ (i.e., there was no convective flow at $X = L^{MI}$ in interstitial medullary).

Continuous flow, solute concentrations, and pressure are required at the junction of the nephron tubules.

$$F_v^{MI}\left(L^{MI}\right) = 0 \tag{10.24}$$

Potential may be discontinuous at the junction of nephron tubules because the diffusion term was omitted in tubules. The boundary conditions, physical constants, cortical interstitial constants, and dimensional parameters were reported by the authors along with their units."

Numerical Procedure

The differential equations [F = (F1, F2, L, FN)] that explain flow, solute conservation, and the Poiseuille equation in the central core model can be generalized as follows:

$$\frac{df}{dx} = F(f, g, x) \tag{10.25}$$

The general form of the algebraic equations [G = (G1, G2, L, GM)] for the flow, solute conservation, and electroneutrality of the central core model:

$$0 = G(f, g, x) \tag{10.26}$$

where $f = f(x) = \left[f_1(x), f_2(x), L, f_N(x)\right]$ are the flow, pressure, and MI of water and solute in tubules, and $g = g(z) = \left[g_1(x), g_2(x), L, g_M(x)\right]$ are Cl^- concentration and electric potential in tubules and MI, water and solute movement in MI, and MI.

The weighted residuals methods for solving boundary value issues include the collocation method. In this method, the user gives a rough solution, and residuals are calculated by putting this solution in place of the governing equations. The residuals are then made to become asymptotically close to zero. The user may supply a rough solution that automatically meets the boundary conditions for the straightforward governing equations. The residuals obtained by substituting the estimated solution in boundary equations for big and complex governing equations should also be made to approach zero.

In their investigation, the governing equations and boundary conditions were converted to nonlinear algebraic equations at inner points and boundary points, respectively, using the Lagrangian interpolation polynomial as an approximation solution. The value of f (or g), thus, at any point, including the interpolation points, let's say X*, is given by

$$f_i\left(x^*\right)=\sum_{j=1}^{p+1} f_{i,j} l_j\left(x^*\right) \qquad \text{for } i=1,L,N \tag{10.27}$$

$$g_i\left(x^*\right)=\sum_{j=1}^{p+1} g_{i,j} l_j\left(x^*\right) \qquad \text{for } i=1,L,M \tag{10.28}$$

where, l_j is the Lagrange polynomial and P is the number of interior (discretization) points.

It should be noted that the method's nomenclature and application depend on the kind and placement of the boundary conditions. For example, if the differential quantity's value (rather than its gradient) is known upstream (at x = 0), the boundary condition is Dirichlet, and the last point (i.e., point number P + 1) is not within the set of collocation points.

Collocation methods are known to have a suboptimal order of accuracy, but selecting the right interior points is essential to achieving the best order possible. Each tubule and medulla were divided into P interior locations (referred to as collocation points) that were picked as the roots of a *P*th degree orthogonal Jacobi polynomial to solve the equations. The Jacobi polynomials play a crucial role in determining the ideal collocation point positions. The domain must be defined inside the interval [0, 1] to satisfy the orthogonality requirement of the Jacobi polynomials. All domains were defined within [0, 1] since the model equations were transformed into dimensionless equations.

By substituting these polynomials [eqs 10.24 and 10.25] in the model equations and boundary equations, we obtain a set of governing equations (nonlinear in nature), which were solved by using algebraic solver software. They used the Broyden method to solve the system that has been described above in this section. Here, double precision is employed for computation.

Results

A successful solution obtained by the orthogonal collocation technique to determine transport of renal electrolyte. It was observed that collocation technique is a stable and fast. Additionally, adding equations for solutes and water at junctions of tubules (which is important in the urine concentrating mechanism) is more exact in **orthogonal collocation method** that was used to solve the model equations in this section using the base-case parameters and boundary conditions to deduce a solution for transport of electrolyte in the outer medulla (renal), see Figure 10.7(A)–(H).

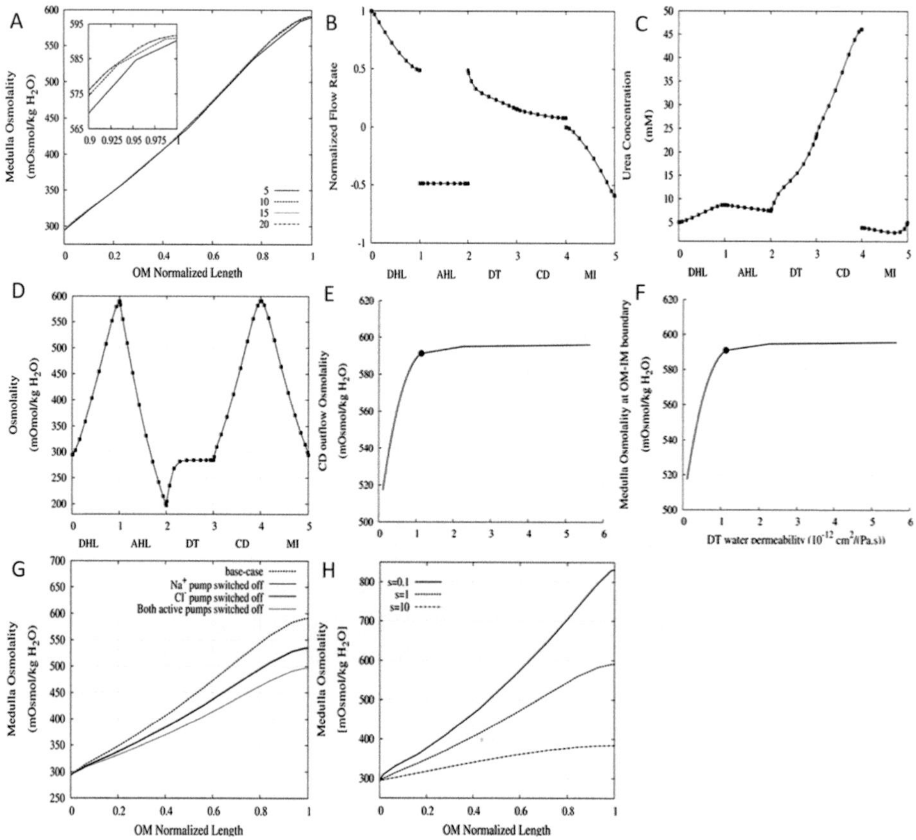

FIGURE 10.7 (A) Line profile of medulla osmolality as function of length of medulla normalized, (B) Line profile of rate of flow normalized along medulla outer and looped nephron, (C) Profile of urea concentrations along the outer medulla and short looped nephron, (D) Short-looped nephron and outer medulla osmolality profiles, DT water permeability's impact on (E) CD outflow osmolality and (F) MI osmolality at OM-IM boundary (marker • for base value), (G) Osmolality profile along MI for different cases such as base-case, **Na**$^+$ pump switched off, **Cl**$^-$ pump switched off, and both active pumps switched off, and (H) Variation in osmolality profile along MI due to countercurrent transport within medulla interstitial.

Conclusions

To resolve the electrolyte transport equations in a central core model of the renal outer medulla, an extremely efficient numerical solution based on the orthogonal collocation approach was developed. The model considered solutes such as urea and 3-ions, along with electrical potential. The results obtained from the study revealed two key findings:

- Careful adjustment of the counter-current exchange between the medulla interstitial and vasa recta is essential in the central core model. Despite the distal tubule being located in the cortex interstitial, it significantly influences backflow in the duct.
- Changes in water permeability or alterations in active pumps in the distal tubule wall have a considerable impact outflow osmolality of the duct and the medulla interstitial osmolality.
- In summary, the presented numerical solution sheds light on the complex interactions between solutes and electrical potential in the renal outer medulla, providing valuable insights into the role of the distal tubule.

10.7 SUMMARY

The urinary system is responsible for maintaining the volume and composition of body fluids within normal ranges, making it an essential part of the excretory process (Table 10.4). While other organs also contribute to excretion, the urinary system remains the primary player in this function. The respiratory system eliminates waste products like carbon dioxide and water through the lungs, while the skin gets rid of wastes via sweat glands. The liver and intestines excrete bile pigments resulting from hemoglobin breakdown. However, the urinary system's role in excretion is crucial, as other organs cannot fully compensate if it fails.

The urinary system performs several important functions, including regulating the excretion of water in urine to maintain appropriate fluid volume. It also controls the concentrations of various electrolytes in body fluids and helps maintain the normal pH of the blood. Additionally, the urinary system contributes to red blood cell production by releasing the hormone erythropoietin and plays a role in controlling blood pressure via renin secretion. Overall, the urinary system's role in maintaining fluid balance, electrolyte levels, and hormonal control is essential for the body's proper functioning.

KEYWORDS

- **analgesic**
- **anus**
- **balloon urethroplasty**
- **benign prostatic hyperplasia**
- **benign prostatic hypertrophy**
- **biopsy**
- **bladder**

REFERENCES

1. Fournier, R. L. *Basic Transport Phenomena in Biomedical Engineering*; CRC Press, 2017.
2. Geankoplis, C. Transport Processes and Separation. In *Process Principles*, 3rd ed.; Prentice Hall, 2003.
3. Geankoplis, C. J. *Transport Processes: Momentum, Heat and Mass*; Allyn and Bacon Series in Engineering, 1983; p 538.
4. Jou, D.; Llebot, J. E., Eds. *Introduction to the Thermodynamics of Biological Processes*; Prentice Hall, 1990.
5. Lightfoot, E. N. *Transport Phenomena and Living Systems*; John Wiley & Sons Inc, 1974.
6. Roselli, R. J.; Diller, K. R. Biological Mass Transport. In *Biotransport: Principles & Applications*, 1st ed.; Springer, 2011; pp 789–1215.
7. Roselli, R. J.; Diller, K. R. Biofluid Transport. In *Biotransport: Principles & Applications*, 1st ed.; Springer, 2011; pp 105–485.
8. Wikipedia. *Urinary System*. https://en.wikipedia.org/wiki/Urinary_system, 2023 (accessed 30 March 2023).
9. CC. *Urinary System Anatomy and Function*. https://www.cincinnatichildrens.org/health/u/urinary-system, 2023 (accessed 30 March 2023).
10. Kelly, C. R.; Landman, J. *The Netter Collection of Medical Illustrations: Urinary System*; Elsevier Health Sciences, 2012.
11. Lose, G.; Colstrup, H.; Saksager, K.; Kristensen, J. K. New Method for Static and Dynamic Measurement of Related Values of Cross-Sectional Area and Pressure in the Female Urethra. *Neurourol. Urodyn.* **1987**, *6* (6), 465–476.
12. NIDDK. *The Urinary Tract & How It Works*. https://www.niddk.nih.gov/health-information/urologic-diseases/urinary-tract-how-it-works#:~:text=The%20urinary%20tract%20is%20the,a%20bladder%2C%20and%20a%20urethra., 2023 (accessed 30 March 2023).
13. EndoText. *The Best Clinical Endocrinology*. https://www.endotext.org/, 2023 (accessed 30 March 2023).
14. Bartley B. *The Male Urethra*. https://www.bartleby.com/lit-hub/anatomy-of-the-human-body/3b-4-the-male-urethra/, 2023 (accessed 30 March 2023).
15. GPNote. *Urinary System*. https://gpnotebook.com/simpleprocess.cfm?querystring=urinary+system, 2023 (accessed 30 March 2023).

16. Hall, J. E.; Hall, M. E. *Guyton and Hall Textbook of Medical Physiology*; Pearson & Prentice Hall, 2020.
17. Martini, F. H. *Fundamentals of Anatomy and Physiology*; Prentice Hall, 2001.
18. MountCastle, V. B. *Medical Physiology*; C. V. Mosby Company, 1974.
19. Glasser, O. *Medical Physics*; Yearbook Publisher, 1944.
20. Skofronich, J. G.; Cameron, J. R.; Grant, R. M. *Physics of the Body*, 2nd ed.; Medical Physics Publishing, 1999.
21. Maton, A.; Hopkins, J.; McLaughlin, C. W.; Johnson, S.; Warner, M. Q.; LaHart, D.; Wright, J. D. *Human Biology and Health*; Prentice Hall, 1993.
22. Seagrave, R. C. *Biomedical Applications of Heat and Mass Transfer*, 1st ed.; The Iowa State University Press, 1971.
23. Tortora, G. J.; Grabowski, R. R. *Principles of Anatomy and Physiology*; Harper Collins, 1996.
24. Truskey, G. A.; Yuan, F.; Katz, D. F. *Transport Phenomena in Biological Systems*; Prentice Hall, 2009.
25. Goyal, M. R. Biofluid Dynamics of Urinary System. In *Biofluid Mechanics of Human Body Systems*, CRC and Apple Academic Press, 2013; pp 283–325.
26. Farahzad, P.; Tewarson, R. P. An Efficient Numerical Method for Solving the Differential Equation of Renal Counterflow Systems. *Comput. Biol. Med.* **1978**, *8* (1), 57–64.
27. Freudenrich, C. *How Your Kidneys Work?* https://health.howstuffworks.com/human-body/systems/kidney-urinary/kidney2.htm, 2023 (accessed 30 March 2023).
28. Layton, H. E.; Pitman, E. B.; Knepper, M. A. A Dynamic Numerical Method for Models of the Urine Concentrating Mechanism. *SIAM J. Appl. Math.* **1995**, *55* (5), 1390–1418.
29. Reddy, J. N.; Gartling, D. K. *The Finite Element Method in Heat Transfer and Fluid Dynamics*; CRC Press, 2010.
30. Saadatmand, M.; Abdekhodaie, M. J.; Pishvaie, M. R.; Farhadi, F. An Efficient Numerical Method and Parametric Study for Electrolyte Transport in the Renal Medulla. *J. Biomed. Sci. Eng.* **2014**, *7* (9), E-article ID 47938. https://www.scirp.org/journal/paperinformation.aspx?paperid=47938 (accessed 31 July 2023).
31. Bronzino, J. D.; Peterson, D. R., Eds. *The Biomedical Engineering Handbook*, Vol. 1–4; CRC Press LLC (Taylor & Francis Group): Boca Raton, 2015; p 5422.
32. Cooney, D. *Biomedical Engineering Principles*; Marcel Dekker, 1976.
33. Enderle, J.; Bronzino, J. *Introduction to Biomedical Engineering*; Academic Press, 2012.
34. Skalak, R.; Chein, S. *Handbook of Bioengineering*; McGraw Hill Book Co, 1986.
35. Sphincter. *Human Anatomy*, 2023. https://www.medfriendly.com/sphincter.html
36. Nielsen, S.; Chou, C.-L.; Marples, D.; Christensen, E. I.; Kishore, B. K.; Knepper, M. A. Vasopressin Increases Water Permeability of Kidney Collecting Duct by Inducing Translocation of Aquaporin-CD Water Channels to Plasma Membrane. *Proc. Natl. Acad. Sci.* **1995**, *92* (4), 1013–1017.
37. Altman, P. *Blood and Other Body Fluids*; Federation of American Societies for Experimental Biology, 1961.
38. Brunzel, N. A. *Fundamental of Urine and Body Fluid Analysis*; Saunders Co., Elsevier, 2018.
39. Abrams, P.; Blaivas, J.; Stanton, S.; Andersen, J.; Fowler, C.; Gerstenberg, T.; Murray, K. Sixth Report on the Standardisation of Terminology of Lower Urinary Tract Function. Procedures Related to Neurophysiological Investigations: Electromyography, Nerve Conduction Studies, Reflex Latencies, Evoked Potentials and Sensory Testing.

The International Continence Society Committee on Standardisation of Terminology, New York, May 1985. *Scand. J. Urol. Nephrol.* **1986**, *20* (3), 161–164.
40. NLM. *Kidney Diseases*. https://medlineplus.gov/(accessed 31 July 2023).
41. Baba, T.; Murabayashi, S.; Tomiyama, T.; Takebe, K. Uncontrolled Hypertension Is Associated with a Rapid Progression of Nephropathy in Type 2 Diabetic Patients with Proteinuria and Preserved Renal Function. *Tohoku J. Exp. Med.* **1990**, *161* (4), 311–318.
42. JHU. *Anatomy of the Urinary System*. https://www.hopkinsmedicine.org/health/wellness-and-prevention/anatomy-of-the-urinary-system, 2023 (accessed 30 July 2023).
43. Tefekli, A.; Cezayirli, F. The History of Urinary Stones: In Parallel with Civilization. *Sci. World J.* **2013**, *2013*.
44. Nahon, I.; Waddington, G.; Dorey, G.; Adams, R. The History of Urologic Surgery: From Reeds to Robotics. *Urol. Nurs.* **2011**, *31* (3).
45. Solomon, B. Diseases, Diagnosis and Treatment. In *The Urinary System Book*, Amazon: Kindle Edition, 2013.
46. Chapple, C. R.; Hillary, C. J.; Patel, A.; MacDiarmid, S. A. *Urodynamics Made Easy*; Elsevier Health Sciences, 2018.
47. Fung, Y.-C. *Biomechanics: Circulation*; Springer-Verlag, 2013.
48. Vignoli, G. *Urodynamics: A Quick Pocket Guide*; Springer, 2016.
49. Balk, E. M.; Rofeberg, V. N.; Adam, G. P.; Kimmel, H. J.; Trikalinos, T. A.; Jeppson, P. C. Pharmacologic and Nonpharmacologic Treatments for Urinary Incontinence in Women: A Systematic Review and Network Meta-Analysis of Clinical Outcomes. *Ann. Intern. Med.* **2019**, *170* (7), 465–479.
50. UCD. *Urinary Incontinence*. https://health.ucdavis.edu/urology/specialties/incontinence/; 2023 (accessed 30 July 2023).
51. Jasper, H. H. Report of the Committee on Methods of Clinical Examination in Electroencephalography: 1957. *Electroencephalogr. Clin. Neurophysiol.* **1958**, *10*, 370–375.
52. Bates, B.; Glen, E.; Griffiths, D.; Melchior, H.; Rowan, D.; Sterling, A.; Zinner, N. R.; Hald, T. Second Report on the Standardization of Terminology of Lower Urinary Tract Function. Procedures Related to the Evaluation of Micturition (Flow Rate, Pressure Measurement, Symbols). *Scand. J. Urol. Nephrol.* **1977**, *11*, 197–199.

CHAPTER 11

Biofluid Dynamics of the Human Digestive System*

MEGH R GOYAL, RAJENDRA BOLDHANE, DANNY GONZALEZ, PUNKAJ GUPTA, and BRIJESH KADAM

11.1 INTRODUCTION

The digestive system prepares the ingested food and liquids for use by millions of body cells.[1] When food is ingested, it cannot directly access the body's cells since it cannot pass through the intestinal walls into the bloodstream in its current form. The digestive system is responsible for altering food both physically and chemically to render it suitable for use while eliminating any nonprocessable waste. The process of digestion relies on the release of exocrine and endocrine substances and the regulated movement of food throughout the digestive tract.[2] The control of digestive processes involves a combination of electrical and hormonal signals originating from the digestive system's own nervous and endocrine systems, as well as from the central nervous system and other endocrine organs like the adrenal gland. Within the body's cells, metabolic reactions initiate, playing a crucial role in generating the necessary energy for sustaining life. However, for these reactions to occur, two essential components are required: oxygen and organic molecules. The respiratory system supplies oxygen, while the digestive system collaborates with the cardiovascular and lymphatic systems to

* This chapter has been modified from the review article that was prepared by students (Rajendra Boldhane, Danny Gonzalez, Punkaj Gupta, and Brijesh Kadam) for the course on Biofluid Mechanics—BME 4994, Course Instructor: Megh R. Goyal, PhD, P.E., Visiting Professor in Biomedical Engineering (2003), Biomedical Engineering Department, Florida International University (FIU), 10555 West Flagler Street, Miami, FL 33174, USA. For details contact at goyalmegh@gmail.com. We acknowledge the contribution by the faculty at FIU for the quality of this chapter.

Biofluid Dynamics of Human Body Systems: Expanded and Revised Edition.
Megh R. Goyal, Arka Bhowmik, & Anamika Chauhan (Eds.)

break down carbohydrates, fats, and proteins into usable energy. Throughout our body, there exists a delicate equilibrium, known as homeostasis, which governs these reactions.[3,4]

This chapter focuses on: (a) the physiology and biomechanics of human digestive system; and (b) secretions of various digestive fluids and chemicals.

11.2 THE HUMAN DIGESTIVE SYSTEM: PHYSIOLOGY AND ANATOMY

The human digestive system, which is a complex organ system crucial for our survival. In this system, food undergoes mechanical breakdown through the actions of chewing and the enzymes found in saliva when it enters the mouth.[5–9] Next, food passes from the pharynx to the epiglottis (a cartilage flap separating the trachea from the esophagus). After this, the food travels down the esophagus, and then the food *bolus* (post-chewed food) moves down the GI tract[2,9] via *peristalsis*. Finally, the food is **digested** in the stomach with the help of stomach acid (HCl).

Following its initial breakdown, the food progresses through the pyloric sphincter of the stomach and subsequently enters the small intestine, where nutrients are absorbed, before continuing into the large intestine. Additionally, numerous auxiliary organs, such as the teeth and tongue, play vital roles in the digestive process.[10–13] Various glandular organs contribute to the digestive process by releasing water, enzymes, and buffers into the digestive tract to facilitate digestion. These auxiliary substances play a crucial role in preparing both organic and inorganic nutrients for absorption. Below, we provide a brief description of the functions of each of these organs[1,3,15–22]:

- **Esophagus:** A simple conduit between the mouth and stomach.
- **Large intestine:** In all animals, water is absorbed. In carnivores, bacterial fermentation takes place, and feces are formed. In herbivores (like horses), the large intestine is huge and of critical importance for the utilization of cellulose.[23]
- **The liver** provides bile salts to the small intestine, which are critical for digestion and the absorption of fats.
- **Mouth:** Foodstuffs are broken down mechanically by chewing, and saliva is added as a lubricant.[5–9]
- **Pancreas** provides digestive enzymes to the small intestine that are critical for the digestion of fats, carbohydrates, and proteins.

- **Small intestine:** Here, the final stages of chemical digestion occur and almost all nutrients are absorbed.
- **Stomach:** Chemical digestion of proteins is initiated, and foodstuffs are reduced to liquid form.
- **Rectum and anus:** The end of the large intestine and the next part of the tract, the rectum, stores the feces. The feces are finally squeezed through a ring of muscle, the anus, and out of our body.

11.3 FUNCTIONS OF THE BIOFLUIDS

11.3.1 SALIVA

Saliva is a protective fluid in the mouth.[5–9] The characteristics of saliva, including its volume, flow rate, and composition, are influenced by factors such as the aroma, taste, and chewing of food. Saliva plays a role in clearing food particles from the mouth and counteracting the effects of acidic plaque. It contains calcium, phosphorus, and fluoride molecules that aid in the remineralization of damaged tooth areas. However, if acid exposure is frequent or prolonged, saliva may struggle to protect tooth enamel from demineralization. A decrease in saliva production raises the risk of dental cavities. Additionally, salivary function can be impacted by radiotherapy used in the treatment of oral cancer.[5–9]

11.3.2 GASTRIC JUICES

- **Gastroferrin** bonds [Fe^{++}] help to prevent the formation of insoluble iron salts and inorganic phosphates.
- **Hydrochloric acid** produces an acidic environment to kill bacteria and activate pepsin. It solubilizes connective tissue.
- **Intrinsic factor** is a glycoprotein that is essential for the normal absorption of vitamin B_{12} in the intestine. Without intrinsic factor, pernicious anemia will develop.
- **Mucus** produces a barrier along the walls of the stomach to protect the stomach from acid and abrasion.
- **Pepsin** is a proteolytic enzyme that is secreted in an inactive form (pepsinogen) and converted by stomach acidity or by autocatalysis to pepsin. It is active at $pH < 5.0$.

11.3.3 BILE

Bile production and recirculation are the main **excretory functions** of the liver. Tumors that obstruct the flow of bile from the liver can also impair other liver functions, which are described below:

- **Catabolic functions** include the detoxification of drugs.
- **Metabolic functions** include the maintenance of glucose (blood sugar) levels.
- **Storage functions** include the storage of sugar (glycogen), fats (triglycerides), iron, copper, and fat-soluble vitamins (A, D, E, and K).
- **Synthetic functions** include the synthesis of serum proteins (such as albumin), blood clotting (coagulation) factors, and complement (a mediator of inflammatory responses).

11.3.4 PANCREATIC JUICE

The exocrine portion of the pancreas releases a variety of enzymes to assist in the digestive process. This includes enzymes like trypsin and chymotrypsin, which initiate protein digestion, a process completed by carboxypeptidase. Nucleases are involved in the breakdown of DNA and RNA. Starch and fats, on the other hand, are digested by amylase and lipase. Goblet cells secrete mucus to provide lubrication and protection to the inner surfaces of the intestines. Within the intestinal mucosa, enterocytes contain digestive enzymes such as peptidases, which break down small peptides into amino acids; disaccharidases like sucrose, maltose, isomaltose, and lactose, which convert disaccharides into monosaccharides; and intestinal lipase, which aids in breaking down neutral fats into glycerol and fatty acids.[7]

11.3.5 SMALL INTESTINE SECRETION

Mucus is a complex of glycoproteins, water, and salts that has the following important functions[2,7]:

- Causes waste to adhere; assists in the formation of the fecal mass.
- Has a low frictional coefficient, so acts as a lubricant.
- It adheres to the luminal surface of the GI tract and food particles.
- It is resistant to the action of digestive enzymes.
- Prevents contact between the GI surface and digestive juices.

The villi in the small intestine quickly reabsorb liquid secretions, creating a fluid movement from the crypts to the villi. This fluid serves as a medium to aid in the absorption of substances from the chyme as it interacts with the villi. Consequently, the principal role of the small intestine is to facilitate the absorption of nutrients and digested products into the bloodstream.[14,26,27]

11.3.6 LARGE INTESTINE SECRETION

Mucus in the large intestine protects the wall against excoriation and provides an adherent medium for holding fecal matter together.[7,23] Additionally, it serves as a protective shield for the intestinal lining against bacterial actions occurring within the fecal matter. The alkaline nature of this secretion acts as a defense mechanism, preventing acids from harming the intestinal wall.

11.4 PHYSICAL PROPERTIES AND COMPOSITIONS OF DIGESTIVE BIOFLUIDS

11.4.1 SALIVARY SECRETION

The physical and chemical properties of digestive biofluids[1,28,29] are shown in **Appendix I** (Table 11.3).[28] The salivary secretion is 97–99.5% water and hypoosmotic. Its osmolarity depends on the nature of the stimulus for the secretion of saliva and on the precise gland. For example, the viscosity of saliva increases from the parotid to the sublingual glands. Normally, saliva has a slightly acidic pH (6.75–7.00). Apart from its digestive function, saliva also contains IgA antibodies[30] and lysozyme, which help to destroy any microorganisms in the oral cavity, including the gums and teeth.

11.4.2 GASTRIC JUICE

Gastric juice is a thin, strongly acidic (pH varying from 1 to 3), and almost colorless liquid secreted by the glands in the lining of the stomach. The essential constituents are the digestive enzymes pepsin and rennin, hydrochloric acid, and mucus.

- Pepsin converts proteins into simpler, more easily absorbed substances.
- The hydrochloric acid provides the acidic environment in which pepsin is most effective.

- Rennin aids in the digestion of milk proteins.
- The mucus secreted by the gastric glands helps protect the stomach lining from the action of gastric juice.

Gastric secretion is triggered by various hormones, chemical compounds, the presence of food in the stomach, and psychological factors like the enticing aroma of a favorite dish. Some specific stomach lining cells produce intrinsic factors, vital for vitamin B12 absorption. Conversely, the suppression of gastric secretion is achieved through pH sensors, the enterogastric reflex (activated by food entering the duodenum), and the release of secretin and cholecystokinin (CCK) in response to acidity in the small intestine.[1,3,16,17,21,22]

11.4.3 BILE

Bile fluid is composed of water, electrolytes, and a range of organic molecules, including bile acids, cholesterol, phospholipids, and bilirubin. It travels through the biliary tract and enters the small intestine as part of its digestive role (as shown in Table 11.1). In adults, the daily production of bile typically falls within the range of 400–800 mL, as indicated in Table 11.2. Despite the substantial quantity of bile acids secreted into the intestine daily, only a relatively small amount is actually lost from the body. This is primarily because around 95% of the bile acids delivered to the duodenum are reabsorbed back into the bloodstream within the ileum.

TABLE 11.1 Composition of Bile.[1]

Description	Units	Bile of liver	Bile of gallbladder
Bile salts	g/dL	1.1	6
Bilirubin	g/dL	0.04	0.3
Cholesterol	g/dL	0.1	0.3–0.9
Fatty acids	g/dL	0.12	0.3–1.2
Lecithin	g/dL	0.04	0.3
Water	g/dL	97.5	92
Na^+	mEq/L	145	130
K^+	mEq/L	5	12
Ca^{++}	mEq/L	5	23
Cl^-	mEq/L	100	25

TABLE 11.2 Daily Secretion of Biofluids.[1]

Biofluid	Rate of flow (mL/min)	Daily volume (mL)
Bile	0.13–0.20	1000
Brunner's gland secretion	0.37–2.50	200
Gastric secretion	0.7–9.5	1500
Large intestinal secretion	–	200
Pancreatic secretion	0.02–5.20	1000
Saliva	0.1–1.8	1000
Small intestine secretion	0.17–0.70	1800

11.4.4 PANCREATIC SECRETION

It is a clear, alkaline secretion of the pancreas containing enzymes that aid in the digestion of proteins, carbohydrates, and fats. There are the following secretions from the pancreas[1]:

- **Bicarbonate secretion:** In an adult, the pancreas secretes an **isosmotic juice** at a daily rate of 1500–3000 mL. Its pH ranges from 8.0 to 8.5, and it contains water, electrolytes, and enzymes for the digestion of fats, carbohydrates, and proteins. The total concentration of the major anions, Cl^- and HCO_3^-, is about 150 mEq/L. At maximum output, as would be observed following a meal, HCO_3^- concentration reaches 150 mEq/L, whereas Cl^- concentration is <50 mEq/L. This ratio is reversed at low flow rates. The high pH of the postprandial juice neutralizes the acidic gastric chyme and raises duodenal contents to an optimal pH for enzymatic digestion. Pancreatic juice[11] also contains Ca^{++} and traces of Mg^{++}, HPO_4^-, and SO_4^{--}.
- **Enzyme secretion:** Proteolytic, lipolytic, carbohydrate-hydrolyzing, and nucleolytic enzymes are secreted by the pancreas. Two of the proteolytic enzymes, trypsin and chymotrypsin, are secreted as zymogens. Both trypsinogen and chymotrypsinogen are activated by enterokinases that are secreted by the mucosa of the proximal intestine. Once trypsinogen is activated into trypsin, it in turn activates further zymogens.

Trypsin and chymotrypsin, categorized as endopeptidases, make up the largest portion of these enzymes. Their primary function is to break the peptide bonds in dietary proteins, leading to the production of oligopeptides and amino acids. Additional enzymes include carboxypeptidases A and B, as well as elastase. In the case of lipolytic enzymes, they are secreted in their active form, with lipase being the primary component. Lipase plays a pivotal role in breaking down triglycerides into diglycerides, monoglycerides, and fatty acids. It operates effectively at the interface of fat droplets, particularly when these droplets are emulsified by bile salts and fatty acids. Bile salts also form molecular groups (micelles) that aid in solubilizing the products of fat breakdown in the watery duodenal fluid, thus facilitating lipase activity.

Colipase, a small cofactor produced by the pancreas, combines with lipase to prevent its inhibition and removal from the fat–water interface caused by bile salts. Colipase additionally lowers the pH of lipase from 8.5 to an optimal value of 6.5. Amylase, another digestive enzyme, is responsible for breaking down starch into maltose, maltotriose, and dextrin. Lastly, nucleolytic enzymes perform the function of breaking the phosphodiester bonds that link nucleotides in nucleic acids.

11.4.5 SMALL INTESTINE SECRETION[18]

Brunner's glands in the duodenum are responsible for producing mucus, which contains a unique glycoprotein specific to the mucus cells in these duodenal glands. This glycoprotein is not found in other parts of the small intestine lacking these glands. The chemical composition of this purified glycoprotein is typical of mucin-type glycoproteins, with its protein content accounting for 35% of its weight. It contains minimal neuraminic acid and only a small amount of sulfate (2%). On the other hand, the crypts of Lieberkühn secrete water and electrolytes. These intestinal secretions, generated by the enterocytes within the crypts, occur at a rate of 1800 mL per day. These secretions mainly consist of extracellular fluid and have a slightly alkaline pH ranging from 7.5 to 8.

11.4.6 LARGE INTESTINE SECRETION

The mucus membrane of the large intestine is rich in crypts of Lieberkühn but notably lacks villi. Furthermore, the epithelial cells in the large intestine do not contain enzymes; instead, they are primarily composed of mucus-secreting cells. Consequently, the major secretion in the large intestine is mucus, which also contains some bicarbonate ions secreted by a small number of non-mucus-secreting epithelial cells. Following its passage through the small intestine, food enters the large intestine, consisting of the cecum and colon.[23]

11.5 FLUID MECHANICS OF SECRETIONS

11.5.1 SECRETION OF SALIVA

Saliva is generated within the salivary glands, as depicted in Figure 11.1. These glands consist of fundamental secretory units called acini, which release a fluid comprising water, electrolytes, mucus, and enzymes. This fluid flows out of the acinus into collecting ducts. Within these ducts, the composition of the secretion undergoes changes. Sodium is largely reabsorbed, potassium is secreted, and substantial amounts of bicarbonate ions are secreted.

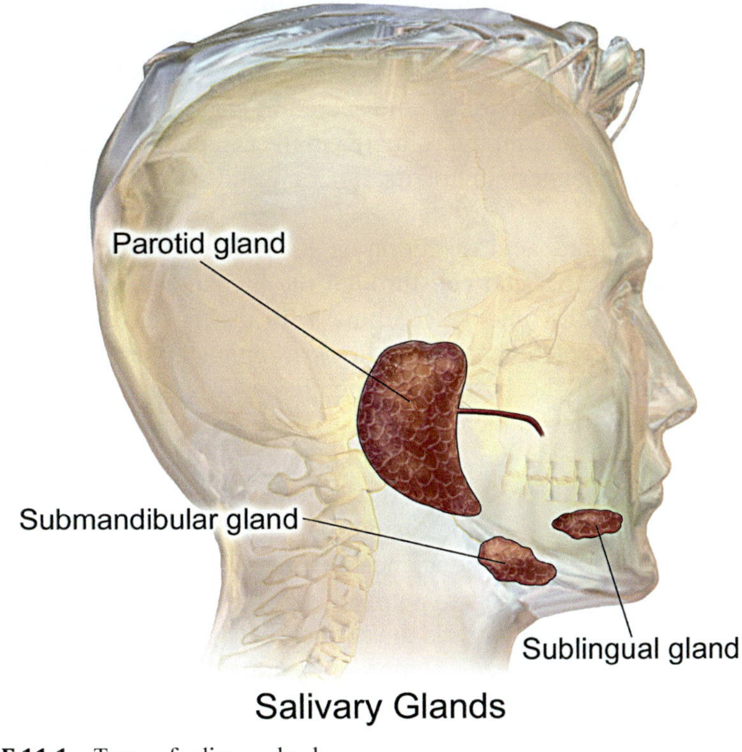

FIGURE 11.1 Types of salivary glands.

Source: Reprinted from Ref. [31]. https://creativecommons.org/licenses/by/3.0/

The secretion of bicarbonate is particularly significant in ruminants. It, along with phosphate, serves as a crucial buffer to counteract the substantial acid production that occurs in the forestomach of these animals. Smaller collecting ducts within the salivary glands merge into larger ducts, ultimately forming a single, large duct that discharges its contents into the oral cavity. Two basic types of acinar epithelial cells exist: (a) serous cells, which secrete a watery fluid essentially devoid of mucus and mucus cells; (b) acini in the parotid glands are almost exclusively of serous type, while those in the sublingual glands are predominantly mucus cells; (c) in the submaxillary glands, it is common to observe acini composed of both serous and mucus epithelial cells.

11.5.2 ACID SECRETION

The parietal cell secretions exhibit a hydrogen ion concentration approximately 3 million times higher than that found in the bloodstream. The secretion of chloride ions goes against both concentration and electrical gradients. Therefore, the parietal cell's capacity to release acid relies on active transport mechanisms. The central actor in acid secretion is an H+/K+ ATPase, often referred to as the "proton pump," situated in the canalicular membrane. This ATPase relies on magnesium for its function and cannot be inhibited by ouabain.

The theory behind acid secretion is as follows: Hydrogen ions are produced within the parietal cell through the breakdown of water. In this process, hydroxyl ions quickly combine with carbon dioxide to produce bicarbonate ions, a reaction facilitated by carbonic anhydrase. Bicarbonate is transported out of the basolateral membrane in exchange for chloride ions. This outward flow of bicarbonate into the bloodstream leads to a slight increase in blood pH, known as the "alkaline tide." This mechanism helps maintain the intracellular pH within the parietal cell.

Concurrently, chloride and potassium ions are transported into the lumen of the canaliculus through conductance channels, and they are essential for acid secretion. Hydrogen ions are pumped out of the cell into the canaliculus in exchange for potassium ions via the action of the proton pump, effectively recycling potassium. The accumulation of osmotically active hydrogen ions in the canaliculus creates an osmotic gradient across the membrane, leading to the outward diffusion of water. As a result, the gastric juice contains 155 mM of HCl, 15 mM of KCl, and a small quantity of NaCl.[31]

11.5.3 SECRETION OF PROTEASES

Pepsinogen, an inert zymogen, is released into the gastric juice by both mucus cells and chief cells. After its secretion, pepsinogen becomes active when exposed to stomach acid, transforming into the active protease known as pepsin. Pepsin plays a significant role in initiating protein digestion within the stomach. Additionally, chief cells also secrete chymosin (rennin), which is a protease that causes milk proteins to coagulate. This coagulation process allows the stomach to retain milk protein in the stomach for a more extended period.

11.5.4 SECRETION OF MUCUS

The predominant type of epithelial cells in the stomach are the mucus cells, which not only line the entire inner surface but also reach into the glands as "mucus neck cells." These cells are responsible for producing mucus that is rich in bicarbonate. This mucus serves the vital function of coating and providing lubrication to the gastric lining. Additionally, it plays a crucial role in safeguarding the epithelium from the damaging effects of acid and various chemical substances.

11.5.5 SECRETION OF HORMONES

The primary hormone released by the gastric epithelium is gastrin, which is a peptide hormone with significant roles in regulating both acid secretion and gastric motility.

11.5.6 SECRETION OF INTRINSIC FACTOR

Intrinsic factor is a glycoprotein secreted by the parietal cells within the gastric mucosa of humans. Its primary role is crucial in the absorption of vitamin B12 (also known as cobalamin) within the intestine. The absence or ineffective utilization of intrinsic factors leads to a condition known as pernicious anemia. Dietary vitamin B12, originally embedded in ingested proteins, is released in the stomach due to the actions of pepsin and stomach acid. It is then swiftly captured by one of two vitamin B12-binding proteins found in gastric juice. These binding proteins exhibit a stronger affinity for vitamin B12 than intrinsic factors do under acidic pH conditions. As the partially digested food travels into the small intestine, pancreatic proteases break down the binding proteins, liberating vitamin B12. This liberated vitamin B12 subsequently binds with intrinsic factors. Importantly, in the ileal mucosa of the small intestine, there are receptors specifically designed for intrinsic factors. These receptors bind with the vitamin B12–intrinsic factor complex, facilitating the absorption of vitamin B12 into the portal bloodstream.

11.5.7 SECRETION OF BILE

- Adult humans produce 400–800 mL of bile daily. The secretion of bile may occur in two stages: Initially, hepatocytes secrete bile into canaliculi, from which it flows into bile ducts. As bile flows through the bile ducts, it is modified by the addition of a watery, bicarbonate-rich secretion from ductal epithelial cells.
- Further modification of bile occurs in the gall bladder. The gall bladder stores and concentrates bile during the fasting state. Typically, bile is concentrated fivefold in the gall bladder by the absorption of water and small electrolytes; virtually all the organic molecules are retained.

The secretion of cholesterol into bile serves as a primary method for removing cholesterol from the body. Cholesterol, in its free form, is nearly insoluble in water-based solutions. However, within bile, it becomes soluble thanks to the presence of bile acids and lipids, such as lecithin. Gallstones, which mainly consist of cholesterol, form due to mechanisms that enable cholesterol to separate and solidify from the solution in bile.

11.5.8 EXOCRINE PANCREATIC SECRETIONS

Pancreatic juice consists of two essential components vital for effective digestion: digestive enzymes and bicarbonate. The digestive enzymes are produced and released by the exocrine acinar cells, while the bicarbonate is secreted by the epithelial cells that line the smaller ducts within the pancreas.

- **Digestive enzymes:** Several proteases are synthesized in the pancreas and secreted into the lumen of the small intestine. The two major pancreatic proteases are trypsin and chymotrypsin, which are synthesized and packaged into secretory vesicles as inactive proenzymes, trypsinogen, and chymotrypsinogen. As one might anticipate, proteases are rather dangerous enzymes to have in cells, and packaging an inactive precursor is a way for the cells to safely handle these enzymes. The secretory vesicles also contain a trypsin inhibitor, which serves as an additional safeguard should some of the trypsinogen be activated to trypsin; following exocytosis, this inhibitor is diluted out and becomes ineffective (the pin is out of the grenade). Once trypsinogen and chymotrypsinogen are released into the lumen of the small intestine, they must be converted into their active forms to digest proteins. Trypsinogen is activated by the enzyme enterokinase, which is embedded in the intestinal mucosa. Once trypsin is formed, it activates chymotrypsinogen. The net result is a rather explosive appearance of active protease once the pancreatic secretions reach the small intestine.[1,16] The major form of dietary fat is triglycerides, or neutral lipids. A triglyceride molecule cannot be directly absorbed across the intestinal mucosa. Rather, it must first be digested into a 2-monoglyceride and two free fatty acids. The enzyme that performs this hydrolysis is pancreatic lipase, which is delivered into the lumen of the gut as a constituent of pancreatic juice. Sufficient quantities of bile salts must also be present in the lumen of the intestine in order for lipase to efficiently digest dietary triglyceride and for the resulting fatty acids and monoglycerides to be absorbed. This means that normal digestion and absorption of dietary fat are critically dependent on secretions from both the pancreas and liver. The major dietary carbohydrate for many species is starch, a storage form of glucose in plants. Amylase is the enzyme that hydrolyzes starch to maltose (a glucose–glucose disaccharide), as well as the trisaccharide, maltotriose, and small branch point fragments called limit dextrins. The major source of amylase in all species is pancreatic secretions, although amylase is also present in the saliva of humans.

- **Bicarbonate and water:** Bicarbonate ions are generated from the formation of carbonic acid from water and carbon dioxide (catalyzed by carbonic anhydrase), which then dissociates to form bicarbonate ions and protons. The bicarbonate ions are pumped out of the cell by secondary active transport into the pancreatic ducts in exchange for chloride ions due to the chloride gradient for the transfer.[33–39] The hydrogen ions produced from the dissociation of water are actively pumped out of the cell in exchange for sodium ions by the Na-H ATPase, and the sodium ions in the cell move out into the lumen following the extrusion of chloride ions down the electrical gradient.

Water follows the ionic movement by osmosis. Thus, the net result is the excretion of an aqueous solution of sodium bicarbonate into the lumen of the duct. This secretion of aqueous bicarbonate solution washes the enzymes from the pancreas into the duodenum, so that the greatest enzyme secretion occurs when there is also bicarbonate secretion.[1,16,17]

11.5.9 SECRETIONS OF SMALL INTESTINE

During the digestive process, significant amounts of water are introduced into the lumen of the small intestine. Most of this water is subsequently reabsorbed within the small intestine. Whether water is being secreted or reabsorbed, its movement across the mucosa is guided by osmotic gradients. In cases of secretion, two distinct mechanisms create an osmotic gradient that attracts water into the intestinal lumen:

- **Increases in luminal osmotic pressure resulting from the influx and digestion of foodstuffs:** The chyme that floods into the intestine from the stomach is typically not hyperosmotic, but as its macromolecular components are digested, the osmolarity of that solution increases dramatically. Starch, for example, is a huge molecule that contributes only a small amount to osmotic pressure, but as it is digested, thousands of molecules of maltose are generated, each of which is as osmotically active as the original starch molecule. Thus, as digestion proceeds luminally, osmolarity increases dramatically, and water is pulled into the lumen. Then, as the osmotically active molecules (maltose, glucose, and amino acids) are absorbed, the osmolarity of the intestinal contents decreases and water can be absorbed.
- **Crypt cells actively secrete electrolytes, leading to water secretion:** The apical or luminal membrane of crypt epithelial cells contains ion channel of immense medical significance: a cyclic AMP-dependent chloride channel known as the cystic fibrosis transmembrane conductance regulator (CFTR). Mutations in the gene for this ion channel result in cystic fibrosis. This channel is responsible for the secretion of water through the following steps: (a) Elevated intracellular concentrations of cAMP in crypt cells activate this channel, resulting in the secretion of chloride ions into the lumen; (b) Accumulation of negatively-charged chloride anions in the crypt creates an electric potential that attracts sodium, pulling it into the lumen across the tight junctions—the net result is the secretion of NaCl; and (c) the secretion

of NaCl into the crypt creates an osmotic gradient across the tight junction that allows water to be drawn into the lumen.

11.5.10 SECRETIONS OF LARGE INTESTINE

Water is absorbed in response to an osmotic gradient, and the mechanism behind this osmotic pressure is quite similar to what occurs in the small intestine. Sodium ions are transported from the intestinal lumen across the epithelial lining. This process is facilitated by the epithelial cells, which have highly active sodium pumps on their basolateral membranes and a means of sodium absorption through their luminal membranes.

Interestingly, the colonic epithelium is even more effective at water absorption compared to the small intestine. The absorption of sodium in the colon is further enhanced by the hormone aldosterone. Chloride ions are absorbed through an exchange process involving bicarbonate ions. This leads to the secretion of bicarbonate ions into the intestinal lumen, which helps neutralize the acids produced during microbial fermentation in the large intestine.

The colonic epithelium is rich in goblet cells, which secrete mucus in response to tactile stimuli from luminal contents and parasympathetic signals from pelvic nerves. The mucus serves as a vital lubricant that protects the epithelium and binds dehydrated ingesta together to form feces.

11.6 FOUR PHASES OF SECRETIONS

11.6.1 CEPHALIC PHASE

The cephalic phase begins with the presentation and ingestion of a meal.[45]

- In the **sight phase,** the smell and taste of food, as well as mechanical stimulation of the oral cavity and swallowing, initiate a number of "long" reflexes, which alter GI activities;
- In the **return phase,** salivation is stimulated by cholinergic reflexes mediated by parasympathetic nerves;
- There is stimulation of primary peristalsis in the esophagus and relaxation of the upper and lower esophageal sphincters. Vagal reflexes inhibit contractile activity in the proximal stomach. This is called **receptive relaxation**. Vagal reflexes initiated during the cephalic phase also stimulate acid secretion by parietal cells in the stomach.[31,46] These "long" reflexes, acting through cholinergic neurons of the enteric nerve

plexus, stimulate secretion of HCl by parietal cells and the secretion of histamine by enterochromaffin-like (ECL) cells. Histamine, in turn, stimulates parietal cell secretion of acid. In addition, vagal reflexes in the antrum stimulate the secretion of the hormone gastrin from G-cells. Elevated levels of circulating gastrin increase acid secretion by direct stimulation of the parietal cells and by stimulating histamine secretion from ECL cells.

- **Return vagal reflexes** also induce a relatively small stimulation of enzyme secretion by the exocrine pancreas.[1,16,22]

During the cephalic phase, the process of digestion of complex carbohydrates is initiated due to the activity of ptyalin secreted in saliva. Little absorption occurs during this phase.

11.6.2 GASTRIC PHASE

The gastric phase begins when food enters the stomach. Distention of the stomach activates stretch receptors, initiating several reflexes, which alter gastric, intestinal, colonic, and pancreatic activities. Gastric acid secretion is stimulated by "long" vago-vagal reflexes and "short" reflexes mediated by the enteric nerve plexus. These reflexes enhance acid secretion by:

- Cholinergic stimulation of parietal cells;
- Stimulation of histamine secretion by ECL cells;
- Stimulation of gastrin secretion by G-cells.

The hormone gastrin plays a pivotal role in stimulating acid secretion, both by directly triggering parietal cells and by prompting the release of histamine from ECL cells. When a meal includes protein, the protein digestion products, such as peptides generated by the combined action of acid and pepsin, further stimulate the secretion of gastrin by G-cells. This, in turn, intensifies acid production. As acid is secreted, the pH within the stomach decreases, which then triggers the release of somatostatin from D-cells in the stomach. Somatostatin acts as a regulator, preventing excessive acid production by inhibiting the secretion of gastrin from G-cells and histamine from ECL cells.

Beyond its role in acid secretion, elevated levels of circulating gastrin, along with long and short reflexes, stimulate gastric motility. Cholinergic reflexes, both long and short, encourage the secretion of pepsinogen from the chief cells. During the gastric phase, gastrin release leads to increased contractions in the ileum and relaxation of the ileocecal sphincter, known as the gastrolienal reflex. Long reflexes can induce mass movements in the distal colon, referred to as the gastrocolic reflex. Vagal reflexes also play a modest role in stimulating enzyme secretion by the exocrine pancreas. Within the stomach, some proteins undergo digestion due to the actions of acid and pepsin, but minimal absorption takes place during this phase.

11.6.3 INTESTINAL PHASE

The intestinal phase commences when chyme enters the small intestine, triggering reflexes and hormonal mechanisms that influence the functioning of the small intestine, pancreas, gallbladder, and stomach. As chyme moves from the stomach into the small intestine, the stretching it causes prompts contractions in the intestines. Additionally, intestinal chyme spurs the exocrine pancreas to secrete its enzymes. When acidic chyme enters the small intestine, it stimulates S-cells in the duodenal mucosa to release the hormone secretin.

Increased levels of circulating secretin stimulate pancreatic ductal cells to secrete larger volumes and more bicarbonate. Digestion products of fats and proteins in the intestinal chyme stimulate the release of CCK from I-cells located in the duodenum and upper jejunum. CCK, in turn, triggers the secretion of enzymes by pancreatic acinar cells. Moreover, CCK induces the contraction of the gallbladder and relaxation of the sphincter of Oddi.

Acidic or hyperosmotic chyme, as well as chyme-containing fats, activate duodenal receptors, leading to a slowdown in the rate of gastric emptying. Although the exact pathways for this inhibition haven't been fully defined, CCK likely plays a role. The presence of acidic or hyperosmotic chyme, along with chyme-containing fats in the small intestine, results in a reduction in the stomach's secretion of acid. This inhibition is partially mediated by neural reflexes and partially by the release of certain hormones collectively known as enterogastrones. Most of the digestion and absorption processes occur while the chyme is within the small intestine.

Regarding organic nutrients, carbohydrates and proteins are initially digested by pancreatic enzymes and then further processed by enzymes located in the brush border membrane of enterocytes. Fats undergo digestion by pancreatic enzymes, with bile salts (acids) secreted by the liver aiding in their digestion and absorption.

The absorption of organic nutrients occurs through various mechanisms, including secondary active transport for substances like glucose, amino acids, and dipeptides; facilitated diffusion for fructose; and simple diffusion for monoglycerides, fatty acids, cholesterol, and fat-soluble vitamins. Apart from organic nutrients, most electrolytes (e.g., sodium, potassium, chloride, and bicarbonate) are absorbed or reabsorbed in the small intestine. This absorption of organic nutrients and electrolytes establishes an osmotic gradient, facilitating the absorption of fluid. Most of the fluid (over 90%) is absorbed or reabsorbed in the small intestine.

11.6.4 INTERDIGESTIVE PHASE

The interdigestive phase corresponds to the fasting period that follows the digestion and absorption of most components from the previous meal, resulting in a relatively empty stomach and small intestine. During this phase, there are brief episodes of vigorous peristaltic contractions, initially starting in the stomach and then progressing to the upper, middle, and finally the distal portions of the small intestine. These intense contractile episodes typically last for only 5–10 min, followed by longer periods of relative quiescence. This recurring pattern repeats at intervals of around 90 min and is referred to as the migrating motility complex (MMC).

The MMC appears to be triggered in the stomach by cyclical fluctuations in the levels of circulating motilin and ceases when food is consumed. In the colon, colonic bacteria break down certain components of the chyme while the remaining electrolytes and fluids are absorbed. Because chyme lingers in the colon for an extended duration, colonic processing generally occurs throughout all four phases of the digestive process.

11.7 REGULATION OF GASTROINTESTINAL SECRETIONS

11.7.1 HORMONE REGULATORS[1,16,17,21,22]

The digestive system possesses its own set of hormonal regulators, which are generated and secreted by cells located within the mucosal lining of the stomach and small intestine. These hormonal regulators play a pivotal role in governing the various functions of the digestive system. They are released into the bloodstream within the digestive tract, then circulate back to the heart through the arteries, and eventually return to the digestive system. There, they stimulate the production of digestive fluids and promote the movement of organs.

- **Gastrin** causes the stomach to produce an acid for dissolving and digesting some foods. It is also necessary for the normal growth of the lining of the stomach, small intestine, and colon.
- **Secretin** causes the pancreas to send out a digestive juice rich in bicarbonate. It stimulates the stomach to produce pepsin that digests protein, and it also stimulates the liver to produce bile.
- **CCK** causes the pancreas to grow and produce the enzymes of pancreatic juice, and it causes the gallbladder to empty.

11.7.2 TYPES OF NERVE REGULATORS

- **Extrinsic** (outside) nerves come to the digestive organs from the unconscious part of the brain or from the spinal cord. They release acetylcholine and adrenaline. (a) Acetylcholine causes the muscles of the digestive organs to squeeze with more force and increase the "push" of food and juice through the digestive tract. Acetylcholine also causes the stomach and pancreas to produce more digestive juice; (b) Adrenaline relaxes the muscles of the stomach and intestine and decreases the flow of blood to these organs.
- **Intrinsic** (inside) nerves make up a dense network embedded in the walls of the esophagus, stomach, small intestine, and colon. The intrinsic nerves are triggered to act when the walls of the hollow organs are stretched by food. They release many substances that speed up or delay the movement of food and the production of juices by the digestive organs.

11.7.3 CHEMICAL EVENTS IN DIGESTION

- **Chewing, churning, and mixing** of food with digestive juices can only separate the molecules from each other. These actions cannot split up molecules. Many nutrient molecules are too large to pass through cell membranes (i.e., they cannot be absorbed).
- Molecules of water, vitamins, and minerals are small enough to be **absorbed** through the intestinal wall and into the bloodstream.
- Proteins, carbohydrates, and fat molecules are too large and must be **broken down** by chemical means.

Chemical digestion of proteins, carbohydrates, and fats is needed for three reasons: "(a) These molecules are too large to pass through a cell membrane; (b)

Fat and some carbohydrates are insoluble in water. Before nutrients can enter the blood, they must be dissolved in water; and (c) The proteins, fats, and carbohydrates that we eat are not usually the same proteins, fats, and carbohydrates that our body uses. The body rearranges these nutrients into usable nutrients."

11.8 FLUID BALANCE IN THE DIGESTIVE SYSTEM

The digestive system resembles a lengthy tube that is open at both ends, allowing it to connect with the external environment. Just like in any part of the human body, it's crucial to maintain a balanced metabolism or homeostasis within the digestive system. Notably, there's a substantial contrast between the input (2000 mL) and the output (150 mL). The organs within the digestive system account for this difference. For instance, saliva, gastric secretions, and secretions from accessory organs contribute approximately 1500 mL. As the digestive process unfolds, there is a continuous absorption of fluids into our body. Among all the components of the system, the intestines and colon are particularly proficient at absorbing biofluids. There exists a direct correlation between the quantity of fluid consumed and the amount of biofluids necessary for the effective operation of an average digestion.

11.8.1 DEHYDRATION DUE TO VOMITING AND DIARRHEA

Dehydration can be caused by fluid loss (through vomiting, diarrhea, or excessive urination), inadequate intake, or a combination of both. The most common cause of dehydration in infants and children is acute gastroenteritis, with its associated vomiting and diarrhea. Dehydration is classified as:

- **Mild dehydration:** a loss of 3–5% of body weight
- **Moderate dehydration** is due to a loss of 6–10% of body weight. Drinking fluids is often sufficient for mild dehydration. It is better to encourage frequent, small amounts of fluid (using a teaspoon or syringe) rather than forcing a child to drink a large amount of fluid at one time. Intravenous fluids and hospitalization may be necessary for moderate to severe dehydration.
- **Severe dehydration** is due to a loss of more than 9–15% of body weight. This is a life-threatening emergency. The majority of cases of acute gastroenteritis are caused by viruses. They tend to be self-limited, resolving on their own after a few days.

11.8.2 GASTROESOPHAGEAL REFLUX DISEASES (GERD), HIATAL HERNIA, AND HEARTBURN[20,48–50]

GERD, which stands for gastroesophageal reflux disease, is a digestive disorder that impacts the lower esophageal sphincter (LES). A significant number of individuals, including pregnant women, experiences symptoms such as heartburn or acid indigestion due to GERD. GER refers to the stomach's contents flowing back into the esophagus. Instances like coughing, vomiting, straining, or sudden physical exertion can elevate abdominal pressure, leading to the development of hiatal hernia. Factors like obesity and pregnancy also contribute to this condition. It's worth noting that many healthy individuals over the age of 50 may have a small hiatal hernia. While often associated with middle age, hiatal hernias can affect individuals of all age groups.[20,48–50]

The treatment of GERD focuses on two main objectives: reducing the frequency of reflux and minimizing damage to the esophageal lining caused by refluxed substances. To achieve this, it is advisable to avoid certain triggers like chocolate, peppermint, fatty foods, coffee, and alcoholic beverages, as these can weaken the LES. Citrus fruits, citrus juices, tomato products, and pepper should also be avoided, as they can irritate an already damaged esophageal lining.

Another strategy is to reduce portion sizes during meals, which can help manage symptoms. Eating dinner at least 2–3 h before bedtime can reduce reflux by allowing stomach acid to decrease and the stomach to partially empty. Professor Goyal, for instance, chooses to have his last meal at 6:00 p.m. For overweight individuals, weight loss often leads to symptom relief. Quitting smoking is essential for reducing GERD symptoms.

Elevating the head of the bed by using 15-cm blocks can be effective in reducing heartburn because it allows gravity to minimize the reflux of stomach contents into the esophagus. Antacids, when taken regularly, can neutralize acid in the esophagus and stomach, providing relief from heartburn. Many people find temporary or partial relief from over-the-counter antacids. However, prolonged use of antacids can result in side effects like diarrhea, alterations in calcium metabolism, and excessive magnesium accumulation in the body. Excessive magnesium can be problematic for individuals with kidney disease. If antacids are required for more than 3 weeks, it's advisable to consult a doctor. In cases of chronic reflux and heartburn, a doctor may prescribe medications such as H-2 blockers (cimetidine, famotidine, nizatidine, and ranitidine) or omeprazole to reduce stomach acid.

11.9 BIOFLUID MECHANICS OF THE DIGESTIVE SYSTEM

The exploration of biofluid mechanics is essential for gaining a comprehensive understanding of the intricacies of the digestive system. It's important to note the significant differences between the external flow around a body and the internal flow within blood vessels. This distinction has paved the way for potential clinical applications, particularly in areas like artificial organs, blood rheology, and diffusion processes.

To understand this phenomenon, model systems have been developed to represent the desired system.[22,27,51,52] It allows for the investigation of variables including geometry, non-Newtonian flow characteristics, and steady/pulsatile flow.[1] Some basic mathematical relations include Hagen–Poiseuille's equation, Bernoulli's equation, Newton's law of viscosity, Reynolds number, Navier–Stokes equation, etc. Hagen–Poiseuille's law[1,34,38,52] for a steady flow is defined as follows (eq 11.1):

$$Q = \frac{\pi \Delta P \left(r^4\right)}{8\mu L} \tag{11.1}$$

This equation reveals that the volume flow rate is (a) directly proportional to the pressure difference between the ends of the tube; (b) inversely proportional to the length of the tube; (c) inversely proportional to the viscosity of the fluid; and (d) directly proportional to the fourth power of the radius of the vessel. In the 1700s, Daniel Bernoulli investigated the forces present in a moving fluid as follows[35,53–56]:

$$\text{Total pressure} = P + 0.5(\rho V^2) + \rho g h = \text{constant} \tag{11.2}$$

Equation 11.2 states that the static pressure in the flow plus one half of the density times the velocity squared plus density times gravity times the height of the liquid is constant throughout the flow path. This equation is valid only for inviscid, incompressible, steady flow, and no heat transfer. Considering the biofluids in the digestive system are inviscid, incompressible, and steady, the equation closely defines the system. Also, the relationship states that the pressure throughout the system is uniform. This may be true for the idealized system, but not in actual situations. The effects of the digestive system peristalsis are leading contributors to the disparity in the equation and reality."[3]

The pumping of fluids through muscular tubes by means of peristaltic waves is important physiologically. Thus, peristalsis plays a significant role in the transport of biofluids in the digestive system. Usha and Rao[31] have completed research on the peristaltic transport of biofluids in elliptic pipes.

They were able to calculate flux through the system by elaborating on flux through a circular tube:

$$Q = \frac{1}{\pi l^2 WT} \iiint \left[c^2 f^2 \left(w^* + W \right) \right] d\xi d\eta dt \tag{11.3}$$

Equation 11.3 is a time-averaged flux of biofluids through an elliptical tube. In this equation, Q = time-averaged flux, l = tube length, W = speed of peristaltic wave, T = wave period, c = constant of elliptical foci, f = friction force, w^*= velocity in the z-direction, and W = speed of peristaltic pump over velocity, wave frame, and time. Therefore, the movement of biofluids through the elliptical tube is dependent on the tube characteristics and the velocity of the peristaltic wave. However, there is an inverse relationship between flux and tube dimensions.[31]

11.10 GENERAL DISCUSSIONS

When we consume foods like bread, meat, and vegetables in forms that the body cannot directly utilize for nourishment, a process of transformation is necessary. This process involves converting our food and beverages into smaller molecules that can be absorbed into the bloodstream and transported to cells throughout the body.

Digestion encompasses several key elements, including the mixing of food, its movement along the digestive tract, and the chemical breakdown of large food molecules into smaller ones. The initial stages of digestion commence in the mouth, where chewing and swallowing initiate the process, and it culminates in the small intestine. The specific chemical processes can vary depending on the type of food being digested.

The digestive system's large, hollow organs are equipped with muscular walls that facilitate their movement. This muscular action serves to propel both food and liquids and facilitates the mixing of contents within each organ.

One common mode of movement in the esophagus, stomach, and intestine is known as peristalsis. Peristalsis resembles the rhythmic motion of ocean waves passing through muscle tissue. The muscle in these organs creates a narrow pathway and then gradually propels the constricted portion along the length of the organ. These wave-like contractions push food and fluids through each hollow organ.

The initial muscle movement required for swallowing can be consciously initiated, but once it begins, it becomes an involuntary process under the control of nerves. The esophagus serves as the passage into which the swallowed food is pushed. At the juncture of the esophagus and stomach, there is a ring-like valve that initially closes off the passage between the two organs. As food approaches this closed ring, the surrounding muscles relax, permitting the food to pass. Subsequently, the food enters the stomach, which has three primary mechanical functions to fulfill[45]:

- The stomach must store the swallowed food and liquid. It requires the muscles of the upper part of the stomach to relax and accept large volumes of swallowed food.
- The stomach mixes up the food, liquid, and digestive juice produced by the stomach. The lower part of the stomach mixes these materials by its muscle action.
- The third task of the stomach is to empty its contents slowly into the small intestine. Several factors affect the emptying of the stomach, including the nature of the food (mainly its fat and protein content) and the degree of muscle action of the emptying stomach and the next organ to receive the contents (the small intestine).

As food undergoes digestion within the small intestine and combines with the secretions from the pancreas, liver, and intestines, the mixture within the intestine is thoroughly blended, and it is propelled forward to facilitate continued digestion. Ultimately, all the digested nutrients are taken up through the walls of the intestine.

The byproducts of this digestive process encompass undigested components of the food, referred to as fiber, and older cells that have been naturally shed from the mucosal lining.[7] These materials are propelled into the colon, where they typically remain for a day or two until they are eliminated from the body through a bowel movement.

A comprehensive understanding of the biomechanics of the digestive tract has significantly contributed to improving the comfort and well-being of patients. For instance, in cases of fat malabsorption within the small intestine, there is an increase in the osmolarity of the fluid within the digestive tract, leading to osmotic diarrhea. This condition can result in dehydration for patients and elevate the osmolarity of their blood or plasma. In severe cases, it can even lead to shock. Therefore, to address this issue, patients are administered parenteral fluids with lower osmolarity levels.

The optimal pH range for most digestive juices falls between 5.0 and 8.0. However, the stomach functions at a notably lower pH level, ranging from 1.5 to 8.0. Additionally, the specific gravity of all digestive biofluids falls within the range of 1.00–1.03, and their freezing point is approximately

0.4–0.5°C. It's worth noting that data regarding osmotic pressure, surface tension, and viscosity are not universally available for all digestive biofluids.

Saliva, as well as the duodenum and pancreas, exhibit elevated levels of sodium and potassium, both exceeding 90 mg/mL. Furthermore, the duodenal and jejunal regions (stomach) have the highest chloride levels, reaching approximately +110 mEq/L. Lipase is found in high concentrations in the pancreas and small intestine, specifically in illegal secretions, with levels ranging from 300 to 2000 units/mL. Lastly, bile demonstrates increased cholesterol levels, at around 120 mg/100 mL.

The ratio of fluid input to output in the digestive system is approximately 40:3, exemplifying a system in a state of homeostasis during the process of digestion. The compositions of digestive fluids, coupled with the equilibrium in volume and secretion rates, work together to maintain a stable digestive system.

11.11 SUMMARY

Exploring the characteristics of biofluids within the digestive system holds significant importance in addressing issues related to alterations in their physical and chemical properties. This knowledge facilitates the diagnosis and treatment of various pathological conditions, rendering it a valuable asset in the medical field. Investigating the physical attributes and flow dynamics of these fluids is crucial for gaining insights into the digestive tract's functioning. Furthermore, comprehending the dynamics of biofluids plays a pivotal role in managing certain digestive tract disorders, such as diarrhea or constipation.[20,48,49]

The consistent characteristics observed in the pH, specific gravity, and freezing point of the biofluids, as indicated by the research findings, underscore the substantial degree of homeostasis within the body's digestive processes. This homeostasis is crucial due to the extensive cooperation required among various organs for the successful functioning of the digestive system. Additionally, the heightened levels of sodium and potassium in saliva, the duodenum, and the pancreas may have a connection to the enzyme levels in the system, particularly those enzymes responsible for carbohydrate digestion. This may relate to the chemical breakdown of carbohydrates. Furthermore, the presence of lipase in the pancreas and small intestine may be attributed to the initial release of lipase from the pancreas. Lastly, the elevated cholesterol levels in bile might provide insights into the chemical composition of this fluid.

It is evident that substantial amounts of fluids are essential to maintain homeostasis, and there is a considerable volume of biofluids involved during digestion. This signifies a cyclical equilibrium in the homeostatic processes of the human body during digestion. Although a relatively small amount of liquid is expelled, a significant volume of biofluids is required to ensure efficient digestion. Further research in these areas is imperative to draw well-founded conclusions. The analysis of biofluid dynamics in digestion could hold great utility in the treatment of diseases, disorders, and drug development. Emerging technologies and novel concepts may be applied to attain a more comprehensive understanding of the human digestive system.

In this chapter, the authors have explored the chemical, physical, and biological properties of biofluids in the human digestive system. The primary aim has been to examine the fundamental aspects of the digestive system, encompassing fluid balance and secretion phases. The digestive system serves as a gateway for nutrients from the external environment to enter the circulatory system. It plays a pivotal role in breaking down various types of food, allowing for their distribution throughout the entire body via the circulatory system.

KEYWORDS

- **abdominal ultrasound**
- **abdominal X-ray**
- **absorption**
- **acute**
- **allergy**
- **anal fissure**
- **anemia**

REFERENCES

1. Hall, J. E.; Hall, M. E. *Guyton and Hall Textbook of Medical Physiology*; Pearson & Prentice Hall, 2020.
2. NLM. *Lower Gastrointestinal Tract*, 2023. https://www.niddk.nih.gov/health-information/digestive-diseases/anatomic-problems-lower-gi-tract/about-lower-gi-tract#:~:text=What%20is%20the%20lower%20GI,large%20intestine%20and%20the%20anus.&text=The%20anus%20is%20a%201,which%20stool%20leaves%20your%20body (accessed 03 April 2023).
3. Martini, F. H. *Fundamentals of Anatomy and Physiology*; Prentice Hall, 2001.
4. Goyal, M. R. Biofluid Dynamics Of Digestive System. In *Biofluid Mechanics of Human Body Systems*, CRC and Apple Academic Press, 2013; pp 327–360.

5. Edgar, W. Saliva: Its Secretion, Composition and Functions. *Br. Dental J.* **1992**, *172* (8), 305–312.
6. Frenkel, E. S.; Ribbeck, K. Salivary Mucins Protect Surfaces from Colonization by Cariogenic Bacteria. *Appl. Environ. Microbiol.* **2015**, *81* (1), 332–338.
7. Jowett, A.; Shrestha, R. Mucosa and Taste Buds of the Human Epiglottis. *J. Anatomy* **1998**, *193* (4), 617–618.
8. Nanci, A. *Ten Cate's Oral Histology-e-book: Development, Structure, and Function*; Elsevier Health Sciences, 2017.
9. Piludu, M.; Lantini, M. S.; Cossu, M.; Piras, M.; Oppenheim, F. G.; Helmerhorst, E. J.; Siqueira, W.; Hand, A. R. Salivary Histatins in Human Deep Posterior Lingual Glands (of von Ebner). *Arch. Oral Biol.* **2006**, *51* (11), 967–973.
10. Kong, F.; Singh, R. P. Disintegration of Solid Foods in Human Stomach. *J. Food Sci.* **2008**, *73* (5), R67–R80.
11. Mazzarello, P.; Harari, M. Left to Digest. *Nature* **2007**, *448*, 753.
12. Nixon, S. E.; Mawer, G. The Digestion and Absorption of Protein in Man: 2. The Form in Which Digested Protein Is Absorbed. *Br. J. Nutr.* **1970**, *24* (1), 241–258.
13. Śródka, A. The Short History of Gastroenterology. *J. Physiol. Pharmacol.* **2003**, *54* (3), 9–2.
14. Sauvant, D.; Bastianelli, D. Digestion, Absorption and Excretion. In: *A Quantitative Biology of the Pig*, Vol. 1; 1999; pp 249–273.
15. Boland, G. W. *Gastrointestinal Imaging: The Requisites*; Elsevier Health Sciences, 2013.
16. Glasser, O. *Medical Physics*; Yearbook Publisher, 1944.
17. Wikipedia. *Human Digestive System*, 2023. https://en.wikipedia.org/wiki/Human_digestive_system (accessed 03 April 2023).
18. Read, N.; Al-Janabi, M.; Holgate, A.; Barber, D.; Edwards, C. Simultaneous Measurement of Gastric Emptying, Small Bowel Residence and Colonic Filling of a Solid Meal by the Use of the Gamma Camera. *Gut* **1986**, *27* (3), 300–308.
19. Shalman, E.; Barak, C.; Dgany, E.; Noskowitcz, H.; Einav, S.; Rosenfeld, M. Pressure-Based Simultaneous CFR and FFR Measurements: Understanding the Physiology of a Stenosed Vessel. *Comput. Biol. Med.* **2001**, *31* (5), 353–363.
20. Thompson, W. G.; Longstreth, G. I.; Drossman, D. A.; Heaton, K.; Irvine, E. J.; Muller-Lissner, S. Functional Bowel Disorders. In: *The Functional Gastrointestinal Disorders. Diagnosis, Pathophysiology and Treatment*; Allen Press, 2000.
21. Tortora, G. J.; Grabowski, R. R. *Principles of Anatomy and Physiology*; Harper Collins, 1996.
22. Wood, J. D. Gastrointestinal Physiology. In: *Medical Physiology: Principles for Clinical Medicine*, 3rd ed.; Lippincott Williams & Wilkins, 2009; pp 463–496.
23. Wikipedia. *Large Intestine*, 2023. https://en.wikipedia.org/wiki/Large_intestine (accessed 03 April 2023).
24. OpenStax. *An overview of digestive system.* https://openstax.org/books/anatomy-and-physiology-2e/pages/23-1-overview-of-the-digestive-system, 2023. (accessed 03/10/2023).
25. Sheng, X. *The Digestive Tract*, 2023. https://www.pinterest.com/pin/442619469603293918/ (accessed 03 April 2023).
26. Waaler, B.; Toska, K. Digestive System's Large and Changing Needs of Blood Supply. *Tidsskrift for den Norske Laegeforening: Tidsskrift for Praktisk Medicin, ny Raekke* **1999**, *119* (5), 664–666.
27. Dzwinel, W.; Boryczko, K.; Yuen, D. A. A Discrete-Particle Model of Blood Dynamics in Capillary Vessels. *J. Colloid Interface Sci.* **2003**, *258* (1), 163–173.

28. Altman, P. *Blood and Other Body Fluids*; Federation of American Societies for Experimental Biology, 1961.
29. Skofronich, J. G.; Cameron, J. R.; Grant, R. M. *Physics of the Body*, 2nd ed.; Medical Physics Publishing, 1999.
30. Fagarasan, S.; Honjo, T. Intestinal IgA Synthesis: Regulation of Front-Line Body Defences. *Nat. Rev. Immunol.* **2003**, *3* (1), 63–72.
31. Usha, S.; Rao, A. R. Peristaltic Transport of a Biofluid in a Pipe of Elliptic Cross Section. *J. Biomech.* **1995**, *28* (1), 45–52.
32. NIDDK. *The Digestive System*, 2023. https://www.niddk.nih.gov/search?s=all&q=digestive+system (accessed 03 April 2023).
33. CancerResearchUK. *Diagram showing the bile ducts in the pancreas.* https://commons.wikimedia.org/wiki/File:Diagram_showing_the_bile_ducts_in_the_pancreas_CRUK_293.svg, 2014. (accessed 03/10/2023).
34. Šarenac, T. M.; Mikov, M. Bile acid synthesis: from nature to the chemical modification and synthesis and their applications as drugs and nutrients. *Frontiers in Pharmacology* **2018**, *9*, 387752.
35. Fung, Y.-C. *Biomechanics: Motion, Flow, Stress, and Growth*; Springer-Verlag, 2013.
36. Geankoplis, C. J. *Transport Processes: Momentum, Heat and Mass*; Allyn and Bacon Series in Engineering, 1983; p 538.
37. Jou, D.; Llebot, J. E., Eds. *Introduction to the Thermodynamics of Biological Processes*; Prentice Hall, 1990.
38. Lightfoot, E. N. *Transport Phenomena and Living Systems*; John Wiley & Sons Inc, 1974.
39. Roselli, R. J.; Diller, K. R. Biofluid Transport. In *Biotransport: Principles & Applications*, 1st ed.; Springer, 2011; pp 105–485.
40. Truskey, G. A.; Yuan, F.; Katz, D. F. *Transport Phenomena in Biological Systems*; Prentice Hall, 2009.
41. Kline, T. L.; Zamir, M.; Ritman, E. L. Relating Function to Branching Geometry: A Micro-CT Study of the Hepatic Artery, Portal Vein, and Biliary Tree. *Cells Tissues Organs* **2011**, *194* (5), 431–442.
42. Medico. *Pancreatic Secretions*, 2023. https://medicoapps.org/pancreatic-secretions-2/ (accessed 03 April 2023).
43. VeterianKey. *Pancreas*, 2023. https://veteriankey.com/pancreas/ (accessed 03 April 2023).
44. Sohma, Y.; Gray, M. A.; Imai, Y.; Argent, B. E. 150 mM HCO3 (-)—How Does the Pancreas Do It? Clues from Computer Modelling of the Duct Cell. *JOP* **2001**, *2* (4 Suppl), 198–202.
45. Bowen, R. *Secretion in the Small Intestine*, 2023. http://www.vivo.colostate.edu/hbooks/pathphys/digestion/smallgut/secretion.html (accessed 03 April 2023).
46. Bowen, R. *Prehension, Mastication, Swallowing*, 2023. http://www.vivo.colostate.edu/hbooks/pathphys/digestion/pregastric/mastication.html (accessed 03 April 2023).
47. Liepsch, D. An Introduction to Biofluid Mechanics—Basic Models and Applications. *J. Biomech.* **2002**, *35* (4), 415–435.
48. Thomas, J. H. Fluid Dynamics of Cerebrospinal Fluid Flow in Perivascular Spaces. *J. R. Soc. Interface* **2019**, *16* (159), 20190572.
49. Cummings, J. H.; Macfarlane, G. T. Collaborative JPEN-Clinical Nutrition Scientific Publications Role of Intestinal Bacteria in Nutrient Metabolism. *J. Parenteral Enteral Nutr.* **1997**, *21* (6), 357–365.
50. Danese, S.; Fiocchi, C. Ulcerative Colitis. *New Engl. J. Med.* **2011**, *365* (18), 1713–1725.

51. Morris, A. M.; Regenbogen, S. E.; Hardiman, K. M.; Hendren, S. Sigmoid Diverticulitis: A Systematic Review. *JAMA* **2014**, *311* (3), 287–297.
52. Reddy, J. N.; Gartling, D. K. *The Finite Element Method in Heat Transfer and Fluid Dynamics*; CRC Press, 2010.
53. Seagrave, R. C. *Biomedical Applications of Heat and Mass Transfer*, 1st ed.; The Iowa State University Press, 1971.
54. Bronzino, J. D.; Peterson, D. R., Eds. *The Biomedical Engineering Handbook*, Vol. 1–4; CRC Press LLC (Taylor & Francis Group): Boca Raton, 2015; pp. 5422.
55. Cooney, D. *Biomedical Engineering Principles*; Marcel Dekker, 1976.
56. Enderle, J.; Bronzino, J. *Introduction to Biomedical Engineering*; Academic Press, 2012.
57. Kutz, M. *Standard Handbook of Biomedical Engineering Design*; McGraw-Hill, 2003.
58. VirginiaEdu. *Poiseuille's Law and the Viscosity of Fluids*. http://galileo.phys.virginia.edu/classes/241L/poise/poise.htm, 2023. (accessed 03 April 2023).
59. Strain, A. J.; Neuberger, J. M. A Bioartificial Liver—State of the Art. *Science* **2002**, *295* (5557), 1005–1009.

APPENDIX I

TABLE 11.3 Physical and Chemical Properties of Digestive Biofluids (Range Values Were Followed by Average Values in Parenthesis).[28]

Property	Units	Saliva	Gastric juice	Bile of gallbladder	Bile of liver	Pancreas secretion	Jejunum secretion	Duodenum secretion	Ileal secretion
Volume	mL/kg weight	8.6–28.6	–	–	–	–	–	–	–
Specific gravity	–	1.01–1.02	1.004–1.01	1.03	1.01	1.005–1.014	–	1.004–1.011	–
pH	–	5.6–7.6	1.49–8.38	6–8 (6)	6.2–8.5 (7.5)	8.6–8.8	5.1–7.1 (6.5)	5.8–7.6	6.1–7.2
Freezing point	°C	0.07–0.34	0.55–0.62	0.56	0.56	0.63	0.40–0.58 (0.5)	–	–
Rate of flow	mL/min	0.1–1.8	0.7–9.5	0.13–0.20	2.6–15	0.02–5.20 (1.0)	0.17–0.70 (0.43)	0.4–2.5 (1.1)	0.3–
Surface tension	dynes/cm	–	–	39–41.3	39–41.3	–	–	–	–
Viscosity	Centipoise	–	–	0.84–2.3	0.84–2.3	–	–	–	–
Osmotic pressure	mOsm/L	–	–	–	–	–	–	–	238
Bicarbonate	mEq/L Mg/10 mL	3–10.7 (6.44) –	– 0–130	8–12 –	40 –	66–127 (70) –	2.0–32 (13) –	4.0–21 (7.8) –	3.9–39.8 (19.8) –
Ca^{++}	mg/100 mL mEq/L	– –	– 1.04–200	– 5–7	– 2–4.5	– 2.2–3.2	5.2–11.6 (10.1) –	12.4 –	– 5.0–12.8 (7.9)
CO_2	vol.%	5.2–9.7	–	–	–	–	–	–	–
Cl^-	mEq/L	8.4–17.7	77.6–159.0	15–30	75–110	54–95 (77)	80–139 (117)	51–133 (86)	101–133 (119)
Copper	µg/100mL mg/100 mL	2–22 –	– –	– 0.06–1.1	– 0.06–1.1	– –	– –	– –	– –

APPENDIX I *(Continued)*

Property	Units	Saliva	Gastric juice	Bile of gallbladder	Bile of liver	Pancreas secretion	Jejunum secretion	Duodenum secretion	Ileal secretion
K^+	mg/100 mL	56–148	–	–	–	–	–	–	–
	mEq/L	–	0.5–32.5 (10)	–	2.6–12	2.6–7.4 (4.6)	–	1–11	5.9–29.3 (11.2)
Na^+	mg/100 mL	8–56	0–116 (60)	–	131–164	–	–	85–143.4	–
	mEq/L					113–153 (141)			105–144 (129)
Nitrogen	mg/100 mL	0.5–9.9	–	–	–	–	45–63 (55)	33–61 (47)	53–86 (69)
	mEq/L	–	0.91–2.18	490	67–92	–	–	–	–
Cholesterol	mg/100 mL	3–15	–	350–930 (630)	80–170 (120)	–	–	0–315 (36)	–
Glucose	mg/100 mL	11.3–28	0.35–1.19	–	–	8.5–18	–	–	–
B1	μg/100mL	0.7	–	–	–	–	–	–	–
B12	μg/100mL	0.15–0.5	–	–	–	–	–	–	–
C	mg/100 mL	0–0.3 (0.1)	0.91–1.05 (0.95)	–	–	–	–	–	–
K	μg/100 mL	1.5	–	–	–	–	–	–	–
Cholinesterase	Units/L	0.23	–	–	–	–	–	–	–
Lipase	Units/L	0.3–2.6 (1.4)	7.0–8.4	–	–	300–2700	3–10 (6)	–	–
	Units/mL		–	–	–	–	–	20–60	–
	Units/100 mL	–					–	–	80–780 (325)/day
Phosphatase	Units/L	2.5–8 (4.2)	–	–	–	–	–	–	–
	Units/100 mL	–	1.04–200	–	–	0.8–12.7	–	10–30	–

CHAPTER 12

Biofluid Dynamics of the Human Maternal–Fetal System*

MEGH R. GOYAL, ANAMIKA CHAUHAN, DELHY ARIAS, VÍCTOR GUERRA, JACKELINE MARTÍNEZ, and ROSAURA DE ZAYAS

12.1 INTRODUCTION

During a certain period in our lives, we have all formed intimate connections with others. However, one of the earliest and most significant close bonds we have experienced occurred during the initial 9 months of our prenatal development with our mothers. It might sound both reasonable and perplexing to acknowledge that our primary relationship commences within the womb. Nevertheless, this connection extends well beyond emotional experiences and sentiments. It is a complex arrangement that facilitates physical interaction between the mother and the developing fetus, particularly involving the exchange of biological fluids.[1] The biological fluids within the human fetal system play essential roles in guaranteeing the optimal growth and well-being of the baby throughout the pregnancy period.

The fetal circulatory system[2–8] differs from the maternal circulatory system in that it contains two openings: "the ductus arteriosus and the foramen ovale." These openings enable a continuous and unobstructed flow of blood throughout the body, efficiently passing through the heart and lungs. The ductus arteriosus establishes a connection between the systemic and

* This chapter has been modified from the review article that was prepared by students (Delhy Arias, Víctor Guerra, Jackeline Martínez and Rosaura De Zayas) for the course on Biofluid Mechanics–BME 4994, Course Instructor: Megh R. Goyal, PhD, P.E., Visiting Professor in Biomedical Engineering (2003), Biomedical Engineering Department, Florida International University (FIU), Miami, FL, USA. For details, contact goyalmegh@gmail.com; We acknowledge the contribution of the faculty of the Biomedical Engineering Department at FIU for the quality of this chapter.

Biofluid Dynamics of Human Body Systems: Expanded and Revised Edition.
Megh R. Goyal, Arka Bhowmik, & Anamika Chauhan (Eds.)

pulmonary arteries, located in proximity to the heart. On the contrary, the foramen ovale is a passage covered by a tissue flap, guiding the circulation from the right side of the heart to the left. Oxygen distribution throughout the body relies on its attachment to hemoglobin, a protein with four heme groups, forming a tetrameric structure. Hemoglobin can accommodate up to four oxygen molecules. In adult hemoglobin, two α-chains and two β-chains constitute the structure, whereas fetal hemoglobin incorporates γ subunits in place of the usual β-subunits.

Changes in pH occur in both maternal and fetal systems due to the fetal utilization of oxygen. The pH level in fetal blood fluctuates based on its specific location within the circulatory system. Moreover, while fundamental principles of fluid equilibrium remain consistent, differences in electrolyte intake arise because neonates, or newborns, require frequent feeding.[9–12]

The fetal environment is mostly made up of the amniotic fluid that provides protection for normal fetal development and shields against fetal distress.[5,13–18] Maintaining the amniotic fluid within a suitable range is vital for promoting the best possible fetal development. The fetus manages the volume of amniotic fluid through actions such as urination, expelling fluid from the lungs, and reducing fluid levels by both swallowing and absorbing it. Additionally, the fetus maintains an acid–base equilibrium, with metabolic byproducts contributing to its acid production. Nonetheless, the placenta assumes responsibility for regulating this balance. Fetal fluids consist of maternal fluids that are transferred either through the amniotic fluid from the uterine wall or via the placenta, where they transition from maternal circulation to fetal circulation.[13]

Deviations in fluid levels can lead to imbalances that have adverse implications for both the well-being of the mother and the fetus. Such irregularities can give rise to a range of circumstances that pose threats to the fetus's life and its regular developmental progression.[19]

Moreover, substantial progress in medical imaging has significantly enriched our understanding of the fetal system. Enhancements in the precision of magnetic resonance imaging (MRI), ultrasound, and Doppler imaging have played a pivotal role in generating distinct visual representations of the fetus. These advancements have enabled the formulation of mathematical formulas to characterize various aspects of the fetus, such as its biofluids. Through these techniques, we can gain a comprehensive and vivid perspective of the fetus and its intricate movements.[20,21]

This section examines the characteristics of different bodily fluids in the maternal system, including blood, vaginal fluid, cervical mucus (CM), and seminal fluid. It also delves into the properties of fetal blood and urine, as well as amniotic fluid. Furthermore, the chapter explores the fetal circulatory and pulmonary systems.

12.2 THE HUMAN FETAL SYSTEM[21–23]

12.2.1 WHERE DO BABIES COME FROM?

Throughout the duration of the baby's development, numerous captivating processes take place. The human fetal system consists of two distinct living entities:

- The fetus relies entirely on the mother for sustenance. Through the placenta, the fetus acquires all essential nutrients while eliminating most of its waste products. The fetus is encompassed within a safeguarding space that creates an environment akin to weightlessness, ensuring optimal temperature conditions. In the maternal–fetal system, diverse biofluids serve crucial roles in various processes.
- The amniotic fluid offers buoyancy for the fetus, serving as a protective buffer against minor shocks and contributing to fetal fluid equilibrium. It is intriguing to observe that fetal blood and maternal blood are distinct and do not mix with each other.

12.2.2 PHYSIOLOGY OF HUMAN FETAL SYSTEM[13, 21–28]

The maternal–fetal system consists of the mother and the fetus, the latter of which resides within the mother's womb. The mother serves as a source of nourishment for the fetus, facilitated by the placenta—a specialized "feto-maternal organ" situated in the uterus. The placenta enables the transfer of essential nutrients and oxygen to the fetus and simultaneously eliminates waste products and carbon dioxide. This exchange occurs through the umbilical cord, linking the fetus's abdomen to the placenta.[19] The chorion is the fetal component of the placenta, and the maternal component is the decidua basalis.[22] The relationship between progesterone production and the gestation period.

- Roughly 50% of the fetal blood flows into the liver, while the remaining 50% enters the ductus venosus (DV) conduit, enabling the blood to circumvent the liver.[21–23]

- The second bypass mechanism, known as the ductus arteriosus, diverts the majority of the blood present in the pulmonary trunk away from the lungs. At this stage, the lungs are contracted and present elevated resistance to blood circulation. This is due to the absence of direct external oxygen intake, as the fetus is not yet receiving oxygen from the air outside the womb.[21–23]
- The third diversion, the foramen ovale, redirects a significant portion of the incoming blood from the right atrium to the left atrium. After birth, the foramen ovale seals shut, resulting in the division of the heart into two distinct pumping chambers.[21–23]

Meconium, which constitutes fetal feces, is continually produced within the gastrointestinal (GI) tract and is expelled through the anus into the amniotic fluid. This substance comprises elements such as remnants from ingested amniotic fluid, mucous, and various waste materials originating from the GI lining and glands.[21–23]

The placental structure facilitates the proximity of maternal and fetal bloodstreams without actual mixing. It operates as a pivotal location for the exchange of oxygen, carbon dioxide, and an array of other particles, ions, and molecules critical for fetal development. Nutrients and various compounds traverse the placenta primarily via diffusion. Simultaneously, waste materials originating from the metabolic processes of the developing fetus are eliminated from the placenta.[17]

Like how carbon dioxide moves by diffusion from fetal blood to maternal blood, various waste products also undergo diffusion across the placental membrane into the maternal blood. This enables these waste products to be eliminated alongside the mother's excreted substances.[17,36]

The trophoblastic phase of nourishment, gradually transitioning into placental sustenance. Initially, a significant portion of fetal nutrition stems from the trophoblast's breakdown and absorption of nutrients originating from the endometrial decidua. Subsequently, the majority of later-stage nourishment is achieved through the diffusion process across the placental membrane.[22]

12.3 SEMINAL FLUID: FUNCTIONS AND COMPOSITION

Semen, also referred to as seminal fluid, is an organic liquid that potentially includes spermatozoa. A schematic representation of semen can be found in Appendix – I. This fluid is generated and originates in the pelvic region from a gland called the seminal vesicle. One of the essential fluids crucial in the process of creating a new offspring is the male's seminal fluid, or "semen." Seminal fluid is a composite mixture of secretions from various glands, each characterized by distinct ionic and nutritional content. Approximately 60% of the seminal fluid is derived from the seminal vesicles, 30% from the prostate gland, while the remaining 5% each originates from the epididymis and the bulbourethral glands.[23] The average ejaculate can be produced from 2 to 6 mL of seminal fluid. The concentration of spermatozoa per milliliter of semen may reach over 50 million.[23]

Semen contains significant components, including buffers aimed at counteracting the acidic conditions within the vaginal environment. This mechanism imparts semen with a slightly alkaline pH exceeding 7. Upon entering the vagina, specialized enzymes initially cause the fluid to coagulate. After a period of 15–30 min, the semen undergoes a process of reverting to a liquid state. Additionally, two primary constituents of semen are water and mucus, serving the purposes of lubrication and furnishing a fluid medium to support spermatozoa.[27] Table 12.1 lists the physical properties of seminal fluid.

TABLE 12.1 Physical Properties of Semen.[3]

Property	**Units**	**Value**
Conductivity	mho	$(88–107) \times 10^{-4}$
Freezing point depression	°C	0.55–0.58
pH	–	7.1–7.5
Specific gravity	–	1.031–1.039
Spermatozoa	10^6/mL	50–100
Volume of ejaculate	mL	2–6"

The seminal fluid transports minerals and nutrients that provide sustenance for the spermatozoa during their journey through the vaginal canal. These nutrients encompass ascorbic acid, citric acid, and fructose. The seminal vesicles, epididymis, and prostate gland are responsible for producing these nourishing elements.[27] Hormones within the seminal fluid fulfill additional roles. For example, prostaglandins, a type of hormone, assist in facilitating the movement of sperm and in contracting the smooth muscle within the vagina. Information regarding the chemical characteristics of semen is provided in Table 12.2.

TABLE 12.2 Chemical Properties of Semen.[2]

Species	Units	Value
Water	% mass	91.8
Calcium	mEq/L	43.0
Chloride	mEq/L	43.0
Sodium	mEq/L	117.0
Potassium	mEq/L	23.0
Zinc	mg/100 mL	14.0
	Acids	
Ascorbic acid	mg/100 mL	12.8
Citric acid	mg/100 mL	480.0
	Carbohydrates	
Fructose	mg/100 mL	290.0
Protein: Creatine	mg/100 mL	20.0"

12.3.1 SEMINAL VESICLES

The seminal vesicles, also referred to as vesicular glands or glandulae vesiculosae, are a pair of uncomplicated tubular glands located posteriorly and inferiorly to the male's urinary bladder. Positioned within the pelvic region, these glands owe their yellowish color to lipofuscin granules derived from deceased epithelial cells. Approximately 50–70% of the seminal fluid in humans originates from the seminal vesicles. However, this component is not expelled in the initial ejaculate portions that are primarily dominated by spermatozoa and zinc-rich prostatic fluid. The excretory duct of each seminal gland opens into the corresponding vas deferens as it enters the prostate gland. The fluid from the seminal vesicles is characterized by its alkaline nature, contributing to the mildly alkaline pH of human semen. This alkalinity aids in counteracting the acidity of the vaginal tract, thus extending the lifespan of sperm. It is worth noting that an acidic ejaculate (pH below 7.2) might be connected to ejaculatory duct obstruction. Furthermore, the seminal vesicles produce a substance that leads to the semen acquiring a sticky or jelly-like consistency after ejaculation. This attribute is thought to play a role in retaining the semen close to the womb.

The dense secretions originating from the seminal vesicles encompass proteins, enzymes, fructose, mucus, vitamin C, flavins, phosphorylcholine, and prostaglandins. The significant fructose levels within these secretions offer a source of nutrient energy for the spermatozoa. While spermatozoa released into the vagina are unlikely to come into contact with the fluid from the seminal vesicles, they directly transition from prostatic fluid to CM. This marks the initial phase of their journey through the female reproductive system. The expulsion of this fluid occurs due to the sympathetic contraction of the muscular is muscle layer.

12.3.2 SPERM[38,39]

Sperm, the smaller and usually mobile male reproductive cell found in most sexually reproducing organisms, is haploid, possessing half the number of chromosomes as other cells in the organism's body. Often equipped with at least one flagellum, sperm serves a vital role in fertilization, where its nucleus combines with the nucleus of the larger egg cell, forming a new organism. In males, the testes generate sperm in substantial quantities to enhance the likelihood of egg fertilization. Human sperm cells consist of distinct sections: a nucleus-containing head, a mitochondria-rich mid-section providing energy, and a flagellum for mobility. During fertilization, the sperm's nucleus and contents are incorporated into the egg's cytoplasm, but the sperm's mitochondria are eliminated and do not persist in the resulting zygote. Because mitochondria have their own DNA, a baby inherits its entire mitochondrial DNA exclusively from the mother. Semen produced by the male reproductive tract functions as a medium, typically containing more than 100 million sperm cells, all aimed at fertilizing the single available egg.

12.3.3 MALE PREEJACULATE

Pre-ejaculate, also referred to as pre-ejaculatory fluid, preseminal fluid, or colloquially as precum, is a clear, colorless, sticky, and thick fluid that is released from the urethra of a man's penis when he becomes sexually aroused. It shares some similarities in composition with semen but possesses notable chemical distinctions. The presence of sperm in this fluid is a subject of debate. While existing research, which is limited due to its focus on a small population of men, has shown either no sperm or very low sperm levels in pre-ejaculate, recent and contrasting individual case studies have suggested higher sperm concentrations.

Pre-ejaculate is thought to serve as both a lubricant and an acid neutralizer. The quantity of pre-ejaculate varies significantly among individuals. Some men may not release any noticeable pre-ejaculate, while others might emit as much as 5 mL. Men who do not typically produce substantial pre-ejaculate can benefit from actively engaging the pubococcygeus muscle, which helps release seminal fluid before the full orgasm occurs.

In rare instances, some men may generate an excessive amount of pre-ejaculate fluid, leading to feelings of embarrassment or irritation. Limited case reports have indicated positive outcomes when these individuals are treated with a 5-α-reductase inhibitor. In such scenarios, doctors might suggest considering the possibility of prostatorrhea (a condition involving the emission of prostatic secretions during straining related to urination or defecation) as part of the differential diagnosis.

- **Origin and composition:** The fluid is released during sexual arousal, masturbation, foreplay, or early stages of sexual activity, preceding full orgasm, and ejaculation of semen. It is primarily produced by the bulbourethral glands ("Cowper's glands"), with contributions from the Littre glands ("mucus-secreting urethral glands"). While pre-ejaculate contains some chemicals found in semen, like "acid phosphatase, certain semen markers such as γ-glutamyl-transferase are entirely absent in pre-ejaculate fluid."
- **Function:** Acidic conditions are unfavorable for sperm survival. Pre-ejaculate serves to neutralize residual urethral acidity from urine, creating a more conducive environment for sperm passage. As the vagina is naturally acidic, depositing pre-ejaculate before semen emission can alter the vaginal environment to support sperm survival. Moreover, pre-ejaculate functions as a lubricant during sexual activity and plays a role in semen coagulation.
- **Health risks:** Research has detected HIV in most pre-ejaculate samples from HIV-positive men. HIV infection can lead to AIDS. The belief that pre-ejaculate might contain sperm capable of causing pregnancy is commonly cited against relying on the withdrawal method as contraception. However, some studies suggest withdrawal could be nearly as effective as condoms in preventing pregnancy. Investigations in Boston and New York City examined pre-ejaculate fluid from HIV-positive and HIV-negative men. Macrophages and CD4 lymphocytes, indicative of HIV presence, were commonly found. Notably, most pre-ejaculate samples lacked sperm. With only limited immobile sperm present, often in the thousands, the probability of pregnancy from pre-ejaculate fluid entering the vagina is very low. The risk of HIV transmission could outweigh the risk

of unplanned pregnancy, emphasizing the use of condoms before penetration of the vagina, mouth (oral sex), or anus. Urinating after ejaculation is widely believed to flush the remaining sperm from the urethra.

- **Religious views:** In Islam, the emission of pre-ejaculate fluid prompts ritual purification, involving washing with water the affected areas, including clothing. While pre-ejaculate fluid calls for ablution before praying, it does not necessitate a ritual purificatory bath such as semen does postejaculation.

12.3.4 SEMEN[38–51]

The seminal plasma found in humans is a complex mixture of diverse organic and inorganic constituents. Its role is to serve as a nourishing and protective medium for spermatozoa during their journey through the female reproductive tract. The components within the seminal plasma aim to counterbalance the challenging conditions of the vaginal environment. An average ejaculation, roughly a teaspoon in volume, contains approximately 150 mg of protein, 11 mg of carbohydrates, 6 mg of fat, and 3 mg of cholesterol, along with 7% potassium and 3% each of copper and zinc. The caloric content of semen is around 15 calories per ejaculation, which translates to about 5–7 calories in a teaspoon of ejaculation. Sperm cells make up about 1% of the semen volume, surrounded by fluids from various reproductive system structures. These fluids nourish and sustain the sperm as they move down the reproductive tract, ensuring their motility and participating in specific chemical processes. During ejaculation, substances from the prostate gland and seminal vesicles dilute the sperm, creating a suitable, mildly alkaline environment.

Around 99% of ejaculate is composed of a variety of elements, including "ascorbic acid (vitamin C), blood-group antigens, cadaverine, calcium, chlorine, cholesterol, choline, citric acid, DNA, enzymes, fructose sugar, glutathione, hyaluronidase, inositol, lactic acid, magnesium, nitrogen, phosphate and bicarbonate buffers, potassium, protein, purine, putrescine, pyrimidine, pyruvic acid, sodium, sorbitol, spermidine, spermine, urea, uric acid, vitamin B12, water, and zinc." A regular, healthy ejaculation contains trace amounts of various vitamins, with the quantity of vitamin C being comparable to that found in an orange.

12.3.4.1 THE COMPONENTS AND CONTRIBUTIONS OF HUMAN SEMEN

The seminal plasma found in humans is a complex mixture of diverse organic and inorganic constituents. Its role is to serve as a nourishing and protective medium for spermatozoa during their journey through the female reproductive tract. The components within the seminal plasma aim to counterbalance the challenging conditions of the vaginal environment. "An average ejaculation, roughly a teaspoon in volume, contains approximately 150 mg of protein, 11 mg of carbohydrates, 6 mg of fat, 3 mg of cholesterol, along with 7% potassium and 3% each of copper and zinc. The caloric content of semen is around 15 calories per ejaculation, which translates to about 5–7 calories in a teaspoon of ejaculate."

Sperm cells make up about 1% of the semen volume, surrounded by fluids from various reproductive system structures. These fluids nourish and sustain the sperm as they move down the reproductive tract, ensuring their motility and participating in specific chemical processes. During ejaculation, substances from the prostate gland and seminal vesicles dilute the sperm, creating a suitable, mildly alkaline environment.

Around 99% of ejaculate is composed of a variety of elements, including "ascorbic acid (vitamin C), blood-group antigens, cadaverine, calcium, chlorine, cholesterol, choline, citric acid, DNA, enzymes, fructose sugar, glutathione, hyaluronidase, inositol, lactic acid, magnesium, nitrogen, phosphate and bicarbonate buffers, potassium, protein, purine, putrescine, pyrimidine, pyruvic acid, sodium, sorbitol, spermidine, spermine, urea, uric acid, vitamin B12, water, and zinc." A regular, healthy ejaculation contains trace amounts of various vitamins, with the quantity of vitamin C being comparable to that found in an orange. The seminal fluid is mostly produced by the accessory genital ducts, the seminal vesicle, the prostate glands, and the bulbourethral glands.

The components and contributions of semen are as follows:[38,39,45,46,50,51]

Gland	%	Components
Bulbo-urethral glands	<1	Galactose, mucus (serve to increase the mobility of sperm cells in the vagina and cervix by creating a less viscous channel for the sperm cells to swim through, and preventing their diffusion out of the semen; contributes to the cohesive jelly-like texture of semen), pre-ejaculate, and sialic acid.
Prostate	25–30	Acid phosphatase, citric acid, fibrinolysin, prostate-specific antigen (PSA), proteolytic enzymes, and zinc (the zinc level is about 135 ± 40 μg/mL for healthy men; zinc serves to help to stabilize the DNA-containing chromatin in the sperm cells; a zinc deficiency may result in lowered fertility because of increased sperm fragility; zinc deficiency can also adversely affect spermatogenesis).

Gland	%	Components
Seminal vesicle	65–75	Amino acids, citrate, enzymes, flavins, fructose (2–5 mg/mL semen, the main energy source of sperm cells, which rely entirely on sugars from the seminal plasma for energy), phosphorylcholine, prostaglandins (involved in suppressing an immune response by the female against the foreign semen), proteins, and vitamin C.
Testes	2–5	Approximately 200–500-million spermatozoa (also called *sperm* or *spermatozoans*), produced in the testes, are released per ejaculation. If a man has undergone a vasectomy, he will have no sperm in the ejaculation.

12.3.4.2 CULTURAL ASPECTS OF SEMEN[42,52,53]

Appendices I–III encompass poetic descriptions of sperm, the process of being born, and the role of mothering, respectively. In various cultures, semen is attributed with distinctive qualities of masculinity. Among certain tribes in Papua New Guinea, such as the Sambia and the Etoro, semen is believed to facilitate sexual maturation in younger individuals. These tribes consider sperm to possess the masculine attributes of tribal elders, and to ensure the transfer of their authority and powers, younger men are required to consume the semen of their elders through acts of fellatio. This practice is observed among both prepubescent and postpubescent males and might be influenced by culturally significant homosexuality within these societies. Similarly, in India, younger men are advised against wasting semen as it is thought to be a vital essence of human bones. Conversations about sex are taboo in Indian culture.

In the realm of Qigong and Chinese medicine, great significance is placed on "jing," a form of sexual energy that is believed to deplete upon ejaculation, leading to masturbation being termed as "energy suicide" within these practices. Qigong theory posits that energy from various pathways or meridians converges in the sexual organs during sexual arousal, with orgasm and ejaculation expelling this energy entirely. The Chinese scientific terms for semen are "jīng yè," denoting the "fluid of essence," and "jīng zǐ," referring to the "basic element of essence."

In Aristotle's perspective, semen is regarded as a residue originating from blood. He explains that sperms result from the excretion of food or are the most perfected component of our nourishment. This connection between food, physical growth, and semen led Aristotle to caution against engaging in sexual activity too early in life, as the nourishment intended for physical development is diverted to semen production. Furthermore, he suggests that the region around the eyes is particularly fertile for seed production, reflecting the effects of sexual indulgence on the eyes and practices implying the connection of seed with the liquid near the eyes.

In the realm of Greek Stoic philosophy, the concept of "*Logos spermatikos*" or the "seminal word" symbolized the principle of active reason that impregnated passive matter. This was akin to the Jewish philosopher Philo's analogy of the Logos as a masculine principle of reason sowing seeds of virtue within the feminine soul. Clement of Alexandria extended this notion by comparing the Logos to physical blood, equating animal semen with foam made from blood. Clement's view aligns with early Christian beliefs, maintaining that seed is not wasted, scattered thoughts, or sown in a manner that prohibits growth (akin to the Law of Conservation).

12.3.4.3 APPEARANCE AND CONSISTENCY OF HUMAN SEMEN[42,52,53]

Semen is an organic fluid of whitish, viscous consistency released from the male reproductive tract. It contains both sperm and fluids known as seminal plasma that assist in preserving the viability of the sperm. Semen is generally clear but may have a white, gray, or occasionally yellowish hue. The presence of blood in the semen can result in a pink or reddish color, termed hematospermia, which may indicate an underlying medical issue.

Following ejaculation, the later portion of the ejaculated semen undergoes coagulation, forming globules. In contrast, the initial part of the ejaculate typically does not coagulate. After a period ranging from 15 to 30 min, an enzyme called "Prostate-specific antigen (PSA)" found in the semen induces the de-coagulation of the coagulated seminal substance. It is believed that the initial clotting aids in retaining the semen within the vagina, while the subsequent liquefaction facilitates the sperm's movement towards the ova.

12.3.4.4 SEMEN QUALITY, QUANTITY AND TASTE[42,52,53]

Semen quality is a measure of the ability of semen to accomplish fertilization. Thus, it is a measure of fertility in a male. It is the sperm in the semen that is the fertile component. The volume of semen ejaculate ranges from 1.3 to 4.99 mL.

"The 1992 World Health Organization (WHO) report describes normal human semen as having a volume of 2 mL (one teaspoon), pH of 7.2–8.0, sperm concentration of 20×10^6 spermatozoa/mL or more, sperm count of 40 $\times 10^6$ spermatozoa per ejaculate or more, and motility of 50% or more with forward progression of 25% or more with rapid progression within 60 min of ejaculation." The actual amount depends on: "(a) the age (younger men usually make more semen); (b) when he last ejaculated, and (c) how long he is aroused before ejaculating; etc."

12.3.4.5 SEMEN STORAGE[54]

Semen has the potential to be preserved using solutions such as the "Illini variable temperature (IVT) diluent," which could maintain its fertility for over 7 days. The IVT diluent consists of a mixture of salts, sugars, antibacterial agents, and CO_2. Alternatively, semen cryopreservation allows for even longer storage periods. In the case of human sperm, the most extended successful storage duration using this technique has been reported as 21 years. Modern times have witnessed the development of sophisticated storage methods.

12.3.4.6 HEALTH BENEFITS[42,52,53]

Semen is antidepressant and contains epinephrine. In addition to its central role in reproduction, the semen may have certain beneficial effects on human health:

- **Antidepressant:** Vaginal absorption of semen could act as an anti-depressant.
- **Cancer prevention:** Seminal plasma might reduce breast cancer by "not less than 50%." This effect is attributed to its glycoprotein and selenium content, with apoptosis being induced by TGF-β.
- **Preeclampsia prevention:** It is hypothesized that substances in semen condition a mother's immune system to accept the "foreign" proteins found in sperm as well as the resulting fetus and placenta, keeping blood pressure low and thereby reducing the risk of preeclampsia. Oral sex and swallowing sperm may help to make a woman's pregnancy safer and more successful, because she is absorbing her partner's antigens.

12.3.4.7 HEALTH RISKS[13,19,49 55,56]

Seminal plasma contains prostaglandin elements that could potentially accelerate the progression of existing cervical cancer. Semen can serve as a carrier for numerous sexually transmitted diseases, including HIV, which causes AIDS. Research by Mathur and Goust demonstrated that "*human bodies produce non-preexisting antibodies in response to sperm. These antibodies mistakenly target native T-lymphocytes as foreign antigens, leading to an immune response where B-lymphocytes attack T-lymphocytes.*"

Semen comprises multiple proteins with potent antimicrobial properties against bacteria, fungi, and viruses. However, these proteins do not exhibit activity against Neisseria gonorrhoeae, a common cause of sexually transmitted disease.

In uncommon instances, individuals may experience allergic reactions to seminal fluids, a condition known as human seminal plasma hypersensitivity. Symptoms can manifest locally or systemically and may include vaginal itching, redness, swelling, or blisters occurring within half an hour of contact. More generalized symptoms such as itching, hives, and even breathing difficulties can also occur. One method to test for sensitivity to human seminal plasma is to use a condom during intercourse. If symptoms subside when a condom is used, semen sensitivity might be present. Mild cases of semen allergy can sometimes be managed through repeated exposure to seminal fluid. In more severe cases, seeking medical advice is crucial, especially if the couple is attempting to conceive; in such instances, artificial insemination might be considered.

Researchers at Utrecht University explored a condition where certain men experience flu-like symptoms, such as fever, runny nose, extreme fatigue, and burning eyes immediately following ejaculation. This condition, referred to as post orgasmic illness syndrome (POIS), arises from an allergy to the individual's own semen. Hypo-sensitization therapy or allergen immunotherapy can help mitigate the effects of this condition.

12.3.4.8 RISKS OF PREGNANCY USING THE WITHDRAWAL METHOD

The withdrawal method is when a man pulls his penis out of the vagina and ejaculates away from the vagina. This method was the original contraceptive before condoms, the pill, etc. were developed and is still very common today. When used correctly, it has a high success rate but does have a higher failure rate than more modern contraceptive methods. Most of these failures may be attributed to the male not pulling out in time; many men (particularly younger, less experienced men) lack self-control and in the heat of the moment leave it too long before pulling out. Under this situation, it is likely that a couple of spurts of semen (containing millions of tiny sperm) enter the vagina. It is also possible for a man to have one or two spurts of semen then a gap of a second or two and then more spurts of semen during an orgasm. In this situation it may appear that the male has ejaculated outside of the vagina, but millions of sperm have been deployed into the vagina. These risks are summarized as follows:

- **High risk:** male does not withdraw in time and shoots inside the vagina; Male misses times his orgasm and a couple of spurts of semen enter the vagina; Male withdraws in time but inserts again without urinating.
- **Medium risk:** male withdraws just in time but close to orgasm; Male withdraws in time but inserts again in the same session after urinating.
- **Low risk:** male withdraws nowhere near his orgasm and has urinated beforehand; Male inserts penis briefly, is nowhere near orgasm and has urinated since his last ejaculation.
- **Abstinence** is the only way that is 100% positive for the prevention of pregnancy. Anytime a male has sexual contact, there is always a chance of pregnancy because no birth control method that is 100% guaranteed to prevent pregnancy.

12.3.4.9 SEMEN INGESTION

Some reasons for human ingestion of human or other semen are "(a) erotic gratification; and (b) physical and spiritual benefits." There is no risk in ingesting the semen of a healthy man. Swallowing semen carries no additional risk other than those inherent in "*fellatio.*" "*Fellatio*" does carry some transmission risk for sexually transmitted diseases, such as "HIV or herpes, especially for people with bleeding gums, gingivitis, or open sores. Even if the semen is cold before the individual ingests it, viruses can stay active for a long period of time once outside the body. Research has suggested that performing unprotected oral sex on a person infected with human papillomavirus (HPV) might increase the risk of oral or throat cancer. The study indicated that 36% of the cancer patients had HPV compared to only 1% of the healthy control group. It is believed that this is due to the transmission of HPV; because this virus has been implicated in the majority of cervical cancers. There are several sexual practices involving the ingestion of semen." They can be done (a) with one or more partners, such as snow-balling, filching, and creampie-eating; or (b) with multiple partners, such as practices of bukkake and gokkun, which originated in Japan. However, the authors of this chapter do not recommend such practices.

12.3.4.10 BLOOD IN THE SEMEN

The occurrence of blood in semen, known as hematospermia, might not be noticeable to the naked eye and might only be observable under a microscope. This condition could result from inflammation, infection, obstruction, or injury within the male reproductive tract, including the urethra, testicles, epididymis, or prostate. Typically, hematospermia resolves on its own without requiring treatment, or it may respond to antibiotics. However, if the issue persists, additional semen analysis and urogenital tests could be needed for further evaluation.

12.3.4.11 SACRED SEMEN

In some preindustrial societies, semen and other body fluids were revered because they were believed to be magical.[39] Throughout history, semen has been attributed with supernatural qualities and effects, often seen as holy or sacred. There were beliefs that drew connections between semen and natural elements. Dew, for instance, was once thought to be a form of rain that fertilized the earth and was metaphorically associated with semen. The Bible employed the term "dew" in this context, referring to virility and vitality.

In ancient times, gemstones were believed to be crystallized drops of divine semen, coagulated after fertilizing the earth. Some cultures, such as ancient China, associated jade with celestial dragon semen. Various plants and flowers were also connected to semen. Dandelion juice's resemblance to human semen led to the belief that it had magical properties for promoting sperm flow. Orchids, with their twin bulbs resembling testicles, were thought to have sprung from the spilled semen of copulating satyrs in Roman mythology.

Primitive myths often linked semen to breast milk in symbolic ways. In Bali's traditions, semen represented a return or refund of the mother's milk in a metaphorical sense. This notion symbolized the reciprocity of nourishment and kindness between husband and wife.

In certain medical philosophies, like traditional Russian medicine and Herbert Nowell's Vital Force theory, semen was viewed as the result of a complex interaction between both men and women, rather than solely a product of the male reproductive system.

Some historical religious and philosophical groups had unique perspectives on semen. Gnostic sects like the Borborites consumed semen as a spiritual practice, while other texts such as the "Pistis Sophia" and "Testimony of Truth" condemned such practices. In the modern context, the St. Priapus Church considers semen consumption a form of worship, attributing it with divine life-giving power. Certain chapters of the Ecclesia Gnostica Catholica include semen consumption in rituals, as seen in Aleister Crowley's Gnostic Mass.

12.3.4.12 RETROGRADE EJACULATION

Throughout history, semen has been attributed with supernatural qualities and effects, often seen as holy or sacred. There were beliefs that drew connections between semen and natural elements. Dew, for instance, was once thought to be a form of rain that fertilized the earth and was metaphorically associated with semen. The Bible employed the term "dew" in this context, referring to virility and vitality.

In ancient times, gemstones were believed to be crystallized drops of divine semen, coagulated after fertilizing the earth. Some cultures, like ancient China, associated jade with celestial dragon semen. Various plants and flowers were also connected to semen. Dandelion juice's resemblance to human semen led to the belief that it had magical properties for promoting sperm flow. Orchids, with their twin bulbs resembling testicles, were thought to have sprung from the spilled semen of copulating satyrs in Roman mythology.

Primitive myths often linked semen to breast milk in symbolic ways. In Bali's traditions, semen represented a return or refund of the mother's milk in a metaphorical sense. This notion symbolized the reciprocity of nourishment and kindness between husband and wife.

In certain medical philosophies, like traditional Russian medicine and Herbert Nowell's Vital Force theory, semen was viewed as the result of a complex interaction between both men and women, rather than solely a product of the male reproductive system.

Some historical religious and philosophical groups had unique perspectives on semen. Gnostic sects like the Borborites consumed semen as a spiritual practice, while other texts like the "Pistis Sophia and Testimony of Truth" condemned such practices. In the modern context, the St. Priapus Church considers semen consumption a form of worship, attributing it to divine life-giving power. Certain chapters of the Ecclesia Gnostica Catholica include semen consumption in rituals, as seen in Aleister Crowley's Gnostic Mass.

12.3.4.13 SPERM COMPETITION[40,57]

Sperm competition refers to the competitive process that takes place between spermatozoa from different males, all attempting to fertilize an egg within a single female. This phenomenon occurs particularly when females engage in promiscuous mating to increase their chances of producing viable offspring. Sperm competition serves as an evolutionary pressure on males, driving the development of adaptations that enhance their odds of reproductive success.

The origin of sperm competition lies in the sexual conflict between males and females. To counter this, males have evolved defensive strategies such as mate-guarding, employing mating plugs, and producing toxic substances in their seminal fluid to discourage females from re-mating. This adaptation helps them better deal with the challenges of sperm competition. Analogously, sperm competition can be likened to owning tickets in a raffle, where a male 's chances of winning (fathering offspring) increase with the number of "tickets" (sperm) he introduces to a female. However, sperm production requires energy, prompting males to produce a specific quantity and size of sperm that optimizes their success in this competition.

The strategy of producing numerous sperm allows males to acquire more "raffle tickets," and this concept aligns with the evolution of anisogamy—the presence of very small sperm due to the trade-off between sperm size and quantity. Another potential adaptation is the evolution of faster-moving sperm that can reach and fertilize the egg more swiftly, giving them an advantage in the race for fertilization.

12.3.4.14 INCREASING SEMEN VOLUME

Males with erectile dysfunction or impotency use natural semen volume pills. Semen volume pills are designed to increase the amount of ejaculate or semen volume during sex. These herbal pills are not approved by "the Food and Drug Administration (FDA)" and fall under the male enhancement category.

12.3.4.15 SEMEN IN ESPIONAGE

When the British Secret Intelligence Service discovered that semen made a good invisible ink, Sir George Mansfield Smith-Cumming noted of his agents that "*Every man (is) his own stylo*."

12.3.4.16 SPERM WASHING

This term describes the process in which individual sperms are separated from the seminal fluid. The sperms are then used in "intrauterine insemination (IUI) or in vitro fertilization (IVF)." Sperm washing is a standard procedure in infertility treatment. Sperm may be "washed" by density gradient centrifugation or by a "direct swim-up" technique that does not involve centrifugation. In normal semen samples, centrifugation causes no DNA damage to spermatozoa than a direct swim-up technique.

12.4 VAGINA SECRETIONS

The female vagina is the passage of the semen on its way to the uterus, secreting fluids for lubrication and protection. In addition to fluid, the vagina contains resident bacteria, whose metabolic rate causes the vagina to have an acidic environment with a pH of 5.0. This environment restricts the growth of many pathogens, and it also decreases the motility of sperm in the vagina.[15]

Some of the fluids in the vagina[21,51,53,58–60] originate from the mucus formed in the cervix. When a woman is pregnant, the cervix forms a mucus plug that serves as a physical barrier between the developing fetus in the uterus and the vagina.[61] A recent study has indicated that cervical plugs contain antimicrobial properties to kill bacteria.[61] Laboratory tests confirmed that samples of mucus plugs contained immune system proteins that help fight infection. Upon labor, the mucus plug breaks down and helps in lubricating the vagina when the baby is ready to be delivered.[61]

12.4.1 HUMAN CERVICAL MUCUS

Sperm competition refers to the competitive process that takes place between spermatozoa from different males, all attempting to fertilize an egg within a single female. This phenomenon occurs particularly when females engage in promiscuous mating to increase their chances of producing viable offspring. Sperm competition serves as an evolutionary pressure on males, driving the development of adaptations that enhance their odds of reproductive success.

The origin of sperm competition lies in the sexual conflict between males and females. To counter this, males have evolved defensive strategies such as mate-guarding, employing mating plugs, and producing toxic substances in their seminal fluid to discourage females from re-mating. This adaptation helps them better deal with the challenges of sperm competition. Analogously, sperm competition can be likened to owning tickets in a raffle, where a male's chances of winning (fathering offspring) increase with the number of "tickets" (sperm) he introduces into a female. However, sperm production requires energy, prompting males to produce a specific quantity and size of sperm that optimizes their success in this competition.

The strategy of producing numerous sperm allows males to acquire more "raffle tickets," and this concept aligns with the evolution of anisogamy—the presence of very small sperm due to the trade-off between sperm size and quantity. Another potential adaptation is the evolution of faster-moving sperm that can reach and fertilize the egg more swiftly, giving them an advantage in the race for fertilization.

12.4.1.1 CERVICAL MUCUS AND PREGNANCY[16,21,51,53,59,60]

Cervical mucus (CM) plays a crucial role in the process of achieving pregnancy. For pregnancy to occur, sperm must traverse the vaginal canal, cervix, and uterus to reach the Fallopian tube, where they can fertilize the egg. This journey is quite a distance for sperm to cover, and CM acts as a conducive medium for sperm to swim forward effectively. Particularly around the time of ovulation, the cervix secretes increased amounts of CM due to glandular activity.

The quality of CM changes during the ovulation period. The CM that is considered "fertile" during ovulation is typically thin, stretchy, and more alkaline compared to other times of the menstrual cycle—often likened to the texture of egg white. This type of mucus facilitates the movement of sperm and provides an environment in which they can function optimally.

In most phases of a woman's menstrual cycle, there is minimal production of cervical fluids. Many hormonal contraception methods primarily work by preventing ovulation; however, they also inhibit the production of fertile types of CM, enhancing their overall effectiveness. Conversely, methods that thin CM may aid in achieving pregnancy. Various fertility awareness methods involve observing changes in the body to estimate periods of fertility and infertility. These observations encompass aspects of CM, such as its sensation at the vulva, its elasticity (known as Spinnbarkeit), its transparency, and the presence of a pattern called "ferning."

To promote pregnancy, some suggest using guaifenesin a few days before ovulation, as it may help improve the quality of CM. Observing CM involves using clean and dry fingers or toilet paper to collect and assess its characteristics. The composition and behavior of CM vary throughout the menstrual cycle and are described as follows:

- **CM before ovulation:** A woman typically experiences little visible cervical fluids. Gradually, more CM will accumulate, but it will be yellow, cloudy, and exhibit a sticky texture. This is nonfertile CM (sometimes referred to as "hostile cervical mucus"). During nonfertile times of the month, one will typically observe a dryness (or scant amount of CM).

- **Near the ovulation date:** The presence of CM will increase markedly due to higher levels of estrogen in the body. First, there will be a moistness or slight tackiness to the mucus, and it will exhibit a white or cream-colored appearance.
- **During ovulation (fertile phase):** For several days around the time of ovulation, "fertile" types of mucus are produced with higher water content. The quantity of cervical fluids will increase greatly and the appearance will resemble raw "egg whites." CM becomes slippery, pliable, stretchable, semitransparent, less acidic, and higher in electrolytes. These electrolytes cause the "ferning" pattern that can be observed in drying mucus under low magnification; as the mucus dries, the salts crystallize, resembling the leaves of a fern. This is the most fertile time of the month for conceiving. During ovulation, there is an increase in cervical fluids and a change in CM texture.
- **After Ovulation (luteal phase):** After the menstrual period ends, the CM becomes sticky, cloudy, thick, and acidic. This "infertile" mucus blocks spermatozoa from entering the uterus. Postovulatory dryness may also ensue because the levels of progesterone rise in the body.
- **During pregnancy:** The cervix is blocked by a special antibacterial mucosal plug that prevents infection. This is somewhat like its state during the infertile portion of the menstrual cycle. The mucus plug comes out as the cervix dilates during labor or shortly before."
- **The fertility dietary supplements** are designed to promote the production of cervical CM during a woman's ovulatory phase and promote the endometrial secretions, which help build the uterine lining for pregnancy. In published studies, the ingredients in a fertility supplement have been demonstrated to:
  - Naturally and safely support the production of cervical fluids.
  - Increase arousal and sexual stimulation.
  - Increase hydration of the mucin (liquidity of mucus).
  - Promote mucus alkalinity and mucosal thinning.
  - Support a healthy endometrial lining for successful implantation of the embryo.

12.4.1.2 ROLE OF CM IN FERTILITY

- Acts as a natural fertility sign: The presence of "fertile" cervical fluids tells a woman that she is nearing ovulation that is her most fertile time of the month.

- Assists in preventing infections in the vagina.
- Depending on the water content that varies during the menstrual cycle, the CM functions as a barrier or a transport medium to spermatozoa: Facilitates passage of sperm through the vagina, cervix and uterus to the Fallopian tube.
- Facilitates sperm movement so that conception can take place.
- Helps to neutralize acidity in the vagina during a woman's fertile time.
- Provides a healthy medium for sperms to survive. It also insulates sperm from the natural acidity in the vagina.
- Provides protection and nutritional support for sperm."

12.4.2 HUMAN FEMALE EJACULATE

Female ejaculation (also known as gushing or squirting) refers to "the expulsion of noticeable amounts of clear fluid by a human female from the paraurethral ducts through and around the urethra during or before an orgasm. The exact source and nature of the fluid continues to be a topic of debate among medical professionals and is related to doubts over the existence of the G-Spot."

- **History:** In the 17th century, François Mauriceau described glands at the urethral meatus that "*pour out great quantities of saline liquor during coition, which increases the heat and enjoyment of women.*" The Dutch anatomist Regnier de Graaf identified the source as the glandular structures and ducts surrounding the urethra; and indicated that the urethra is lined by a thin membrane. In the lower part, near the outlet of the urinary passage, this membrane is pierced by large ducts (called lacunae) through which pituito-serous matter occasionally discharges in considerable quantities. Between this very thin membrane and the fleshy fibers, there is a whitish membranous substance about one finger-breadth thick, which completely surrounds the urethral canal. The substance is called the "female prostate or corpus glandulosum." The function of the female prostatae is to generate a pituito-serous juice, which makes women more libidinous with its pungency and saltiness and lubricates the sexual parts during coitus. This fluid "which rushes out with such impetus during venereal combat or libidinous imagining" is derived from a number of sources: vagina, urinary tract, cervix and uterus. In 1918 A.D., Almeda Sperry wrote to Emma Goldman about the "*rhythmic spurt of the love juices.*" Female ejaculation is mentioned as normal in "TH Van de Velde's Ideal Marriage: Its Physiology and

Technique (1926)." It appears that a layman believes that something is forcibly propelled or extruded from the woman's body in orgasm. In describing sexual relations among Trukese Micronesians, Gladwin, and Sarason indicated that female orgasm is commonly signaled by urination. Among the Batoro, older women teach the younger women "*kachapati*" (spraying the wall) at puberty. The discussion entered popular culture in 1982 with the publication of the bestselling book, "*The G-Spot and Other Recent Discoveries About Human Sexuality by Ladas, Whipple, and Perry.*" This book discusses female ejaculation and brings the issue back into discussions of women's sexuality both in the medical community and among the general public.

- **Research:** A significant portion of research focusing on the nature of this fluid revolves around the determination of whether it constitutes urine or contains urine, as outlined by WHO guidelines. Challenges arise in the collection of specimens and concerns about potential contamination. Given that the area of interest involves paraurethral glands, achieving a complete separation of secretions from urine proves difficult, particularly considering the possibility of retrograde ejaculation into the urethra towards the bladder. Notably, the most reliable data available originates from studies where women abstain from sexual activity, and their own urine serves as a control for pre- and postorgasm samples. Researchers have attempted to incorporate chemicals excreted in urine to identify any urinary contamination. Further methodological complexities encompass variations in fluid composition across the menstrual cycle and age-related shifts in the biochemical profile of para-urethral tissues. The chosen markers' sensitivity and specificity also contribute to methodological considerations. Addressing critical queries involves understanding the origin of the fluid production and its specific composition. Scientific investigations carried out in the 1980s and later established that the produced fluid is distinct from urine, although certain similarities persist, such as the presence of alkalinity. Recent studies focusing on women experiencing ejaculation have not identified any evidence of urological issues, implying that ejaculation and coital incontinence are distinct physiological conditions. However, they might not always be clearly distinguishable in individual cases. To ensure accurate medical assessment, it becomes essential for healthcare professionals to ascertain whether any reported ejaculation is associated with incontinence in women, thereby avoiding unnecessary interventions. Furthermore, it is crucial for physicians to differentiate between orgasmic ejaculation and vaginal discharges, as the latter might necessitate further investigation and treatment. In specific instances, pinpointing the exact source of reported discharge might not be apparent without additional inquiry.

- **The nature and source of female ejaculate:** Research in this field has predominantly focused on disproving the notion that female ejaculate is urine, utilizing measurements of various substances such as urea, creatinine, prostatic acid phosphatase (PAP), PSA, glucose, and fructose levels. A study conducted in 2007 involving two women employed ultrasound, endoscopy, and biochemical analysis of the fluid in question. This ejaculation was then compared to the women's preorgasmic urine samples, as well as published data on male ejaculate. The study revealed that the ejaculate had higher levels of PSA, PAP, and glucose, but lower levels of creatinine compared to urine. PSA levels were comparable to those found in males. The volume of expelled fluid varies greatly, ranging from 1 to 5 mL, though reports of volumes as high as 473 mL have been documented. However, certain practical concerns arise regarding the reported volume of female ejaculate. Given that this fluid needs to be accommodated within the pelvis, with the urinary bladder being the largest available reservoir, it is important to note that the actual volume of paraurethral tissue is quite limited. In contrast, male ejaculate varies from 0.2 to 6.6 mL, with a maximum of 13 mL. This suggests that claims of larger ejaculate amounts are likely to include at least some urine. Nevertheless, Skene's glands can discharge as much as 30–50 mL in a span of 30–50 s. One approach to differentiate between urine and ejaculate involves using methylene blue to detect any urinary component. Biochemical and immune–histochemical methods have identified PAP and PSA in paraurethral tissues, indicating that the ejaculate probably originates from the ducts within these tissues in a manner analogous to males. Another shared marker in the prostate/paraurethral tissue between both sexes is human protein-1. While PSA is present in urine and postorgasmic samples show elevated levels compared to preorgasmic samples, simultaneous collection of ejaculates consistently demonstrated the presence of PSA in both urine and ejaculate, albeit in higher concentrations in the ejaculate than in the urine.
- **Functions:** The physiological function of the purported liquid is unknown. However, the physicians claim that it may have an antimicrobial function, protecting it from urinary tract infections.

- **Social significance:** The societal implications surrounding the understanding of female ejaculation have been substantial, regardless of the specific details involved. Within the realm of female healthcare, these popular accounts have carried significant weight. In certain cultures, such as India, women have felt a sense of embarrassment or have steered clear of sexual intimacy due to misconceptions that they have wet the bed, prior to comprehending the concept of ejaculation. This has led to the suppression of sexual climax and a reluctance to openly discuss this topic. Furthermore, an excessive focus on ejaculation has led to performance anxiety during sexual encounters. Contemporary literature on women's health underscores the vast variability in the amount of fluid produced, which may go unnoticed, and the occurrence of ejaculation with or without vaginal stimulation. It also highlights that ejaculation can either accompany orgasm or be a result of intense sexual pleasure and that orgasm itself may occur without ejaculation. Certain proponents view female ejaculation as a fundamental aspect of female creativity, describing it as a birthright.
- **Legal implications:** Chemical indicators such as PSA or PAP found in the female reproductive system have been utilized as evidence in rape trials. In more recent times, the understanding of these markers originating from females has led to instances of acquittal based on forensic evidence. Notably, the British Board of Film Classification has prohibited films that are alleged to depict female ejaculation, citing guidance from medical experts that female ejaculation does not exist and that such depictions are considered to show urine. This perspective is linked to the notion that urinating during sexual activity is considered obscene under English Law. Similarly, in Australia, proposals for Internet censorship have adopted a similar stance, banning explicit websites featuring female ejaculation.

12.4.3 HUMAN VAGINAL FLUID[16,21,51 53,58–60]

Vaginal fluid, also referred to as vaginal lubricating fluid or vaginal secretion, serves a crucial role in a woman's reproductive health. Just before the onset of menstruation during puberty, girls may notice a white stain in their underwear due to vaginal secretion, which is a normal occurrence. Both vaginal secretion and CM are produced to maintain vaginal moisture, facilitating comfortable sexual intimacy. Over time, the volume of secretion decreases, and the presence of blood becomes more apparent. This blood signifies the implantation of the egg cell into the uterine wall and is a natural process in a woman's reproductive cycle. Vaginal lubrication is consistently present, with production significantly increasing during a woman's sexual arousal as a preparation for sexual intercourse. This lubrication is essential to prevent discomfort or pain during intercourse, and in cases of inadequate natural lubrication, artificial lubricants may be used. Unlike glands found in other parts of the body, the vagina does not possess glands to produce lubrication. Instead, the lubrication relies on various mechanisms. While plasma seepage resulting from vascular engorgement in the vaginal walls is the main source of lubrication, the Bartholin's glands, situated slightly below and to the left and right of the vaginal opening, also contribute by secreting mucus. Vaginal dryness is a condition where this natural lubrication is insufficient.

- **Nature of vaginal fluid:** The vagina maintains an acidic pH level, which acts as a protective measure against infections. This acidic environment is established by naturally occurring bacteria. Typically, the pH of vaginal fluid ranges from 3.8 to 4.5, whereas male semen has a pH of 7.2–8.0. The vagina maintains its health by producing secretions that help cleanse and regulate itself. These secretions are referred to as normal vaginal discharge. Disrupting the delicate balance of these secretions can create a favorable environment for infections to develop. The vaginal lubricating fluid contains various components such as water, pyridine, squalene, urea, acetic acid, lactic acid, complex alcohols, glycols, ketones, and aldehydes. Normal vaginal discharge can have different appearances, including being clear, cloudy white, or yellowish when dry on clothing. It might also contain small white flecks and occasionally appear thin and stringy. Changes in normal discharge can occur due to factors such as sexual arousal, menstrual cycle phase, emotional stress, nutritional status, pregnancy, medication use (including birth control pills), and genetic factors.
- **Vaginal dryness:** Inadequate lubrication or vaginal dryness can lead to dyspareunia, a form of sexual pain disorder. Vaginal dryness may stem from insufficient arousal or stimulation, as well as hormonal shifts triggered by menopause (which might lead to atrophic vaginitis), pregnancy, or breastfeeding. Contraceptive creams and foams can also cause irritation and dryness, while anxiety and fear regarding sexual activity can exacerbate the issue.

  Various factors can hinder vaginal lubrication, including the use of over-the-counter antihistamines, life events such as pregnancy, breastfeeding, menopause, aging, or medical conditions such as diabetes. Medications with anticholinergic or sympathomimetic effects can dry out the vaginal tissues. These medications are commonly used to treat allergies, cardiovascular problems, psychiatric conditions, and other medical issues. Oral contraceptives may also influence vaginal lubrication, either increasing or decreasing it. In older women, diminished estrogen levels can lead to reduced vaginal lubrication and increased dryness.

- **Artificial lubricants:** Using artificial lubricants can alleviate the pain associated with sexual intercourse for women, but it does not address the root cause of vaginal dryness. Personal lubricants can be applied to either the vaginal opening or the penis to supplement the natural lubrication produced and minimize discomfort or pain. Occasionally, a vaginal suppository might be inserted before engaging in intercourse. It is important to note that oil-based lubricants can weaken latex and diminish the effectiveness of condoms, latex gloves, or dental dams for contraception and protection against sexually transmitted infections. Hence, water- or silicone-based lubricants are frequently preferred alternatives.
- **Dry sex:** This practice involves a woman engaging in sexual intercourse after deliberately reducing or removing vaginal lubrication. The idea behind this practice is that a dry vagina is believed to be tighter than a lubricated one, which some believe enhances sexual pleasure for the man. However, aside from causing discomfort and pain for the woman, this practice could potentially elevate the chances of transmitting sexually transmitted infections to both partners. Notably, the absence of lubrication can lead to vaginal tissue lacerations, which in turn increases the risk of HIV transmission and other infections.
- **The effects of the menstrual cycle:** An increase in vaginal moisture and the presence of clear discharge can be observed around the middle of the menstrual cycle. The vaginal pH balance undergoes variations throughout the cycle, with the least acidity occurring in the days leading up to and during menstruation. As a result, infections tend to be more prevalent during this period.
- **Signs of abnormal discharge:** Normal physiological vaginal discharge typically appears clear, creamy, or slightly yellowish in color. Any alterations in the color or quantity of discharge might indicate a vaginal infection. Many women will experience a vaginal

infection at some point in their lives. Common symptoms of a vaginal infection include discharge accompanied by itching, irritation, or a rash; persistent and increased discharge; discomfort or burning during urination; the presence of white, clumpy discharge resembling cottage cheese; and the occurrence of gray/white or yellow/green discharge with an unpleasant odor. Vaginal infections can also coincide with bacterial vaginosis, which results from an overgrowth of bacteria upsetting the natural vaginal balance. Bacterial vaginosis often recurs. It is advisable to seek medical advice if there is a sudden increase in discharge, a change in discharge color from yellow-white to brown, red, or green, or if the discharge becomes malodorous.

12.5 BIOFLUIDS IN HUMAN FETAL SYSTEM

Saliva The formation of the fetus begins with the introduction of seminal fluid into the vagina. Seminal fluid is composed of spermatozoa, hormones (such as "prostaglandins"), and sperm nutrients (such as "fructose"). Seminal prostaglandins may function to "*influence the female response to the disposition of semen in the female tract.*"[62]

Also, the blood in this system consists of two entities, namely: the maternal blood and the fetal blood. There are no blood vessels that connect the fetus to the mother. Instead, fetal red blood cells (RBCs) travel to vascular capillaries that "sit in puddles (called intervillous spaces)." These spaces contain maternal blood that bathes the lining of the "placental capillaries." Oxygen and nutrients diffuse through the villi and into the fetal capillaries. The fetal tissue that is bathed with the maternal blood uses only the nutrients the fetus requires to grow.[17] Furthermore, the amniotic fluid is composed of secretions from the umbilical cord,[52,53,58,63–67] old skin cells of the fetus, placenta residue, and fetal urine.[4] The amniotic fluid protects the fetus, regulates fetal body temperature, gives the fetus space to move and develop as it grows, and allows the lungs to fully develop.[16,18,36]

The fetal urine is also an integral part of the amniotic fluid.[16,18,36] It is not urine as one may think, because the waste products of the fetal urine are transported through the placenta and secreted with the urine of the mother. Fetal urine increases the volume of amniotic fluid in the womb. To keep the volume of amniotic fluid stable, the fetus swallows fetal urine to reduce it. Table 12.3 describes the functions of the biofluids.

12.6 THE FETAL AND MATERNAL BLOOD: FUNCTIONS AND CHARACTERISTICS

The blood characteristics of the fetus and the mother have similarities and differences.[3,10,14,50,68] For example, "the blood content of the mother goes through some changes during pregnancy, and the circulatory system differs for both the mother and the fetus. The main difference in the fetal circulatory system is that the lungs are collapsed and nonfunctional, as well as the digestive system. The placenta delivers the respiratory and nutritional needs to the fetus. The umbilical vein (UV) provides the nutrients for the fetus from the placenta."[23]

TABLE 12.3 Biofluids in the Fetal: Types and Functions.[3,10,14,17,50,63]

Fluid	Functions	Fluid	Functions
Amniotic[13]	1. Prevents trauma to the fetus. 2. Permits free movement within the cavity. 3. Stabilizes temperature. 4. Mediates hydrostatic forces during labor.	Semen[23]	1. Contains sperm for reproduction.
Blood[23]	1. Conveying nutrients and oxygen. 2. Carrying wastes and CO_2. 3. Carrying hormone and regulatory agents. 4. Homeostatic role: a. Thermoregulation. b. Buffering capacity. 5. Transporting numeral agents and cells.	Urine[24]	1. Waste disposal: a. Remove waste substances like urea, uric acid and creatine. b. Excrete foreign compounds and nonnutritive materials like drugs and food additives

In addition, there are some extra connections in the heart of a fetus. The purpose of these connections is to direct the blood away from the lungs.[9] When the blood enters the right atrium, about a quarter of it flows through the foramen ovale, enters the left atrium, and skips the pulmonary circuit. About 90% of the blood that goes into the right ventricle is pumped out and enters the ductus arteriosus.[23] The components of blood are slightly different from the mother and the fetus. Table 12.4 shows the composition of normal blood of a mother. Table 12.5 indicates variations of substances in maternal blood during pregnancy.[3] Table 12.6 shows the composition of the fetal blood.

The temperature of blood is approximately 38°C. Additionally, blood is about five times more viscous than water, primarily due to the presence of various components such as leukocytes, erythrocytes (RBCs), proteins, and ions. Several elements in maternal blood experience minor fluctuations over the course of pregnancy. Instances of substances that undergo changes during pregnancy include iron, cholesterol, and albumin.

TABLE 12.4 Composition of Normal Blood.[23]

Whole	46–63% Plasma
blood	37–54% Formed Elements
Blood plasma	92% Water
	7% Proteins:
	>60% Albumins
	35% Globulins
	4% Fibrinogen
	<1% Regulatory proteins.
	1% Other solutes:
	Ions (Na, Cl, K, etc.)
	Organic Nutrients
	Organic Wastes
Formed elements	99.9% Erythrocytes
	0.1% Platelets and leukocytes

TABLE 12.5 Variations of Substances in Maternal Blood During the Pregnancy.[3]

Gestation age in (Months)	Hemoglobin (%)	Vitamin E (mg/100 mL)
Whole blood		
2	–	–
3	87.08	0.66
4	84.80	–
5	84.29	–
6	81.88	0.75
7	77.21	–
8	77.49	–
9	78.41	0.88

TABLE 12.5 *(Continued)*

	Serum, mg/100 mL			
Gestation age, (months)	**Iron**	**Cholesterol**	**Albumin**	**Lipids**
2	144.0	4360	200	688
3	155.1	4430	152	653
4	151.4	4180	189	694
5	133.6	3950	207	745
6	104.3	3790	211	737
7	79.9	3740	239	900
8	94.5	3690	266	964
9	96.5	3520	257	1018

TABLE 12.6 Chemical composition of Fetal Blood.[3]

Component	Vessel	Units	Value
	Whole Blood		
Erythrocytes	UV	mm^{-3}	4.6×10^6
Leukocyte	Capillary	mm^{-3}	24.9×10^3
Platelets	UV	mm^{-3}	227×10^3
O_2 saturation	UV	%	65
	Umbilical artery	%	30
O_2 pressure	UV	mmHg	30.0
	Umbilical artery	mmHg	18.0
CO_2 dissolved	UV	mM/L	1.3
	Umbilical artery	mM/L	1.4
CO_2 pressure	UV	mmHg	42.1
	Umbilical artery	mmHg	46.4
	Plasma		
Osmotic pressure	UV	mOsm/KgH_2O	292.7
pH (38°C)	UV	–	7.35
Calcium	UV	mg/100 mL	11.3
Chloride	UV	mEq/100 mL	11.29
Potassium	UV	mEq/100 mL	4.4
Sodium	UV	mEq/100 mL	140.2
Albumin	UV	mg/100 mL	4.16
Cholesterol	UV	mg/100 mL	34.0
Fructose	UV	mg/100 mL	4.2

Table 12.5 outlines the changes observed in specific components of maternal blood throughout pregnancy. The variations in these components appear to lack a consistent pattern and do not exhibit a linear relationship. Nonetheless, certain elements in the blood do exhibit some form of correlation. Notably, there is a negative correlation observed among hemoglobin percentage, iron concentration, and albumin concentration during pregnancy. This suggests that a small proportion of these three blood components contribute to the developing fetus's blood composition. Additionally, vitamin E and lipid concentrations demonstrate a positive correlation during pregnancy. Furthermore, fetal blood showcases the presence of surface antigens within its erythrocytes. These antigens are determined by genes inherited from both parents, potentially leading to a different blood type in the fetus compared to either parent.[23]

TABLE 12.7 Physical Properties and Chemical Composition of Amniotic Fluid.[3]

Physical properties	**Units**	**Value**
Volume	mL	400–1200
Appearance	–	Clear
Osmotic pressure	mOsm	265.5
P_{CO2}	mmHg	33–55
Specific gravity	ratio	1.006–1.008
	Chemical composition	
Water	%	98.5
Calcium	mg/100 mL	7.2
Chloride	mg/100 mL	365
Fructose	mg/100 mL	1–2
Glucose	mg/100 mL	4
Iron	mg/100 mL	0.07
Potassium	mg/100 mL	19
Sodium	mg/100 mL	290
Urea	mg/100 mL	20
Uric acid	mg/100 mL	4.4

Blood is a viscoelastic fluid.[3,10,14,50,63,68,69] The changes in viscosity and elasticity are due to changes in the arrangement, orientation, and stretching of RBCs. Viscosity is related to the energy dissipated during flow primarily due to sliding and deformation of RBC and RBC aggregates (for more details, see Chapters 1 and 7 in this book). Elasticity is related to the energy stored during flow due to the orientation and deformation of RBC. The viscoelastic profile of normal human blood can be divided into three regions: low, medium, and high shear rate.[3] From Newton's law of viscosity, viscosity can be determined by the ratio of shear stress to strain rate of the fluid as shown in eqs 12.1 and 12.2.

$$\text{“}\mu = (\tau)/\ (du/dy) \tag{12.1}$$

$$\mu = (F/A)/\ (du/dy) \tag{12.2}$$

where μ = viscosity, u = velocity of the plate moving over the liquid, y = distance between two plates, τ = shear stress = F/A, A = cross-sectional area, and F = frictional force."[13]

12.7 THE AMNIOTIC FLUIDS: FUNCTIONS AND COMPOSITION

A fetus develops inside a fluid-filled sac called the amniotic sac that contains amniotic fluid.[23] It contains a small amount of nutrients and urine from the fetus. The pH of the fluid is slightly alkaline ranging from 7.12 to 7.38, and it is slightly acidic ranging from 6.49 to 7.12 in the latter part of pregnancy.[23] This occurs because of a buildup of uric acid and other wastes in the fluid. Table 12.7 lists the physical and chemical composition of the amniotic fluid.

The amniotic fluid provides a safe environment for the developing fetus during pregnancy. It also absorbs any shock from any light stimulus. Since the main component of this fluid is water, many of the physical properties of the amniotic fluid are like water. The volume of the fluid can range from 400–1200 mL depending on the gestational age in pregnancy.[3]

12.8 PHYSIOLOGY OF THE FETAL CIRCULATORY SYSTEM[13,22–26,28]

The fetal circulation initiates within the placenta, which is connected to the fetus through the umbilical cord. This cord is divided into veins and arteries. Blood flow is directed into the placenta through the umbilical arteries, which then branch into the internal iliac arteries. Meanwhile, the fetus receives oxygen and essential nutrients through the UVs. Figure 12.1 illustrates the fetal circulatory system in humans, demonstrating the pathway of the UV carrying oxygen-rich blood into the developing liver. A fraction of this blood enters the inferior vena cava, contributing to approximately 65–75% of the total venous return to the heart.[14]

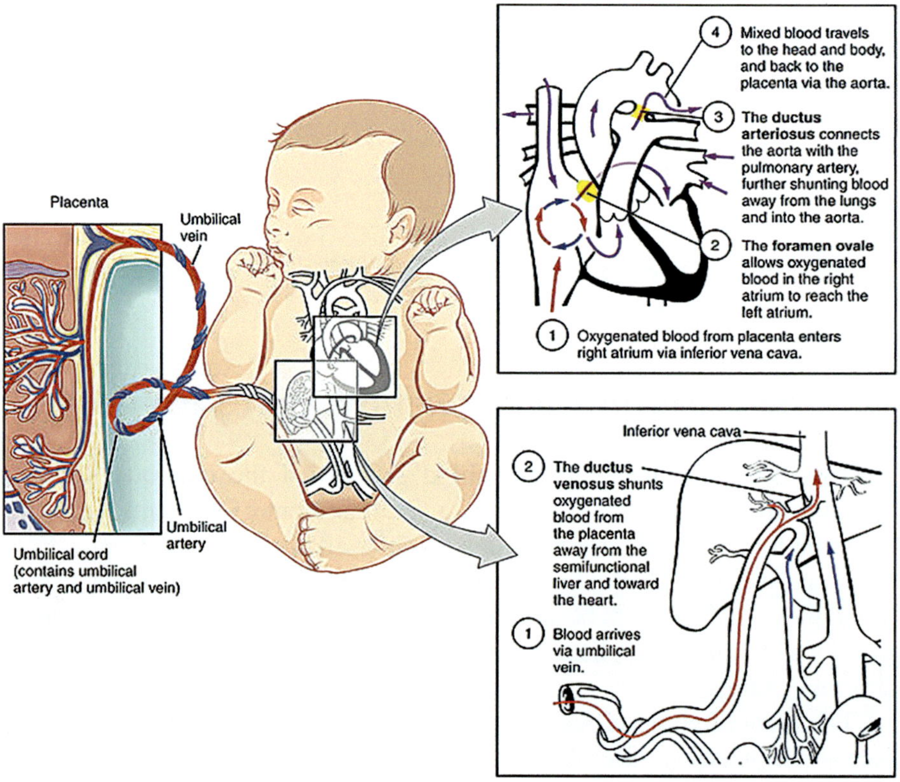

FIGURE 12.1 Fetal circulation.

Source: Reprinted from Ref. [64]. https://creativecommons.org/licenses/by/3.0/

Starting from the inferior vena cava, the blood proceeds into the foramen ovale, positioned between the inferior vena cava and the left atrium. Once within the right atrium, it then moves through the right ventricle. At this point, around one-third of the blood is directed to the lungs, while the remaining portion enters the ductus arteriosus. The blood subsequently travels through the pulmonary arteries and enters the aorta, facilitating the distribution of oxygen-rich blood throughout the fetus's body. Deoxygenated blood and waste materials exit via the umbilical arteries and pass through the umbilical cord before reaching the placenta.[22]

The estimated cardiac output of the human fetus (553 mL/kg/min^{-1}) is higher than that of a sheep (450 mL/kg/min^{-1}). In addition, the right and left ventricular outputs are more similar in the human compared with the sheep. The ratio of the right-to-left ventricular outputs decreases with advancing gestation (from 1.3 at 15 weeks to 1.1 at 40 weeks). These data are consistent with the fact that the larger human brain requires a higher left ventricular output than the brain of the sheep.

12.9 MATHEMATICAL MODELING: THE FETAL CIRCULATORY AND PULMONARY SYSTEMS[9,11,12]

In the fetal system, the lungs are collapsed, due to the fact that the fetus is floating inside a liquid atmosphere and cannot breathe air. Therefore, the fetus must obtain its oxygen and nutrients through the placenta diffusing into the fetal circulation. The fetal circulation can be described using eqs 12.3–12.6 of an uncontrolled circulation that lacks the control mechanisms needed to regulate circulation:

$$Q_R = K_R P_{sv} \tag{12.3}$$

$$Q_L = K_L P_{pv} \tag{12.4}$$

$$Q_P = (P_{pa} - P_{pv})/R_P \tag{12.5}$$

$$Q_s = (P_{sa} - P_{sv})/R_S \tag{12.6}$$

$$Q_d = (P_{pa} - P_{sa})/R_d \tag{12.7}$$

where Q_R = the right-heart output, Q_L = the left-heart output, Q_P = the pulmonary flow, Q_S = the systemic flow, Q_d = the flow through the ductus arteriosus, K_R = the pump coefficient for the right side, K_L = the pump coefficient for the left side, R_p = the pulmonary resistance, R_S = the systemic resistance, P_{sv} = the systemic venous pressure, P_{sa} = the systemic arterial pressure, P_{pv} = the pulmonary venous pressure, P_{pa} = the pulmonary arterial pressure, and Q_f = the flow through the foramen ovale, and Q_p is small through the collapsed lungs. The six distinct flows of the fetal circulation are shown in Figure 12.2.

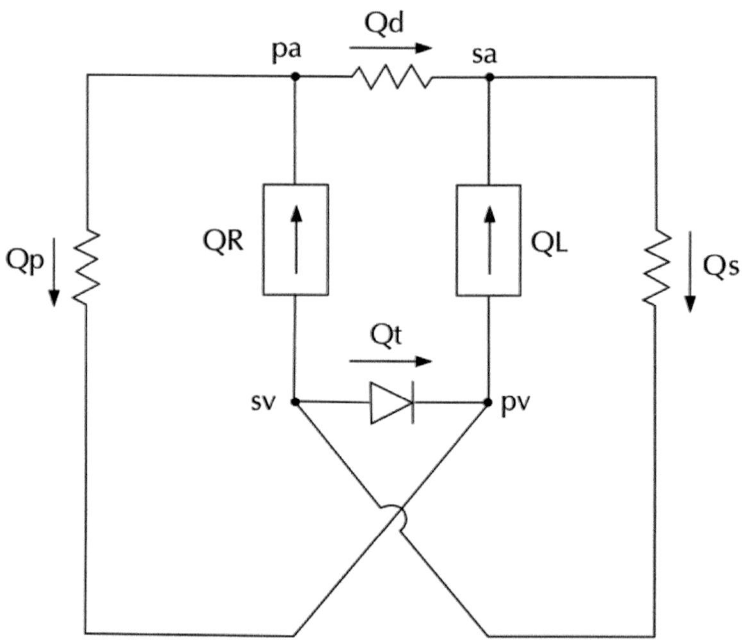

FIGURE 12.2 The fetal circulation composed of six distinct flows for Q_d is greater than zero.

The ductus arteriosus and the foramen ovale play vital roles during the changes in fetal circulation before and after birth. The flow through the ductus is defined as a simple resistance given by eq 12.7. The foramen is modeled as an "ideal valve" using eqs 12.8 and 12.9:

$$\text{Open: } Q_f > 0 \text{ and } P_{sv} = P_{pv} \tag{12.8}$$

$$\text{Closed: } Q_f = 0 \text{ and } P_{sv} <= P_{pv} \tag{12.9}$$

To shunt blood away from the high-resistance fetal lungs, the arteries and atrial shunts work to balance the flow through both sides of the heart. The arterial shunt is the ductus arteriosus, and the atrial shunt is the foramen ovale. Furthermore, $R_p > R_s$ is due to collapsed lungs and blood flow through the umbilical cord. Also, the pump coefficients, K, on both sides of the heart are nearly equal before birth and after birth as shown in eq 12.10. Before birth, the ductus is open and the ductus resistance equals to zero yielding eq 12.11. The foramen ovale is also assumed to be open before birth giving eq 12.12.

$$K_R = K_L = K \tag{12.10}$$

$$P_{sa} = P_{pa} = P_a \tag{12.11}$$

$$P_{sv} = P_{pv} = P_{v} \tag{12.12}$$

Both sides of the heart have common venous and arterial pressures. Combining eqs 12.3–12.12 will yield the eqs 12.13 and 12.14 for systemic flow and pulmonary flow (scheme of placental circulation). The foramen ovale will continue to be open as long as we have $Q_f > 0$ and $R_p > R_s$.

$$Q_s/Q = 1 + (Q_d/Q) = 1 + (R_p - R_s)/(Rp + Rs) = 2 \times R_p/(R_p + R_s) \tag{12.13}$$

$$Q_p/Q = 1 - (Q_d/Q) = 1 - (R_p - R_s)/(R_p + R_s) = 2 \times R_s/(R_p + R_s) \tag{12.14}$$

When the baby takes its first breath, the lungs inflate with air causing the reduction of flow resistance through the lungs. Pulmonary resistance falls and systemic resistance rises abruptly, indicating the effects of entering a gaseous environment, resulting in $R_p < R_s$ and $Q_f < 0$.

Moreover, the closure of the foramen ovale is triggered by heightened pressure in the left atrium due to an escalated blood flow. This pressure surpasses that in the inferior vena cava and right atrium, which results from a decline in pulmonary resistance (R_p). Consequently, a significant volume of blood moves through the lungs into the left atrium, while blood flow to the right atrium diminishes. Additionally, resistance to the output of the left ventricle increases. This shift in the pressure gradient across the atria leads to the sealing of the foramen ovale valve. As for the ductus arteriosus, it typically closes within 2 days after birth.[5,9,11,63]

The analysis of fetal venous system velocity has emerged as a method for assessing the health of an unborn baby. Under normal circumstances, the blood flow within the UV remains consistently smooth. However, cases of fetal distress, impending asphyxia, cardiac anomalies, severe growth restriction, and congestive heart disease have exhibited a pulsating velocity pattern in the UV. This phenomenon was initially interpreted as a negative prognostic indicator. It's worth noting that pulsations in UV velocity are also observed in pregnancies prior to the 13-week mark and even in some healthy late-stage pregnancies. These pulsations, seen both in compromised fetuses and in the early stages of healthy pregnancies, manifest as minor deviations linked to the cardiac cycle.[73]

The DV serves as a fetal link connecting the intra-abdominal section of the UV to the inferior vena cava, enabling the direct introduction of oxygenated blood into the fetal heart. Within this setup, the fetal heart generates pressure and flow waves that move counter to the blood flow in the DV. This interaction results in pulsations within the DV velocity pattern, as depicted in Figure 12.3. The researchers then applied the law of conservation of mass to derive eqs 12.15 and 12.16. These equations were subsequently solved by the researchers using frequency domain techniques.[73] The researchers suggested that UV pulsations occur in healthy fetuses because the umbilical cord was either hypercoiled or compressed.[73]

$$\text{“}d(V)/dt = C(uv)d[p(uv)]/dt = Qb - Qa - Q \quad (12.15)$$

$$Q(t) = Qb - Qa + (\Sigma)\ [Qn \times e^{(\omega)n{*}t}] \quad (12.16)$$

where $(\omega)_n$ = nth harmonic frequency, (Σ) = summation for $n = 1$ to N, V = the UV volume, C = compliance, and Q = flow.

In the UV study, data can be collected using ultrasound and the Doppler effect. When an ultrasound wave is backscattered from a target that is moving relative to the wave source, the reflected wave has a different frequency than the transmitted wave, known as the Doppler effect.

The Doppler frequency shift is given by the eq (12.17). In eq 12.17, the frequency shift (F_d) is equal to the received frequency (F_1) minus the transmitted frequency (Fo); and $[(-2V\cos\alpha/c) \times Fo]$ describes blood cells moving with a velocity V. In these equations, c is the velocity of sound in the respective tissue, and α is the angle between the ultrasound beam and the direction of flow.[65] Since we cannot have direct contact with a fetus, it is used to determine the flow of blood in the fetus.

$$F_d = F_1 - Fo = [(-2V\cos\alpha/c) \times Fo] \quad (12.17)$$

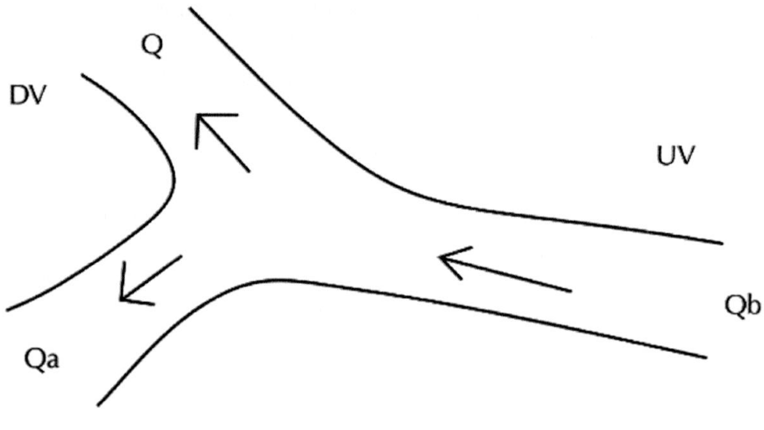

FIGURE 12.3 Direction of flows in the fetal umbilical vein and ductus venosus: Q_b = Flow in the umbilical vein (UV) before the split, Q_a = Flow after the split, Q = Flow in the ductus venosus (DV)

12.10 SATURATION OF HEMOGLOBIN VERSUS PARTIAL PRESSURE OF OXYGEN

A comparative oxygen–dissociation curves for maternal hemoglobin and fetal hemoglobin, demonstrating that the curve for fetal hemoglobin is shifted to the left of that for maternal hemoglobin. This means that at the low P_{O2} levels in fetal blood, the fetal hemoglobin can carry 20–50% more oxygen than the maternal hemoglobin.[22]

The fetal hemoglobin has a higher affinity for oxygen than adult hemoglobin and can carry more oxygen per RBC adult hemoglobin.[23] The oxygen affinity of hemoglobin is lowered by 2,3-bisphosphoglycerate (BPG) available in erythrocytes. BPG binds to adult hemoglobin thus prohibiting oxygen molecules from binding to hemoglobin at certain places. Fetal hemoglobin is structured with γ subunits rather than the β-subunits found in adult hemoglobin. Therefore, it has a lower affinity to BPG and can bind more oxygen molecules.[23] Under equal oxygen pressure conditions, fetal blood can carry a greater amount of oxygen compared to maternal blood, which depicts the relationship between hemoglobin saturation and the partial pressure of oxygen for both fetal and maternal hemoglobin. The transfer of oxygen from the mother to the fetus is influenced by numerous factors, including intervillous blood flow, fetal placental blood flow, oxygen tension in maternal and fetal arterial blood, the oxygen affinity of maternal and fetal blood, hemoglobin concentration in both maternal and fetal blood, maternal and fetal blood pH, carbon dioxide tension (Bohr effect), placental diffusing capacity, placental vascular geometry, maternal-to-fetal blood flow ratio in exchange areas, shunting around exchange sites, and placental oxygen consumption.

Given that the fetus is enveloped by the amniotic cavity filled with fluid, pressures must be considered relative to amniotic fluid pressure. When uterine contractions are absent, this amniotic fluid pressure remains relatively constant. Fetal systemic arterial blood pressure is notably lower than that of the mother, averaging around 55 mmHg (systolic/diastolic approximately 70/45 mmHg). The right ventricular pressure slightly surpasses the left ventricular pressure by 1–2 mmHg. Pulmonary arterial pressure aligns with systemic arterial pressure. The right atrium experiences slightly higher pressure (3 mmHg) compared to the left atrium (2 mmHg), facilitating right-to-left blood flow across the foramen ovale. Systemic venous bed pressure is akin to that of the right atrium, except for the UV, where pressure measures approximately 8–10 mmHg.

Early in gestation, systemic blood pressures are somewhat lower, evident in premature newborns having lower blood pressure than infants. For instance, at 30 weeks' gestation, mean arterial blood pressure is about 35 mmHg. Despite the fetal mean PO_2 being lower, fetal blood possesses the capacity to transport nearly as much oxygen as maternal blood.

Fetal heart rate analysis is crucial in evaluating fetal oxygenation adequacy, making understanding its rate and regulation essential for obstetricians. Prior to labor, the average heart rate in an unmedicated fetus is around 140 beats/min. In earlier stages of pregnancy, this rate is slightly higher but not significantly so. At 20 weeks' gestation, the average fetal heart rate reaches 155 beats/min, decreasing to 144 beats/min by 30 weeks' gestation. Normal fetuses can exhibit variations of up to 20 beats/min above or below these values. Like an adult heart, a fetal heart possesses its own intrinsic pacemaker activity, leading to rhythmic contractions. The sinoatrial (SA) node, located in the right atrial wall, demonstrates the fastest rate of contraction, establishing the normal heart rate. The atrium follows with the next fastest pacemaker rate, and the ventricle exhibits a slower beating rate than the SA node or atrium. In cases of complete or partial heart block in a fetus, rates below the norm can occur. A fetus with complete heart block typically maintains a rate of approximately 60 beats/min.

12.11 PH OF MATERNAL AND FETAL BLOOD

Table 12.8 highlights the distinction between maternal and fetal pH levels. Maternal blood is extracted from the antecubital vein, whereas fetal blood is collected from two different sources: the UV and the umbilical artery. Fetal pH is compared to maternal pH at these sites under normal health conditions. Notably, fetal pH is lower compared to maternal pH when considered as independent entities. During the transfer of oxygen from maternal blood to fetal blood, a notable shift in pH occurs. Specifically, the pH of maternal placental blood decreases, while the pH of fetal placental blood increases. This shift is influenced by the fact that fetal hemoglobin exhibits a higher affinity for oxygen in contrast to maternal hemoglobin. Consequently, fetal hemoglobin must effectively extract the necessary oxygen essential for the fetus's complete development.

Oxygen binding to hemoglobin intensifies as pH rises and is released as pH decreases. As maternal pH decreases, the oxygen–dissociation curve for maternal hemoglobin shifts to the right, facilitating oxygen release. Fetal hemoglobin capitalizes on this scenario by binding oxygen molecules as fetal pH increases, leading to a corresponding rightward shift in the fetal oxygen–dissociation curve.[10,23,56]

TABLE 12.8 pH of Maternal and Fetal Blood.[3]

"Source of blood	pH Normal (range)
Maternal antecubital vein	7.28
Fetal UV	7.32 (7.30–7.34)
Fetal umbilical artery	7.24–7.32)"

Umbilical cord blood gas analysis serves to assess the fetal acid–base status during the delivery process. The placenta facilitates the circulation of umbilical venous blood, characterized by higher pH, PO_2, base excess, and lower PCO_2 in comparison to umbilical arteries. Variations in acid–base parameters acquired from umbilical cord, venous, and arterial blood samples at birth reflect both fetal metabolism and the mingling of venous and arterial blood within the fetal circulation.

Significant discrepancies between arterial and venous umbilical cord pH readings are often attributed to prenatal cardiac failure. Interestingly, research conducted at the University of Connecticut Health Center indicates that alterations in pH are not contingent upon the gestational age.[22,66]

12.12 EXCRETION AND RETENTION OF FLUID IN THE FETAL

Normally, fluid undergoes filtration from the plasma as it traverses the walls of fetal capillaries, entering the interstitial space. This process in the capillaries is balanced by the removal of an equivalent volume of interstitial fluid by the lymphatic system, subsequently rejoining the circulatory system. When the filtration occurring in the capillaries surpasses the fluid removed by the lymphatic system, the interstitial space expands. The fetal interstitial space exhibits a compliance of 45 mL/mmHg/kg of body weight. This elasticity allows for the absorption of substantial fluid quantities into the interstitial space while resulting in only minor increases in interstitial fluid pressure.[13] The rate of fluid filtration across the capillary wall is described by the classical Startling eq 12.18

$$J_v = CFC \times (\Delta P - \sigma \times \Delta\pi) \tag{12.18}$$

where CFC is the capillary filtration coefficient; ΔP and $\Delta\pi$ are the transcapillary hydrostatic and protein osmotic pressure gradients, respectively; σ is the reflection coefficient (a factor that characterizes the effectiveness of the plasma and interstitial proteins at generating osmotic pressures).

The value of "σ" ranges from a value of zero for large pores that do not restrict protein movement to a value of 1 for pores that allow water but inhibit proteins from passing through the pore. The capillary filtration will only change if there is a change in either the filtration coefficient or in the reflection coefficient. No changes will occur due to changes in hydrostatic pressures or protein concentration."

Apart from managing fluid volume, the lymphatic system also plays a role in reintroducing dissolved plasma proteins back into the circulation. The purpose of the lymphatic system is to prevent an excessive buildup of fluid in the tissues by facilitating its return. By bringing back proteins, it diminishes the concentration of proteins within the interstitial space. This mechanism collectively lowers the osmotic pressure caused by proteins in the interstitial space, resulting in a reduction of the volume of fluid that is filtered across capillaries.[13]

The movement of fluids in and out of the fetus is substantial. Research conducted on sheep fetuses has demonstrated that a 3-kg ovine fetus eliminates around 25–30% of its body weight per day (approximately 900 mL/day) through urination in the later stages of gestation. Additionally, it ingests slightly less fluid (around 700 mL/day) through swallowing. During the final third of gestation, the fetal lungs of sheep secrete 10% of the fetus's body weight per day (equivalent to 300 mL/day). Out of this volume, only 1–2 mL/day is necessary to facilitate lung expansion as they develop, while the remaining portion of about 300 mL/day exits the lungs via the trachea. Roughly half of this fluid is swallowed as it exits the trachea, while the remaining portion enters the amniotic fluid.[13]

When considering the fluid dynamics, the combined processes of swallowing and placental filtration contribute approximately 725 mL/day of fluid intake into the fetal body. Conversely, the total fluid loss from the fetal body due to renal, lung, and head secretions amounts to around 1225 mL/day. The difference between these values, which is about 500 mL/day (calculated as 1225–725), corresponds to the volume of fluid that moves directly from the amniotic sac into the fetal blood and crosses the fetal side of the placenta. This phenomenon is referred to as intra-membranous flow. The specifics of this fluid movement vary depending on the species. For instance, in primates, there exists an extensive network of fetal capillaries positioned just beneath the fetal surface of the placenta. This arrangement provides the necessary capillary surface area for the exchange of water and solutes through intra-membranous channels.[13]

The quantity of fluid ingested by the fetus increases as pregnancy advances and is influenced by the availability of fluid. While a decrease in amniotic fluid doesn't impact the frequency of electromechanical events associated with swallowing, it does lead to a reduction in the overall volume swallowed. Fetal swallowing plays a critical role in upholding a normal balance between fetal and amniotic fluid. Fluid found within the fetal lungs originates from blood that traverses lung capillaries and is subsequently either swallowed or reabsorbed from the amniotic sac through intra-membranous routes before being returned to the fetal circulation. The fetal lung fluid introduces solutes and chloride into the amniotic fluid.

Amniotic fluid dynamics rely on an ongoing cycle encompassing fetal kidney, lung, and head secretions. Changes occurring in any of these flows can have repercussions on the amniotic fluid volume. Toward the latter portion of pregnancy, fetal urine significantly contributes to the composition of amniotic fluid. An estimated rate of fetal urine production over the course of pregnancy. In the later stages of gestation, the bladder fills approximately every 20–30 min and is then emptied through active contractions of the bladder wall during urination. Urine is expelled through the urethra and released into the amniotic fluid. During the final stages of pregnancy, fetal urine production averages around 0.2 mL/min/kg of fetal body weight, amounting to about 900 mL/day near term.[13]

The rate of fetal urine production changes with fetal behavioral state, and there is a 24-hour rhythm flow that may be related to maternal eating and drinking. Fetal urine production also varies with the volume of amniotic fluid.[13] Uterine blood flow through the maternal circulation of the placenta also increases, causing an increase in the mother's metabolism and cardiac output. Blood volume also increases during pregnancy due to the increase of aldosterone and estrogen.[22]

The quantity of fluid ingested by the fetus increases as pregnancy advances and is influenced by the availability of fluid. While a decrease in amniotic fluid doesn't impact the frequency of electromechanical events associated with swallowing, it does lead to a reduction in the overall volume swallowed. Fetal swallowing plays a critical role in upholding a normal balance between fetal and amniotic fluid. Fluid found within the fetal lungs originates from blood that traverses lung capillaries and is subsequently either swallowed or reabsorbed from the amniotic sac through intra-membranous routes before being returned to the fetal circulation. The fetal lung fluid introduces solutes and chloride into the amniotic fluid.

Amniotic fluid dynamics rely on an ongoing cycle encompassing fetal kidney, lung, and head secretions. Changes occurring in any of these flows can have repercussions on the amniotic fluid volume. Toward the latter portion of pregnancy, fetal urine significantly contributes to the composition of amniotic fluid. An estimated rate of fetal urine production over the course of pregnancy. In the later stages of gestation, the bladder fills approximately every 20–30 min and is then emptied through active contractions of the bladder wall during urination. Urine is expelled through the urethra and released into the amniotic fluid. During the final stages of pregnancy, fetal urine production averages around 0.2 mL/min/kg of fetal body weight, amounting to about 900 mL/day near term. It is not responsive to changes in respiratory gas tensions. The UBF is determined by eq (12.19):

$$\textit{Uterine blood flow} = [\textit{uterine artery pressure} - \textit{uterine venous pressure}] \div [\textit{uterine vascular resistance}] \quad (12.19)$$

Hence, any factor affecting any of the three values on the right side of this equation will alter UBF. Causes of decreased uterine blood flow are: "Uterine contractions; Hypertonus; Abruptio placentae; Tetanic contraction; Overstimulation with oxytocin; Hypotension; Hypertension; Sympathetic block; Hypovolemic shock; Supine hypotensive syndrome; Preeclampsia; Vasoconstrictors, endogenous and exogenous; Sympathetic discharge; Adrenal medullary activity; Most sympathomimetics (α-adrenergic effects); Exception is ephedrine (primarily β-adrenergic effect); and Strenuous exercise."

Using ultrasonographic methods, the measurement of umbilical blood flow in an undisturbed fetus is approximately 360 mL/min (or 120 mm/min/kg). This value is notably lower than that observed in sheep, where it averages around 200 mm/min per kg. Within the umbilical blood flow, roughly 40% is derived from the combined ventricular output. Of this, about 20% is shunted, meaning it doesn't engage in maternal blood exchange within sheep. This shunted blood is either transported through specific vascular routes within the fetal aspect of the placenta or remains distant from maternal blood to facilitate exchange.

The impact of acute moderate hypoxia doesn't significantly affect umbilical blood flow, but severe hypoxia leads to its reduction. Although the umbilical cord lacks innervation, administration of catecholamines or acute cord occlusion results in decreased umbilical blood flow. In cases of chronic decrease, no known methods exist to substantially increase umbilical flow. However, specific fetal heart rate patterns, such as variable decelerations, have been attributed to temporary umbilical cord compression during labor.

Modifying maternal position, such as adopting a lateral or Trendelenburg posture, and replenishing amniotic fluid using amnioinfusion techniques can sometimes alleviate these patterns, implying relief from cord compression through these interventions.

The total blood volume within the fetus encompasses the blood volumes present in the fetal body, umbilical cord, and the fetal part of the placenta. In a healthy fetus, the circulating blood volume undergoes exponential growth as gestation advances. Nevertheless, the average blood volume in the fetus remains relatively stable at about 113 mL per kg and doesn't show substantial alterations with the progression of gestational age. The fetus possesses the capacity to maintain its blood volume effectively, achieved through both rapid and widespread fluid movements across capillaries. These fluid movements are characterized by their swiftness, as the CFC is approximately five times higher than that observed in adults. Additionally, they can span a broad area due to the interstitial space being about 10 times more compliant in comparison to adults.[13]

12.13 CARBON DIOXIDE AND MAINTENANCE OF ACID–BASE BALANCE

The fetus has different requirements for the maintenance of fluid and acid–base balance than full-grown adults. It obtains water, organic nutrients, electrolytes, and other needs from the mother's bloodstream. About 75% of a newborn's body weight is water. In addition, the fetus spends all nine months of pregnancy under virtual weightless conditions floating in the amniotic fluid.[23]

The metabolism of the fetus, like an adult, results in production of acids.[65] The fetus body breaks down glucose by cellular respiration to produce energy. One of the byproducts of cellular respiration is the carbon dioxide produced in the body and is converted to carbonic acid in blood by an enzyme in erythrocytes by the reaction in eq 12.19.

$$CO_2 + H_2O <-> H_2CO_3 <-> H+ + HCO_3^- \tag{12.19}$$

Carbonic Acid formation is dependent on the partial pressure of carbon dioxide in the blood. If there is an increase in carbon dioxide, the reaction will shift to the left, causing more carbonic acid to be produced.[23] As the fetal blood passes through the placenta, the bicarbonate ion joins with an erythrocyte to gain a hydrogen ion to dissociate into carbon dioxide and water.[65] The carbon dioxide formed then diffuses rapidly through the placenta and into the mother's blood stream.

Carbon dioxide exhibits an even greater ease of passage across the placenta than oxygen. The factors governing oxygen transfer also largely apply to carbon dioxide transport. The rate of blood flow, rather than diffusion resistance, is the limiting factor for carbon dioxide transfer. Under normal undisturbed conditions, the carbon dioxide tension in fetal blood is approximately 40 mmHg. It's noteworthy that maternal arterial carbon dioxide tension is around 34 mmHg, with pregnant women often experiencing a state of compensated respiratory alkalosis. Fetal blood pH remains close to 7.4, and the bicarbonate concentration is like that of maternal blood.

Bicarbonate and fixed acids, in contrast, cross the placenta at a much slower rate than carbon dioxide. This process takes hours rather than seconds for equilibrium. Instances like respiratory acidosis can arise in the fetus when there is an acute compromise in blood flow, whether uterine or umbilical. This is often observed during severe variable decelerations associated with uterine contractions, particularly during the second stage of labor. While these acid–base alterations are typically swiftly rectified following the contraction's cessation and the deceleration, persistent oxygen deficiency can lead to decreased fetal oxygen consumption, redistribution of blood flow, and a reliance on anaerobic metabolism to meet energy demands, albeit with reduced efficiency. This anaerobic metabolism results in the production of lactate, contributing to metabolic acidosis. Furthermore, combined respiratory acidosis can occur due to carbon dioxide retention. Unlike carbon dioxide, lactate elimination from the fetus is gradual.

Substantial quantities of fetal-produced carbon dioxide can be effectively eliminated through maternal respiration, even at a regular rate. Fetal metabolism can generate various types of acids in two principal ways: First, the consumption of nonsulfur containing amino acids leads to uric acid formation, and second, incomplete carbohydrate and fatty acid combustion yields lactic and keto acids. These acids are expelled through the placental diffusion route and maternal kidneys.[65]

Maintaining the amniotic fluid within a normal range is crucial for proper fetal development. The average amniotic fluid volume progresses from about 20 mL at 10 weeks of gestation to approximately 800 mL at 32–34 weeks. Subsequently, minimal changes occur until the approach of labor, at which point the volume starts to decrease. It's worth noting that there exists a broad span of amniotic fluid volume deemed healthy for fetuses. For instance, the maximum value encompasses a range of 300–2000 mL within a 95% confidence interval at 32 weeks of gestation.[13]

12.14 TRANSPORT OF BIOFLUIDS BETWEEN THE MOTHER AND FETUS[1,12,20,63,69,74–80]

All fetal fluids are eventually derived from maternal fluids. There are two routes for fluid entry into the fetal compartment: (a) Transfer from the uterine wall across the chorion and amnion and into the amniotic fluid, with subsequent absorption by the fetus; (b) Transfer from the maternal to fetal circulation across the placenta. The mechanism by which the substances are transferred across the placenta is simple diffusion, an exchange of molecules across a membrane due to random molecular movements with a net transfer from the highest to the lowest concentration.[3] The factors that influence the diffusion rate are described by Fick's law in eq 12.20.

$$\text{“Diffusion rate} = [(D \times A)\,(C_1 - C_2)]/t \qquad (12.20)$$

where D = diffusion coefficient; A = area of exchange; $C_1 - C_2$ = concentration gradient across the membrane, and t = membrane thickness. The diffusion coefficient is a factor that describes the ease with which a particular substance crosses a particular membrane."[57,81] "Table 12.9 indicates the factors that affect placental transfer. The same factors that influence the delivery of the placenta on the maternal side are operative in removing materials into the fetal circulation.[57] Amniotic fluid volume in normal fetuses is a function of gestational age showing a regression line at a 95% population confidence interval.[13]"

TABLE 12.9 Factors Affecting the Placental Transfer of a Diffusible Substance.[13,17,57,81]

Maternal	Placental	Fetal
1. The amount delivered to the placenta: a. Blood concentration; exogenous and endogenous supplies; homeostatic mechanisms. b. Flow-rate in intervillous space; Hemodynamic factors in mother; local circulatory factors. 2. "Shunting" and arteriovenous mixing in intervillous space.	1. Area of the diffusing membrane. 2. Diffusion pressure: Difference in concentration in intervillous space and in villous capillary. 3. Diffusion resistance: a. Characteristics of transferred. material: size, charge, polarity, etc. b. Characteristics of membrane: physicochemical composition; thickness.	1. Blood concentration. 2. Hemodynamic factors: systemic and local. 3. Flow characteristics: pool, cross-current, etc."

12.15 ABNORMALITIES VERSUS ALTERNATIONS IN BIOFLUID CONCENTRATIONS

Fluctuations in fluid levels within both the maternal and fetal environments can give rise to various conditions and health issues that affect the normal progression of fetal development and the overall well-being of the mother. These irregularities are particularly evident in the composition of amniotic fluid and the blood of both the mother and the fetus. Medical practitioners commonly evaluate the depth of amniotic fluid across four divisions within the uterus and aggregate these measurements to calculate the amniotic fluid index (AFI). AFI measurements below 5 cm are indicative of oligohydramnios in pregnant women, while a cumulative fluid level surpassing 25 cm indicates the presence of polyhydramnios.[19]

A study performed with 2081 high-risk pregnancy human patients showed that 72 (3.5%) of the patients displayed evidence of polyhydramnios when using the AFI to assess the amniotic fluid. It was observed that there was an increase in fetal macrosomia, premature births, prenatal morbidity, and fetal anomalies.[83] The chronic hypertension in the maternal blood leads to a preeclampsia. A study of pregnant chronic hypertensive patients at "King–Drew Medical Center (KDMC)" reported that there was a decrease in the birth weight of newborns with chronic hypertensive mothers.[55]

12.16 ADVANCES IN FETAL BIOFLUID MECHANICS

There have been numerous advances in the biomechanics of fetal biofluids. The umbilical cord is being stored for stem research.[54] A notable advancement in medical imaging is the utilization of "MRI." This technology employs the principles of nuclear magnetic resonance to create detailed images within the body. Furthermore, the incorporation of cutting-edge technology has significantly improved the clarity and precision of these images. MRI offers exceptional differentiation between various soft tissues.[20] Ultrasound represents an acoustic wave that is propagated through a medium, generating a mechanical disturbance within that medium. By employing Doppler's principle, it can capture real-time images and offer insights into blood flow dynamics. This technology is extensively utilized to generate high-quality images of the fetus and to assess blood circulation.

An ultrasound image of a fetal heart is displayed, revealing the discernible partitions between the heart's chambers. As technological advancements continue, the resolution of these images becomes notably enhanced.[82]

Doppler velocity recordings are employed to visualize blood flow, enabling the presentation of three-dimensional images. These imaging methods offer significant utility in observing the fetus, given its complete confinement within the mother's body. The progress made in understanding fetal fluid dynamics owes much to advancements in imaging technologies. Presently, our knowledge about the functionalities of the fetal system has expanded because of enhanced imaging capabilities.

12.16.1 NIPPLE ASPIRATE FLUID (NAF)

12.16.1.1 BREAST ANATOMY AND NAF COLLECTION

The period from the ninth week of development to delivery marks the fetal phase, during which the male and female gonads undergo separation. In comparison to the embryonic stage, the fetus 's circulatory system becomes more efficient and specialized. It features three shunts—the DV, foramen ovale, and ductus arteriosus—that enable it to bypass the liver and pulmonary circuit until after birth. During this phase, the brain continues to develop, the body elongates, the skeletal structure solidifies, and facial characteristics undergo changes. The developing fetus engages in movements, blinking, practicing sucking, and circulating amniotic fluid within the womb.

The fetus is reliant on maternal organ systems for its growth and development, with the placenta and amniotic fluid playing crucial roles. The placenta serves as a specialized organ connecting the fetus to the mother, responsible for transporting essential nutrients, oxygen, and other substances crucial for fetal growth. In male fetuses, the bipotential gonads develop into the testes and accompanying epididymis, while the Müllerian ducts regress. The Wolffian ducts form the vas deferens, and the cloaca gives rise to the urethra and rectum. In female fetuses, the bipotential gonads transform into ovaries, the Wolffian ducts degenerate, and the Müllerian ducts form the uterus and uterine tubes. The cloaca undergoes division and development to create the vagina, urethra, and rectum.

Amniotic fluid offers protection to the fetus, acts as a reservoir for swallowing, and provides space for movement. Despite the fetal dependency on the maternal system, there is no direct connection between maternal and fetal blood vessels. Consequently, the circulatory systems of the mother and fetus remain distinct, and their blood does not mix. The ongoing exchange of various fluids within the maternal–fetal system is crucial for promoting the healthy growth and survival of the fetus.

12.16.2 CLINICAL UTILITY OF RISK ASSESSMENT BREAST DISEASES

NAF examination offers a valuable means of detecting benign breast conditions and the associated increase in relative risk, even before any abnormalities become detectable through imaging or palpation. Dr. George Papanicolaou, renowned for developing the "pap smear" for cervical cancer, highlighted that nonpregnant, nonlactating adult women secrete fluid into their breast milk ducts. This fluid remains trapped due to the closure of nipple orifices caused by smooth muscle contraction, dried secretions, and keratinized epithelium. Within this fluid, a variety of cell types are present, including normal, atypical, and even malignant breast epithelial cells. It also contains molecular biomarkers for diagnostic purposes, such as associated proteins, complex lipids, RNA, and DNA.

The utilization of NAF for identifying women at high risk of breast cancer is not a new concept. In 1958, Dr. George Papanicolaou introduced a method of extracting fluid from breast milk ducts through suction to analyze cell samples, like his technique for detecting abnormal cells in cervical samples (the "pap test"). This approach demonstrated the potential to identify abnormal cells within NAF from the breast ducts. Numerous clinical studies spanning up to 25 years have consistently shown that benign breast diseases, particularly those with atypia, significantly elevate the risk of developing breast cancer. On average, these studies indicate that the presence of atypia indicates a more than fourfold (400%) increase in relative risk of breast cancer compared to women who do not produce this fluid.

12.17 SUMMARY

The period from the ninth week of development to delivery marks the fetal phase, during which the male and female gonads undergo separation. In comparison to the embryonic stage, the fetus's circulatory system becomes more efficient and specialized. It features three shunts—the DV, foramen ovale, and ductus arteriosus—that enable it to bypass the liver and pulmonary circuit until after birth. During this phase, the brain continues to develop, the body elongates, the skeletal structure solidifies, and facial characteristics undergo changes. The developing fetus engages in movements, blinking, practicing sucking, and circulating amniotic fluid within the womb.

The fetus is reliant on maternal organ systems for its growth and development, with the placenta and amniotic fluid playing crucial roles. The placenta serves as a specialized organ connecting the fetus to the mother, responsible for transporting essential nutrients, oxygen, and other substances crucial for fetal growth. In male fetuses, the bipotential gonads develop into the testes and accompanying epididymis, while the Müllerian ducts regress. The Wolffian ducts form the vas deferens, and the cloaca gives rise to the urethra and rectum. In female fetuses, the bipotential gonads transform into ovaries, the Wolffian ducts degenerate, and the Müllerian ducts form the uterus and uterine tubes. The cloaca undergoes division and development to create the vagina, urethra, and rectum.

Amniotic fluid offers protection to the fetus, acts as a reservoir for swallowing, and provides space for movement. Despite the fetal dependency on the maternal system, there is no direct connection between maternal and fetal blood vessels. Consequently, the circulatory systems of the mother and fetus remain distinct, and their blood does not mix. The ongoing exchange of various fluids within the maternal–fetal system is crucial for promoting the healthy growth and survival of the fetus.

KEYWORDS

- **amnion**
- **amniotic fluid**
- **amniotic fluid index**
- **blood**
- **buffer**
- **chorion**
- **ductus arteriosus**

REFERENCES

1. Powls, R. *The Human Body and Why It Works*; Prentice-Hall Inc., 1985.
2. Aaronson, P. I. *The Cardiovascular System at a Glance*; Blackwell Science, 1999.
3. Altman, P. *Blood and Other Body Fluids*; Federation of American Societies for Experimental Biology, 1961.

4. Beckmann, C. R.; Herbert, W.; Laube, D.; Ling, F.; Smith, R. *Obstetrics and gynecology*; Lippincott Williams & Wilkins, 2013.
5. Berne, R. M. Cardiovascular Physiology. *Ann. Rev. Physiol.* **1981**, *43* (1), 357–358.
6. Fung, Y.-C. *Biomechanics: Circulation*; Springer-Verlag, 2013.
7. Greenfield, A. D. The Circulation Through the Skin. In *Handbook of Physiology: Circulation*, Vol. 2; American Physiology Society, 1963; pp 1325–1351.
8. Wood, N. Aspects of Fluid Dynamics Applied to the Larger Arteries. *J. Theor. Biol.* **1999**, *199* (2), 137–161.
9. Hoppensteadt, F. C.; Peskin, C. S. *Modeling and Simulation in Medicine and the Life Sciences*; Springer Science & Business Media, 2012.
10. Nelson, D. L.; Lehninger, A. L.; Cox, M. M. *Lehninger Principles of Biochemistry*; Macmillan, 2008.
11. Reddy, J. N.; Gartling, D. K. *The Finite Element Method in Heat Transfer and Fluid Dynamics*; CRC Press, 2010.
12. Truskey, G. A.; Yuan, F.; Katz, D. F. *Transport Phenomena in Biological Systems*; Prentice Hall, 2009.
13. Harding, R.; Bocking, A. D. *Fetal Growth and Development*; Cambridge University Press, 2001.
14. CHKD. *Blood Circulation in the Fetus and Newborn* [Online], 2023. Children's Hospital of The King's Daughters. https://www.chkd.org/ (accessed Apr 4, 2023).
15. de Kroon, R. R.; de Baat, T.; Senger, S.; Van Weissenbruch, M. M. Amniotic Fluid: A Perspective on Promising Advances in the Prevention and Treatment of Necrotizing Enterocolitis. *Front. Pediatr.* **2022**, *10*.
16. GYNOB. *Amniotic Fluid: What is this Stuff, Anyway?* [Online], 2023. https://www.gynob.com/amniotic.htm (accessed Apr 4, 2023).
17. GYNOB. *The Placenta* [Online], 2023. https://www.gynob.com/placenta.htm (accessed Apr 4, 2023).
18. MedlinePlus. *Amniotic fluid* [Online], 2023. https://medlineplus.gov/ency/article/002220.htm (accessed Apr 4, 2023).
19. BDF. *Birth Defects* [Online], 2023. https://www.marchofdimes.org (accessed Apr 4, 2023).
20. Enderle, J.; Bronzino, J. *Introduction to Biomedical Engineering*; Academic Press, 2012.
21. Paladine, H. L.; Desai, U. A. Vaginitis: Diagnosis and Treatment. *Am. Fam. Physician* **2018**, *97* (5), 321–329.
22. Hall, J. E.; Hall, M. E. *Guyton and Hall Textbook of Medical Physiology*; Pearson & Prentice Hall, 2020.
23. Martini, F. H. *Fundamentals of Anatomy and Physiology*; Prentice Hall, 2001.
24. WOM. *Renal Functions* [Online], 2023. https://umanitoba.ca/medicine/medicine/department-physiology-pathophysiology (accessed Apr 4, 2023).
25. Jou, D.; Llebot, J. E.; Eds. *Introduction to the Thermodynamics of Biological Processes*; Prentice Hall, 1990.
26. MountCastle, V. B. *Medical Physiology*; C. V. Mosby Company, 1974.
27. Silverthorn, D. U. *Human Physiology*; Jones & Bartlett Publishers, 2015.
28. Tortora, G. J.; Grabowski, R. R. *Principles of Anatomy and Physiology*; Harper Collins, 1996.
29. Sciencia58. *Schematic representation of a sagittal cross section through the human female genital and pelvic organs.* https://commons.wikimedia.org/wiki/File:Female_reproductive_system_lateral_1.png, 2023. (accessed 12/10/2023).

30. Sciencia58. *Schematic representation of a sagittal cross section through the male pelvic and genital organs in humans.* https://commons.wikimedia.org/wiki/File:Male_anatomy_1.png, 2023. (accessed 12/10/2023).
31. Quizlet. *The Menstrual Cycle* [Online], 2023. https://quizlet.com/gb/323119080/the-menstrual-cycle-diagram/ (accessed Apr 4, 2023).
32. SMCH. *Anatomy: Fetus in Utero* [Online], 2023. Standford Medicine Children Health. https://www.stanfordchildrens.org/en/topic/default?id=anatomy-fetus-in-utero-85-P01189 (accessed Apr 4, 2023).
33. Wallen, K.; Lloyd, E. A. Female Sexual Arousal: Genital Anatomy and Orgasm in Intercourse. *Horm. Behav.* **2011**, *59* (5), 780–792.
34. Wikipedia. *Pre-ejaculate* [Online], 2023. https://en.wikipedia.org/wiki/Pre-ejaculate (accessed Apr 4, 2023).
35. Terzo, L. *The Fertility and Cervical Fluid Connection* [Online], 2023. https://www.drlaurieterzo.com/dr-laurie-terzo-blog/2019/7/30/9dic6x3s6r5kbbbsqdjt2o909fewfu (accessed Apr 4, 2023).
36. Gest, T. *Placenta and Extraembryonic Membranes* [Online], 1999. https://www.med.umich.edu/lrc/coursepages/m1/embryology/embryo/06placenta.htm (accessed Apr 4, 2023).
37. EndoText. *Endocrinology of Pregnancy* [Online], 2023. https://www.ncbi.nlm.nih.gov/books/NBK278962/figure/endocrin-preg.F4/ (accessed Apr 4, 2023).
38. Bonde, J. P. E.; Ernst, E.; Jensen, T. K.; Hjollund, N. H. I.; Kolstad, H.; Scheike, T.; Giwercman, A.; Skakkebæk, N. E.; Henriksen, T. B.; Olsen, J. Relation Between Semen Quality and Fertility: A Population-Based Study of 430 First-Pregnancy Planners. *Lancet* **1998**, *352* (9135), 1172–1177.
39. Dening, S. Sex in Ancient Civilizations. In *The Mythology of Sex*; Macmillan, 1996.
40. Gallup G. G. Jr.; Burch, R. L. Semen Displacement as a Sperm Competition Strategy in Humans. *Evol. Psychol.* **2004**, *2* (1), 147470490400200105.
41. Harvey, C. Relation Between the Volume and Fructose Content of Human Semen. *Nature* **1948**, *162* (4125), 812–813.
42. Wikipedia. *Simple Description of a Human Spermatozoon with Special Details of the Head* [Online], 2023. https://en.wikipedia.org/wiki/Sperm (accessed Apr 4, 2023).
43. ndubuisi, v. *Male Sperm Poem* [Online], 2023. https://www.poemhunter.com/poem/male-sperm/. (accessed Apr 4, 2023).
44. Mann, T. *The Biochemistry of Semen*; John Wiley & Sons, 1954.
45. Owen, D. H.; Katz, D. F. A Review of the Physical and Chemical Properties of Human Semen and the Formulation of a Semen Simulant. *J. Androl.* **2005**, *26* (4), 459–469.
46. Poiani, A. Complexity of Seminal Fluid: A Review. *Behav. Ecol. Sociobiol.* **2006**, *60*, 289–310.
47. Rosenthal, S. *The Gynecological Sourcebook*; McGraw-Hill Professional, 2003.
48. Salam, A. P.; Horby, P. W. The Breadth of Viruses in Human Semen. *Emerg. Infect. Dis.* **2017**, *23* (11), 1922.
49. Shanis, B.; Check, J.; Baker, A. Transmission of Sexually Transmitted Diseases by Donor Semen. *Arch. Androl.* **1989**, *23* (3), 249–257.
50. Skofronich, J. G.; Cameron, J. R.; Grant, R. M. *Physics of the Body*, 2nd ed.; Medical Physics Publishing, 1999.
51. WHO. *Laboratory Manual for the Examination of Human Semen and Semen–Cervical Mucus Interaction*; Cambridge University Press, 2023.

52. Wikipedia. *Umbilical Cord* [Online], 2023. https://en.wikipedia.org/wiki/Umbilical_cord (accessed Apr 4, 2023).
53. Wikipedia. *Vaginal Discharge* [Online], 2023. https://en.wikipedia.org/wiki/Vaginal_discharge (accessed Apr 4, 2023).
54. Watson, P. The Potential Impact of Sperm Encapsulation Technology on the Importance of Timing of Artificial Insemination: A Perspective in the Light of Published Work. *Reprod. Fertil. Dev.* **1993**, *5* (6), 691–699.
55. Hon, E. H.; Fukushima, T.; Park, J. M.; Davidson, E. C.; Huerta, E. Cutaneous Pressure Pulse Contours Associated with Chronic Hypertension in Pregnancy. *J. Matern.-Fetal Med.* **1995**, *4* (2), 71–80.
56. Ohning, B. L.; Aslam, M.; Bissinger, R. L. *Neonatal Resuscitation* [Online], 2021. Medscape. https://emedicine.medscape.com/article/977002-overview (accessed Apr 4, 2023).
57. Baker, R. R.; Bellis, M. A. Human Sperm Competition: Ejaculate Adjustment by Males and the Function of Masturbation. *Anim. Behav.* **1993**, *46* (5), 861–885.
58. Hacker, N. F.; Gambone, J. C.; Hobel, C. J. *Hacker & Moore's Essentials of Obstetrics and Gynecology*; Elsevier Health Sciences, 2015.
59. Rice, A.; ElWerdany, M.; Hadoura, E.; Mahmood, T. Vaginal Discharge. *Obstet. Gynaecol. Reprod. Med.* **2016**, *26* (11), 317–323.
60. Wathne, B.; Holst, E.; Hovelius, B.; Mårdh, P.-A. Vaginal Discharge-Comparison of Clinical, Laboratory and Microbiological Findings. *Acta Obstet. Gynecol. Scand.* **1994**, *73* (10), 802–808.
61. CC. *Mucus Plug* [Online], 2023. https://my.clevelandclinic.org/health/symptoms/21606-mucus-plug (accessed Apr 4, 2023).
62. Kjeldsberg, C. R.; Joseph, A. K. *Body Fluids*; American Society of Clinical Pathologists, 1986.
63. Goyal, M. R. Biofluid Dynamics of Maternal Fetal System. In *Biofluid Mechanics of Human Body Systems*; CRC and Apple Academic Press, 2013; pp 361–412.
64. Openstax. *The fetal circulatory system includes three shunts to divert blood from undeveloped and partially functioning organs, as well as blood supply to and from the placenta.* https://commons.wikimedia.org/wiki/File:2916_Fetal_Circulatory_System-02.jpg, 2013. (accessed 03/10/2023).
65. Resnik, R.; Creasy, R. K.; Iams, J. D.; Lockwood, C. J.; Moore, T.; Greene, M. F. *Creasy and Resnik's Maternal-Fetal Medicine: Principles and Practice*; Elsevier Health Sciences, 2008.
66. Egan, J. F.; Vintzileos, A. M.; Campbell, W. A.; Rodis, J. F.; McLean, D. A.; Fleming, A. D.; Fleming, A. D. Arteriovenous Cord Blood pH Discordancy in a High-Risk Population and Its Clinical Significance. *J. Matern.-Fetal Med.* **1992**, *1* (1), 39–44.
67. Kiserud, T.; Acharya, G. The Fetal Circulation. *Prenat. Diagn.* **2004**, *24* (13), 1049–1059.
68. Latrobe. *Role of Viscosity* [Online], 2023. https://www.latrobe.edu.au/ (accessed Apr 4, 2023).
69. Bronzino, J. D.; Peterson, D. R.; Eds. *The Biomedical Engineering Handbook*, Vols. 1–4; CRC Press LLC (Taylor & Francis Group): Boca Raton, FL, 2015; p 5422.
70. OBGYNKey. *Amniotic Fluid Volume* [Online], 2023. Obgyn Key. https://obgynkey.com/amniotic-fluid-volume/ (accessed Apr 4, 2023).
71. BrainKart. *The Fetal Circulation* [Online], 2023. https://www.brainkart.com/article/The-Fetal-Circulation_25509/ (accessed Apr 4, 2023).
72. Parer, J. T. Fetal Circulation. *The Global Library of Women's Medicine*; 2008. DOI: 10.3843/GLOWM.10194.

73. Hellevik, L.; Stergiopulos, N.; Kiserud, T.; Rabben, S.; Eik-Nes, S.; Irgens, F. A Mathematical Model of Umbilical Venous Pulsation. *J. Biomech.* **2000**, *33* (9), 1123–1130.
74. Cooney, D. *Biomedical Engineering Principles*; Marcel Dekker, 1976.
75. Fournier, R. L. *Basic Transport Phenomena in Biomedical Engineering*; CRC Press, 2017.
76. Geankoplis, C. J. *Transport Processes: Momentum, Heat and Mass*. Allyn and Bacon Series in Engineering, 1983; p 538.
77. Lightfoot, E. N. *Transport Phenomena and Living Systems*; John Wiley & Sons Inc, 1974.
78. Roselli, R. J.; Diller, K. R. Biofluid Transport. In *Biotransport: Principles & Applications*, 1st ed.; Springer, 2011; pp 105–485.
79. Seagrave, R. C. Biomedical Applications of Heat and Mass Transfer, 1st ed.; The Iowa State University Press, 1971.
80. Skalak, R.; Chein, S. *Handbook of Bioengineering*; McGraw Hill Book Co, 1986.
81. Seashore, M. R. Human Growth: Vol. 1, Principles and Prenatal Growth; Vol. 2, Postnatal Growth. *Yale J. Biol. Med.* **1979**, *52* (4), 416.
82. BMUS. *Fetal Heart: Scenes from the History of Ultrasound* [Online], 2023. The British Medical Ultrasound Society. https://www.bmus.org/. (accessed Apr 4, 2023).
83. Phelan, J.; Park, Y.; Ahn, M.; Rutherford, S. Polyhydramnios and Perinatal Outcome. *J. Perinatol.* **1990**, *10* (4), 347–350.
84. Anim, J. T. Breast. In *Pathology of Diseases*; Jaypee Digital, 2000; p 24.
85. Mancall, E. L.; Brock, D. G. *Gray's Clinical Neuroanatomy: The Anatomic Basis for Clinical Neuroscience*; Elsevier Health Sciences, 2011.
86. Chigama, B. E. *Letter To My Unborn Child* [Online], 2023. https://www.poetryinternational.com/en/poets-poems/poems/poem/103-27203_LETTER-TO-MY-UNBORN-CHILD (accessed Apr 4, 2023).
87. O'Donohue, J. *For A Mother-To-Be: Poetry For Pregnancy and Motherhood* [Online], 2023. www.yogaformodernlife.com/blog/poetry-for-pregnancy(accessed Apr 4, 2023).

APPENDIX I

A Simple Pictorial Description of a Human Spermatozoon with Special Details of Head

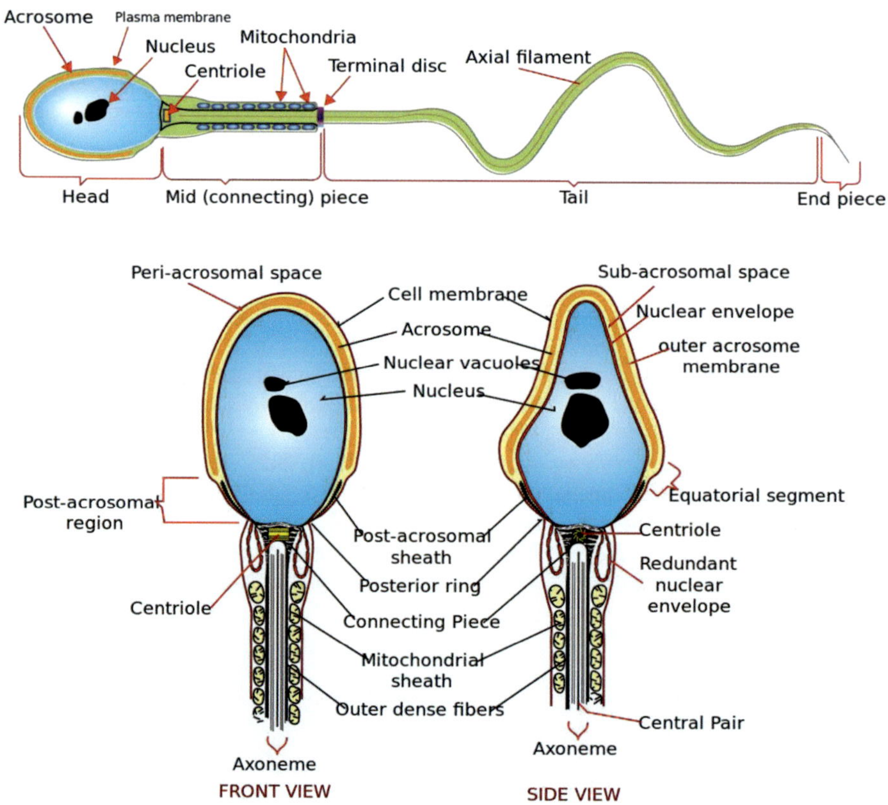

FIGURE 12.4 Simple description of a human spermatozoon with special details of the head.
Source: Public domain.

APPENDIX II

Male Sperm Poem by Poet Virginus Ndubuisi[43]

"Male sperm is like,
Super glue,
It holds both jeans and,
Slices.
It is exorbitant and,
Hard to see,
But is an odorant,
Object that attract the,
Environment.
It is like a magnet,
It magnets both iron,
And metals.

It scares both the adult,
And teenage girls away,
Like masquerade with a Cain.

Oh! Yee,
Adama masquerade,
Tell the Izaga masquerade that his,
Sperm is dangerous,
Tell the Otangburugbu masquerade,
That his sperm is bad,
But,
Tell the Ijele masquerade
That his sperm is,
Good.
Tell also the
Egwumawu teams that,
Their sperms are bad,

And forget not to,
Intimate the,
Adama masquerade that she,
Needs more sperm."

APPENDIX III

Letter to My Unborn Child[86]

"Child:

when I get the courage
when I am brave, strong and ready
to bring you into a world with too many
clenched fists, too many open palms
too many potholes, too many ZESAless*
nights, too many frolicking leaders
I want to make sure
you are brave, strong and ready too
for the furrowed ride
for now I will keep you safe
wrapped in my scared womb
only when I am ready
will I birth you freely
like the way this poetry flows
from my tongue
When you get here
I mean if you get here
I imagine the joy of your small
cute smiles lighting up my day
your tiny hands curling
into little adorable fists
incapable of hurting any one
then slowly opening like
the petals of a blooming rose
I imagine in your sleep
you will dream of happy things
I will watch you smile, stretch your tiny
limbs
saying 'when I grow up I wanna be . . .
better than mummy'
but child you are not to say that aloud
especially with me around
like the words flowing from my tongue
You will wear many names of your choice
I want to give you that freedom from birth
to make your own choices beginning with
your name
without me breathing down my failed
dreams
on you

Child:

I want you to be proud in your skin
so comfortable no one can convince you
otherwise
be weary of brain-pickers I would say
those who will pick on your brains with
sjamboks
like they did on the backs of grandma in the
cotton fields
you will be gifted with brawn but child that
does not mean
you are to be a slave
and when you are old like these locks
tying my world together, at thirteen
I want your world to be open
to limitless possibility
and I want you to be brave
just like me when I will you into this world
to labour for your own happiness
to strive to cut the fences, prejudices
around the skin you will unashamedly be
proud of
child I seek you to find all-weather wings
I seek to find a heart as warm
I want you to find love
give love, above all, I want you to be you
for now I will keep you safe
wrapped in my scared womb
safe from the stale promise of democracy
safe from misfired bullets of hatred
pelting my ears from the radio relentlessly
safe from circus governments of disunity
pawning our rights and freedoms
like zhing-zhong products at the flea market
only when I am ready
only when I am ready
will I birth you freely
like the way this poetry flows from my
tongue

© 2011,"

APPENDIX IV

For A Mother-To-Be (Adapted from O'Donohue, 2023)[87]

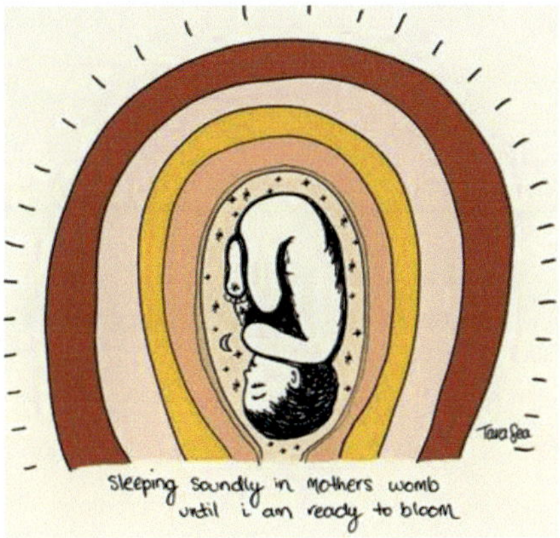

"Nothing could have prepared your heart to
open like this.

From beyond the skies and the stars
This echo arrived inside of you and started to
pulse with life
Each beat a tiny act of growth,
Traversing all our ancient shapes,
On its way home to itself.

Once it began, you were no longer your own.
A new, more courageous you, offering itself
In a new way to a presence you can sense
But you have not seen or known.

It has made you feel alone
In a way you never knew before;
Everyone else sees only from the outside
What you feel and feed
With every fibre of your being.

Never have you travelled farther inward
Where words and thoughts become half-light
unable to reach the fund of brightness
Strengthening inside the night of your womb.

Like some primeval moon,
Your soul brightens
The tides of essence
That flow to your child.

You know your life has changed forever,
For in all the days and years to come,
Distance will never be able to cut you off
From the one you now carry
For nine months under your heart.

May you be blessed with quiet confidence
That destiny will guide you and mind you.

May the emerging spirit of your child
Imbibe encouragement and joy
From the continuous music of your heart,
So that it can grow with ease,
Respectant of wonder and welcome when its form is fully filled

And it makes it journey out
To see you and settle at last
Relieved and glad in your arms."

CHAPTER 13

Kinetics of Drug Transport in the Human Body Systems*

MEGH R. GOYAL, MAHWISH AHMED, NELLY ARBOLEDA, JOSE CESAR, and RAFAEL OLIVER-VIDAUD

13.1 INTRODUCTION

Drug delivery encompasses research into the development of novel technologies and delivery systems that are able to overcome biological barriers. Using biology to target or trigger drug delivery certainly opens new possibilities in drug treatment. The fundamental knowledge allows the design of delivery systems for improved drug delivery and enables access to drugs targeted at various body systems. Additionally, models have been established that allow reliable correlations of in vitro work with in vivo studies to ensure the successful transfer from the laboratory to human.

Several ways are used to effectively deliver the drugs in the human body system. The conventional methods in use today include transdermal drug delivery, oral delivery, intranasal, and ocular delivery. One of the requirements for these conventional methods is to deposit the drug at the target site by overcoming the degradative processes of our body. Also, the drug must be in uniform concentration when delivered to the desired area.

However, it is difficult for the drug to reach the desired therapeutic blood levels in a short period. Some of the delivery systems also suffer from

*This chapter has been modified from the review article that was prepared by students [Mahwish Ahmed, Nelly Arboleda, Jose Cesar, and Rafael Oliver-Vidaud] for the course on Biofluid Mechanics BME—4994, Course Instructor: Megh R. Goyal, PhD, P.E., Visiting Professor in Biomedical Engineering (2003), Biomedical Engineering Department, Florida International University (FIU), 10555 West Flagler Street, Miami, FL 33174, USA. For details, contact at goyalmegh@gmail.com. We acknowledge the contribution by Dr Anthony McGoron and the faculty in Biomedical Engineering at FIU.

Biofluid Dynamics of Human Body Systems: Expanded and Revised Edition.
Megh R. Goyal, Arka Bhowmik, & Anamika Chauhan (Eds.)

serious side effects, and the requirement of increased dosage to achieve the successful results. On the other hand, there are alternative methods, such as implants that require less volume of the drug with lower side effects, and easy delivery and patient comfort. An extensive amount of research is being conducted to find new methods of successful drug delivery in every area of the health industry. The research focuses on eliminating or at least minimizing the side effects of powerful drugs. Also, these advances are still developing unique therapeutic advantages and easy delivery for the patients. **Caution:** The information in this chapter does not replace a visit to a doctor's office. One must consult the physician or a specialist.

This chapter is focused on the functions, physical and chemical properties,[1] composition of biofluids, and fluid balance. The development of a new drug involves more than the synthesis of a substance that has a particular effect on the body.

13.2 PHYSIOLOGY OF THE HUMAN LYMPHATIC SYSTEM

The cells, tissues, and organs of the lymphatic system[2–8] play a central role in the body's defenses against a variety of pathogens, or disease-causing organisms. Lymphocytes provide an immune response to specific threats to the body. Immunity is the ability to resist infection and disease through activation of specific defenses.

The lymphatic system includes a network of lymphatic vessels. A series of lymphoid organs are connected to the lymphatic vessels. The lymphatic system produces, maintains, and distributes lymphocytes. This system also helps to maintain blood volume and eliminate local variations in the composition of the interstitial fluid. Lymph flows along a network of lymphatic vessels, the smallest of which are the lymphatic capillaries. The lymphatic vessels empty into the thoracic duct and the right lymphatic duct. The three classes of lymphocytes are: T cells (thymus-dependent), B cells (bone marrow derived), and NK cells (natural killer).

Lymphoid tissues are connective tissues dominated by lymphocytes. In a lymphoid nodule, the lymphocytes are densely packed into an area of loose connective tissue. Important lymph organs include the lymph node, the thymus, and the spleen. Lymphoid tissues and organs are distributed in areas especially vulnerable to injury or invasion. These defenses are either nonspecific defenses, which protect without distinguishing one threat from another, or specific defenses, which only protect against particular threats.[9] Table 13.1 shows the chemical composition of the lymph.

13.3 FUNCTIONS OF BIOFLUIDS

One of the most vital and abundant fluids in the human body is water. Body water is responsible for the distribution of various nutrients, waste products, and gases for human cells to be functional (see Chapter 1 in this book). In our body, most of the water lies in the intracellular compartment of the body. The extracellular fluid is divided into three compartments: the interstitial fluid, plasma, and lymph. The cells are surrounded by this fluid, which acts as a carrier of nutrients and waste materials. If the water content of the body were to change, cellular activities would be jeopardized.[2]

Lymphatic fluid (see Chapter 7 in this book) is colorless and opaque. It travels through the lymphatic vessel and connects the lymph nodes, which are lymphatic organs. The main functions of lymph are: (a) the return of protein and fluid from the ISF to the circulation to maintain a low interstitial fluid protein concentration; and (b) maintaining the oncotic pressure gradient across the capillary membrane. For instance, edema will occur if ISF oncotic pressure is not kept low. Lymph plays a role in the absorption and transport of fat from the small intestine; immunological role; lymph glands and circulation of immune cells such as lymphocytes and dendritic cells; and removal of bacteria.

TABLE 13.1 Chemical Composition of Lymph[1,15] (Range of Values is Shown in the Brackets).

Constituent	No. of tests	Units	Thoracic duct lymph	Blood plasma or serum
		Electrolytes		
Calcium	5	mEq/L	4.2 (3.4–5.6)	5 (4.3–5.9)
	1	mEq/L	4.4	4.8
Chloride	5	mEq/L	98 (87–103)	96 (94–98)
	1	mEq/L	97	98
Phosphorus	5	mEq/L	2.4 (2–3.6)	2.5 (2.1–3.5)
Inorganic P	1	mg/100 mL	2.7	2.9
Potassium	5	mEq/L	4.7 (3.9–5.6)	5(4.1–5.9)
	1	mEq/L	3.3	4.7
Sodium	5	mEq/L	(118–132)	(113–135)
	1	mEq/L	138	142
		Nitrogenous substances		
Protein, total	5	g/100 mL	4.8 (2.9–7.3)	7.08 (5.4–9.4)
	1	g/100 mL	(2.8–3.6)	6.0
Albumin	5	g/100 mL	2.3 (1.5–2.7)	2.8 (2–3.5)
	1	g/100 mL	(1.6–2.5)	3.5

TABLE 13.1 *(Continued)*

Constituent	No. of tests	Units	Thoracic duct lymph	Blood plasma or serum
Globulin	5	g/100 mL	2.6 (1.5–4.8)	4.2 (3.1–7)
α_1	1	g/100 mL	(1.2–1.2)	2.5
α_2	1	g/100 mL	0.26	0.34
β_1	1	g/100 mL	0.25	0.62
β_2	1	g/100 mL	0.25	0.39
γ	1	g/100 mL	0.15	0.29
	1	g/100 mL	0.78	1.46
Bilirubin	1	mg/100 mL	0.8	1
Creatinine	4	mg/100 mL	3 (0.8–8.9)	3.1 (0.8–9)
Uric acid	5	mg/100 mL	5 (1.7–10.8)	5.1 (1.6–10.9)
	1	mg/100 mL	4.1	4.2
Nitrogen:	5	mg/100 mL	47 (13–139)	49 (15.8–141)
nonprotein	1	mg/100 mL	23	29
		Lipids and carbohydrates		
Cholesterol:				
Total	5	mg/100 mL	72 (34–106)	119 (83–167)
Free	1	mg/100 mL	60	207
	5	mg/100 mL	34 (15–51)	39 (28–56)
	1	mg/100 mL	36	57
Fat, total	1	g/100 mL	1.1	–
Glucose	1	mg/100 mL	136	117
		Enzymes		
Aldolase	1	Units/mL	10	6
Phosphatase:				
Acid	1	Units/mL	0.7	0.3
prostatic				0
Alkaline	1	Units/mL	1.9	2.3
Transaminase	1	Units/mL	12	13
GOT				

The blood provides food and oxygen to the human body's organs and tissues. It is often referred to as a "river of life" because it collectively flows through the blood vessels, pumping blood from the heart. It is made up of blood cells, such as[13] RBC, WBC, platelets, and fragments of cells.

13.4 COMPOSITION OF THE BIOFLUIDS

To maintain a healthy body, both the volume of the fluid and the composition of the fluids should remain constant. Blood is made up of erythrocytes in plasma, which account for almost half the volume of the blood (see Chapter 1 in this book). Even though white cells and platelets are part of blood, their proportion is smaller to affect any flow properties. Plasma is yellow in color because there are proteins dissolved in the watery liquid. Plasma is made up both of water and salts. In addition, plasma contains vitamins, nutrients, minerals, and waste products. Blood also contains gases, such as carbon dioxide and water.[14]

The composition of blood is similar to that of extracellular fluid, except that it contains cells, albumin, and other fractions. Any abnormalities in these fractions can cause infections or diseases. Albumin is fractioned from plasma and helps in eliminating any traces of the hepatitis virus in the whole blood, and often used for the treatment of hypokalemia. Globulin is made up of three types: alpha, beta, and gamma. Gamma globulin is elevated in liver disease, chronic infections, and collagen infections. However, a decrease in gamma globulin is observed in patients suffering from malnutrition. This particular globulin is very beneficial in treating certain infections because it provides protection against them.

RBCs cause certain diseases if the quantity is altered. Anemia is caused by a lack of red blood cells. It results from abnormalities in the bone marrow. In addition, in hemoglobin, the respiratory protein is at low levels. When the bone marrow quantity is increased, hemolytic jaundice results. Fibrinogen forms blood clots that seal leaks after an injury.[2]

Lymph closely resembles plasma. However, it contains very low concentrations of "suspended proteins," water, and electrolytes (see Table 13.1). The lymphatic route has proven to be beneficial for drug delivery and the dissemination of deadly cancers. Thus, a thorough knowledge of the lymphatic system is necessary to understand the targeted drug delivery.[14]

13.5 PHYSICAL PROPERTIES OF BIOFLUIDS[2, 4,15–19]

In an average adult, body water accounts for 60% in a male and 50% in a female. In both genders, it makes up 99% of the fluid volume outside the cells. The intracellular fluid compartment (IFC) contains two-thirds of the total body water, and the extracellular fluid compartment comprises one-thirds of water, and this volume must be maintained for survival. If any alterations occur with the volume of water in the intracellular compartment, an individual's health is in danger.[2,14,16,17] The physical properties of blood are different for each type of blood cell. For the whole blood, the pH varies for each age group.[19,20] From 2–3 years of age, arterial blood has a pH of 7.35. From 3–18 years of age, it varies from 7.38–7.40 depending on the individuals; the pH of the venous blood is about 7.31. For a person more than 70 years old, the pH ranges from 6.2 to 8.5.

The relative viscosity of blood is approximately 4.71 at 20°C and 3.00 at 37°C. The specific gravity at 4°C is about 1.058 for an adult and 1.0638 for an infant. The blood has a specific heat of 0.96 g-cal and a surface tension of approximately 55.5–61.2 dynes/cm. At 4°C, the erythrocyte value of the blood is 1.09 and the specific heat is 0.77 g-cal.[2,14,16,17]

The pH for erythrocytes is about 7.4. The plasma blood has a pH of approximately 7.30–7.40. Compared to both the whole blood and the erythrocyte, the specific gravity is low and is about 1.0273. However, the specific heat of the plasma is higher compared to the whole blood and the erythrocyte and is about 0.94 g-cal. The surface tension of the plasma for a child is 52 dynes/cm. The relative viscosity of the plasma is approximately 1.86 at 13–22°C, 1.89 at 14–23°C, and 1.32 at 37°C.[15,21,22]

For a white adult male at 22–28°C: plasma volume is 41.1 mL/kg of body weight, erythrocyte volume is 28.0 mL/kg body weight, and whole blood volume is 69.1 mL/kg body weight. For a white adult female at 22–28°C: plasma volume is 43.1 mL/kg body weight, erythrocyte volume is 23.4 mL/kg body weight, and whole blood volume is 66.5 mL/kg body weight.[15,21,22]

13.6 MECHANISMS OF DRUG DELIVERY[9,18,23–37]

Drugs can enter the body in two possible ways: (a) Through the enteral route, which refers to drugs that are given via the gastrointestinal (GI) tract. The enteral route concerns absorption via one or more of the following components of the GI tract: the buccal cavity, sublingually, gastrically, intestinally, and rectally. Once the drug is absorbed from the GI tract, it enters the bloodstream and is further distributed throughout the body. However, it is important to notice that in some cases (i.e., drugs absorbed by the stomach, intestines, colon, and upper rectum), the drug does not enter the circulatory system directly. Instead, it enters the liver first before entering the circulatory system. This is important since drugs entering this way suffer extensive degradation due to enzymes present in the liver. (b) The second route is called the parenteral route, which is the name given to all other routes (Table 13.2).

As seen, the ways a drug enters the body are varied, and depending on them, the concentration of the drug with respect to time will vary. For instance, an intravenous injection will show a peak in concentration almost immediately,

TABLE 13.2 Parenteral Routes of Drug Administration.[18]

Route	Results
Inhalation	Mouth/nose, pharynx, trachea, bronchi, bronchioles, alveolar sacs, and alveoli
Intra-arterial	Introduced directly into the artery, regional drug delivery
Intradermal	Within the dermis usually a local effect
Intramuscular	Within the muscle
Intraperitoneal	Within the peritoneal cavity
Intrathecal	Cerebrospinal fluid and subarachnoid space
Intravenous (IV)	Introduced directly into the venous circulation
Percutaneous	Topical treatment applied to the skin
Subcutaneous	Beneath the epidermal and dermal skin layers
Vaginal	Within the vagina

while a pill might take longer to show a peak in plasma concentration as it has to be absorbed first. Moreover, one of the most interesting systems is controlled-release drug delivery, in which the drug reaches a steady-state plasma concentration after a short period of time. Therefore, the method of drug administration will depend on the type of action that is desired from the drug under consideration.

Most controlled-release systems fall into one of three categories: diffusion, solvent-activated release, and polymeric degradation (Figures 13.1A–C). The diffusion and solvent-activated systems will be explained briefly in this chapter, and the polymeric degradation systems will be explained in more detail due to their present importance in current research.

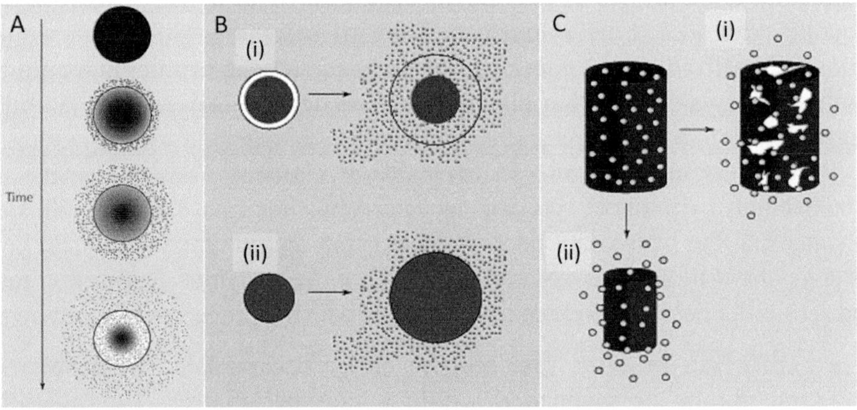

FIGURE 13.1 (A) Drug delivery from a typical matrix drug delivery system, (B) solvent-activated systems of drug delivery, and (C) drug delivery from (a) bulk-eroding and (b) surface-eroding biodegradable systems.

Polymeric degradation is important to observe that one of the most remarkable and useful features of a polymer's swelling ability manifests itself when that swelling can be triggered by a change in the environment surrounding the delivery system. Depending on the polymer, the environmental change can involve pH, temperature, or ionic strength, and the system can either shrink or swell upon a change in any of these environmental factors.

For most of the polymers used for this purpose, the structural changes are reversible and repeatable with additional changes in the external environment. Furthermore, polymeric degradation (biodegradation) is one of the most researched methods of drug release. In this case, as with the diffusion method, the drug is contained within a polymer membrane or matrix. The polymer is designed to degrade and release the drug at a specific location in the body. As the polymer degrades, the drug is freed and made available to the body. Degradation occurs via three major mechanisms:

- **Mechanism A:** Water-soluble polymers are made insoluble by cross-linking them together. When the cross-links are broken at some point in the body, the polymer will dissolve.
- **Mechanism B:** Water-insoluble polymers are made soluble by hydrolysis or ionization of side groups.
- **Mechanism C:** Insoluble polymers are cleaved into soluble monomers.

Many degradation systems use a combination of these mechanisms. Most matrix degradation systems currently in use are known as **bulk degradation.** This means that degradation occurs throughout the polymer structure in a rather random fashion. The problem with bulk degradation is that the erosion rate is dependent on the volume of the matrix rather than its thickness. Furthermore, the rate of release is rather unpredictable, and entire dose dumping can often occur. These problems would be eliminated by a system that displays surface degradation rather than bulk degradation.

Researchers are developing such polymer systems. These polymers are usually highly hydrophobic yet contain water-labile linkages. Thus, the diffusion of water into the matrix and internal degradation are minimized. In these new systems, the drug release rate is proportional to the polymer degradation rate, and zero-order degradation kinetics is possible with the proper surface geometry.

- **Diffusion system:** The drug is either encapsulated in a polymer membrane or suspended within a polymer matrix. Typically, the course of events is as follows: water diffuses into the membrane or matrix, the drug dissolves, and finally the dissolved drug diffuses out of the polymer. In a membrane system, the diffusion of water through the polymer is the rate-determining step. Furthermore, the rate of release is constant and proportional to the initial concentration of the drug. In a matrix system, however, the rate of release depends on the amount of drug present at a particular time; therefore, the rate of release is time-dependent.

- **Solvent-activated systems:** The most common mechanism employs a semipermeable membrane (reservoir) containing a small, laser-drilled hole(s). Within the membrane, there is a high concentration of an osmotic agent, either the drug itself or a salt, which causes water to enter through the membrane. The drug is then forced out through the hole because of the increased pressure. Drug release proceeds at a constant rate in solvent-activated systems.

The question arises as to how the degradation of the polymer may be induced in specific parts of the body. As mentioned above, the location of the release depends on the nature and purpose of the drug. A common application of these principles is transdermal drug delivery, which is the transport of drugs through the skin. There are two kinds of devices used for this purpose, the semisolids and the patches. However, the therapeutic effects of semisolids are inferior to those of patches. Furthermore, there are two types of transdermal patches: the reservoir type and the matrix type. The reservoir type uses a rate-limiting membrane to control the rate of drug delivery to the skin, whereas the matrix type uses the stratum corneum (one of the skin layers) as a rate-limiting membrane to control the rate of drug delivery.

One of the mechanisms for transdermal drug delivery is electroporation, that is, the creation of hydrophilic pores by increasing the transmembrane voltage. The creation process involves a transition from spontaneously occurring hydrophobic pores to hydrophilic pores. A hydrophilic pore is believed to dominate most electroporation phenomena. Furthermore, it can be modeled as a leaky, microscopic capacitor whose parallel resistance for small radii is extremely large.[16]

13.7 KINETICS OF DRUG DELIVERY[13,17,18,22,26,38–49]

Mathematical models can describe the transport of drugs after the controlled delivery. These models can predict (a) the depth or distance of the drug and (b) the local concentration of the drug as a function of time and location.[16] To make a mathematical analysis of drug delivery systems, it is necessary to consider factors such as[18]: (1) blood perfusion rate; (2) capillary permeability; (3) drug biological affinity; (4) metabolism of drugs; and (5) renal excretion.

It is important to consider that most analytical methods measure the total drug concentration (bound and unbound to proteins) in the plasma. The total plasma concentration (C_{total}, mg/L) can be easily measured.

In eq 13.1 for the apparent distribution volume ($V_{apparent}$ in L), A_{body} is the total amount of drug in the body in milligrams. However, eq 13.1 may give values different from the *true distribution volume* (V) due to protein binding. To analyze the effects of such protein binding, more accurate models are required.[16]

In eq 13.2 for total plasma concentration, C is the concentration of the true-distributed drug and C_{BP} is the concentration of the drug bound to plasma proteins. C_{BP} and C_{BT} (concentration of drug bound to tissue protein) are related to the total plasma concentration by the binding fractions x and y, where x is defined as the fraction of the total drug that is bound to plasma protein and y is the fraction of the total drug in the tissue region that is bound to tissue proteins.[18]

The law of conservation of mass yields eq 13.3. Solving eq 13.3 for C and letting $C_{BP}/C = x/(1 - x)$, $C_{BT}/C = y/(1 - y)$, and $V_{tissue} = (V - V_{plasma})$, yields eq 13.4. Combining eqs 13.1 and 13.4 will yield eq 13.5. Note that for $x = y = 0$ in eq 13.5, the apparent distribution volume equals the true distribution volume. After the drug has been administered, it begins to be absorbed by the body. The major site of drug metabolism is in the liver [24: 120]. The drug breaks down and is carried by several enzymes that follow the Michaelis–Menten rate model (eq 13.6). When the drug concentration is much smaller than K_m, eq 13.6 reduces to eq 13.7.

Once the drug has been degraded, some of it will be eliminated by the kidneys. The elimination of drugs by the kidneys is called renal clearance, which is a volume of plasma that is totally cleared of the drug per unit time. A simple model of renal clearance (CL_{renal}) is given by an unsteady mass balance of the drug in the body's apparent distribution volume (eq 13.8).

Equation 13.9 is obtained by integrating eq 13.8. If Q_{urine} is the volumetric urine formation rate and C_{urine} is the concentration of the drug in the urine, then another mass balance (eq 13.11) is obtained. Furthermore, plasma clearance (CL_{plasma}) is the sum of all drug elimination processes in the body. It includes metabolism, kidneys, and other secondary processes (such as sweating). Each process has its own clearance value and rate constant, which are given by eq 13.12. Therefore, the plasma clearance is given by the total first-order elimination rate constant, k_{te} (eq 13.13).

Now, taking into consideration all elimination processes, the change in drug concentration with time[18] is given by eq 13.14. Consequently, the half-life (which is the time taken for the drug concentration to be reduced by one half) is given in eq 13.15.

$$V_{\text{apparent}} = A_{\text{body}}/C_{\text{total}} \tag{13.1}$$

$$C_{\text{total}} = C + C_{\text{BP}} \tag{13.2}$$

$$A_{\text{body}} = VC + C_{\text{BP}}V_{\text{plasma}} + C_{\text{BT}}V_{\text{tissue}} \tag{13.3}$$

$$C = A_{\text{body}}/\,[V + x/(1-x)V_p + y/(1-y)\,(V - V_p)] \tag{13.4}$$

$$V_{\text{apparent}} = (1-x)V + xV_{\text{plasma}} + y[(1-x)/(1-y)]^{*}\,(V - V_{\text{plasma}}) \tag{13.5}$$

$$r_{\text{metabolic}} = V_{\text{max}}C/(K_m + C) \tag{13.6}$$

$$r_{\text{metabolic}} = k_{\text{metabolic}}C, \tag{13.7}$$

where $k_{\text{metabolic}}$ is the first-order rate constant for the metabolic reactions.

$$V_{\text{apparent}}\,(dC/dt) = -\,CL_{\text{renal}}\,{}^{*}C \tag{13.8}$$

$$C(t) = C_0 e^{-(\text{CLrenal/Vapparent})t} = C_0 e^{-\text{Krenal}^{*}\text{t}}, \tag{13.9}$$

where K_{renal} is the apparent distribution volume given by eq 13.10 and CL_{renal} is renal clearance.

$$K_{\text{renal}} = CL_{\text{renal}}/V_{\text{apparent}} \tag{13.10}$$

$$CL_{\text{renal}}{}^{*}C = Q_{\text{urine}}{}^{*}C_{\text{urine}} \tag{13.11}$$

$$k_i = CL_i/V_{\text{apparent}}, \tag{13.12}$$

where i = renal, metabolic, sweat, bile, respiration, or feces.

$$k_{\text{te}} = \sum k_i = (1/V_{\text{apparent}})\sum CL_i = [CL_{\text{plasma}}/V_{\text{apparent}}] \tag{13.13}$$

$$C(t) = C_0 e^{-\text{ktet}} \tag{13.14}$$

$$t_{1/2} = [0.693/k_{\text{te}}] = [0.693V_{\text{apparent}}/CL_{\text{plasma}}] \tag{13.15}$$

13.7.1 SINGLE COMPARTMENT MODEL

An intravenous injection of drug can be modeled with eqs 13.16–13.18, where: C_0 is the initial drug concentration; D is the dose; $AUC^{0\rightarrow\infty}$ is the area under the curve C versus time. Equation 13.17 is integrated from 0 to ∞.

Using the equation for change in concentration with respect to time yields eq 13.18 for $AUC^{0\rightarrow\infty}$.

$$C_0 = D/V_{\text{apparent}} \tag{13.16}$$

$$AUC^{0\rightarrow\infty} = \int C(t)\, dt \tag{13.17}$$

$$AUC^{0\rightarrow\infty} = C_0/k_{te} = D/(V_{\text{apparent}})(k_{te})` = D/(CL_{\text{plasma}}) \tag{13.18}$$

The accumulation of drug in the urine is described in eq 13.19, where M_{urine} is the mass of drug (unchanged) in the urine. By substituting eq 13.14 into eq 13.19 and integrating, we get eq 13.20. For $t\rightarrow\infty$, eq 13.20 shows the total amount of drug collected in the urine.

Equations 13.22–13.24 describe the controlled-release drug delivery system for a single compartment model. For I_0 = drug-infusion rate (mg/min), the unsteady mass balance is given in eq 13.22. For no initial drug present, integrating eq 13.22 yields eq 13.23. For long periods of time, eq 13.23 reduces to a steady-state plasma drug concentration, as shown in eq 13.24.

$$(dM_{\text{urine}}/dt) = k_{\text{renal}} V_{\text{apparent}} C \tag{13.19}$$

$$M_{\text{urine}}(t) = [(k_{\text{renal}} C_0 V_{\text{apparent}})/k_{te}]\,(1 - e^{-\text{ktet}}) \tag{13.20}$$

$$M_{\text{urine}} = k_{\text{renal}} C_0 V_{\text{apparent}}/k_{\text{te}} = D(k_{\text{renal}}/k_{\text{te}}) = AUC^{0\rightarrow\infty}\, CL_{\text{renal}} \tag{13.21}$$

$$V_{\text{apparent}}(dC/dt) = I_0 - Cl_{\text{plasma}} C = I_0 - V_{\text{apparent}} k_{\text{te}} C \tag{13.22}$$

$$C(t) = [(I_0/(k_{\text{te}} V_{\text{apparent}})]\,(1 - e^{-\text{ktet}}) \tag{13.23}$$

$$C_{ss} = I_0/(k_{\text{te}} V_{\text{apparent}}) = I_0/CL_{\text{plasma}} \tag{13.24}$$

13.7.2 QUANTITATIVE MODEL FOR ELECTROPORATION OF ARTIFICIAL PLANAR BILAYER MEMBRANES

The free energy of pore formation is shown in eq 13.25. The change in specific capacitance is given by eq 13.26. Equation 13.27 is for a quantitative description of the critical pore radius at maximum (W_p). The potential drop near a pore's mouth is due to a spreading resistance, R_{spd}, within the bathing electrolyte[16,18] that is given in eq 13.28. Equation 13.29 is for the internal pore resistance. Equation 13.30 is obtained by applying the voltage division principle to both resistances (eqs 13.28 and 13.29).

The flux of pores in pore radius space is described by eq 13.31, and the equation for the rupture is shown in eq 13.32. In eqs 13.25–13.34, U_m is the spatially averaged transmembrane voltage; r_p is the radius of the pore; $C_0 = \epsilon_m/h_m$ is the capacitance per area of a pore-free lipid bilayer of thickness, h_m; $\epsilon_w = K_w\epsilon_0$ = the permittivity of water and $\epsilon_m = K_l\epsilon_0$ = the permittivity of water lipid, respectively (we used $K_l = 2$ and $K_w = 80$); σ_e is the bulk electrolyte conductivity; n is the pore distribution function; A_m is the planar membrane area; and c_p is the number of pores per area.

Equation 13.33 is a circuit differential equation and describes the behavior of the interaction of the membrane and its pore population with its environment during a rectangular pulse. When the pulse stops, the corresponding eq 13.34 is obtained by simplifying eq 13.33.

$$W_p(r, U_m) = 2\pi\gamma r_p - \pi\Gamma r_p^2 - 0.5C_pU_m^2r_p^2 \tag{13.25}$$

$$C_{lw} = C_0\,[(\epsilon_w/\epsilon_m) - 1] \tag{13.26}$$

$$R_{p,c} = \gamma/(\Gamma + 0.5C_{lw}U_m^2) \text{ and } \Delta W_{p,max} = \pi\gamma^2/(\Gamma + 0.5C_{lw}U_m^2) \tag{13.27}$$

$$R_{spd} = 1/\,(2\sigma_e r_p) \tag{13.28}$$

$$R_{int} = h_m/(\pi r_p^2\sigma_p) \tag{13.29}$$

$$U_{m,p} = U_m\,[R_{int}/(R_{int} + R_{spd})] \tag{13.30}$$

$$J_p = -D_p\,[(\delta n/\delta r_p) + (n/kT)\,(\delta\Delta W_p/\delta r_p)] \tag{13.31}$$

$$\tau = [(kT)^{3/2}/(4\pi c_pA_mD_p\gamma(\Gamma + 0.5C_{lw}U_m^2)]^* \\ \exp\,[\pi\gamma^2/(kT(\Gamma + 0.5C_{lw}U_m^2))] \tag{13.32}$$

$$C_m(dU_m/dt) = I_pR_N/(R_E + R_N) - U_m[G_m(t) + 1/(R_E + R_N)], \\ \text{for } 0<t<t_{pulse} \tag{13.33}$$

$$C_m(dU_m/dt) = -(U_m/R_m)\ for\ t>t_{pulse} \tag{13.34}$$

13.8 TRANSDERMAL DRUG DELIVERY

Transdermal drug delivery is the delivery of drugs through the skin. This method is superior compared to traditional drug delivery systems. After its application, it keeps the drug concentration constant in tissues and in blood for a long period of time. It is less likely to be delivered as an overdose or underdose, termination is easily possible, and the patient can easily avoid intravenous therapy. The patches, a form of transdermal drug delivery, are more convenient and painless than injections. These patches achieve "pseudo-zero-order release kinetics" through the rate-controlling membranes.[13,22,39]

However, there is a major obstacle in transdermal drug delivery. One of the most important functions of the skin is to prevent any foreign or potent substances from entering the body and stop any component loss from the inside of the body. To analyze the problem a drug encounters from the skin to reach the circulatory system, it is necessary to know the physiology of the skin.

13.8.1 PHYSIOLOGY OF THE SKIN

The skin consists of three layers (see Figure 13.2):[3–7,20] (a) The **stratum corneum** is the outermost layer and the barrier and is about 10–50 μm thick. It is made up of dead cells that are closely packed together. Both the intercellular phase and the intracellular phase consist of proteins. However, lipids are only present in the intracellular phase. Also, there are sweat glands and hair follicles in the stratum corneum. (b) The epidermis is a layer comprising rapidly dividing basal cells. (c) The dermis is the innermost and noncellular layer.

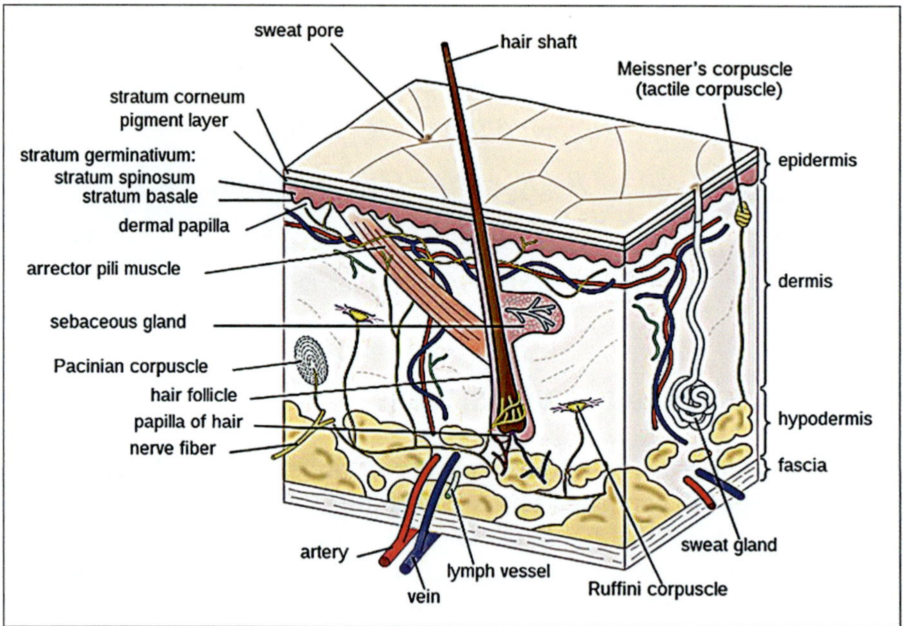

FIGURE 13.2 Layers of the skin.

Source: Reprinted from Ref. [51] https://creativecommons.org/licenses/by-sa/4.0/deed.en)

Even though extensive research has been conducted on transdermal drug delivery, progress to bring those products into use is very slow because researchers cannot find many drugs that would overcome the low permeability of the skin. Also, one has to consider that skin is extremely sensitive to any forced drug or skin permeation enhancers and can result in skin irritations that can also slow the efforts of the researchers in developing new transdermal drug delivery systems.[50]

13.8.2 EXAMPLES OF TRANSDERMAL DELIVERY

Some of the applications today and the products for future use in transdermal delivery are hormones, cardiovascular drugs, analgesics, antihistamines, and central nervous system drugs. The female reproductive hormones (such as estradiol and progesterone) can be delivered transdermally.

To treat hypertension and angina, the switch to transdermal delivery from ocular delivery seems beneficial in reducing blood pressure. Chronic pains can be treated transdermally by transferring analgesics. When antihistamines are delivered transdermally, the common side effects (such as drowsiness) can be reduced; however, these can cause skin irritations or hypersensitization.

There are some drugs that diffuse through almost impermeable layers of the skin because of their low molecular weight and are effective in small doses (0.1–15 mg per day). The transdermal patches are used to deliver nicotine, estrogen, scopolamine, fentanyl, and nitroglycerine. For easy transdermal delivery to the bloodstream, some properties of a drug are: low molecular weight; soluble both in oil and water; and effective enough at a low dosage to diffuse through difficult areas of the skin. There are other drugs that can be forced to diffuse through the skin, but only with the help of additional enhancers. Some of the current enhancers in transdermal delivery involve electricity for iontophoresis, ultrasound for phono- or sonophoresis, magnetism, and heat.[14]

Experiments are being conducted to deliver large molecules (such as poly-1-lysines) through the skin, enhanced by ultrasound at 20 kHz frequency. The experiment by researchers at Ultra-Sonic Technologies, LLC, showed that high molecular weight compounds were transported only when the ultrasound was turned on, suggesting that the use of the ultrasound at the set frequency was the only way to diffuse the molecule across the skin.

Another experiment was carried out by the same researchers to examine the structural effects on the skin due to the transportation of microparticles, varying from 1–25 μm in size. As a result, the 20 kHz ultrasound created micropores in the stratum corneum that helped in delivering both larger and tiny molecules transdermally.[18]

A simpler method is to use heat as an enhancer for transdermal delivery. It is effective because it increases "skin permeability, body fluid circulation, blood vessel wall permeability, rate-limiting membrane permeability, and drug solubility." Heat increases the kinetic energy of the drug molecules and any other biomolecules located in the cell membrane and helps in the absorption of the drug (see Chapter 5 in this book). An example to elaborate on the importance of heat as an enhancer is the increase in skin permeability of lipophilic drugs due to temperature changes, even 5°C, which change the lipid structure.[2,16,18] The effectiveness of heat as an enhancer can also be demonstrated by discussing the situation of the cancer patients taking morphine and fentanyl. Over an extended period of time, they need an increased dosage at times of unbearable pain. Thus, heat can be applied to the patch that decreases the time for the drug to reach the blood as skin permeability is increased, skin blood flow is accelerated, and the drug is released faster from its reservoir, providing a proficient delivery. All of these mechanisms contribute highly to making heat-induced enhancers one of the most effective ways of transdermal delivery of drugs, especially high molecular weight drugs, to the bloodstream.[18]

13.9 ORAL DRUG DELIVERY

Oral drug delivery is an easy, simple, painless, and cost-effective way of administering drugs. However, this method may upset the stomach and cause metabolism-related problems associated with taking medicines. In addition, some drugs are too large to be delivered by mouth or contain an electrical charge that inhibits them from passing through the small intestine in the body and eventually to the circulatory system. Another problem with oral-administered drugs is that acids and enzymes in the body may decompose them before they can reach the target. An economic problem associated with oral drug delivery is that some large molecules have high production costs, and as a consequence, their delivery is also high. Also, there are variations in the dosage amount for each patient of the same drug.

Around the 1960s, the concept of controlled drug delivery was formed. It focused on drug safety and a longer lifetime of the drug in the system. Currently, this system produces both controlled and sustained drug release.

In oral drug delivery, the polymers are released by the regulation of diffusion, degradation, and swelling. This system is beneficial to the drug industry today because it reduces toxicity due to low variations in serum concentrations and lowers the amount and frequency of doses administered. However, the drawback to this system is that it takes more time to achieve the desired therapeutic blood concentrations and is an expensive system.

Research is being conducted to improve the widely used drugs' ability to deliver increased amounts over longer periods. Acetaminophen, a widely used pain reliever, is found in Tylenol, Midol, Benadryl, Excedrin, and Lortab. A research group at Stanford analyzed the changes brought by the extended-release version of Tylenol. For everyday use, Tylenol is known to have different pills in different dosages. Besides the regular-strength tablet, the extra-strength contains 500 mg of acetaminophen. There is an extended-release version with an increase in the drug amount to 650 mg, from which 325 mg dissolves rapidly in the body, and the other 325 mg continues to be delivered over a period of 8 h. Some methods have been proposed to enhance this oral delivery: (a) Firstly, there can be multiple layers of drugs in a pill, which can increase the time period of drug delivery. (b) Secondly, the drug can be placed in water as tiny granules or in a capsule, and there would be a "constant rate" maintained for the delivery. The possible method of delivering the drug in liquid can be beneficial compared to swallowing large tablets. Thus, by analyzing the proposed methods, acetaminophen can be delivered for an extended period of time and in a greater amount.

13.10 INTERNASAL DRUG DELIVERY

Since ancient times until today, in Eastern India, in the form of the Ayurvedic system, nasal drug delivery has been recognized. In the last decade, since proteins and peptides are available with advanced biotechnology, there has been a tremendous amount of research conducted on this particular delivery system. However, there are some restrictions on this delivery because not all drugs can be administered nasally. The chemical properties of a particular drug need to be analyzed before it is delivered through the nasal mucosa. Some drugs require additional permeation enhancers, which carry their own side effects.

Compared to orally delivered drugs, certain drugs that can be delivered through the nose have benefits such as lower dose requirements, minimal side effects, and the achievement of the desired therapeutic blood levels. It should be considered that these benefits are not applicable to all of the drugs delivered nasally, and sometimes none of these benefits are shown by a certain drug. However, the advantage of easy and painless delivery is notable enough for further research on this system.

Some conditions must be considered for nasal delivery, such as: it should be delivered "parenterally," in an unconventional form, absorbed poorly or slowly, and produce side effects when delivered through the mouth.

13.10.1 EXAMPLES OF INTRANASAL DRUG DELIVERY SYSTEM

A total of 93 patients with nasal surgery were selected and given Tobispray, a nasal aerosol that controlled the bleeding caused during the surgery. The treatment was successful at a rate of approximately 95%. Xylometazoline, a nasal spray, is widely used to relieve congestion due to coryza. There is still some research being conducted to further evaluate this spray to relieve congestion associated with respiratory infections and coryza. Another example is atropine sulfate to treat rhinorrhea.[54] This nasally delivered drug produced no side effects, such as dry mouth or visual disturbance, commonly associated with oral administration. To treat bronchoconstrictive disease, Ipratropium, in aerosol packaging, is successfully used. In addition to drugs, there are devices being used for intranasal delivery. These are the mechanical pump and the pressurized aerosol system. Both are widely used, but the mechanical pump is easier to use than the pressurized aerosol system.

13.11 OCULAR DRUG DELIVERY

Ocular drug delivery is the application of solutions, ointments, and suspensions to the cornea of the eye or the space between the eyeball and the eyelid. As soon as the drops are applied to the cornea, they are quickly washed away when they are thinned with tears. This is the reason why the drops are frequently applied, as there is a brief contact between the drops and the eye. In comparison, suspensions can stay in the eye for a long time but can cause irritation, tears, and the loss of the applied dose. The most easily stored and remaining for a longer time are ointments, but they can blur the vision of the patient and can attach to the epithelial cells, causing interference with the usual base.

A drug is administered ocularly by a "pulse entry" and later by a drop in drug concentration. To achieve the best results, the eye drops may either be applied to a greater pulse, or applied repeatedly at certain time intervals. Because ocular delivery often requires multiple applications, there are new methods being developed to improve the delivery system. These methods have produced a greater lifetime of the solution or ointment in the eye; however, they also may carry side effects.

13.12 IMPLANTS IN DRUG DELIVERY

When conventional drug delivery methods (such as transdermal, ocular, internasal, and oral) fail to provide a therapeutic drug concentration to the desired site, implants are used to overcome these limitations. This method is designed particularly for drugs that face problems in oral delivery. These systems deliver drugs without pain and transport the drugs and fluids into the bloodstream via implants and without any needle insertions. Insulin, steroids, chemotherapy, antibiotics, analgesics, and heparin are being delivered by this system. A major advantage is that the implants are located under the skin without an opening in the skin, thus reducing the chance of infection. However, there is still some negative feedback for accepting this system as a possible candidate.

Two approaches are in consideration today: (a) The first is to implant electrical pumps, where the drug is refilled into the reservoir. This approach is beneficial in regulating the pump rate; however, this device is very large, and surgery can cause infections. (b) The second method is designed to be safe and control the erosion rate. As a result, an accurate release rate can be maintained by the implanted device. However, it is difficult to achieve "zero-order surface erosion" because it can cause some structural changes in the polymers.

Polymers used in medicines are used for two main purposes: for chronic periods and for transient use. Polymers for chronic periods are for cardiovascular and plastic surgery, orthopedics, and otolaryngology purposes. The crucial information needed for the polymer to be fully functional is its chemical composition and its fabrication method.

Another alternative drug delivery method is the DAD (drug administration device). The titanium housing contains a 20-cm refillable reservoir, an electronic control module, an integrated battery, and a peristaltic drive pump for drug delivery. The DAD contains an audible system that alerts patients and physicians to low battery power.

13.13 DRUG DOSAGES FOR CONDITIONS/DISEASES DUE TO IMBALANCE OF BIOFLUIDS

In reality, some changes in pressure, physical properties, temperature, and volume of a body fluid can be caused by many different unrelated conditions. The treatments are usually directed at the original disease and only secondarily at the manifestations (syndrome). This creates a situation where many different treatments can be used to correct the same type of fluid imbalance depending on the original cause, and conversely, the same treatment may be used to correct imbalances of different body fluids if the original cause remains the same. One must consult a physician or a specialist.[55]

For example, in the respiratory system, alveolar interstitial edema appears as a consequence of heart pump failure, and the treatment is directed mainly at improving heart function. The same interstitial edema can appear in malnutrition and liver failure because, among other reasons, there is a reduction in albumin concentration, rendering the blood unable to maintain adequate osmotic pressure. In this last case, the treatment will be completely different.

A field in drug delivery is now finding cell markers to deliver drugs with increased affinity for specific body tissues, making them less dependent on blood levels and consequently less toxic to other systems that do not need to be targeted. Most drugs are absorbed in one or more of the following ways:

- **Passive diffusion,** as in the case of transdermal medications like nitroglycerin, the delivered dose varies linearly with the area of contact. It can be used to treat high blood pressure because organic nitrates are vasodilators.[18,56]
- **Active uptake,** as is the case of radioactive iodine to treat thyroid disease.[4,18,20]
- **Intestinal absorption,** as in most oral medications like prednisone, dexamethasone, and oral penicillin.[18]
- **Volume of distribution** depends on the physical properties of the drug; most drugs work this way by achieving a concentration in various body fluids that allow the drug to reach them with varying degrees of success.
- Drugs can be administered **intrathecally,** as in the case of spinal anesthetics.[56]

13.14 CURRENT RESEARCH

Factors that stimulate and suppress GH secretion under physiological conditions. The current research is mainly focused on two aspects of the field: traditional drug delivery, stem cells, and tissue engineering.[21] The main goal of traditional drug delivery research is to do away with needles. Thousands of people with diabetes and other diseases depend on needles because the treatments are based on large protein molecules that must be delivered intravenously to avoid being devoured in the stomach.

Nevertheless, new research is concentrating on ways to get those macromolecules past the digestive tract. This research looks for less invasive and more efficient ways to deliver therapies, such as patches, inhalers, ultrasound, MRI, nanotechnology, and of course, pills (which still seem to be the overwhelming preference of patients).

However, more than half of today's medical problems cannot be treated with drugs. For instance, if somebody is dying of liver failure, there's no drug to treat them. The only way to treat the patient is by transplanting the liver from another person. To approach this problem, the field of tissue engineering came into the picture, which involves delivering cells to the body, not just drugs. This principle starts with the fact that cells are inherently intelligent; if we can provide a polymer support system, they will organize themselves on this scaffolding to create new tissue. The scaffolding eventually degrades, leaving only the living cells. In this way, scientists believe they can grow new cartilage, bone, skin, and eventually entire organs. Retrometabolic design of drug delivery is a new area of research.[9,21,23,43] Following is a look into past and future achievements:

- Some common examples include a tissue engineering triumph, where researchers grew a human ear in the lab, attached it to the back of a mouse, and watched it thrive. Or the case of a boy who was born without any bones or cartilage on his left side, so scientists grew on him a new chest from his own cells. These experiments were merely the first frontier in the field of tissue engineering, and they hint at an incredible future.
- Recently, scientists made a polymer scaffold that mimics the spinal cord. They "created" paraplegic rats by placing a defect in their spinal cord that hindered the use of their hind legs. The researchers placed neuronal stem cells on the polymer scaffolds and implanted them in rats. After a while, the rats with the implants could actually support their own weight. The results do not translate into a total cure; however, they are certainly deemed a step in the right direction toward helping the patients.
- The knot that ties itself is a new form of suture. Sometimes surgeons do operations in areas of the body that are hard to access, making it almost impossible to tie a suture. To help them, scientists have manipulated tissue engineering polymers to make materials that can change shape under changing conditions, such as the change from room temperature to body temperature. This technology may lead to a suture that can be tied loosely and then placed in the body, where it automatically

tightens itself. These materials and principles could also be used to make new blood vessels and cardiovascular stints that can be placed through tiny incisions and then expand to their proper shape.

- An insulin pill is in current development. The reason for an insulin pill failure is that insulin is a large protein molecule that is digested in the stomach. Researchers, however, are creating a pill that survives stomach acids and carries insulin safely to the bloodstream. Insulin patches and continuous monitoring of blood glucose levels are being used.
- The pharmacy on a chip is another goal. Researchers are developing a microchip that can be implanted under the skin to deliver drugs on cue. The chip has tiny reservoirs that can hold different types of medicine as well as varying doses of the same medicine. It can be programmed to release drugs at specific time intervals, and it could change the way we think about medical recording as every time one takes a drug, it could actually transmit that information from the chip to the computer at home, to the doctor's office, or hospital.
- Self-destructing sensors that could prevent diseases: To this end, scientists are developing nanoparticle sensors and MRI therapies[59–61] with tiny particles as a single atom that will recognize compounds, such as glucose and cholesterol, whose overproduction may signal a disease. The particles will then trigger a mechanism that tells a system (like the pharmacy on a chip) to release another compound to deal with the chemical imbalance. These nanoparticles are biodegradable, and they will self-destruct after 2 or 3 days.

Transport of drugs to BBB is a challenging task[9,30,40,46,62–70] for the researchers. Dr. Anthony McGoron at Florida International University, Miami, FL, is currently working on an in vitro Blood-Brain Barrier (BBB) model. His research objectives are:

1. Development of the in vitro BBB model.[62,71]
2. Validation of the BBB model.
3. Reversible alterations in the BBB permeability.
4. Design of peptide-conjugated doxorubicins.
5. Modeling of the doxorubicin transport across the in vitro BBB model.

Furthermore, the research is concentrating on a novel nanoparticulate delivery system, which incorporates the drug into polybutylcyanoacrylate nanoparticles (200–300 nm).

13.15 SUMMARY

Drug delivery systems are important areas of study against diseases that use knowledge of biology to treat targeted cells. However, the use of biology requires a thorough knowledge of different human body systems. It is also important to have a good understanding of the fluids involved in all the systems because these act as mediums through which the drugs will eventually be administered. Moreover, the analysis of the behavior is imperative not only because an imbalance between these might produce sickness but also to know how these will react in a certain drug delivery system. For instance, a fluid imbalance might produce edema, which is an accumulation of water in the tissues and is also a circulatory problem. Hence, a diuretic drug should be able to eliminate the excess water from the kidneys. This simple example summarizes the importance of knowledge and new research in this area.

Recent developments in new drug delivery systems offer promise for the future. The research to administer drugs with minimal side effects and dosage amounts can improve the drug delivery methods currently in use. The transdermal system is efficient, except that certain drugs are only permeable through the human skin. However, enhancers are being used to increase the permeability. The oral drug delivery method is the most convenient way of drug administration; however, it takes a sufficient amount of time for results and can cause side effects. Nasal drug delivery is selective to particular drugs and shows a decrease in side effects and is a painless method of consumption. Aside from the traditional delivery methods, implants are extremely powerful in drug delivery because they offer controlled drug rates over an extended time period and solutions for treating many chronic diseases. Moreover, researchers can rely on several principles or mechanisms of action of drug delivery systems and may apply them to develop new ways. An important example is the controlled-release system. It is important to know the electrical properties of the different cells in the body to mathematically model the behavior of cells and also to be able to control that behavior. The transport across membranes is mostly driven by electric and ion gradients. An important application of these principles is electroporation. Finally, research is continuously being done to improve the existing drug delivery systems and come up with new, less invasive methods. In addition, simple mathematical models for describing the kinetics of the drug are presented.

KEYWORDS

- **biodegradation**
- **biological barrier**
- **bulk degradation**
- **diffusion**
- **drug delivery**
- **drug delivery mechanism**

REFERENCES

1. Altman, P. *Blood and Other Body Fluids*; Federation of American Societies for Experimental Biology, 1961.
2. Burke, S. R. *The Compostion and Function of Body Fluids*; The C.V. Mosby, 1972.
3. Greenfield, A. D. The Circulation Through the Skin. In *Handbook of Physiology: Circulation*, Vol. 2; American Physiology Society, 1963; pp 1325–1351.
4. Hall, J. E.; Hall, M. E. *Guyton and Hall Textbook of Medical Physiology*; Pearson & Prentice Hall, 2020.
5. Harrison, T. R. Principles of Internal Medicine. *Academic Medicine* **1950**, *25* (6), 458.
6. Martini, F. H. *Fundamentals of Anatomy and Physiology*; Prentice Hall, 2001.
7. Rhoades, R. *Human Physiology*; Thomson Learning, 2003.
8. Sherwood, L. *Human Physiology: From Cells to Systems*; Cengage Learning, 2010.
9. Wikipedia. *Drug Delivery to the Brain* [Online], 2023. https://en.wikipedia.org/wiki/Drug_delivery_to_the_brain. (accessed Apr 3, 2023).
10. Hur, S.; Kim, J.; Ratnam, L.; Itkin, M. Lymphatic intervention, the frontline of modern lymphatic medicine: part I. history, anatomy, physiology, and diagnostic imaging of the lymphatic system. *Korean Journal of Radiology* **2023**, *24* (2), 95.
11. Sharma, A.; Kakkar, A. Designing Dendrimer and Miktoarm Polymer based Multi-Tasking Nanocarriers for Efficient Medical Therapy. *Molecules* **2015**, *20* (9), 16987–17015.
12. CIVHC. *Presciption Drug Supply Chain and Fund Flow* [Online], 2023. https://www.civhc.org/wp-content/uploads/2022/05/Drug-Rebate-Flow-Chart-Graphic.png (accessed Apr 3, 2023).
13. Ashauer, R. Toxicokinetic–Toxicodynamic Modelling in an Individual based Context—Consequences of Parameter Variability. *Ecol. Modell.* **2010**, *221* (9), 1325–1328.
14. Pitt, W. G.; Husseini, G. A.; Staples, B. J. Ultrasonic Drug Delivery–A General Review. *Expert Opin. Drug Deliv.* **2004**, *1* (1), 37–56.
15. Dittmer, D. S. *Blood and Other Body Fluids*; Federation of American Societies for Experimental Biology, 1961.
16. Bronzino, J. D.; Peterson, D. R.; Eds. *The Biomedical Engineering Handbook*, Vols. 1–4; CRC Press LLC (Taylor & Francis Group): Boca Raton, FL, 2015; p 5422.
17. Cooney, D. *Biomedical Engineering Principles*; Marcel Dekker, 1976.

18. Fournier, R. L. *Basic Transport Phenomena in Biomedical Engineering*; CRC Press, 2017.
19. Goyal, M. R. Kinetics of Drug Transport in the Human Body. In *Biofluid Mechanics of Human Body Systems*; CRC and Apple Academic Press, 2013; pp 413–440.
20. Glasser, O. *Medical Physics*; Yearbook Publisher, 1944.
21. Bodor, N.; Buchwald, P. *Retrometabolic Drug Design and Targeting*; Wiley Online Library, 2012.
22. Bonate, P. L. Recommended Reading in Population Pharmacokinetic Pharmacodynamics. *AAPS J.* **2005**, *7* (2), E363–E373.
23. Abraham, D. J.; Myers, M. *Burger's Medicinal Chemistry, Drug Discovery and Development*; John Wiley & Sons, 2021.
24. Jain, K. K. An Overview of Drug Delivery Systems. *Drug Deliv. Syst.* **2020**, 1–54.
25. Wikipedia. *Drug Delivery* [Online], 2023. https://en.wikipedia.org/wiki/Drug_delivery#:~:text=Common%20routes%20of%20administration%20include,deliver%20medications%20through%20other%20routes (accessed Apr 3, 2023).
26. Li, J.; Zeng, M.; Shan, H.; Tong, C. Microneedle Patches as Drug and Vaccine Delivery Platform. *Curr. Med. Chem.* **2017**, *24* (22), 2413–2422.
27. Rao, K. M.; Rao, K. S. K.; Ha, C.-S. Functional Stimuli-Responsive Polymeric Network Nanogels as Cargo Systems for Targeted Drug Delivery and Gene Delivery in Cancer Cells. In *Design of Nanostructures for Theranostics Applications*; Elsevier, 2018; pp 243–275.
28. Mishell D, Jr. Pharmacokinetics of Depot Medroxyprogesterone Acetate Contraception. *J. Reprod. Med.* **1996**, *41* (5 Suppl), 381-390.
29. Nahler, G.; Nahler, G. *Dictionary of Pharmaceutical Medicine*; Springer, 1994.
30. Park, K. Controlled Drug Delivery Systems: Past Forward and Future Back. *J. Control. Release* **2014**, *190*, 3–8.
31. Schoenmaker, L.; Witzigmann, D.; Kulkarni, J. A.; Verbeke, R.; Kersten, G.; Jiskoot, W.; Crommelin, D. J. mRNA-lipid Nanoparticle COVID-19 Vaccines: Structure and Stability. *Int. J. Pharm.* **2021**, *601*, 120586.
32. Singh, A. P.; Biswas, A.; Shukla, A.; Maiti, P. Targeted Therapy in Chronic Diseases Using Nanomaterial-based Drug Delivery Vehicles. *Signal Transduct. Target. Ther.* **2019**, *4* (1), 33.
33. Stewart, P.; Tuor, U. Blood-eye Barriers in the Rat: Correlation of Ultrastructure with Function. *J. Comp. Neurol.* **1994**, *340* (4), 566–576.
34. Strohl, W. R. Current Progress in Innovative Engineered Antibodies. *Protein Cell* **2018**, *9* (1), 86–120.
35. Tamai, I.; Tsuji, A. Drug Delivery through the Blood-Brain Barrier. *Adv. Drug Deliv. Rev.* **1996**, *19* (3), 401–424.
36. Tekade, R. K.; Maheshwari, R.; Soni, N.; Tekade, M.; Chougule, M. B. Nanotechnology for the Development of Nanomedicine. In *Nanotechnology-Based Approaches for Targeting and Delivery of Drugs and Genes*; Elsevier, 2017; pp 3–61.
37. Teleanu, D. M.; Chircov, C.; Grumezescu, A. M.; Volceanov, A.; Teleanu, R. I. Blood-Brain Delivery Methods Using Nanotechnology. *Pharmaceutics* **2018**, *10* (4), 269.
38. Archilla, J. F. R.; Palmero, F.; Lemos, M. C.; Sánchez-Rey, B.; Casado-Pascual, J. Nonlinear Systems: Nonlinear Phenomena in Biology. In *Optics and Condensed Matter (Understanding Complex Systems)*, 1st ed.; Vol. 1; Springer, 2018; p 359.

39. Thevis, M.; Opfermann, G.; Schänzer, W. High Speed Determination of Beta-Receptor Blocking Agents in Human Urine by Liquid Chromatography/Tandem Mass Spectrometry. *Biomed. Chromatogr.* **2001**, *15* (6), 393–402.
40. Geankoplis, C. J. *Mass Transport Phenomena*; Holt, Rinehart and Winston, 1972; p 495.
41. Hanley, H. J. Transport Phenomena in Fluids. *J. Macromol. Sci. Part B: Phys.* **1969**, *3* (2), 365–365.
42. Wikipedia. *Pharmacokinetics* [Online], 2023. https://en.wikipedia.org/wiki/Pharmacokinetics (accessed Apr 3, 2023).
43. Wikipedia. *Retrometabolic Drug Design* [Online], 2023. https://en.wikipedia.org/wiki/Retrometabolic_drug_design (accessed Apr 3, 2023).
44. Nordberg, M.; Duffus, J.; Templeton, D. M. Glossary of Terms Used in Toxicokinetics (IUPAC Recommendations 2003). *Pure Appl. Chem.* **2004**, *76* (5), 1033–1082.
45. Peattie, R. A.; Fisher, R. J.; Bronzino, J. D.; Peterson, D. R. *Transport Phenomena in Biomedical Engineering: Principles and Practices*, 1st ed.; CRC Press, 2012; p 197.
46. Tuomanen, E. Breaching the Blood-Brain Barrier. *Sci. Am.* **1993**, *268* (2), 80–84.
47. Tiwari, G.; Tiwari, R.; Sriwastawa, B.; Bhati, L.; Pandey, S.; Pandey, P.; Bannerjee, S. K. Drug Delivery Systems: An Updated Review. *Int. J. Pharm. Investig.* **2012**, *2* (1), 2.
48. Tosun, I. *Modeling in Transport Phenomena: A Conceptual Approach*; Elsevier, 2007.
49. Truskey, G. A.; Yuan, F.; Katz, D. F. *Transport Phenomena in Biological Systems*; Prentice Hall, 2009.
50. Zaid Alkilani, A.; McCrudden, M. T.; Donnelly, R. F. Transdermal Drug Delivery: Innovative Pharmaceutical Developments Based on Disruption of the Barrier Properties of the Stratum Corneum. *Pharmaceutics* **2015**, *7* (4), 438–470.
51. Paharmacy180. *Plasma Drug Concentration-Time Profile* [Online], 2023. https://www.pharmacy180.com/article/plasma-drug-concentration-time-profile-2506/ (accessed Apr 3, 2023).
52. Padalkar, A. N.; Shahi, S. R.; Thube, M. Microparticles: An Approach for Betterment of Drug Delivery System. *Int J Pharm Res Dev* **2011**, *3* (1), 99–115.
53. Takata, K.; Takai, H.; Yoshizaki, Y.; Nagata, T.; Kawahara, K.; Yoshida, Y.; Kuzuya, A.; Ohya, Y. Peptide Drug Release Behavior from Biodegradable Temperature-Responsive Injectable Hydrogels Exhibiting Irreversible Gelation. *Gels* **2017**, *3* (4), 38.
54. Bodor, N.; Buchwald, P. Corticosteroid Design for the Treatment of Asthma: Structural Insights and the Therapeutic Potential of Soft Corticosteroids. *Curr. Pharm. Des.* **2006**, *12* (25), 3241–3260.
55. PDR. *Physicians' Desk Reference PDR* [Online], 2023. https://www.pdr.net/ (accessed Apr 3, 2023).
56. R&M. *Pharmaceutical Drug Delivery Market Forecast to 2027—COVID-19 Impact and Global Analysis by Route of Administration; Application; End User, and Geography* [Online], 2023. https://www.researchandmarkets.com/reports/5067620/pharmaceutical-drug-delivery-market-forecast-to (accessed 2020).
57. EndoText. *Normal Physiology of Growth Hormone in Adults* [Online], 2023. https://www.ncbi.nlm.nih.gov/books/NBK279056/ (accessed Apr 3, 2023).
58. CDC. *Asthma* [Online], 2023. https://www.cdc.gov/dotw/asthma/index.html (accessed Apr 3, 2023).
59. Arruebo, M.; Fernández-Pacheco, R.; Ibarra, M. R.; Santamaría, J. Magnetic Nanoparticles for Drug Delivery. *Nano Today* **2007**, *2* (3), 22–32.

60. Wikipedia. *Magnetic Drug Delivery* [Online], 2023. https://en.wikipedia.org/wiki/Magnetic_drug_delivery (accessed Apr 3, 2023).
61. Ohtake, M.; Umemura, M.; Sato, I.; Akimoto, T.; Oda, K.; Nagasako, A.; Kim, J.-H.; Fujita, T.; Yokoyama, U.; Nakayama, T. Hyperthermia and Chemotherapy Using Fe (Salen) Nanoparticles Might Impact Glioblastoma Treatment. *Sci. Rep.* **2017**, *7* (1), 42783.
62. McGoron, A. In-Vitro Blood-Brain Barrier Model for Drug Delivery. In *Personal Communication*; Associate Dean (Academic Affairs), College of Engineering, Florida International University (FIU), 2013.
63. Moore, J. E. *Personal Communication*; Biomedical Engineering Institute, Florida International University: Miami, FL, 2003.
64. Begley, D. J. The Blood-Brain Barrier: Principles for Targeting Peptides and Drugs to the Central Nervous System. *J. Pharm. Pharmacol.* **1996**, *48* (2), 136–146.
65. Bodor, N.; Buchwald, P. Recent Advances in the Brain Targeting of Neuropharmaceuticals by Chemical Delivery Systems. *Adv. Drug Deliv. Rev.* **1999**, *36* (2–3), 229–254.
66. Crone, C.; Lassen, N. A. *Capillary Permeability: The Transfer of Molecules and Ions Between Capillary Blood and Tissue*; 1970.
67. Dadparvar, M.; Wagner, S.; Wien, S.; Kufleitner, J.; Worek, F.; von Briesen, H.; Kreuter, J. HI 6 Human Serum Albumin Nanoparticles—Development and Transport Over an in Vitro Blood–Brain Barrier Model. *Toxicol. Lett.* **2011**, *206* (1), 60–66.
68. Miller, G. *Breaking Down Barriers*; American Association for the Advancement of Science, 2002.
69. Neuwelt, E. A. *Implications of the Blood-Brain Barrier and Its Manipulation: Volume 1 Basic Science Aspects*; Springer Science & Business Media, 2013.
70. Pardridge, W. M. Recent Developments in Peptide Drug Delivery to the Brain. *Pharmacol. Toxicol.* **1992**, *71* (1), 3–10.
71. Ranade, V. V.; Cannon, J. B. *Drug Delivery Systems*; CRC Press, 2011.

CHAPTER 14

Instrumentation and Measurements of the Human Body Fluids*

MEGH R. GOYAL, ANAMIKA CHAUHAN, BAYRON ORTIZ, XAVIER PIÑERO, and JOEL BLANCO

14.1 INTRODUCTION

Various instruments are available for measuring fluid properties in our body, each with its unique characteristics (such as: capacity, accuracy, error, precision, linearity, durability, simplicity of operation, output, calibration, and cost). Medical instrumentation aims to measure physiological variables like electrical activity, fluid flow rate and volume, blood pressure, and more. This allows specialists to focus on specific properties that should be known by them, but it can lead to some redundancy in the quantities being measured due to the overlapping nature of these techniques across different bodily systems.

This chapter focuses on the instruments and equipment's that are available in the market to measure the fluid property in a human body.[1–5] The reader is directed to *Medical Instrumentation*, 3rd Edition[6] for more in-depth research.[7,36]

*This chapter has been modified from the review article prepared by the students (Bayron Ortiz, Xavier Piñero, and Joel Blanco) of Prof. Goyal for the course on Fluid Mechanics, INGE 4015. Course instructor: Megh R. Goyal, Ph.D., P.E., Retired Professor in Biomedical Engineering, General Engineering Department, University of Puerto Rico—Mayaguez Campus; Anamika Chauhan, Assistant Professor (Nutrition). For details, contact at <goyalmegh@gmail.com>; Prof. Goyal appreciates for providing advice, and photos of instruments and tools in the office of Physicians (MDs) in Mayaguez—Puerto Rico, namely, Jesus M. Roman-Velez; Ricardo Santaella; Gilbert Baez; Norberto Baez; Carlos Cardona—Perez; Norma Santiago-Gabrielini; and Luis Vazquez. Authors recognize significant contribution in this field by Prentice, William, *Therapeutic Modalities in Rehabilitation* (Third Edition, 2005, p. 608), New York: McGraw-Hill Medical Publishing Division, The McGraw-Hill Companies.[36]

Biofluid Dynamics of Human Body Systems: Expanded and Revised Edition.
Megh R. Goyal, Arka Bhowmik, & Anamika Chauhan (Eds.)

14.2 HUMAN EYE

14.2.1 THE SCHIRMER STRIP TEST

The Schirmer test is utilized to assess tear production in the eyes (Figure 14.1A). During the test, a small filter paper strip measuring 35 × 5 mm is placed at the junction of the lower eyelid's middle and lateral thirds. For 5 min, the patient is told to look ahead while using regular blinking, and quantity of tears soaked by the strip is determined. If the wetting of the paper is less than 10 mm in 5 min, it indicates a deficiency in tear production, known as tear deficient dry eye. One significant drawback of the Schirmer test is its potential for variability in results when conducted by different doctors or at different times. Although many clinicians use anesthetic eye drops to numb the eye before the test, the National Eye Institute workshop recommends avoiding their use. For evaluating reflex tear secretion, the Schirmer II test can be performed by first irritating the nasal mucosa with a cotton-tipped applicator. On the other hand, the Phenol red thread test is an invasive method that involves using a cotton thread infused with phenol red dye. The dye changes color from yellow to red when exposed to tears. After 15 s, the length of the thread that has changed color indicates the wetting length (in mm) by tears. The normal range for wetting length is between 9 and 20 mm, and a wetting value below 9 mm is suggestive of dry eyes.

14.2.2 THE NON-CONTACT TONOMETER

The non-contact tonometer operates based on the time interval taken for a puff of air to flatten the cornea precisely. The device measures this interval in milliseconds, starting from the moment the air puff is released until the cornea reaches the flattened state, where the timing device stops. Soft eyes are flattened more quickly than hard eyes, resulting in shorter time intervals for soft eyes and longer intervals for hard eyes.

14.2.3 THE APPLANATION TONOMETER

The pressure within an eye, an artery or the pregnant uterus can be measured from outside the walls with an applanation tonometer (Figure 14.1B). When the device flattens, a section of the wall under pressure to keep the wall flat is equal to the pressure within the sphere or cylinder. The pressure is calculated as the force applied divided by the area of the inner disc. This device is also called a toco dynamometer or guard-ring toco graph or a cardio toco graph to record the heart rate.

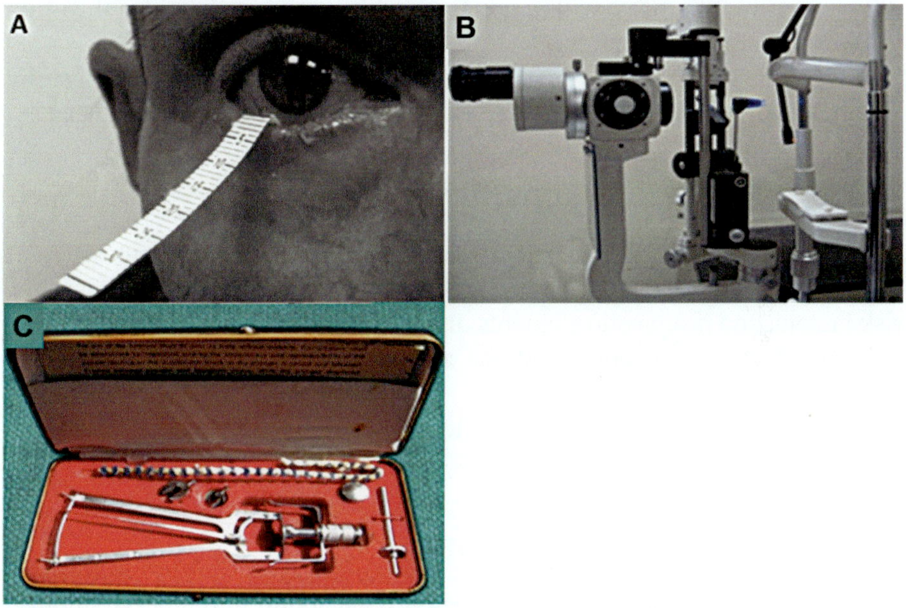

FIGURE 14.1 (A) Red thread test (Reprinted from 8 Image by Rocha-Cabrera et al, 2015. https://creativecommons.org/licenses/by-nc/4.0/), (B) Applanation tonometer attached to slit-lamp (Reprinted from 9 Image by Maheshwari et al, 2012. https://creativecommons.org/licenses/by/3.0/), and (C) Schiotz tonometer (Reprinted from 9 Image by Maheshwari et al, 2012. https://creativecommons.org/licenses/by/3.0/).

14.2.4 INDENTATION (SCHIOTZ) TONOMETER (FIGURE 14.1C)

Once a numbing drop is applied to the eye's surface, the doctor places the instrument directly on the cornea and gauges the eye's pressure by measuring the indentation caused by a specific weight. However, this test is less precise compared to applanation tonometry and is not as frequently employed by ophthalmologists and optometrists in modern practice.

14.2.5 PANOPTIC OPHTHALMOSCOPE

- Facilitates easy entry into small pupils that are not dilated.
- Offers a view of the fundus that is noticeably larger than that of a standard ophthalmoscope in an undilated eye.

- Provides a 25° field-of-view, which is five times wider than the standard 5° field-of-view.
- Enhances magnification by 26% compared to the standard scope.
- Improves comfort with a greater working distance between the practitioner and the patient.
- Every 3.5 V battery handle and wall transformer already in use by Welch Allyn are compatible with the new, reasonably priced head.

14.2.6 THE BIOMICROSCOPY (UBM)

Anatomical aspects, physio-pathological processes, and different types of glaucoma can be evaluated with UBM. This helps in differential diagnosis, risk factors, and prognostic elements. In addition, UBM after filtering surgery helps to detect early complications and to understand the poor surgical outcome.

The ophthalmoscope provides a clean white light for optimum examination of the eyeball.

14.3 THE HUMAN EAR: OTOSCOPE

The human ear converts mechanical energy via sound into a nerve impulse that travels to the brain. Due to the complicated nature of sound waves, this enables us to sense the pitch, loudness, and timbre of sounds.

In an otoscope, on/off switch with a pocket clip allows to investigate the ear with either hand. A full set of apertures (main beam, small beam, semicircle, green filter, and a fixation star) allows complete inspection.

14.4 RESPIRATORY SYSTEM

During the breathing process, we inhale oxygen and exhale carbon dioxide. The non-invasive method to measure carbon monoxide levels during breathing is by a spirometer. Smokers can see with the Breath CO how high their CO levels are, which is deadly.

14.4.1 THE SPIROSTAR DX

The SpiroStar spirometer incorporates a heuristic system algorithm to ensure system stability and consistently deliver reliable results throughout the session. It is compatible with Spiro2000, offering user-friendly operation, flexible parameter selection, and various report formats to choose from. Technicians have the option to customize the reports to meet specific requirements. The session results can be easily transferred to central information systems through a local area network or mobile software. To account for variations in the operating environment, SpiroStar is equipped with an automatic temperature sensor that continuously monitors and adjusts as needed. Additionally, SpiroStar utilizes a disposable flow transducer.

14.4.2 THE BRONCHOSCOPE

This device enables examination of the tracheobronchial tree's interior as well as endo-bronchial diagnostic and therapeutic procedures, including the removal of foreign objects and collection of samples for culture and biopsy.

14.4.3 THE INCENTIVE SPIROMETER

The incentive spirometer is employed post-surgery to maintain clear and active lungs by expelling mucus effectively. The recommended steps for using the incentive spirometer are as follows:

- If possible, sit on the edge of the bed or sit up as much as you can while in bed.
- Place the incentive spirometer on its side.
- Insert the mouthpiece and firmly close your lips around it.
- By taking a slow, deep breath, raise the piston toward the top of the column. The colorful outline area and the coach indicator should line up.
- Hold your breath for at least 5 s, letting the piston return to the column's base.
- After a brief break, repeat steps one through five at least 10 times each hour you're awake.
- To represent your finest effort, aim to place the yellow indicator on the left side of the spirometer. Use this as a goal during each repetition.
- After completing each set of ten deep breaths, practice coughing to ensure the lungs are clear.
- Once you can get out of bed, walk in the hallway and cough effectively. You may discontinue using the incentive spirometer unless instructed otherwise by the nurse.
- Following these steps can help increase lung volume during inhalation, reduce mucus or phlegm, and prevent serious lung infections.

14.4.4 BODY PLETHYSMOGRAPH

The patient is requested to inhale or exhale to a predetermined volume during body plethysmography, usually at functional residual capacity (FRC). Once the desired volume is reached, a shutter is dropped across the breathing tube, creating a closed system. The patient then exhales forcefully against the sealed shutter, expanding their chest capacity and releasing pressure in their lungs. This expansion of the chest slightly reduces the volume of the box.

Boyle's Law ($P_1V_1 = P_2V_2$) is used to calculate the initial volume of air in the lungs when the shutter was closed. At constant temperature, the equation calls for beginning pressure (P_1), volume (V_1), and end pressure (P_2), volume (V_2). To determine the change in chest volume, the first step is to solve for the box's volume during the respiratory effort. The change in chest volume is determined by the difference between this volume and the box's original volume. Boyle's Law can be used to calculate the unknown volume, which corresponds to the initial volume of gas present in the lungs at the moment the shutter closed. The shutter typically closes at the conclusion of a typical exhale or at FRC.

14.4.5 THE PULSE OXYMETER

This allows continuous measurements of arterial oxygen saturation (%Sp_{O2}) and pulse rate (beats/min). In general, the sensor is placed on a finger.

14.5 CIRCULATORY SYSTEM

Maintaining an appropriate blood pressure level is crucial to ensure sufficient blood flow throughout the body. Blood pressure plays a vital role in enabling a person to transition quickly from a lying down position without experiencing blood pooling in the legs. The regulation of normal blood pressure relies on factors such as heart contractions, arterial wall elasticity, blood volume, and the resistance of blood vessels to blood flow.

14.5.1 MEASUREMENT OF BLOOD PRESSURE

Blood pressure is measured by determining two points during the heart's activity: (1) the highest point when the heart contracts to push blood into circulation (systole); and (2) the lowest point when the heart relaxes to fill with blood returning from circulation (diastole). The instrument used for this measurement is a sphygmomanometer, which records blood pressure in millimeters of mercury (mmHg). This process involves wrapping an inflatable rubber cuff around the upper arm and inflating it by squeezing a connected rubber bulb. The Korotkoff sounds, which are distinct sounds heard via stethoscope placed above the artery below the cuff, are observed during the measurement.

When the first Korotkoff sound is heard, it indicates the systolic blood pressure, which is then noted on the manometer. The pressure in the cuff is gradually reduced, and the Korotkoff sounds continue until the cuff pressure is no longer enough to fully occlude the vessel during the length of cardiac cycle. The point where the Korotkoff sounds disappear below this pressure is logged as a diastolic pressure.

14.5.2 AUTOMATIC INDIRECT METHOD

Most automatic and semi-automatic devices for blood pressure measurement utilize a pressure transducer connected to the cuff of the sphygmomanometer, along with a microphone placed beneath the cuff. These devices also include a standard physiological system to record both cuff pressure and Korotkoff sounds. In semi-automatic devices, the recording is interpreted by a technician. Fully automated devices, such as the electrophygmo-manometer PE-300, use a signal detecting circuit to identify the occurrence of the first and last Korotkoff sounds. The device then retains and displays the cuff pressure readings corresponding to these points automatically. This method is designed to work in conjunction with an occluding cuff, microphone, or pulse transducer, along with a recorder for the automatic measurement of indirect systolic and diastolic blood pressure. The automatic detection of Korotkoff sounds and the cessation at the diastolic value are indicated by a microphone and light flash at the systolic pressure.

Mean arterial pressure (MAP) can be calculated using a simple formula: MAP = 1/3 (systolic–diastolic) + diastolic. It represents a weighted average of systolic and diastolic pressure. MAP falls approximately one-third of the way between the diastolic low and the systolic peak. The oscillometric method used in the Dinamap is employed to determine MAP. Within the enclosure of the device, a solid-state pressure transducer senses pressure pulsations within a cuff bladder. As the cuff pressure decreases and passes through the mean arterial blood pressure, pressure oscillations increase in amplitude and reach their maximum.

14.5.3 DIRECT MEASUREMENTS

Direct blood pressure readings are much more accurate than indirect ones and provide a continuous readout and recording of the blood pressure waveform. Typically, percutaneous insertion, catheterization (vessel cutting), or transducer implantation in a blood vessel or the heart are used to directly measure blood pressure. A local anesthetic is applied close to the invasion site in the case of percutaneous insertion. A hollow needle is then inserted at a small angle toward the blood vessel after it has been momentarily blocked. After the needle has been placed correctly, a catheter is hooked on the hollow needle, frequently with the use of a guide. The needle and guide are eliminated once the catheter has been correctly inserted into the blood artery.

14.5.4 CATHETERIZATION

A key diagnostic method for examining the heart and other circulatory system components is catheterization. Through a superficial vein or artery, a catheter—a lengthy tube—is introduced into the heart and major blood arteries. Blood pressure measurements using a catheter can be performed in two ways:

- A sterile saline solution is introduced into the catheter, allowing the fluid pressure to be transmitted to a transducer located outside the body.
- The transducer is inserted into the catheter and advanced to the specific point where the pressure needs to be measured.

Heart sounds can be detected by instruments such as: manual stethoscope, electronic stethoscope, phono-cardiogram, vibro-cardiograph, and apex cardiograph. For volume changes in different parts of our body, physician may use plethysmograph, pseudo-plethysmographs, photoelectric plethysmograph, impedance plethysmograph electrode system, reoencephalography (RECG), oculo-pneumo plethysmograph, and recording suction ophthalmic dynamometer.

14.5.5 ELECTROCARDIOGRAPHY (ECG)

Electrocardiography is a graphic display of the electrical impulses to control muscular contraction and relaxation of the heart. The ECG is obtained by attaching electrodes to various parts of the body to get the electrical signals. The P-wave, a deflection brought on by current coming from the atrium, the

QRS complex, which depicts the electrical activity moving into the ventricles, and the T-wave, which occurs when the ventricles reset themselves, are the most noticeable components of the ECG. A doctor can treat patients by identifying the cause of irregular heart rhythms by monitoring their ECG.

The 12-lead ECG uses electrodes attached to all four limbs and six positions on the chest wall to examine the heart rhythm from different viewpoints. One lead provides a one-dimensional view of heart rhythm, and multiple leads allow a three-dimensional view. The six electrodes are placed at the extremities (usually the wrists and ankles). The six limb leads from four electrodes are accomplished by varying the signal orientation between the four extremities. The six following chest leads result from the six electrodes pasted outer wall of chest, and these are known as leads I, II, III, aVR, aVL, and aVF.

- Lead I results from designating the left-arm electrode as a positive electrode and the right-arm electrode as a negative electrode.
- Lead II results from designating the left-leg electrode as positive and the right-arm electrode as negative.
- Lead III results from designating the left arm as negative and the left leg as a positive electrode. To produce the ECG signal, the voltage difference between the two electrodes is compared.
- The aVR lead designates the right arm as the positive electrode.
- The aVL lead designates the left arm as the positive electrode, and the aVF lead designates the left leg as the positive electrode.

The chest leads (precordial leads) are unipolar leads designated as V1, V2, V3, V4, V5, and V6; when properly placed, V1 and V2 are above the right ventricle, V3 and V4 are over the intraventricular septum; and V5 and V6 are lateral leads for left lateral view of the heart. Today, the six chest electrodes are manufactured in a continuous strip.

Each 1-mm box on the ECG paper represents 0.04 s, and each 5-mm box represents 0.2 s. By simple extrapolation, five of the 5-mm boxes are equivalent to 1 s. ECG papers or "rhythm strips" generally have "tic" marks at the top of the paper, outside the grid representing 3 s. The vertical axis uses the same boxes, with the scale being the signal amplitude. Standard ECG recordings use 10-mm to represent 1 millivolt.

14.5.6 PRESSURE TRANSDUCER

This device consists of a catheter filled with a saline solution and a pressure sensor. One end of the catheter is inserted into the lumen of an artery and the other end is connected to the pressure sensor. Four-wire strain gauge (Wheatstone Bridge) displacement sensors measure the deflection of the diaphragm.

14.5.7 ULTRASOUND

Ultrasound is a versatile diagnostic tool used to examine various body parts such as the arterial system, heart, pancreas, peritoneal cavity, urinary system, ovaries, gall bladder, venous system, brain, spinal cord, fetus during pregnancy, and cysts, among others. A transducer is employed to transmit sound waves and receive the echoes produced. To enhance sound transmission, a jelly-like substance is applied to the skin. This technique is beneficial for physicians since fluid conducts sound waves effectively. The behavior of ultrasound is influenced by the frequency of sonic energy and the density and mechanical compliance of the medium it travels through. The speed of sound propagation in a medium varies depending on its density, elastic properties, and temperature.

It should be noted that ultrasound cannot be used to examine calcified tissues, as they absorb most of the ultrasound beam, rendering bones or lungs unsuitable for this technique.

14.5.8 ULTRASONIC IMAGING

The most common application of ultrasound is for non-invasive, diagnostic imaging of interior organs. Pulsed ultrasound or pulsed Doppler mode are frequently used by the imaging system. After amplification, the received information is displayed in one of the following modes:

- **M-scan display:** In this mode, each transmitted pulse triggers the oscilloscope sweep, and echoes appear as dots with brightness proportional to their intensity. The transducer remains stationary, and the movement of the dots along the sweep indicates the movement of received targets.
- **B-scan display:** This mode presents a two-dimensional image of a stationary organ or body structure. The transducer is moved with respect to the body, while the vertical deflection of the oscilloscope or chart paper corresponds to the transducer's movement.
- **A-scan display:** Here, each transmitted pulse triggers the oscilloscope sweep, and vertical deflections on the trace represent the returning echoes. In some cases, the transducer is kept stationary, and any movement of echoes along the transducer indicates moving targets. An example of an A-scan display is an echoencephalogram.

14.5.9 DOPPLER EFFECT

According to Doppler's principle, if a sound source moves away from an observer while maintaining a constant pitch, the sound appears to be lower in pitch. Doppler ultrasound is employed to measure the flow of liquids, such as blood flow. The amount of frequency shift of the reflected wave is calculated using the formula: $f = [(2V)/\lambda]$, where V is the velocity of the interface, and λ is the wavelength of the transmitted ultrasound. Ultrasonic Doppler equipment can detect and assess the characteristics of blood flow in arteries and veins. The ultrasound beam is directed at the vessel of interest while the sound source or ultrasonic transducer is in contact with the outer skin layer. The echoes are generated by the movement of red blood cells within the vessel.

In a simple Doppler instrument, the transmitter continuously sends a sinusoidal electrical signal of a specific frequency to the ultrasonic transducer. The transducer converts this electrical energy into ultrasonic energy, which propagates into the medium. Echo signals, resulting from reflections and scattering, return to the transducer, creating an electrical signal that is amplified by the receiver. The amplified signal is then passed through a demodulator and low-pass filter, allowing only low-frequency Doppler shift signals to be extracted at the output.

14.5.10 PULSED DOPPLER

Pulsed Doppler technology allows for the detection of moving interfaces and scatters within a well-defined sample volume. Anywhere along, the axis of the ultrasonic beam can be changed to change the sample volume's position. In this technique, the ultrasonic transducer is excited by a short-duration burst from a gated transmitter. The echo signals, resulting from scattering and reflection, are received by the same transducer, amplified in the receiver, and then processed in the demodulator. The Doppler signal in a pulsed Doppler instrument is derived from the changes in phase of the echo signal between consecutive pulses. This phase change occurs due to the movement of a target either moving toward or receding from the ultrasonic transducer. The process involves emitting a transmit pulse from the transducer, and subsequently, echoes are detected from reflectors in the medium, resulting in a sequence of single-pulse echoes.

Duplex Ultrasound

The duplex ultrasound instrument integrates real-time B-mode scanning with built-in Doppler capabilities. In typical applications, the B-mode image obtained helps pinpoint areas for Doppler examination of blood flow. Certain duplex instruments even allow operators to indicate the direction of blood flow relative to the ultrasound beam. For determining flow velocity based on the frequency of the Doppler signal, the angle between the flow vector and ultrasound beam is needed.

Echocardiography is a valuable tool for studying various cardiac conditions like congenital heart diseases, coronary artery diseases, cardiac tumors, and other disorders. It involves obtaining an echocardiogram, which is derived from duplex ultrasound, providing measurements of blood flow direction, volume, and velocity. The technique yields colored images through the reflection of ultrasound waves, enabling determination of the direction and rate of blood flow. Blind Doppler ultrasound is also used, generating a graph to analyze wave-length, amplitude, and rhythm for recording the heartbeat.

Echocardiology can identify conditions such as calcific aortic stenosis, pulmonary valve stenosis, mitral valve stenosis, left atrial myxoma, and other heart diseases. When selecting a transducer for echocardiography, factors to consider include the type of tissue being studied, the anatomical area involved, the patient's physical health, the type of study being conducted, and the depth of the structure.

Ultrasonic Devices

Most of the devices in the ultrasound techniques employ principles based on Doppler or Duplex. Examples of these devices are:

- **ACUSON 128XP/10 95'** uses Doppler to detect the pulse of the patient at the desired location. And Duplex is used to obtain an image of this location. With a colored image, the technician can determine the direction of the blood flow, the speed, and flow volume.
- **ADVANCED ACUSON ASPEN** uses the same principle as ACUSON 128XP, but is more versatile. Here, all studies (images or videos) can be stored in a memory up to one month.

14.5.11 ECHOCARDIOGRAM (ECG)

The Echocardiogram (ECG) is a valuable diagnostic tool for identifying various heart problems. During the test, a transducer is placed against the chest, emitting ultrasound waves that echo from different parts of the heart. This data is processed by computer software to produce an image of the heart that is presented on a screen and captured on paper or videotape. The echocardiogram study typically combines three techniques:

- M-mode echo, which produces an image useful for accurately measuring the size of heart chambers.
- Two-dimensional (2-D) echoes, which depict moving "slices" of the heart and provide visual representations of the heart's form and motion.
- Doppler echo, used to assess blood flow through the heart. The signals are displayed as black-and-white tracings or color images on the screen.

Echocardiograms are sometimes performed during or after an exercise test. They provide valuable information about the heart, including:

- The finding of fluid surrounding the heart, blood clots, lumps within the heart, and unusual holes between the heart chambers.
- Measuring the thickness of the heart muscle and the size of the heart chambers.
- Evaluation of the motion and form of the heart's valves to spot leaky or constricted valves.
- Determining if the heart is pumping strongly or weakly, and whether the various cardiac regions are pumping equally.

14.6 DIALYSIS

Dialysis is a process by which free blood toxins and excess fluids are removed from the blood in the kidney. Two main methods are used in dialysis: hemodialysis and peritoneal dialysis.

14.6.1 HEMODIALYSIS

Toxins are taken out of the blood by a filter. The dialyzer functions as an alternative kidney. Two compartments of the dialyzer are divided by a semi-permeable membrane. Only particles of specific size can cross through this membrane. A solution called dialysate circulates to one side of a membrane and the blood flows from the other side. Dialysate is composed of water, dextrose (sugar), and elements similar to the normal fluids of the body without toxins. In the diffusion process across a semi permeable membrane, particles move along the concentration gradient until the two solutions reach equilibrium.

In the dialysis, fresh dialysate flows through the dialyzer and the waste from the blood turns toward it until a vast proportion of waste products have been evacuated from the blood. The osmosis is like the diffusion, but instead of evacuating waste products, it evacuates excess blood. The dialysis technician can regulate the number of fluids to be extracted during dialysis. After a few hours of dialysis, the blood is relatively purified and the excessive fluids and waste are removed.

Care during Hemodialysis

There must be a substantial aperture to permit blood passage from the body to the dialyzer. To create this opening, A small surgery is performed to create this opening that is usually made in an arm or leg. Two different sorts of entrances exist:

- Temporary entrance includes fistulas, such as: the intra-clavicula, the Femur and the internal Jugular. The intra-clavicula is a temporally catheter placed in the intra-clavicular vein, exactly underneath the bone of the neck. In the Femur, the catheter is placed temporally in the vein in the ingle. It is only used during the hospitalization. In the internal Jugular, the catheter is placed in the great vein of the neck. This catheter can be used for the external dialysis during several weeks up to months. The catheters are covered with clean and dry bandages.
- The most common permanent entrances are fistulas and grafts. A fistula is formed under the skin sewing a vein to an artery. This causes the vein to expand by increasing the volume of blood that flows through it.

14.6.2 PERITONEAL DIALYSIS

The peritoneal dialysis uses peritoneal membrane (also called the peritoneum). In order to obtain the entrance in the peritoneum, a flexible tube is surgically implanted in the lower abdomen. Since the peritoneum is rich in tiny glasses of blood, it continuously provides blood that is filtered by osmosis and diffusion. The toxins in the blood are replaced periodically. The time space defines how long the dialysis solution remains in the peritoneal cavity during the peritoneal dialysis. Axial and centrifugal blood flow pumps are essential elements of these machines.[20]

14.6.3 MEASUREMENTS OF BLOOD FLOW AND CARDIAC OUTPUT

The blood flow (L/min or mL/min) meters employ following physical principles:

- Electromagnetic induction (magnetic blood flow meter): These flow meters measure the mean velocity of the blood stream.
- Indicator dilution: Measurement of blood temperature and injection of cool saline are performed with a specially designed single catheter. The 10 ml volume cold saline is normally injected, and the circulating blood is rewarmed rapidly and no measurable recirculation occurs.

The device used for this technique is called the Swan-Ganz catheter with four separate lumens:

- One of the lumens is used to inject the cooled saline.
- The second lumen contains two thin wire leads close to the top of the catheter.
- The third lumen ends at the catheter tip to measure the blood pressure.
- To inflate a small rubber balloon at the tip of the catheter, the fourth lumen of the catheter is used. The balloon is inflated and the returning venous blood carries the catheter until its tip is positioned in the pulmonary artery once the catheter has been inserted into the vein.

14.6.4 NONINVASIVE BLOOD GLUCOSE MONITORING

The technology employed in continuous glucose monitoring includes various methods such as near infrared light (near-IR) spectroscopy, far infrared radiation (far-IR) spectroscopy, radio wave impedance, optical rotation of polarized light, fluid extraction from skin (reverse iontophoresis), and interstitial fluid harvesting.

- Fluid extraction from the skin (reverse iontophoresis) is capable of continuously measuring blood glucose levels without requiring much effort from the patient. In reverse iontophoresis, a small current inducted in skin, and fluid is extracted. This current draws out salt, which carries water, and subsequently, glucose. The glucose concentration in this extracted fluid can be measured and is proportional to the blood glucose level. The GlucoWatch utilizes reverse iontophoresis to extract fluid from the skin and includes a programmable alarm to indicate high or low panic values.

- Interstitial fluid harvesting involves measuring interstitial fluid from the skin with minimal skin trauma and sensation. This process includes fluid extraction from the skin using methods like penetration with a 29-gauge needle, disruption with chemicals, or microporation with a laser. Research demonstrates that while the level of serum glucose is steady, the concentration of glucose in interstitial fluid is comparable to that of serum glucose.

14.7 HUMAN REPRODUCTIVE SYSTEM

14.7.1 ULTRASOUND

An ultrasonic scan is non-invasive and frequently used to examine breast tissue, stomach, and pelvic organs. The ultrasound and ultrasonography are used to determine the age and sex of the fetus, screen for prenatal anomalies (such spinal bifida), and screen fetal development. Every 3 months (each trimester) throughout pregnancy, it could be necessary to get an ultrasound scan.

14.7.2 SPERM COUNT

Sperm count is a measure of male fertility and refers to the quantity of sperm found in an ejaculate. The test assesses various aspects of sperm production and release. In approximately one out of every 20 samples, no sperm are observed, which can result from either obstructive issues or a failure in sperm production.

14.7.3 RIGID SCAN

The Rigid Scan will read the blood pressure of the patient during the night and the ring will contract or expand depending on the penis activity during the night. This helps the specialist to determine if the circulation in the penis is a normal one, or if the patient is important.

14.8 HUMAN DIGESTIVE SYSTEM

14.8.1 COLONOSCOPY

The procedure known as a colonoscopy includes looking within the colon screening the lower portion of the big bowel with a sigmoid scope. The patient consumes only liquids for one to two days prior to the treatment, then takes a laxative and one or more enemas to achieve full bowel emptying. Sedatives or tranquilizers may also be administered. The colonoscopy typically lasts 30–60 min and can be performed as an outpatient procedure, although it is usually done in a hospital setting. During the procedure, the patient lies on their side with knees flexed, and a colonoscope is inserted into the intestine. Air may be injected to expand the folds of the colon, providing a clear view of the entire surface area. The colonoscope is slowly withdrawn, allowing the doctor to examine the lining for any lesions or polyps. Special attachments on the scope enable the removal of small polyps or the collection of biopsy samples if necessary.

14.8.2 UPPER GI ENDOSCOPY

Upper GI endoscopy is a procedure conducted by a gastroenterologist to diagnose and address issues related to the upper digestive system. It proves beneficial in evaluating and diagnosing various problems like difficulty swallowing, stomach or abdominal pain, and conditions such as bleeding, ulcers, and tumors. The endoscopy process involves several steps, including:

- Carefully guiding the endoscope down the esophagus after being inserted via the mouth.
- Advancing the endoscope until it reaches the stomach.
- Close examination of the stomach to identify any potential problem areas.
- If any suspicious findings are detected, the doctor may take a sample for biopsy.

14.8.3 GASTROSCOPY

Gastroscopy enables a direct visualization of the esophagus, stomach, and duodenum using an endoscopic tube. It is employed to investigate the cause of bleeding, perform biopsies, diagnose inflammatory diseases, tumors, ulcers, and structural abnormalities. The patient is given a sedative or tranquilizer and told to fast for 6–12 h prior to the surgery. Local anesthetic is sprayed into the mouth and throat. Then, the mouth, esophagus, stomach, and duodenum are all traversed via a flexible fiber-optic tube. Although, there is a minor coincidental that organs or other tissues could be perforated, this is a relatively uncommon complication of the treatment.

14.8.4 ENDOSCOPIC RETROGRADE CHOLANGIO PANCREATOGRAPHY (ERCP)

Endoscopic retrograde cholangio pancreatography (ERCP) is a valuable procedure used for diagnosing and treating issues causing jaundice or abdominal pain. Other imaging techniques involving X-rays or sound waves may complement the information obtained from ERCP. The ERCP process involves the following steps:

- The mouth is used to implant an endoscope, which then travels via the stomach, esophagus, and duodenum.
- A small tube is threaded into the duct.
- A dye is injected backward through the ducts, enabling the taking of X-rays.
- The specialist can then remove any stones blocking the duct.

Before the ERCP test, patients recommended to fast 12 h and are typically given a sedative or tranquilizer. Local anesthetics in spray or gargle form are used to prevent gagging, and a mouth guard may be inserted to protect the teeth. The ERCP procedure takes about 60 min.

- The endoscope is inserted through the mouth, and the patient is instructed to swallow, helping pass it down into the esophagus.
- Guided by a fluoroscope, the doctor continues passing the endoscope down into the stomach and duodenum, then injects a drug to relax the duodenum.
- The contrast medium is injected through the endoscope, and a series of X-rays is taken.
- Additional X-rays may be taken from different positions after removing the scope.

14.8.5 INTRAVENOUS CHOLANGIOGRAPHY

Cholecystography is a medical procedure that uses contrast X-rays to detect gallstones, blockages, or abnormalities in the gallbladder and bile ducts. Alternatively, a CT scan can be used instead of conventional X-rays. Before the test, the patient consumes a high-fat meal the night before and only consumes water or fat-free liquids thereafter. A laxative may be given to prevent shadows caused by a full bowel during the X-rays. The test usually lasts about 2 to 2 1/2 h if there are no obstructions in the bile ducts. However, if there are obstructions, it may take 4–24 h for the contrast medium to accumulate in the gallbladder. The procedure starts with preliminary X-rays while the patient lies on a tilting X-ray table. Later, the contrast medium is injected intravenously, and additional X-ray images acquired from variable angles.

14.8.6 PERCUTANEOUS TRANSHEPATIC CHOLANGIOGRAPHY

This is a contrast X-ray procedure aimed at diagnosing the underlying causes of jaundice by examining the bile ducts. Before the test, the patient is instructed to fast for about 8 h. During the procedure:

- The patient lies on a tilting X-ray table, and a local anesthetic is applied to the skin covering the liver and its surrounding capsule.
- Using fluoroscopic guidance, a long, flexible needle is inserted into the liver to locate a dilated bile duct, through which the contrast medium is administered.
- Next, the table is rotated through a variety of angles while a sequence of X-rays are obtained.

14.8.7 CHOLECYSTOGRAPHY

This is a contrast X-ray examination specifically designed to identify gallstones, inflammatory diseases, and tumors in the gallbladder. The procedure involves the following steps:

- On the day before the test, at noon, the patient consumes a meal with a normal amount of fat.
- This is followed by a fat-free dinner, which causes the contrast agent-containing bile to accumulate in the gallbladder.
- After dinner, at 5-min intervals, the patient takes a series of pills containing the contrast dye.
- This outpatient procedure includes a series of X-ray images taken while the patient is positioned in different ways.
- After the initial pictures, the patient may be given a high-fat meal or a pill to produce the same effect, and then asked to wait for 15–20 min. Following this, another set of X-ray images is taken.

14.8.8 DUODENOGRAPHY

Duodenography is a diagnostic procedure used to obtain X-ray images of the duodenum and pancreas for the detection of tumors or lesions. Before the test, the patient must fast for 6–12 h. During the procedure:

- The patient is seated, and a long, flexible tube known as a catheter is inserted through the nose and carefully guided into the stomach.
- Next, the patient lies down, besides through the help of a screen inside fluoroscope, the doctor continues to advance the catheter into the duodenum.
- To relax the duodenum, a hormone or drug is administered through injection or IV, followed by the introduction of a contrast medium through the catheter while X-rays are taken from various angles.
- Later, some of the barium is withdrawn, and air is injected through the catheter. Additional X-rays are taken before removing the catheter.

14.8.9 BARIUM X-RAY

This X-ray may help in directing attention to an area of possible abnormality. The endoscopy permits a more accurate view of the upper digestive system.

14.8.10 ESOPHAGOGRAPHY

Although this test is occasionally done alone, when the doctor is reasonably sure that the problem is confined to the esophagus, it is usually done as part of the upper GI series.

14.8.11 GASTRIC SECRETION STUDIES

This procedure is employed to collect and analyze samples of gastric juices secreted by the stomach, aiming to diagnose the causes of epigastric pain or gain further insights into suspected ulcers. The process involves the following steps:

- Typically, the patient is provided with a liquid meal, followed by a 12-h fasting period during which liquids and smoking are limited used for 8 h beforehand the test.
- To stimulate the production of gastric juices, the test may include consuming food, drinking, or chewing. The duration of the test can vary, ranging from 2 to 12 h, depending on the specific objective.

- The doctor inserts a long, flexible tube through one of the patient's nostrils, passing it down the throat and into the stomach. The patient is then asked to swallow until the tube reaches the stomach.
- Gastric juices are subsequently aspirated and collected through the tube over several hours. The test can be uncomfortable and may cause initial gagging when the tube is inserted, as well as some irritation in the nostril or throat afterward.

14.8.12 LIVER SCAN

A non-invasive method is being employed in this isotope study to assess the liver's composition and functionality, aiding in the detection of conditions like cirrhosis, abscesses, or cystic growths. The procedure carries minimal risk, with an extremely remote possibility of isotope overdose. There is no requirement for specific preparation before the test. Typically, a structural scan takes approximately 45–60 min, while each function scan lasts around 30 min, and they are conducted in a series, with the final one scheduled after 24 h. Usually, this surgery can be done without hospitalization. The isotope is administered through an injection, and when necessary, a blocking agent may be used.

14.8.13 LOWER GI SERIES

Like the upper GI Series, this test, also known as a barium enema, utilizes barium as a contrast medium to obtain fluoroscopic views of the large intestine, aiding in the detection of polyps, tumors, gastroenteritis, telescoping bowel, irritable colon, or other causes of lower abdominal pain, as well as pus and blood, mucus in the stool.

- Before the test, the patient may be advised to fast or follow a liquid diet, and laxatives or enemas may be given to empty the bowels.
- The barium enema is administered while the patient is laying on their side on an inclined X-ray table. A tube is inserted into the rectum to deliver the solution.
- The table is rotated to ensure complete coating of the large intestine with barium, and X-ray images are captured.
- The patient is then instructed to use a bathroom or given a bedpan to eliminate as much barium as they can.

- Additional X-ray images are taken to examine the remaining thin film on the mucous lining of the intestine.
- In some cases, a double contrast study follows, involving careful injection of air into the colon to provide better contrast in imaging film and detect potential polyps.

14.8.14 PANCREAS SCAN

This scan is employed to examine the pancreas and evaluate its structure and function, aiding in the detection of conditions like cancer, cysts, or infections.

- Before the test, the patient follows a high-protein diet for several days and then fasts for 8 to 12 h, except for a glass of skim milk a few hours before the procedure.
- Another high-protein liquid is administered during the scan.
- The pancreatic scan can be done either as an inpatient or an outpatient procedure, and it normally takes 1–2 h.
- Occasionally, the isotope used in the scan may cause nausea or vomiting.

14.8.15 PROCTOSIGMOIDISCOPY

A proctoscope test, also referred to as proctoscopy, is performed to inspect the lower region of large intestine and rectum, aiming to identify conditions like hemorrhoids, polyps, abscesses, and the cause of bleeding. It is commonly used as a cancer screening tool especially in patients with age greater than 40.

- Before the test, the patient may be instructed to adhere to a liquid diet or fast for a few hours, and an enema or suppository may also be given.
- During the examination, the patient will be positioned in the knee-chest position, and the doctor will begin with a digital rectal exam using a gloved finger, followed by inserting the proctoscope.
- Comparable to a colonoscopy, the process enables a complete evaluation of the rectum and lower intestine.

14.8.16 SIALOGRAPHY

This test, often conducted alongside an upper GI series, involves a fluoroscopic examination (also known as barium milkshake or barium swallow) of the esophagus, stomach, and small intestine.

- Before the test, the patient may be advised to follow a low-residue diet for 2 or 3 days, and they will need to refrain from eating or smoking for 6–12 h prior to the exam.
- The procedure commences with the patient swallowing a "barium milkshake," a sweetened and flavored substance with a chalky taste that contains barium sulfate, serving as a contrast medium to highlight the upper digestive tract. Sometimes, a more liquid radiopaque material may be used instead.
- The patient is securely strapped to a tilting X-ray table, starting in a vertical standing position. During the test, the table is tilted at various angles to spread the contrast medium and obtain different views on the fluoroscope.
- Pressure may be applied to the patient's abdomen to further distribute the contrast medium.

If the test includes a small bowel exam, the patient will need to wait several hours for the barium to reach the intestine. They may be allowed to leave and return later or spend the time reading.

14.8.17 SMALL BOWEL EXAMINATION

This test, often performed in conjunction with an upper GI series, involves a fluoroscopic examination (also referred to as barium milkshake or barium swallow) of the esophagus, stomach, and small intestine.

- Prior to the test, the patient may be advised to adhere to a low-residue diet for 2 or 3 days, and they will be instructed to abstain from eating or smoking for 6–12 h before the exam.
- The procedure starts with the patient ingesting a "barium milkshake," a sweetened and flavored substance with a chalky taste that contains barium sulfate, used as a contrast medium to highlight the upper digestive tract. In some cases, a more liquid radiopaque material may be used instead.
- The patient is securely fastened to a tilting X-ray table, beginning in a vertical standing position. During the test, the table is tilted at various angles to distribute the contrast medium and obtain different views on the fluoroscope.
- Pressure may be applied to the patient's abdomen to further spread the contrast medium.

If the test includes a small bowel exam, the patient will need to wait several hours for the barium to reach the intestine. They may be allowed to leave and return later or spend the time reading.

14.9 THE NERVOUS SYSTEM

The central control system for our body, the human nervous system, is extremely important. The central nervous system, the peripheral nervous system, and the autonomic nervous system make up its three main subsystems. There are several methods available to evaluate the electrical transport of the nervous system in real time.

14.9.1 OPEN MAGNETIC RESONANCE IMAGING (MRI) EQUIPMENT

Magnetic resonance imaging (MRI) is a sophisticated medical imaging technique that produces detailed body images using the principles of nuclear magnetic resonance. It can create precise cross-sectional images of various body parts from multiple angles and directions in a relatively short time. This technology takes advantage of the presence of small biological magnets in the body. By stimulating the body with radio waves, the MRI alters the orientation of protons, then detects the electromagnetic signals emitted by the body at a specific frequency to create internal organ images.

MRI is especially effective in diagnosing brain and central nervous system disorders, as it can differentiate between various soft tissues in both healthy and diseased states. It offers advanced procedures, including high-resolution imaging, diverse vascular imaging techniques, and breath-hold imaging for the abdomen and chest. OPART® is an advanced MRI system that incorporates cutting-edge gradient, RF, and computer technology, enabling specialized clinical applications like MR Cholangio-Pancreatography, MR Myelography, and MR Urography. The open MRI system allows for examinations that may not be possible in closed systems, and it features an intuitive user interface for routine examinations.

14.9.2 CAT SCAN

X-rays and computer technology are used in a computerized tomography (CT) scan to produce three-dimensional images of any portion of the body.

The CT scanner consists of an X-ray source that emits beams passing through the organ and detected on the other side. X-ray emissions are captured from various perspectives by revolving the source and detector 360° around the patient. The computer then interprets this data and generates cross-sectional images displayed on a monitor, providing detailed views of soft tissues, blood vessels, muscles, and the brain.

CT scans are valuable for diagnosing brain injuries, detecting brain bleeding, and identify the tumor geometric characteristics. CT angiography produces detailed images of small blood vessels and hollow structures like the colon and airways of the lungs. Contrast agents may be used to enhance the images, either through injection or ingestion. Toshiba CT systems offer expanded clinical applications to address the specific needs of cardiology, trauma care, surgery, pediatrics, and oncology. They provide high-performance CT angiographies with multi-slice, half-second acquisition. Aspire CT technology ensures precision in scanning and intervention. Cardiac gating allows for cost-effective and rapid coronary artery screening. The advanced CAT Aquilion is designed for half-second, multi-slice scanning, providing detailed images with increased anatomical detail for electroencephalograms (EEG). The Aquilion CAT can withstand the g-forces generated during fast scanning. It can also be combined with Aspire CI for improved CT interventional applications. EEG can identify normal brain patterns, such as the alpha and theta waves, and is used to diagnose conditions like epilepsy and brain damage. EEG has helped categorize four stages of sleep. In a clinical electroencephalograph, the readout is a multichannel pen recorder with a pen for each channel, typically operating at 30 or 60 mm/s. Some electroencephalographs can interface with analog tape recorders for signal recording and playback.

14.9.3 THE ELECTRODE SHOCK

Individuals suffering from diabetic foot pain, sprains, fractures in the feet, ankles, and other pediatric conditions can now receive treatment at home using an electrode sock. This sock provides stimulation to the entire foot and ankle, promoting increased blood flow, which in turn reduces swelling and associated pain.

14.9.4 TRANSCUTANEOUS ELECTRICAL NERVE STIMULATOR (TENS)

Transcutaneous electrical nerve stimulation (TENS) delivers electrical impulses through lead wires and surface electrodes to the nerves beneath the skin. The stimulator converts the battery's direct current into stimulation pulses. Most stimulators have adjustable settings to control the intensity of stimulation by managing voltage, current, and pulse duration for each pulse. Specific sites on the body are targeted with electrodes for pain treatment. The current passes through the electrodes and into the skin, stimulating particular nerve pathways, resulting in a tingling or massaging sensation that helps reduce the perception of pain. When used as instructed, TENS is a safe, non-invasive, and drug-free approach to pain management.

14.9.5 MICROCURRENT ELECTRICAL NEUROMUSCULAR STIMULATOR (MENS)

Microcurrent electrical nerve stimulation (MENS) utilizes a low voltage current ranging from 1 to 1000 μA, while most TENS devices typically have 1–80 mA. Applying microcurrent to an injured area assists in readjusting the body's electrical current, enhancing ATP production, which in turn promotes improved healing and faster recovery. Additionally, it helps to alleviate perceived pain.

14.9.6 ELECTRICAL MUSCLE STIMULATOR (EMS)

- Preserving and enhancing range of motion in situations where the normal range of motion is reduced due to factors like fractures with subsequent immobilization, surgical procedures, or arthroscopy, especially in the shoulders, knees, and back.
- Using electrical stimulation to induce muscle contractions, thereby promoting muscle building and strengthening, which aids in preventing muscle wasting due to disuse.
- Alleviating muscle spasms.
- Facilitating muscle retraining and improving localized blood circulation.
- Electrical Muscle Stimulation (EMS) helps boost blood flow, thereby aiding in the prevention of venous thrombosis.

14.9.7 HIGH VOLTAGE PULSED GALVANIC STIMULATION (HVPGS)

High-voltage pulsed galvanic stimulation, or HVPGS, is used to promote wound healing, lessen discomfort and edema, treat carpal tunnel syndrome, and take care of diabetic foot issues. Sensory, motor, and pain-conducting nerve fibers are effectively stimulated by its special combination of very short pulse duration and high peak current, which allows for reasonably painless stimulation. With just minor thermal and electrochemical consequences, the HVPGS current passes through the skin.

14.9.8 INTERFERENTIAL GENERATOR

The delivery of therapeutic frequencies to tissues using interferential electrical stimulation is efficient. Interferential stimulators use a set carrier frequency of 4000 Hz/s and an adjustable frequency of 4001–4400 Hz/s, unlike traditional TENS and neuromuscular stimulators. These frequencies combine to provide the desired signal frequency (Interference frequency) when they are heterodyned. At the intersection of the electrodes, interferential stimulation is focused and penetrates both the skin's surface and deeper tissues. Skin resistance is eighty (80) times less at 4000 Hz (interferential unit) than at 50 Hz (in the TENS range). In treating patients with severe pain, increasing bone healing in delayed and nonunion fractures, stimulating deep skeletal muscle in venous insufficiency, and decreasing patient activity of specific cervical and lumbosacral sympathetic ganglia, interferential current is very effective.

14.9.9 CLINICAL ELECTROMYOGRAPHY (EMG)

The two primary parts of clinical electromyography are needle electromyography (needle EMG) and nerve conduction studies (NCS) (Figure 14.7C-E). It functions as a diagnostic instrument to assess the peripheral nervous system.

A noninvasive stimulator delivers short electrical impulses to a peripheral nerve through the skin during NCS tests. The impulse is then transmitted via the nerve, and the reaction is captured by electrodes positioned far from the stimulation point. Since the latency, time taken for stimulus to get to the recording electrodes, can be precisely determined, the transmission velocity may also be calculated. Nerves can be evaluated on both a motor and sensory level. Faster electrical impulse transmission occurs in healthy nerves than in unhealthy ones. In the same session, NCS and needle EMG are frequently carried out in tandem because they are complementary for the majority of diseases. The suggested circumstances for carrying out an EMG study are shown in Table 14.1.

- Carpal tunnel syndrome (CTS) is often frequent diagnosis observed in EMG results. Electrophysiological features include: prolonged median nerve motor latency at the wrist ("terminal" or "distal" latency); normal median motor conduction velocity in the forearm; and slow median sensory conduction velocity across the wrist.
- Ulnar neuropathy: Key EMG findings for elbow involvement are: slowed ulnar motor conduction velocity across the elbow; conduction hinder in the elbow region; and needle EMG evidence of denervation in ulnar nerve- innervated muscles.

TABLE 14.1 The conditions for EMG.

Condition	Yield of EMG	EMG features	Comment
CTS	91–98%	+ Median n. TL, + Median SCV, nl ulnar NCS	Most common entrapment neuropathy
Myopathy	Varies*	Small-short MUP	Polymyositis is most common
Polyneuropathy	>80%	Diffusely slow	Diabetes most common etiology in USA
Radiculopathy	80–100%	Denervation in paraspinals and myotome	EMG not positive before 7–20 days
Ulnar neuropathy (elbow)	73–91%	Slow MCV across elbow, CB	Wrist entrapment is less frequent

*The yield of EMG varies with the etiology and severity of the myopathy.

Abbreviations: n, nerve; TL, terminal latency; NCS, nerve conduction studies; SCV, sensory conduction velocity; MCV, motor conduction velocity; CB, conduction block; MUP, motor unit potentials.

- Peripheral neuropathy: EMG features of peripheral neuropathy consist of: generalized slowing of nerve conduction velocity in sensory and/ or motor nerves, usually more prominent in the lower limbs; sensory nerves are often affected earlier than motor nerves; and needle EMG evidence of denervation in distal muscles may not always be present and is not always necessary for diagnosis. NCS results in diagnosis in about 80% of cases.
- Radiculopathies: EMG features include: needle EMG evidence of denervation affecting paraspinal muscles; and denervation affecting limb muscles with the same root innervation (myotome). The diagnostic yield is at least 80%.

- Myopathy: The timing of symptom onset is crucial, as needle EMG evidence of denervation takes at least 7–10 days to develop following nerve injury. Therefore, a negative needle EMG examination conducted within 10 days of symptom onset should be repeated later. To ensure enough time has passed for detectable needle EMG findings, a 2-week interval is recommended.

14.10 THE URINARY SYSTEM

By removing water and waste from the blood, the kidneys play a crucial part in the urinary system. These waste materials are eliminated by the kidneys in the form of urine. An adult bladder holds about 475 ml of urine and excretes from 0.7 to 2 L of urine daily. The sweat has the same composition as the urine.

14.10.1 THE CYSTOSCOPE

In a medical procedure called a cystoscopy, a specialized endoscope with a tiny camera is used to view inside urethra as well as bladder. This test serves several purposes, including diagnosing and evaluating urinary tract disorders, checking for bladder or urethral cancer, diagnosing enlarged prostate in men, determining the cause of pain during urination, and diagnosing recurrent bladder infections. There are two types of cystoscopies:

- The standard rigid cystoscope requires the patient to lie in the lithotomy position.
- The flexible cystoscope, on the other hand, is easier to insert and does not require the lithotomy position for insertion.

Before the procedure, the urethra is cleansed, and a local anesthetic is applied. The cystoscope is then inserted through the urethra into the bladder. A water or saline solution is used to fill the bladder during the examination. Additionally, a small specimen can be obtained through the cystoscope for further examination if needed.

14.10.2 ENDOSCOPE

Endoscopy is a method of examining and inspecting the interior of body cavities using a device called an endoscope. This device utilizes fiber optics and powerful lenses to provide illumination and visualization inside a joint. The endoscope consists of two fiber optic lines—one for carrying light into the body cavity (light fiber) and the other for transmitting the reflected light on to viewing lens (image fiber). Additionally, there is a separate port that allows for drug administration, suction, and irrigation. Various tools, including forceps, scissors, brushes, snares, and baskets, can also be introduced through this port for the purposes of tissue excision, sampling, or other diagnostic and therapeutic procedures. The type of anesthesia used during endoscopy will depend on the technique.

Endoscope can be used in conjunction with cameras or video recorders to document images of the joint's interior or to record the entire endoscopic procedure. Modern endoscopes now come with digital capabilities, enabling manipulation and enhancement of video images. A rigid endoscope is commonly employed for arthroscopy, where it is used to remove torn cartilage from the joint's bearing surfaces. Biopsy, involving the sampling of tissue for pathological testing, can also be performed under the guidance of an endoscope.

14.10.3 NEEDLES AND SYRINGES

Needles come in many different sizes and diameters, or gauges (Figure 14.8 F–H). Needles also have different bevels on the tip for better performance. Syringes are available in many sizes and shapes. Many are specialized or used for just one thing. These can be disposable syringes or re-usable syringes. The syringe is used to draw blood from our body.

14.11 THE ENDOCRINE SYSTEM

The radio-immuno analysis is used for diagnosis of the endocrine diseases. It is based on the usage of specific antibodies for a specific hormone. A urine analysis is also used to quantify parameters in the diagnosis of glycosuria or concentration of glucose in urine. It can also determine the quantity of hormones excreted by the urine that gives an idea of the functioning of the endocrine glands. The X-rays are used: to observe the endocrine glands; and to see the shape and size of any disorder.

14.12 MICROSCOPE

The microscope allows for larger views for examination and analysis. The microscope may be static, or dynamic. An optical or light microscope, which has magnifying powers of 1 to 10 magnitudes and resolutions of roughly 0.01 mm, is the most used form of microscope. The compound optical microscope has magnifying of 2.5 to 1000 magnitudes and resolutions of 0.01 to 0.0002mm. The electron microscope uses a beam of electrons and has a magnifying power of 1000 to 1,000,000 magnitudes with resolutions from 0.001 to 0.00000001mm. The acoustic microscope or scanning tunneling microscope uses other physical effects to view individual molecules and atoms.

14.13 PHYSICAL THERAPY

14.13.1 HIGH VOLTAGE PULSED GALVANIC (HVPG) STIMULATION

In order to treat and offer therapeutic relief from the following conditions, HVPG stimulation is used: to calm muscular spasms; to ease arthritis pain; to increase local blood circulation; and ensure or bring movement variability. It is also used in edema reduction, reducing motor spasticity, preventing or retarding disuse atrophy, and managing post traumatic or post-surgical pain conditions, and it is an ideal therapy for carpal tunnel syndrome. There are two types of HVPG: portable galvanic stimulator and pulsed-Galvanic stimulator.

14.13.2 ELECTRICAL STIMULATING CURRENT

Electrical stimulating units utilize wavelengths with frequencies ranging from 1 to 2000 Hz. These stimulators offer the flexibility to adjust the frequency output to achieve the desired physiological response. By releasing β-endorphin at lower frequencies or by stimulating cutaneous sensory nerves at high frequencies, the stimulating currents can effectively control pain. They can cause muscle contraction, relaxation, or tetany depending on the frequency and kind of current (alternating or direct). They also cause a net flow of ions, which results in a chemical change, in the case of direct current.

14.13.3 INFRARED MODALITIES

Inducing localized or occasionally extensive warmth or cooling of the surface tissues is accomplished using these methods. They can either increase or reduce circulation depending on whether heat or cold is applied. Ice at a temperature of 0 °C and infrared bulbs at over 3000 °C are examples of infrared modalities.

As for ultraviolet therapy, its physiological effects are primarily confined to the cutaneous layers of the skin, penetrating to a maximum depth of approximately one mm. Sports therapists seldom use ultraviolet therapy to treat cutaneous lesions.

14.13.4 SHORTWAVE AND MICROWAVE DIATHERMY

Diathermies are characterized by high-frequency currents that oscillate at beyond million cycles/second. Electrotherapeutic devices utilizing shortwave and microwave technology come with preset frequencies and wavelengths, which cannot be adjusted. Microwave units have shorter wavelengths than shortwave diathermy units and are set at wavelengths of 33 or 12 cm. Shortwave diathermy units are fixed at either 13.56 MHz or 27.12 MHz. Microwaves can penetrate slightly deeper than shortwave diathermy.

14.13.5 THE ACOUSTIC SPECTRUM AND ULTRASOUND

Ultrasound devices generate energy that belongs to the acoustic spectrum rather than the electromagnetic spectrum. The frequencies of ultrasound waves typically vary from 700,000 to 1,000,000 cycles per second. At a frequency of 1 megacycle, approximately 50% of the energy produced can pierce to a deepness near to 5 cm. Therefore, ultrasound is a preferred option for achieving deeper penetration into tissues.

14.13.6 ELECTRICAL TREATMENTS

Electricity can be beneficial in treating injuries as it elicits responses in all biological tissues. The type and extent of response depend on the tissue type, its characteristics, and the nature of the current used. Electrical currents are used by therapists for a variety of purposes, including as stimulating muscles or nerves to contract muscles, relieving pain by stimulating sensory nerves, and generating electrical fields in biological tissues to speed up the healing process. Ultrasound devices, for instance, operate within the acoustic spectrum, with frequencies ranging from 700,000 to 1,000,000 cycles per second. At 1 megacycle, about 50% of the energy can penetrate to a depth of approximately 5 cm, making it a suitable choice for deeper tissue penetration.

Electrical currents have substantial impacts on muscles and nerves. The stimulus's intensity and duration must be high enough to exceed the membrane's fundamental threshold for excitation in order for it to be effective. It alters the membrane, leading to depolarization and resulting in an action potential. Various clinical responses can occur when electrical systems are applied to soft tissues, inducing ion movement and generating action potentials. Electrical muscle stimulation can cause muscle contraction through motor nerve stimulation or muscle membrane depolarization. The use of alternating current (AC) and direct current (DC) has different effects on nerve stimulation and chemical changes in the body.

Electrically produced muscular contractions can be used to strengthen muscles, delay atrophy, and perform muscle reeducation. The positioning of the electrodes is essential in the therapy regimens intended to stimulate sensory nerves and alleviate pain. Proper electrode placement, either around the painful area, over specific dermatomes, or myotomes, close to the corresponding spinal cord segment, or near superficial peripheral nerves, can effectively reduce or eliminate pain. Combining different electrode placement systems or using crossing patterns can also yield positive results. Systematic and organized electrode placement evaluation is essential for the success of the treatment.

14.13.7 PROTOCOL HYPER STIMULATION AND B-ENDORPHIN

In this approach, a specific point stimulation setup is required, which involves using a large dispersive pad along with a small pad or handheld probe as the point electrode. The electrode placement is carefully chosen, and the point electrode is applied to the area of interest. The intensity of the electrical current is gradually increased until the patient perceives it. The patient is then asked to indicate any changes in their sense of intensity while the probe is moved about the area. Once the location with the maximum intensity perception is identified, the current intensity is increased to the highest tolerable level.

It is possible to combine TENS with strong point stimulation. To keep the patient at ease, TENS treatments are modified as necessary. When used frequently, strong point stimulation offers great pain relief for a predetermined amount of time and can improve overall pain suppression. Daily, severe point stimulation, though, might eventually cause central nervous system adaptation, which would lessen the impact of this kind of stimulation.

14.13.8 CLINICAL USES OF LOW-VOLTAGE UNINTERRUPTED DIRECT CURRENT

The therapeutic benefits of this therapy are attributed to the polar and vasomotor effects, resulting in acid reactions around the positive pole and alkaline reactions at the negative pole. Additionally, there is a vasomotor effect observed for skin, leading to increased blood flow between the electrodes and through the treated area.

- Promotion of wound healing: Treated ulcers exhibit faster healing rates compared to untreated ulcers. For the initial three days, the negative electrode should be placed at the wound site, while the positive electrode is positioned approximately 25 cm away from the wound.
- Promotion of fracture healing: Fracture healing may be expedited by applying direct current through the fracture site.
- Iontophoresis: In order to apply direct current, two different methods are used: (1) using an active pad over gauze soaked in an ion-containing solution, and (2) using an immersion technique where the active electrode is suspended in a container of the solution and the area to be treated is submerged in the container.

14.13.9 THERMO THERAPY

The term "thermotherapy" refers to the employment of heat therapies to raise tissue temperature. Cryotherapy, often known as cold therapy, works best during the initial phases of the wound's recovery. The step of injury therapy known as reconditioning can continue to use cold applications. Any cryotherapy or thermotherapy method that transfers heat through water is referred to as hydrotherapy. The following physiologic effects of cold have been documented:

- Reduced blood flow, leukocyte and phagocyte delivery, edema formation, and accumulation.
- A reduction in the local temperature, perhaps to a significant depth.
- A reduction in venous and lymphatic outflow.
- A slower metabolism.
- A reduction in muscular excitability.
- Reduced nerve conduction velocity; decreased muscle spindle depolarization.
- A decrease in muscle viscosity.
- Significant anesthetic effects.
- Capillary and arteriole vasoconstriction.

Clinical Use of the Infrared Energy

The effects of infrared radiation are generally superficial and have a direct impact on the cutaneous blood vessels and nerve receptors. Infrared energy has the ability to penetrate more than 1 cm deep. The hypothalamus is stimulated to increase blood flow to the underlying tissues when this energy is absorbed by the skin for a long enough time to raise the temperature of the blood in circulation. On the other hand, the skin's ability to absorb cold can reduce blood flow in the treated area via a related mechanism. Therefore, it is advised to choose modalities like diathermy or ultrasound if the primary goal is to increase tissue temperature and blood flow to deeper tissues. Conversely, the use of ice or cold is the most appropriate method if the main objective is to lower tissue temperature and inhibit blood flow to an injured location.

Cryotherapy

Cryotherapy involves the use of cold or ice to treat acute and subacute injuries and to reduce discomfort during athletic reconditioning and rehabilitation. Ice packs, cold whirlpools, ice massage, commercial chemical cold spray, and contrast baths are a few examples of cryotherapy techniques. When used, cryotherapy frequently results in a three- to four-stage feeling.

Ice massage can be administered by a therapist or self-applied by the individual. It is particularly beneficial when stretching activities are involved. There are five types of cold treatments:

- Athlete position: Depending on the area to be treated, the individual can be positioned side-lying, prone, supine, hook-lying, or sitting.
- Self-treatment: When athletes can comfortably access the area to be treated on their own, this method is appropriate.
- Circular motion: Ice massage is applied in a circular pattern, with each successive stroke covering half of the previous stroke.
- Longitudinal strokes: Each stroke of an ice massage is applied longitudinally, overlaying 50% of the stroke before it.
- Peripheral coverage: Use a regular pattern of circular and longitudinal strokes to apply ice for 15–20 min.

Physiologic responses: (1) The four stages of a cold are cold, stinging, burning, and numbness. (2) The skin becomes reddened (erythematous) due to blanching or a lack of blood in the capillary bed.

Commercial (cold) hydrocollator treatment includes—Athlete position: Depending on the location to be treated, side laying, prone, supine, hooking, or sitting; To maintain proper positioning of the cold pack during treatment, the patient only needs to remain still; The skirt needs to be fitted with a cold pack; To prevent the loss of cold, the cold pack should be covered with a towel. A 20-min timer should also be set.

The cold whirlpool is recommended for acute and subacute conditions where it is wanted to exercise the injured pan while receiving a cold treatment. Treatment consists of: Body segment must be submerged; water temperature should be set at 65 to 80 °F for entire body immersion; treatment duration should be 5–15 min. Temperature should be set at 50 to 60 °F.

Although they don't offer sufficient deep penetration, cold sprays (such those containing fluori-methane or ethyl chloride) do aid in reducing muscle spasm. The recommended course of treatment is from distal to proximal, and the area to be treated should be sprayed before being stretched. Use a fast jet stream spray or stroking motion; there should be no frosting and only light coating; Acupressure may be combined with cold sprays; treatment duration needs to be determined for each body section.

Treatments for subacute swelling, gravity-dependent Contras edema, and vasodilation include contrast baths. Alternating hot and cold immersions are used as part of the treatment, which should last at least 20 min. Five 1-min cold immersions and five 3-min warm immersions should make up each treatment.

Ice immersion: The patient should be positioned comfortably and the wet area should be chosen (where water spills are not a problem). Similar to the contrast bath, the immersion should be maintained until the desired outcomes are obtained. The container needs to be big enough to accommodate the movement of the body segment if cryokinetics is used in the therapy. Similar to a cold whirlpool, ice immersion can cause body segments to move in a gravity-dependent manner.

Cryokinetics

Cryokinetics is a method that involves integrating different cryotherapy treatments with exercise. The goal of cryokinetics is to gradually restore normal range of motion through controlled activities after completely analgesicing the damaged area. The exercises are continued until pain sensation reappears, at which point the injured area is iced again for several minutes until it becomes numb, and the exercise is resumed. This series is repetitive 4–5 times.

Thermotherapy

Superficial heat applications are preferred for addressing discomfort in areas such as the neck, back, low back, and pelvic regions. The main objective of thermotherapy is to enhance blood flow and elevate muscle temperature to promote analgesia, improve cellular nutrition, reduce edema, and eliminate inflammatory byproducts and metabolites.

Hydrotherapy

Hydrothermal therapy encompasses techniques like hot baths, saunas, and wraps. Its healing properties are derived from the mechanical and thermal effects it induces. By utilizing the body's response to hot and cold stimuli, prolonged heat application, water pressure, and sensory experiences, hydrotherapy aims to invigorate and relax. The practice is employed to invigorate the body, improve digestion, circulation, and the immune system, as well as alleviate pain. Additionally, hydrothermal therapy has a calming effect on the lungs, heart, stomach, and endocrine system through stimulation of nerve reflexes along the spinal cord.

Sauna Bath

Both saunas and steam rooms have advantages for the body. While the moist air of a steam bath has a pleasing effect on the respiratory system, saunas quickly remove pollutants through the skin. Both increase heart rate, promote blood flow, and modulate the immunological system. They increase mucosal secretions in the respiratory system, encourage hormone synthesis, and open airways, which eases breathing resistance. These treatments also control the vegetative system, promote relaxation, and enhance one's view on life. Children as young as two or three are considered safe to use saunas.

Full and Partial Immersion Baths

- Cold arm bath: Fill a basin with cold water, submerging the arm several inches above the elbow.
- Cold foot bath: Place the feet in a footbath with cold water that is at least calf-deep. If you have chilly feet, extremely high blood pressure, an irritable bladder, a urinary tract infection, diabetes, or a vascular occlusion, you should not use this medication.
- Cold packs: Wrap the affected area in a cloth that has been covered with cooled cataplasm. As an alternative, you might use a plastic bag filled with crushed ice that you would apply for a minute before removing.
- Herbal bath: Combine a handful of each of the following herbs: valerian, lavender, linden, chamomile, hops, and burdock root to

make a calming bath. To your bathwater, add this combination, and soak for 30 min.

- Paraffin bath: An efficient technique for applying localized heat.
- Rising temperature arm bath: Similar to the rising temperature foot bath. Follow it with a cold arm douche and then rest for half an hour.
- Rising temperature footbath: Immerse the feet in a foot bath with water at body temperature. Gradually add hot water to reach a final temperature of 103°F. The procedure should last 10–15 min and can be done daily.
- Rising temperature hip bath: Fill a tub with a hand's breadth of tepid water and gradually add hot water until it reaches the navel. Maintain a final temperature of 103–104 °F. Afterward, proceed to bed. 15–30 min should be spent in the bath, but no more than 3 times per week.
- Sitz bath: Usually consumed as a sitz bath that is either warm, rising in temperature, or chilly. Before taking a sitz bath, the feet should be warmed in a warm foot bath. The body's exposed areas should be covered. While a warm or rising-temperature sitz bath is suitable for bladder problems, an irritable bladder, inflammation or infection of the prostate, or pregnancy preparation, a cold sitz bath is beneficial for hemorrhoids or anus inflammation.
- Warm packs: Apply a wrapping towel to the affected area after soaking it in a hot herbal infusion or stew.

14.13.10 INFRARED LAMPS

Infrared lamp therapy generates a rise in temperature in the superficial tissues without direct contact with the body. The radiant heat produced has a limited penetration of less than 1 mm. To enhance the heating effects, a warm moist towel is placed over the specific body segment being treated, while the rest of the body is covered with dry towels. There are two types of infrared generators: luminous and nonluminous. Luminous generators include incandescent bodies and lamps with tungsten and carbon filaments.

14.13.11 FLUID THERAPY

The fluido-therapy device uses a suspended air stream to offer therapeutic advantages for healing and rehabilitation. It combines pressure oscillations, levitation, sensory stimulation for desensitization, application of heat, and massage. This dry, organic medium, in contrast to water, is kind to the skin and doesn't produce temperature shocks. Unlike aqueous or paraffin heat transfer methods, it enables higher treatment temperatures. Clinical fluidotherapy treatments have been effective in lowering pain and accelerating wound, acute injury, edema, and blood flow difficulties. The hand or foot is submerged in the fluidotherapy unit after protective toweling has been placed on the body segment's interface and the patient has been made comfortable. The suggested treatment time ranges from 15 to 20 min.

14.13.12 ULTRASOUND

Ultrasound is a valuable tool for therapists in their efforts to restore athletes to their pre-injury condition. Therapeutic ultrasound units used on the musculoskeletal system typically operate within a frequency range of 800,000 to 1,000,000 Hz. The effectiveness of ultrasonic energy in biological tissue is influenced by factors such as tissue absorption and density characteristics.

- Ultrasound and underwater therapy can effectively treat irregular surfaces, such as the hand. During the treatment, the sound head is moved in a circular motion and does not need to be in direct contact with the body part. The applicator should be positioned approximately ½ inch away from the body part, and the intensity setting remains the same as for regular application.
- Pulsed ultrasound: Pulsed beams, as opposed to continuous beams, allow for rest periods to facilitate cooling.
- Simultaneous ultrasound and electrical stimulation units combine the use of ultrasound with either low-voltage AC or high-voltage DC.
- Coupling agents: The coupling medium used should possess good lubricating properties, enabling smooth gliding of the applicator against the skin without significant friction. Commercial gels, mineral oil, and de-gassed water are all viable options for coupling agents, with no significant differences between them.

14.13.13 INTERFERENTIAL THERAPY

Interferential Therapy is a method that utilizes low-frequency electrical stimulation to target muscle and nerve tissues without causing pain. It involves the use of two medium-frequency currents passing through the tissues simultaneously, resulting in interference and the creation of a low-frequency beat frequency. Different tissues respond optimally to specific stimulation bands, which can be estimated based on factors such as conduction velocity, latency, and refractory period. For example, motor nerves are stimulated at 10–50 Hz, while nociceptive fibers respond well to 90–150 Hz (with 130Hz being particularly effective).

The clinical application of Interferential Therapy is guided by the above data and a deep understanding of how stimulated tissues behave physiologically. There are four main clinical applications for Interferential Therapy: pain relief, muscle stimulation, increased blood flow, and reduction of edema.

14.14 SUMMARY

Biomedical engineering has made significant contributions through the advancement of biomedical instrumentation. These developments have produced a variety of biomedical devices and clinical methods, including electronic patient monitoring, critical care medicine, and assistive technology for the disabled. This chapter specifically focuses on different types of instruments used to measure fluid properties. For instance, the sphygmomanometer is utilized to measure blood pressure, while the Schirmer strip quantifies total tear secretion. This chapter complements the contents of Chapter 15, titled "Medical Imaging of Human Body Fluids: Principles and Diagnostics," authored by Bhowmik et al., included in this book.

KEYWORDS

- **acoustic spectrum**
- **active electrode**
- **amplitude**
- **anesthesia**
- **biomedical instruments**
- **bronchoscope**
- **carpal tunnel syndrome**

REFERENCES

1. Goyal, M. R. Instrumentation and Measurements of Body Fluids. In *Biofluid Mechanics of Human Body Systems*, 1st ed.; CRC and Apple Academic Press, 2013; pp 441–520.
2. Geddes, L. A.; Baker, L. E. *Principles of Applied Biomedical Instrumentation*, 3rd ed.; John Wiley & Sons, 1991; p 992.
3. Bronzino, J. D.; Peterson, D. R., Eds. *The Biomedical Engineering Handbook*, Vo. 1–4; CRC Press LLC (Taylor & Francis Group): Boca Raton, 2015; p 5422.
4. Cobbold, R. S. C. *Transducers for Biomedical Measurements: Principles & Applications*; Wiley Interscience, p 486.
5. Goyal, M. R. Proceedings of Congress on: Biofluid Dynamics of the Human Body Systems, 2003. http://faculty.eng.fiu.edu/~mgoyal; (accessed 31 July 2023).
6. Webster, J. G. *Medical Instrumentation*, 3rd ed.; John Wiley & Sons, 1998; p 736.
7. Anandanatarajan, R. *Biomedical Instrumentation and Measurements*, 2nd ed.; PHI Learning Pvt. Ltd., 2010; p 324.
8. Rocha-Cabrera, P.; Reyes-Batlle, M.; Martín-Navarro, C. M.; Dorta-Gorrín, A.; López-Arencibia, A.; Sifaoui, I.; Martínez-Carretero, E.; Piñero, J. E.; Martín-Barrera, F.; Valladares, B. Detection of Acanthamoeba on the ocular surface in a Spanish population using the Schirmer strip test: pathogenic potential, molecular classification and evaluation of the sensitivity to chlorhexidine and voriconazole of the isolated Acanthamoeba strains. Journal of Medical Microbiology **2015**, *64* (8), 849–853.
9. Maheshwari, R.; Choudhari, N. S.; Singh, M. D. Tonometry and Care of Tonometers. Journal of Current Glaucoma Practice **2012**, *6* (3), 124.
10. Lowe, J.; Cleland, C. R.; Mgaya, E.; Furahini, G.; Gilbert, C. E.; Burton, M. J.; Philippin, H. The arclight ophthalmoscope: a reliable low-cost alternative to the standard direct ophthalmoscope. *Journal of ophthalmology* **2015**, 2015.
11. Seppänen, T. M.; Alho, O.-P.; Seppänen, T. Dynamic changes in heart rate variability and nasal airflow resistance during nasal allergen provocation test. *Journal of Healthcare Engineering* **2016**, 2016.
12. Fung, A. G.; Tan, L. D.; Duong, T. N.; Schivo, M.; Littlefield, L.; Delplanque, J. P.; Davis, C. E.; Kenyon, N. J. Design and benchmark testing for open architecture reconfigurable mobile spirometer and exhaled breath monitor with GPS and data telemetry. *Diagnostics* **2019**, *9* (3), 100.
13. Xenogiannis, I.; Tajti, P.; Hall, A. B.; Alaswad, K.; Rinfret, S.; Nicholson, W.; Karmpaliotis, D.; Mashayekhi, K.; Furkalo, S.; Cavalcante, J. L. Update on cardiac catheterization in patients with prior coronary artery bypass graft surgery. *JACC: Cardiovascular Interventions* **2019**, *12* (17), 1635–1649.
14. Drew, C.; Hughes, D. Characteristics of the Vitalograph Spirometer. *Thorax* **1969**, *24* (6), 703–706.
15. Elgendi, M.; Eskofier, B.; Dokos, S.; Abbott, D. Revisiting QRS detection methodologies for portable, wearable, battery-operated, and wireless ECG systems. *PloS one* **2014**, *9* (1), e84018.
16. Qiu, L.; Chuttani, R.; Pleskow, D. K.; Turzhitsky, V.; Khan, U.; Zakharov, Y. N.; Zhang, L.; Berzin, T. M.; Yee, E. U.; Sawhney, M. S. Multispectral light scattering endoscopic imaging of esophageal precancer. *Light: Science & Applications* **2018**, *7* (4), 17174-17174.

17. Xenogiannis, I.; Tajti, P.; Hall, A. B.; Alaswad, K.; Rinfret, S.; Nicholson, W.; Karmpaliotis, D.; Mashayekhi, K.; Furkalo, S.; Cavalcante, J. L. Update on Cardiac Catheterization in Patients with Prior Coronary Artery Bypass Graft Surgery. *JACC: Cardiovasc. Intervent.* **2019**, *12* (17), 1635–1649.
18. Elgendi, M.; Eskofier, B.; Dokos, S.; Abbott, D. Revisiting QRS Detection Methodologies for Portable, Wearable, Battery-Operated, and Wireless ECG Systems. *PloS One* **2014**, *9* (1), e84018.
19. Srinivas, N.; Babu, A. V.; Rajak, M.; Ahmed, S. M. Identification of Cardiac Arrhythmia with Respect to ECG Signal by Neural Networks and Genetic Programming. *Can. J. Artif. Intell.* **2013**, *2* (2), 479–486.
20. Wang, Y.; Shen, P.; Zheng, M.; Fu, P.; Liu, L.; Wang, J.; Yuan, L. Influence of Impeller Speed Patterns on Hemodynamic Characteristics and Hemolysis of the Blood Pump. *Appl. Sci.* **2019**, *9* (21), 4689.
21. Polaschegg, H.-D. Hemodialysis Machine Technology: A Global Overview. *Expert Rev. Med. Devices* **2010**, *7* (6), 793–810.
22. McDonald Jr, H. P. An Automatic Peritoneal Dialysis Machine: Preliminary Report. *Am. Soc. Artif. Intern. Organs* **1965**, *11* (1), 83–85.
23. Ahmed, R.; Santhirakumar, K.; Butt, H.; Yetisen, A. K. Colonoscopy Technologies for Diagnostics and Drug Delivery. *Med. Devices Sens.* **2019**, *2* (3–4), E-article ID 10041.
24. Qiu, L.; Chuttani, R.; Pleskow, D. K.; Turzhitsky, V.; Khan, U.; Zakharov, Y. N.; Zhang, L.; Berzin, T. M.; Yee, E. U.; Sawhney, M. S. Multispectral Light Scattering Endoscopic Imaging of Esophageal Precancer. *Light: Sci. Appl.* **2018**, *7* (4), 17174–17174.
25. Lee, S.-H.; Park, Y.-K.; Cho, S.-M.; Kang, J.-K.; Lee, D.-J. Technical Skills and Training of Upper Gastrointestinal Endoscopy for New Beginners. *World J. Gastroenterol.* **2015**, *21* (3), 759–768.
26. Siddiqui, Z.; Singh, P.; Kushwaha, S.; Srivastava, R. MRI and Fear of Confined Space: A Cause and Effect Relationship. *Int. J. Contemp. Med. Surg. Radiol.* **2017**, *2* (1), 19–24.
27. Bess, D. W.; Roberge, E. A. Battlefield Teleradiology. *Curr. Trauma Rep.* **2016**, *2*, 173–180.
28. Costa, Á.; Hortal, E.; Iáñez, E.; Azorín, J. M. A Supplementary System for a Brain-Machine Interface Based on Jaw Artifacts for the Bidimensional Control of a Robotic Arm. *PloS One* **2014**, *9* (11), e112352.
29. Ragnerius, A.; Widelund, F. Automated Foot-Strike Pattern Recognition Using a Sock with Textile Piezoelectric Sensors. Chalmers University of Technology, Gothenburg, Sweden, 2016. https://odr.chalmers.se/server/api/core/bitstreams/98283e85-30b0-40d2-b43d-7868246ad1ea/content (accessed 31 July 2023).
30. Matran-Fernandez, A.; Rodríguez Martínez, I. J.; Poli, R.; Cipriani, C.; Citi, L. SEEDS, Simultaneous Recordings of High-Density EMG and Finger Joint Angles During Multiple Hand Movements. *Sci. Data* **2019**, *6* (1), 186.
31. Wang, W.-H.; Hung, Y.-C.; Hsu, S. P.; Lin, C.-F.; Chen, H.-H.; Shih, Y.-H.; Lee, C.-C. Endoscopic Hematoma Evacuation in Patients with Spontaneous Supratentorial Intracerebral Hemorrhage. *J. Chin. Med. Assoc.* **2015**, *78* (2), 101–107.
32. Taslakian, B.; Ingber, R.; Aaltonen, E.; Horn, J.; Hickey, R. Interventional Radiology Suite: A Primer for Trainees. *J. Clin. Med.* **2019**, *8* (9), 1347.
33. Karpuz, M.; Özer, A. Y. Syringes as Medical Devices. *FABAD J. Pharm. Sci* **2016**, *41*, 27–37.

34. Szekeres, M.; MacDermid, J. C.; Birmingham, T.; Grewal, R.; Lalone, E. The Effect of Therapeutic Whirlpool and Hot Packs on Hand Volume During Rehabilitation After Distal Radius Fracture: A Blinded Randomized Controlled Trial. *Hand* **2017**, *12* (3), 265–271.
35. Blank, V.; Wiegand, J.; Keim, V.; Karlas, T. Evaluation of a Novel Tomographic Ultrasound Device for Abdominal Examinations. *PloS One* **2019**, *14* (6), e0218754.
36. Prentice, William, *Therapeutic Modalities in Rehabilitation*, 3rd ed.; McGraw-Hill Medical Publishing Division, The McGraw-Hill Companies: New York, 2005, p. 608.

CHAPTER 15

Medical Imaging of the Human Body Fluids: Principles and Diagnostics

ARKA BHOWMIK, MANPREET SINGH, and RUPESH SINGH

15.1 INTRODUCTION

Medical imaging is a specialized discipline of medical physics, radiology, medical informatics, and biomedical engineering that requires seeing the operating principle of image acquisition systems, image reconstruction, image analysis, and its applications in diagnostics and treatment. We need medical imaging to quantify (a) the disease state in the human body, (b) the structural changes or damages, and (c) the functional changes due to diseases and treatment procedure. In medical imaging techniques, electromagnetic and mechanical energy/sound wave provides an approach to interact, penetrate, and image tissue. In fact, the strength of wave energy defines the image quality. In practice, a compromise is often made between the strength of energy (or dose) and quality of the image since many forms of energy can induce irreversible damage to a patient. Hence, safety should be of paramount importance. Imaging systems can be broadly classified based on their ability to resolve the details of the tissue and/or organs. For instance, electron, optical and acoustic microscopes can provide details as small as a few nanometers to micrometers, often employed based on their resolution and the ability to resolve small structural changes. On the other hand, imaging methods such as MRI and CT can resolve both structural and functional details in the range of 100–500 μm with the aid of an appropriate contrasting agent, but PET, ultrasound, and optical imaging can resolve image details of the order of a few millimeters.

It is common practice today to use imaging techniques for accurate detection of structural and functional changes in a medical condition on different

Biofluid Dynamics of Human Body Systems: Expanded and Revised Edition.
Megh R. Goyal, Arka Bhowmik, & Anamika Chauhan (Eds.)

physical scales. These physical scales are nano- (molecules), micron- (cell), and millimeter (tissue) level, since the inherent property of biological sample varies both spatially and temporarily at these scales due to disease and treatment. This enables a better understanding of the underlying symptoms and enhances treatment by specifically targeting.

This chapter briefly illustrates the operating principle of common medical imaging systems and their application in body fluid imaging. A methodical elaboration of working principle, inherent physics, image construction, and applications involving different medical imaging systems is presented here. The chapter is well-suited for medical physics and biomedical engineering students or researchers looking for a basic understanding of medical imaging devices and its application in imaging/measuring the changes in biofluid with particular emphasis in disease diagnosis.

15.2 ESSENTIAL MEASURING INSTRUMENTS FOR BIOFLUID DYNAMICS

Biofluid dynamics is a specialized field of biomedical engineering, suitable for predicting the flow behavior of body fluid and its interrelationship with diseases. The instruments for measurement of biofluid dynamics can be classified into two categories: (a) direct measurement devices; and (b) indirect measurement devices. Here, the direct measurement refers to quantitative measurement of flow behavior and indirect measurement refers to a qualitative measurement of flow behavior. For instance, direct measurement devices are used for predicting variation in regional blood flow in the arteries, veins, and tissue in milliliters per minute, or in perfusion index or by image contrast ranges. For example, Doppler ultrasound, positron emission tomography (PET), optical coherence tomography (OCT), dynamic contrast-enhanced magnetic resonance imaging (MRI), invasive catheter flow sensors, and other optical methods. The indirect measurement devices, on the other hand, are used for predicting variation in regional blood flow by observing the blood pressure measurement, optical-reflectance-based flow measurement, electrical impulse measurement, sound or beating rhythm, and thermal diffusion-based flow measurements. For instance, blood pressure measurement using a sphygmomanometer or impulse charts using electrocardiogram (ECG) indirectly measures the blood flow associated with cardiovascular flow problems. A sphygmomanometer measures the change in blood pressure or pressure force exerted on the vessel walls of brachial artery due to increase (hypertension) or decrease (hypotension) in blood flow. Albeit, an ECG electrode measures the electrical impulses due to pressure waves generated by contraction and relaxation of different body parts due to a change in blood flow associated with the beating of the heart. Higher impulses reflect more blood flow revealing the strain on the heart (hypertension), and chronically, lower impulse indicates deprived blood flow to the cells (hypertension). One important take away is that any physical quantity can be evaluated using basic energy signatures, such as light, pressure, sound, temperature, and electrical signals.

In this chapter, we will discuss some of the essential devices regularly used in clinical or research settings for direct measurement of biofluid behavior.

15.3 COMPUTED TOMOGRAPHY (CT)

CT is a medical imaging modality widely used as an alternative to X-ray or mammography. The word computed refers to the use of computer to construct images, and the word tomography refers to the acquired images in the form of slices. The standard X-ray on the other hand, only provides the projection image of the body. Both CT and X-ray use X-rays as a source of energy. When an X-ray beam traverses through the body, it undergoes attenuation within the body. The level of attenuation depends on the density of the medium in the path of the rays. The remnant transmitted rays from the medium traverses to the detector or the film which gives rise to image contrast. Here, the medium could be soft tissue, bones, body fluid, or contrasting agents. The density of the medium determines the transmitted rays to the detector. For example, bones are denser than soft tissue and therefore attenuate more compared with soft tissue. Similarly, fibro glandular tissue and calcifications are denser than normal breast tissue thus appears distinctly in the mammogram. The limitations of X-ray modality are: (1) difficulty in reading due to overlapped features, and (2) ionizing radiation is damaging to tissue. CT, on the other hand, enables a better 3D visualization of features that are not resolvable in X-ray images. The working principle of CT images is completely different from standard 2D images as discussed in the next subsection.

15.3.1 WORKING PRINCIPLE OF CT SCANNER

CT scan is used to obtain 2D images of the body cross-section not same as the transmitted planner X-ray images of the body. The major component in a CT scan is the reconstruction approach (discussed later in detail). The scanner consists of a circular rotating frame with X-ray source on one side and arc-shaped detector array on the opposite side (Figure 15.1A). The patient is located at the center of the circular frame. During imaging, the circular frame rotates around the patient's body. During each rotation angle, thin collimated fan beam of X-ray passes through body parts (in the direct path of the rays) and the attenuated rays at different angles were collected by detector arrays. Here, the entire frame rotates at different angles and rays from the same body section are collected by a detector array at different angles. In other words, the same section of the body was irradiated or probed from different angles before moving to the next section of the body. Therefore, CT scan exposure is often greater than a single exposure of X-rays. Upon receiving the attenuated signals from different rotation angle, the scanner uses a computer or computer algorithm to reconstruct the body cross-section by mathematical inversion technique also referred to the backpropagation method (discussed later in this chapter).

The central hypothesis is: if we know the amount of X-ray irradiated from the source and the amount attenuated after passing through the body, we can determine the medium density by mathematical inversion technique. The reconstruction of medium density led to a 2D image of a body *cross-section*.

15.3.2 COMPONENTS OF CT SCANNER

A typical CT scanner is a complex electromechanical system that includes a variety of parts for X-ray generation, attenuated signal acquisition and image construction. The essential components for signal acquisition are X-ray production, filtration, collimator, cooling system and detector array that are packed on a rotating frame around the patient's body (also referred to as gantry). The image construction is performed by accompanying workstations using specialized and synchronized software capable of performing image reconstruction simultaneously during image acquisition by the scanner. These workstations also serve as an operator console and viewing console for physicians. In the gantry, the source of X-ray and arc-shaped detector array are attached to the same circular frame separated by 180 degrees (Figure 15.1B) so that the detector array rotates by the same angle as the source.

15.3.2.1 X-RAY SOURCE

The X-ray source is a vacuum tube, wherein X-rays are produced when beams of accelerating electrons emitted from the thermionic cathode (negatively charged) bombarded on a rotating anode (positively charged). The electrons are emitted from the metallic surface of the cathode due to excessive heating of the cathode by applying electric current. These free electrons are then focused and accelerated toward the positively charged anode due to potential difference and due to a focusing cup around cathode that is having a negative charge. Upon bombardment of accelerating electrons in the anode, the electrons lose energy. The loss of electron kinetic energy led to an increase in the temperature of the anode and production of X-rays. In fact, the ray production can be explained by understanding the interaction of electrons with the atomic structure of a node (discussed in the next subsection). To avoid melting of anode due to heat, the anode is configured to rotate during the entire process using an induction motor. The rotation of anode ensures that the electron does not bombard the same location for a longer time and dissipation of heat. The entire X-ray tube is enclosed in a radio-opaque lead protective housing and is surrounded by circulating oil within the housing. The circulating oil is used to dissipate the heat and protective housing is used to restrict unwarranted radiation in all directions. Only a small portion of X-ray radiation is allowed to pass through a radio translucent window.

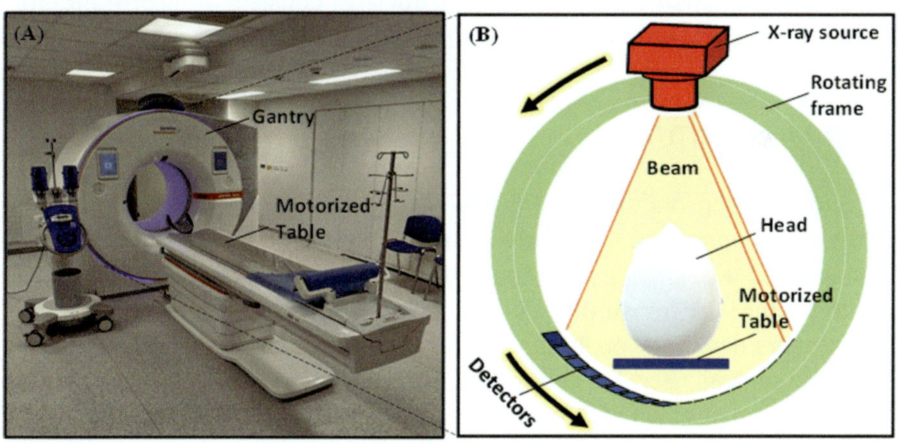

FIGURE 15.1 CT scanner overview. (A) Complete scanner unit. (B) Simplified model of source and detector assembly within the gantry.

Source: Reprinted from Ref. [1]. https://creativecommons.org/licenses/by-sa/4.0/

15.3.2.2 FILTRATION

Filtration modulates the dose of radiation to patient by only allowing the wavelength that contributes to CT imaging. This is crucial since X-rays produced by source are not monochromatic (i.e., polychromatic) and ideally the longer wavelength or low energy (soft) X-rays contribute to tissue absorption compared with shorter wavelength. Essentially, the filters attenuate these longer wavelengths that can increase patient radiation dose during imaging. These filters are thin metal sheets made of copper, aluminum, and tin placed in the path of X-ray radiation after radio translucent window. The filters also make the X-ray beam energy to be more uniform and improve the image quality (reduces the artifacts). In addition, filters can be of different shape and thickness to improve or resolve finer details of the body section.

15.3.2.3 COLLIMATOR

Collimators are specially designed lead shutters that shape the X-ray beam to be irradiated on the body section. These collimators are placed after the filters which restrict the X-ray beams to a defined region of interest (ROI). In the case of CT scanner, the collimators role is to narrow and widen the slice thickness of the section of body. However, the slice thickness also depends on the detector positioning. The collimator also consists of laser light units to define accurately the ROI over the patient's body.

15.3.2.4 DETECTORS

Detectors are sensors that are arranged in a grid array around the body in an arc shape to capture the transmitted X-rays from the body. This detector array is attached to the same circular frame as the X-ray source but on the opposite end to capture the transmitted X-rays passed from the patient body section. The sensors/detectors on both ends of the arc-shaped array is called the reference sensors used for calibration purpose since these reference sensors capture the unattenuated X-ray intensities. The individual detector in the array can be either made of xenon-filled ionization channels or scintillator.

15.3.2.5 XENON-FILLED DETECTOR

The xenon-filled ionization chambers are composed of two parallel metallic electrodes separated from the pressurized xenon gas. The interaction of X-rays with the xenon gas between the electrodes produces pairs of electron–ion. These charged particles travel toward the electrodes by force of attraction by establishing a small potential difference across the electrodes resulting in the generation of current. The generated current/analog signals are directly proportional to the number of photon–gas interaction. These analog signals are amplified and transformed to digital signals using analog-to-digital (A/D) converters. The digital signals were gathered simultaneously from all detectors in the detector array and passed to reconstruction pipeline for the reconstruction of body slices.

15.3.2.6 SCINTILLATOR

Scintillator, on the other hand, is made up of special crystals or solid-state materials, and when interacting with the X-rays produces photons. This solid-state material is mounted on a photodiode to receive photons from the crystal and generate electric signals by photoelectric effect. Next, the analog signals from each detector in the array were converted to digital signals simultaneously using A/D converter that was fed into the reconstruction pipeline for the reconstruction of slice image. Scintillators are dense materials as compared with xenon gas which ensures 100% X-ray interaction and energy conversion.

15.3.2.7 COOLING SYSTEM

All the above components of CT system in the gantry unit have energy conversion units and moving parts which lead to unwanted heat loss. This heat needed to be dissipated from the gantry to avoid damage to the CT unit during its operation. A CT scanner or its gantry may use cooling systems like blowers or heat exchangers to remove excess heat generated by the operating units.

15.3.3 PRODUCTION OF X-RAY

The interaction of speeding electrons with the anode's atomic structures leads to the production of X-rays. Primarily, two scenarios may exist that could lead to production of X-rays either by excitation or by Columbic interaction. Excitation is direct collision-based interaction during which speeding electrons either force the electrons in atomic orbits to a higher energy level or completely splits the electrons out of the orbit. In either scenario, there exists a change in the energy of the speeding electrons and orbital electrons which leads to the production of X-rays. The columbic interactions on the other hand could be viewed as an indirect interaction (i.e., deflection or deceleration due to the charged particle's electric field) of speeding electrons with a nucleus and orbiting electrons of atoms that leads to a loss of incident electron energy resulting in X-ray production. For instance, deceleration of speeding electrons while traversing the nearby nucleus of the atom and deflection of speeding electrons by orbital electrons of anode's atoms. In particular, the inelastic scattering (or deflection and deceleration) of speeding electrons due to nucleus contributes to most of the X-ray production used for diagnostic purpose also referred to as Bremsstrahlung radiation or characteristic radiation. If all the speeding electrons undergo the same nature of interaction with anode, then the resulting X-ray would be monochromatic or same energy level. However, the interaction leads to X-ray production of varying energy levels due to different forms of interaction (combination of both columbic and excitation). Hence, the X-rays generated by X-ray tube are polychromatic in nature.

15.3.4 INTERACTION OF X-RAY WITH TISSUE

As X-ray photons traverse through the tissue, it undergoes a loss of energy due to interaction with varying density of the materials or tissue. This loss of energy is due to absorption and scattering of traversing X-rays. The absorption of X-ray energy contributes to patient dose and scattering of X-ray photons contributes to imaging. However, excessive scattering also degrades the image contrast due to increase in scattering noise. In essence, material density plays a vital role in the degree of interaction with X-rays. This density variation is defined by the term attenuation coefficient in the units per cm. This property tells us the degree of incoming radiation attenuated while traversing through a material. For instance, the denser material will interact more with photons and hence display higher attenuation coefficient compared with less dense materials. Mathematically, it can be written as follows:

$$\begin{aligned} &\mu_{eff} = \mu_{scat} + \mu_{abs} \\ &\text{or, } \mu_{eff} = \left(\mu_{Rayleigh} + \mu_{Compton}\right) + \mu_{abs} \end{aligned} \tag{15.1}$$

where μ is the representing attenuation coefficient, subscript *scat*, *abs*, *Rayleigh,* and *Compton* refer to scattering, absorption, Rayleigh scattering, and Compton scattering, respectively.

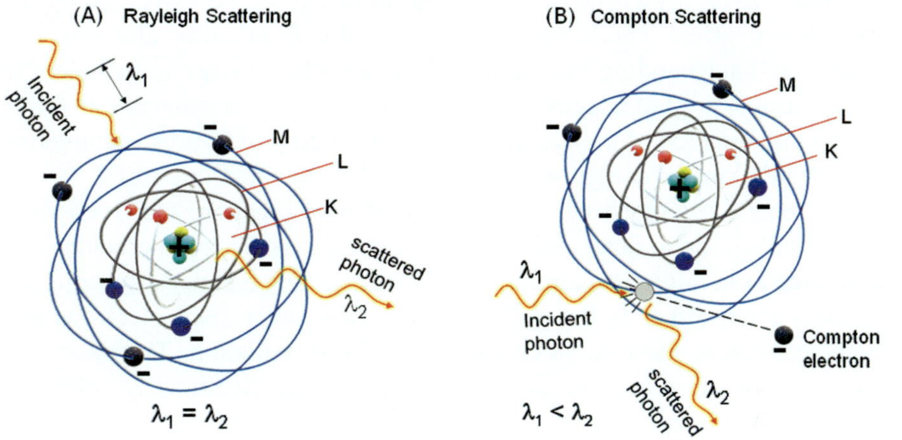

FIGURE 15.2 Schematic of scattering of X-ray in the tissue. (A) Rayleigh scattering of incident X-ray. (B) Compton scattering of incident X-ray.

Rayleigh scattering can be explained by the nonionizing interaction of low energy (15–30 KeV) X-ray photon with the tissue atom, during which electric field of the photons induces electrons of tissue atoms to undergo harmonic oscillation. These oscillating atoms emanate a second X-ray photons in all directions having the same wavelength (*see* Figure 15.2A). This type of scattering is also referred to as coherent scattering that includes 5–10% interaction with the tissue for diagnostic X-ray energy range.[2,3]

Compton scattering can be explained by the ionizing process of X-ray photons interaction with the electrons at outer shell of tissue atoms, during which the incident photons are energetic enough to split the electrons from the shell and ionize the atom. It is important to note, the energy absorbed by electrons positioned at the atom's exterior is less than the incident photon energy due to which the remnant photon is deflected by an angle from the incident direction (see Figure 15.2B). The ejected electron is the by-product of the ionizing process which may lose its energy in the medium of further interaction or may lead to further ionization. The remnant photon deflected by the tissue atoms loses its energy while traversing through the tissue by multiscattering and/or absorption before reaching to the detector. The electron density determines the probability of Compton scattering, whereas the incident photon energy has less effect.[2] For incident energy level in a diagnostic application (15–150 KeV),[3] Compton scattering often contributes to noise in the acquired image slices since attenuation of X-ray in the tissue is relatively small compared with incident radiation. Finally, the attenuation of incident X-ray by virtue of rays being absorbed can be explained by the Photoelectric effect. The X-ray energy is initially absorbed by electrons in a packed shell surrounding the nucleus. This absorbed energy is sufficient to split electrons (or photoelectrons) from the inner shells causing an ionization process and creation of the vacancy. Here, the energy of X-rays is converted to electron's kinetic energy which is equivalent to subtraction of $E_{\text{x-ray}}$ and binding energy of electrons. As a result of hole creation, electrons from higher shells jump to inner shells to fill the vacancies. This jump can be in a single step or in multiple quantum states (in cascading manner). Importantly, the extra energy that is the difference between outer shell electrons and photoelectrons is released as characteristic X-ray radiation. The energy of secondary radiation and the kinetic energy of photoelectrons is quite low in the diagnostic applications that get absorbed and attenuated within the tissue. The probability of photoelectric interaction typically has an inverse relationship with the incident energy of X-ray. The probability drops at higher incident energy of X-ray with maximum probability being observed energy equivalent to K-shell electrons. Figure 15.3 depicts a diagram of attenuation of incident energy of X-ray within the tissue by virtue of scattering and absorption. A simplified expression for the transmitted X-ray from the tissue that reaches the detector can be given by the Beer–Lambert Law as follows:

$$I_{X-ray,tr} = I_{X-ray,in} \exp\left(-\mu_{eff} \Delta x\right) \tag{15.2}$$

wherein I is the ray intensity or energy, Δx is the length of the unit voxel of the tissue in the direction of the ray, and subscripts tr and in represents the transmitted and incident rays.

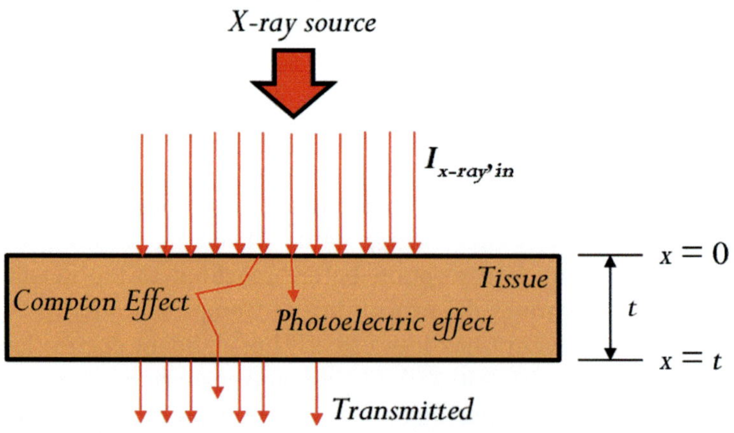

FIGURE 15.3 An illustration of interaction of incident X-ray within the tissue.

15.3.5 CT SCANNER IMAGE RECONSTRUCTION

There are many operating parameters, which are needed to be set in a CT scanner for image acquisition and to obtain a good quality image. In general, these parameters are selected by the operator or provided with proprietary software. These parameters can be incident X-ray energy (KeV), rotation time and speed, slice thickness, reconstruction method or kernels, and many more, which are adjusted by the software and/or by the operator depending on the protocol (i.e., with or without contrasting agents). It is important to realize that in certain procedures contrasting agents could be administered to visualize the finer details of the ROI. The contrasting agents are specially made fluids that can provide better contrast than surrounding tissue at the target site. The image acquisition unit is composed of multiple algorithms for preprocessing the raw data, calibration correction of raw data, image reconstruction, and image post-processing. More details on the calibration process and preprocessing algorithms can be found in the literature.[2,3] Here, we will elaborate the key mathematical algorithms involved in the image reconstruction process. The image reconstruction can be performed in three different ways (a) matrix inversion, (b) back-projection method, and (c) iterative method. To understand each approach, we need to understand the forward projection of raw signals received by the detectors. The forward projection is the normalization of measured data from the detector and its logarithmic conversion. The projection data can be mathematically expressed from eq (15.2) as follows:

$$P_j = \ell n\left[\frac{I_{X-ray,tr}}{I_{X-ray,in}}\right] = \left(\mu_{eff}\right)\Delta x$$
$$\text{or, } P_j = \ell n\left[\frac{I_{tr}}{I_{in}}\right] = \left(\mu_1 + \mu_2 + \ldots\ldots + \mu_n\right)\Delta x \tag{15.3}$$

where P_j represents projection data in a *jth* ray direction, I_{tr} and I_{in} are the transmitted and unattenuated X-ray reaching the detector, μ is the attenuation coefficient in the direction of traversing ray, and Δx is the unit element of the voxel. Figure 15.4 illustrates the variation of X-ray intensity in an elliptical tissue with lesion about the radial distance of the arc-shaped detector array.

To compute the projection data (P_j) in each rotation angle between $j = 0$ to 360, the attenuated rays reaching the detector after traversing through the tissue are normalized by the reference intensity I_0 or I_{in}. The reference intensities are the intensities on both end edges of the detector array which receives X-rays unattenuated by the tissue as can be seen from Figure 15.4.

The normalized data are further subjected to logarithmic conversion for generating the forward projection data which is typically represented by lines in 2D graph for a thin slice. It is important to note that the attenuation of X-ray in the air is also compensated by calibrating the detector gain in detector array. For many practical situations, degradation of detector sensitivity and damaged sensors is compensated by algorithms or by periodical maintenance of the scanner. From eq (15.3), projection can also be defined as a linear combination of attenuation coefficient along the path of traversing X-ray. The mathematical basis of forward projection of an object to incoming ray can be given Radon transformation.

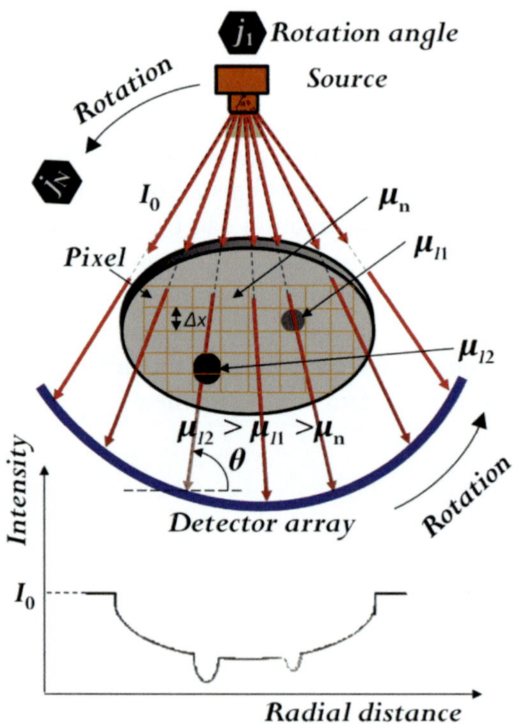

FIGURE 15.4 Illustration of X-ray intensity captured by detector passing through the elliptical tissue and through the air.

In Figure 15.4, the intensity arriving to the detectors in both sides of the detector arc not interacting with the tissue is the reference intensity or incident intensity (I_0 or I_{in}). The elliptical tissue can be divided into small voxels of unit length Δx. The tissue contains two lesions having higher attenuation coefficients resulting to decrease in the magnitude of intensity arriving to the detector as can be seen from one-dimensional magnitude plot.

This transformation is the projection of the intensity attenuated by an object along a radial line oriented at a specific angle (see Figure 15.5).

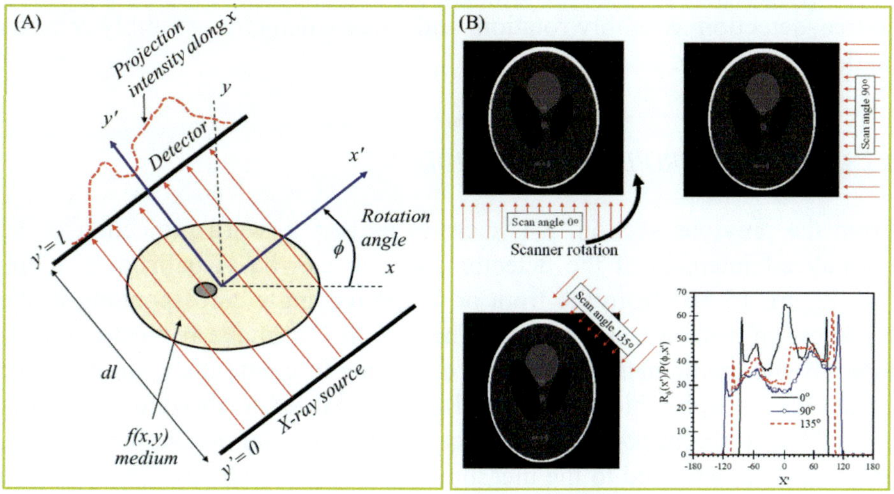

FIGURE 15.5 (A) An illustration of the X-ray intensity line integral about the path $y' = 0$ to l representing forward projection data using Radon transformation. (B) An example of parallel beam projection of Shepp-Logan head phantom and projection data obtained from different orientation angles using Radon transformation. It should be noted that the phantom or object is stationary, but the source–detector assembly is rotating causing variation of projected value since the line integral of attenuation coefficient varies with the path (l) and direction (φ) (refer Eq. 15.4).

It should be noted that eq (15.4) is same as eq (15.3) in the integral form. In other words, Radon transformation provides us with line integral of attenuation or pixel value of the medium defined by $f(x',y')$ at a particular orientation of the source–detector assembly.

$$R_\phi(x') = P(x',\phi) = \int_0^l f(x',y')\,dl \tag{15.4}$$

Note, $f(x',y')$ within will vary with each rotation angle as well as line integral or projection value. To account the rotation about the centroid of the gantry, we can use the trigonometric relation for rotation axis to translate eq (15.4) to a generalized expression as follows:

$$R_\phi = P(\phi) = \int_0^l f\left(x\cos\phi + y\sin\phi, -x\cos\phi + y\cos\phi\right)dl \tag{15.5}$$

where f is referred to as a bounded region defined by its attenuation property, l is the ray propagation distance from source to detector, x and y is the reference coordinate system, x' and y' is the coordinate system after the source–detection assembly rotation, and source–detector assembly rotation angle is given as ϕ.

15.3.5.1 BACK-PROJECTION METHOD

From the previous section, it is clear to us that forward projection is the normalized intensity at the detector end that can be visualized as a line (*see* Figure 15.5B) from the front or as 2D rectangle corresponding to the thickness of a slice from the top. The key in CT image reconstruction is to apply the mathematical inversion technique to create the unknown slice properties through which X-rays had passed that led to forward projection. As indicated, the back-projection requires an inversion technique to create the slices that have led to the measured normalized intensity. This problem is mathematically ill-posed, means no unique value of properties (such as attenuation coefficient) can be predicted that has led to measure normalized intensity. However, in CT imaging identifying the exact property of tissue voxels is of less interest. Instead, the objective in imaging is to generate good quality images.

Figure 15.6 illustrates the method of back-projection for simplified 2×2 tissue block. Let us assume, the pixel value of the tissue block is unknown to the imager as shown in Figure 15.6(A). Initially, X-rays are passed at different scan angles and the projected/attenuated intensity is collected by the detector. Next, these projected intensity values at different scan angles is back-projected to fill the pixel value as shown in Figure 15.6(B). To simplify the concept of back-projection, one can mentally visualize the method as back-projecting the measured forward projected intensity by the detectors from different scan angles. In one sentence, smear the "forward projection", that is, intensity back and add the contribution from each scan angle at overlapping pixels until the original image is reconstructed as shown in Figure 15.6(C). In Figure 16.6(C), one can clearly visualize back-projecting from all scan angles (0°, 45°, 90°, and 135°) led to the same picture as the original. Mathematically, this approach of back-projection is also referred to as an inverse Radon transform.

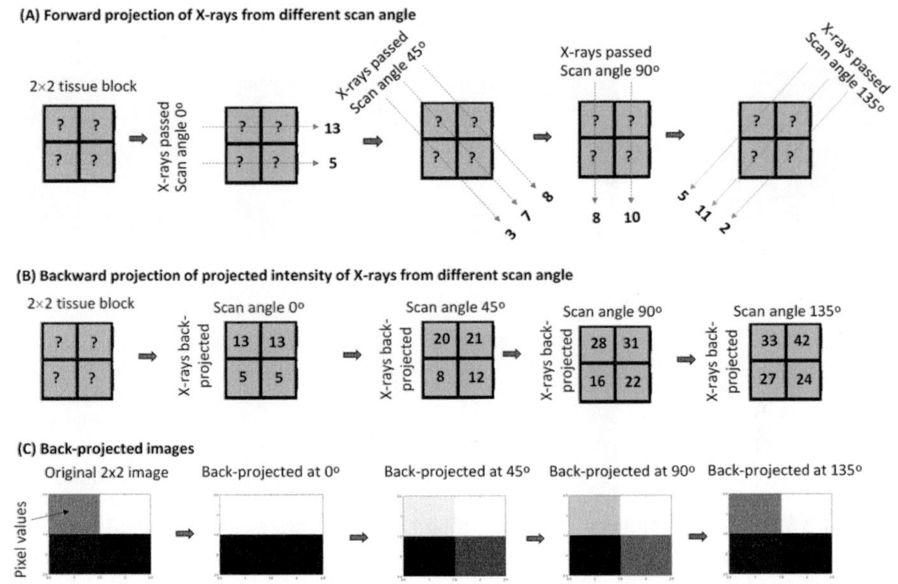

FIGURE 15.6 An illustration of the back-projection method works in CT image reconstruction. (A) Forward projection of X-ray intensity at different scan angles. (B) Backward projection of forward intensity at different scan angles. (C) An example of backward projection image reconstruction.

$$f_{recons}(x,y)=\sum_{j=1}^{N}P(x',\phi_j)\,d\phi \tag{15.6}$$

where *j* is the index for rotation angle of source–detector assembly and other symbols has usually meaning discussed earlier.

Eq (15.6) is simply the addition of projected intensity in the pixel space with each scan angle denoted by index *j*. Note, this is a simplified approach of back-projection which often led to image artifacts, noise, degradation of resolution, and sharpness of edges after reconstruction. Figure 15.7 illustrates the back-projected images of Shepp-Logan head phantom generated by simple back-projection method.

It can be realized from Figure 15.7, as we increase the number scan angle for the total scan of 180° the back-projection led to better image construction. However, using simple back-projection, we can only obtain a diffused image of phantom since the pixels with no pixel values outside the phantom in the original image is also filled with certain value during back-projection. To overcome these artifacts during back-projection, an advanced form of back-projection method is used which is also known as filtered back-projection.

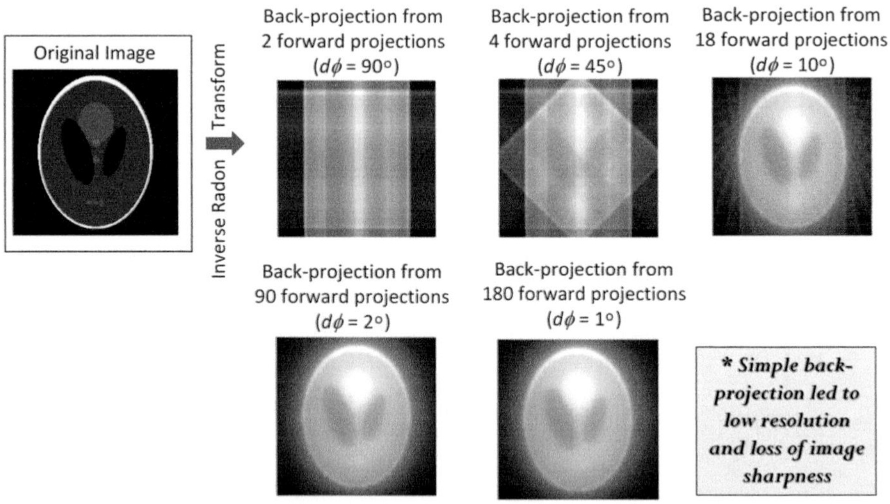

FIGURE 15.7 A visual illustration of simplified X-ray back-projection of Shepp-Logan head phantom. Here, back-projected images are shown for total scan angle of 180° done at angular discretization of 90°, 45°, 10°, 2°, and 1°.

The filtered back-projection employs a spatial filtering (a mathematical operation) method to the forward projection at each scan angle before employing the inverse Radon transform (or back-projection). This significantly reduces the noise and clutter in back-projected images. This spatial filtering can be processed in the spatial domain and frequency domain. In spatial domain, the filtering of projected intensity is convolutional process (integration of functions) during which a spatial kernel/function $h(x')$ is shifted over projected intensity $P(x')$ along axis x' (see Figure 15.5(A)) and add its contribution to $P(x')$. This process resulted in a synthetic projected intensity $P'(x')$.

$$P'(x',\phi) = P(x',\phi) * h(x') \tag{15.7}$$

The mathematical application of the convolutional process in the spatial domain is computationally intensive; hence, the filtering is often processed in special domain by using Fast Fourier transform of projected intensity as $F\,[P\,(x')]$ and filtering function as $F[h(x')]$ along the x' Here, F is defining a Fourier transform operator. In frequency domain, the convolutional process is simplified to multiplication which is computationally faster. Next, an inverse Fourier transform is employed on the multiplication of $F^{-1}[F[P(x')]F[h(x')]]$ to revert back the filtered projected intensity in frequency domain to spatial domain, that is, $P'(x', \varphi)$ followed by applying a Radon transformation on the filtered projection intensities similar to eq (15.6). The commonly used spatial kernel/filter functions *h*(*x*') are Ram-Lak filter, low-pass cosine, and Hamming filters.

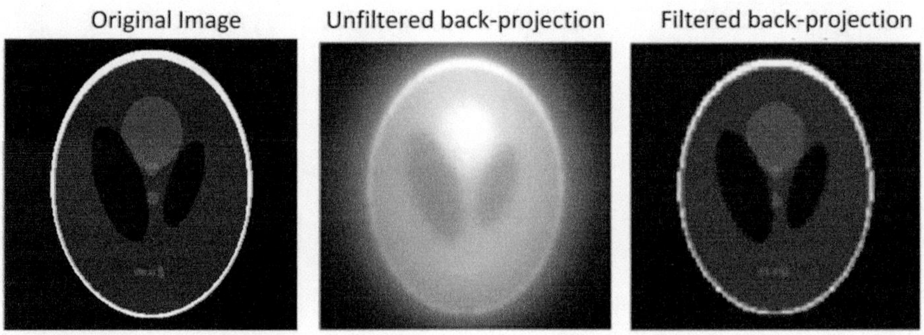

FIGURE 15.8 A comparison between unfiltered and filtered back-projection of a head phantom known as Shepp-Logan.

Figure 15.8 depicts a comparison between normal back-projection and filtered back-projection for the head phantom called Shepp-Logan.

After reconstruction of the images using back-projection or iterative algorithm, the pixel value of represented maps is given in CT number (CT_0) which is non-dimensional number. The unit of the CT number given by Hounsfield unit (HU).

$$CT_0 = \left(\frac{\mu_0 - \mu_{H_2O}}{\mu_{H_2O}} \right) \times 1000 \tag{15.8}$$

where CT_0 is CT number and μ_0 attenuation coefficient of tissue. Here, μ_{H2O} is the attenuation coefficient of water used for normalizing the tissue attenuation coefficient. The reconstructed images have a CT number of pixels varying from +3000 to −1000 and needed proprietary standard windowing functions to convert the CT number range to 0–255 numeric value to form a grayscale image.

15.3.5.2 MATRIX INVERSION METHOD

Another conventional reconstruction method is a matrix inversion process. Let us consider a 2×2 tissue block (*see* Figure 15.9), which is scanned at 0° and 90°.

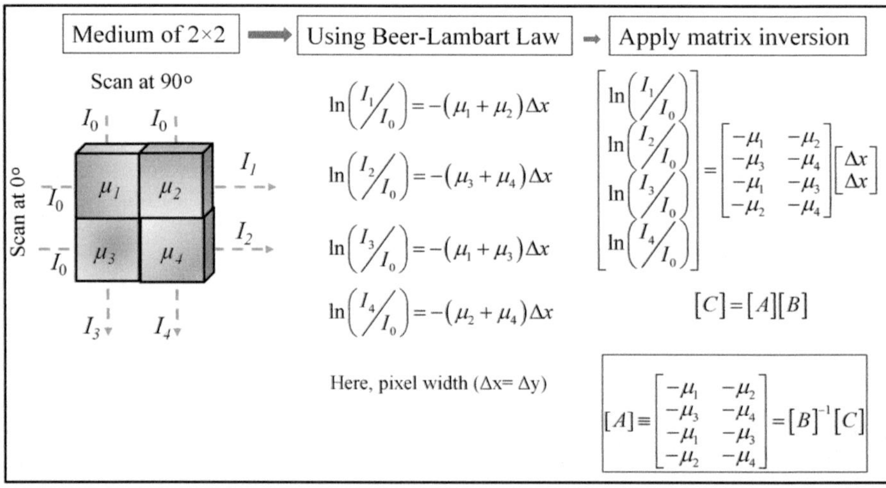

FIGURE 15.9 An illustration of matrix inversion process for reconstruction of CT images.

The incident X-ray intensity (I_0) is known to us, and the attenuated intensities I_1, I_2, I_3, and I_4 are measured by the detector. Using Beer-Lambert law (Eq. 3), we can compute the projected signal by taking the natural log of the ratio (I_j/I_0) in each scan angle. Hence, the reconstruction process involves finding the attenuation coefficients μ_1, μ_2, μ_3, and μ_4. This is possible by arranging the coefficients of equations in a matrix format and taking an inverse of the matrix as shown in Figure 15.9 to compute the unknown coefficients or attenuation properties. It is important to note that matrix inversion is only possible if the grid spacing Δx and Δy is approximated. Ideally, the unit voxel width and height are approximated to be a small value. However, actual tissue has $N \times M$ voxels in the spatial domain, and applying the matrix inversion could lead to solving many equations which is a computationally intensive process and could take longer time to reconstruct. This process can be achieved by today's computer at a faster rate, but in the year 1967 (first CT invented), matrix inversion method for image reconstruction was not a feasible solution. Hence, in practice, image reconstruction in CT was carried out either using backpropagation or iterative algorithms.

15.3.5.3 ITERATIVE METHOD

Image reconstruction can also be performed by iterative methods, also referred to as iterative reconstruction algorithms. These reconstruction algorithms involve minimizing the measured and predicted projection values at each scan angle by assuming a guess value of the attenuation coefficients for each pixel/voxel of the tissue domain.

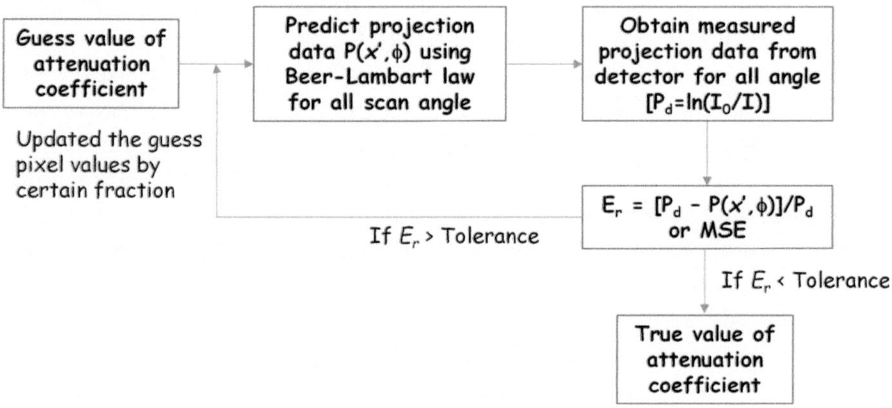

FIGURE 15.10 An iteration algorithm flowchart for reconstruction of CT images.

Figure 15.10 depicts a simplified flowchart of iterative algorithms for each/all scan angles, wherein an initial guess value of attenuation coefficient is assigned for all voxels and a projection data are computed using Beer-Lambert law (Eq. 3). Next, the predicted projection data based on guess value is compared with measured projection by a detector (P_d) by defining an error function (E_R) or adding the error from each scan angle using the mean squared error (MSE). If the MSE for all scan angle is less than the predefined tolerance than the iteration will terminate, else the algorithm will proceed to the next iteration by updating the guess value of attenuation coefficients by certain fraction and repeat the same loop until MSE achieve the termination condition. In this way, the guess value of attenuation coefficient is refined to achieve pixel values which reflect the best value for each voxel in the CT reconstructed images. Iterative algorithms are particularly useful in the case of CT reconstruction of tissue where filtered back-projection was unable to reduce the artifacts.

15.3.6 CT ANGIOGRAPHY AND PERFUSION IMAGING OF HUMAN BODY FLUID

CT angiography (CTA) and perfusion imaging are specialized techniques to visualize the related structural changes associated with the blood flow and measure the changes in the blood flow in a particular region of the imaged site. CTA is ideally used for identifying structural changes in the vessels whereas CT perfusion is used to quantify the functional changes in blood microcirculation in the tissue over a short time. These methods use radio-opaque contrasting agents to visualize and compute the change in the blood circulation (often appear white in the image). CTA is widely used to visualize the blockage in blood flow or extent of aneurysms in major arteries, blood clots, vascular malformation, tumor, vessel rupture, gastrointestinal organ's interior, biliary organs, and urinary tract. CTA only requires venous access, and contrast agent is administered through a vein in the arm. Depending on imaging sites, other ways of administration of contrasting agents are by mouth (i.e., oral contrast), through anus (i.e., rectal contrast), or injected into a joint using a needle. CTA is also referred to as contrast-enhanced CT wherein finer details of vascular anatomy or other fluids along with tissue are highlighted in order to have better visualization of the investigated site. In order to obtain a detailed image of the vascular structure or highlighted organs, often algorithms are used to subtract the pre-contrast images from the contrast-enhanced images. In fact, maximum intensity projection of individual slices produces excellent CTA images. CT perfusion imaging predicts the blood perfusion (or volumetric flow of blood) in the tissue of a particular organ. CT perfusion is widely used for evaluating brain strokes, evaluating brain tissue at risk after stroke, and assess treatment response of brain tumors. During CT perfusion an overall high radiation dose is employed, since images of the same organ are collected over time to determine the changes in contrasting agent's flow pattern in the organ. The duration of the acquisition depends on the number of slices and scanning duration for each slice. In practice, the same slices are imaged for 1–2 s to evaluate the dynamic change in the distribution of the contrasting agent in the tissue. It uses additional software to generate parametric maps of tissue and obtain parameters like blood volume, tissue perfusion, etc. In order to create the parametric map of blood perfusion/contrast agent in the tissue, the scanner takes images of the organ before and after contrast agents reach the designated organ. This procedure is performed to extract the blood flux value to the organ of interest in HU/time.[3] Depending on the type of test and body parts to be probed, the contrasting agents in CTA or CT perfusion may differ. The contrast agent commonly used during CTA or CT perfusion of vascular structures is an iodine-based solution, whereas for gastrointestinal tracts commonly used contrasting agents are barium or iodine solution. Contrasting agents administered by mouth or through rectum cause less side effects to patients, but patients may experience side effects such as allergic reaction or kidney problems.

15.4 NUCLEAR MEDICINE IMAGING (NMI)

Nuclear medicine imaging (NMI) is a medical imaging method widely known as molecular imaging, which maps the gamma radiation emitted by different body parts after a radiotracer also referred to as "radiopharmaceuticals" is administered to patients. These radiotracers are introduced to the body either via oral/nasal administration or via subsurface injection to cutaneous bloodstream. Radiotracers can be either biochemical conjugates of standard metabolic compounds/carrier molecules (such as carbohydrate/sugar) with radioactive isotope or they can be target-specific radiopharmaceutical compounds to target a specific organ/molecule found in the organ. One part of radiotracer is radioactive isotope which emits gamma rays due to unstable nucleus, and another part is a molecule that is metabolized by tissue or binds to tissue. The necessity of NMI arises from the fact that a pathology-specific biochemical alteration in tissue often takes place at an early stage. The binding of radiotracer in the target tissue and time-dependent increase (high-intensity region in the reconstructed image) or decrease (low-intensity region in the reconstructed image) in the accumulation of radiotracer enable identification of early indicators and localization. In other words, the accumulation would increase in particular tissue having abnormality and is also time-dependent. Unlike X-ray/CT imaging method, NMI generates an image based on the gamma (γ) energy emitted by the body due to decay of radioactive isotopes in the radiotracer attached to the tissue instead of transmitting X-rays for imaging. NMI unable to provide a structural image of the body instead it can only generate multiparametric maps of body function. Hence, it is also referred to as a functional imaging method. Depending on applications, NMI enable mapping functional parameters, such as glucose metabolism, blood flow and perfusion, tissue oxygenation, and receptor binding. These functional parametric maps were created based on the time-dependent distribution of radiotracers to the body sites or by quantifying the changes in the uptake of the radiotracer by targeted tissue over time. For instance, an elevated glucose metabolism and perfusion in the tissue can be an early indication of cancer or pre-cancer which can be identified by NMI. These changes are visually quantifiable by an increase or decrease in intensity in the image such that higher intensity indicates more accumulation in the tissue and less intensity indicates less accumulation in the tissue. Often, NMI (functional image) is fused with planar X-ray or CT (structural images) to pin-point the explainable changes in a particular organ.

The benefits of NMI are: (1) NMI is capable to identify a disease in its earliest stage, (2) NMI can differentiate between benign and malignant lesions and reduce unnecessary biopsy, (3) NMI can provide useful diagnostic and treatment details of many diseases. The limitations are: (1) NMI uses a tiny quantity of radioactive molecules (in nanogram) and has very low radiation risk, (2) ionizing radiation is damaging to tissue, (c) rare and mild allergic reaction is possible, and (d) restricted during pregnancies. NMI can be performed in three ways (a) planar imaging, (b) Single-photon emission computed tomography (SPECT) and (c) PET combined with CT imaging. The working principle of SPECT and PET methods differs from the nature of emission of gamma rays because of the decay of radiotracers discussed below.

15.4.1 WORKING PRINCIPLE OF SPECT AND PET IMAGING

Figure 15.11(A) depicts a complete SPECT setup with dual-head gamma camera, wherein the patient table is moved in z-direction to scan the whole body. The dual-head gamma camera rotates around each section of the patient's body to capture emission images at different angles for slice reconstruction. The planar (2D) nuclear imaging/planar scintigraphy configuration is rather like planar X-ray imaging (Figure 15.11B), with the only exception is that in nuclear imaging, no external X-ray radiation source is used. Instead, the image source is the γradiation emitted by the radiotracer while traversing through different body parts. It is important to note that any radioactive isotope in radiotracer has a certain decay period while traversing or attached tissue. Naturally, the scan or imaging is performed within the decay period after administration of radiotracer. Also, the attenuated gamma rays reaching the detector differ with time. These gamma rays are detected by gamma camera and the scans can be either static or dynamic in nature. In static imaging, the distribution of radiotracer in the body is static such as bone imaging or dimercaptosuccinic acid (DMSA) renal imaging. Whereas in dynamic imaging, the imaging is performed at a faster rate since the distribution of radiotracer in the body changes with time such as time-dependent radio-renogram and hepatobiliary iminodiacetic acid scan (HIDA) for gallbladder clearance. The most SPECT scanner can also be used for planar scintigraphy. However, planar scintigraphy suffers from similar problems as planar X-rays overlapping image features in the acquired images. In contrary to planar imaging, SPECT and PET on the other hand is 3D scanning method which is capable to construct cross-sectional images or slices similar to CT scan discussed earlier. However, the source of energy in SPECT and PET imaging is gamma rays originated due to decay of unstable nuclei of the radioisotope instead of the external X-ray source. SPECT uses gamma cameras rotated about a contour around the body that captures emitted γ rays by the body (see Figure 15.11C). The scanner rotation can be 180° around the body or 360° in an elliptical manner like human body contour. The term single photon in SPECT originates from the fact that the γrays captured by the camera are in the single-photon direction due to emission of gamma photons by excited/unstable nucleus or radioisotope in a particular direction. For instance, ^{12}C has 6 neutrons and 6 protons, resulting in stable nucleus whereas ^{14}C has 8 neutrons and 6 protons resulting to an unstable nucleus.

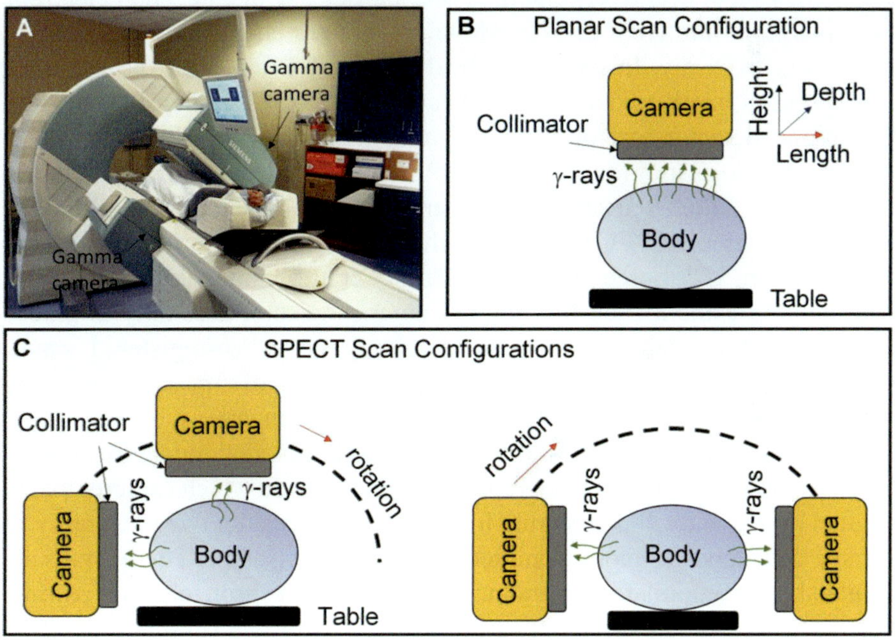

FIGURE 15.11 Illustration for nuclear imaging scanning configuration. (A) Complete SPECT scanner unit (B) Scan configuration for planar imaging wherein the camera head moves in the transverse direction to obtain a planar image. (C) Scan configuration for SPECT tomographic method wherein single- or dual-head gamma camera can be placed in L-shape or at 180° apart and the duo-camera rotates around the body in an elliptical contour of the body.

Source: (A) Reprinted from Ref. [4]. https://creativecommons.org/licenses/by-sa/4.0/

The nucleus with too many neutrons aims to reach a stable configuration in terms of energy, but one can visualize as the number of protons and neutrons. This is achieved by converting or decaying the excess neutrons to proton plus negatively charged highly energetic particle called beta-minus particle (β^-) along with other anti-particles. The minus beta-particle is simultaneously ejected along with the gamma photon from the nucleus leaving the proton in the nucleus. The emission of a gamma photon from the nucleus is to balance the excess beta-negative decay energy which the beta-particle and anti-particle did not carry. The emitted γ rays were collected by a gamma camera in a series of angle around the patient.

Finally, the acquired data from all angles were reconstructed similar to CT (either using back-projection or iteration method) to obtain emission tomography (i.e., functional image of body slice). It is important to note that the interaction of negatively changed beta-particles with outer shell electrons can change its direction or slow down its kinetic energy that can lead to generation of X-ray. In case of hybrid SPECT/CT imaging, the scanning configuration involves dual-camera heads for SPECT imaging followed by CT source–detector assembly for CT imaging. The section of the body is translated (using a motorized table) in SPECT and CT imaging field of view in series and an automated software registration/fuse the anatomical CT image with the functional SPECT image for the same body section. In yet another configuration, CT source–detector assembly could be made operable along with gamma camera such that the CT source–detector assembly is placed between gamma cameras to perform simultaneous co-registered scans. Hybrid SPECT/CT also improves the performance of functional imaging since information collected from CT helps in attenuation correction and scattering in SPECT images and accurately delineate anatomical locations.

Like SPECT, PET is a tomographic imaging method which captures γ-rays emitted by the body and construct images of the slices. PET uses an array of gamma detectors that capture emitted γ rays by radioisotope. These γ rays are emitted by the annihilation phenomenon due to the positron–electron pairing. For instance, ^{11}C isotope has 6 protons and 5 neutrons (unstable nucleus) and tends to attain stable configuration by beta-positive decay. This decay is attained by converting excess proton into a neutron and emission of beta-plus particle (also referred to as positron). These positrons are positively charged and has the same mass as electrons. The interaction of positron particles with electrons lead to annihilation of the positron–electron pair. The annihilation process releases the energy in the form of γ-rays in two opposite directions. The emitted γ rays were recorded by two gamma cameras at 180° around the patient (Figure 15.12). The projected data are reconstructed similar to CT to obtain PET of functional changes in the body. PET imaging is often accompanied by CT imaging in order to perform attenuation and scatter correction and to detail structural information from overlapping PET and CT images. The commonly used hybrid PET/CT detectors are made possible in the same gantry by arranging PET camera configuration and CT source–detector configuration positioned in series (see Figure 15.12A). Here, the patient's body is translated inside the gantry using the motorized table and automated software to co-register the image sections for slice overlap. PET/CT is commonly used in most clinical settings instead of standalone PET imaging.

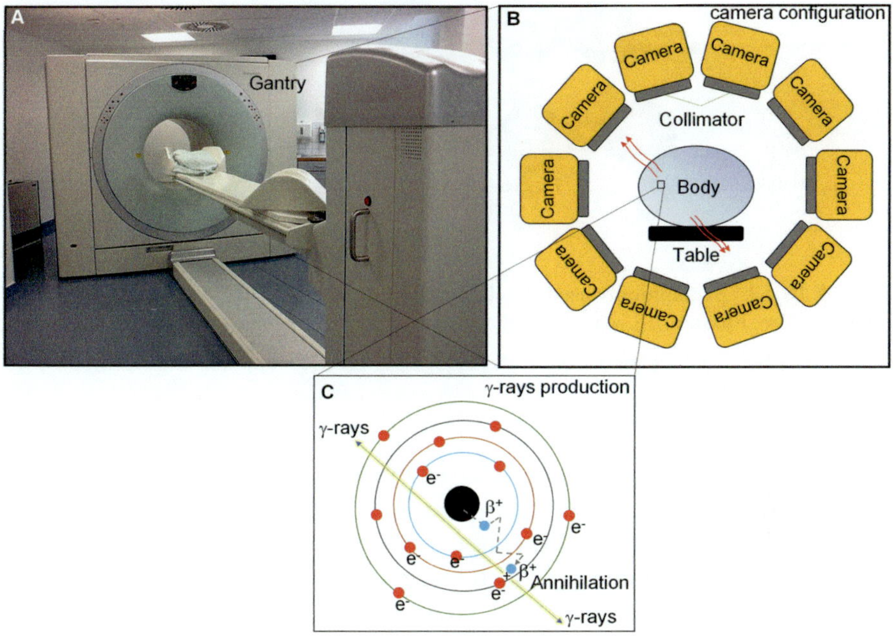

FIGURE 15.12 Illustration for PET scanner configuration. (A) Complete PET/CT scanner unit. (B) Scan configuration for PET tomographic method wherein multiple gamma camera can be placed that captures gamma rays in 180° apart; (C) Inset representing the gamma ray production in PET imaging (wherein β+ is positron and e- is electrons) due to decay of radioisotopes.

Source: (A) Reprinted from Ref. [5]. https://creativecommons.org/licenses/by-sa/4.0/

15.4.2 RADIOTRACERS USED IN NUCLEAR IMAGING

Radiotracers in nuclear imaging is a combination of radioisotopes and chemical compounds as shown in Figure 15.13. The radioisotope has an unstable nucleus that undergoes radioactive decay (for instance, beta-minus and beta-plus decay discussed earlier) within the nucleus which led to the release of γ-rays. The chemical compounds on the other hand are molecules that bind with a tissue or specific molecule or metabolic molecules, which decide the uptake of radiotracer in the body and imaging site.

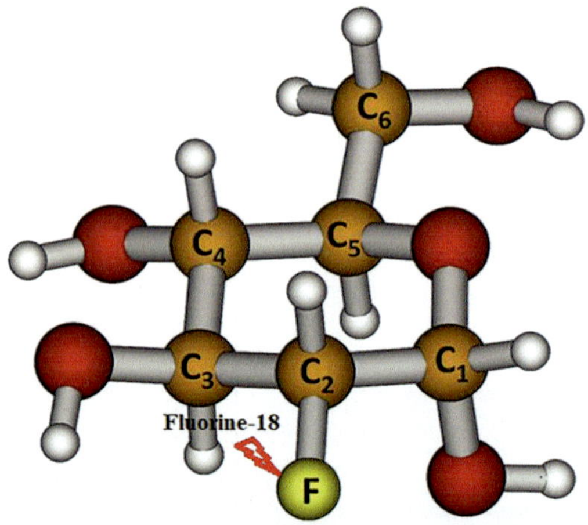

FIGURE 15.13 A commonly used metabolic radiotracer fluorodeoxyglucose (FDG) in PET imaging. Here, Fluorine-18 is a radioactive isotope attached to glucose carrier molecule.

Source: Modified and adapted from Ref. [29]. https://creativecommons.org/licenses/by/4.0/deed.en

The decay of radioisotope is referred to as the continuous break down of unstable nucleus till a stable state is reached. It is important to note that the decay of the protons or neutrons in the nucleus changes its atomic number and mass. The new stable state is different from the previous unstable state; hence, decay or disintegration of nucleus particles gives rise to the new element. The stability of a nucleus somewhat depends on the relationship between atomic number (number of protons in the nucleus of an atom) and atomic mass (total mass of an atom also depends on number of protons and neutrons in the nucleus). Atoms with atomic mass less than 50 has a stable nucleus when the number of protons and neutrons is equal, whereas in atoms with atomic mass greater than 50, the stability can be possible at slightly greater than one in terms of the ratio of the number of neutrons to protons.[2] The decay of radioactive material can be expressed by the number of disintegrations of nucleus with time as

$$A = -\frac{dN}{dt} = \lambda N \tag{15.9}$$

where N is the number of radioactive atoms, A is the radioactivity, and λ is the decay constant. The negative sign indicates that the decay leads to a decrease in the number of radioactive atoms. Radioactivity A is measured in curies (Ci). eq (15.9) can be further simplified by taking integration as

$$N_f = N_i \exp(-\lambda t) \tag{15.10}$$

where N_f is the number of radioactive atoms after time t, N_i is the initial number of radioactive atoms ($> N_f$). It should be noted that the decay constant gives us the sense of decay rate of radioactive elements to the stable configuration (or also enables us to compute the length of imaging). Some radioisotopes decay rate is shorter, and others are longer. Alternatively, half-life of radioactive elements can also be used to quantify the timespan of radioactive elements to reach half of its radioactive life as $(N_f = N_i/2^n)$ where superscript n describes the number of half-lives elapsed. The half-life ($t_{1/2}$) can be obtained as ($= \ln 2/\lambda$), which allow us to identify the time dependence of the radioactivity. However, while calculating the time dependence in nuclear medicine, the biological residence time of radiotracer in tissue also needed to be considered for computation of effective time for imaging the organ. The biological residence time of radiotracer is the time radiotracer uptake and excretion by tissue as part of the cells natural process. The patient's dose is also an important criterion to consider, which is the amount of ionization that takes place within the tissue due to the attenuation of λ photons or alpha particles ejected from the nucleus. It is important to know that the choice of a radionuclide or amount of tracer molecules should be in a manner such that the patient's dose is low and the half-life is sufficient to carry out imaging.

The radioisotopes are produced for nuclear imaging in a separate facility by artificial (or man-made) means using particle accelerators (such as cyclotron), nuclear reactors, and radionuclide generators. In brief, the cyclotron is nuclear collider or bombardment vacuum chamber that enables collision of accelerating particles (i.e., protons, deuterium, or alpha particles) with the nucleus of an atom to create radioisotopes, such as Technetium-99m, Fluorine-18, Gallium-67, and Thallium-201. In a nuclear reactor, the radioisotopes are produced by two ways (a) as a by-product of nuclear fission (process of disintegrating nucleus of an atom into two smaller nuclei) and (b) neutron activation method. The fission reaction is possible either spontaneously or by the absorption of neutrons in the nucleus. The absorption of neutrons by primary radioactive fuel leads to by-products (or fission fragments). These fission fragments (unprocessed radioisotopes) from the nuclear reactor are further processed or chemically separated before using for medical imaging. The commonly obtained fission-based by-products used in nuclear imaging are molybdenum-99, iodine-131, and xenon-133.

It important to note that the radiation activity (Ci/gram) of fission produced radioisotopes is high which makes them favorable radioisotopes for the creation of high labeling efficiency (i.e., higher imaging performance with less injected material) radiotracers after combining with carrier molecules. The neutron activation method on the other hand uses stable target materials having stable nuclei inserted in the reactor core by ports that capture the neutrons produced by nuclear fission reaction and converts to radioisotope. Most radioisotopes produced by neutron activation has impurities that are hard to separate and the presence of unwanted materials in the capturing target rods lowers the radiation activity. Few radioisotopes or daughter isotopes produced by the neutron activation method can be purified by chemically or physically separated from the target rods. Radionuclide generators are lead made containers that hold slow decaying parent radioactive isotopes kept in an adsorbed form in porous materials, which is later extracted chemically to daughter radioisotopes used for NMI purpose in clinic. The decay rate of daughter radioisotopes is commonly relatively fast and the parent radioactive material stored in the sealed generator. These parent radioisotopes are stocked in the generator from separate facility or supply which resolves the issue of developing radioisotopes in in-house facilities. The daughter isotopes are generated in the same container by passing eluting solvents. The eluted daughter isotopes were finally collected in a vial for imaging use. The methods involved with the production radionuclides are particularly needed for nuclide short-lived radio activity. These types of short-lived radioactive nuclide need to be generated close by, so it may be transported quickly.

15.4.3 INTERACTION OF GAMMA RAYS WITH TISSUE

Gamma rays produced by the decay of radioisotopes lead to mainly three types of indirect ionization (or interaction) in the tissue. At low energies of gamma rays, photoelectric effect is dominant wherein the secondary X-ray emitted due to photoelectric effect when gamma rays interact with tissue atoms. At moderate energies of gamma rays, the Compton effect is dominant, wherein X-rays are generated due to the Compton effect. At relatively higher gamma energies (above a certain threshold of electron–positron pair production), gamma interaction with tissue leads to pair production. The electron and positrons further lose its kinetic energies by interaction or even attain total annihilation with the release of secondary gamma rays as in PET imaging. Either secondary X-rays or gamma rays are produced as a consequence gamma interaction with tissue and the interaction with atoms can be the combination of Rayleigh, Compton, photoelectric and pairing effects. The photoelectric effect and Compton effect have been discussed earlier in CT imaging. The attenuation of gamma photon takes place in a similar fashion in X-rays which depends on the amount photons absorbed and scattered (or removed) in the tissue (expressed by eq 15.2). However, the attenuation coefficient in eq 15.2 can be expressed for gamma rays as

$$\mu_{eff} = \mu_{Rayleigh} + \mu_{Compton} + \mu_{photoelectric} + \mu_{pair-production} \tag{15.11}$$

15.4.4 GAMMA DETECTOR OF SPECT/PET SCANNER

The gamma camera/detector in SPECT and PET scanner is the only important components that detect the emitted γ-rays from the body into measurable digital signals. These detectors are composed of multiple components (such as collimator, gamma detectors, and a photodiode) coupled together into a sensing unit. Each sensing unit is arranged in a 2D array to construct a 2-D projected data (width is equivalent to slice thickness) of the emitted gamma rays. The design of gamma camera is very similar to that of the CT detector except the material used for detection is different (being a scintillator crystal or gas-filled ionization chamber). It is also important to note that the scintillator crystal for SPECT and PET are different, since the γ-rays produced by positron–electron annihilation in PET scan has a greater photon count/energy than γ-rays generated by beta-minus decay in SPECT scan.

The gamma rays are initially filtered through parallel-hole thin lead sheet-based collimators that allows rays parallel to the holes or rays less than 90 angle from the inner surface of the hole. The collimation process ensures the removal of scattered rays and reduces/improves the image contrast-to-noise ratio. To resolve image pixels with high resolution, the length of the parallel sheet should be greater than the distance between source and collimator. One can imagine the collimator as an optical lens such that the focal length and distance from the lens to the detector array of the camera plays a vital role in developing a better spatial resolution in images. Also, spatial resolution depends on the distance of the source (radiotracer accumulated deep inside the body) from collimator. Naturally, the region/source close to the surface of the body will manifest as high resolution in acquired image data compared with that of deep organ. Yet, other design of the collimator (or lenses) can be used for improving resolution such as pin-hole collimators are often used to magnify images, converging collimators are used for creating large image of small organs and diverging collimators are used to image large organs.

The collimator is coupled with detectors (either gas-filled ionization chambers or scintillator crystals) similar to a CT scanner (discussed earlier) except for the gas or scintillator crystal used in SPECT/PET scanner has a different composition compared with the CT scanner. In brief, the gas in gas-filled detectors is subjected to ionization process after the interaction of high energy gamma rays with the gas. Due to ionization, the charged particles/ions are attracted toward the electrodes that encompass the gas. Thus, establishing a small potential difference across the electrodes. This led to the generation of current. The generated current is amplified and transformed to digital signals using analog-to-digital (A/D) converters. The digital signals were gathered simultaneously from all detector units in the 2D detector array and passed to the reconstruction pipeline for reconstruction of body slices. On the other, scintillator-based sensing unit is made up of special crystals or solid-state materials (designed with optimal optical thickness) when interact with the γ-rays produce visible or ultraviolet rays. This solid-state material is mounted on optical sensors (photodiode of photo-multiplier tubes) to receive rays/photons from the crystal and generate electric signals by photoelectric effect. Next, the analog signals from each detector in the array were converted to digital signals simultaneously using A/D converter that was fed into the reconstruction pipeline for reconstruction slice image. Interested readers can dive in these literatures[2,3] for better understanding of the detection components of the gamma camera.

15.4.5 IMAGE RECONSTRUCTION IN SPECT AND PET/CT

The image reconstruction or 3D volume rendering in SPECT and PET scan is performed like that of CT image reconstruction discussed earlier, that is, using filtered back-projection and iterative algorithms. As we are aware of the section on CT scan, reconstruction can be performed based on measured projected data from the body section when a well-collimated X-ray source passed through the tissue. However, one of the major challenges in SPECT or PET reconstruction is that the projected data contains a large amount of scattered gamma photons which is significantly greater than that of the CT scanner. Clearly, scattering effect reduces the image quality of reconstructed images. The primary reason for enhanced scattering is that the source is not tightly collimated as in the CT scanner, since in SPECT or PET, the gamma radiation is emitted from the body in all directions (no external source illuminates the tissue). To improve the image quality and reconstruction process, special mathematical operations are used to correct the scattering and attenuation before or after back-projection or iteration-based reconstruction of tissue slices from the projected data.

In fact, various proprietor methods are implemented in SPECT and PET scanners for performing scattering and attenuation corrections. Attenuation correction primarily corrects the measured attenuation coefficient or image voxel value in SPECT or PET scanner by either removing corrected fraction based on a standard reference. This standard reference could be obtained by placing (a) radioactive source between body and detector to obtain the attenuation of source originated gamma rays before and after radiotracer introduced to the body, and (b) by placing a CT source–detector assembly in the same gantry to obtain approximate coefficient value that would be possible if X-ray source equivalent to SPECT/PET gamma photon energy illuminates the tissue. In fact, the later method is commonly used for attenuation correction in SPECT/CT and PET/CT scanning system. The scattering correction on the other hand is similar to removing unwanted scattered rays in the projected data. This type of correction is possible in the SPECT scanner by considering dual-energy window method wherein one window is placed at γ-intensity peak that encompasses contribution from scattered as well as un-scattered γ-rays and another sub-window placed slightly offset of the primary window which has encompassed only scattered γ-rays.

The fraction of scattered γ-rays in sub-window is removed from the total γ-photon count in the main window to obtain corrected projection data. The filtered back-protection algorithm and iteration algorithm is possible of using similar concepts as in the CT scanner. In reality, the filtering kernel differs from the CT reconstruction method. It is also important to know that the optical correction approach as well as mathematical treatment for reconstruction is distinct in SPECT and PET scanning system. This difference is primarily due to the fact that in SPECT scanner, the gamma ray emission is acquired in the single-photon direction whereas in the PET, gamma rays emitted are acquired in two opposite directions.

15.4.6 COMPARTMENTAL ANALYSIS IN NUCLEAR MEDICINE

Compartmental analysis is a mathematical method to construct computational models of pharmacokinetics of radionuclide concentration in the whole body or multiple subsystems of the body referred to as compartments. The term pharmacokinetics refers to quantification of the change in drug (e.g., radioactive sugar molecule) disposition in the body with time, and the change of concentration due to distribution, metabolism, and excretion of the tracer substance. Indeed, the conventional experimental approaches can be performed to quantify overall change in tracer molecule concentration in blood and urine over time. An approach widely useful to determine a change in the drug concentration for the whole body or sampled parts. However, experimental methods do not provide information about the kinetics in the individual parts since it is impractical to sample deep body parts. Therefore, a compartment analysis for PET systems is needed because a PET imager only displays the measured gamma ray intensity due to change in disposition of a tracer molecule but fails to provide any information about individual organs or whole-body metabolism and tracer concentration. The following section briefly discusses the steps involved in pharmacokinetic modeling of the PET system. For better understanding, the readers are encouraged to read the referred literature.[6,7]

15.4.6.1 STEPS IN PHARMACOKINETICS MODELING

Step I: Essential assumptions

- Amount of tracee (*substance associated with mass transfer, i.e., blood, plasma, and urine*) is conserved.
- Radioactive decay of tracer (^{18}FDG-Fluorodeoxyglucose) is conserved.
- Amount of tracee within the compartment remains uniform.
- Exchange of tracee between two compartments is constant.

Step II: Essential definitions

- Any physiological system that is divided into a number of interconnected subdivision blocks having a mass of well-mixed, homogeneous substance that behaves uniformly is referred to as compartment.
- Compartment can be defined for any anatomical, physiological, chemical, or physical subdivision of a system.
- The exchange of mass across the interconnected compartments are conserved. The exchange phenomenon may lead to an increase or

decrease in mass in compartments which is mathematically represented by a set of ordinary differential equations (ODE). These ODEs are deduced from the principle of mass conservation applicable to each compartment.

$$\begin{bmatrix}\text{Change in mass stored}\\ \text{in } \textit{compartment} \text{ during}\\ \text{time } \Delta\text{t}\end{bmatrix} = \begin{bmatrix}\text{Mass flow in } \textit{Compartment}\\ -\text{Mass flow out of } \textit{Compartment}\end{bmatrix} + \begin{bmatrix}\text{Rate of change of mass}\\ \text{due to species consumption}\\ \text{or accumulation}\end{bmatrix} \quad (15.12)$$

- The rate at which mass of tracer element leave/enter one compartment (or consumed/unconsumed by other compartment) lead to decrease or increase in concentration of tracer element. This rate is defined by rate constants or exchange coefficients.
- The initial concentration of ^{18}FDG which is injected into the blood system (C_{blood}) is also referred to as Input Function (IF). After injection, the tracer element (^{18}FDG) in tracee–I (blood) which distributes the tracer element to organs. Next, within the organ, the tracee–II (plasma) distributes the tracer element within the interstitial space and exchange from the interstitial space to intracellular region of the cell. Within the cell, the glucose molecule of the FDG undergoes phosphorylation (ATP generation), also referred to as metabolism. The cell contains phosphorylated FDG tracer, and the interstitial space contains free FDG tracer (see Figure 15.14, a representative exchange across cell).

Step III: Model development

The concentration of FDG in blood (C_{blood}/C_b) is represented by the input/output function (or source/sink), the concentration of FDG in interstitial space (C_1) is represented by one compartment and concentration of FDG in cell (C_2) is represented by another compartment. Hence, two-compartment models are shown in Figure 15.14B.

The rate constants k_1, k_2, k_3, k_4 are related to systems metabolism (or mass accumulation and removal) that are unknown for each compartment.

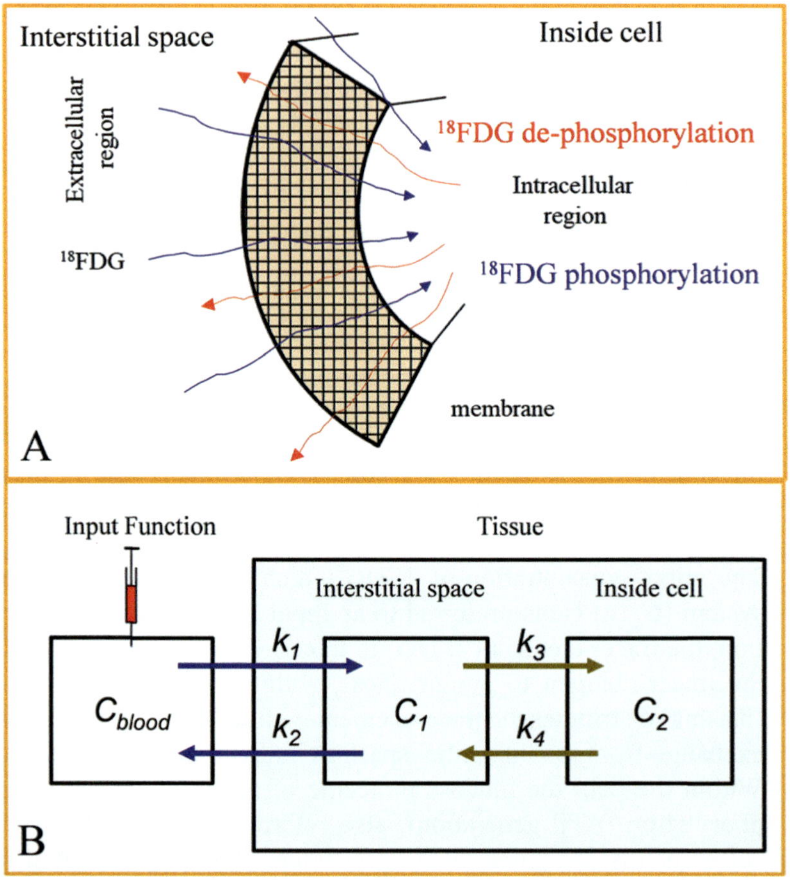

FIGURE 15.14 (A) An illustration of ^{18}FDG transport from extracellular space to intracellular space via cell membrane. Note that ^{18}FDG undergoes phosphorylation in intracellular space and ^{18}FDG de-phosphorylation solute transport back to extracellular space. (B) A simplified two-compartment model of the exchange process (compartment 1 being extracellular space and compartment 2 being inside cell).

- In compartment analysis, diffusive effect is neglected. The movement of tracee (plasma) from one compartment to another compartment is modeled by the removal rate or accumulation rate of tracer elements. Therefore, from eq (15.12)

$$\begin{bmatrix} \text{Change in mass stored} \\ \text{in } \textit{compartment} \text{ during} \\ \text{time } \Delta t \end{bmatrix} = \begin{bmatrix} \text{Rate of change of mass} \\ \text{due to species consumption} \\ \text{or accumulation} \end{bmatrix} \qquad (15.13)$$

$$\text{Now,}\quad \begin{bmatrix}\text{Mass accumulated}\\ \text{in } \textit{compartment}\end{bmatrix} = \left[+\dot{r}_A \Delta t\right] = k_i c_A \Delta t$$

$$\text{or,}\quad \begin{bmatrix}\text{Mass removal}\\ \text{from } \textit{compartment}\end{bmatrix} = \left[-\dot{r}_A \Delta t\right] = -k_i c_A \Delta t$$

where k_i is the rate constant (in per sec) also known as fractional turnover rate, and c_A is the concentration of ^{18}FDG (in mass/volume).

$$\text{and,}\quad \begin{bmatrix}\text{Change in mass stored}\\ \text{in } \textit{compartment} \text{ during}\\ \text{time } \Delta \text{t}\end{bmatrix} = \left[\Delta c_A\right]$$

From eq (15.13), we can obtain form of any compartment as

$$\begin{aligned} &\Delta c_A = \pm\left(k_i c_A\right)\Delta t \\ &\text{or, } \frac{\Delta c_A}{\Delta t} = \pm\left(k_i c_A\right) \\ &\text{or, } \lim_{\Delta t \to 0} \frac{\Delta c_A}{\Delta t} = \frac{dc_A}{dt} = \pm\left(k_i c_A\right) \end{aligned} \tag{15.14}$$

where i is 1, 2,...,n no of exchanges.

$$\text{or, } \frac{dc_A}{dt} = \sum_{i=1}^{n} k_i c_A$$

Using the expression, we can obtain two-compartment model as

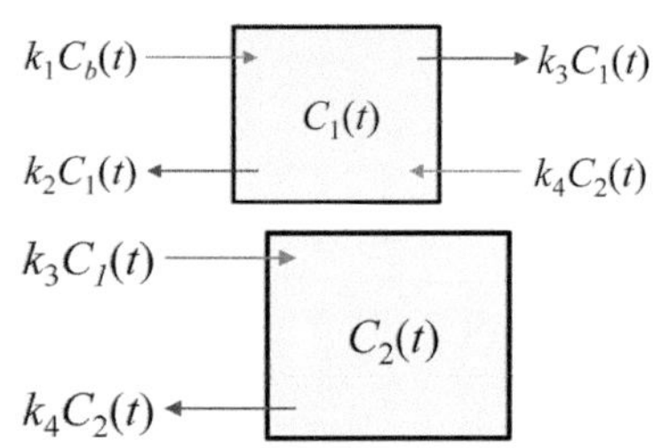

$$\frac{dc_1}{dt} = +k_1 c_b - \left(k_2 + k_3\right) c_1\left(t\right) + k_4 c_2\left(t\right) \tag{15.15}$$

$$\frac{dc_2}{dt} = +k_3 c_1\left(t\right) - k_4 c_2\left(t\right) \tag{15.16}$$

- Now, we have a *system of linear ODE* for concentration at a constant rate.
- These ODE is a first-order derivative of time, which is also referred to as an *initial value problem*. The solution of such ODE depends on the *initial conditions.*

$$\begin{aligned} c_1\left(t\right)\Big|_{t=0} &= 0 \\ c_2\left(t\right)\Big|_{t=0} &= 0 \end{aligned} \tag{15.17}$$

- The solution of system of equation subject to initial condition can be evaluated using numerical techniques (*finite difference method and matrix inversion*) and by the analytical approach (*integration*).
- With such compartmental equation setup, now the *objective* is to predict the *unknown coefficients* which are indicative *metabolism of the system*.

Step IV(A): Analytical solution of the models

The analytical method involves the integration to determine the concentration of 18-FDG in compartment 1 and compartment 2 using a set of ODE's (Eq. 15 and 16) and initial value (Eq. 17). We also know $C_b(t) = c_1(t) + c_2(t)$ and substituting in eq (15.16)

$$\frac{dc_2}{dt} = +k_3\left(c_b - c_2\right) - k_4 c_2(t)$$
$$\text{or, } \frac{dc_2}{dt} + pc_2 = k_3 c_b \qquad \text{where, } p = k_3 + k_4 \tag{15.18}$$

Next, multiply with e^{pt} both sides as

$$\frac{dc_2}{dt}\cdot e^{pt} + pc_2 \cdot e^{pt} = k_3 c_b \cdot e^{pt}$$
$$\text{or, } d\left[c_2 \cdot e^{pt}\right] = k_3 c_b \cdot e^{pt} dt$$
$$\text{or, } \int_0^t d\left[c_2 \cdot e^{pt}\right] = k_3 c_b \int_0^t e^{pt} dt$$
$$\text{or, } e^{pt}\left[c_2(t) - c_2(0)\right] = \frac{k_3 c_b}{p}\left[e^{pt} - 1\right]$$

Substituting initial condition $c_2(t)|_{t=0} = 0$, we get

$$c_2(t) = \frac{k_3 c_b}{p}\left(1 - e^{-pt}\right) \tag{15.19}$$

Next, we substitute the expression of c_2 in eq (15.15) and get

$$\frac{dc_1}{dt} = +k_1 c_b - \left(k_2 + k_3\right)c_1(t) + \frac{k_3 k_4 c_b}{p}\left(1 - e^{-pt}\right)$$
$$\text{or, } \frac{dc_1}{dt} + qc_1 = k_1 c_b + \frac{k_3 k_4 c_b}{p}\left(1 - e^{-pt}\right) \qquad \text{where, } q = k_2 + k_3$$

Next, multiply both sides of the above equation with e^{qt} and integrate.

$$\int_0^t d\left[c_1 \cdot e^{qt}\right] = k_1 c_b \int_0^t e^{qt}\,dt + \frac{k_3 k_4 c_b}{p}\left[\int_0^t e^{qt}\,dt - \int_0^t e^{(q-p)t}\,dt\right]$$
$$\text{or, } c_1(t) = \frac{k_1 c_b}{q}\left(1-e^{-qt}\right) + \frac{k_3 k_4 c_b}{pq}\left(1-e^{-qt}\right) - \frac{k_3 k_4 c_b}{p(q-p)}\left(e^{-pt}-e^{-qt}\right) \tag{15.20}$$

Step IV(B): Numerical solution of the models

We can also use a numerical method such as finite difference formulation to discretize the time derivative and to btain two algebraic equations. Here, we show backward differencing scheme for the time derivative of Eq. 15 and 16, respectively.

$$\left(\frac{c_1^k - c_1^{k-1}}{\Delta t}\right) = k_1 c_b - \left(k_2 + k_3\right)c_1^k + k_4 c_2^k$$
$$\text{or, } \left(1 + k_2\Delta t + k_3\Delta t\right)c_1^k - \left(k_4\Delta t\right)c_2^k = \left(k_1\Delta t\right)c_b + c_1^{k-1} \tag{15.21}$$

and

$$\left(\frac{c_2^k - c_2^{k-1}}{\Delta t}\right) = k_3 c_2^k - k_4 c_2^k$$
$$\text{or, } -\left(k_3\Delta t\right)c_1^k + \left(1 + k_4\Delta t\right)c_2^k = c_2^{k-1} \tag{15.22}$$

where the subscript *k* represents the index for time evolution. The solution for the above set of equations can be deduced using matrix inversion as

$$[A][X] = [B]$$

$$\begin{bmatrix} \left(1 + k_2\Delta t + k_3\Delta t\right) & -\left(k_4\Delta t\right) \\ -\left(k_3\Delta t\right) & \left(1 + k_4\Delta t\right) \end{bmatrix}\begin{bmatrix} c_1^k \\ c_2^k \end{bmatrix} = \begin{bmatrix} \left(k_1\Delta t\right)c_b + c_1^{k-1} \\ c_2^{k-1} \end{bmatrix} \tag{15.23}$$

$$\begin{bmatrix} c_1^k \\ c_2^k \end{bmatrix} = [A]^{-1}[B]$$

Step V: Implement an optimization method to determine unknown rate constants

We now have a set of equations deduced from analytical/numerical in step IV, such that the concentration (or relative concentration) of compartment 1 and compartment 2 is known. Note that the relative concentration of compartments

1 and 2 can be evaluated by measuring the change in radiotracer gamma intensities from the ROI in the acquired images.

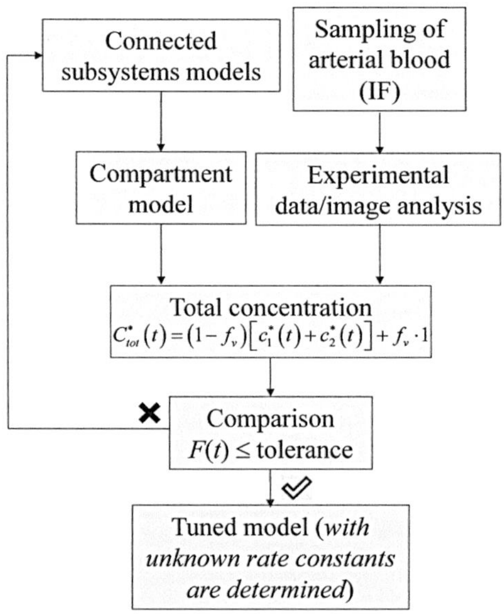

$$C_{total}(t)=(1-f_v)\left[c_1(t)+c_2(t)\right]+f_v c_b(t)$$

$$\text{or, } \frac{C_{total}(t)}{c_b(t)}=(1-f_v)\left[\frac{c_1(t)}{c_b(t)}+\frac{c_2(t)}{c_b(t)}\right]+f_v\cdot 1$$

$$\text{or, } C^*_{tot}(t)=(1-f_v)\left[c_1^*(t)+c_2^*(t)\right]+f_v\cdot 1$$

Note: the concentrations here are relative quantities, which can be related to changes in the intensities of gamma rays in the images for the ROI as compared with the image of the region where the drug is injected. However, the rate constants or fractional turnover rate (k) are not known to us. This parameter defines the rate at which the radioactive glucose molecule (18-FDG) is consumed by the cells due to metabolism. Thus, the rate constant is directly related to the metabolism of the 18-FDG molecule. We will construct an objective function to optimize the model unknowns by fitting with measured experimental outputs.

$$F(t)=\left\|C^*_{tot,\exp}(t)-C^*_{tot,num/anal}(t)\right\|$$

$$\text{or, } F(t)=\sqrt{\left\{C^*_{tot,\exp}(t)\right\}^2-\left\{C^*_{tot,num/anal}(t)\right\}^2} \quad (15.24)$$

where, $F(t)=$ Objective function, and $\|\bullet\|=$ Euclidean L^2 norm

Next, we can iteratively vary the input rate constants and f_v of the models until $F(t) \leq$ tolerance. The fitting methods that could be used are the least square curve fitting method, maximum likelihood, and Monte Carol Simulation.

15.4.7 PRINCIPLE OF PERFUSION AND VENTILATION IMAGING IN NUCLEAR MEDICINE

Perfusion imaging in nuclear medicine measures the flow of radiotracer with the body fluid and evaluates the function of the organ based on the uptake of the radiotracer by tissue. The body fluid under consideration is blood for myocardial, renal, lung, and cerebral perfusion tests. Whereas the inhaled air is another fluid used for lung function test by the inhalation of radiotracer to lung. Perfusion or ventilation testing quantifies (a) spatiotemporal changes in blood flow and (b) uptake of radiotracer (in inhaled air or in the blood) by specific tissue with and without abnormality. Myocardial perfusion test is performed to identify the function of heart muscle by comparing (a) difference in the amount of radiotracer absorbed by healthy muscle than that of the damaged muscle (radiotracers are made specifically to bind with chemical released by damaged muscle), and (b) observed image changes (cold-spots) in image caused by insufficient blood flow or muscle activity or defects during rest and after induced stress.

Muscle with abundant radiotracer will appear in high-intensity region in acquired images compared with muscle lacking rich concentration of radiotracer. Apart from the absorption of radiotracer by muscle, the variation of blood flow to the heart can also be identified in the image as cold or hot spots. Similarly, renal scintigraphy (or perfusion test) is performed to quantify renal blood perfusion or artery narrowing, kidney blockage or obstruction to urine flow, renal cortical tissue function, identifying renal mass being benign or malignant. The common approach in renal perfusion or functional test is to compare the images before and after the administration of radiotracer. In general, images are captured after a wait period post-administration of radiotracer depending on the decay period of radioisotopes in a time-dynamic manner. The lung ventilation–perfusion (VQ) scan tests two different functions (a) breathing (or ventilation) and (b) circulation (perfusion) capability of an entire lung. During the ventilation scan, the radiotracer is inhaled in a gas form with a mask to quantify pulmonary air transport whereas during perfusion scan radiotracer is injected into a vein to quantify pulmonary blood circulation. The areas where radiotracer accumulates in greater concentration appear as hotspots and areas without a flow of air or blood appear as cold-spots. The lung VQ test is normally performed to identify blood clots, small masses and quantify treatment response.

In cerebral perfusion scan, the transport of radiotracer in blood in different compartments of the brain is imaged either in static or time-dynamic maps. The tissue which is well perfused in the brain will appear as hot spots especially the area of enhanced oxygen and glucose metabolism (malignant lesion), and tissue not perfused well due to injury appears as cold-spots. Brain perfusion imaging can be performed in various medical conditions, such as epilepsy, dementia, ischemic stroke, carotid blockages, brain benign, and malignant tumor.

15.5 MAGNETIC RESONANCE IMAGING (MRI)

MRI is a specialized medical imaging equipment (Figure 15.15A) that maps the electromagnetic signal intensity due to change in the net magnetization of nucleus in the human body. These nuclei are at the core of every atom in the human body. However, most MRI procedures (except MR spectroscopy) make use of only the hydrogen nucleus of generating signal intensity or imaging purpose since hydrogen atoms are having largest nuclear magnetic moment (or field strength) and are the most abundant atoms in the human body (in biological molecules, fat, and water molecules).[3] It is also worth noting that other nuclei such as isotopes of phosphorous, sodium, and carbon can be an option for generating MR signals in MRI with the specific purpose of identifying certain metabolites or biomarkers. However, the MR signals generated by other isotopes are weaker than hydrogen nuclei. Another very important criterion for generating nuclear magnetic signals are that the nucleus should not have an equal number of neutrons and protons. In the case of balanced nucleus (i.e., same number of protons and neutrons), nuclear magnetic resonance (NMR) effect would be absent because of non-zero spin of the nucleus. Nuclear magnetic moment is defined by a magnetic force vector indicating magnitude and direction. A large group of nuclei ($\sim 10^{19}$) together presents a net magnetic moment used for the purpose of imaging in MRI. In body, nucleus is having spins, and its strength and directions of spin also influence the net forces (i.e., electric, and magnetic field). However, nucleus spin and magnetization stem from its quantum properties, and to date, there is no unique theory for the same.

One can recognize that each nucleus in our body has some spin and behaves like a tiny bar magnet. Still, the net magnetic field generated by all nuclei spinning in the body sums up to be zero due to its random orientations.

Hence, the human body has no net magnetization in the initial resting state. Once the body is placed in MRI's strong magnetic field, the hydrogen nuclei/protons were forced to align parallel or anti-parallel with its respective angular momentum along with additional setup of precession or wobbling about the magnetic field, resulting in some net magnetization of the body in longitudinal direction since nuclei/protons are no longer randomly oriented longitudinally. Next, an external oscillating magnetic field also referred to as radiofrequency (RF) pulse is applied in transverse direction which modifies the spin configuration of hydrogen nuclei/protons (i.e., modifies the net longitudinal magnetization). In other words, the transverse RF pulse establishes a net transverse magnetization due to phase coherence in the precession of transverse magnetic field of protons, which induces a voltage or signal intensity in gradient coils according to Faraday's law of induction. The measured signal intensity or voltage in gradient coils is used for 3D image construction. Having a simplified understanding of the MRI's working, it is also important to know that MRI truly originated from its forerunner NMR. NMR is a spectroscopic method widely used since the 1940s in chemistry and biochemistry research, that exploits the magnetic properties of protons and neutrons of the nucleus (associated with nuclear spin and charge distribution).

In NMR, a magnetic field which is strong enough to enable selective absorption of energy by nuclei in the sample due to resonance effect and later releases the energy that is unique to specific nuclei. The same idea was borrowed (in the early 1970s) from NMR to use the magnetic properties of protons with the help of body-size magnets and technology to scan or image the whole human body. In fact, the term was changed to MRI for clinical applications. MRI is a highly sensitive method for disease diagnosis because it allows better contrast for soft tissue differences compared with X-ray and CT, and it does not use nonionizing radiation. With the increase in MRI's magnetic field strength and use of contrasting agents, MRI can perform advanced imaging method like MR spectroscopy and functional imaging.[2,3] However, it is important to know that increasing the field strength of MRI for better image is not always a preferable solution since it also increases the specific absorption rate (SAR) absorbed by the human body.

15.5.1 WORKING PRINCIPLE OF MRI

It can be realized from the above discussion that the MRI examination can be performed in five steps (a) placing the patient in a strong magnetic field, (b) sent a radio frequency pulse for excitation, (c) radio frequency pulse turned off, (d) map-induced signals in gradient coils, and (e) reconstructs images. Figure 15.14(B) depicts some of the inner components of MR machines (discussed in detail later in this chapter). The physics of magnetic resonance and signal detection is interesting and exciting. In fact, the working on MRI is entirely dependent on the hydrogen nucleus/protons dynamics under strong magnetic field. It is important to consider these dynamics constituent particle in a magnetic field to fully comprehend the different MR protocols (such as T_1 and T_2 weighted images) discussed later. Hydrogen is one of the lightest and most abundant atoms in the universe and in the human body. It consists of one proton and one electron (no neutron) which are found in water and biomolecules.

Nucleus in hydrogen has one proton (nucleus is unbalanced) and are referred to as proton hereafter. Proton in the atom has a positive charge and magnetic moment that stems from its quantum property. These protons possess a rotation around its own axis also referred to as spin. As the spinning proton with its own magnetic moment placed in a magnetic field which is external to its own field (i.e., the body is positioned in a magnetic field B_0 as in Figure 15.15C), the spinning protons in the body align with the external magnetic field in parallel and anti-parallel orientations by flipping their poles. In addition, these protons start wobbling about the external field like a top. In other words, these protons in an external field spin about their own axis and wobble about the external field (refer the wobbling magnetic force of a proton represented by a vector in inset of Figure 15.15C). This wobbling rotation of the proton is also referred to as precession.

It is important to know that this precession speed (number of precessions per second or precession frequency) in the MR magnetic field is not constant, it varies with the element/proton type and magnitude of the field. The stronger field induces higher precession frequency. The precession frequency (also referred to as Larmor frequency) can be computed using an equation called Larmor equation.[2]

$$\omega_0 = \gamma B_0$$
$$\text{or, } f_0 = \frac{\gamma}{2\pi} B_0 \tag{15.25}$$

where f_0 is the precession frequency (in MHz), B_0 is the strength of the magnetic field given in Tesla (T) and $\gamma/2\pi$ is the so-called gyromagnetic ratio. The symbol γ here should not be confused with gamma rays discussed earlier. The value of the gyromagnetic ratio differs for different elements. For hydrogen (protons), the value of $\gamma/2\pi$ is 42.58 MHz/T.

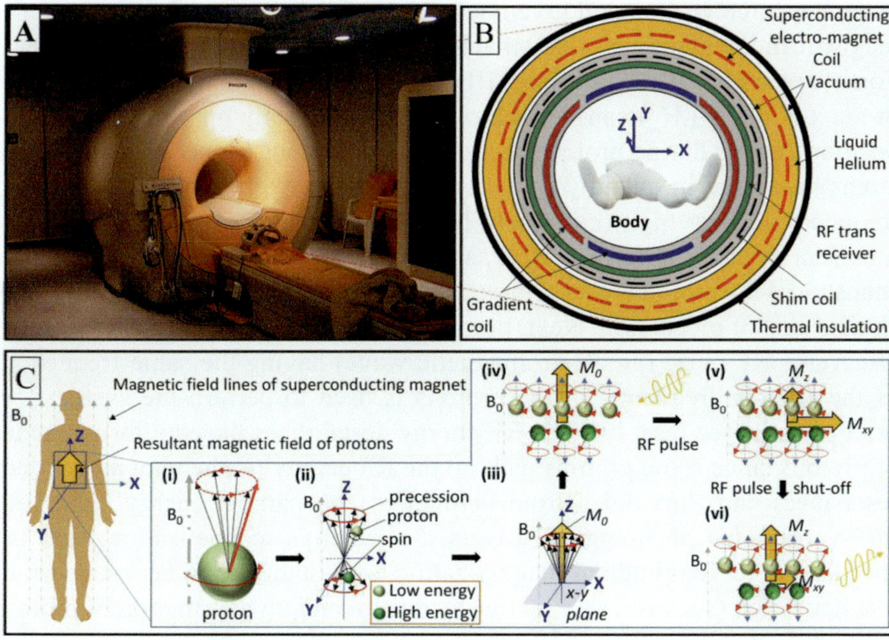

FIGURE 15.15 Illustration for MRI scan. (A) Complete 3T MRI scanner unit with dedicated body part specific receiver coil. (B) MRI scanner cross-section consist of main superconducting electromagnet, magnetic gradient coils and RF transmitting coil. (C) Patient body under strong magnetic field induces a resultant longitudinal field about the same direction as the magnetic field lines. Inset (i) precession of single proton with its magnetic moments in each angular position, (ii) precession of parallel and anti-parallel protons in the primary magnetic field (black arrow represents magnetic moments or force vectors), (iii) net magnetization in longitudinal direction in equilibrium state, (iv) applying RF pulse (frequency equivalent to Larmor frequency) of the processing protons lead to resonance effects, (v) Shrinking of magnetization in longitudinal direction and establishes new magnetization in x–y plane, and (vi) RF pulse shut-off causing the new magnetization in x–y plane to shrink and longitudinal magnetization to grow back (a process of relaxation) with the release of MR signals.

Source: (A) Reprinted from Ref. [8]. https://creativecommons.org/licenses/by-sa/4.0/

Note: the arrows in inset (i–vi) of Figure 15.15(C) represent the magnetic force (or magnetic moments) exerted by individual protons. In an equilibrium state after turning on external magnet, there are protons spinning in parallel and in anti-parallel directions with many magnetic moments/vectors acting either upward and/or downward in each angular position of precession (see inset II). This gives rise to two magnetic components for each angular position of individual protons (components in z-direction and in x–y plane). Due to parallel and anti-parallel precession, protons with the component of magnetic vectors exactly opposite to each other lead to cancelation of some magnetic force vectors and give rise to sum of the vectors of each proton's magnetic moments as a resultant magnetization (M_0) in the longitudinal direction of external field, B_0 (see inset III). However, magnetic field components in the x–y plane (M_{xy}) cancel each other due to exactly opposite horizontal position vectors during proton precession. The protons (or nucleus/hydrogen atoms) are large in quantity (> 10^{15} protons per voxel) and the precession frequency is very high (extremely high speed) which is quite difficult to represent in pictorial format. Here, a simplified image is shown as a timely snapshot of the proton precession such that all protons are assumed to behave similar axis of precession. Next, perpendicular to the main magnetic field, a transverse RF pulse (an electromagnetic wave) having the same frequency as the Larmor frequency of the protons is used to perturb the system for brief periods (see inset iv). The RF energy destabilizes the equilibrium state of protons since some protons pick up the RF energy (a phenomenon called resonance) and flips poles from parallel to anti-parallel energy state and vice-versa. This phenomenon affects the patient's magnetization, that is, shrinks the net longitudinal magnetization and establishes a net transversal magnetization (see inset v). Previously, due to random distribution of proton in the main magnetic field B_0, the vector sum of all transversal magnetization (M_{xy}) is canceled in the transverse direction. Due to the resonance effect, all protons are no longer precessing in transverse direction randomly. In fact, a group of protons which has picked up RF energy are in phase (synched or phase coherence) have their magnetic moment in the same direction at the same time. This results in a sum of magnetic vectors pointing in the transverse direction due to a group of precessing protons pointing in one side. This newly established transversal magnetization also precesses with the precession frequency in x–y plane about z-axis. The precession of the net transverse magnetic field about the patient's body induces voltage or signal intensity in gradient coils according to Faraday's law of induction. This induced voltage or MR signals thus originated from each voxel and are used for image reconstruction. After a brief period, the RF pulse is set to zero which causes the whole system of protons to slowly return to their initial equilibrium energy state. In fact, the measured MR signals are also time-dependent which means that the originated signal intensity from each voxel varies with time due to the changing net magnetization in both orthogonal directions, that is, time-dynamic shrinking of transversal magnetization of the system (also referred to as transversal relaxation) and simultaneous increase in longitudinal magnetization of the system (also referred to as longitudinal relaxation).

15.5.2 INTERACTION OF MAGNETIC FIELD WITH PROTONS

As discussed above, the MRI acquisition process deals with two net magnetization, a longitudinal magnetic field (M_z) or parallel to applied strong field B_0 and a rotating transversal magnetization (M_{xy}) perpendicular to z-axis in x–y plane. At equilibrium state, the longitudinal magnetization is maximized and the vector sum of all transversal magnetization M_{xy} is zero. As proton system is perturbed by RF pulse, some of the M_z tipped into transversal plane M_{xy} and the signal is picked up by the gradient coils due to rotating M_{xy}. As protons in voxel absorb RF energy, the magnetization evolves due to flipping of protons energy state and in-phase synchronization. This perturbation process can be best understood by quantum mechanical models.

According to quantum mechanics model, the RF pulse energy can be assumed as energy packets (quanta) such that when energy delivered by RF is equivalent to E which is the energy difference between two energy level of hydrogen proton in the magnetic field, the M_z magnetization flips from positive ($+M_z$) to zero ($M_z = 0$ but $+M_{xy}$) and negative ($-M_z$) depending on pulse time and amplitude. Note, this quantum energy levels in magnetic field determines the probability of protons being in anti-parallel/parallel spins. The complete reversal of original equilibrium state is also possible by releasing quantum of energy which is only possible at certain flip angle of 720°, and thus maintaining energy conservation. Note, the newly established transversal magnetization (M_{xy}) is not equal to zero because due to resonance, the selected group of protons are precessing in phase and setup a $+M_{xy}$. Additionally, the flip angle (also referred to as displacement angle) of proton system can be controlled by varying the RF pulse amplitude and duration since the displacement angle is linearly related to the production of RF pulse amplitude and time. In practice, the RF pulse amplitude and duration is varied to achieve a desired flip angle of protons, a technique commonly used to carry out fast MRI or to improve tissue image contrast, since tissue composition also varies with the speed of relaxation. The common flip angle (or displacement angle) is 90° and 180°, although smaller angles (< 90°) are also implemented for fast MRI. One of the benefits of faster MRI or smaller flip angle or short RF pulse is that it can still achieve higher magnetization in transverse direction. For instance, 45 (displacement angle can create 70% magnetization in the transverse direction (since sin 45 °is 0.707) compared with 100% magnetization in 90 (displacement angle (sin 90 ° is 1.0).

Finally, the MRI imaging protocols depend on the relaxation characteristics of protons in the tissue, spatial localization, and pulse sequences. These factors were briefly touched upon in the following illustration. It is important to note that the tissue energy relaxation in transverse and longitudinal direction occurs simultaneously. However, both relaxations do not take place at the same rate. It should be noted that MR signals are only collected during transverse relaxation by gradient coils. However, MR signals for longitudinal direction are indirectly evaluated from transverse relaxation using a modification in pulse sequences. Different sequences of RF pulse are applied to tissue to create varying MRI imaging protocols, such as T_1-weighted (T_1W), T_2-weighted (T_2W), proton density (PD) weighting, and fluid attenuated inversion recovery (FAIR) method.

15.5.2.1 T_2 RELAXATION OR SPIN-SPIN RELAXATION

T_2 is the time constant value associated with proton relaxation and evolution of newly formed transverse magnetization (M_{xy}) just after the RF pulse or delay time between two repetitive pulses. Typically, relaxation can be viewed as protons affinity to return back to its initial equilibrium state (i.e., stabilized spin configurations) over time by shrinking the transverse magnetization back to zero and growing the longitudinal magnetization back to maximum. However, transverse magnetization (M_{xy}) keeps on precessing with precession frequency, thus induces a frequency-dependent electrical signal in gradient coils. These frequency-dependent signals are used for image construction and varies with time. The decay is primarily due to a decrease in the amplitude of M_{xy} which is loss of in-phase synchronization (also referred to as loss of phase coherence). Mathematically, the decay of transverse magnetic field is given as[3]

$$M_{xy}\left(t_f\right) = M_{xy}\left(t_{in}\right)\exp\left(-\Delta t/T_2\right) \tag{15.26}$$

where $M_{xy}(t_f)$ is transverse magnetization at any relaxation time t_f, $M_{xy}(t_{in})$ is the maximum transverse magnetization at initial time after 90° flip RF pulse, T_2 is time constant and Δt is ($= t_f - t_{in}$). It is clear from eq (15.26) that the slope is exponentially decaying which depends on the magnitude of $\Delta t/T_2$. In all practical situations, the restriction of free motion of water bound molecules or hydration layers in the tissue or solvents varies with the T_2 value and to the ratio t/T_2.

The rigid and bigger molecules restrict the free motion of water molecules (e.g., in bones) which lead to increase in the T_2 value due to efficient transfer of energy of excited spinning proton to other less excited spinning protons (spin-spin interaction) and consequently, $-\Delta t/T_2$ would be a smaller quantity, meaning that the decay will be much faster. On the other hand, tissue with less constraint on molecular motion such as cerebral spinal fluid would lead to low T_2 value and longer decay period. In other words, different organs or tissue has a different composition of water and molecules which in turn varies with the decay time of transversal magnetization. Hence, the spatial images also lead to contrast between tissue groups. The T_2 value also increases due to inhomogeneity of main magnetic field B_0. In practice, the magnetic field B_0 lines generated by superconducting magnet of MRI is not completely homogeneous over the body section (i.e., the gradient in a magnetic field with a maximum value at the center of the body that reduces away from the center). The T_2 also varies due to the presence of MR contrasting agents in the tissues.

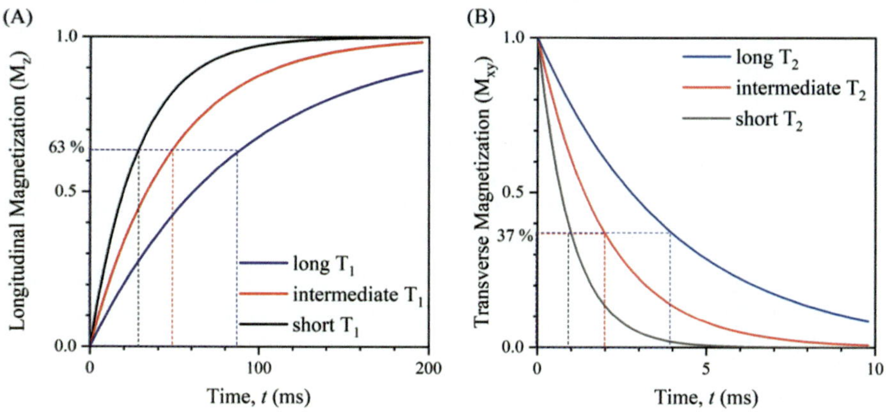

FIGURE 15.16 T_1 and T_2 relaxation curves in tissue. (A) T_1 relaxation curves for different tissue. (B) T_2 relaxation curves for different tissue. Note that both T_1 and T_2 are the time constants defined by 63% and 37% evolution in longitudinal and transverse magnetization, respectively. Tissues with longer T_2 losses transverse magnetization (or relaxes) faster compared with tissue with shorter T_2. Whereas tissue with longer T_1 gains (restores) longitudinal magnetization faster compared with tissue with shorter T_1. Time taken for recovery of longitudinal magnetization is much longer than that for the time required for transverse relaxation.

To remember the concept, the tissue with longer T_2 value leads to dark pixels due to low signal (or faster transverse relaxation) and shorter T_2 values lead to bright pixel (or slow transverse relaxation) in T_2-weighted images (see Figure 15.16).

15.5.2.2 T_1 RELAXATION OR SPIN-LATTICE RELAXATION

T_1 is a time constant value associated with the evolution of longitudinal magnetization M_z. The M_z evolves simultaneously as transverse magnetization (M_{xy}) shrinks just after the RF pulse. However, the growth of longitudinal magnetization (M_z) to a maximum value as in the equilibrium state takes place slowly as compared with transverse relaxation (see Figure 15.16). In this process, the energy associated with flipping of protons to its original equilibrium state is released back to the lattice or surrounding molecular structure via collision, rotation, and electromagnetic interaction. The T_1 relaxation is also referred to as spin-lattice relaxation. Mathematically, the regrowth of the longitudinal magnetic field is given as[3]

$$M_z(t_f) = M_z(t_{in})\left[1 - \exp(-\Delta t/T_1)\right] \tag{15.27}$$

where $M_z(t_f)$ is longitudinal magnetization at any relaxation time t_f, $M_z(t_{in})$ is the longitudinal magnetization at the initial time after 90° flip RF pulse, and T_1 is the time constant and Δt is $(= t_f - t_{in})$. It is important note that at t_{in} the longitudinal magnetization (M_z) is zero just after initial 90° flip RF pulse, therefore from eq 15.27 M_z exponential increases with time and the speed of recovery depends on the [-exp(-$\Delta t/T_1$)] or on the value of T_1. Tissues with shorter T_1 values lead to bright pixels and longer T_1 values lead to dark pixel in T_1-weighted images. The rate of energy released back to lattice or composite biological molecules (i.e., the viscous hydration layer surrounding the molecules in the tissue or freely moving water molecules) depends on the "tumbling" frequencies of molecules relative to that of the Larmor frequency. T_1 relaxation is dependent on field magnitude of the external field (B_0), that means the larger the magnitude lengthier the T_1 time. Here, "tumbling" frequencies are referred to the frequencies at which molecules are rotating in magnetic field B_0 though different from Larmor frequency. These molecules are bound viscous molecules (such as a hydration layer) or unbound/lossy molecules (such as bulk water or lipid hydrogens). For molecules made of tightly bound viscous hydration layer and freely moving bulk water, the "tumbling" frequencies do not exactly overlap with Larmor frequency. This resulted in constrained energy dissipation (T_1 value short) and exponential growth of M_z is longer. Whereas molecules such as triglycerides and moderately bound hydration layers lead to partial overlap of molecular "tumbling" frequency with Larmor frequency. Hence, takes relatively short time to M_z growth (as T_1 value is large).[3]

In practice, image quality or contrast can be improved by a varying sequence of RF pulse, MR field strength and administering contrasting agents. Especially, contrasting agent gadolinium chelated with macromolecules establishes a spin-lattice sink that decreases the T_1 value for local tissue and results in longer time to equilibrium value.[3]

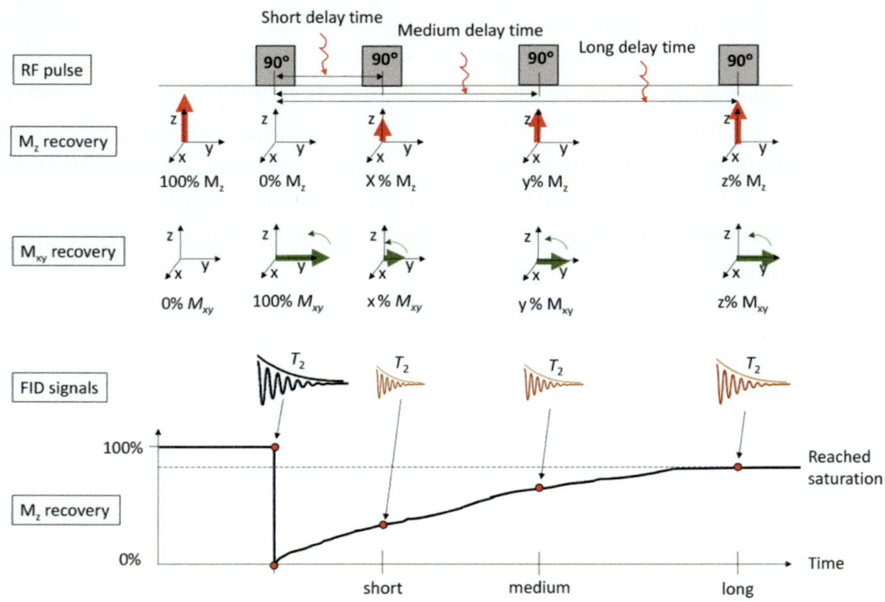

FIGURE 15.17 Longitudinal relaxation (M_z) for proton system or sample can be created by the measured amplitude of free induction decay (FID) signals associated with repeated perturbation of the sample by 90° pulse after different delay time from equilibrium. Row 1 depicts the transverse RF pulse schedule with varying delay time. Rows 2 and 3 depict the M_z and M_{xy} magnitude before and after 90° RF pulse, respectively. Row 4 depicts induced FID signals measured by coil just after 90° pulse. The measured FID signal shows maximum amplitude during pulse that gradually decays due to no pulse representing T_2 relaxation process. Row 5 depicts the constructed T_1 relaxation curve using measured FID signals.

Source: Modified and adapted from Ref. [3]

It is important to note that the direction of external magnetic field (B_0) is same as longitudinal magnetization (M_z), and is not directly measurable, as $M_z <<< B_0$. In fact, a longitudinal magnetization for samples is created from measurable transverse magnetization signals by applying specialized pulse sequence (i.e., repeated perturbation) to a proton system with different pulse delay from an equilibrium condition (see Figure 15.17). When the proton system is at equilibrium M_z is having maximum value (but M_{xy} was zero). Next, applying a 90 (RF pulse from the transverse direction causes M_z to tip-off to xy plane having maximum M_{xy} (but M_z becomes zero). This change is measured by peak amplitude of FID signal by the coil.

After a short delay period, another 90 (RF pulse from the transverse direction perturbed the proton system. However, within this first delay period only x-fraction of M_z is recovered since the rate of longitudinal relaxation is much slower than the transverse relaxation of M_{xy}. Thus, applying 90 (RF pulse after first short delay time makes only the x-fraction of the proton system to be perturbed again, which sets up another x% transverse magnetization. This change is again measured by peak amplitude of FID signal. The above process is repeated with different delay time from an equilibrium condition to measure the variation in peak FID amplitudes at different time points until the system of protons saturates (means no more change in peak FID signal takes place). The peak amplitude data points are fitted using an exponential equation to estimate the T_1 value.

The above curves represent signals from single voxel or mean of signals. The magnitude of the signals received by gradient coils is spatial temporal in nature since different voxels in the human body have different PD and varying distribution of contrasting agents which effects the acquired image contrast. In addition, the interplay of T_1 and T_2 relaxation times (or time to repetition and time to echo) plays a vital role in image contrast.

TR represents time to repetition that means delay in subsequent RF pulses. It is the duration during which T_1-recovery as well as T_2-decay take place. Time of echo (TE) is the time between excitation of 90° RF pulse and appearance of the peak amplitude of an induced echo. Echo is a repetition of the FID signal in the coil, primarily induced to separate the overlap of FID signals with that of the duration of energy deposition in the system after initial RF pulse. This is done by inverting transverse magnetization by applying a 180 (pulse at time TE/2 after initial 90 ° pulse, which leads to an echo at a later time.

15.5.2.3 SPIN ECHO (SE) CONTRAST WEIGHTING

Spin echo sequence of pulse is the process of initial excitation of a proton system by a 90 RF pulse to produce FID, after which another 180° refocusing pulse is induced for inception of echo later (*see* Figure 15.18A). The primary reason to produce an echo later is to avoid FID amplitude overlap between energy deposition and release. Introducing a 180° refocusing pulse inverts the transverse magnetization or spin system and produces a phase coherence (or echo) at time TE (see Figure 15.18B).

It can be realized that in a reference frame that is rotating, by inverting magnetization its vector originates in the mirror side of the transverse field.

Figure 15.18C depicts a conventional SE pulse sequence, that is, measurement of the amplitude of echo/FID by applying to proton system subsequent 90°and 180°pulses followed by repetition of the same pulse order after significant delay/relaxation time. Here, the TR value is long enough to attain a significant T_2 decay since TR value plays a vital role on the relaxation curve. However, shortening the TR value by introducing subsequent 180° refocusing pulse such that TR is the time between two repetitive 180° refocusing pulse would reduce the T_2 decay or relaxation.

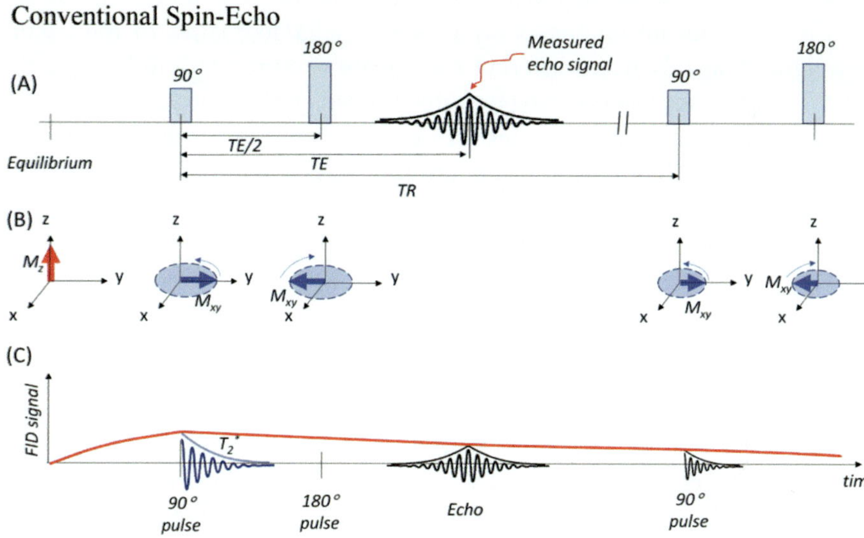

FIGURE 15.18 (A) Conventional spin echo (SE) pulse sequence and echo generation. (B) Transverse magnetization evolution due to 180° focusing pulse. (C) Decay of the FID signal (or actual T2 relaxation curve) with time due to repetitive perturbation.

Source: Modified and adapted from Ref. [3]

Weighted image in MRI is referred to artificially induced modification into tissue intrinsic parameters (i.e., PD, T_1 and T_2 characteristics) to acquire signals that can lead to significant image contrast. Image contrast can be defined by measuring signal intensity difference between voxels. One of the reasons to improve the image contrast is to differentiate between different tissue or pathology. In MRI, the time required for stationary tissue or pathology to relax varies with their molecular composition, PD and free movement, and many other intrinsic factors discussed earlier. In other words, both intrinsic tissue parameters (i.e., PD, T_1 and T_2) and pulse timing parameters (i.e., TE and TR) can affect the image contrast among different tissues. In fact, by varying the time (i.e., TE/TR) of pulse, the contrast dependence can be weighted toward T_1, PD, and T_2 characteristics of the tissue.

15.5.2.4 PROTON DENSITY (PD) CONTRAST-WEIGHTED IMAGE

PD-weighted image refers to image constructed, or signal measured by modifying the pulse timing parameters (i.e., TR and TE) with a chief weighting toward the signal generated due to a number of magnetized protons in the tissue, with no emphasis on T_1 and T_2 characteristics of the tissue. PD-weighted signals from different tissues are acquired by minimizing the T_1 and T_2 differences among tissues, which means only relative difference in the number of magnetized protons will produce contrast in the PD contrast images. The T_1 differences in tissue are reduced by the longer TR value to attain substantial recovery of M_z for all tissues. Whereas T_2 differences in tissue are reduced by the shorter TE value to lower T_2 decay for all tissues.

15.5.2.5 T1-WEIGHTED CONTRAST IMAGE

T1-weighted image refers to image constructed, or signal measured by modifying the pulse timing parameters with a chief weighting toward increasing the relative difference between T_1 characteristics of the tissues, with no emphasis on PD and T_2 characteristics of tissue. T_1-weighted signals from different tissues are acquired by maximizing the relative difference between longitudinal magnetization (M_z) while recovering to equilibrium. This can be achieved by short TR to maximize the relative difference between recovering longitudinal magnetization and short TE to minimize the relative difference among tissues during T_2 decay.

15.5.2.6 T2-WEIGHTED CONTRAST IMAGE

T2-weighted image refers to image constructed, or signal measured by modifying the pulse timing parameters with a chief weighting toward increasing the relative difference between T_2 characteristics of the tissues, with no emphasis on PD and T_1 characteristics of tissue. T_2-weighted signals from different tissues are acquired by maximizing the relative difference between transverse magnetization (M_{xy}) while recovering to equilibrium. This can be achieved by relatively long TR to minimize the relative difference between recovering longitudinal magnetization (M_z) and by long TE to maximize the relative difference among tissues during T_2 decay.

15.5.3 ESSENTIAL COMPONENTS OF MRI SCANNER

MRI scanner is made up of complex engineering components like any other type of scanner discussed earlier. There are many components in the scanner to perform its task efficiently (see Figure 15.15B). We will identify some of the essential components and elaborate upon them.

15.5.3.1 SUPERCONDUCTIVE ELECTROMAGNET

The chief component in an MRI is the superconducting electromagnet to create a powerful magnetic field in the bore. Electromagnet creates a magnetic field by allowing current flow through the coils. The design of the superconducting electromagnetic coil depends on the criteria, such as temporal field strength, stability, field strength, and magnetic field homogeneity in the bore. The coil is designed in a manner to create a nearly homogeneous magnetic field along the z-direction of the bore. It is important to note that high magnetic field (> 0.5–1 Tesla) can only be achieved if the electromagnetic coil achieves the property of a metal called superconductivity (no loss or resistance to current flow). 1 Tesla is equal to 10,000 Gauss. Superconductivity is achieved by some metals or alloys at extremely low temperatures. In MRI, superconductivity is achieved in electromagnet coils by maintaining liquid helium at temperature less than 4°K in the cryogenic containment compartment (see Figure 15.15B).

15.5.3.2 SHIM COIL

Shim coils operate on magnetic field and introduces magnetic field homogeneity over the bore volume. Shim coil introduces minimal variation magnetic flux density.

15.5.3.3 RADIOFREQUENCY (RF) TRANS-RECEIVER COIL

RF trans-receiver coil transmits energy for perturbation of magnetized protons in the body as well as receive inducted signals due to relaxation of perturbed protons in the body.

15.5.3.4 GRADIENT COILS

Orthogonally faced gradient coil pattern (*see* Figure 15.19) on cylindrical shell or etched coil pattern on the conducting cylindrical shell are used in the innermost compartment of the MRI (see Figure 15.15B) to introduce linear variation in the magnetic field strength inside the bore or patient body. As current flows through these gradient coils, it induces a secondary magnetic field that brings predictable variation in the magnetic field's magnitude (B_0). Thus, causes a spatial gradient in precession or resonance frequency of the protons since main magnetic field (B_0) is altered over the tissue volume. The superposition of magnetic field generated by any coil and main magnetic field (B_0) would not affect the B_0 at the middle of the coil also known as the null space of the bore. In this space, field's magnitude remains unchanged (i.e., constant protons frequency of precession). However, the magnetic field either increases or decreases from the null space linearly introducing gradient and varying the precession frequency of protons. If we consider the frame of reference to be rotational in nature, there will be an incremental change in frequency of the protons symmetrically about the null defined by a frequency bandwidth. It should be noted that the superimposition of magnetic field generated by all three gradient coils simultaneously with the main field would lead to an oblique gradient from the null space not in axial, sagittal, and coronal plane. When the RF pulse frequency bandwidth (variable frequencies) are used to perturb proton system that are resonating within the frequency bandwidth established by a particular gradient coil, only protons system at a particular position or slices that are having same frequency as interacting RF pulse frequency will resonate and able to generate signal at the specific frequency with phase information. In fact, using a specialized method (discussed later), phase shift is brought into effect during relaxation of rotating transverse magnetization of proton system such that the proton system precess in the same frequency but in a different phase. Hence, the position of protons system or protons in unit voxels is captured by both frequency and phase in k-space. The k-space data store spatial information in frequency and phase. These spatially encoded data are used later to construct 2D images of the body slice (discussed later).

15.5.3.5 THERMAL INSULATION

Thermal insulation and vacuum separation (see Figure 15.15B) are used to maintain the cooling of the superconducting coil during operation.

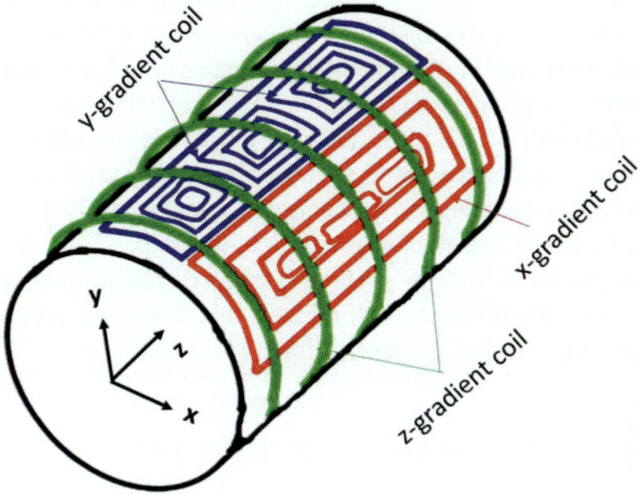

FIGURE 15.19 A representative cylindrical shell with gradient coils mounted on the surface of the cylindrical shell. x- and y- gradient coils are orthogonal to each other and z- gradient coil is mounted along the length of the bore. Net gradient or gradient direction can be achieved by superimposition of magnetic field generated in orthogonal axes and the main field. The gradient direction in a particular axis can be altered by reversing the flow of current through the coil.

15.5.4 IMAGE RECONSTRUCTION IN MRI SCANNER

Image reconstruction in MRI from acquired signals is significantly different from earlier diagnostic devices such as CT and NMI. The primary reason being the signals measured from individual voxel in the MRI slice are fluctuating/sinusoidal in nature and the spatial information (or tissue projection data) is embedded in the frequency and phase differences in the signals. To accurately capture the spatial information, MRI specialist applies well-timed three gradients (slice selection, phase-encoded, and frequency-encoded gradients (FEG)) between the pulse sequences (e.g., SE sequence) to capture MR signals in k-space (two-dimensional frequency versus phase data). The acquired two-dimensional data in k-space is converted using inverse Fourier transform to reconstruct the spatial two-dimensional contrast images.

15.5.4.1 SLICE SELECTION GRADIENT (SSG)

As indicated earlier, the proton precession frequencies in tissue volume increase or decrease from null space in z-direction due to the gradient established in the main magnetic field (B_0) by z- gradient coil. The pulse generated by the RF trans-receiver is not capable of selectively energizing a particular section of tissue (or proton system). However, under the state of axial or z-gradient field in the tissue volume, applying SSG simultaneously with RF pulse ensures narrow frequency bandwidth of RF pulse (known amplitude and duration) to only perturb the proton system of tissue slice having similar precession frequency bandwidth as RF pulse. Altering the central frequency of RF shifts the proton system slice location. The thickness of slice also depends on the frequency bandwidth of pulses as well as gradient strength in axial direction. Narrow frequency bandwidth perturbs protons in a thin slice and broad frequency bandwidth perturbs protons in a thick slice.

15.5.4.2 FREQUENCY-ENCODED GRADIENT (FEG)

The FEG is the readout gradient applied orthogonally to z-axis (or SSG) in x-direction during the evolution and decay of echo or transverse magnetization. Due to applied FEG, the net changes in frequency of protons within the selected slice varies from maximum positive frequency to maximum negative frequency (creates a frequency gradient) with zero-frequency at null space (i.e., $x = 0$). The measured complex signals by the receiver coil due to decay of echo contains local variation in x-direction of the slice (or position of protons in the x-direction), which can be amplified, digitized, and signal processed (applying transforms) to create a projection of slice in x-direction. Indeed, it is possible to gather large numbers of projections by incremental rotation of FEG direction about the slice and then use filtered back-projection to reconstruct a two-dimensional image of a slice by superimposition of projection data from each angle. This approach of reconstruction of MR image is not applicable in today's MRI system due to associated motion-related artifacts. Instead, the spatial variation in both x- and y- axis are obtained by applying phase shift to RF pulse in the y-axis.

15.5.4.3 PHASE-ENCODED GRADIENT (PEG)

PEG is applied to the proton system within the slice after the SSG and before FEG to introduce a small phase shift in the complex frequency-dependent signal measured by receiver coil. After PEG, a small shift in the phase of precession frequency of protons is introduced in the orthogonal direction (i.e., y-axis) from null space. That means a phase gradient is established to the precession frequency of protons in positive and negative y-direction of the slice (i.e., incremental positive change in phase to negative change in phase), but no change in phase at null space (i.e., y = 0). The incremental gradient in the phase of the proton frequency in y-direction of the tissue slice is implemented to avail different view angle of the volume. In other words, PEG is applied to the slice to determine the position of proton in y-axis and FEG is applied to determine the position of proton in x-axis.

15.5.4.4 TWO-DIMENSIONAL IMAGE DATA RECONSTRUCTION

The frequency and phase information of evolving transverse magnetization is transformed into a two-dimensional spatial image of the slice using an inverse Fourier transform. The entire process of the image reconstruction is detailed in Figure 15.20 which elaborates the two-dimensional image acquisition of tissue slice using spin echo pulse sequences.

Figure 15.20 depicts the excitation of a proton system in the tissue volume with spin echo pulse sequence along with SSG, PEG, and FEG for two-dimensional image acquisition from measured composite signals by the receiving coil. Here, at first, RF pulse with certain frequency bandwidth is simultaneously applied with SSG which enables selection or excitation of a slab of tissue protons along z-axis having same frequency range as RF pulse. The thickness of the proton slice is chosen by the frequency bandwidth. After excitation of protons in the slab by a 90° pulse, net maximum transverse magnetization (M_{xy}) is produced by flipping of protons. Next, briefly a PEG is applied to the precessing protons to cause a small shift in the phase of the precession frequency of those protons in the orthogonal direction to avail different view angles of the tissue volume (useful for image reconstruction). The application of PEG leads to phase difference in the composite signal originated due to decay of transverse magnetization. Next, a refocusing 180RF pulse introduced with SSG to limit the range of frequency for same slice and inverts the net transverse magnetization which would lead to appearance of echo at a later stage. While echo signals are evolving, a FEG is introduced in logical x-direction, that is, orthogonal to SSG (z-) and PEG (y-) direction, thus extracting spatial changes in frequency of evolving complex signals with its phase (i.e., corresponding amplitudes for positions) that are simultaneously stored in k-space. Data or amplitude of signals were stored in k-space matrix with row (k_y) and column (k_x) specifies the phase and frequency variation in the measured signal, respectively.

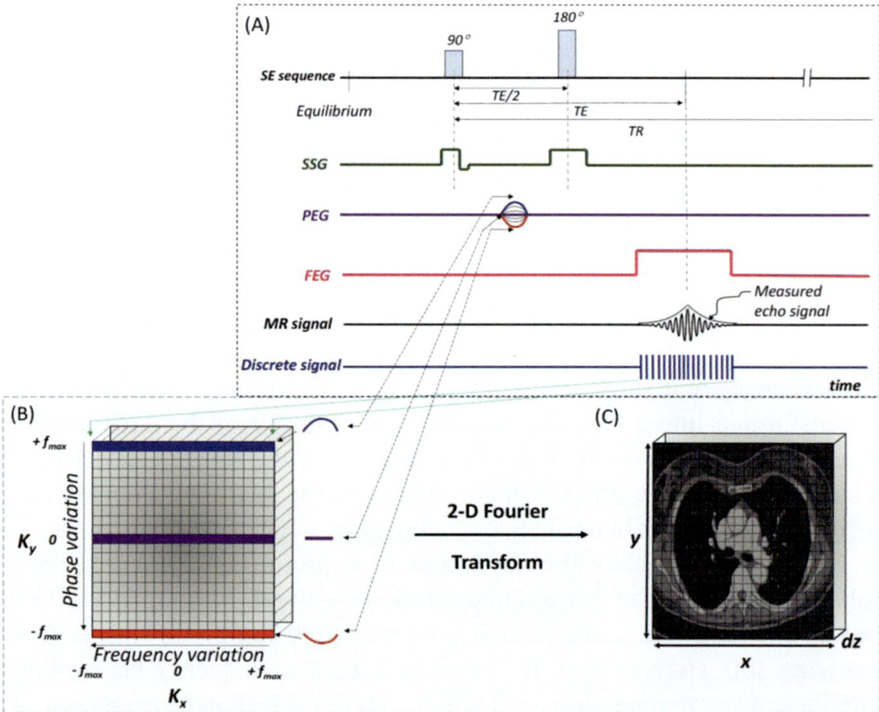

FIGURE 15.20 MR data acquisition in *k*-space matrix and transformation of positional data of 2-D spatial images. (A) Pulse sequence for spin echo approach with SSG applied along with RF pulse to excite protons in a particular slice, followed by application of PEG to introduce a brief phase shift in precession frequency of protons, followed by FEG applied at TE to store the complex signal data with its frequency variations and phase information. (B) Storing the amplitude of signal each row one-by-one in k-space with variation of frequency in x-direction and corresponding phase shift in y-direction. (C) Transformation of k-space data to spatial image data.

Source: Modified and adapted from Ref. [3]

The matrix of k-space is filled in rows by (a) one-dimensional Fourier transformation of acquired digital data of the complex signal to discrete frequency which allows the mapping of signal amplitude corresponding to a frequency along k_x axis, and (b) the PEG-induced phase variation in signal allows mapping the corresponding amplitudes along k_y axis in two-dimensional space.[3] And for each repetition of pulse sequence after TR, a small increment in PEG from earlier repetition is introduced to modify the view angle, although keeping SSG and FEG same. This automatically ensures filling of signal amplitudes of protons in their respective phase position in k-space of the specific slice. Here, phase shifting by PEG does a similar function of rotating detector in CT system, since shifting the phase enables different view angle of the same sample in MRI. Finally, all rows are populated by taking an inverse of Fourier transform in 2D space which decodes the frequency-domain variations for each of the columns of k-space to produce the spatial domain representation.[3]

15.5.5 PRINCIPLE AND APPLICATION OF FUNCTIONAL MRI (FMRI)

fMRI is a time-dependent brain MRI which is used to determine the function of the human brain. It primarily maps local variation in blood oxygenation or small blood flow changes in brain tissue which reflect the brain neuronal activity and associated damaging tissue conditions. The fMRI data involve a statistical map of brain responses toward imagined mental or perpetual functions performed by patients while in the scanner. The region of brain activity (i.e., blood oxygenation) is reflected by relative hyper- or hypo- regions in color maps on structural images of the regular MRI. Interestingly, the activation area or activity of neurons shows surplus oxygen transport in the area of the brain by artery, which leads to higher concentration of oxyhemoglobin in the activation area and variation in the paramagnetic property of the tissue. This variation in the magnetic property of the tissue is measured as the blood oxygenation level-dependent (BOLD) signals in fMRI.

There are two essential features of BOLD fMRI that should be considered during data acquisition: (a) the neuronal activity in the brain lasts for a few seconds but the associated hemodynamic changes last longer, and (b) hemodynamic response can be approximated as a linear time-invariant system.[9] These functional images of the brain are particularly used by neurologist or neurosurgeons (a) for the diagnosis of stroke, trauma, and brain degenerative diseases (e.g., Alzheimer), (b) to monitor the treatment efficacy, and (c) to determine the course of brain surgery. fMRI is also used to determine brain sections that are responsible for speech, thought, sensation, and movement. The analysis of fMRI data is complex in nature which demands various data processing techniques and statistical analysis to map the brain function. This book briefly introduced the technical basis for fMRI and some of its applications. Readers can find a detailed understanding of fMRI approaches and various sophisticated data analysis in the linked literature.[9,10]

15.5.6 PRINCIPLE OF MR PERFUSION, ANGIOGRAPHY, DIFFUSION, AND SPECTROSCOPY IMAGING

Apart from BOLD approach discussed above, measuring blood perfusion and blood flow, water diffusion, and spectral signature of tissue biochemistry is also referred to a class of functional imaging in MRI since these features allow clinicians to noninvasively predict the physiology and tissue function. The MR techniques that highlight physiology, such as gross blood flow, tissue ultrastructure, and biochemistry include MR perfusion or angiography, diffusion, and spectroscopy methods, respectively. In this section, we briefly touched upon all these techniques one by one. Whereas, their applications in clinical understanding of disease pathophysiology or therapeutic response are discussed in the next section.

15.5.6.1 MR PERFUSION

Perfusion imaging in MRI measures the flow of extrinsic (e.g., gadolinium) and intrinsic (e.g., water molecules) contrast agents in small vessel–tissue composite and evaluate the dynamic function of the organ based on the uptake of the contrast agent by the tissue. Perfusion imaging involves sophisticated manipulation of pulse sequence parameters, or magnetization/spin label of blood water *w.r.t.* non-flowing tissue water, or kinetic/compartment analysis of contrast to quantify spatiotemporal variations in blood flow, blood volume, vascular permeability, mean transit time, etc., in the vessel–tissue composite with or without abnormality. The time-dynamic variation is quantified based on a change in the image contrast indicating the underlying variation in the volume of contrast agents in the voxels, since there exists a rate of change of flux of blood in the composite voxels due to any abnormal changes. Detailed application-based variation in perfusion MRI methods can be found in the suggested literature for further reading.[11]

15.5.6.2 MR ANGIOGRAPHY (MRA)

MRA involves monitoring the variation in extrinsic or intrinsic contrast agent in large vessels, an important difference compared with MR perfusion imaging of small vessel-tissue composite. Perfusion is a term associated with the combination of the flow of blood and the diffusion of its constituents in microscale vessel–tissue junctures. However, angiography quantifies the bulk flow of blood in the large vessel with or without obstruction to flow. Apart from the variation in image contrast due to volumetric changes in contrast agents (such as gadolinium or blood water) in large vessels, it can exploit slice and phase encoding to predict the velocity of blood in the large blood vessels. Here, the data acquisition focuses on (a) increasing the contrast of large vessel compared with background tissue and (b) reducing signal loss due to time-of-flight effects by sophisticated manipulation of pulse sequence and magnetic parameters.[12]

15.5.6.3 DIFFUSION MRI

Diffusion MRI focuses on imaging diffusion of water molecules or low-viscosity molecules in tissue voxels. This method is sensitive to the associated restriction of the movement of water molecules in tissue. It uses special MR pulse sequences along with timely application of diffusion-encoded gradient in three orthogonal directions to map direction-dependent diffusion of water molecule in tissue from the measured signals, since diffusion of water in tissue encounters restriction due to macromolecules, membranous structure, viscous hydration layers, and fibers. This is supported by the fact that the decay of magnetization (i.e., relaxation of proton signals) is not only affected by T_1 and T_2 relaxation of tissue but also by self-diffusion of low-viscosity molecules (or proton system) in local magnetic field gradient.[13] In inhomogeneous magnetic gradient, the diffusing species or the proton system is not stationary. Thus, increasing the gradient time for diffusing species having higher diffusivity would lead to greater spread of proton system from its original spin location which contributes to the decay of magnetization (based on Einstein's relation between displacement and diffusion time).[13] Thus, applying diffusion-encoded gradient between the RF pulse sequences allows capturing signals that encompass information about proton system diffusion. Especially, a diffusion-weighted image (DWI) can be mapped by specialized software by weighting toward increasing the relative difference between diffusion characteristics of the tissues, with no emphasis on T_1 and T_2 characteristics of tissue. In DWI, the signal attenuation is directly related to net displacement of proton system (or low-viscosity molecules) over the applied diffusion gradient. Tissue structure (normal or abnormal) with restriction to molecular diffusion would appear hypointense as compared with less restricted sections in DWI images. A discussion of commonly used RF sequence for diffusion imaging can be found in Ref. [13].

15.5.6.4 MR SPECTROSCOPY (MRS)

MRS is an approach like Fourier transform NMR used in chemical analysis. NMR is a highly sensitive method to determine the presence or absence of certain biological and chemical compounds based on measured signals appearing at a particular frequency since nuclear resonance frequency is influenced by the chemical and structural composition of the compounds. MRS on the other hand, extracts the hydrogen protons signature or other isotopes in tissue to predict the presence and absence of biological molecules in the tissue. This method is particularly useful for noninvasive confirmation of abnormal condition based on the acquired signature of biological molecules in specific regions of tissue.

In practice, MRS of isotopes (such as ^{31}P, sodium-23, carbon-13) compared with hydrogen protons demands higher magnetic field strength (> 1.5 Tesla) to properly resolve signals from the noise, since these protons have a relatively low gyromagnetic ratio compared with hydrogen protons. Hence, hydrogen proton MRS is widely used in clinical settings as it can achieve a high-resolution spectrum of tissue chemical compounds within short time and with lower magnetic field strength widely used in clinical settings. The high sensitivity of proton MRS is attributed to the higher gyromagnetic ratio of hydrogen proton, abundance in the tissue metabolites and favorable magnetic relaxation time. The commonly known metabolites (or biological molecules) that proton MRS can resolve are N-acetyl aspartate (NAA), choline, Creatine (Cr), lactate, lipids, etc. A detailed discussion of MRS techniques is available in Ref. [14] for further reading.

15.5.7 APPLICATION OF MR PERFUSION, ANGIOGRAPHY, DIFFUSION, AND SPECTROSCOPY IMAGING

MR perfusion is a clinical tool widely used for evaluating cerebral ischemia (or stroke) and occlusion of carotid supply. It can provide qualitative or quantitative information on the changes in cerebral blood flow due to obstructive and non-obstructive ischemia. It is also useful for the evaluation of tumor severity in the brain which is reflected by the level of perfusion and necrosis, or for treatment efficacy of drugs especially for agents that suppress vascular growth in neoplasm. MR perfusion is also applicable in evaluating psychiatric conditions, such as dementia and migraine headaches.[15] It is also useful for the measurement of volume of cerebral fluid and blood flow due to the administration of psychiatric drug.[16] MR angiography on the other hand, is used for the identification of aneurysm in the aorta of chest and abdomen, plaque formation in the carotid artery, image guidance for stent placement/evaluation and for coronary treatment planning.

Diffusion-weighted MRI or DWI are popularly used for mapping water molecule movement or restriction in the brain, breast, and prostate lesions. DWI is highly sensitive to restriction of water diffusion in the brain and has clinical application in the diagnosis or prognosis of ischemia and treatment planning for severe stroke. Diffusion MRI with specialized methods (tractography) can map patterns of three-dimensional axonal structure of the brain. MRS in clinical practice allows chemical comparison of normal and abnormal tissue (e.g., brain) noninvasively. It is also useful in quantifying the changes in the brain tissue due to stroke and epilepsy. MRS could provide information about the presence or concentration of some important metabolites found in neurons, cell membranes and brain matter, such as abundance of lactate in brain MRS highlights anaerobic metabolism, a sign for irreversible ischemia and stroke. MRS also enables grading of gliomas based on the abundance of metabolites since higher grade of glioma is related to an increase in choline, lipids, and lactate and decrease in the concentration of NAA and creatinine.

15.6 ULTRASOUND (US) IMAGING

Ultrasound (US) in medical imaging is used to differentiate the soft or rigid structure characteristics based on the associated changes in the acoustic properties of the tissue. The acoustic properties of tissue or abnormalities differs due to its rigidity, composition, conformity, and flow which allows medical practitioners to identify the medical condition, such as stones, malignant and benign, fatty tissue, intra-tumoral calcifications, and blood velocity. The traditional US imaging techniques are sensitive in capturing sound signals from reflecting structures, which is also capable of accurately measuring the depth of echo-producing structures, volume measurement, and reconstruction of 2D/3D tomographic images. For example, in traditional US imaging, a rigid body such as stone is differentiated from soft tumor based on its rigidity and image features. A stone is a rigid structure which has the bright echogenic appearance due to strong reflection, and a non-echogenic shadow behind the structure since sound cannot penetrate stone. Similarly, calcifications are also notified in US imaging by its bright echogenic appearance. The pathological significance of calcifications is related to abnormal calcium speckles in soft tissue often related to metastatic lesion mineral imbalance. On the other hand, specialized methods, such as color Doppler

US imager (image blood flow) and elastography (measures the elasticity of tissue) can provide essential information about blood velocity fluctuation (or spectral data) and modulus of elasticity, respectively, to differentiate malignant and benign tumors.[17] In addition, a comet tail time-dependent twinkling feature in color Doppler images is a prominent sign of small calcifications.[18] In clinical practice, US imaging is used for noninvasive identification of structure and blood flow (or functional characteristics) based on measured reflected signals of back scattered sound waves, which acts as a useful tool to image sound features of abnormality that are not clear in other radiographic images (X-ray radiation and magnetic properties) discussed earlier.

In human subjects, abdominal ultrasound is performed to dynamically capture the abdominal tissues/organs, monitoring the heartbeat of fetus through Doppler fetal heart rate monitors or to visualize the fetus in pregnancy using fetal ultrasound, to image the blood flowing through any blood vessels or organs using Doppler ultrasound, to image ocular structures using ophthalmic ultrasound, to determine the volume of urine (*see* Chapter 10, §10.5) and to see the heart through Echocardiogram (*see* Chapter 14, §14.5.11), or to perform ultrasound-guided biopsies, or to assess the fragility of bones using bone sonometry, or in the detection of malignant lesions in breast tissue using breast ultrasound etc.

15.6.1 WORKING PRINCIPLE OF US IMAGING

Ultrasound (US) imaging also referred to as sonography is a noninvasive diagnostic tool to obtain structural and functional information of the targeted ROI. It is used to probe the organs inside the body and to determine the velocity of moving fluid, primarily blood and urine using the reflected high-frequency sound waves. Unlike X-ray imaging, it is based on nonionizing radiation where a transducer probe is placed directly in contact with the superficial skin layer (epidermis) after applying a very thin layer of gel that allows very smooth transmission of ultrasound waves into the body relative to the acoustic impedance (product of tissue density and the transmission velocity of sound) of each tissue. The common operating procedure involves application of gel to the skin tissue before ultrasound scanning. It serves two purposes, firstly, it avoids the formation of air pockets and secondly, it avoids the occurrence of refractive index mismatch between the transducer and skin layer.

Ultrasound waves are primarily sound waves non-detectable to human ears with frequencies higher than 20,000 Hz (cycles/sec) and velocity of approximately 1540 m/s in the soft tissues. The ultrasound frequencies typically lie within the range of 3 and 10 MHz. There are four important parameters (namely wavelength- λ (in m), frequency - f (in Hz), velocity - c (in m/s), and amplitude or intensity) describing the wave through a relationship:

$$\lambda = \frac{c}{f}$$

15.6.2 ESSENTIAL COMPONENTS OF US IMAGER

The basic components of the US imaging system (see Figure 15.21), essentially include a transducer, transmitter pulse generator, amplifiers, a computer/CPU (central processing unit), a monitor, and a program to store and process the images.

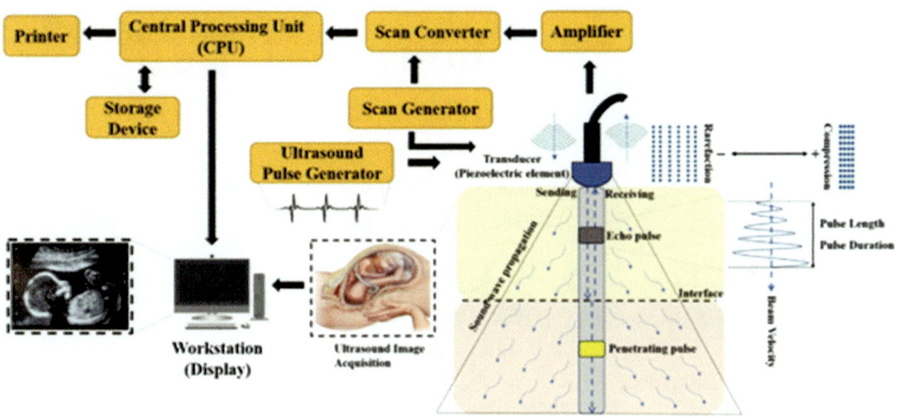

FIGURE 15.21 Components of an ultrasound imaging system.

CPU is primarily responsible for providing the required inputs to a transducer and visualize it on the monitor. CPU receives an electronic input from the transducer to construct an image and extract quantitative features from the obtained images and displayed on a visualization unit. First ultrasound waves are produced by transducer which also receives the reflected ultrasound echoes. In principle, transducers comprise of special ceramic piezoelectric crystals that vibrate on exposure to small electrical currents, thereby generate the beam of sound waves that transmits through the soft tissues. A small electric current is also produced by piezoelectric crystals when these crystals are deformed by reflected sound waves. The conversion of electrical signal into mechanical vibration or vice-versa is done through an inverter which is leaned upon the body to emit an ultrasonic beam focused through lenses and ultrasonic mirrors.

15.6.3 INTERACTION OF SOUND WITH TISSUE

The ultrasound waves are primarily used for disease diagnosis or therapy purposes. Here, a cone beam of sound waves is transmitted through the biological tissue(s) to monitor the inside activities along the path of the beam. The macroscopic processes of interactions are dependent on the intensity of sound wave, also depends on the tissue physiological properties through which the sound beam traverses. Figure 15.22 illustrates the interaction of ultrasound beam at the interface of two different media.

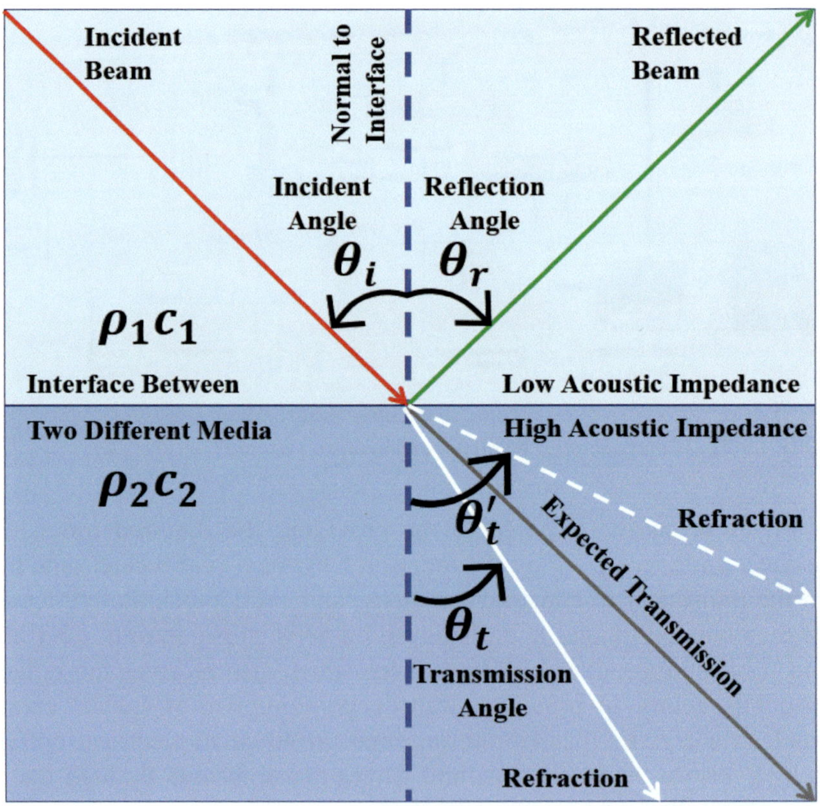

FIGURE 15.22 Illustration of an ultrasound beam interaction between two different media.

The ultrasound beam may show reflection or transmission (refraction) at tissue boundaries into the media with different acoustic impedance (Figure 15.22). A non-perpendicular ultrasound beam is shown with angle of incident beam, θ_i equals the angle of reflected fraction of beam, θ_r. A portion of sound wave transmitted shows refraction in sound transmitting tissue at refraction angles (θ_t and θ_t') less than or greater than the incident ultrasound beam incident angle. The refraction angle of transmitted wave, $\theta_t' >$ incident angle, θ_i whenever the velocity of sound in the media 2, c_2 is greater than the velocity of sound in media 1, c_1. Also, the transmitted refraction angle, θ_t is smaller than the incident beam angle, θ_i whenever the velocity of sound in the media 2, c_2 is smaller than the velocity of sound in media 1, c_1.

The expansion/compression of the medium enables the propagation of sound waves across a tissue. As summarized in Figure 15.23, the interaction of ultrasound waves through the biological tissues leads to reflection, scattering, refraction, and attenuation (energy absorption).

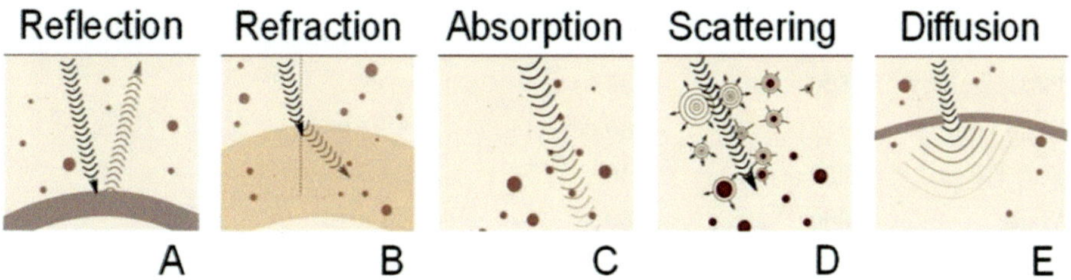

FIGURE 15.23 Demonstration of interaction of sound wave with physical medium.

Reflection in the tissue (Figure 15.23) is denoted in (a), difference in acoustic impedance at the interface leads to the refraction of sound waves (b). The absorption (c) and sound scattering produced by moving medium (d) lead to a decrease in the amplitude of sound wave. The diffraction phenomenon (e) is also noticed due to spreading of the sound waves. The scattering and reflected sound waves are the main signals received by the transducer of ultrasound system further used for image reconstruction. (Source: Adapted from Cloutier et al., 2021).[19]

The acoustic impedance mismatch will incur when the ultrasound wave traverses through two mediaof different densities (see Figure 15.22). This biological tissue interface where these two media change is known as acoustic boundaries. Larger density differences cause large change of acoustic impedance resulting in the reflection at the tissue boundaries. For example, the soft tissues located closer to the air-filled lungs show a greater difference in acoustic impedance to reflect all the incident ultrasound beams. Likewise, an acoustic boundary exists between urine and bladder wall due to difference in acoustic impedance value. The acoustic impedance for air, water, and biological tissues are summarized in Table 15.1. Among biological tissues, bone absorbs maximum ultrasound energy than soft tissues.

TABLE 15.1 Acoustic Impedance of Biological Tissues of Interest.

Biological tissue	Density [kg/m^3]	Velocity of sound [m/s]	Acoustic impedance (Density [kg/m^3] × Velocity of sound[m/s])
Air	1.2	330	0.0004×10^6
Blood	1058	1560	1.65×10^6
Fat	924	1450	1.34×10^6
Kidney	1041	1565	1.63×10^6
Liver	1061	1555	1.65×10^6
Lung	300	600	0.18×10^6
Muscle	1068	1600	1.71×10^6
PZT	7500	4000	–
Skull bone	1912	4080	7.8×10^6
Soft Tissue	1050	1540	–
Water	1000	1480	1.48×10^6

At the boundaries/interfaces reflection takes place due to mismatch in impedance of sound (product of density (in kg/m^3) and speed of sound (in m/s)) of two interacting media causing the ultrasound waves to deflect toward the transducer. There are four major factors that are responsible for reflection, that is, incident angle, impedance mismatch, the boundary angle and width. The incident angle is the angle between beam and plane of the tissue. In order to occur reflection, the incident angle should be $< 90°$. If the wavelength of ultrasound wave is larger than width of tissue boundary (width $< \lambda$), the ultrasound wave will show no reflection. Hence, tissue boundaries exhibiting smooth and greater width than the ultrasound wave (width $> \lambda$) will result in significant amount of reflection of signal, that is, act as mirror/specular reflector.

Scattering of ultrasound (non-specular reflection) will occur when the boundary width < wavelength. This phenomenon is a cause of reflection of refraction which can be seen whenever there is presence of larger amount of smaller tissue boundaries (red blood cells having width 20 times < ultrasound wavelength) causing scattering or diffusion of ultrasound beam. Typically, signal received in transducer typically 100–1000 times less than the transmitted signal. In other words, whenever the reflecting acoustic boundary is uneven and less than beam diameter, it causes the incoming beam to reflect in many different directions.

Refraction can be seen whenever there is deflection of ultrasound signal in the direction of sound propagation (*see* Figure 15.22). Refraction may result in double-image artifacts and contributes beam intensity loss. When beam frequency of sound remains unchanged, causing the wavelength to change while ultrasound beam traverses from one medium to another at different velocities (*see* above equation and Figure 15.22).

Absorption is the process of conversion of acoustic energy into a different form of energy, mostly thermal energy (heat energy). The loss of sound energy during absorption gives rise to "noise." The extent of absorption of ultrasound beam is affected by three main variables, that is, the viscosity of the medium, the relaxation time of the medium and the frequency of beam. The absorption phenomenon of ultrasound beam increases with the increased viscosity of the medium. Higher viscosity suggests greater frictional forces allowing more heat generation from the vibrating particles of the medium as they move across each other.

Relaxation time suggests the time taken by vibrating particles to come back to their original mean position within the medium following incurred displacement caused by an ultrasound pulse. When a relaxation time is shorter, the vibrating particles can reach to their original mean position in a very short span of time before the next ultrasound beam pulse encounter the same particles; however, longer relaxation time suggests that the particles are in contact with another ultrasound pulse before they are in fully relaxed position, thereby causing higher absorption of ultrasound and more heat dissipation. Higher frequency implies that the medium particles move at fast pace, thus generating more frictional heat by viscous drag and relaxation process. This can be inferred that with the increased beam frequency causes more absorption of ultrasound beam. The significance of ultrasound absorption is that maximum beam is reduced, thus less propagation and intensity to probe the biological structures (tissues) beyond the absorbing media.

Attenuation refers to any propagation losses (reflection, refraction, scattering, and absorption) causing reduction of intensity of sound while propagating through a medium. Beam attenuation depends not only on the properties of the propagating medium but also on the frequency of ultrasound beam. The half-value thickness (HVT) of sound is widely used for quantifying attenuation in a specified medium so that within the propagating medium, the intensity become one half of the original value. Hence, the amplitude (or intensity) of ultrasound wave(s) decreases when it passes denser biological structures. Here, the ultrasound wave follows a consistent single path without being deflected. In nutshell, Attenuation is a frequency-dependent behavior where lower frequency ultrasound has a lower resolution resulting in better penetration depth while higher frequency ultrasound has higher resolution; however, exhibits a lower penetration depth. The transducer frequency of 1 MHz achieves a penetration depth of 30 cm, 5 MHz of 6 cm, 20 MHz of 1.5 cm. Therefore, transducer frequency and acoustic impedance mismatch affects attenuation. Ultrasonic half value of ultrasonic thickness for different biological tissues is summarized in Table 2 for two different frequencies of 2 MHz and 5 MHz, respectively.

TABLE 15.2 Ultrasonic Half-Value Thickness (HVT) at Two Different Frequencies of 2 MHz and 5 MHz.

Biological tissue domain	Ultrasonic half-value thickness (cm)	
	2 MHz	**5 MHz**
Air	0.06	0.01
Blood	8.5	3.0
Bone	0.1	0.04
Brain	2.0	1.0
Liver	1.5	0.5
Muscle	0.75	0.3
Soft tissue (average)	2.1	0.86
Water	340	54

15.6.4 IMAGE CONSTRUCTION IN US IMAGER

Ultrasound imaging is the reflection mode imaging wherein sound reflected by anatomical structures were used to construct an image. When two tissues of different densities are positioned next to each other, a mismatch of acoustic impedance (product of tissue density and transmission velocity of sound) is created and resulting in the reflection of sound waves. There are two types of reflection that can occur during ultrasound beam interaction with the size of the boundary and irregularities within the reflectors.

Reflection is specular if the boundary is very smooth and greater than beam diameter, whereas reflection is not specular if acoustic boundary is irregular. The greater the intensity of sound wave reflection toward transducer, greater would be the acoustic mismatch/difference in the density of adjoining tissues and hence, brighter appearance in obtained image. The amplitude or strength of sound signal and time taken by the sound wave to traverse through the body essentially generates an image. The scanner calculates the distance between tissue boundary and the transducer to generate two-dimensional images of the ROI. The US imaging is mostly advised for imaging soft tissues and with limited usage for bones- and structures filled with gas. The reasoning behind limited usage is that the transmission of sound waves is very poor in air and air-filled structures and rapid transmission in the bone tissue. This causes a large mismatch of acoustic impedance at bone and air-filled structures with the soft tissues causes the reflection of sound waves. However, these larger reflections decrease the penetration of sound waves into deeper tissues, thereby may result in imaging artifacts.

15.6.5 PRINCIPLE OF US IMAGING

There are four different diagnostic modes of anatomical imaging using ultrasound that are currently used in medical imaging to examine arterial or venous system, urinary system, heart, pancreas, ovaries, spinal cord, etc. The different modes of sound waves are:

- **A-mode:** It is also referred to as amplitude modulation mode. It is a 1-d scan wherein the amplitude of the backscattered echoes from tissue boundaries is plotted as a function of depth using a transducer that scans A-line throughout the body. It is based on the pulse-echo principle. A-mode is most often used in the field of ophthalmology (echoophthalmoscopy) to examine the distance between different regions of the eye and/or the detachment of cornea. It is also used to detect abnormalities within the cranial cavity (echoencephalography) of the head region by delivering high-frequency sound pulses and is well adapted for emergency examination of patients with brain hemorrhage.
- **B-mode:** In B-mode or Brightness modulation mode ultrasound, a linear array of transducer elements scans a plane throughout the body to produce a two-dimensional gray-scale image of cross-section of a stationary or moving biological tissue structure. In this type of scanning, each A-line in the image cross-section represents an A-mode scan with signal brightness proportional to the backscattered echo. In

other words, it can be inferred that a more intense echo produces a brighter image. It has wide acceptance to measure the dimensions of cardiac chambers and in assessing the valvular structure and functions.

- **M-mode:** In M-mode or Motion mode ultrasound (Time Motion mode), a rapid sequence of B-mode ultrasound scans follows a sequential order to measure the range of motion of boundaries moving relative to the probe thereby capturing the movement of various biological tissues within the human body. Its time resolution is dependent on the time taken by the reflected echoes from the deepest biological tissue to return to the transducer. M-mode scanning is often used to detect motion of heart valves. This is analogous to recording of a video in an ultrasound.
- **Doppler imaging mode:** Doppler mode of imaging is a noninvasive approach to visualize the velocities of moving biological tissues. There is an apparent frequency variation of sound that transducer received due to the motion of sound source from sound detector which results in Doppler effect. It is often used for computing the velocity of blood in blood vessel, especially for lumen narrowing or in the areas of stenosis.

This chapter briefly focuses on some of the commonly known ultrasound imaging methods. Readers can look in the previous Chapter 14 for more details on Doppler effect (see § 14.5.9), pulsed Doppler and Duplex ultrasound (see § 14.5.10) methods and echocardiogram (§ 14.5.11).

15.6.6 APPLICATION OF US AND US IN BLOOD FLOW STUDIES

Ultrasound (US) imaging has numerous applications in the field of veterinary and human medicine. In preclinical (veterinary medicine) and clinical studies (human medicine), ultrasound imaging has been employed by researchers/physicians to assess specifically the animal cardiac structures/functions or several medical conditions of human subjects. In animal studies, the spatial and temporal resolution allows capturing accurate images of rat/mouse hearts, kidneys, or abnormal tumor masses, etc. In humans, it is used for the diagnosis of vascular diseased conditions, obstetrics and pregnancy, nephrology, musculoskeletal, cancer, urology, gynecology, gastroenterology, cardiology, pulmonology, abdomen pains, ultrasonic-guided biopsies, etc. Several vascular disease conditions, such as arterial blockages (Atherosclerosis), formation of blood clots (deep vein thrombosis), arterial spasm (Raynaud's phenomenon), brain aneurysms (a balloon-like bulging in the arteries), arterial compression, or swollen and twisted veins of the legs (varicose veins), structure of the heart, identifying narrowed or blocked arteries in the neck region (atherosclerotic carotid artery stenosis) and in arms and legs (peripheral artery disease).

Fetal ultrasound is the commonly used procedure during pregnancy to monitor the development of embryo or fetus activities (prenatal visualization) in real-time. In fetal ultrasound, three-dimensional ultrasound in motion (also known as four-dimensional ultrasound) allows the visualization of fetus and its movement in realistic time which may help to see if the umbilical cord is wrapping around the neck of the baby and to avoid any problems or complicacies associated with stillbirth. It can also be used to check the number of fetuses (multiple pregnancy) and heartbeat at different stages of pregnancies. Fetal ultrasounds can detect any growth-related abnormalities within 18–22 weeks of gestation period. Apart from the neonates and fetuses, other potential applications involve monitoring placenta and its location in relation to the cervix and umbilical blood flow in pregnant women. It can also be employed for endovenous (vein) ablation and imaging gallbladder and renal system. Other clinical applications of ultrasound involve transabdominal imaging of testicles in men and uterus and ovaries in women.

15.7 OPTICAL COHERENCE TOMOGRAPHY

15.7.1 INTRODUCTION

OCT being a optical modality implements the principles of interferometry. Since its development in the early 90s, OCT has now become an essential examination tool in ophthalmology across the world. OCT, also known as optical sectioning, provides a cross-sectional view of anterior and posterior eye structure or any optically accessible tissue (e.g., skin). OCT imaging in ophthalmology has now been proven to provide early imaging biomarkers that can help physicians manage the progression of diseases. OCT is also finding applications in dermatology with capability to noninvasively see several layers of skin and vasculature. An increasingly promising application of OCT is to visualize fluid (e.g., blood, sub-retinal fluid, etc.). Visualizing blood in the vessels is referred as angiography. In ophthalmology, cardiology, and radiology, angiography is typically done with the help of dyes/contrast agents mixed and flowing with the blood. OCT, however, gives a dye/contrast agent-free angiography, known as OCT angiography (OCTA).

15.7.2 MODEL OF LIGHT-TISSUE INTERACTION IN OCT

The OCT imaging is based on the principles of low-coherence interferometry. Interferometry is a technique in which the reflected light from a surface contains information that can be extracted by superimposing reference arm light. The overlapping of these two reflecting lights gives interfering fringes, which have intensity and depth information based on birefringence and movement of the tissue. The interferometer component of OCT splits the light source into two: reference arm and one for the sample arm. The backscattered light from the sample arm is then compared with the reference arm light. In OCT imaging, we particularly use low-coherence light source, whose beam is then split into sample and reference arms (Figure 15.24).[20–22]

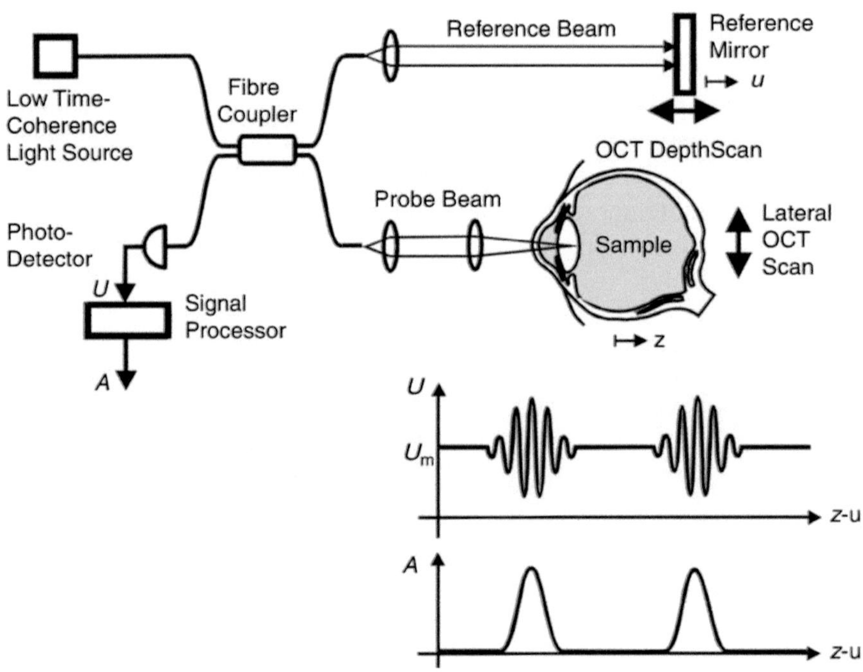

FIGURE 15.24 Basic components and working principle of OCT system. The light splits into two paths, reference arm, and sample (probe) arm. Reflected depth-encoded light from the sample is then compared with the reference arm light (interferometer) and detected using photodetector.

Source: Reprinted from Ref. [21]. https://creativecommons.org/licenses/by-sa/4.0/.

The light reflected from the sample arm has depth-encoded (depth profile) information (causes layers/surfaces of different refractive indices), and when combined with reference arm light forms the interference patterns. A depth profile at one location is called an A-scan, while scanning the A-scan in second dimension is called a B-scan OCT. Obtaining B-scan in third dimension gives a 3-D OCT volume scan. OCT imager can use visible to near-infrared light to produce a tradeoff between varying penetration, high spatial resolution, and signal-to-noise ratio.

15.7.3 TIME-DOMAIN OCT IMAGING

The OCT system typically has an interferometer and a broadband light source. The light is split into sample and reference arms and combined back in a coupler. The optical arrangement, type, patterns of scans, variation in arm's length, etc., decides the subtype of OCT system. When the reference mirror is moved in time to obtain depth profile, the system is called a time-domain OCT or TD-OCT. Time-domain OCT system was developed in early 90s. So, the time-domain OCT involves mechanical movement of the reference arm mirror, making it slow in scanning. The mechanical movement of the mirror is also limited by course steps of motors, leading to low resolution in imaging. Time domain OCT is being replaced with faster and high-resolution spectral or frequency domain and swept source OCT.

15.7.4 SPECTRAL-DOMAIN OR FREQUENCY-DOMAIN OCT IMAGING

The frequency-domain OCT is scheme where the depth-encoded broadband spectrum is read using spatial detectors, for example, a spectrometer. Instead of moving the reference mirror (as in time-domain OCT), the spectrometer analyses the spectra of reflected broadband light.

The spectrum is separated using diffraction grating, which divides separated different wavelengths of white broadband light into narrow band components. The spectral-domain OCT or SD-OCT does not have moving mirrors and can result in up to 50—100 times faster scanning compared with time-domain OCT scanning. The SD-OCT also has a very high signal-to-noise ratio compared with TD-OCT. This increased speed of scanning can be utilized for imaging moving objects (e.g., blood), as will be discussed in Doppler OCT and OCTA. Typically, the light source is a broadband near-infrared super luminescent diode (SLD) (Figure 15.25). Since the interferogram contains information from all depth layers, a Fourier transform is required to extract individual depth profile. For this reason, spectral-domain OCT is also known as Fourier domain OCT (FD-OCT).[20–22]

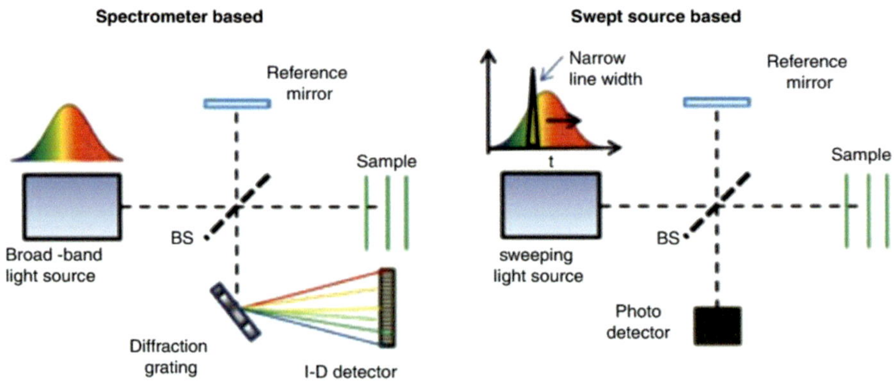

FIGURE 15.25 The basic setup and working principles of (left) spectrometer-based spectral or Fourier domain OCT and (right) swept source OCT.

Source: Reprinted from Ref. [21]. https://creativecommons.org/licenses/by-sa/4.0/

15.7.5 SWEPT SOURCE OCT IMAGING

As indicated in Figure 15.25, the working principle of swept source OCT is very similar to SD-OCT or FD-OCT, other than the broadband light source, typically an SLD, is replaced by sweeping narrow width light source (right inset in Figure 15.25). This rapid sweeping of narrow line width light ranging over a broad wavelength range removes the need of diffraction grating. Each reflecting wavelength from the sample is overlapped with reference light to form fringes and detected using high-speed photodetectors.[21] SS-OCT offers better visualization of deeper tissues (decreased sensitivity roll-off and improved signal-to-noise ratio) compared with SD-OCT. Both SD-OCT and SS-OCT have established themselves in clinical applications such as ophthalmology. SD-OCT use is widespread compared with SS-OCT, mainly because of more affordable commercial availability.

15.7.6 PRINCIPLE OF DOPPLER OCT AND OCT ANGIOGRAPHY

Angiography, in general, is accompanied by a dye and/or contrast agent for visualization of blood vessels (hence the term "angio-"), blood flow, and other bodily fluid in motion. One example of angiography is an X-ray of blood vessels with a radiopaque contrast media injection in the blood. Another example of angiography is the introduction of sodium fluorescein for retinal vessel angiography using the scanning laser ophthalmoscope. Typically, the angiography procedure is invasive, due to contrast agent injections of due to imaging modality being ionizing (e.g., X-ray). Traditional angiography is also a time-consuming procedure, which can be cumbersome for the patients. Some dyes and contrast agents can have adverse effects on the patients, such as anaphylactic reactions.

The OCTA uses physics and optical properties of moving objects, such as blood cells, for obtaining a dye-free and contrast agent-free, noninvasive angiography. Especially the red blood cells, when moving in the vessels, result in high variance OCT signal, which differs in intensity and phase (also known as Doppler OCT) compared with a stationary tissue. The repeated B-scan at one location helps remove stationary tissue signals from moving cell signals, and the in-face image created with volumetric B-scans of OCTA gives the angiography image.[21] OCTA, although being successful in optically easily accessible and superficial tissues angiography (e.g., ophthalmology and dermatology), yet has limitations in terms of deep-seated tissue imaging, mainly because of high optical attenuation.

15.7.7 APPLICATION OF OCT IN VASCULAR DISORDER AND MALFORMATION

In ophthalmology, and in some recent studies, in dermatology, OCTA has been used to obtain clinically meaningful information, which is otherwise difficult to visualize. In ophthalmology, the retinal and choroidal microvasculature structure and function can be assessed with high resolution using OCTA. The OCTA is a noninvasive and non-dye-based, is quick and easy to perform without losing the quality of imaging data. In the retina, blood and fluid leakage can have serious consequences such as vision loss and blindness. Diabetic retinopathy and neovascular age-related macular degeneration are two leading causes of blood and fluid leakage led retinal degenerative diseases and can be assessed in early phase using OCTA.[23–25]

The OCTA can capture dermal and subdermal microvasculature with high resolution indicating potential clinical application in dermal and subdermal vascular diagnosis and pathology.[26,27] Some recent animal studies using OCTA were performed for the detection of vascular microenvironment in brain tumor in a xenograft mouse model of glioblastoma.[28] In summary, despite its limitation of deep tissue imaging, at present, OCTA provides several benefits compared with traditional angiography when imaging vascular disorders and malformations.

15.8 NUMERICAL PROBLEMS

- **Problem 1:** The linear attenuation coefficient of a gadolinium-based phosphor used for the detection of X-rays is 560 cm^{-1} at an X-ray incident energy of 150 keV. What percentage of X-rays are detected by phosphor layers of 100, 250, and 500 μm thickness (use Eq. 2)? What are the tradeoffs in terms of spatial resolution?
- **Problem 2:** Calculate the projection value (P_j), using measured signals S_1, S_2, and S_3 produced by each crystal. The value of μ_{tissue} is 0.5 cm^{-1}, μ_{bone} is 1 cm^{-1}, and $\mu_{crystal}$ is 2 cm^{-1}. The dimension of the tissue is depicted in Figure 15.26.
- **Problem 3:** Using eqs (15.9) and (15.10) calculate the following:
  a) In a sample of 1000 atoms, if 50 atoms decay in 5 s, what is the radioactivity measured in mCi, of the sample?
  b) In order to produce a level of radioactivity of 1 mCi, how many nuclei of ^{99m}Tc ($\lambda = 3.22 \times 10^{-5}$ s^{-1}) must be present? What mass of the radionuclide does this correspond to? (Avogadro's number is 6.02×10^{23}).
  c) A radioactive sample of ^{99m}Tc contains 10 mCi activity at 9 am. What is the radioactivity of the sample at 3 p.m. on the same day?

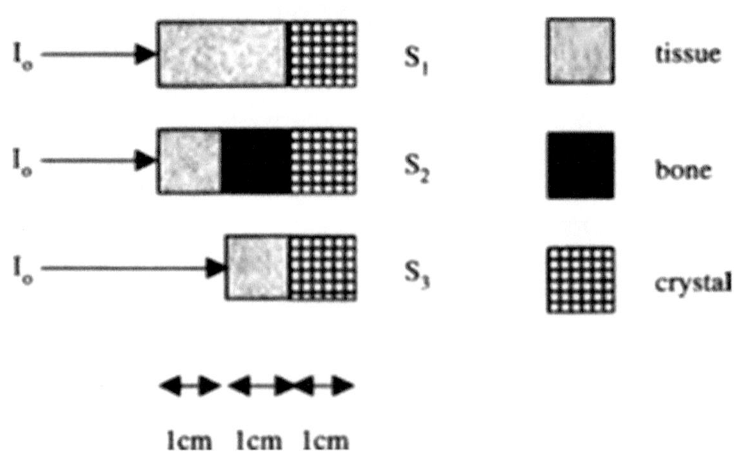

FIGURE 15.26 Illustration for Problem 2.

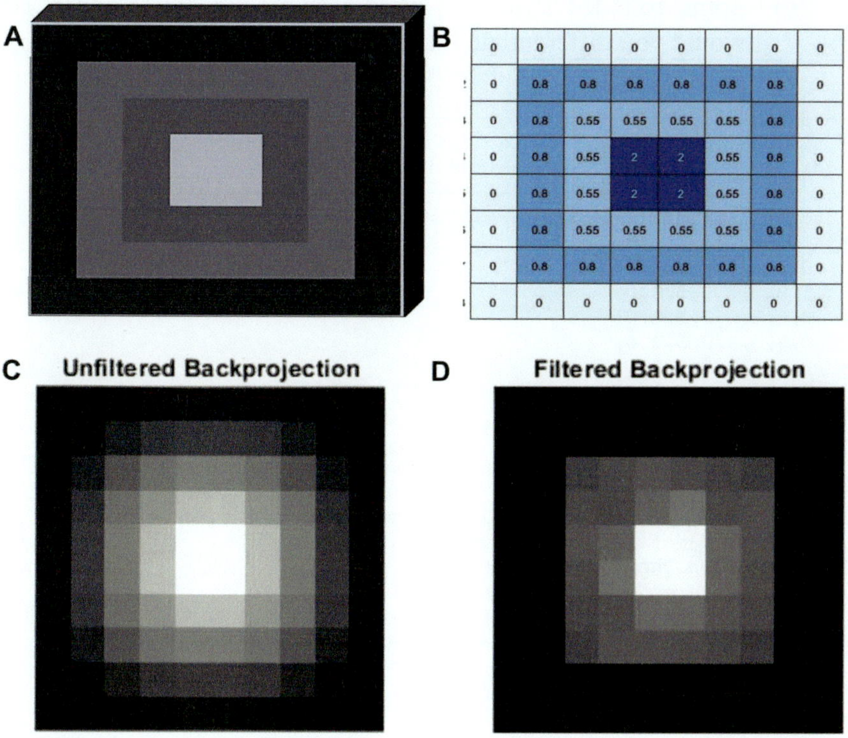

FIGURE 15.27 (A) Schematic of original tissue slice. (B) Attenuation coefficient values of tissue slice, (C and D) unfiltered and filtered back-projection image from 180 forward projection data, respectively.

- **Problem 4:** Using the provided thin slice of tissue phantom with its effective attenuation properties shown in Figure 15.27 (A and B), reconstruct the unfiltered (Figure 15.27C) and filtered (Figure 15.27D) back-projection images using inverse Radon transformation from 180° forward projections (assume no property variation along slice thickness). Use the sample MATLAB script (Figure 15.28).
- **Problem 5:** Rob was experiencing some pain in his upper right abdomen. Upon suggestion of the doctor, Rob was asked to get a US scan. As a trainee intern in the diagnostic center, your job is to perform a visual diagnosis using US scan. How will you arrive at solving each of the following:
  a) What is the probe you will choose and why?
  b) How will you differentiate between hepatic vessels and hepatic ducts?

c) Some rounded structures were seen in the gall bladder. How would you differentiate if these are polyps or stones?

```
%-----------------------------------------------------
% This script allow to create a reconstruction images
% (Using Radon Transformation and its Inverse)
%-----------------------------------------------------
% Step1: Create a Phantom
Phantom = [[0,0,0,0,0,0,0,0;
    0,8,8,8,8,8,8,0;
    0,8,5,5,5,5,8,0;
    0,8,5,2,2,5,8,0;
    0,8,5,2,2,5,8,0;
    0,8,5,5,5,5,8,0;
    0,8,8,8,8,8,8,0;
    0,0,0,0,0,0,0,0]];
%
% Step 2: Obtain the forward projection using Radon transformation
%
theta=0:179; % projection angles (theta= 0 to 179 at dtheta = 1)
% Forward projection using function "radon"
[R,xp] = radon(Phantom,theta);
% Here, R referes to forward projection (Pj) data
%       xp is the radial coordinates
% ----- Display the transformed data (by sinogram) ---------
figure
imshow(R,[],'Xdata',theta,'Ydata',xp,'InitialMagnification','fit')
xlabel('\theta (degrees)')
ylabel('x''')
colormap(gca,hot), colorbar
%
%
% Step 3: Obtain the backprojection images using inverse
%         Radon transformation (a) Use both filtered/unfiltered
%
I_unfilt = iradon(R,theta,'linear','none');% unfiltered backprojection
I_filt = iradon(R,theta);                  % filtered backprojection
%----Display reconstructed images after backprojection------
figure
subplot(1,2,1)
imshow(I_unfilt,[])
title('Unfiltered Backprojection')
subplot(1,2,2)
imshow(I_filt,[])
title('Filtered Backprojection')
```

FIGURE 15.28 Sample MATLAB code for Problem 4.

- **Problem 6:** Between time-domain and frequency-domain OCT which one would you prefer and why? (Hint: Speed of image acquisition)
- **Problem 7:** Two swept source OCT systems are available. System A uses 900 nm IR laser with a bandwidth of 10 nm, and System B uses

1300 nm IR laser with a bandwidth of 100 nm. Which one would you prefer to use for imaging the skin and why? (Hint: Scattering effect influences the image resolution).

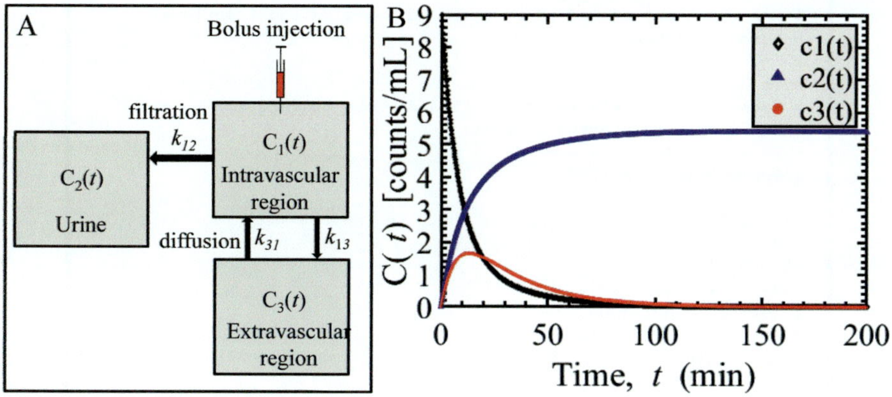

FIGURE 15.29 (A) Schematic of multi-compartment system of renal clearance, (B) concentration curve for each compartment.

- **Problem 8:** Consider the case of renal clearance of radioactive drug (see Figure 15.29A). The clearance of a drug by renal/kidney is identified as complete filtration of the drug from plasma in the kidney which gets accumulated in the form of urine in the bladder. The multi-compartment system consists of an intravascular compartment connected irreversibly with urine compartment, and an extravascular compartment exchanging the tracer in the plasma (Tracee) with compartment 1. At time t=0, $c_1(0)$ = 9 counts/mL concentration of tracer equivalent to total dose. At time t=0, $c_2(0)$ = 0 and $c_3(0)$ =0 concentration of tracer equivalent to zero. If the unknown rate constants are k_{12}= 0.05 min^{-1}, k_{13} = 0.04 min^{-1} or k_{31} =0.06 min^{-1}. (1) Using the basic principle of mass balance in Section 15.4.6, obtain three ODE for the three-compartment model. (2) Compute the concentration curve (Figure 15.29B) by numerically solving the ODE's using a MATLAB script given in Figure 15.30.

15.9 SUMMARY

This chapter introduces essential medical devices widely used in clinical settings for imaging human pathology and biofluid dynamics. Each section

```
%---------------------------------------------
% PROGRAM for multi-compartment system
%---------------------------------------------
clear
clc
%
% Represent c1, c2 and c3 by symbolic function and fractional
% turnover rates
%
syms u(t) v(t) x(t)
% Note u can be referred to c1(t) and v can be referred to c2(t)
% and x can be referred to c3(t)
%-------------------------
% KNOWN RATE COSNTANTS
%----------------------
k12= 0.05;
k13 = 0.04;
k31 =0.06;
%-------------------
% INITIAL CONDITIONS
%-------------------
cond1 = u(0) == 9;
cond2 = v(0) == 0;
cond3 = x(0) == 0;
conds = [cond1; cond2; cond3];
%--------------------------------
% DEFINE THE DIFFERENTIAL EQUATION
%--------------------------------
ode1=diff(u)==-(k12+k31)*u+k13*x;
ode2=diff(v)==k12*u;
ode3=diff(x)==-k31*x+k13*u;
odes=[ode1;ode2;ode3];
%---------------------------
% SOLVE system of ODES
%--------------------------
[uSol(t), vSol(t), xSol(t)] = dsolve(odes,conds);
%
t=0;
tmin=0;
tmax=200;
dt=0.2;
n=(tmax-tmin)/dt;
% Preassign the matrix values
data=zeros(n+1,4);
%
for i=1:n+1
    c1=subs(uSol,t);
    c2=subs(vSol,t);
    c3=subs(xSol,t);
    data(i,1)=t;
    data(i,2)=c1;
    data(i,3)=c2;
    data(i,4)=c3;
    t=t+dt;
end
```

FIGURE 15.30 Sample MATLAB code for Problem 8.

is divided by imaging devices, and its specific sections were designed to impart initial concentration on the physics and image acquisition methods followed by its application in structural and functional imaging of human diseases. The dynamics of biofluids especially blood in human physiology and pathophysiology is a source of local changes in the function of tissues. In other words, this dynamic nature of blood is essential for the human system to live and function. Therefore, monitoring the dynamic nature of blood in clinical settings provides us the essential clue for underlying pathology or treatment efficacy. In vast literature, there are numerous techniques that harness the electromagnetic and sound energy to probe the biofluid dynamics or pathology. However, we restricted our discussion throughout the chapter to essential few approaches and their application for brevity.

KEYWORDS

- **computed tomography**
- **diffusion-weighted MRI**
- **functional imaging**
- **single-photon emission computed tomography**
- **time-domain OCT**
- **ventilation imaging**
- **X-rays**

REFERENCES

1. Vendiš, T. *Modern CT Scanner Located at the Lochotín University Hospital in Pilsen, Czech Republic*.wikipedia.org,2022.https://en.wikipedia.org/wiki/CT_scan#/media/File:Modern%C3%AD_v%C3%BDpo%C4%8Detn%C3%AD_tomografie_s_p%C5%99%C3%ADmo_digit%C3%A1ln%C3%AD_detekc%C3%AD_rentgenov%C3%A9ho_z%C3%A1%C5%99en%C3%AD.jpg (accessed 18 March 2023).
2. Webb, A. G. *Introduction to Biomedical Imaging*; John Wiley & Sons, 2017.
3. Bushberg, J. T.; Seibert, J. A.; Leidholdt, E. M.; Boone, J. M. *The Essential Physics of Medical Imaging*; Lippincott Williams & Wilkins, 2012.
4. Ytrottier. *Siemens Single-Photon Emission Computed Tomography Machine in Operation, Doing a Total Body Bone Scan at the Credit Valley Hospital, Ontario, Canada*. wikipedia.org, 2012. https://commons.wikimedia.org/wiki/File:SPECT_CT.JPG (accessed 18 March 2023).

5. Brudersohn. *Combined Apparatus for Positron Emission Tomography (PET) and X-ray Computer Tomography (CT), Siemens Biograph*. wikipedia.org, 2012. https://commons.wikimedia.org/wiki/File:PET-CT_Siemens_Biograph01.jpg (accessed 18 March 2023).
6. Goris, M. L. *Nuclear Medicine Applications and Their Mathematical Basis*; World Scientific Publishing Company, 2011.
7. WikiBooks. *Basic Physics of Nuclear Medicine/Dynamic Studies in Nuclear Medicine*. 2022. https://en.wikibooks.org/wiki/Basic_Physics_of_Nuclear_Medicine/Dynamic_Studies_in_Nuclear_Medicine#Compartmental_Analysis (accessed 18 March 2023).
8. Kasuga, H. *Modern High Field Clinical MRI Scanner (3T Achieva, Philips, Netherlands)*. wikipedia.org, 2006. https://commons.wikimedia.org/wiki/File:Modern_3T_MRI.JPG (accessed 18 March 2023).
9. Poldrack, R. A.; Mumford, J. A.; Nichols, T. E. *Handbook of Functional MRI Data Analysis*; Cambridge University Press, 2011.
10. Stroman, P. W. *Essentials of Functional MRI*; CRC Press, 2016.
11. Gillard, J. H.; Waldman, A. D.; Barker, P. B. *Clinical MR Neuroimaging: Diffusion, Perfusion and Spectroscopy*; Cambridge University Press, 2004.
12. Rubin, G. D.; Rofsky, N. M. *CT and MR angiography: comprehensive vascular assessment*; Lippincott Williams & Wilkins, 2009.
13. Jones, D. K. Fundamentals of Diffusion MR Imaging. In *Clinical MR Neuroimaging: Diffusion, Perfusion and Spectroscopy*; Gillard, J. H., Waldman, A. D., Barker, P. B., Eds.; Cambridge University Press, 2004; pp 54–85.
14. Barker, P. B. Fundamentals of MR Spectroscopy. In *Clinical MR Neuroimaging: Diffusion, Perfusion and Spectroscopy*, Gillard, J. H., Waldman, A. D., Barker, P. B., Eds.; Cambridge University Press, 2004; pp 7–26.
15. Lev, M. H.; Rosen, B. R. Clinical Applications of Intracranial Perfusion MR Imaging. *Neuroimaging Clin. N. Am.* **1999**, *9* (2), 309–331.
16. Petrella, J. R.; Provenzale, J. M. MR Perfusion Imaging of the Brain: Techniques and Applications. *AJR Am. J. Roentgenol.* **2000**, *175* (1), 207–219. DOI: 10.2214/ajr.175.1.1750207.
17. Yan, F.; Song, Z.; Du, M.; Klibanov, A. L. Ultrasound Molecular Imaging for Differentiation of Benign and Malignant Tumors in Patients. *Quant. Imaging Med. Surg.* **2018**, *8* (11), 1078–1083. DOI: 10.21037/qims.2018.12.08.
18. Yu, M. H.; Kim, Y. J.; Park, H. S.; Jung, S. I.; Jeon, H. J. Imaging Patterns of Intratumoral Calcification in the Abdominopelvic Cavity. *Korean J. Radiol.* **2017**, *18* (2), 323–335. DOI: 10.3348/kjr.2017.18.2.323.
19. Cloutier, G.; Destrempes, F.; Yu, F.; Tang, A. Quantitative Ultrasound Imaging of Soft Biological Tissues: A Primer for Radiologists and Medical Physicists. *Insights Imaging* **2021**, *12* (1), 1–20.
20. Popescu, D. P.; Choo-Smith, L. P.; Flueraru, C.; Mao, Y.; Chang, S.; Disano, J.; Sherif, S.; Sowa, M. G. Optical Coherence Tomography: Fundamental Principles, Instrumental Designs and Biomedical Applications. *Biophys. Rev.* **2011**, *3* (3), 155. DOI: 10.1007/s12551-011-0054-7 From NLM.
21. Aumann, S. et al. *Optical Coherence Tomography (OCT): Principle and Technical Realization*; Springer, 2019. DOI: 10.1007/978-3-030-16638-0_3.
22. Fercher, A. F.; Drexler, W.; Hitzenberger, C. K.; Lasser, T. Optical Coherence Tomography—Principles and Applications. *Rep. Progress Phys.* **2003**, *66* (2), 239–303. DOI: 10.1088/0034-4885/66/2/204.

23. Sun, Z.; Yang, D.; Tang, Z.; Ng, D. S.; Cheung, C. Y. Optical Coherence Tomography Angiography in Diabetic Retinopathy: An Updated Review. *Eye* **2021**, *35* (1), 149–161. DOI: 10.1038/s41433-020-01233-y.
24. Bae, K.; Kim, H. J.; Shin, Y. K.; Kang, S. W. Predictors of Neovascular Activity During Neovascular Age-Related Macular Degeneration Treatment Based on Optical Coherence Tomography Angiography. *Sci. Rep.* **2019**, *9* (1), 19240. DOI: 10.1038/s41598-019-55871-8 From NLM.
25. Laíns, I.; Wang, J. C.; Cui, Y.; Katz, R.; Vingopoulos, F.; Staurenghi, G.; Vavvas, D. G.; Miller, J. W.; Miller, J. B. Retinal Applications of Swept Source Optical Coherence Tomography (OCT) and Optical Coherence Tomography Angiography (OCTA). *Progress Retinal Eye Res.* **2021**, *84*, 100951. DOI: https://doi.org/10.1016/j.preteyeres.2021.100951.
26. Men, S.; Wong, J. M.; Welch, E. J.; Xu, J.; Song, S.; Deegan, A. J.; Ravichander, A.; Casavant, B.; Berthier, E.; Wang, R. K. OCT-Based Angiography of Human Dermal Microvascular Reactions to Local Stimuli: Implications for Increasing Capillary Blood Collection Volumes. *Lasers Surg. Med.* **2018**, *50* (9), 908–916. DOI: https://doi.org/10.1002/lsm.22944.
27. Deegan, A. J.; Talebi-Liasi, F.; Song, S.; Li, Y.; Xu, J.; Men, S.; Shinohara, M. M.; Flowers, M. E.; Lee, S. J.; Wang, R. K. Optical Coherence Tomography Angiography of Normal Skin and Inflammatory Dermatologic Conditions. *Lasers Surg. Med.* **2018**, *50* (3), 183–193. DOI: https://doi.org/10.1002/lsm.22788 (accessed 27 July 2022).
28. Andleeb, F.; Katta, N.; Gruslova, A.; Muralidharan, B.; Estrada, A.; McElroy, A. B.; Ullah, H.; Brenner, A. J.; Milner, T. E. Differentiation of Brain Tumor Microvasculature From Normal Vessels Using Optical Coherence Angiography. *Lasers Surg. Med.* **2021**, *53* (10), 1386–1394. DOI: https://doi.org/10.1002/lsm.23446 (accessed 26 July 2022).
29. Arthur-Baidoo, E.; Ončák, M.; Denifl, S. Electron attachment to fluorodeoxyglucose: Dissociation dynamics in a molecule of near-zero electron affinity. *The Journal of Chemical Physics* **2022**, *157* (7), 074301. DOI: https://doi.org/10.1063/5.0101726

Index

C

D

I

O

P

W

X

For Product Safety Concerns and Information please contact our
EU representative GPSR@taylorandfrancis.com Taylor & Francis
Verlag GmbH, Kaufingerstraße 24, 80331 München, Germany